Advanced Non-Thermal Power Generation Systems

Generally, sources for power generation are broken down into two categories: thermal and non-thermal. Thermal sources for power generation include combustion, geothermal, solar, nuclear, and waste heat, which essentially provide heat as a means for power generation. This book examines non-thermal (mechanical, electrochemical, nanoscale self-powered, and hybrid) sources of power generation and emphasizes recent advances in distributed power generation systems.

Key Features

- Details recent advances made in wind power, including onshore, offshore, fixed and floating platform, and air wind energy systems, and offers detailed assessments of progress
- Covers advances in generation of hydropower, exploring dam hydropower, novel wave energy converters, and novel systems and turbines for hydrokinetic energy conversion to power
- Examines all types of fuel cells and their multi-functional roles, along with hybrid fuel cell systems in complete detail
- Explores advances in the development of self-powered nanogenerators for use in portable, wearable, and implantable power electronics
- Focuses on technologies with the best commercial possibilities and provides perspectives on future challenges that need to be solved

This book will be of value to all researchers in academia, industry, and government interested in pursuing power generation technologies and seeking a comprehensive understanding of available and emerging non-thermal power generation sources. Readers who are interested in learning about thermal power generation sources can find it in the author's companion text *Advanced Power Generation Systems: Thermal Sources* (2023).

Sustainable Energy Strategies

by Yatish T. Shah

Chemical Energy from Natural and Synthetic Gas

Thermal Energy: Sources, Recovery, and Applications

Modular Systems for Energy and Fuel Recovery and Conversion

Modular Systems for Energy Usage Management

Hybrid Power: Generation, Storage, and Grids

Hybrid Energy Systems: Strategy for Industrial Decarbonization

CO2 Capture, Utilization, and Sequestration Strategies

Advanced Power Generation Systems: Thermal Sources

Advanced Non-Thermal Power Generation Systems

Other related books by Yatish T. Shah

Biofuels and Bioenergy: Processes and Technologies

Water for energy and Fuel Production

Energy and Fuel Systems integration

For more information on this series, please visit: https://www.routledge.com/Sustainable-Energy-Strategies/book-series/CRCSES

Advanced Non-Thermal Power Generation Systems

Yatish T. Shah

CRC Press is an imprint of the
Taylor & Francis Group, an informa business

First edition published 2024
by CRC Press
2385 Executive Center Drive, Suite 320, Boca Raton, FL 33431

and by CRC Press
4 Park Square, Milton Park, Abingdon, Oxon, OX14 4RN

CRC Press is an imprint of Taylor & Francis Group, LLC

ISBN: 978-1-032-55281-1 (hbk)
ISBN: 978-1-032-55284-2 (pbk)
ISBN: 978-1-003-42990-6 (ebk)

DOI: 10.1201/9781003429906

Typeset in Times
by codeMantra

This book is dedicated to my three sons James, Jonathan and Keith

Contents

Sustainable Energy Strategies Series Preface

While fossil fuels (coal, oil and gas) were the dominant sources of energy during the last century, since the beginning of the twenty-first century an exclusive dependence on fossil fuels is believed to be a non-sustainable strategy due to (1) their environmental impacts, (2) their non-renewable nature, and (3) their dependence on the local politics of the major providers. The world has also recognized that there are in fact ten sources of energy: coal, oil, gas, biomass, waste, nuclear, solar, geothermal, wind, and water. These can generate our required chemical/biological, mechanical, electrical, and thermal energy needs. A new paradigm has been to explore greater roles of renewable and nuclear energy in the energy mix to make energy supply more sustainable and environmentally friendly. The adopted strategy has been to replace fossil energy by renewable and nuclear energy as rapidly as possible. While fossil energy still remains dominant in the energy mix, by itself, it cannot be a sustainable source of energy for the long future.

Along with exploring all ten sources of energy, sustainable energy strategies must consider five parameters: (1) availability of raw materials and accessibility of product market, (2) safety and environmental protection associated with the energy system, (3) technical viability of the energy system on the commercial scale, (4) affordable economics, and (5) market potential of a given energy option in the changing global environment. There are numerous examples substantiating the importance of each of these parameters for energy sustainability. For example, biomass or waste may not be easily available for a large-scale power system making a very large-scale biomass/waste power system (like a coal or natural gas power plant) unsustainable. Similarly, an electrical grid to transfer power to a remote area or onshore needs from a remote offshore operation may not be possible. Concerns of safety and environmental protection (due to emissions of carbon dioxide) limit the use of nuclear and coal-driven power plants. Many energy systems can be successful at laboratory or pilot scales, but may not be workable at commercial scales. Hydrogen production using a thermochemical cycle is one example. Many energy systems are as yet economically prohibitive. The devices to generate electricity from heat such as thermoelectric and thermophotovoltaic systems are still very expensive for commercial use. Large-scale solar and wind energy systems require huge upfront capital investments which may not be possible in some parts of the world. Finally, energy systems cannot be viable without market potential for the product. Gasoline production systems were not viable until the internal combustion engine for the automobile was invented. Power generation from wind or solar energy requires guaranteed markets for electricity. Thus, these five parameters collectively form a framework for sustainable energy strategies.

It should also be noted that the sustainability of a given energy system can change with time. For example, coal-fueled power plants became unsustainable due to their impact on the environment. These power plants are now being replaced by gas-driven power plants. New technology and new market forces can also change sustainability of the energy system. For example, successful commercial developments of fuel cells and electric cars can make the use of internal combustion engines redundant in the vehicle industry. While an energy system can become unsustainable due to changes in parameters, outlined above, over time, it can regain sustainability by adopting strategies to address the changes in these five parameters. New energy systems must consider long-term sustainability with changing world dynamics and possibilities of new energy options.

Sustainable energy strategies must also consider the location of the energy system. On the one hand, fossil and nuclear energy are high-density energies and they are best suited for centralized operations in an urban area, while on the other hand, renewable energies are of low density and they are well suited for distributed operations in rural and remote areas. Solar energy may be less affordable in locations far away from the equator. Offshore wind energy may not be sustainable if the distance from shore is too great for energy transport. Sustainable strategies for one country may be quite different from another depending on their resource (raw material) availability and local market potential. The current transformation from fossil energy to green energy is often prohibited by required infrastructure and the total cost of transformation. Local politics and social acceptance also play an important role. Nuclear energy is more acceptable in France than in any other country.

Sustainable energy strategies can also be size dependent. Biomass and waste can serve local communities well at a smaller scale. As mentioned before, the large-scale plants can be unsustainable because of limitations on raw materials. New energy devices that operate well at micro- and nanoscales may not be possible on a large scale. In recent years, nanotechnology has significantly affected the energy industry. New developments in nanotechnology should also be a part of sustainable energy strategies. While larger nuclear plants are considered to be the most cost-effective for power generation in an urban environment, smaller modular nuclear reactors can be the more sustainable choice for distributed cogeneration processes. Recent advances in thermoelectric generators due to advances in nanomaterials are an example of a size-dependent sustainable energy strategy. A modular approach for energy systems is more sustainable at smaller scale than for a very large scale. Generally, a modular approach is not considered as a sustainable strategy for a very large, centralized energy system.

Finally, choosing a sustainable energy system is a game of options. New options are created by either improving the existing system or creating an innovative option through new ideas and their commercial development. For example, a coal-driven power plant can be made more sustainable by using very cost-effective carbon capture technologies. Since sustainability is time, location, and size dependent, sustainable strategies should follow local needs and markets. In short, sustainable energy strategies must consider all ten sources and a framework of five stated parameters under which they can be made workable for local conditions. A revolution in technology (like nuclear fusion) can, however, have global and local impacts on sustainable energy strategies.

The CRC Press Series on Sustainable Energy Strategies will focus on novel ideas that will promote different energy sources sustainable for long term within the framework of the five parameters outlined above. Strategies can include both improvement in existing technologies and the development of new technologies.

Yatish T. Shah,
Series Editor

Preface

In general, sources for power generation are divided into two categories; thermal and non-thermal. As described in my previous book (*Advanced Power Generation Systems: Thermal Sources*, CRC Press, 2023), thermal sources for power generation include combustion (particularly of fossil fuels, biomass, waste, hydrogen, etc.), geothermal, solar, nuclear, and waste heat. Essentially, these are the sources which provide heat as a means for power generation. Heat can be converted into power in a number of different ways such as thermodynamic cycles, photovoltaics, thermo-photovoltaics, thermoelectricity, etc. These technologies are covered in detail in my previous book. Combustion technology (particularly using fossil fuels) is the biggest source of carbon emission and needs to be either used with carbon capture or utilization or by replacing fossil fuels with biomass, hydrogen, etc. In recent years, significant progress has been made in the use of geothermal and solar energy for the utility-scale power production. Nuclear energy is yet to capture public acceptance in many parts of the world. Geothermal energy, while dispatchable, cannot be easily accessed in all parts of the world. Solar energy, while accessible, is intermittent and requires storage device or another source of energy to make it dispatchable. The use of solar energy, both thermal and photovoltaics, is the fastest growing industry both at the utility scale and at the distributed (such as residential, industrial, etc.) level. The scientific advancements in all of these thermal sources for power generation have been fully described in *Advanced Power Generation Systems: Thermal Sources*, my first book on advanced power generation systems.

This book is the second part of my previous book on advanced power generation systems and it deals with non-thermal, such as mechanical (wind and hydro), electro chemical, nanoscale self-powered, hybrid, etc., sources of power generation. It also emphasizes the recent improvements in distributed power generation systems. Energy industry is going through numerous paradigm changes. The centralized utility-scale power production is being increasingly replaced by distributed power generation systems that can be used for residential and commercial purposes and for the transportation industry. Utility-scale grid is increasingly connected to localized microgrids or even nanogrids. Small-scale distributed power generation appears to be more suitable for some renewable sources. The use of mechanical and electrochemical energy for this purpose is gaining more attention.

While capturing of wind energy has been in existence since the 18th century, modern methods to harness wind power began in 1990. Wind power, that is, the kinetic energy of the wind, is a renewable energy source that is used mainly for the production of electrical power, among others. The global wind resources (land and near-shore) are estimated to be 72 TW, which is seven times the world's electricity demand and five times the world's energy demand. In contrast to the global pollution issues and the significant cost of fossil fuels, wind seems to be a clean, affordable, and inexhaustible source of energy, and its exploitation can meet directly both the global demand for renewable and clean energy and the need to secure new energy sources. Fundamentally, wind power can be generated by four different methods: onshore fixed turbine, offshore fixed turbine, offshore floating turbine, and airborne wind energy. This book examines recent advancements in all four methods of wind power generation.

Currently, land-based wind power is the most developed and matured. However, offshore wind farms seem to be advantageous due to the enormous energy potential associated with large continuous areas and the stronger winds that imply greater power generation, even though they have higher initial investment, operation, and maintenance costs. Offshore wind farms can be further divided into near-shore wind farms (where the tower is connected to the bottom of the sea) and floating wind farms (where the wind tower is connected to a floating base on the water). A floating wind turbine has the most potential for wind power production. Airborne wind energy is at the early stages of development. Among other changes in modern wind power, in 1993, asynchronous generator in wind turbines was replaced with synchronous generator. Recent advancements in land-based wind power also include advanced control mechanisms, advanced energy conversion systems, better understanding of operational and maintenance issues, among others, which are addressed in this book.

In both the USA and globally, hydropower is the largest renewable resource in the energy mix and certainly the largest source of renewable electricity. There are two fundamental ways power can be generated using hydro energy: (1) by hydro-potential energy using dam for power generation and storage; these dams can be very large or very small; and (2) using hydrokinetic energy associated with ocean waves, currents, tidal waves, river flow, etc. There are also other ways to generate power from ocean such as ocean thermal gradient, salinity gradient, etc., but these are not as dominant and not as yet fully developed. While growth in the use of hydroelectricity (at least the traditional type—generated by very large dams) has slowed to near zero in the USA, many other countries in both the developed and developing world are pushing ahead with major projects to obtain energy from non-power dams (NPD) as well as large, small, and micro-hydropower projects. New advancements in hydropower include variable speed hydropower and digitalization of hydropower industry. Advanced electrical systems, drive trains, and converters to handle transient hydropower operations are also being developed. Innovations in pumped storage hydropower (PSH) are also pursued. New methods to capture hydrokinetic energy from rivers, streams, tidal waves, and ocean waves are being developed. This book examines these and many other issues on hydropower in detail.

The third major source of non-thermal power generation is fuel cells. Electrochemical processes such as source for power generation or energy storage are very flexible and versatile, and they are rapidly progressing both at stationary and mobile levels. The use of fuel cells for distributed and stagnant power generation is attracting more attention. Unlike combustion processes using fossil fuels, fuel cells use hydrogen as a source of fuel, which results in very low level or negligible carbon emission. Fuel cells can be roughly divided into low-temperature (ca. <200°C) and mid- to high-temperature (ca. >450°C) fuel cells. Alkaline fuel cell (AFC), polymer electrolyte membrane fuel cell (PEMFC, also known as proton exchange membrane fuel cell), direct methanol fuel cell (DMFC), and microbial fuel cell (MFC) are typical low-temperature fuel cells. Phosphoric acid fuel cell (PAFC), molten carbonate fuel cell (MCFC), direct carbon fuel cell (DCFC), and solid oxide fuel cell (SOFC) belong to the mid- to high-temperature fuel cell class. In general, low-temperature fuel cells (AFC, MFC, DMFC, and PEMFC) feature a quicker start-up, which makes them more suitable for portable applications. PEMFCs have recently gained momentum for the application in transportation and as small portable power sources. However, AFC, DMFC, and PEMFC require pure hydrogen (minimum 99.999%) and consequently an external fuel processor, which increases the complexity and cost and decreases the overall efficiency. They also require a higher loading of the precious metal catalysts. In low-temperature PEMFCs and DMFCs, the high cost originates from the expensive Nafion® membrane and Pt catalyst for electrodes. Poor durability of these cells is due to the degradation of membrane and catalyst, as well as the instability of catalyst support. In parallel to the development of classic fuel cells, a new and promising type of fuel cell, the microbial fuel cell (MFC), or biological fuel cell, is currently under intensive research. MFC produces electricity from microbially catalyzed anodic oxidation processes. The greatest potential of MFC lies in the use of wastewater as fuel, which allows simultaneous waste treatment and energy recovery. However, MFC still severely suffers from the short active lifetimes (typically 8 hours to 7 days) and limited power generation, which make it far from being used in practical applications.

In contrast to low-temperature fuel cells, PAFC, DCFC, MCFC, and SOFC are more flexible because they can reform various fuels (methanol, ethanol, natural gas, gasoline, etc.) inside the cells to produce hydrogen while offering advantages for stationary applications, especially for cogeneration. They are also less prone to catalyst *poisoning* by carbon monoxide and carbon dioxide. However, their slower start-up limits them to more stationary applications. In recent years, the use of molten carbonate as electrolyte has been found to be very formidable for high-temperature fuel cell. High-temperature fuel cells use solid electrolytes that could make the operation and maintenance of fuel cells easier. However, their commercialization is still hampered by high cost, poor durability issues, and operability problems that are directly linked to severe material challenges and systems issues. For example, in SOFC, high cost is a result of the high operating temperature, which also results in the use of expensive interconnect and sealing materials. SOFC durability is mainly affected by the microstructure decline, carbon deposition, and sulfur poisoning to catalysts. While both MCFC and SOFC present several challenges, in recent years, significant advances are made toward their commercialization potential. This book covers advances made in all low-, medium-, and high-temperature fuel cells in detail.

This book also covers in detail the recent advances made for the multi-functional roles of several fuel cells such as MCFC, MFC, and regenerative fuel cell. MCFC captures CO_2 and produces hydrogen and heat along with the generation of power. MFC treats waste materials like wastewater and other biological waste fluids while generating power. Regenerative fuel cells can generate power and produce hydrogen by electrolysis. These multi-functional roles of fuel cells are described in detail in this book. Fuel cell can also be operated in a hybrid mode with several combinations of NGSS-MCFC, SOFC-PEMFC, fuel cell with other renewable energy sources such as wind, hydro, solar, etc. Such hybrid combinations improve its energy efficiency, increase durability, and increase lifetime. These issues are discussed in detail in this book.

Fuel cells are currently used in numerous static and mobile applications. With more advancements in technology, it is predicted that fuel cell will find a major application in the future technology of cars and the vehicle industry. Fuel cell is ideally suited to replace ICE in the not-so-distant future.

One of the fastest growing industries in recent years is the portable, wearable, and implantable power electronics. This industry is largely distributed in nature and requires low power levels but for a large number and continuously increasing devices. With the advent of cell phones, mini laptop computers, wearable and implantable, medical and environmental devices and sensors, the growth in small-scale power electronics is enormous, and it is likely to grow even faster with the emphasis on Internet of Things and digitalization of almost all industries at small scale. In the past, a major method of supplying power to these devices was the use of battery or supercapacitors. Batteries are, however, heavy, require constant recharging, and are non-implantable. Medical and environmental devices and sensors, require micro-level power, and they operate at the scale where the use of battery may not be possible. One of the most novel advancements made in the power industry is the development of technologies such as TENG, PENG, and PyENG, TEG, solar cell, etc. to self-harvest the energy required for small-scale portable, wearable, and implantable devices. These technologies use triboelectricity, piezoelectricity, pyroelectricity, thermoelectricity, etc. as sources for energy. These technologies can operate at very small and flexible scales and function either individually or in a variety of hybrid forms. They use both mechanical and thermal sources of energy. Recent advancements made in the development of nanogenerators using concepts of TENG, PENG, etc. in individual and hybrid forms are also discussed in detail in this book.

Thus, wind, hydro, fuel cell, and self-powered nanogenerators along with their hybrids are the major sources of non-thermal power generation systems. In this book, we examine in detail advancements made in these sources of power generation. This book will particularly emphasize the technologies that have the best commercial possibilities and give generally accepted perspectives on future work that needs to be done in each technology. Similar to my previous book, this book should be very helpful to all researchers in academia, industry, and government who are interested in pursuing power generation technologies.

Author

Yatish T. Shah received his BSc in chemical engineering from the University of Michigan, Ann Arbor, and MS and ScD in chemical engineering from the Massachusetts Institute of Technology, Cambridge, USA. He has more than 40 years of academic and industrial experience in energy-related areas. He was chairman of the Department of Chemical and Petroleum Engineering at the University of Pittsburgh, Pennsylvania; dean of the College of Engineering at the University of Tulsa, Oklahoma, and Drexel University, Philadelphia; chief research officer at Clemson University, South Carolina; and provost at Missouri University of Science and Technology, Rolla, the University of Central Missouri, Warrensburg, and Norfolk State University, Virginia. He was also a visiting scholar at the University of Cambridge (UK) and a visiting professor at the University of California, Berkley, and Institut fur Technische Chemie der Universitat Erlangen, Nurnberg, Germany. He has written 14 books related to energy, most of which are under the *Sustainable Energy Strategies* book series (published by Taylor & Francis), for which he is the editor. He has also published more than 250 refereed reviews, book chapters, and research technical publications in the areas of energy, environment, and reaction engineering. He is an active consultant to numerous industries and government organizations in energy areas.

1 Introduction

1.1 WHAT ARE NONTHERMAL SOURCES FOR POWER GENERATION?

Generally, sources for power generation are divided into two categories: thermal and nonthermal. As shown in my previous book [1], thermal sources for power generation include combustion (of fossil fuels, biomass, waste, hydrogen, etc.), geothermal, solar, nuclear, and waste heat, and their appropriate hybrid combinations. Essentially, these are the sources that provide heat as a means of power generation. Heat can be converted to power in many different ways such as various thermodynamic cycles, photovoltaics (PV), thermo-PV, and thermoelectricity. These technologies are examined in detail in my previous book [1]. Combustion technology (particularly using fossil fuels) is the biggest source of carbon emission and needs to be either used with carbon capture or utilization or replacing fossil fuel raw materials with biomass, hydrogen, etc. In recent years, significant progress has been made in the use of geothermal and solar energy for utility-scale power production. Nuclear energy is yet to capture public acceptance in many parts of the world. Geothermal energy, while dispatchable, is so far not accessible in all parts of the world. More accessible technologies for enhanced or advanced geothermal systems are not yet fully developed so they can be used in any part of the world. Solar energy, while accessible, is intermittent and requires a storage device or another source of energy to make it dispatchable. The use of solar energy, both thermal and PV, is the fastest-growing industry both at the utility scale and at the distributed (like residential and industrial) level. Advances in all of these thermal sources for power generation are fully described in my previous book. The hybrid combination of solar and wind, which is very popular in recent days, was also covered in several previous books [1–3].

The energy industry is going through numerous paradigm changes. Centralized utility-scale power production is being increasingly replaced by distributed power generation that can be used for residential and commercial purposes and the transportation industry. Utility-scale grid is increasingly connected to localized microgrids or even nano-grids. Small-scale distributed power generation appears to be more suitable for some renewable and nuclear sources. The use of mechanical and electrochemical energy for power generation is capturing more attention. Wind energy is one of the fastest-growing industries. Novel distributed methods to capture different forms of hydro-energy (from rivers, ocean waves, tidal waves and currents, ocean salinity, and thermal gradients) are rapidly advancing. The use of large-scale hydro-dam for energy storage is still the most prominent. Since hydro-energy is dispatchable, its combination with solar and wind energy in hybrid forms is rapidly expanding. Both wind energy and hydro-energy are mechanical (nonthermal) sources of energy. They are described in full detail in this book.

DOI: 10.1201/9781003429906-1

Electrochemical processes as a source for power generation or energy storage are very flexible and versatile, and they are rapidly progressing at both stationary and mobile levels. The use of a fuel cell (FC) for distributed and stagnant power generation is capturing more and more attention. Unlike combustion processes using fossil fuels, a FC uses hydrogen as a source of fuel, which results in very low level or negligible carbon emission. The FC can be operated at low or high temperatures and can use a multitude of fuels including fossil fuels and biofuels to provide hydrogen it needs to generate power. FCs can be combined with other methods of power generation to reduce carbon emission, improve energy efficiency, and increase their own durability through a variety of combined hybrid systems. Some of the FCs provide multiple functions. For example, molten carbonate fuel cell (MCFC) can not only generate power but also capture carbon dioxide (CO_2) and produce hydrogen and heat. Microbial FC can be used for wastewater treatment along with power generation. Two major current issues with the FC are its cost and durability. Significant progress has been made to address these issues. FCs are currently used in numerous static and mobile applications. With more advances in technology, it is predicted that FC will find a major application in the future transportation industry replacing internal combustion engine for power generation. This book illustrates advances made in both low- and high-temperature FCs and their applications in great detail. The use of electrochemical processes in hydrogen and FC industry has a very bright future. The successful commercialization of FC will radically change the power industry.

Along with the nonthermal power generation technologies mentioned above, another fastest-growing industry is portable, wearable, and implantable power electronics. This industry is largely distributed in nature and requires low power levels for a large number of rapidly increasing devices. With the advent of cell phones, mini-laptop computers, and wearable and implantable medical and environmental devices and sensors, the growth in small-scale power electronics will be extensive and it is likely to grow even faster with the emphasis on the Internet of things and digitalization of almost all industries at small scale. In the past, a major method of supplying power to these devices was the use of battery or supercapacitors. Batteries are, however, heavy, require constant recharging, and are generally non-implantable. Medical and environmental devices and sensors require micro-level power, and they operate at a scale where the use of battery may not be possible. One of the most novel advances made in the power industry is the development of technologies such as triboelectric nanogenerator (TENG), piezoelectric nanogenerator (PENG), pyroelectric nanogenerator (PyENG), thermoelectric generator (TEG), and solar cell to self-harvest energy required for small-scale portable, wearable, and implantable devices. These technologies use triboelectricity, piezoelectricity, pyroelectricity, thermoelectricity, etc., as sources of energy. The technologies can operate at very small and flexible scales and function either individually or in a variety of hybrid forms. They use both mechanical and thermal sources of energy. Recent advances made in the development of nanogenerators (NGs) using the concepts of TENG, PENG, etc., in individual and hybrid forms are discussed in detail in this book.

Thus, wind, hydro-, FC, and self-powered NGs along with their hybrids are the major sources of nonthermal power generation systems. In this book, we examine the advances made in these sources of power generation in detail. This book will particularly emphasize the technologies that have the best commercial possibilities and give generally accepted perspectives on future work that needs to be done to make each nonthermal source of power generation more successful.

1.2 WIND POWER

While wind power has been the fastest-growing renewable energy industry over the last decade, going forward, advances in power generation from wind energy face a multitude of challenges. Future advances in wind power will occur in three distinct directions. First, significant advances will be made to make conventional land wind power more efficient, less costly, and more durable for changing needs of wind power for the utility grid. Wind energy is already an affordable and significant contributor to US electricity; it provided 8.4% of electricity production in 2020 [4], is now the largest US source of renewable energy [4], and accounted for 47.4% of new electricity capacity commissioned in the USA in 2020. There are challenges, however, to the continued deployment of traditional wind technology. In the next decade, with significant growth, many of the best wind resources on land will be developed, and the trend toward larger rotors and taller towers has made transportation logistics challenging. The advances will include more digitalization of wind power system, more use of variable-speed wind power, more sophisticated electrical and control systems, and better technologies for the insertion of intermittent wind power at a larger scale in the grid. Advances will also include methods to capture more efficient low-velocity wind for power, which will enhance the region for wind power capture. Advances in capturing wind energy require a better understanding of atmospheric flow physics. These include the impact of atmospheric turbulence on the performance and loads of wind turbines. To achieve optimal wind turbine performance and reliability, the industry will require better characterization of turbulence and its effects under the wide range of atmospheric conditions in which wind plants are expected to generate power continuously and reliably. Furthermore, wakes, or regions of slower and more turbulent air downwind of wind turbines, need to be better understood, as does the impact on local climates that large-scale deployment of wind energy may introduce.

There are also new challenges in understanding advanced wind turbine system dynamics and materials. The size and flexibility of modern wind turbines have pushed design beyond where assumptions and modeling tools were first established, which creates unprecedented risks. Researchers lack the experimental data at the large scale necessary to validate the models and materials used to develop innovative solutions for future wind energy systems. While modern wind turbines have become by far the largest rotating machines on earth, a renewed interest in small wind turbines is fostering energy transition and smart grid development. Small machines have traditionally not received the same level of design refinement as their larger counterparts, resulting in lower efficiency, lower capacity factors, and therefore a higher cost of energy. The future of air wind systems depends on an advanced understanding of small-scale wind turbine dynamics.

Wind energy is generally captured by so-called wind farms. Managing the airflow through wind power plants is a complex challenge but offers opportunities to evolve optimal plant design, enhance production, lower maintenance costs, and provide the controllability demanded by the larger energy system. A grid dominated by wind energy and solar power will impose system needs that will also challenge how we approach the design of individual turbines, wind power plants, hybrid power plants, and the grid itself.

Advances in wind energy also require better implementation of digital technology and a better understanding of related environmental and social issues. A future in which digitalization has made data accessible in the right places and at appropriate times has many valuable outcomes, but significant technical and cultural impediments must be resolved before this aspirational goal for wind energy can be achieved. Environmental research must define the wildlife and habitat impacts of large-scale deployment in collaboration with the engineering of wind turbines and plants. The social aspects of how wind plants interact with the communities in which they are built and the communities that are served by low-cost clean electricity need to be addressed. Solutions will need to evolve beyond assessments of acceptance to include engagement in planning and design processes and different ownership structures to embrace the transition as a shared task among members of society.

The second major direction for advances in the wind power industry will be offshore wind energy, in particular, capturing wind power over deep waters. Offshore wind speeds tend to be faster than on land. Small increases in wind speed yield large increases in energy production: A turbine in a 15-mph wind can generate twice as much energy as a turbine in a 12-mph wind. Faster offshore wind speeds mean much more energy can be generated. Offshore wind speeds tend to be steadier than on land. Wind speed and directions are more predictable on the water than on land. A steadier supply of wind means a more reliable source of energy. Many coastal areas have very high-energy needs. Half of the USA' population lives in coastal areas, with concentrations in major coastal cities. Building offshore wind farms in these areas can help meet those energy needs from nearby sources. Offshore wind farms have many of the same advantages as land-based wind farms—they provide renewable energy; they do not consume water; they provide a domestic energy source; they create jobs; and they do not emit environmental pollutants or greenhouse gases (GHGs).

Offshore wind farms can be expensive and difficult to build and maintain. In particular, it is very hard to build robust and secure wind farms in water deeper than around 200 feet (~60 m), or over half a football field's length. Although coastal waters off the east coast of the USA are relatively shallow, almost all of the potential wind energy resources off the west coast are in waters exceeding this depth. Floating wind turbines are beginning to overcome this challenge. Wave action, and even very high winds, particularly during heavy storms or hurricanes, can damage wind turbines. The production and installation of power cables under the seafloor to transmit electricity back to land can be very expensive. The effects of offshore wind farms on marine animals and birds are not fully understood. Offshore wind farms built within view of the coastline (up to 26 miles offshore, depending on viewing conditions) may be unpopular among local residents and may affect tourism and property values.

The design, manufacture, and operation of offshore wind assets have their own set of challenges including corrosion, fatigue, erosion, lightning strikes, and biofouling. Addressing these challenges and maintaining the operational availability of offshore wind turbines will become increasingly important as the reliance on offshore wind energy grows. As offshore wind farms move to increase water depths and aim to operate with larger turbines, foundation designs will be needed to adapt accordingly. In recent years, this has led to both increased monopile sizes and an increased interest in the use of jacket structures, both of which present new manufacturing challenges. There is a drive for cost reduction through high-throughput fabrication, and for that, advanced manufacturing methods are required.

The aggressive marine environment means that the monopile foundations are subject to both internal and external corrosions. Internal corrosion can be exacerbated by a limited exchange of trapped seawater, while intermittent electrolytic contact can cause extensive corrosion at the splash and tidal zone. Added to this are concerns surrounding microbial-induced corrosion and biofouling. There is also a requirement for high-visibility coatings at the transition piece. However, conventional paint systems can suffer from both damage and ultraviolet (UV) degradation, requiring expensive maintenance. There are challenges relating to fatigue, including the effect of loading during the initial piling operations and the cyclic loading of the structure from wind and waves. These fatigue difficulties can be intensified by the composition of the seabed and any developing biofouling, which increases the hydrodynamic load and creates challenges in routine inspection and maintenance.

Significant improvements in efficiency and reductions in cost have been achieved through the use of larger turbine blades, with the next generation of composite blade structures expected to be over 100 m in length. However, the move to ever larger blades can also create logistical barriers. Manufacturers face challenges with transporting the blades to installation sites and are considering segmented blade designs, which can be bonded on-site before final installation. The proposed increased size of turbine blades is limited by weight, meaning that lighter materials such as thermoplastic foams and alternative composites are being considered. Lighter blades allow for easier installation and repair and improve performance. However, there are inherent difficulties with composite manufactures, such as the misalignment of fibers and inconsistent resin distribution, which can lead to lowered fatigue strength.

The impact of fatigue on turbine blades is an ongoing challenge, with each blade being subjected to more than 100 million loading cycles for its lifetime. The cyclic loading of the blades is also worsened by leading edge erosion and ice buildup. Leading edge erosion is caused by the repeated impact of rain, ice, and particulate matter, which leads to a loss of aerodynamic efficiency and can compromise the structural integrity of the blades, leading to water ingress and UV damage. Even a small amount of leading edge erosion can result in a ~5% drop in annual energy production.

The increased height of turbines and the span of blades both raise the risk of lightning strikes and the cost of repair. Lightning strikes can result in the loss of turbine blades and damage to electrical systems. Although there are existing lightning strike protection systems, failure can still occur due to moisture ingress, the detachment of diverter strips, and the erosion of blade surfaces, among other factors. The issues outlined above for larger blades and taller towers apply to both onshore and offshore wind turbines.

As the number—and size—of wind farms increases and their distance from the shore grows, so too do the challenges associated with connecting wind farms to the grid. Some of the main challenges that are facing the industry related to offshore wind farm export cables are (1) supply chain constraints for the manufacturing of export cables, (2) routing the export cables to an onshore point of interconnection, (3) protecting the export cables, and (4) minimizing the environmental impact of installed cables.

There are two types of basic technology for offshore wind transmission: high-voltage alternating current (HVAC) and high-voltage direct current (HVDC). HVDC is the preferred option for larger offshore wind projects, especially as they move farther away from the shore. This is because HVDC cables are capable of longer transmission distances with lower losses compared with HVAC cables. The insurance for protecting export cables is very expensive. Mechanically protecting submarine cables can assist with reducing damage from anchors, fishing gear, dredging, and more. This requires a detailed geophysical and geotechnical surveying of the cable route, along with a cable burial risk analysis. If cable burial is not possible, then other means such as articulated concrete mattresses or rock dumping are employed. The submarine cable route may also need to cross over other linear infrastructure assets such as communications cables, other power cables, or pipelines. In these instances, it is important to engage with the asset owners as early as possible to develop cable crossing requirements and agreements.

Submarine cable monitoring systems can identify preventive maintenance requirements or serve as an early warning of potential cable damage. There are several monitoring systems, and teams can also use a remotely operated vehicle (ROV) to inspect the cables. The primary environmental concerns with offshore wind transmission are the impact on the seafloor, sensitive coastal environments, and other marine life. There is also concern with potential conflict with other ocean uses. Installing electric transmission cables typically involves cable burial or cable protection methods. When burying cables in soft sediments, teams often employ water jetting methods to lay and bury cables approximately 2 m deep. Offshore wind projects also must consider avoiding sensitive habitats wherever possible.

Cable installation must also account for other ocean uses such as fishing interests, sand borrow areas, artificial reefs, and navigation channels. Burial must be at a sufficient depth in or around these areas to avoid direct contact and potential impacts. The use of HVDC cable systems also reduces the number of cables as compared to HVAC systems required for a given amount of capacity. This helps to reduce the overall impact on the environment.

The third major direction for advances in wind energy is the development of airborne wind power. Airborne wind technology replaces the support structure with a lightweight tether, reducing mass by around 90% [4], which may lead to lower lifecycle emissions and lower visual footprint. The tether allows airborne wind to harvest wind at higher altitudes, which can be stronger and more consistent, and to adjust its flight path to (1) find the optimal height for power production, (2) control its influence on neighboring airborne devices, and (3) mitigate airspace use conflicts or viewshed concerns—including landing if needed. Airborne wind can be deployed or redeployed quickly and lowered to the ground for maintenance, potentially opening new markets and improving technician safety. The lower mass inputs, easier logistics, and potentially higher-capacity factor (fraction of time that the system produces the rated amount of power) may lead to a reduced cost of energy compared with traditional wind technology in certain regions or markets.

The International Renewable Energy Agency called AWE a potential "game changer" and ranked the technology third most important below next-generation offshore turbines and floating foundations. The projected commercialization timeline for airborne wind is vague, spanning nearly a decade from 2024 to 2033. AWE technology is still relatively immature, and much remains to be done. AWE technologies can be classified into two basic categories: *Ground-Gen* and *Fly-Gen* systems. In *Ground-Gen,* AWE systems generate electricity by flying in figure-eight or circular crosswind trajectories while spiraling upward and reeling out the tether that connects the kite to a winch-generator unit on the ground station. *In Fly-Gen,* AWE systems fly onboard turbine–generator units that are connected to an airframe along figure-eight or circular crosswind trajectories defined by the tether length. The potential for reduced levelized cost of energy (LCOE) has not been validated given that the first commercial units were planned for deployment by SkySails Power in 2021 [4]. The reliability and availability of airborne wind energy system (AWES) over many weeks, months, or years have not been demonstrated. Further, if any failure occurs, the device cannot stop midair and wait for help; it must be fail-safe, returning safely to the ground and avoiding any personnel or property nearby. Robust automatic launch and recovery and adequate protection against extreme weather (e.g., high winds, gusts, or lightning) must be developed for utility-scale deployment. There are regulatory and siting concerns related to noise, wildlife impacts, radar mitigation, airspace use, and grid compliance, which will need to be studied and addressed with the appropriate stakeholders. AWES must be more than simply cost-competitive; if the energy harvesting characteristics of mature AWES are not differentiated from traditional wind turbines in some way, the sector may not achieve significant market penetration [4]. None of these challenges are viewed as insurmountable, but they must be addressed for the successful application of the technology. AWE is an incredibly diverse field of technologies with each having its own strengths and weaknesses. Since the sector is still rapidly evolving, a combination of AWES architectures or an entirely new architecture may emerge that enhances a benefit or mitigates a challenge attributed to airborne wind.

The three directions of wind power growth show an increasing path of high-risk high-reward direction as one moves from land-based wind power to offshore wind power to airborne wind power. The level of maturity of technology also decreases in three directions. However, each presents its own set of opportunities and challenges. Besides these three directions of advances in wind power, in the future, wind power will be more and more hybridized with other sources of energy and energy storage, particularly as its role in the grid increases. Hybridization with solar power is already occurring at an increasing rate as outlined in my previous books [1–3]. Wind energy can also be effectively hybridized with hydropower and FCs as shown later in this book.

1.3 HYDROPOWER

Just like wind power, the growth in hydropower is manifested in three different ways. **First,** the large-scale dam-based hydro-energy for generation and storage of power will continue to grow throughout the world, despite more in developing countries and some other parts of the world then the USA. Large-scale dam hydropower and pumped storage hydropower (PSH) compose a significant fraction of renewable generation and electrical storage in the USA. The USA has approximately 101 gigawatts (GW) of nameplate hydropower capacity, including 80 GW of conventional hydropower and 21 GW of pumped storage. Hydroelectric generation in the USA represented 41% of all renewable energy generation and 6.8% of the total generation in the USA in 2018. At least 40% of the hydropower capacity is composed of PSH and "peaking" hydropower plants, which can store water to produce electricity at times of greatest need and value. PSH represents 95% of grid-scale energy storage in the USA. While total hydropower capacity appears to be growing due to retrofits, values from large-scale hydropower resources appear to be flat for at least the last 20 years. Proportionately, hydropower resources are decreasing in dominance. In 2019, annual wind generation surpassed annual hydroelectric generation. Battery storage growth domestically is growing rapidly, while only one pumped storage facility has been built in the last 15 years.

These static annual generation and total capacity values for hydropower resources, however, do not reflect the dynamic contributions of this sector to electric grid reliability, economic efficiency, and resilience. The economic and financial context for hydropower resources has undergone significant changes in recent years due to changing power grid conditions, such as low natural gas prices and low load growth leading to a reduction in energy prices. The rapid growth of renewable energy resources has led to the displacement of some baseload-scheduled energy generators while putting a greater emphasis on ancillary services. Further changes are on the horizon—technical, technological, sociopolitical, and market structures—that can substantially change the operational requirements of the power grid. The varying degrees to which these changes eventually manifest will affect the value drivers for hydropower resources differently. Evaluating the changing value landscape for hydropower resources and understanding the ability of different resources to provide power system services will enable prudent decisions regarding changes in operating paradigms and capital investments.

The new operating paradigms are creating opportunities well suited for hydropower installations to provide valued power system services. The flexibility of hydropower plants is already being used in many regions of the country to provide services needed to integrate renewables and maintain the reliability of grid operations. The opportunities, however, may also present additional costs, such as accelerated machine wear and tear due to frequent cycling and start–stop operations. This implies that asset management and reinvestment programs need to consider changing operational paradigms in order to ensure prudent long-term investments. These changing conditions also require new research into technology innovation, data development, analytical tools, and operational strategies to preserve and enable important hydropower capabilities and contributions into the future.

Various government and industrial organizations are focused on the review of the current hydropower operations landscape. The value of a resource can be discerned from the composition of the portfolio of services (energy, capacity, and ancillary services) it provides and the relative value of each of those services. This task illustrated the recent trends in the provision of grid services by hydropower resources based on a comprehensive, data-driven analysis of hydropower operations in various markets across the country. These trends shed some light on the impacts of further changes on the horizon relative to hydropower operations and value. The work also identified and estimated the value of grid services provided by hydropower that are not currently monetized, such as inertia.

Efforts are also being made to examine hydropower capabilities and operations in future grid states. The ability of hydropower resources to provide value to the power system will require a comprehensive understanding of the resources' technical and technological capabilities, costs, and constraints. Hydropower's capabilities to provide valued grid services, and the factors influencing how these capabilities vary unit-to-unit and plant-to-plant, are qualitatively and quantitatively different. This task was designed to analyze the capabilities and constraints that affect a hydropower facility's ability to provide various grid services, both now and in the future.

Hydropower operations are changing in many parts of the country because of changing grid conditions. The changes, however, differ regionally based on the prevailing value drivers, such as changing arbitrage patterns due to the increasing penetration of solar resources. In some regions, these changes manifest as new market opportunities, such as the Western Energy Imbalance Market (EIM), which is designed to better incorporate the penetration of variable renewable energy (VRE). However, in some parts of the country, such as the Pacific Northwest, hydropower resources continue to operate primarily in load-following mode. Even with these changing conditions, the capacity factor for conventional hydropower resources across the USA has remained relatively consistent through the years, between 35% and 45%.

Hydropower generators are important contributors to grid reliability. Even in a changing power system, hydropower continues to be a significant contributor to system reliability through inertial and primary frequency responses, reactive power support, and black-start capabilities. Approximately 40% of units maintained and tested for providing black start in the USA are hydropower turbines, even though hydropower makes up only approximately 10% of overall US generating capacity. In addition, in California independent system operator (CAISO), hydropower resources have been observed to contribute up to 25% of the total regulation reserve (up and down) requirements, as well as up to 60% of the total spinning reserve requirements even though hydropower constitutes approximately 15% of installed capacity. Not all generators currently have these capabilities; the ability of inverter-based resources is currently being evaluated in laboratory and field demonstrations, and other traditional generators that supply these services may retire from service in the future, which may increase the demand for certain reliability services from hydropower.

There is wide variation in hydropower plant conditions and capabilities to provide grid services. Hydropower's contributions to the grid are multifaceted, in that hydropower may serve several roles in a generating stack. At least 40% of hydropower resources, by capacity, comprise pumped storage and "peaking" hydropower plants that can store water to produce electricity at times of greatest need and value, and at least 18% comprise run-of-river plants, which may have some operational flexibility but typically cannot impound and store additional water beyond inflows. Even within a given resource class, i.e., peaking or run-of-river, the ability to provide grid services depends on the site-specific electromechanical (physical) attributes, which are in turn governed by the hydrological and geological conditions. For a given plant, these capabilities will naturally vary over seasons and water years (wet/average/dry). In addition, institutional factors such as existing contracts and Federal Energy Regulatory Commission (FERC) licenses determine a resource's ability to provide grid services. In many cases, non-power services such as flood control or environmental flows govern the ability of a hydropower plant to supply energy. The value of these services is not always accounted for in the overall value of hydropower. Additionally, the value of these non-power services is locational, based on stakeholder perspectives and the valuation methodologies that are employed.

Traditional economics for hydropower plants may not provide stable revenue in the future. Conventional value streams, such as energy and ancillary services prices, are exhibiting declining trends in some parts of the country, and these changes have affected hydropower resources adversely. This trend is evident in conventional hydropower plants in the Northeast and in PSH plants in the Midwest. Energy generation remains the primary source of revenue for many hydropower plants, and while most hydropower is technically capable of providing ancillary services, this provision often includes opportunity costs associated with reduced capacity. New market mechanisms are emerging that could compensate for hydropower flexibility. While not all services that hydropower provides are currently monetized, new markets for grid services are emerging that can offer alternative revenue streams.

Just like advances in wind power, advances in hydropower will also occur in three different directions. First, in the near future, significant advances in large-scale hydropower will occur in many different ways in order to adjust to the new reality described above. The role of hydropower in balancing the grid with intermittent renewable power will require variable-speed hydropower, which can follow variable load requirements. Advanced control mechanisms to handle the transient nature of hydropower operation will become more important. New and innovative drive trains, converters, and other electrical components will be needed to satisfy changing needs. Most importantly, digitalization of hydropower will be a new reality to increase its efficiency, decrease cost for maintenance and operation, and optimize performance. New innovations in PSH will also take a front seat. New applications of large-scale hydropower include the use of hybrid hydropower with a combination of floating PV power with hydropower in many different ways and a combination of wind power with hydropower. In both cases, a combination of intermittent power with dispatchable power renders more grid stability. Significant theoretical work for hybrid hydropower is also being pursued.

The second direction for growth in hydropower will be energy recovered from freshwater river and streams. This can be done in two ways: first from the development of small and microscale hydropower using "run-of-river" or small stream dams or impediments and second by harnessing hydrokinetic energy from river tidal waves and kinetic energy from flowing river and streams. In the USA, at least 18% is composed of run-of-river plants, which can only produce electricity when water is flowing. *Small hydro* is the development of hydroelectric power on a scale suitable for local community and industry, or to contribute to distributed generation in a regional electricity grid. Small hydro-projects may be built in isolated areas that would be uneconomic to serve from a national electricity grid, or in areas where a national grid does not exist. The use of the term "small hydro" varies considerably around the world, and the maximum limit is usually somewhere between 10 and 30 megawatts (MW). While a minimum limit is not usually set, the US National Hydropower Association specifies a minimum limit of 5 MW. In California, hydroelectric generating stations with a maximum capacity of less than 30 MW are classified as small and are eligible for inclusion in the state's renewable portfolio standard, while hydroelectric generating stations with a higher capacity are classified as large and are not considered renewable. The "small hydro" description may be stretched up to 50 MW in the USA, Canada, and China. In India, hydro-projects up to 25 MW station capacities have been categorized as small hydropower (SHP) projects. Small hydro can be further subdivided into mini-hydro, usually defined as 100–1,000 kilowatts (kW), and micro-hydro, which is 5–100 kW. Micro-hydro is usually the application of hydroelectric power sized for smaller communities, single families, or small enterprises. The smallest installations are pico-hydro, below 5 kW. Since small hydro-projects usually have correspondingly small civil construction work and little or no reservoir, they are seen as having a relatively low environmental impact compared with large hydro. Even though the output capacity is small, capacity can be increased by an array or modular installation.

The kinetic energy of flowing water from a river or stream can also be harvested and converted to power by a suitable hydrokinetic system. Such a system is easy to transport and relocate due to the small size of the plant. Moreover, the system can be installed along the riverside either mooring to a fixed structure or on a floating pontoon. The system will require different types of turbines and generators. Despite its benefits, enormous research efforts are still necessary to improve hydrokinetic technology, especially for energy conversion applications. Areas of research that require further attention include (but are not limited to) turbine selection and enhancement, assessment studies, energy conversion efficiency, and environmental impacts. Hydrokinetic systems continue to receive significant attention from researchers in order to improve the technology, reduce the barriers to implementation, gain further insights, and understand the limitations of the technology.

The third direction of growth in hydropower is capturing various forms of energy from the ocean normally known as marine energy and converting them into power. This includes the energy carried by ocean waves, tides, salinity, and ocean temperature differences. The movement of water in the world's oceans creates a vast store of kinetic energy, or energy in motion. Some of this energy can be harnessed to generate electricity to power homes, transport, and industries. The term marine energy encompasses both wave power, i.e., power from surface waves, and tidal power, i.e., obtained from the kinetic energy of large bodies of moving water. Offshore wind power is not a form of marine energy, as wind power is derived from the wind, even if the wind turbines are placed over water.

The oceans have a tremendous amount of energy and are close to many if not most concentrated populations. Ocean energy has the potential of providing a substantial amount of new renewable energy around the world. Electricity generation from marine technologies increased by an estimated 16% in 2018 and by an estimated 13% in 2019. The global potential of marine energy is outlined in Table 1.1

Ocean power can be generated by salinity gradients. At the mouth of rivers where freshwater mixes with salt water, the energy associated with the salinity gradient

TABLE 1.1
Global Potential of Marine Power [5]

Global Potential

Form	Annual Generation
Tidal energy	>300 TWh
Marine current power	>800 TWh
Osmotic power salinity gradient	2,000 TWh
Ocean thermal energy thermal gradient	10,000 TWh
Wave energy	8,000–80,000 TWh

Source: International Energy Agency-Ocean Energy System (IEA-OES), Annual Report 2007 [6].

can be harnessed using the pressure-retarded reverse osmosis process and associated conversion technologies. Another system is based on using freshwater upwelling through a turbine immersed in seawater, and one involving electrochemical reactions is also in development. Significant research took place from 1975 to 1985 and gave various results regarding the economy of pressurized reverse osmosis (PRO) and reverse electrodialysis (RED) plants. It is important to note that small-scale investigations into salinity power production take place in countries such as Japan, Israel, the USA, Norway, and the Netherlands.

Ocean thermal energy conversion (OTEC) uses the ocean thermal gradient between cooler deep and warmer shallow or surface seawaters to run a heat engine and produce useful work, usually in the form of electricity. OTEC can operate with a very high capacity factor and so can operate in baseload mode. Among ocean energy sources, OTEC is one of the continuously available renewable energy resources that could contribute to the baseload power supply. The resource potential for OTEC is considered to be much larger than for other ocean energy forms. Up to 88,000 TWh/yr of power could be generated from OTEC without affecting the ocean's thermal structure.

Systems may be either closed cycle or open cycle. Closed-cycle OTEC uses working fluids that are typically thought of as refrigerants such as ammonia or R-134a. The most commonly used heat cycle for OTEC to date is the Rankine cycle, using a low-pressure turbine. Open-cycle engines use vapor from the seawater itself as the working fluid. OTEC can also supply quantities of cold water as a byproduct.

Tidal power or *tidal energy* is harnessed by converting energy from tides into useful forms of power, mainly electricity using various methods. Although not yet widely used, tidal energy has the potential for future electricity generation. Tidal forces result from periodic variations in gravitational attraction exerted by celestial bodies. These forces create corresponding motions or currents in the world's oceans. This results in periodic changes in sea levels, varying as the earth rotates. These changes are highly regular and predictable, due to the consistent pattern of the earth's rotation and the Moon's orbit around the earth. Among sources of renewable energy, tidal energy has traditionally suffered from relatively high cost and limited availability of sites with sufficiently high tidal ranges or flow velocities, thus constricting its total availability. However, many recent technological developments and improvements, in both design (e.g., dynamic tidal power (DTP), and tidal lagoons) and turbine technology (e.g., new axial turbines and cross-flow turbines), indicate that the total availability of tidal power may be much higher than previously assumed and that economic and environmental costs may be brought down to competitive levels. Tidal power can be classified into four generating methods.

Tidal stream generators make use of the kinetic energy of moving water to power turbines, in a similar way to wind turbines that use the wind to power turbines. Some tidal generators can be built into the structures of existing bridges or are entirely submersed, thus avoiding concerns over esthetics or visual impact. Land constrictions such as straits or inlets can create high velocities at specific sites, which can be captured using turbines. These turbines can be horizontal, vertical, open, or ducted. Figure 1.1 illustrates the world's first commercial-scale and grid-connected tidal stream generator—SeaGen—in Strangford Lough.

Tidal barrages use potential energy in the difference in height (or hydraulic head) between high and low tides. When using tidal barrages to generate power, the potential energy from a tide is seized through the strategic placement of specialized dams. When the sea level rises and the tide begins to come in, the temporary increase in tidal power is channeled into a large basin behind the dam, holding a large amount of potential energy. With the receding tide, this energy is then converted into

FIGURE 1.1 World's first commercial-scale and grid-connected tidal stream generator—SeaGen—in Strangford Lough. The strong wake shows the power of the tidal current [7].

mechanical energy as the water is released through large turbines that create electrical power through the use of generators. Barrages are essentially dams across the full width of a tidal estuary.

DTP is a theoretical technology that would exploit an interaction between potential and kinetic energies in tidal flows. It proposes that very long dams (e.g., 30–50 km length) be built from coasts straight out into the sea or ocean, without enclosing an area. Tidal phase differences are introduced across the dam, leading to a significant water-level differential in shallow coastal seas—featuring strong coast-parallel oscillating tidal currents such as those found in the UK, China, and Korea. Induced tides (TDP) could extend the geographic viability of a new hydro-atmospheric concept lunar pulse drum (LPD) discovered by a Devon innovator in which a tidal "water piston" pushes or pulls a metered jet of air to a rotary air actuator and generator. The principle was demonstrated at London Bridge in June 2019.

A tidal lagoon is a new tidal energy design option to construct circular retaining walls embedded with turbines that can capture the potential energy of tides. The created reservoirs are similar to those of tidal barrages, except that the location is artificial and does not contain a preexisting ecosystem. The lagoons can also be in the double (or triple) format without pumping or with pumping that will flatten out the power output. The pumping power could be provided by excess grid demand renewable energy from, for example, wind turbines or solar PV arrays. Excess renewable energy rather than being curtailed could be used and stored for a later period of time. Geographically dispersed tidal lagoons with a time delay between peak productions would also flatten out peak production providing near baseload production at a higher cost than other alternatives such as district heating renewable energy storage.

Historically, *tide mills* have been used both in Europe and on the Atlantic Coast of North America. The incoming water was contained in large storage ponds, and as the tide goes out, it turns waterwheels that use mechanical power to mill grain. The world's first large-scale tidal power plant was France's Rance Tidal Power Station, which became operational in 1966. It was the largest tidal power station in terms of output until Sihwa Lake Tidal Power Station opened in South Korea in August 2011. The Sihwa Station uses sea wall defense barriers complete with ten turbines generating 254 MW.

Wave power is the capture of energy of wind waves to do useful work—for example, electricity generation, water desalination, or pumping water. A machine that exploits wave power is a *wave energy converter* (WEC). Waves are generated by wind passing over the sea's surface. As long as the waves propagate slower than the wind speed just above, energy is transferred from the wind to the waves. Air pressure differences between the windward and leeward sides of a wave crest and surface friction from the wind cause shear stress and wave growth.

Wave power is distinct from tidal power, which captures the energy of the current caused by the gravitational pull of the Sun and Moon. Other forces can create currents, including breaking waves, wind, the Coriolis effect, cabbeling, and temperature and salinity differences. As of 2022, wave power is not widely employed for commercial applications, after a long series of trial projects. In 2000, the world's first commercial wave power device, the Islay Land Installed Marine Power Energy Transmitter (LIMPET), was installed on the coast of Islay in Scotland and connected to the National Grid. In 2008, the first experimental multi-generator wave farm was opened in Portugal at the Aguçadoura Wave Park. WECs can be classified based on their working principle. Novel devices to capture energy from the ocean are constantly being pursued. Power from tidal and wave energy has a good chance of being commercially successful. This book examines recent advances made toward all three directions of future growth in hydropower.

1.4 FUEL CELLS FOR POWER GENERATION

According to many experts, we may soon find ourselves using FCs to generate electrical power for all sorts of objects we use every day. A FC is a device that uses a source of fuel, such as hydrogen, and an oxidant to create electricity from an electrochemical process. Much like the batteries that are found under the hoods of automobiles or in flashlights, a FC converts chemical energy to electrical energy All FCs have the same basic configuration, an electrolyte and two electrodes, but there are different types of FCs, based mainly on what kind of electrolyte they use. Many combinations of fuel and oxidant are also possible. The fuel could be diesel or methanol, while air, chlorine, or chlorine dioxide may serve as oxidants. Most FCs in use today, however, use hydrogen and oxygen as the chemicals. FC can be operated at low, medium, or high temperatures. FCs have three main applications: transportation, portable uses, and stationary installations.

In the future, FCs could power our cars, with hydrogen replacing the petroleum fuel that is used in most vehicles today. Many vehicle manufacturers are actively researching and developing transportation FC technologies. Stationary FCs are the largest, most powerful FCs. They are designed to provide a clean, reliable source of on-site power to hospitals, banks, airports, military bases, schools, and homes. FCs can power almost any portable device or machine that uses batteries. Unlike a typical battery, which eventually goes dead, a FC continues to produce energy as long as fuel and oxidant are supplied. Laptop computers, cellular phones, video recorders, and hearing aids could be powered by portable FCs. FCs have strong benefits over conventional combustion-based technologies currently used in many power plants and cars. They produce much smaller quantities of GHGs and none of the air pollutants that create smog and cause health problems. If pure hydrogen is used as a fuel, FCs emit only heat and water as byproducts. Hydrogen-powered FCs are also far more energy efficient than traditional combustion technologies.

Global electricity systems are currently witnessing a paradigm shift from traditional centralized to distributed generation technologies One method to provide distributed energy is to use microgrids, which are focused on local needs. With appropriate use of energy sources, microgrids can provide energy needs and security along with environmental compatibility [1–3]. Microgrids are best served by the use of diversified electrical energy production resources beyond the current solar, wind, hydro, biomass, diesel, and battery technologies. Interestingly, FC systems are considered promising energy resources based on being clean, pollution-free, and efficient, including their potential to store higher calorific value, in the hydrogen form, compared with the chemical energy that may be stored using most other materials [1–3], and the capability to supply energy for a relatively long time. Since microgrids are often used for distributed energy, the use of FC technologies for microgrid system applications has gained serious interest. This is often done in conjunction with other renewable sources since FC can also provide energy storage capabilities.

FCs carry many benefits. Gasoline- and diesel-powered vehicles emit GHGs, mostly CO_2, which contribute to climate change. FC vehicles (FCVs) powered by pure hydrogen emit no tailpipe GHGs, only heat and water. Producing hydrogen for FCVs can generate GHGs, depending on the production method. Still, it generates much fewer GHGs than conventional gasoline and diesel vehicles. FCVs could reduce our dependence on foreign oil since hydrogen can be derived from domestic sources. These sources include natural gas and coal, as well as renewable sources such as water, biogas, and agricultural waste, which would make our economy less dependent on other countries and less vulnerable to oil price shocks. Highway vehicles emit much of the air pollutants that contribute to smog and harmful particulates in the USA. FCVs powered by pure hydrogen emit no harmful pollutants. Producing hydrogen from fossil fuels causes some pollutants, but much less than the amount generated by conventional vehicles. FCs are highly efficient, particularly when utilized in cogeneration mode, and FCs can attain over 80% energy efficiency. They provide reliable quality of power as long as they do not degrade over time. They are silent, significantly lighter, and more compact.

The development of FC to a commercial level, however, faces several challenges. Only a few models are now available for sale or lease, and availability is limited to areas with hydrogen fueling stations, mostly in California. Several challenges must be overcome before FCVs will be a successful, competitive alternative for consumers.

FCVs are currently more expensive than conventional vehicles and hybrids. They are expensive to manufacture due to the high cost of catalysts (platinum). Their costs, however, have decreased significantly and are approaching Department of Energy's (DOE's) goal for 2020 (see Figure 1.2). There is a lack of infrastructure to support the distribution of hydrogen. A lot of the currently available FC technology is in the

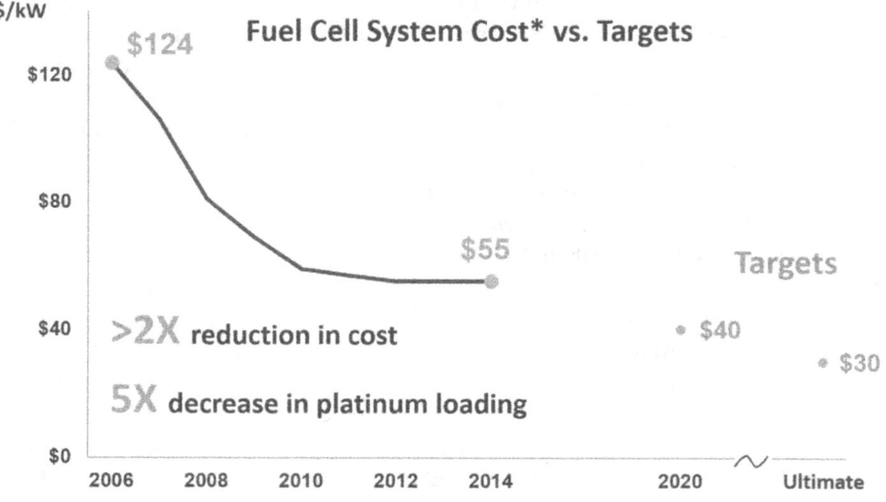

FIGURE 1.2 United States Department of Energy (USDOE), Fuel Cell Technologies Office, Fuel Cell Technologies Office Accomplishments and Progress [8,9].

prototype stage and not yet fully validated. Hydrogen is expensive to produce and not widely available. Car makers must continue to lower costs, especially for the FC stack and hydrogen storage, for FCVs to compete with conventional vehicles. The current infrastructure for producing and getting hydrogen to consumers cannot yet support the widespread adoption of FCVs. In 2013, Hydrogen-USA (H2USA)—a partnership program between DOE, other federal agencies, automakers, state governments, academic institutions and additional stakeholders—was launched as a public–private partnership between DOE and other federal agencies, automakers, state government, academic institutions, and additional stakeholders. Its goal is to coordinate research and identify cost-effective solutions for deploying hydrogen infrastructure. By the end of 2015, more than 50 public stations were available, mostly in California. This is an important first step for making hydrogen available to consumers. FC systems are not yet as durable as internal combustion engines, especially in some temperature and humidity ranges. On-road FC stack durability is currently about half of what is needed for commercialization. Durability has increased substantially over the past few years from 29,000 to 75,000 miles, but experts believe that FCVs must achieve a 150,000-mile expected lifetime to compete with gasoline vehicles. FCVs must be embraced by consumers before their benefits can be realized. As with any new vehicle technology, consumers may have concerns about the dependability and safety of these vehicles when they first hit the market. Furthermore, consumers must become familiar with a new kind of fuel. Public education can accelerate this process.

A FC provides electricity via an electrical circuit with a DC load. Problems arise when FCs are manufactured. The electrochemical reaction needs a consistent area of contact, while normally FCs have a very small area of contact between the electrolyte, the sites of the electrode, and the flows of reactant gases. Moreover, the geometric distances between the electrodes introduce resistance in the FC operation, reducing the production of electricity. Therefore, to address these problems, FCs have been manufactured with improved design and new approaches to tackle these issues. Among the common solutions, there is the adoption of porous electrodes and a thin electrolyte, to reduce the electrical resistance. The porous structure allows interaction among gas, ions, and electrolyte molecules to occur more effectively, improving the electrochemistry of the cell. In this way, the contact area is maximized, guaranteeing better performance, efficiency, and current production.

By considering a hydrogen/oxygen FC operation, hydrogen flows and reacts at the anode (negative electrode), while oxygen flows and reacts at the cathode (positive electrode). Through an electrochemical reaction, hydrogen is split into an electron and a proton, producing electricity for a given load and generating the harmless byproduct of water. The use of platinum catalysts for electrodes is problematic, and future research needs to find alternate cheaper materials.

There is also the need to have several policy changes that can facilitate the commercialization of this technology, i.e., standardization, safety codes and analysis, and regulations and best practices for hydrogen production, distribution, and dispensing. As mentioned above, the major disadvantage of the FC is that the technology currently presents a more expensive capital expenditure than other forms of power conversion. Once this barrier is overcome, due to the economy of scale and the adoption of cost reduction actions, FCs will eventually become a dominant and efficient solution for energy conversion. By considering the current state of the art, FCs operate with an electric efficiency of about 40%–50% and overall efficiency in cogeneration assets (production of combined heat and power) of more than 80%. Their performance is indeed higher if compared to combined heat and power (CHP) internal combustion engines. FCs have no moving parts, such as pistons, and for specific FC types, most of the components are entirely made of solids, which simplifies the manufacturing process. Depending on the FC type and the supplied fuel, the emissions can vary, but fall below the existing standards of emissions. Generally, a FC system emits "<1 ppm of NO_x, 4 ppm of CO, and <1 ppm of reactive organic gases." All of these features make FC technology an attractive and efficient solution in different energy sectors.

There are several types of FCs, according to the technology adopted and the operating parameters, and the final application depends on the FC chosen type and configuration.

According to market experts, such as 6W research, the global market of stationary FCs is going to achieve high levels of growth in the coming years. In 2018, the world FC market was characterized by a marked growth in fuel cell shipments: The overall shipment units increased to 57,500, with a total power of 240 MW. Compared to the situation in 2017, the market registered a 5% increase in shipments and an 8% increase in installed power. A better picture is noticeable if the 2018 market is compared to the 2012 situation: FC shipments in 2012 were 125 MW, which means in 2018 the market has lived an increase of 92% compared to 2012.

According to "Research and Market, 2019," the stationary FC market will grow up to USD 5.08 billion by 2030, with a compound annual growth rate (CAGR) increase of 3.9%. The total installed capacity is forecast to grow from 220 to 612 MW, and the market will be led by the USA in North America and Japan (and China soon) in Asia.

As reported in the "Solid Oxide Fuel Cell-Global Market Outlook (2017–2026)," solid oxide fuel cell (SOFC) technology had an important market share in 2017, accounting for USD 389.21 million, and it is forecasted to increase up to USD 1,356.51 million by 2026, with a CAGR of 14.9%. In North America, CAGR (compound annual rate of growth) is expected to grow by more than 13% by 2023; in Europe and Asia-Pacific CAGR will be 15% and in South America and MEA CAGR will be 13% during the same period of time. With a lower market share, but with an important trend, direct methanol fuel cells have been valued at USD 137 million in 2018, expected to grow up to USD 367 million by 2025.

Supportive government policies, the economy of scale, and technology improvements are the main drivers of this important growth. Several prototypes, experimental projects, and proof-of-concept studies are being executed and validated, allowing a deeper understanding of the technology's performance to be obtained in real operating conditions. The FC equipment adopted and installed up to 2018 is reported to have a rated power mostly following within the range of 0.5–400 kW. Within these projects, related to FC systems applied to stationary applications, different configurations, and end-user applications have been tested, namely as "backup power supplies, power generation for remote locations, stand-alone power stations for one or more consumers, and distributed generation for buildings and cogeneration."

Chapter 4 of this book examines recent advances in low-temperature (alkaline FC, direct methanol FC, and proton-exchange membrane FC), medium temperature phosphoric acid fuel cell (PAFC), and high-temperature direct carbon and solid oxide FCs.

Depending on the FC type, FC companies are consolidated in different countries, and depending on the size of applications, large-scale FC stationary installations or micro-CHP fuel-cell-based installations, the predominant market, in terms of FC adoption, can differ. Most of the big players are located in Europe, Japan, and the USA. For large-scale stationary applications, three main technologies (MCFC, SOFC, and PAFC) are manufactured and mostly adopted within the USA, with a specific reference to bigger sizes for large-scale FC stationary installations. Japan and Europe have lower installations of large-scale FC stationary installations, which are commonly based on PAFC and MCFC technology, while in South Korea, PAFC and MCFC are the most installed technologies. For residential micro-CHP applications, Europe and Japan are leading the market, due to ad hoc-aimed subsidies and programs.

Europe has installed more than 4,100 FC units for combined heat and power applications, due to three main projects and actions: Callux, PACE (pathway to a competitive European fuel cell micro-CHP market program), and ene.field. The three programs have been key actions for the technology rolling out. Only in the ene.field program, 603 PEMFC micro-CHP units have been installed, and there have been 403 installations of SOFC. In Germany, an incentive program, namely KfW, is supporting the micro-CHP early market, at different levels: 5700 EUR as a fixed amount for a new FC, and other additional amounts and flat rate supplement. For every 100 watts of electrical power started, another EUR 450 is added, up to EUR 6750. When used in CHP mode, a subsidy is paid for each kilowatt-hour of electricity produced: EUR 0.04 per kilowatt-hour for electricity that is consumed and EUR 0.08 for electricity that is fed into the grid. The program aims to provide funding for the installations of about 60,000 CHP units by 2022. The current cost of FCs in Europe for micro-combined heat and power production is about EUR 10,000/kW, with more than 2,000 micro-CHP fuel-cell-based installations adopted on the field, and another 2,800 planned by 2021. The largest stationary FC power plant currently operating in Europe is 1.4 MW.

Japan is the main leader in fuel-cell-based micro-combined heat and power unit installations, with the ENE-FARM program. They have been able to decrease the price per sale to USD 7,000/unit for PEM and USD 8,800/unit for SOFC. The overall installations can be counted for 360,000 units in 2020: almost 62% of them are PEMFCs, and 38% are SOFCs. The program supported also subsidizes 50% of the cost—USD 730/unit for SOFC. However, there are no more subsidies for PEMFC, since the commercial price is now competitive without additional financial support.

Asia has been the more active area for installing FC units, especially for commercial micro-CHP applications. This is particularly applicable to Japan in the last 5 years, which has seen an increase of almost 30% in 2018 (55,500 installed units) compared to 2014. Most of the US market is based on SOFC (300 MW installed), with subsidies between 600 and 1,200 EUR/kW (NG or biogas), and a price per sale of 10,000 USD/kW. PEMFC large installations in the USA are still not common (10 MW) compared with other technologies. Europe counts 1.8 MWs of SOFC systems and 1.5 MWs of PEMFC, with EUR 34 million available under the Horizon 2020 program for stationary FCs. Korea has 1.5 MWs of PEMFC systems installed, which may be because the Hyundai Nexo Stack is also used in stationary applications. Subsidies for demonstration projects are helping these innovative technologies spread. Japan has 2.5 MW of PEMFC installed, and research and development are considered key actions since USD 300–400 million is available for R&D on stationary FC. For other FC technologies, MCFC tech is the main tech applied to large stationary applications.

Chapter 4 examines recent advances made in all low-, medium-, and high-temperature FCs.

1.5 MULTIFUNCTIONAL AND HYBRID FUEL CELLS

Besides generating power, several types of FCs provide multiple functions. For example, molten carbonate FC can capture CO_2 and produce hydrogen and heat besides producing power. Exxon in partnership with FuelCell Energy Company is developing a hybrid combination of natural gas combine cycle-molten carbonate fuel cell (NGCC-MCFC), which can capture CO_2 coming out of NGCC and generate additional power. MCFC can also generate hydrogen and heat. Another multifunctional FC is a regenerative FC, which can operate both as an electrolyzer to generate hydrogen or as a FC to generate power. National renewable energy laboratory (NREL) first reported [10] that in the Technology management Inc. (TMI) reversible FC—electrolyzer system employs a high-temperature solid oxide-based electrochemical process to produce electricity from common hydrocarbon fuels (e.g., natural gas, propane, and bio-derived fuel) and hydrogen. In electrolyzer mode, the reversible system uses electricity and thermal energy to convert pure water into fuel (hydrogen and oxygen). TMI's reversible system uses the waste thermal energy produced during electricity generation mode to achieve high system efficiency during electrolysis mode, ultimately lowering product life cycle costs for the combined system. To further increase system efficiency, TMI has adopted a "passive" cell design, which minimizes the balance of plant components. More recently, Wang et al. [11] presented a review of unitized reversible FC involving proton-exchange membrane FC. Such a cell is getting more industrial acceptance. Finally, a multifunctional microbial FC can treat wastewater besides generating power. These multifunctional FCs are finding increasing acceptance in both the energy and environmental industries, and they would be commercialized in not distant future.

The FC can also be easily combined with other FCs or other sources of energy to form a formidable hybrid system. Hybrid systems are often formed to reduce environmental pollution, increase efficiency, or increase the life of the individual FC. They can also enhance the durability of the power system. Hybrid power systems (HPSs) assure continuous power supply to end users. These systems consist of more than one energy source such as wind–diesel, wind–PV, hydro–solar, wind–solar–FC, FC–micro-turbine, and wind–PV–diesel, with and without battery backup. According to the report on the global HPS market (Zion Market Research, 2019), the market size for hybrid systems was US$477.71 million in 2017 and is expected to touch US$836.92 million by 2024. This simply means that HPSs are being employed globally though at a slow pace. HPSs are reliable, efficient, low cost, and have minimal GHG emissions. Taylor [12] presented a comprehensive overview of off-grid applications (such as health clinics, schools, water pumping, meteorological and communication towers, water purification, rural telephony, home systems, grain grinding, carpentry, refrigeration, and lighting) of hybrid systems around the globe. The hybrid systems include a large number of configurations based on wind, solar PV, biomass, micro-hydro, FC, and diesel generation with or without battery backup. They are prevalent in many countries such as Russia, Kazakhstan, Nepal, India, Bangladesh, Indonesia, the Philippines, China, Korea, Mongolia, Morocco, Egypt, Ghana, 14 Southern African development community (SADC) Countries, South Africa, Mozambique, Mexico, Central America, Chile, Argentina, Brazil, and the Dominican Republic [12,13].

Oil-rich Saudi Arabia is also diversifying its existing energy mix by supplementing it through renewable sources of energy such as solar and wind for grid-connected and off-grid power systems. Saudi Arabia observes high intensities of solar radiations throughout the year and long hours of sunshine durations, and it is available in all parts of the country. Additionally, the nation has a good wind power resources distributed in different regions of the country. These two resources can be combined with diesel and battery backup option for developing HPS in remotely located areas, which are being supplied power through diesel-only system in the present framework. There are more than 40 locations where only diesel power systems are deployed and could be supplemented through wind, solar power, and FC systems. The deployment of HPSs will help in reducing the GHG emissions in these localities, tend to reduce health and electricity bills, and will create new jobs.

HPSs are efficient, economical, and reliable off-grid power systems and assure continuous power supply to end users. These systems are getting popular among remotely located communities in developing countries, especially in Asia and Africa. The applications of such hybrid systems include buildings, individual houses, villages, islands, hotels and resorts, communication and meteorological towers, schools and clinics in villages, alpine huts, industrial fencing, and water desalination, and recently, they are being studied for space applications. In the context of electricity generation in oil- and gas-producing countries, it has become implicit to develop new, clean, and renewable sources of energy to supplement fossil fuel-based power generation. With this single aim, one can hit two targets, the conservation of fixed fossil fuel resources and combating adverse climatic changes to safeguard our planet. In the regional context, solar PV, solar thermal, wind power, geothermal, and hydropower are alternative sources for power mitigation. Of these renewables, wind, solar PV, diesel, and energy storage in hybrid combinations are the possible ways to supply continuous energy for all sizes of applications.

Hybrid nonthermal sources are becoming increasingly important. As shown in my earlier books [1–3], solar–wind combination provides some stability and dispatchability for power generation from two intermittent sources. Wind energy can also be combined with hydropower and FC to provide stable power generation for grid usage. A novel combination of hydropower and solar power has captured worldwide attention. As mentioned earlier, laying down solar cells on water near hydrodam can provide stable power generation with increasing efficiency and durability. Hydro-dam in this setup can also be used as a storage device. FC can be combined with numerous other renewable and fossil fuel-based sources to either increase power generation, improve efficiency, reduce CO_2 emission, or increase lifetime. A hybrid SOFC-PEMFC is gaining significant popularity because it expands the range of FC applicability in many different ways. FC in many different hybrid combinations can also serve as an energy storage device along with power generation device.

Chapter 5 examines in detail various types of multifunctional and hybrid FC systems.

1.6 SELF-POWERED ELECTROCHEMICAL SYSTEMS AND NANOGENERATORS

During the last two decades, the portable electronic industry has exploded. Portable, wearable, and implantable electronics (WIEs) have experienced a period of rapid development and are more and more important and attractive to the public in general. They have a wide range of applications, which include consumer industry, military applications, and medical and environmental industries. The current technology to supply power to portable electronics mainly relies on rechargeable batteries, but for the near future, micro/nano-systems will be widely used in health monitoring, infrastructure and environmental monitoring, Internet of things, and defense technologies; the traditional batteries may not meet or may not be the choice as power sources for the following reasons. First, with the increasing shrinkage in size, the size of the total micro/nano-systems could be largely dominated by the size of the battery rather than the devices. Second, the number and density of micro/nano-systems to be used for sensor network could be large; thus, replacing batteries for these mobile devices becomes challenging and even impractical. Lastly, the power needed to drive a micro/nano-system is rather small, in the range of micro- to milli-Watt range. To meet these technological challenges, the researchers have proposed self-powering nanotechnology aiming at harvesting energy from the environment to power the micro/nano-system-based sensor network.

A *self-powered dynamic system* is defined as a dynamic system (see Figure 1.3) powered by its own excessive kinetic energy, renewable energy, or a combination of both. A particular area of work is the concept of fully or partially self-powered dynamic systems requiring zero or reduced external energy inputs. The exploited technologies are particularly associated with self-powered sensors, regenerative actuators, human-powered devices, and dynamic systems powered by renewable resources (e.g., solar-powered airships) as self-sustained systems. Various strategies can be employed to improve the design of a self-powered system, and among them,

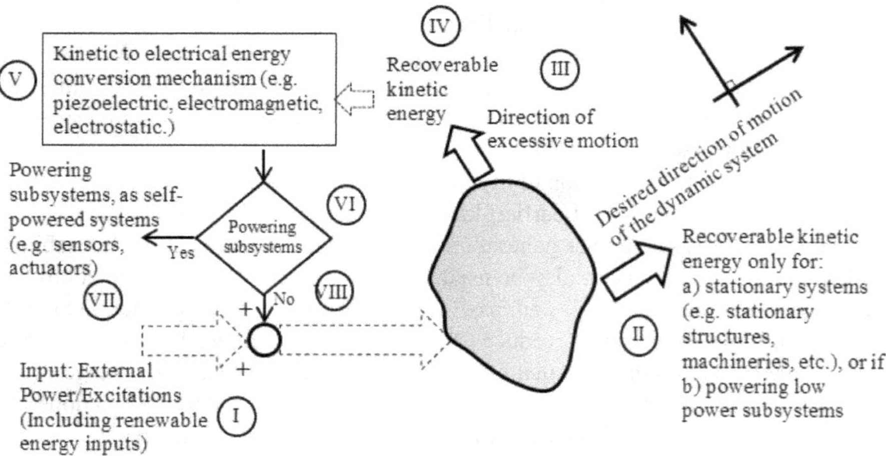

FIGURE 1.3 Concept of self-powered dynamic systems [14].

adopting a bioinspired design is investigated to demonstrate the advantage of biomimetics in improving power density.

The concept of "self-powered dynamic systems" in Figure 1.3 is described as follows [14].

I. Input power (e.g., fuel energy powering a vehicle engine or propulsion system) or input excitation (e.g., vibration excitation to a structure) to the system. The source of this input energy can be renewable energy source (e.g., solar power to a dynamic system).
II. The kinetic energy in the direction of motion of a dynamic system is only recovered if the system is stationary (e.g., a bridge structure), or the recoverable energy is negligible in comparison with the power required for motion (e.g., a low-powered sensor).
III. The movement of the dynamic system perpendicular to the desired direction of the motion is usually the wasted kinetic energy in the system (e.g., the vertical motion of an automobile suspension is wasted as heat energy in the shock absorbers, or the vibration of an aircraft wing is converted into heat energy through structural damping).
IV. The vertical movement of the dynamic system is a source of recoverable kinetic energy.
V. The recoverable kinetic energy can be converted to electrical energy through an energy conversion mechanism such as an electromagnetic scheme (e.g., replacing the viscous damper of a car shock absorber with a regenerative actuator), piezoelectric (e.g., embedding piezoelectric material in aircraft wings), or electrostatic (e.g., vibration of a micro-cantilever in a micro-electromechanical system [MEMS] sensor).
VI. The recovered electrical power can be stored or used as a power source.
VII. The recovered electrical energy can power subsystems of the dynamic system such as sensors and actuators.
VIII. The recovered electrical power can be realized as an input to the dynamic system itself.

Such self-powered schemes are particularly beneficial in the development of self-powered sensors and self-powered actuators by employing energy harvesting techniques, where kinetic energy is converted to electrical energy through piezoelectric, electromagnetic, or electrostatic electromechanical mechanisms. Developing a self-powered sensor eliminates the use of an external source of power such as a battery and therefore can be considered a self-sustained system. A self-sustained system does not require maintenance (e.g., replacing the battery of the sensor at the end of the battery life). This is particularly beneficial in remote sensing and applications in hostile or inaccessible environments.

Ever since Wang and his group demonstrated the first NGs using piezoelectric nanowires for converting mechanical energy into electricity [15], a great interest has been excited worldwide in developing various approaches for energy harvesting. A key idea presented in the 2006 paper is self-powered nanotechnology, aiming at powering nanodevices/nano-systems using the energy harvested from the environment in which the systems are supposed to operate. NGs can harvest triboelectric, piezoelectric, pyroelectric, thermoelectric, and solar energy individually and in hybrid forms. NGs can be very small scale, wearable, and implantable.

In the past decades, WIEs have experienced a period of rapid development and are more and more important and attractive to the public. Nowadays, WIEs have penetrated every aspect of our lives, making people's lifestyles more efficient and convenient. As the core components of WIEs, practical mobile sensors require integration ability with accessories and fabrics such as bracelets, watches, eyeglasses, necklace, or implantation into the human body. However, their developments still need to overcome lots of challenges, such as how to reduce weight and size. Furthermore, for implantable electronics especially, flexibility is particularly necessary.

Another great limitation is the power supply. For full functions, most current practical mobile electronics still require outer power supplies usually batteries to provide power. The battery occupies most of the volume and weight, and periodic replacement of it will lead to electronic waste, physical burden, and financial strain on patients. Therefore, self-powered systems are imperative to practical WIEs. Our bodies contain a variety of energy, including chemical, thermal, and mechanical energy, among which mechanical energy is the most abundant. For the purpose of utilizing this mechanical energy, a NG, which can transform mechanical energy into electric energy, was invented by Wang in 2006 and has made remarkable progress in recent years. Typically, NGs can be divided into three types based on electricity generation mode, namely PENG, TENG [16], and PyENG. They have been widely used as micro-nano-energy or blue energy harvesters and self-powered sensors. Considering that more and more implantable and wearable electronic sensors have been employed by humans, developing NG-based technology is extremely attractive. The developments in PENG, TENG, PyENG, and other hybrid NGs will be the subjects of intense research for future nonthermal power generation [15,16]. This subject is extensively discussed in Chapter 6.

1.7 ORGANIZATION OF THIS BOOK

This book is organized into five additional chapters. Chapter 2 examines advances made in power generation from wind energy. Chapter 3 examines advances made in the harnessing of power from a variety of hydro-energy. Chapter 4 examines low-temperature and high-temperature FCs. Chapter 5 examines multifunctional and hybrid FCs. Finally, Chapter 6 examines advances made in a variety of self-powered NGs, which can be used for portable, wearable, and implantable power electronic devices.

REFERENCES

1. Shah, Y. T., *Advanced Power Generation Systems-Thermal Sources*. CRC Press, Taylor and Francis, New York, 2023.
2. Shah, Y. T., *Hybrid Power- Generations, Storage and Grids*. CRC Press, Taylor and Francis, New York, 2020.
3. Shah, Y. T., *Hybrid Energy Systems- Strategy for Decarbonization*. CRC Press, Taylor and Francis, New York, 2021.

4. U.S. Department of Energy, *Challenges and Opportunities for Airborne Wind Energy in United States*, a report by Department of Energy for Congress, Washington, DC, 2021.
5. Marine energy, 2013. Wikipedia, The free encyclopedia, last edited 28 June 2023. https://en.wikipedia.org/wiki/Marine_energy.
6. Bard, J., *Ocean Energy Systems Implementing Agreement An International Collaborative Programme*. International Energy Agency, Paris, France, 2007, p. 5.
7. Tidal power, 2023. Wikipedia, The free encyclopedia, last edited 12 August 2023. https://en.wikipedia.org/wiki/Tidal_power.
8. U.S. Department of Energy, *Hydrogen and Fuel Cell Technologies Office Accomplishments and Progress*. U.S. Department of Energy, Hydrogen and Fuel Cell Technologies Office, Washington, DC, 2023.
9. U.S. Department of Energy, *FY 2018 Progress Report for the DOE Hydrogen and Fuel Cells Program*. DOE/GO-102019-5156. U.S. Department of Energy, Washington, DC, 2019.
10. Milliken, C. E., Ruhl, R. C., Low cost, high efficiency reversible fuel cell systems. In *Proceedings of the 2002 U.S. DOE Hydrogen Program Review*. NREL/CP-610-32405. NREL, Golden, CO, 2002.
11. Wang, Y., Leung, D.Y., Xuan, J., Wang, H., A review on unitized regenerative fuel cell technologies, Part-A: Unitized regenerative proton exchange membrane fuel cells. *Renewable Sustainable Energy Rev* 2016;65:961–977.
12. Taylor, R., *Renewable Hybrid System Applications around the World NREL – NETL – DOE Natural Gas/Renewable Energy Hybrids Workshops*, 21–22 August, 2001.
13. Rehman, S., Hybrid power systems – Sizes, efficiencies, and economics. *Energy Explor* 2021;39(1):3–43. DOI: 10.1177/0144598720965022
14. Self powered dynamic systems, 2023. Wikipedia, The free encyclopedia, last edited 8 July 2023. https://en.wikipedia.org/wiki/Self-powered_dynamic_systems.
15. Wang, W., Song, J., Piezoelectric nanogenerators based on zinc oxide nanowire arrays. *Science* 2006;312(5771):242–246. DOI: 10.1126/science.1124005
16. Wang, Z., Wu, W., Nanotechnology-enabled energy harvesting for self-powered micro-/nanosystems. *Angew Chem Int Ed* 2012;51:2–24

2 Advanced Wind Power Systems

2.1 INTRODUCTION

While capturing wind energy has been in existence since the 18th century, modern methods of wind power began in 1990. Wind power, i.e., the kinetic energy of the wind, is a renewable energy source that is used among others mainly for the production of electrical power. The global wind resources (land and nearshore) are estimated to be 72 TW (terrawatts), which is seven times the world's electricity demand and five times the world's energy demand. In contrast to the global pollution issues and the significant cost of fossil fuels, the wind seems to be a clean, affordable, and inexhaustible source of energy and its exploitation can directly meet both the global demand for renewable and clean energy and the need to secure new energy sources. Fundamentally, wind energy can be generated by four different methods: onshore, offshore fixed turbine, offshore floating turbine, and airborne wind energy (AWE). This chapter examines recent advances in all four methods of wind power generation. At present time, land wind energy is most developed and matured. However, offshore wind farms seem to be advantageous due to the enormous energy potential associated with the large continuous areas and the stronger winds that imply greater power generation, even though they have higher initial investment, operation, and maintenance costs. Offshore wind farms can be further divided into nearshore wind farm where the tower is connected to the bottom of the sea and floating wind farm where the wind tower is connected to a floating base on the water. As shown later, a floating wind turbine has the most potential for wind power production. AWE is in the early stages of development. Among other notable changes in modern wind power, in 1993, asynchronous generator in wind turbines started slowly changing to synchronous generator [1,2].

Wind power has grown more than 50-fold over the last 20 years [3]. Globally, an unprecedented 108 gigawatts (GW) of onshore wind was installed in 2020. China and the USA made up 79% of this global deployment. An offshore wind capacity of 6 GW was added, half of which was commissioned in China, and almost all the rest was in the European Union and the UK. Overall, 1,592 TWh of electricity was generated from wind installations in 2020. The International Energy Agency (IEA) study showed that the best close-to-shore offshore wind sites globally could provide almost 36,000 TWh of electricity per year, which is very close to the global electricity demand projected for 2040. However, several challenges will have to be overcome for this enormous potential to be successfully exploited. Government policies will continue to be critical in determining the future of offshore wind [3].

The net-zero emission target by 2050 requires wind power to achieve 8,000 TWh by 2030. This translates into 18% per year growth from 2021 to 2030. It is

DOI: 10.1201/9781003429906-2

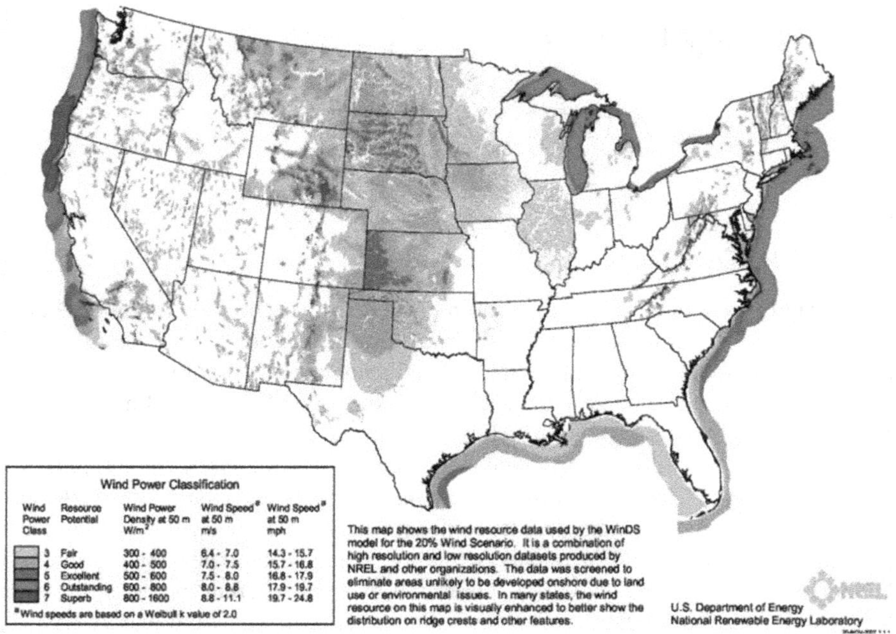

FIGURE 2.1 Wind resource potential at 50 m on land and offshore [4,5].

also necessary to raise annual capacity additions to 310 GW of onshore wind and 80 GW of offshore wind. These lofty targets require cost reductions and technology improvements for offshore wind and facilitating permitting for onshore wind. In the USA, developers have proposed multiple offshore projects in four different states (Maryland, New York [Long Island], New Jersey, and North Carolina). New growth markets for offshore wind are constantly emerging in the USA, China, Taipei, and Japan. In the USA, the land-based and offshore wind resource has been estimated to be sufficient to supply the electrical energy needs of the entire country several times over. The Midwest region, from Texas to North Dakota, is particularly rich in wind energy resources, as illustrated in Figure 2.1. Along with solar energy, wind energy has the best potential for growth to meet the required renewable energy needs of the world.

2.2 ADVANCES IN THE WIND TURBINES AND ASSOCIATED HARDWARE

The amount of power that can be harvested from wind depends on the size of the turbine and rotor, the height of the tower, and the length of its blades. The output is proportional to the dimensions of the rotor and the cube of the wind speed. Theoretically, when wind speed doubles, wind power potential increases by a factor

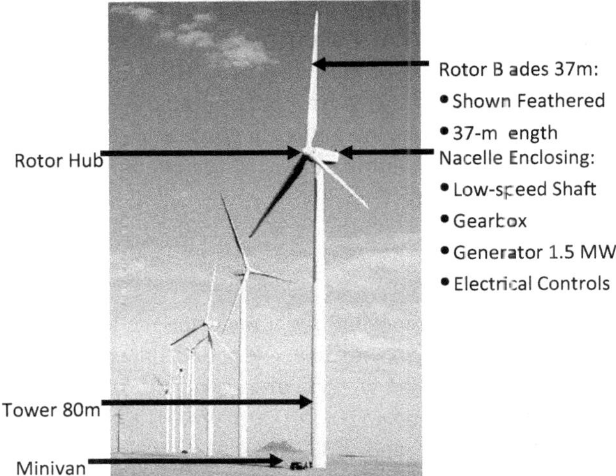

FIGURE 2.2 A modern 1.5-MW wind turbine installed at a wind farm (photo by Mark Rumsey, Sandia National Laboratories) [4,5,7].

of eight. Wind turbine capacity has increased from 0.05 megawatts (MWs) in 1985 to 2 MW onshore and 3–5 MW (megawatts) offshore at present time. During the same time, the rotor diameter has increased from 15 m to about 150 m. Commercially available wind turbines at present time have reached 8 MW capacity, with rotor diameters of up to 164 m.

Modern wind turbines deployed throughout the world today have three-bladed rotors with typical diameters of 70 to 80 m mounted atop 60- to 80-m towers as shown in Figure 2.2. The turbine power output is controlled by rotating the blades on their long axis as the blades spin about the rotor hub, which is referred to as "controlling the blade pitch." The turbine is pointed into the wind by rotating the nacelle about the tower, which is called "yaw control."(The yaw control mechanism comprises a motor and drive. The main purpose of this arrangement is to move the nacelle and blades according to the wind direction. It enables the wind turbine to capture the maximum available wind. During the nacelle movement, a fair chance of cable twisting occurs inside the tower.) Almost all modern turbines operate with the rotor positioned as an "upwind rotor." Wind sensors on the nacelle tell the yaw controller where to point the turbine and, when combined with sensors on the generator and drive train, tell the blade pitch controller to regulate the power output and rotor speed to prevent overloading structural components. A turbine will generally start producing power in winds of about 5.4 m/s (12 mph) and reach maximum power output at about 12.5–13.4 m/s (28–30 mph). The turbine will "feather the blades" (pitch them to stop power production and rotation) at about 26.8 m/s (60 mph).

While the amount of energy in the wind available for extraction by the turbine increases with the cube of wind speed, a turbine can only capture a portion of this cubic increase in energy due to limitations posed by the rated power of the rotor. Over the years, the height and size of wind turbines have increased to capture the more energetic winds at higher elevations. For land-based turbines, size is not expected to grow as dramatically in the future as it has in the past. In general situation, many turbine designers do not expect land-based turbines to become much larger than about 100–120m in diameter, with corresponding power outputs of about 3–5 MW. While larger sizes are physically possible, the logistical constraints of transporting the components over the highway and obtaining cranes large enough to lift the components are potential barriers [4–7].

2.2.1 Growth in Wind Turbine

In a wind turbine, the rotational mechanism transforms the kinetic energy of the air stream into electric power. A wind turbine is divided into two types based on the nature of its blades. The first type is vertical-axis turbines (VAWTs), and the second type is horizontal-axis turbines (HAWTs). In VAWTs, the movement of blades is normal to the earth's surface, and in HAWTs, the movement of blades is alongside with earth's surface. Although the function of both types of blades is the same, they have different configurations. The normal placing of blades in VAWTs makes them operative in all directions. So, this is also called omnidirectional, and hence, this type of turbine can generate electricity even if winds reach it from any angle. These are mostly useful for low-power projects and are also very much effective in unsettled wind speed. In HAWTs, blades are placed alongside the axis due to which this type of turbine gets less air intensity captured by its blades. This type of turbine, however, can generate a large amount of electrical power because its blades produce more movement in comparison with VAWTs. These are used for generating large amount of power.

There is also a further subclassification of turbines. The VAWT turbines are divided into two types: One is Savonius and the second is Darrieus. The advantages of the Savonius turbine are its large torque, low air resistance, and more movement of blades. Darrieus turbines have blades in aerodynamic shapes so that they can produce large wind speed. It can bear up to 220 km/hour wind speed. Darrieus turbine is further subdivided as Darrieus, Darrieus H, and helicoidal based on the shape of blades, which are oval, H-shaped, and helix-like, respectively. The disadvantage of the Darrieus turbine is its low efficiency.

Over the past 20years, average wind turbine ratings have grown almost linearly (see Figure 2.3). Each new generation of wind turbines (roughly every 5years) has grown along the linear curve and has achieved reductions in the life cycle cost of energy (COE). The long-term drive to develop larger turbines stems from a desire to take advantage of wind shear by placing rotors in the higher, more energetic winds at a greater elevation above ground (wind speed increases with height above the ground). This is a major reason why the capacity factor of wind turbines has increased over time [8]. However, there are constraints to this continued growth to larger sizes as in general it costs more to build a larger turbine. The primary argument for a size limit for wind turbines is based on the "square–cube law," which states that "as a wind turbine rotor increases in size, its energy output increases as the rotor swept area (the diameter squared), while the volume of material, and therefore its mass and cost, increases as the cube of the diameter." In other words, at some size

FIGURE 2.3 Development path and size growth of wind turbines [4,5,7].

the cost for a larger turbine will grow faster than the resulting energy output revenue, making scaling a losing economic game. Engineers have successfully skirted this law by changing the design rules by increasing size and removing material or by using material more efficiently to trim weight and cost [4,5,7,9–11].

Breakthroughs in control technology have allowed general electric (GE) to use a 120-m rotor on its GE 2.5–120 turbine, the company's largest. While a previous focus of the industry was increasing the total nameplate capacity of wind turbines, the focus has shifted to the capacity factor of the turbine, which helps keep energy costs low by providing the most possible power. Ten years ago, the turbine was at about 25% capacity factor. Today, it is over 50%. The improvement in capacity factor and the COE allow wind turbines to go into more and more locations where the wind is at a lower speed. One of the deciding forces so far for increasing capacity factors has been an increase in the size of the rotors used on wind turbines. GE's predominant turbine in the USA, which has a 1.6 MW capacity, currently comes with a 100-m rotor, compared with a 70-m rotor in the past. Increasing the size of the turbine rotors creates new challenges for manufacturers, however. Rotors scale poorly with size, so the cost can go up faster than the revenue generated by the increased capacity factor.

Turbine rotors are affected by two different forces: torque, which turns the rotors and creates energy, and thrust, which pushes against the turbine. Dealing with thrust can be difficult when designing a rotor. Engineering breakthroughs in turbine controls have allowed for the handling of the additional thrust generated by wind. To some extent, the controls used on turbines are similar to antilock braking systems on cars. The way the turbine is controlled and shut down, along with how it responds to wind gusts, allows for a bigger rotor on a turbine. A larger area of the rotor gets more energy at a lower wind speed. Alstom has utilized this principle and increased its ECO100, 3-MW turbine with a 100-m rotor to about 122-m rotor with a larger turbine [4–7,12].

While the focus on increasing the power produced by wind turbines may be on the capacity factor, it is important to make sure wind turbines are operational and available. The availability of wind turbines with larger capacity has increased over time. Today, GE has 22,000 wind turbines with an average production-based availability of around 98%. This is achieved by improving the individual components used in turbines, both electronics and gearboxes. For gearboxes, GE has combined the manufacturing processes and design processes so they are designing components that can be reliably manufactured. In addition, the company does highly accelerated lifetime testing on all its gearbox designs to validate the design on all new gearboxes. GE is looking at ways to harden gears and different types of bearings and bearing configurations.

According to Caduff et al.'s study [11], the bigger the wind turbine, the greener the produced electricity. Of course, a bigger wind turbine means mainly producing more power more efficiently. During the last decade, the wind turbine size has more than doubled. In 2005, the maximum operational wind turbine was 3.6 MW and now it is 8 MW. The largest operational wind turbine models based on capacity and their manufacturers are the following: Vestas V164-8 MW (with 164 m rotor diameter), Enercon E-126 7.5 MW, Samsung S7.0-171 7 MW, Mitsubishi Heavy Industries (MHI) SeaAngel 7 MW, Repower 6M 6.2 MW, Siemens SWT-6.0-150, Alstom Haliade 6.0 MW, Sinovel SL6000 6.0 MW, Areva M5000 5.0 MW, and Gamesa G5 MW [4,6,7,9,11,12]. The largest operational onshore wind farm in the world is Gansu Wind Farm in China. It has a current capacity of power of 6,800 MW and a target capacity of 20,000 MW by 2020. The first world's offshore wind power plant was installed in Vindeby (Denmark) in 1991 and had a total capacity of 4.95 MW (11 X 0.45 MW—bonus wind turbines). At this moment, the largest offshore wind farm in the world (and the largest wind farm in Europe by MW capacity) is the London Array. It is a 630 MW wind farm consisting of 175 Siemens Wind Power SWT-3.6 turbines and located 20 km off the Kent Coast in the UK. In February 2016, Dong Energy announced the Hornsea Project, a 1.2 GW offshore plant, which was made up of 7-MW turbines and will occupy more than 400 km^2, situated about 120 km off the Yorkshire Coast in the UK. As of 2020, among the European Union (EU) members, Germany remains the country with the largest installed capacity (45 GW), followed by Spain (23 GW), the UK (14 GW), and France (10 GW) [4,9,13].

Land transportation represents another potential limiting factor for wind turbine growth. Cost-effective transportation can be achieved by remaining within standard over-the-road trailer dimensions of 4.1 m high by 2.6 m wide. Weights should remain under 80,000 lbs gross vehicle weight (GVW) corresponding to a cargo weight of about 42,000 lbs. Loads that exceed 4.83 m in height will trigger expensive utility and police assistance. These dimension limits have the most impact on the base diameter of wind turbine towers. Rail transportation is even more dimensionally limited. Overall widths should remain within 3.4 m, while heights are limited to 4.0 m. Limitations are driven by tunnel and overpass widths and heights. Transportation weights are less of an issue in rail transportation with a GVW of 360,000 lbs possible [5].

The practical size of wind turbines is also affected by other constraints. Crane requirements are quite stringent because of the large nacelle mass in combination with the height of the lift and the required boom extension. As the height of the lift to install the rotor and nacelle on the tower increases, the number of available cranes with the capability to make this lift becomes fairly limited. Other limiting factors are that cranes with large lifting capacities are difficult to transport, require large crews, and therefore have a high operation, mobilization, and demobilization costs [4,5,14]. The National Renewable Energy Laboratory (NREL) report points out that another factor that will affect the design of turbines of the future, particularly on the Great Plains, is the low-level (or nocturnal) jet. The nocturnal jet is a poorly understood phenomenon that occurs at nighttime as cooling allows the winds to stabilize into parallel layers flowing at different velocities. Kelvin–Helmholtz (K-H) instabilities can form at the boundaries between these parallel layers (Figure 2.4). These instabilities can rapidly grow into large-scale coherent vortices that can reach the ground and cause gust loads on turbine rotors. While the nocturnal jet can occur anywhere, it is most prevalent over areas of the Great Plains, regions that represent some of the best class 4 wind resources. While the low-level jet will not limit wind growth, it is clear that wind turbine designers must clearly understand the implications of the extremely coherent gusts as they place ever larger rotors on taller towers.

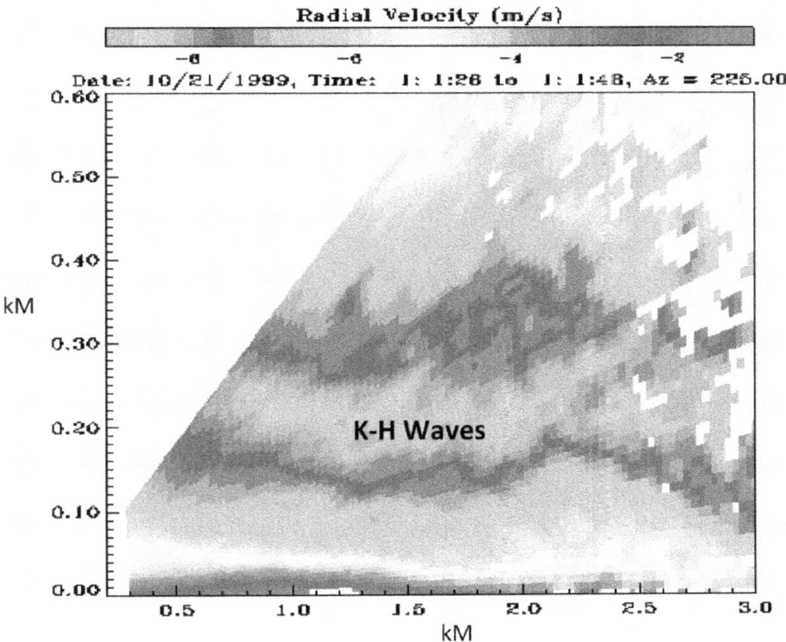

FIGURE 2.4 Kelvin–Helmholtz (K-H) instabilities can form at the boundaries between these parallel layers [4,5,7].

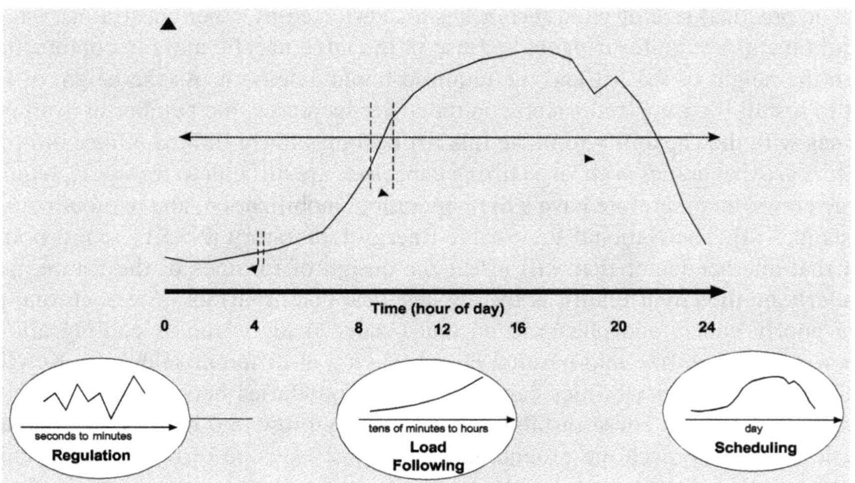

FIGURE 2.5 Requirements of power plant operators deal with three different timescales: regulation—being seconds to minutes; load [4,5,7].

The USA possesses more than 8,000 GW of wind resources that could be harnessed to produce electricity at a reasonable cost if transmission expenditures are excluded. Considering some elements of the transmission required to access these resources, the NREL report [4,5,7] showed that more than 600 GW of potential wind capacity is available for \$60–\$100/MWh, a range that is likely to decrease as more installment experiences are gained. The NREL report also indicated that an abundance of modestly priced wind energy is available in the USA, even with limited transmission access.

There are other concerns as well in capturing high levels of wind power. Wind turbines do not follow a predictable pattern of power generation, except in rare locations (e.g., in California) driven strongly by local weather patterns and landforms. The intermittent nature of wind power can adversely affect grid stability. The power plant operators have to simultaneously deal with three different timescales: regulation—being seconds to minutes; load following—being tens of minutes to hours; and scheduling—being daily (Figure 2.5). The timing associated with wind power generation has been difficult to predict until recently. The fear that the output of large wind power plants will plummet to zero in a matter of minutes or seconds has hampered many major developments, but recent research is shedding a brighter light on the ability of wind power plants to provide a more grid-friendly connection. The use of hybrid power and energy storage has also helped with grid balancing.

2.2.2 Progress on Rotor and Blades

Manufacturers are perfecting how to capture more wind energy with longer blades in both onshore and offshore applications. Ten years ago, the length of the rotor blade, from vertex to tip, measured 30–40 m long, about 115 ft. Today's blades, at 140 m

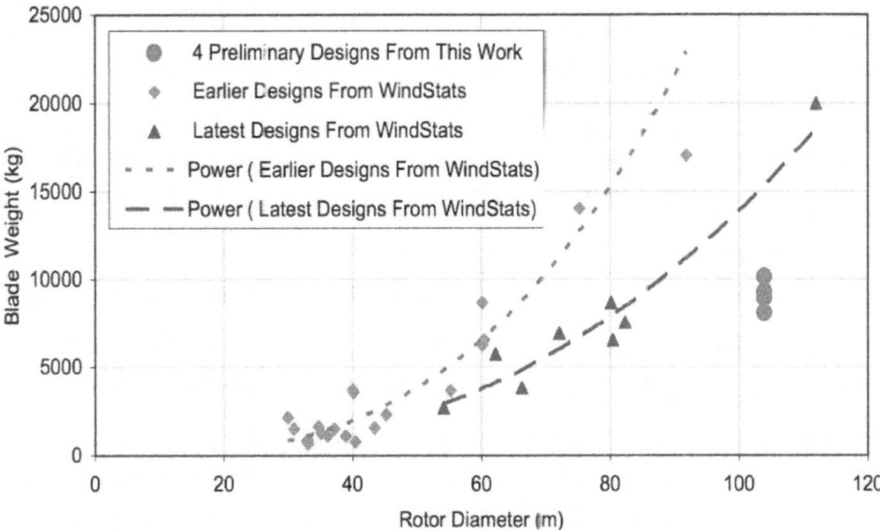

FIGURE 2.6 WindPACT study results indicating the lowering of growth in blade weight due to the introduction of new technology [4,5,7].

(460 ft) from tip to tip, have blade span that now equals the wingspan of the average passenger airplane. The greater the radius of the rotor blades, the more wind can access and, in turn, the greater the torque needed to power the electrical generators. Blade size, along with turbine height, will only continue to grow in the coming years. Bigger blade size and taller turbines make for stronger production capacity. For comparison, the average turbine from one decade back had the capacity to produce 1.5 MW of power. Since then, they have grown in capacity. The average nameplate of newly installed land-based wind turbines in the United States in 2019 was 2.55 MW. The capacity of turbine numbers will continue to rise in the future [4–7,10–12].

Studies have shown that, in recent years, blade mass has been scaling at roughly an exponent of 2.3 instead of the expected 3, as shown by the WindPACT blade scaling study. The WindPACT study shows how successive generations of blade design have moved off the cubic weight growth curve to keep weight down as illustrated in Figure 2.6. The circular points in this figure relate to the results reported by Thresher et al. [7]. If advanced research and development were to provide even better design methods, as well as new materials and manufacturing methods that allowed the entire turbine to scale as the diameter squared, then it would be possible to continue to innovate around this limit to size. The desire to increase the size of machines is also driven by the deployment of wind in offshore environments. Since the cost of installing marine foundations is significantly larger than the cost of onshore foundations, it is important to be able to use larger rotor diameters in offshore wind farm, which can capture larger wind energy [4,5,7,15].

Nanotechnology and nanomaterials have significantly contributed to blade technology. Using composite materials based on carbon nanotubes, lightweight and high-strength rotor blades can be developed. Carbon nanotube composites can be used in wind turbines, which can provide excellent conductivity. The wind turbine lifetime can be increased by using nano-paints, the weight can be reduced by using fiberglass, and the efficiency can be increased by coating. An epoxy containing carbon nanotubes is being used to make windmill blades. Stronger and lower-weight blades are made possible by the use of nanotube-filled epoxy. The resulting longer blades increase the amount of electricity generated by each windmill. Researchers have also developed lubricants using inorganic buckyballs that significantly reduce friction [4,5,7,15–17].

To reduce energy costs and optimize energy production, researchers from the NREL and Sandia National Laboratories (SNL) are working together to design 206-m rotors for land-based turbines [7]. With larger blades, more kinetic energy from the wind can be harnessed and used to generate electricity. However, transportation and manufacturing costs will rise in response to increased material mass and demand. The NREL and SNL have also come up with a way to decrease the stiffness of the blades and for the turbine to hold more upcoming wind, regardless of speed [7]. The downward rotor configuration lowers the stiffness requirement because the wind pushes the blades away from the tower, which ensures that lighter blades keep safe clearance with the tower. Carver's Wind Blade Division and Sandia National Laboratories are developing a wind turbine blade, which could potentially capture 12% more wind than regular turbines. The blade has a small, curved tip, unlike other turbines, which can optimize wind capture. Changing the shape of traditional turbine blades permits lighter blades to be made. This approach allows for aeroelastic stability, but at the cost of increased complexity for manufacturing and control. Researchers from the NREL are currently working on designing lighter blades and optimizing spar cap placements on the blades without increasing the thickness. This would minimize blade mass and increase the strength [4,5,7,12].

The advantage of a curved rotor blade compared to a flat blade is that lift forces allow the blade tips of a wind turbine to move faster than the wind that is moving and generating more power and higher efficiencies. As a result, lift-based wind turbine blades are becoming more common now. Also, homemade polyvinyl chloride (PVC) wind turbine blades can be cut from standard-sized drainage pipes that have the curved shape already built-in giving them the best blade shape, but curved blades also suffer from drag along their length, which tries to stop the motion of the blade. We can reduce this drag force by bending or twisting the blade and also tapering it along its length producing the most efficient wind turbine blade design. To increase the wind turbine blade efficiency, the rotor blades need to have an aerodynamic profile to create lift and rotate the turbine, but curved aerofoil type blades are more difficult to make but offer better performance and higher rotational speeds, making them ideal for electrical energy generation. We can improve the aerodynamics and efficiency even more using twisted, tapered propeller-type rotor blades. Twisting the blade changes the wind angle along the blade, and the combined effect of twisting and tapering the blade along its length improves the angle of attack increasing speed and efficiency while reducing drag. Also, tapered blades are stronger and lighter than straight blades as the bending stress is reduced.

Larger blades may be twirling up high, but those huge blades cause transportation problems on the ground. Their size and shape require specialized trucking routes, permit applications, and the proper vehicles, equipment, and technicians. Even if they move along highways and interstates in the dead of night, the huge equipment can still cause traffic bottlenecks. The idea is to make the blades easier to ship by creating them of segmented pieces, which can later be fitted together. The idea of segmented blade is getting a warm response from the industry. The blades arrive in two pieces—each 70 to 80 meters long—and could be fit together on-site. Some groups are experimenting with new composites such as carbon fiber or other light materials that could make on-site assembly easier. The Department of Energy (DOE) recently awarded GE Renewable Energy, Oakridge National Laboratory, and the National Renewable Energy Lab $6.7 million to develop an additive manufacturing process for new high-performance blade designs for future large rotors. The object is to make them with recyclable materials for onshore and offshore deployment [4,5,7,18,19]. Concepts such as on-site manufacturing and segmented blades are also being explored to help reduce transportation costs. It may be possible to segment molds and move them into temporary buildings close to the site of a major wind installation so that the blades can be made close to or at the wind farm site.

Originally, the damage to the wind turbines was detected by manual methods such as the use of a camera and telephoto lens. These methods were, however, not very effective since damages were only detected when they reached critical states. SNL, International Climbing Machines, and Dolphitech are working together to develop an autonomous robot that can vertically attach to a wind turbine, move, and detect any external or internal problems with onboard cameras and its "phased-array ultrasonic imaging" autonomously [4,5,7]. This method allows on-time detection of damage to minimize maintenance costs and turbine downtime, which can improve turbine lifespan and efficiency. Autonomous robot inspectors can be used for both onshore and offshore wind turbines and can detect any abnormalities on wind turbines from any size of blades. SNL is also working on equipping drones with infrared cameras to detect damages through thermal imaging. This process involves exposing the blades to the sunlight and then covering them in shades. When there is no sunlight shining on the blades, the heat on the blades diffuses inside without harming anything. However, damaged areas prevent heat from diffusing inward, which leaves the surface hot, and these hot spots on the infrared camera show damage.

Wind turbines can freeze in cold weather as noted recently in Texas. Typically, wind turbines employed in colder regions are equipped with deicing devices and integrated heating to protect crucial turbine components, such as the pitch and yaw motors, gearbox, and battery, from extremely low temperatures. Specialized cold weather and anti-icing technologies are designed to prevent ice accumulation on turbine blades, as well as the detection and removal of ice in unavoidable circumstances. Ice buildup on the blades of wind turbines can add weight, change the aerodynamics of the blades, and severely hamper their performance, which can throw the spinning blades out of balance or prevent blade rotation entirely. The Lac Alfred Wind Farm located near Amqui, Quebec, suffered significant turbine downtime due to ice accumulation, which led them to implement Wicetec Oy's Ice Prevention System (WIPS). This system makes use of integrated, carbon-based electrical heaters to quickly heat the surface of turbine blades to a controlled temperature when ice is detected [17,20]. This method is, however, expensive and has not been justified for Caribou Wind Farm in New Brunswick, Canada, for all of its wind turbine blades. The lack of significant icing conditions has resulted in Caribou Wind Farm turning to other deicing options, such as electrically heated tiles, usage of black paint on blades to absorb ultraviolet (UV) energy, and helicopter-applied coatings [4,5,7]. These methods can also be adopted for infrequent cold weather spells in Texas.

Wind turbine blades also have an end-of-life problem. In 2017, researchers at the University of Cambridge Institute for Manufacturing found that waste produced by used wind turbine blades could hit an estimated 43.4 million metric tons by 2050. With those types of projected numbers, more wind turbine makers are looking at ways to recycle their products when they are no longer viable. Wind turbine blades are made from composites, which are difficult to recycle and end up in landfill when their service life ends. The blades will be shredded at a Veolia North America (VNA) site in Missouri and then used as a replacement for coal, sand, and clay at cement manufacturing facilities across the USA. Using wind turbine blades—that are primarily made of fiberglass—to replace raw materials for cement manufacturing, reduces the amount of coal, sand, and minerals that are needed to produce cement.

In 2015 and 2016, a Spanish start-up company suggested vortex bladeless wind turbines. Instead of using large rotating blades, the company had the idea to harvest energy from vibrations that result from wind passing over a cylindrical structure. These vibrations are called vortex-induced vibrations (VIVs), and the phenomenon itself is known as vortex shedding. It is the same principle that makes flags move and swirl when it is windy. The resulting vortex bladeless technology consists of vertical masts, or straw-like structures, that generate energy using an alternator to convert vibrations created by the masts into electricity [4,5,7]. The technology generated both excitement and doubt. Some believed it was not feasible, and others were encouraged by its potential advantages over traditional wind turbines—lighter weight, no large rotating parts, reduced noise, lower manufacturing and maintenance costs, and reduced interference with birds. Vortex bladeless prototyped small models and produced promotional videos, but the technology was not commercialized. However, there has been renewed interest in the technology. Recent market and technical research shows promise for the technology, if it can be commercialized. Vortex bladeless is now testing a 2.75-m-tall prototype, the Vortex Tacoma, which they hope will be their first commercial product. The company emphasizes its technology's potential role in small-scale, distributed generation. A utility-scale product is not on the near-term horizon [4,5,7].

The conversion of aerodynamic power to mechanical torque is perhaps the most unique aspect of wind turbine operation. The ability to withstand changing aerodynamic loads and effectively extract power from the wind over a 20-year lifespan makes wind turbines cost-effective. As machines grow larger and larger, rotors must increase accordingly. A 1.5-MW turbine has a rotor of approximately 70 m in diameter. A 3-MW rotor is roughly 99 m in diameter. While the blade length only increases by 41%, the blade root flap bending moment can increase by levels approaching 160%, depending upon design [4,5,7,18]. These load increases demand significant innovation in design. Several approaches that include either alleviation

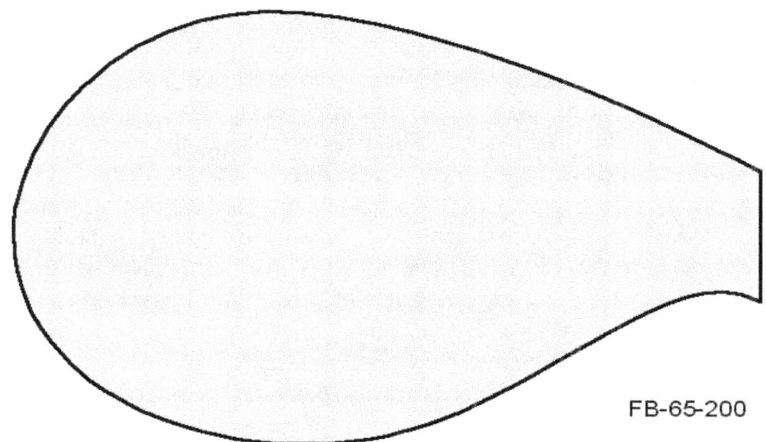

FIGURE 2.7 A flat-back thick airfoil shape with high structural efficiency and easier highway transport [4,5,7].

of these load levels or creating load-resistant designs have been examined. Carbon fibers have been incorporated into wind turbine blades for some years to provide localized strength, but the future will see lighter stiffer blades incorporating large volumes of carbon fabrics presenting options for both tolerating higher blade loads and reducing overall machine loads by reducing rotor weight. Blade planforms are being developed that incorporate unique inboard airfoils using truncated airfoil shapes. These planforms provide increased thickness-to-chord ratios, providing a greater structural cross section to react to outboard loads.

In general, thin flat structures such as airfoils are very inefficient at carrying structural loads. The trick is to make a thick and structurally efficient blade airfoil shape that does not give up much in aerodynamic performance. Figure 2.7 illustrates such an airfoil shape, which is used near the root of the blade where the flatwise bending loads are the highest, and better structure pays the highest return, but where aerodynamic performance is less important. An additional benefit is that blades with a shorter chord at the root are more easily transported over the highway due to the width and height restrictions. Another approach to increasing blade length while restraining weight and cost growth is to reduce the fatigue loading on the blade. There can be a big payoff in this approach because the approximate rule of thumb for fiberglass blades is that a 10% reduction in cyclic stress can provide about an order of magnitude increase in fatigue life. Blade fatigue loads can be reduced by controlling the blade's aerodynamic response to turbulent wind inputs by actively flying the blade using the pitch control system of the turbine. This approach is being explored using modern state space control strategies so that future turbines can take advantage of this innovation [4,5,7,15–18,21,22].

An elegant concept is to build passive means of reducing loads directly into the blade structure. By carefully tailoring the structural properties of the blade using the unique attributes of composite materials, the blade can be built in a way that couples the bending deformation of the blade to twisting deformation. This is referred to as flap–pitch, or bend–twist, coupling and allows the outer portion of the blade to twist as it bends. This is accomplished by designing the internal structure of the blade, or orienting the fiberglass and carbon plies within the composite layups, in such a way as to make the blade twist as it is bent. This twisting changes the angle of attack over much of the blade. If properly designed, this change in the angle of attack will reduce the lift as wind gusts begin to load the blade and therefore passively reduce the fatigue loads. Another approach to achieving pitch–flap coupling is to build the blade in a curved shape so that the aerodynamic load fluctuations apply a twisting movement to the blade, which will vary the angle of attack. These new blade designs are complex and must be developed, tested, and optimized so as not to adversely impact energy production or result in unstable vibrations [4,5,7,15–18,21,22].

Active designs that can reduce loads are also being explored. The concept of aeroelastic coupling, the change in the shape of the blade as it bends under load, offers possibilities for load alleviation. Flap pitch, also called bend–twist, coupling allows the outer portion of the blade to twist as it bends (Figure 2.8). This is accomplished

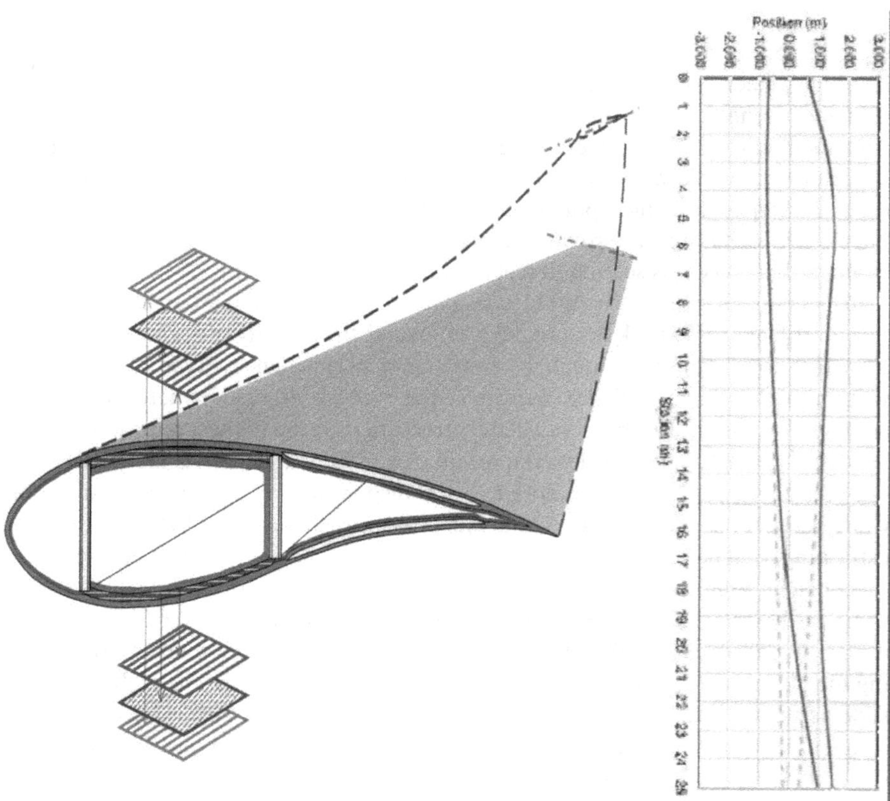

FIGURE 2.8 A twist–flap coupled blade design to alleviate fatigue loads (on the left with material coupling and on the right with a curved blade) [4,5,7,15–18,21,22].

by designing the internal structure of the blade, or orienting the fiberglass and carbon plus within the layups, in such a way as to make the blade flex in pitch as it bends. This twisting therefore changes the angle of attack over much of the blade. If properly designed, this change in the angle of attack will reduce lift and therefore reduce loads. Such designs offer a method of almost instantaneously tailoring blade profiles to alleviate loads without placing additional demands on the blade pitch system. The design must be developed, tested, and optimized so as not to adversely impact energy production.

In September 2017, pricing for electricity from two new offshore wind farms in the UK came in at $76.34 (USD)/MWh—less than gas and new nuclear in the UK. One year later, in August 2018, the 800 MW Vineyard Wind Project off of Massachusetts published pricing for phase 1 (400 MW) starting at $74/MWh (with a 2.5% increase each year thereafter for 20 years). Phase 2 (400 MW) will start at $65/MWh. Part of the reason for these very low prices is the extremely long blades planned for these sites—rotor diameters were 164–180 m (538–590 ft). Each blade was nearly the length of a football field. These longer blades are key for enabling turbines to have large capacities—8 to 10 MW each [4,5,7,15–18,21,22].

The longer blades and higher towers mean more energy produced per turbine. Fewer turbines are required to produce the same amount of energy, which lowers overall project costs through reduced capital, installation, and maintenance expenses. Also, capacity factors for offshore wind projects are 60% or greater—a half to a third better than onshore wind and equivalent to some fossil fuel plants; much of the wind blows during periods of peak loads. Even larger offshore turbines are on the horizon. GE's 12 MW Haliade-X is the largest turbine planned to date, with a 220-m rotor diameter and an overall height of 260 m. With an estimated 63% capacity factor, its output will be 67 GWh annually. This means that a single turbine will generate enough electricity to power more than 6,200 average US homes every year. The swept area of the blades is 38,000 m^2 (410,000 ft^2)—equivalent to seven US football fields. While GE has not made the cost public, industry analysts suggest that the Haliade-X could cost $14 M USD per unit. While the gargantuan size enables increased energy production over existing models, the very long blades present design challenges, such as increased torque on turbine components [4,5,7,9,15–18,21–23].

2.2.3 Advances in Tower Height

Advancements in blade and tower construction technologies are enabling taller towers. Besides capturing more wind energy, taller towers could create wind development opportunities in new areas. Current onshore tower heights (or hub heights) are typically between 80 m (262 ft) and 100 m (328 ft) tall. In this range, successful wind farms in the "wind belt" (the US Midwest and Texas) often achieve capacity factors of 40% and higher, but if hub heights were closer to 140 m (460 ft), wind farms in the Southeast with lower wind speeds could also provide similar capacity factors. In rough terms, each meter of increase in the height of towers can add 0.5%–1% to a turbine's annual energy production (AEP), so being able to access higher wind speeds with taller towers may make some previously poor wind sites economically viable. Using wind speed data, NREL mapped the potential wind resource across the

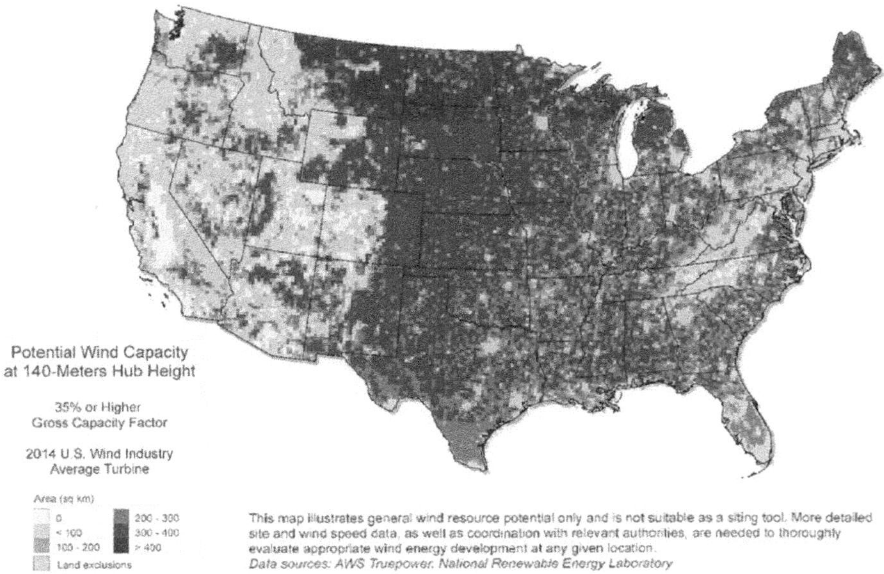

FIGURE 2.9 Potential wind capacity at 140-m hub height [4,5,7].

USA at 140 m (see Figure 2.9). The blue shading indicates which areas could support 35% or greater capacity factors. The darker the area, the denser the concentration of potential wind sites at 140-m hub heights [4,5,7,10,12,13].

NREL study indicates that for Alabama, 140 m hub heights will allow more than 100,000 km^2 area with potential for a 30% or greater capacity factor, and about 60,000 km^2 have the potential for a 40% capacity factor. Today, a 40% capacity factor would be an excellent result for a farm in the wind belt. However, making turbine towers taller introduces some design, cost, and permitting challenges, and just because a site may show new-generation potential at a high elevation does not automatically mean it is a suitable wind farm site. Sites must also undergo airspace evaluations, state and local permitting, and wildlife impact assessments. For example, the Federal Aviation Administration (FAA) requires advance notice for any proposed structure 200 ft or more above ground, so that they can evaluate its impact on aircraft safety and navigation. All proposed structures above 499 ft are automatically subject to an obstruction evaluation [4,5,7].

The cost impact of extremely large cranes and the transport premiums for large tower sections is driving the exploration of novel tower design approaches. Several concepts are under development or being proposed that would eliminate the need for cranes for very high, heavy lifts. One concept is the telescoping or self-erecting tower. This concept allows assembly of the nacelle and rotor at close to ground level and then utilizes hydraulics to jack the tower and nacelle to their operating height. Other self-erecting designs look at lifting dollies or tower climbing cranes that use tower-mounted tracks to lift the nacelle and rotor to the top of the tower. These concepts have the added advantage of being able to bring the nacelle to the ground for major overhauls that would otherwise require expensive cranes [4,5,7].

The materials for turbine tower construction are also the subject of intense research. In the USA, turbine towers are made primarily from tubular steel built on a concrete base, but some manufacturers are experimenting with building towers exclusively from concrete for durability. The concrete pieces that would make up the tower would be fabricated on-site so that huge chunks of concrete are not moving through difficult-to-navigate parts of the country. Currently, the width of the base cannot exceed 4.5 m because of transportation limitations, which limit the height of the turbine. Raising turbines higher calls for a mobile casting or manufacturing facility to be available on location. In another DOE-funded effort, GE Renewable Energy, cement manufacturer LafargeHolcim, and COBOD International are developing optimized three dimensional (3D)-printed concrete bases. COBOD manufactured a 10-m-tall 3D-printed concrete base in 2019 using its large-scale construction 3D printing technology. The goal is to double today's 100-m-tall transportation-based limitation. Printing a pedestal on-site would avoid that issue, and the taller structure would increase energy production. Concrete also provides possible cost reductions that may be realized by limiting the size of steel tower sections. Composites are an additional material under study for wind turbine towers. Small Business Innovation Research Projects are exploring several advanced tower concepts using composite materials and novel structures, both of which may lend themselves to on-site manufacturing and ease of transport. Increases in machine size have several effects on turbine erection [4,5,7,12].

Assuming a tower can be segmented and erected in sections, the nacelle mass appears to be the limiting factor in turbine erection. Projected masses for nacelles in the 2.5–3.5 MW range run from 295,000 to 430,000 lbs. Lifting this mass requires a crane with a much greater lifting capacity than expected because of the extreme nacelle height. Crane costs for placing a 2.5-MW turbine on a 110-m tower are expected to run $40,000–$50,000. While crane costs for erecting a 5-MW turbine on a 156-m tower would run as high as $138,000. Reductions in tower height and turbine tower head mass would significantly reduce these costs. These masses and costs are based on machine scaling projections and spreading the mobilization and demobilization costs over a 50 MW wind farm. Other limiting factors are that cranes with this lifting capacity are rare, are difficult to transport, and require large crews, and they have very large mobilization and demobilization costs. In addition, moving cranes between turbine erection sites is also extremely costly and time-consuming [4,5,7,12].

NREL [4,5,7] points out that the FAA review process also includes a review from the Department of Defense (DOD), base commanders, the Department of Homeland Security (DHS), and other federal agencies to ensure proposed wind farms will not impact military operations, national security, or radar communications. Unlike FAA height rules, DOD review is not tied to a specific height or distance, since operations can vary widely from base to base. Area infrastructure such as roads, bridges, and overpasses can limit the size of tower segments that can be transported to a site. To address this, some original equipment manufacturers (OEMs) have built slip form cast prestressed concrete towers on-site. One such project is taking place at Iowa

State University, where researchers recently completed an 18-month study to design a better 140-m tower using a technology they call Hexcrete. According to Iowa State University (2017), "The basic idea of Hexcrete is that it's assembled from precast panels and columns made with high-strength or ultra-high-performance concrete. Those panels and columns can be cast in sizes that are easy to load on trucks. They are tied together on-site by cables to form hexagon-shaped cells. A crane can stack the cells to form towers as high as 140 m." The prototype Hexcrete tower segment has passed multiple tests that simulate fatigue and operational loads from a Siemens 2.3-MW turbine operation. Anticipated costs, derived from calculations using NREL models, show a 10%–18% levelized cost reduction for a 120- to 140-m Hexcrete tower over today's conventional 80-m tower. In another project, conducted in 2016 at the Technische Universität Wien (TU Wien) in Vienna, Austria, researchers designed and tested an innovative double-wall concrete tower construction method. Rectangular, double-wall components are transported to the site and attached together there to form octagonal tower segments. The segments are stacked on top of each other, and then, the gap in the hollow wall is filled with concrete and reinforcing steel. This method results in a stable, solid concrete tower while avoiding much of the transportation costs associated with delivering solid tower segments to a site.

Wind turbines and their towers are simply getting larger and taller. GE Renewable Energy's Haliade-X offshore wind turbine will generate 12 MW, 13 MW, or 14 MW, depending on the model. Unveiled in 2020, a prototype is being built in the harbor of Rotterdam, the Netherlands. It is about one-third more powerful than the largest wind turbine already in service. The turning diameter of its rotor is 220 m, longer than two American football fields. At 248 m tall, the new towers will far exceed the reach of today's tallest wind turbines. It should be, however, noted that not all installations are better served by sending their turbines higher. A taller tower requires lower wind speeds, but its erection requires proper geological ground conditions.

NREL has carried out extensive studies on advances in wind turbine, blade, and tower designs and associated control, operational, and maintenance issues. Some of the important conclusions regarding tower design resulting from these studies can be summarized as follows [1–7,9–29].

1. Wind resource quality improves significantly with height above ground. Over large portions of the country, our mesoscale resource data indicate an increase in annual average wind speed of 0.5–1.0 meters per second (m/s) when moving from 80 to 110 m and 1.0 to 1.5 m/s when moving from 80 to 160 m.
2. Wind speed differences translate to sizable capacity factor improvements. Although the observed variance is broad, median capacity factor gains with higher hub heights are estimated at approximately 2 to 4 percentage points when going from 80 to 110 m and an additional 2 to 4 percentage points when going from 110 to 140 m. Between 140 and 160 m, median capacity factor gains are approximately 1 percentage point. Relatively larger gains occur east of the Rocky Mountains, with the greatest gains sprinkled throughout the Heartland, the Midwest, and the Northeast.

3. Based on first-order cost estimates informed by current technology, the most wind-rich regions of the country generally show an economic preference for the lowest considered tower height; higher hub heights (e.g., 110 and 140 m) are often preferred in more moderate wind speed regions. This result is consistent with industry experience to date.
4. Higher nameplate and lower specific power turbines (e.g., 150–175 W/m^2) also show a general economic preference for the lowest considered tower height; however, these larger turbines require tower heights of at least 110 m. Tower heights of 140 m and in some cases 160 m tend to be preferred in more moderate wind speed areas.
5. The highest nameplate capacity turbine we considered (4.5 MW) has a relatively greater preference for 140-m hub heights than similar 3-MW-class turbines. This observation is driven by the proportionally lower cost associated with taller towers and tall tower installations in dollars per kilowatt ($/kW) for larger turbines and indicates that turbine scaling (which offers additional cost-saving potential) and taller tower deployment are likely to occur in parallel.
6. Future tower innovations could make higher hub heights more attractive. In a tower cost-bounding scenario, where NREL applies a fixed $200/kW tower cost for each turbine at all hub heights, NREL sees an economic preference for 160-m tower heights in 70%–90% of sites, depending on the specific turbine configuration.
7. Reducing the cost of realizing taller towers is critical to capturing the value of higher wind speeds at higher aboveground levels and for increasing the viability of wind power in all regions of the country.
8. Additional factors that could impact tower height include blade tip clearance requirements, balance-of-station costs, turbine nameplate capacity, and specific power. Turbines with higher specific power ratings experience more energy gain for a given change in wind resource. Larger wind turbines tend to have an economic advantage for tall tower applications and offer additional cost-saving potential in balance-of-station and turbine-level economies of scale. Ultimately, wind turbine design reflects an optimization across an array of potential criteria; focusing on tower height alone may result in suboptimal outcomes.
9. When pursuing higher tower heights, a system-level incremental capital cost of less than $500/kW for low specific power turbines and potentially as low as $200/kW, particularly for higher specific power turbine configurations, could support a levelized cost-of-energy (LCOE) reduction across much of the country and might also push less-energetic wind resource regions further along the path to economic competitiveness. Depending on the specific focus regions and turbine configurations under consideration, variance from this general guidance could be merited.
10. To realize taller wind turbine towers, an array of potential concepts remain in play. These concepts rely on various materials spanning rolled tubular steel (currently the most widely used option), concrete, and lattice steel, for space frame designs, as well as hybrid designs that use a combination of

concepts. Although there are clear advantages and disadvantages to each known concept, the future design of tall wind turbine towers remains to be determined. At the same time, NREL examination suggests that tubular towers can continue to be viable at higher aboveground heights, particularly with continued advances in control technology that allow for reliable use of soft–soft designs. Tower erection strategies and innovation may also be a determining factor in the viability of future tall tower concepts.

NREL suggests that future research needs should include activities that quantify and ultimately reduce the uncertainty of the wind resource data, particularly at higher aboveground levels. More robust cost assessments and analysis including more sensitivities and evaluation of specific technology opportunities and alternative turbine configurations would also be valuable. Further information on the innovative solutions to capture value from taller towers would be very beneficial. Cost optimization that focuses on breakeven cost—computed at the system or total CapEx level—would be desirable. The value of the metric should also consider the LCOE that might be required to support the economic deployment of wind energy in a given region.

2.3 ADVANCES IN THE WIND ENERGY CONVERSION TECHNOLOGIES AND ASSOCIATED ELECTRICAL COMPONENTS

The conversion of wind energy into electrical energy is carried out through a generator, coupled to the wind turbine. The kinetic energy of the wind is converted into rotational motion, which in turn is coupled with a generator. A wind power conversion method (WPCM) considers the transformation of energy in two steps. In the first step, the energy of wind in the shape of air float rotates the turbine due to which the kinetic energy of wind is converted into mechanical power. In the second step, the mechanical strength of turbine rotates the generator, which converts mechanical strength into electrical energy. So, the essential components of a wind strength conversion framework are a turbine, a gearbox, an alternator, and power converter configurations [26–28]. Sometimes, a transformer is also required for grid connection. A strength converter is placed between the grid and the generator. There are numerous viable technical configurations of wind strength conversion systems. In some configurations, gear arrangement is present, and in some, it may not be. The same happens with the power converter and some without it. The power output can be either DC or AC.

The growing desire to allow variable-speed operation and produce high-quality power is being met by a new generation of power electronics and circuit designs. The advent of medium voltage/high amperage power electronics with high-speed switching rates, such as insulated gate bipolar transistors (IGBTs), has allowed a revolution in the operation of wind turbines. However, wind turbine power converter designs present a unique problem to circuit designers. Wind turbines generally operate far below their power ratings (30%–40%) because of the unpredictable wind velocities at most wind sites. This means that generators and power converters that are designed for normal industrial applications and achieve their highest performance at rated power are inefficient for wind turbine operation. New designs are being explored that aim to achieve their highest performance at below-rated power or have a relatively flat performance curve. Other designs are being explored that employ novel circuit designs to reduce the number of switches, such as matrix converters (MCs). Yet, other designs explore the potential advantages of novel switch materials, such as silicon carbide that may allow higher voltages, currents, and operating temperatures in smaller packages. For the wind turbine for offshore operation, a design approach is needed to provide extremely high reliability. Power converters, computer controls, and other power components must be extremely robust or incorporate redundancy to reduce nuisance faults that are much more difficult to deal with on remote offshore installations.

Enhancing wind power generation with wind energy conversion system (WECS) technologies has relied for many years on the common trend of maximizing electricity generation whereby continual installations of wind power grid infrastructures are mandatory. For instance, this trend involves the deployment of many wind farms across vast areas with the intention to capture wind resources over broader geographic ranges. Furthermore, various methods of design engineering have been implemented to enhance WECS technologies. Accordingly, increasing the radius of the swept area of wind turbine blades for extracting energy from a larger volume of air was one of the methods revealed to enhance WECS design at the component level [30]. The increasing electric power generation, however, requires that wind conversion technologies meet certain criteria. This means that power maximization demands should be considered in association with the number of costs and time needed to develop and use WECS technologies, in addition to the efficiency and reliability of the approach employed [31]. A physical prototyping-based engineering approach for harnessing maximum wind power from wind sources is not appropriate. For example, enlarging the radius of the swept area of turbine blades to increase power production in particular requires much time and high material costs. Besides, the installation of wind farm infrastructure requires large land resources, which bring another challenge to the process of power generation. Furthermore, the physical prototyping-based design approach relies on the textual specification, which is ambiguous to analyze, and testing and validation processes could lead to erroneous results, which are not reversible. It also poses complications for designing and implementing robust power management systems for WECSs.

Nowadays, the fundamental goal of enhancing WECSs is to broaden the scales of wind energy extraction from varying wind speed ranges for significantly maximizing electricity generation with remarkably decreasing costs. This principle of enhancing wind energy conversion should be met by ensuring the safety and integration of WECS technologies such as generators, power electronic converters (PECs), and grids. In recent years, WECS technologies have promisingly improved and this has enabled to maximize wind power generation at fewer costs [32,33]. While this improvement is still continuing, maximizing power efficiency is complex because it depends on wind generators, PECs, control systems, environmental conditions, etc. Present research indicates that the implementation of efficient power management systems for electric generators and PECs can have a significant impact in increasing wind energy harvesting and hence enhancing WECS efficiencies [34,35]. Such a management system should also take into account power system disturbances such as the grid frequency variation, the real power disturbance, and the voltage flicker at the buses of the power grid, which are the results of an intermittence characteristic of wind speed [36–39].

Straight interfacing of wind energy conversion system to the power grid is generally not feasible due to mismatching of magnitude and frequency; hence, proficient power converters are needed for connecting wind energy conversion system to the power grid. Earlier, best-investigated choice of power converters was the AC-DC-AC converter [40–44], but recently the best choice of the power converter is the MC [45–50]. The AC-DC-AC converters are also called traditional converters or back-to-back (BTB) converters. The merit of traditional converters is their high-power concentration and moderately low cost, but the presence of bulky DC link increases its dimensions so their heaviness and volume also increases, which leads to their untimely breakdown [49,51]. Another serious drawback of traditional converters is their sensitivity to communications signals and unwanted signals due to which there may be fault in the system.

A WECS has different PECs, which act like an imperative function for converting wind generator output power in changeable voltage and changeable frequency form and finally sending it to the preset voltage and preset frequency grid system. A distinctive arrangement of PECs consists of an alternating current to direct current converter at the generator side and a direct current to alternating current converter at the grid side interconnected with the help of a component called DC link. This component can be capacitive in voltage source converter (VSC) and inductive in current source converter (CSC) [52]. The advancement of technology in electronics at the power level is doing an excellent job in enhancement of the efficiency of turbines and in designing efficient generators used in WECS. The different configurations of power conversion include (1) noncontrollable AC to DC converter, (2) noncontrollable AC to DC converter with direct current boost, (3) phase controlled converter, (4) entirely controllable pulse-width modulation (PWM) AC to DC converter, (5) matrix converters, (6) three-phase to three-phase matrix converter, (7) sparse matrix converter, and (8) Z-source matrix converter [26–28,53].

Of these converters, due to many benefits of MC researchers are promoting their utility in wind energy systems. The MC has enormous advantages over BTB converters. The main advantages are superior voltage magnification, swift control mechanism, absence of a DC link capacitor, enormously quick transient response, and no commutation problems [26–28]. Likewise, it has a provision of the smooth control system, which can change frequency, phase angle, voltage, and input power factor as per requirement. In WECS, the energy surge is toward the power grid; hence, the MC, which has energy surge in both directions, is not so much beneficial in all applications. So, here another topology of MC is used, which has a little bit of change in configuration, and is called an indirect MC. Also, in MC there is no requirement for extra inactive circuit parts or transformers to enhance voltage level.

There are, however, some primary constraints of MCs. Those are the limit in amplification of voltage, the intricacy of the managed circuit, and the large amount of energy electronic gadgets. These have resulted in a totally wide scope of research on novel configurations of strength converters, novel switching techniques, novel management techniques, and gentle computation for one to make the MC extra talented and extra trustworthy in assessment to the traditional converters. The modern-day research trends appear to be on control algorithms of MCs so that maximum performance can be accomplished.

Wind energy conversion technologies can be differentiated based on some criteria, and their performances can differ accordingly. For instance, there are two main types of modern wind turbines: HAWTs and VAWTs. The majority of large-scale wind turbines are horizontal axis due to their greater efficiency and power output. Moreover, WECS technologies can also be classified as constant-speed and variable-speed technologies. Based on this classification criterion, various types and topologies of wind generator technologies have been introduced for generating electricity from wind resources. The constant-speed-based squirrel cage induction generator (SCIG) and variable-speed-based generator technologies such as doubly fed induction generator (DFIG), permanent magnet synchronous generator (PMSG), and excited synchronous generator (EESG) are among the most prominent in the modern wind farm industry.

There are two different drive systems for converting rotational power into electrical power: indirect and direct drive. An indirect drive system uses a gearbox to increase the rotational speed of the shaft that drives the generator. The indirect drive configuration often uses an asynchronous SCIG, wound rotor induction generator (WRIG), or DFIG. The direct drive does not use a gearbox because there is a full-scale power converter with multipole generators: EESG or PMSG. There are also configurations that use gearboxes with a full-scale power converter, and thus, bigger and more expensive generators are unnecessary. These indirect drive types use smaller generators: wound rotor synchronous generator (WRSG), PMSG, and asynchronous SCIG. The configuration that uses EESG is the most reliable, powerful, and expensive configuration. Wind turbines with DFIG configuration are the most commonly offered type by the major manufacturers. The trend is toward full-scale power converter configurations with multipole PMSG, because they reduce losses and they are lighter than types with EESG [54].

As mentioned earlier, the most recent WECSs generally depend on variable-speed generator technologies because of their outstanding efficiencies and wider possibility for future enhancement. In recent days, DFIG- and PMSG-based variable-speed WECS technologies are closely competed in the global wind energy commercial market [27]. Furthermore, the performance of WECSs also relies on the type of mechanical linkage between wind turbine and generator shaft: gearbox, and direct drive technologies. For instance, among the leading variable-speed technologies, multiple- and single-gearbox systems with DFIGs are usually characterized to have low dynamic performance and high-energy harvesting efficiency per cost, whereas single-gearbox and direct drive systems with PMSGs have high dynamic capability and superior power efficiency, but the PMSG-based WECSs are generally costly [55–57]. Yet, even though DFIG WECS has been recently reported to have better cumulative advantages, future trends of research studies indicated that PMSG WECS could become the leading choice for wind farm application as its operation is smoothly compatible with the extended voltage and power scales of its electrical conversion components [58,59]. Hence, the optimization of the electrical components of PMSG-based WECS is one of the major themes of recent and future research studies in the field of wind power engineering [60]. At the same time, EESG-based WECS is currently under continuous research studies for the better enhancement of its design efficiency in terms of cost, size, and weight though it is relatively less popular due to its cumulatively compromised performance in wind energy harvesting [56,61].

Moreover, PECs have a huge impact on the overall performance of grid-connected WECS technologies. Among these technologies, the two levels (2L)—CSC [62,63] and VSC [58,64] topologies, in BTB configurations were conventionally employed in small- and medium-scale wind farms for the last several decades; they were usually compatible with DFIG-based WECS technology. Here, one of the main drawbacks of DFIG-based WECS is that it does not maintain operational compatibility with power converters of increasing power and voltage capacities [65]. Nevertheless, besides its considerable cost advantages, this technology is largely suitable for the vast application in small- and medium-scale onshore wind generation, particularly with BTB 2L-VSC [66,67]. However, modular multicell converter (MMC) [68–70] has been under continuous physical design development with different voltage capacities and is recently being considered the state of the art, particularly for application in the PMSG-based wind farm industry with large-scale electricity production. The main attractive feature of PMSG-based WECS design is that the capacity of its PEC can be scalable to increasing voltage levels, which makes the application of this technology highly desirable for multi-mega-scale offshore wind energy deployment though its higher cost is still the major impediment. Several additional designs of converter technologies were also proposed in the multiple studies for applications in the wind power industries. These include diode clamped converter (DCC) [67,71], neutral-point-clamped (NPC) converter [72,73], and active neutral-point-clamped (ANPC) converter [74,75], and they were introduced to be employed in wind farms of large-scale power capacities that are mainly based on PMSG systems. Similar studies also indicated that these converters are yet to be sufficiently matured for smooth practical applications in recent wind farm trends. Hence, more improvements in converters' limitations with operation and maintenance (O&M) costs, weight, size, and power conversion capability are needed.

2.3.1 Advances in Electrical Components

Wind generation of electricity places an unusual set of requirements on electrical systems. Most applications for electrical drives are aimed at using electricity to produce torque, rather than using torque to produce electricity. Applications that generate electricity from torque usually operate at constant rated power. Wind turbines, however, must generate at all power levels and spend a substantial amount of time at low-power levels. Unlike most electrical machines, wind generators must operate at the highest possible aerodynamic and electrical efficiencies in the low-power/low-wind region to squeeze every kilowatt-hour out of the available energy. Traditional electrical machines and power electronics disappoint because in most motor applications, there is power to spare and efficiency is less important in this low-power region. For wind systems, it is not critical for the generation system to be efficient in

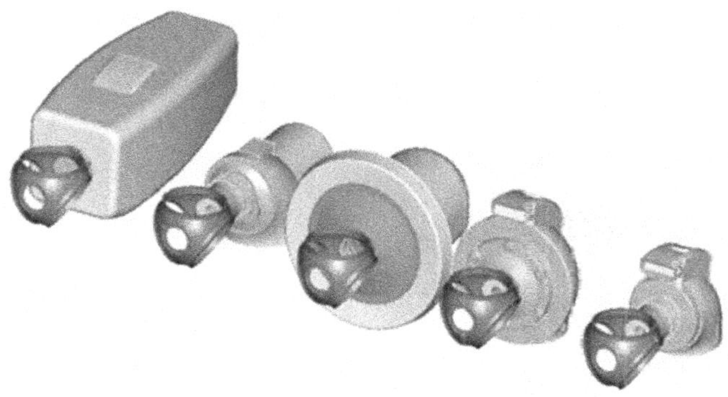

FIGURE 2.10 Unique drivetrain designs under development to reduce size and cost [4,5].

above-rated winds where the rotor lets energy flow through to keep the power down to the rated level. Therefore, wind systems can afford inefficiencies at high power while they require maximum efficiency at low power—just the opposite of almost all other electrical applications in existence [4,5,7,24,25].

Converting torque to electrical power has historically been achieved using a speed-increasing gearbox and an induction generator. Many current MW-scale turbines use a three-stage gearbox consisting of varying arrangements of planetary gears and parallel shafts. Generators are either squirrel cage induction or wound rotor induction, with some newer machines using the doubly fed induction design for variable speed, in which the rotor's variable frequency electrical output is fed into the collection system through a solid-state power converter. Full-power conversion and synchronous machines are drawing interest due to their fault-ride-through and other grid-supported capacities.

Due to fleet-wide gearbox maintenance issues and related failures with some past designs, it has become standard practice to perform extensive dynamometer testing of new gearbox configurations to prove durability and reliability before introducing them into serial production. The long-term reliability of the current generation of MW-scale drive trains has not yet been fully verified with long-term real-world operating experience. There is a broad consensus that wind turbine drive train technology will evolve significantly in the next several years. Several unique designs are under development to reduce drive train weight and cost while improving reliability (see Figure 2.10). One approach for improving reliability is to build a direct drive generator (DDG) that eliminates the complexity of the gearbox. The trade-off is that the slowly rotating generator must have a high pole count and be large in diameter. Depending on the design, the generator can be in the range of 4–10 m in diameter and can be quite heavy [4,5,7,24,25].

The decrease in cost and increase in availability of rare earth PMs are expected to significantly affect the size and cost of future permanent magnet generator designs. Permanent magnet designs tend to be quite compact and lightweight and reduce

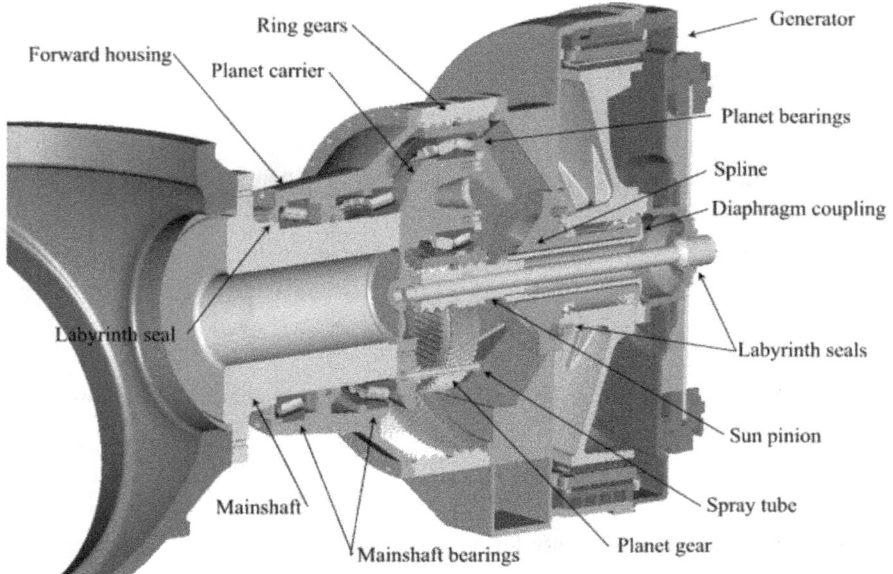

FIGURE 2.11 Single-stage drivetrain uses a low-speed generator [4,5].

electrical losses in the windings. A 1.5-MW DDG using rare earth PMs has been studied, and a prototype has been built. This design uses 56 poles and is only 4 m in diameter, versus 10 m for a wound rotor design. This machine has undergone testing at NREL's National Wind Technology Center.

A hybrid of the direct drive approach that offers promise for future large-scale designs is the single-stage drive using a low-speed generator (see Figure 2.11). This allows the use of a low-speed generator that is significantly smaller than a comparable direct drive design. The WindPACT drive train project has developed a prototype for such a drive train. This design uses a single-stage planetary drive operating at a gearbox ratio of 9.16:1. This gearbox drives a 190 RPM, 72-pole, permanent magnet generator. This approach, which reduces the diameter of a 1.5-MW generator to 2 m, was fabricated and also tested on the dynamometer at NREL's National Wind Technology Center.

Another approach that offers promise for reduced size, weight, and cost is the distributed drive train. This concept is based on splitting the drive path from the rotor to drive several parallel generators. Studies have shown that by distributing the rotor torque on the bull gear over many parallel secondary pinions, a significant size and weight reduction is achieved. In 2006, Clipper Windpower developed a 2.5 MW prototype (Figure 2.12), which incorporates this approach and is currently in the new 2.5-MW Liberty turbine. Enercon had developed a MW-scale DDG with a wound rotor salient pole design with 80 magnetic poles. However, wound rotor DDGs for growing MW-scale machines are 10 m and more in diameter, making their transport over land problematical unless techniques such as segmented rotor designs are employed [4,5,7].

FIGURE 2.12 Clipper Wind Power's new 2.5-MW turbine incorporates an innovative distributed drivetrain [4,5].

The decrease in cost and increase in availability of rare earth PMs are expected to significantly affect the cost of future generator designs. Neodymium–iron–boron, the PM material of choice, has a density close to steel with a magnetic flux density close to that of an electromagnet. However, this magnetic flux density is achieved without the additional mass, size, and cost of the copper windings required for an electromagnet. This allows the creation of smaller generators. A preliminary design for a 1.5-MW DDG using rare earth PMs has been studied under the WindPACT Project at the US DOE's NREL. This design uses 56 poles and is only 4 m in diameter versus 10 m for a wound rotor design [4,5].

2.4 ADVANCED CONTROL METHODOLOGIES FOR WIND TURBINE AND WIND ENERGY CONVERSION SYSTEM

Advanced control methodologies, as applied to wind turbines, are designed primarily to reduce the COE in one of two ways: increase energy capture or reduce mechanical loads. The overall wind energy control system is also designed to maximize and stabilize wind energy conversion into power fed to the grid. To increase energy capture, researchers are exploring a technique called "model-referenced adaptive control." This is a control technique in which the controller constantly adapts itself in an attempt to get the real plant (the turbine) to behave like a plant model. In this way, the turbine controller can adapt itself to perform optimally despite rotor variability, blade erosion, and site-specific parameters. In an attempt to reduce turbine mechanical loading, a technique called "full-state feedback" with "disturbance accommodating control" (DAC) and periodic control has been implemented. In such a system, many turbine states are fed back through a control loop and a DAC controller with

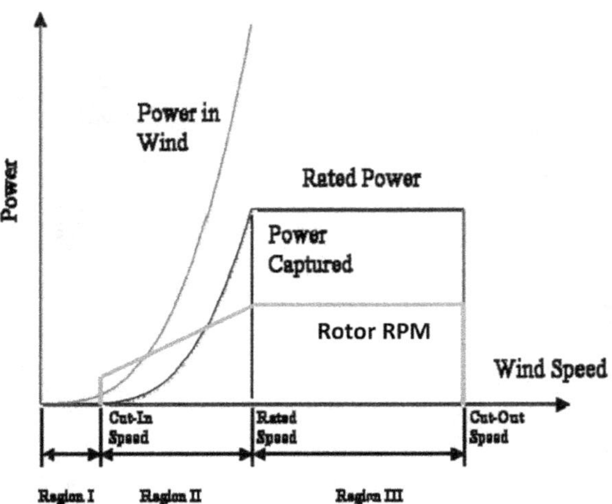

FIGURE 2.13 A typical power output versus wind speed curve [3,5,24].

time-varying gains to decide what controls should be activated. A DAC controller has the advantage of allowing disturbances (such as turbulent eddies) to pass through the system while tightly controlling parameters of interest, such as rotor speed and blade loads. Employing time-varying gains that are synchronized to the azimuthal position of each blade allows the advantage of taking into account turbine parameters that vary during a rotor cycle such as yaw inertia and gravity loads. It is believed that such a system will show a significant reduction in turbine rotor, shaft, and tower loads [4,5,7,76,77].

Today's controllers integrate the signals from dozens of sensors to control rotor speed, blade pitch angle, generator torque, and power conversion voltage and phase. The controller is also responsible for critical safety decisions, such as shutting down the turbine when extreme conditions are realized. Today, most turbines operate at variable speed, and the control system regulates the rotor speed to obtain peak efficiency in fluctuating winds by continuously updating the rotor speed and generator loading to maximize power and reduce drive train transient torque loads (see Figure 2.13). Operating variable speed requires the use of power converters to make the generated power match the grid frequency. The power converter also enables turbines to deliver fault-ride-through protection, voltage control, and dynamic reactive power support to the grid [4,5,7,9,76,77].

Advanced wind turbine controls can reduce the loads on wind turbine components while capturing more wind energy and converting it into electricity. NREL is researching new control methodologies for both land-based wind turbines and offshore wind turbines. At the National Wind Technology Center, researchers design, implement, and test advanced wind turbine controls to maximize energy extraction and reduce structural dynamic loads. These control designs are based on linear models of the turbine that are simulated using specialized modeling software. The resulting advanced control algorithms are field tested on the National Wind Technology

Center (NWTC)'s Controls Advanced Research Turbines (CARTs). Researchers are also studying blade pitch and generator torque and employing advanced sensors to optimize power capture and reduce wind turbine loads. Among its advanced research turbines, the NWTC features two 600-kW CARTs that test new control schemes and equipment for reducing loads on wind turbine components. The CARTs are equipped with light detection and ranging systems to provide some "look-ahead" capability for the turbines, allowing them to adjust to changing wind conditions using feedforward controls.

In addition, NREL's Simulator for Wind Farm Applications (SOWFA) employs computational fluid dynamics to allow users to investigate wind power plant performance under a full range of atmospheric conditions and terrain types. The tool allows researchers and wind power plant designers to examine and minimize the impact of turbine wakes on overall plant performance, either by judiciously locating the wind turbines or by turning some turbines slightly out of the wind to redirect their wakes. SOWFA has demonstrated that wind turbine wakes can reduce the energy production of downstream wind turbines, impacting the overall performance of a wind power plant. Studies show that by coordinating turbine controls to minimize wake effects, the overall wind power plant output could be increased by 4%–5% [4,5,7,76,77].

The intermittent nature of wind energy significantly affects the dynamic behavior and transient stability of power systems [78]. This implies that one of the significant challenges of WECS is to provide power smoothing during transient fault conditions. One approach for power smoothing is to use energy storage system (ESS) devices that include batteries, flywheel, and compressed air storage [79,80]. However, this approach adds high extra costs to WECSs. An alternate approach is to build a virtual system [81–84], which can be built through the implementation of various automated control strategies, which would ultimately regulate different operating parameters of WECSs. Recently, a dual objective—control technique—was presented to reduce the torque ripples of the turbine shaft by implementing the frequency separation principle [85,86]. In addition, the real current control method [87], the generator torque control strategy [88], real and reactive power control [89], and independent pitch control [90] were employed to streamline the generator output power. Moreover, the kinetic energy optimization-based inertial control strategies [84,91] were demonstrated through simulation to be identified as outstandingly outperforming the virtual power smoothing approach.

The condition monitoring systems, which comprise combinations of sensors and signal processing equipment, are used to monitor the status of major operating components. Monitoring may be online (and thus provide instantaneous feedback of condition) and offline (the data are collected at regular time intervals using measurement systems that are not integrated into the wind turbine). A detailed view of a wind turbine condition monitoring system typically presents the following data: wind turbine data (wind speed, active and reactive power, yaw angle, and command, operational, and fault status), electrical and mechanical data, meteorological data, grid data, and statistical data. Different monitoring techniques in wind turbines are used to prevent damage or failure of components. The most common condition monitoring techniques that are used include process parameters, performance monitoring, vibration analysis, acoustic emission, ultrasonic testing (UT), oil analysis, strain measurement, electrical effects, shock pulse method, radiographic inspection, and thermography [76,77].

Reliable, efficient, and effective control strategies are required to be designed in WECSs to maintain the systems' comprehensive performance. In this regard, hybrid [92] and model-based predictive [93] control design strategies have been highly recommended in recent studies due to their robust performances that would enable them to circumvent the nonlinear and unpredictable characteristics of WECS operations. Hybrid control strategies were demonstrated in [94] as being designed by combining hard control that includes proportional integral derivative (PID), sliding mode control (SMC), and adaptive control and soft control that involves fuzzy logic control (FLC), neural network control (NNC), and genetic algorithm (GA) to make use of the cumulative advantages of hard and soft control strategies by reducing the control complexity of the systems in improving efficiency and dynamic stability. In practical applications, hybrid design strategies could optimize the systems by alleviating the respective limitations of PID, SMC, FLC, NNC, etc., and by fusing their respective advantages. In some recent studies [95,96], hard and soft combinations were characterized as more efficient strategies.

In some other works [97–99], model-based predictive control (MBPC) was also demonstrated to have appealing features, which can be utilized to develop efficient and cost-effective power smoothing system. In general, the ultimate goal of implementing these strategies (including hybrid and MBPC) is to establish stringent power control systems that eventually meet advanced operation requirements (power reliability, fault ride through [FRT] capability, maximum power production, and overall cost optimization) for WECS technologies. Moreover, these control design strategies can enhance WECS technologies by reducing their overall design complexities and thereby achieving rapid dynamic and transient responses.

Model-based design approach (MBD) has also been used [100–103] for evaluating WECS control system designs based on the proposed control design strategies. In the typical case, a WECS control system model can be simulated, tested, and preliminarily validated based on a model predictive algorithm and by using MATLAB®/SIMULINK software platform, external target computer, and controller. MBD methodology was generally considered by large study projects [104,105] as a compelling approach as it could ultimately enable to achieve the efficient and reliable wind energy conversions, which will result in more energy transferred to the electrical power systems without needing to build complex infrastructures, and for the similar scales of energy extracted from the wind resources.

2.5 PERSPECTIVES ON MAINTENANCE AND OPERATIONAL ISSUES

While wind turbines are carefully designed with a good selection of materials, rough environmental conditions and high dynamic loading of the components demand regular maintenance. Common causes of wind turbine failure can be component failure, control system failure, high wind, waves, lightning, grid failure, icing, etc. NREL study indicates that in the past, the three most failing rates have happened in blades/pitch, electric system, and control system [4,5,7,106–108]. To achieve the best possible power production efficiency of the installed wind turbine, a high-reliability level should be reached. High reliability can be achieved by understanding and minimizing the failures of the system. The reliability of a wind turbine does not depend exclusively on the reliability of its subsystems, but also on external, indirect factors such as the maintenance strategy used, the existence of spare parts, the time needed to repair, the existence of control systems of the subsystems, the type of wind turbine related to the operational environment (onshore, nearshore, and offshore), and the training of the personnel. Maintenance planning and identification of components that are prone to failures could make maintenance more efficient and lead to a reduction in failure events. The maintenance strategy should take into account reliability improvement, reduction in maintenance cost, and enhanced condition monitoring.

There are numerous operational issues with wind power. High up-front investment cost, high levelized cost of wind energy, the complicated and long permitted procedures of a wind power plant in many countries, together with the long planning phase (including environmental, engineering, feasibility, and site-specific studies) are required pose formidable challenges. However, the high feed-in tariffs and the policies of subsidies that most of the governments follow to promote wind energy make wind a competitor with the existing conventional energy sources that help. The intermittent nature of wind power is the biggest problem for its use since it causes problem of grid balancing. The use of a storage device or another dispatchable source to balance the grid adds significantly to the overall cost. Wind forecasts are important to avoid extreme situations. Finally, a modern wind turbine is designed to work for an estimated time of 20 years. After this point, it has to be either renewed or recycled. While steel, copper, and aluminum used in wind turbine are fully recyclable, the glass-reinforced plastics (based on polyester or epoxy) used in the rotor blades have proven challenging to recycle. Options for blade material waste treatment are mechanical recycling (which is a labor-intensive process and uses the produced material as a filler in artificial wood, cement, or asphalt production), incineration, pyrolysis, and landfill, which is the worst option [4,5,7,106–108].

2.6 COSTS AND THEIR POSSIBLE IMPROVEMENTS

The balance of the wind farm station consists of turbine foundations, the wind farm electrical collection system, wind farm power conditioning equipment, supervisory control, and data acquisition systems, access and service roads, maintenance buildings, service equipment, and engineering permits. The combination contributes about 20% to the installed cost of a wind farm. The cost-of-energy metric remains the principal technology indicator, incorporating the key elements of capital cost, efficiency, reliability, and durability. The unsubsidized cost of wind-generated electricity ranges from about \$0.5 to \$0.085/kWh for projects completed in 2006 [24]. O&M costs have dropped significantly since the 1980s due to improved designs and increased quality. Reference [24] presents data that show O&M expenses are a significant portion of the total system COE. O&M costs with the latest generation of turbines have been reported O&M costs below \$0.1/kWh. The fraction of time during which the equipment is ready to operate (commonly known as availability) is now over 95% and often reported to exceed 98%.

The LCOE is the price of electricity required for a project where revenues would equal costs, including making a return on the capital invested equal to the discount rate. According to the International Renewable Energy Agency (IRENA) [13] for 2014, the LCOE range was for onshore projects of 0.06–0.12 USD/kWh and for offshore projects of 0.10–0.21 USD/kWh. The regional weighted average installed costs in 2014 for onshore wind range from around 1.280 USD to 2.290 USD/kW. The average installed cost ranges in 2014 for offshore wind projects were 2.700–5.070 USD/kW. The variable O&M costs for onshore wind projects in 2014 for the USA ranged from 0.01 to 0.024 USD/kWh, a range similar to many countries in Europe [4,5,7,10,14].

The DOE Wind Program has conducted cost studies under the WindPACT Project that identified many areas where technology advances would result in changes to the capital cost, annual energy production, reliability, O&M, and balance of the station. Many of these potential improvements, summarized in Table 2.1, would have significant impacts on annual energy production and capital cost. Table 2.1 also includes the manufacturing learning–curve effect generated by several doublings of turbine manufacturing output over the coming years. The learning–curve effect on capital cost reduction is assumed to range from zero in a worst-case scenario to the historic level in a best-case scenario, with the most likely outcome halfway in between. The most likely scenario is a sizeable increase in capacity factor with a modest drop in capital cost (over the 2002 levels of each). While by 2006, capacity factor was already increased from just over 30% to almost 35%, another modest increase in capacity factor is still possible. While the capital cost was slightly increased from 2002 to 2006, due to technological development, a 10% capital cost reduction when compare to 2002 baseline is still possible. Table 2.1 describes possible technological improvements and associated cost reductions in various technical areas of the wind energy system.

2.7 OFFSHORE WIND ENERGY

In the middle of the 1990s, strong wind in the high sea propelled the interest in building offshore wind turbines. While the cost and technical difficulties in building offshore wind turbine are higher than onshore, higher wind speed (25% greater than onshore), lower frame of tower, ability to face greater speed, and less roughness in sea level resulted in the pursuit of offshore wind energy. The first country to build offshore wind turbine successfully in the middle of the sea was Denmark in 1991. From 1991 to 1997, Denmark, the Netherlands, and Sweden performed a couple of prototype-based wind turbine operations in the deep sea [109–111]. Requiring less land area, less noise pollution, and no greenhouse gas emission with an addition of great environmental protection, offshore wind energy was getting popular among the European countries. By the end of 20th century, around 50 countries were running wind turbines generating approximately 17,500 MW of electricity. From the start of the 21st century, within 6 years, 21 offshore plants had been built in different countries namely Denmark, Sweden, Ireland, Germany, and Netherlands [109–111]. China constructed 59 wind

TABLE 2.1
Areas of Potential Technology Improvement and Cost Reduction [4,5,7]

Technical Area	Potential Advances	Cost Increments (Best/Expected/Least, Percent): Annual Energy Production	Cost Increments (Best/Expected/Least, Percent): Turbine Capital Cost
Advanced tower concepts	• Taller towers in difficult locations • New materials and/or processes • Advanced structures/foundations • Self-erecting, initial, or for service	+11/+11/+11	+8/+12/+20
Advanced (enlarged) rotors	• Advanced materials • Improved structural aero-design • Active controls • Passive controls • Higher tip speed/lower acoustics	+35/+25/+10	−6/−3/+3
Reduced energy losses and improved availability	• Reduced blade soiling losses • Damage tolerant sensors • Robust control systems • Prognostic maintenance	+7/+5/0	0/0/0
Drivetrain (gearboxes, generators, and power electronics)	• Fewer gear stages or direct drive • Medium/low-speed generators • Distributed gearbox topologies • Permanent magnet generators • Medium voltage equipment • Advanced gear tooth profiles • New circuit topologies • New semiconductor devices • New materials (GaAs, SiC)	−8/+4/0	−11/−6/+1
Manufacturing and learning curve	• Sustained, incremental design and process improvements • Large-scale manufacturing • Reduced design loads	0/0/0	−27/−13/−3
Total		+61/+45/+21	−36/10/+21

farms by 2005 including 1,883 turbine generator. The Northern side of the world has good wind energy, but the icing on the turbine blade is the main obstacle. On average, 20% of power loss occurred annually due to this problem. In 2011, Virk, Homola, and Nicklasson from Narvik University College carried out a numerical study on a HAWT (5 MW) about atmospheric ice accretion. The result of this study indicated that both the blade size and relative section velocity of the blade affected ice growth. Near the root section, the icing was less. In the blade section, from center to top, a significant change in icing was noticed with the variation in atmospheric temperature. Furthermore, the result also proved that the icing could be managed by optimizing geometric design parameters [110,112–115]. As shown in Figure 2.14, interest in

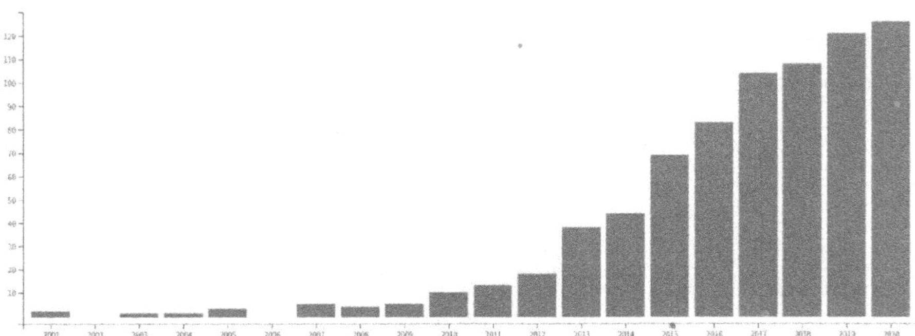

FIGURE 2.14 Number of published papers in design optimization of offshore wind turbine (2000–2020) [112].

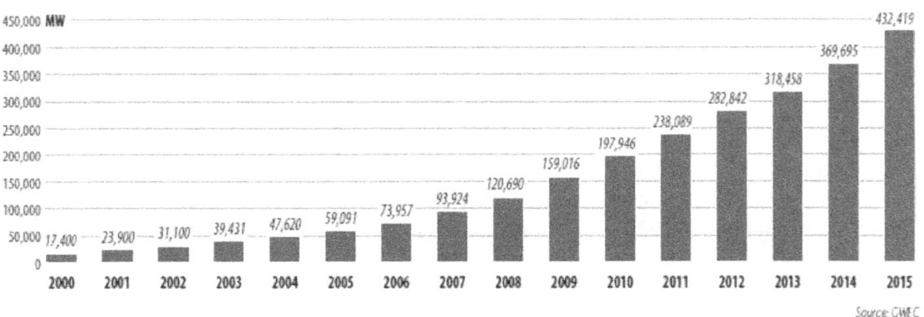

FIGURE 2.15 Global cumulative installed wind capacity (2000–2015) [119].

offshore wind turbine increased significantly over the last decade, which is in line with growth in wind energy as shown in Figure 2.15.

In recent years, micro-wind energy has made a lot of impact due to its low-speed operation along with low-power applications resulting in cost reduction and simplicity. For example, Massimiliano, Marcello, and Gianpaolo from Italy designed a simple, effective, and low-cost micro-wind energy conversion system run by a PMSG and boost-converted and voltage-oriented controller. The system performed a reliable operation at maximum power. It also showed a promising quick response with variable wind speeds. It had low losses and almost zero reactive power exchanged with the power grid. In comparison with a conventional wind turbine of the same maximum power range and with the same average wind speed, it could generate twice the energy produced by the generator [110,112–116]. Recent years have seen a boom in offshore deployments in the Baltic Sea and North Sea. While the marine environments would seem similar, it is believed that weather extremes along the Northern Atlantic Coast may present additional challenges for developers. An improved understanding of wind and wave loading and bottom conditions in these novel environments is critical for future development. The challenging marine environment also makes routine access to machines less than routine.

A recent study shows that the USA has the second highest offshore wind resource outside of the European Union [109,112,117]. The European experience has already demonstrated a range of methods for deploying wind energy in shallow waters (<30 m) using driven piers or tripod foundations. To optimize the cost expenditures for these foundations, the largest rotor diameters and ratings possible must be considered. GE Windpower developed a 3.6-MW turbine for offshore deployment, and models of this machine are being installed in European waters. The challenge for the US shallow water environments appears to be a lack of well-documented knowledge about the potential wind, wave, and ice loadings in desirable sites. Preliminary studies show that significantly greater resources can be accessed if turbines can be deployed on floating platforms in relatively deepwater (up to 200 m). These installations, however, require inexpensive floating or tension leg platforms (TLPs) with affordable and reliable anchoring systems.

US offshore wind energy resources are abundant, indigenous, and broadly dispersed among the most expensive and highly constrained electric load centers. The US DOE's Energy Information Agency shows that the 28 states in the contiguous 48 states with a coastal boundary use 78% of the nation's electricity [118]. Nineteen offshore wind projects now operate in Europe with an installed capacity of 900 MW. All installations have been in water depths less than 22 m. Although some projects have been hampered by construction overruns and higher-than-expected maintenance, projections show strong offshore growth in many EU markets. In the USA, approximately 10 offshore projects are being considered. Proposed locations span both state and federal waters and total more than 2,400 MW. The current shallow water offshore wind turbine is basically an upgraded version of the standard land-based turbine with some system redesigns to account for ocean conditions. These modifications include structural upgrades to the tower to address the added loading from waves, pressurized nacelles, and environmental controls to prevent corrosive sea air from degrading critical drive train and electrical components, and personnel access platforms to facilitate maintenance and provide emergency shelter [109–117,119–121].

To minimize expensive servicing, offshore turbines may be equipped with enhanced condition monitoring systems, automatic bearing lubrication systems, onboard service cranes, and oil temperature regulation systems, all of which exceed the standard for land-based designs. Today's offshore turbines range from 3 to 5 MW in size and typically have three blades, operate with a horizontal-axis upwind rotor, and are nominally 80–126 m in diameter. Tower heights offshore are lower than land-based turbines because wind shear profiles are less steep, tempering the energy capture gains sought with increased elevation. As shown in Figure 2.16, offshore foundations differ substantially from land-based turbines.

The baseline offshore technology is deployed in arrays using monopiles at water depths of about 20 m. Monopiles are large steel tubes with a wall thickness of up to 60 mm and a diameter of 6 m. The embedment depth will vary with soil type, but a typical North Sea installation will require a pile that is embedded 25–30 m below the mud line that extends above the surface to a transition piece with a leveled and grouted flange on which the tower is fastened. Mobilization of the infrastructure and logistical support for a large offshore wind farm is a significant portion of the system

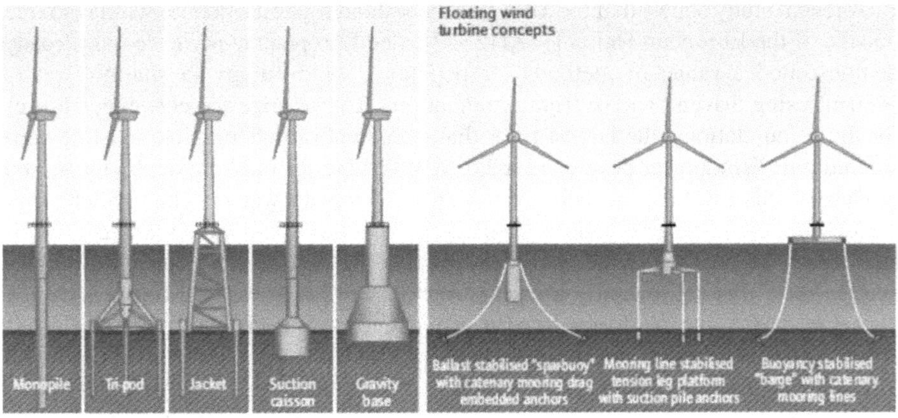

FIGURE 2.16 Fixed bottom foundations and floating offshore concepts [119].

cost. Current estimates indicate that the COE from offshore wind plants is above $0.10/kWh and that the O&M costs are also higher than for land-based turbines due to the difficulty of accessing turbines during storm conditions.

2.7.1 Types of Offshore Wind Turbine Foundations

There are different types of foundations that are used for offshore wind turbines (Figure 2.16). The two main categories are the fixed bottom (for depths up to 50 m) and the floating concepts (for deeper waters). Floating platforms for deeper water are described in detail in the next section. The monopole is the simplest and cheapest type of foundation that is commonly used in the North Sea into the sea bed to a depth of 0–25 m. A gravity-based foundation is normally fabricated in reinforced concrete with ballast, and it uses its own weight to anchor to the seabed at a water depth between 0 and 25 m. An alternative to these two is the suction caisson, in which by removing the water from a caisson onto which the turbine is situated the suction force thus created is used to promote easy installation. A tripod structure is a relatively lightweight three-legged steel structure, in which the frame is submerged in the water, providing good stability and stiffness, and it is used for depths between 20 and 50 m. The jacket is often used in water depths of about 20–50 m and usually consists of a three- or four-legged lattice structure made from tubular steel. Three main types of fixed bed structures are further described below.

2.7.1.1 Monopile Substructure

In the last few decades, the most popular and widely adopted modern offshore wind foundation system has been the monopile foundation. Nearly 81% of all existing European offshore wind turbines consisted of a monopole foundation by the end of 2016 [122,123], such as Horns Rev 1–3, the Anholt Projects in Denmark, the London Array Project in the UK, and the DanTysk Project in Germany. The monopile is generally used in areas with relatively shallow water depth (<40 m). The typical diameter of the steel tubular section is 3–6 m, and the length is 20–50 m, and up to 1,000 tons [124,125]. Depending on seabed characteristics, total applied load, and design criteria, 40%–50% of the steel tubular section is inserted into the seabed to provide resistance by the surrounding soil along the embedded length. A monopile is generally manufactured onshore and then transported to the operation location for installation by pile driving or drilling. Because no seabed preparation is necessary, the installation can generally be achieved within 24 hours [126–128].

2.7.1.2 Tripod Substructure

For larger turbines and deeper water up to 50 m, a tripod, an extension of the monopile, is generally used [129]. Tripods consist of three-legged tripod bases connected to a large diameter central steel tubular section and the seabed. These three piles are embedded 10–20 m into the seabed to provide significant resistance for better stability performance and stiffness of the entire offshore wind turbine substructure [124,130], depending on the special equipment required for driving or drilling. The typical installation of a tripod offshore wind turbine up to 700 tons generally takes 2–3 days [126,131]. Similar to the monopile, the installation of a tripod foundation does not require seabed preparation. However, due to heavier foundations tripod construction and maintenance costs can be higher than those of other base types. In addition, erosion protection is required for the tripod in locations where bottom currents are significant or where sediment is easily eroded. Examples of tripod foundation wind farms are Alpha Ventus, Trianel Windpark Borkum I, and Global Tech I.

2.7.1.3 Jacket Substructure

For deeper-water oil and gas platforms up to 60 m, a jacket or braced frame substructure is generally used [132,133]. The jacket structure is composed of a small-diameter lattice truss. This lattice truss structure is connected by three or four tubular legs that are driven into the seabed. The jacket substructure can be installed down to depths of 10–60 m, and some can be extended to 80 m [129]. The general installation of the jacket substructure can be completed in 3 days. The main advantage of a jacket substructure includes that it is particularly suitable for severe offshore conditions, as truss components offer higher resistance to prevailing ocean waves and current flow in comparison with monopile or tripod structures, and can adjust their application range with geometrical variations without altering the stiffness of the whole structure [134]. The main disadvantage of the jacket substructure is the higher installation and construction cost, and it is always used as a transitional water substructure [124,135]. Due to erosion, the jacket structure's joints generally require long maintenance downtime periods to sustain structural integrity. Some deeper-water wind farms use jacket foundations, for example, Beatrice and Thornton Bank Phases II and III.

2.7.1.4 Gravity-Based Structure

Besides three main fixed foundations for off-shore wind turbines described above two other foundations have been derived from off-shore oil and gas drilling foundations. A *gravity-based structure* (*GBS*) is a support structure held in place by gravity, most notably in offshore oil platforms. These structures are often constructed in fjords due to their protected area and sufficient depth. Before deployment, a study of the seabed must

FIGURE 2.17 Construction of the bases of new windmills for the Thornton Bank Offshore Wind Farm, in Oostende, Belgium [136].

be done to ensure it can withstand the vertical load from the structure. It is then constructed with steel-reinforced concrete into tanks or cells, some of which are used to control buoyancy. When construction is complete, the structure is towed to its intended location. Early deployments of offshore wind power turbines used these structures. As of 2010, 14 of the world's offshore wind farms had some of their turbines supported by GBSs. The deepest registered offshore wind farm with GBSs is Thornton Bank 1, Belgium, with a depth of up to 27.5 m (see Figure 2.17). Newer generations of wind turbines are much larger and deployed in deeper waters, and GBSs are no longer considered competitive in comparison with support structures such as floating moored.

2.7.1.5 Suction Caissons Structure

Suction caissons (also referred to as *suction anchors*, *suction piles*, or *suction buckets*) are a form of fixed platform anchor in the form of an open-bottomed tube embedded in the sediment and sealed at the top while in use so that lifting forces generate a pressure differential that holds the caisson down. They have many advantages over conventional offshore foundations, mainly being quicker to install than deep foundation piles and being easier to remove during decommissioning. Suction caissons are now extensively used worldwide for anchoring large offshore installations, such as oil platforms, offshore drillings, and accommodation platforms to the seafloor at great depths. In recent years, suction caissons have also seen usage for offshore wind turbines in shallower waters [137].

Oil and gas recovery at great depth could have been a very difficult task without the suction anchor technology, which was developed and used for the first time in the North Sea 30 years ago. The use of suction caissons/anchors has now become common practice worldwide. Statistics from 2002 revealed that 485 suction caissons had been installed in more than 50 different localities around the world, in depths of about 2,000 m. Suction caissons have been installed in most of the deepwater oil-producing areas around the world: the North Sea, Gulf of Mexico, offshore West Africa, offshore Brazil, West of Shetland, South China Sea, Adriatic Sea, and Timor Sea. No reliable statistics have been produced after 2002, but the use of suction caissons is still rising.

A suction caisson can effectively be described as an inverted bucket that is embedded in marine sediment. Attachment to the sea bed is achieved either through pushing or by creating a negative pressure inside the caisson skirt by pumping water out of the caisson; both of these techniques have the effect of securing the caisson to the sea bed. The foundation can also be rapidly removed by reversing the installation process and pumping water into the caisson to create an overpressure.

The concept of suction technology was developed for projects where gravity loading is not sufficient for pressing foundation skirts into the ground. The technology was also developed for anchors subject to large tension forces due to waves and stormy weather. The suction caisson technology functions very well in a seabed with soft clays or other low-strength sediments. The suction caissons are in many cases easier to install than piles, which must be driven (hammered) into the ground with a pile driver. Mooring lines are usually attached to the side of the suction caisson at the optimal load attachment point, which must be calculated for each caisson. Once installed, the caisson acts much like a short rigid pile and is capable of resisting both lateral and axial loads. Limit equilibrium methods or 3D finite element analysis are used to calculate the holding capacity (Figure 2.18).

The advantages and disadvantages of the three most common substructure types used for fixed offshore wind turbines as outlined by Chen and Kim [112] are illustrated in Table 2.2. In recent years, the GBS is not as widely used. A suction caisson is largely used for shallow water oil and gas drilling.

There are three logical pathways (Figure 2.19) representing progressive levels of complexity and development that will lead to cost reductions and greater offshore deployment potential. The first path is to lower costs and remove deployment barriers for shallow water technology in water depths of 0–30 m. The second path is transitional depth technology, which is needed for depths where current technology no longer works. This technology deals mostly with substructures that are adapted from existing offshore oil and gas practices. Transitional depths are defined as 30–60 m. The third path is to develop technology for deepwater, defined by depths between 60 and 900 m. This technology will probably use floating systems, which require more R&D to design turbines that are lighter and can survive the added tower motion on anchored, buoyant platforms. The ultimate vision for offshore wind energy is that it

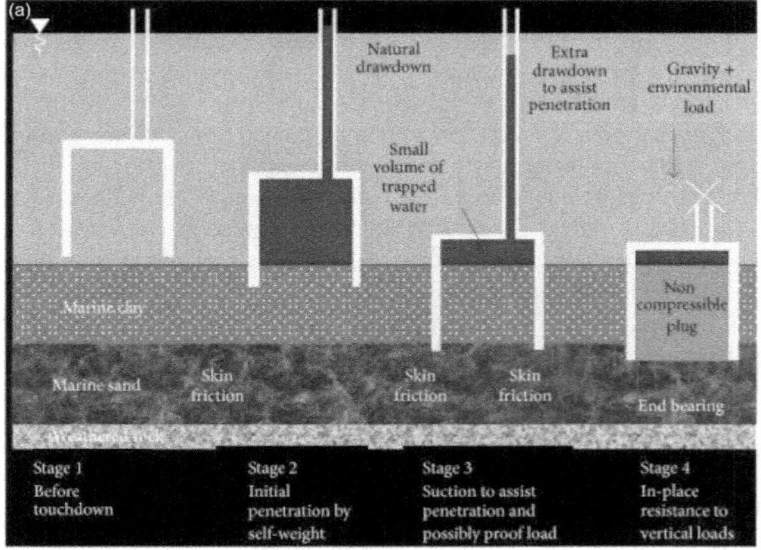

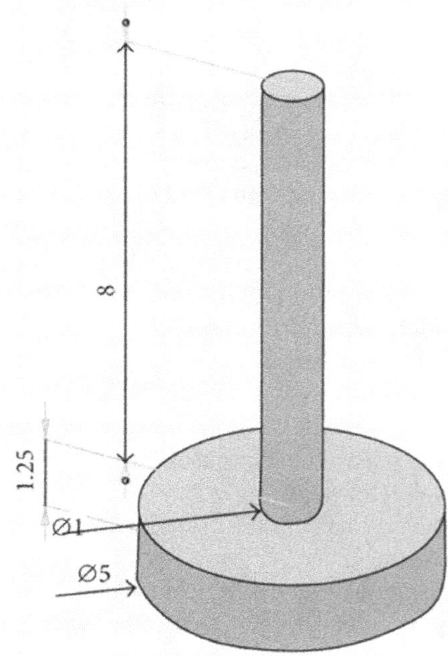

FIGURE 2.18 (a) Suction caisson installation [137]. (b) Suction caisson. Chart with dimension. Designed for supporting structures in deepwaters. The dimension for the top suctions caisson is 8 m × ø1 m and for the bottom suctions caisson is 1 m × ø5 m [137].

TABLE 2.2
Advantages and Disadvantages of Three Main Fixed Offshore Wind Turbine Structures [112]

Monopile

Advantages

1. Work well in sand and gravel soils. No need for seabed preparation.
2. Have a simple design that installs quickly.
3. Adaptable for shallow and deeper installations of various sizes.
4. Cost-effective for installations up to 40 m.

Disadvantages

1. Cost and risks associated with fabrication, installation, and transport increase for larger monopiles required at deeper installations where hydrodynamic loads are an issue.
2. Installation noise can disorient, injure, or kill marine life-sensitive or pressure waves. This includes humpback whales, loggerhead turtles, and manatees.
3. Wind, wave, and seismic loading can negatively affect monopile foundations. This can cause early fatigue damage to the structure if it is not accounted for during installation.

Tripod

Advantages

1. The seabed site does not need advanced preparation before installation.
2. Well-suited for locations where stiff clays or medium-to-dense sands are present and can be used in softer soils too.
3. Become an economical choice for installations at 45 m or more.
4. Provides extra stability to the wind turbine.

Disadvantages

1. Scour/erosion protection may be needed around the tripod base in locations where bottom currents are significant or where sediment is easily eroded.
2. Tripod construction and maintenance costs can be higher than other base types

Jacket

Advantages

1. Can be installed using piles or suction caissons in stiff clays or medium-to-dense sands.
2. Soft-oil installations are possible with longer pile lengths that significantly increase friction resistance.
3. The larger surface area of the lattice configuration may provide an artificial reef location, providing a new habitat for local species.
4. Economical choice using straightforward manufacturing methods.
5. Can be moved by barge.

Disadvantages

1. May allow invasive species to establish and spread.
2. Changes to local water patterns may be detrimental to native marine ecosystems.
3. Higher installation and construction cost.
4. Installations using pile drivers can create underwater noise that may injure or kill some marine life.
5. North Sea installation of jacket foundations has reported ongoing grout joint issues, requiring long maintenance downtime periods to sustain structural integrity.

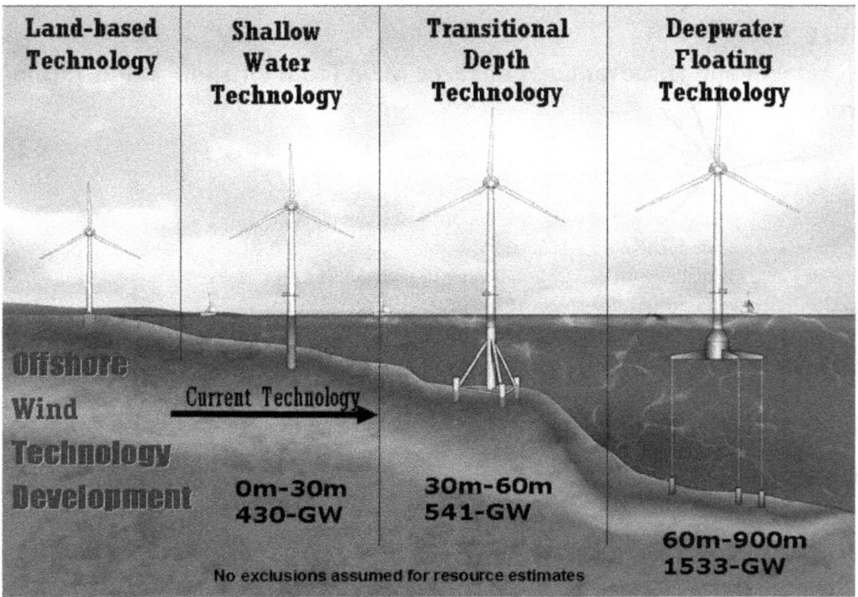

FIGURE 2.19 Offshore development pathways [114]

would open up major areas of the outer continental shelf to wind energy development. This would require the use of deepwater floating platforms that could be mass-produced and assembled in dry docks and then floated out and anchored without extensive assembly at sea. Deepwater technology also avoids the need for long-distance transmission because wind farms can be located much closer to load centers.

New offshore technologies will be required to grow wind turbines to 5 to 10 MW sizes or greater. These technologies may include lightweight composite materials and composite manufacturing, lightweight drive train, modular pole DDGs, hybrid space frame towers, and large gearbox and bearing designs that are tolerant of slower speeds and large scale. The cost of control systems and sensors that monitor and diagnose turbine status and health will not grow substantially as turbine size increases, and high reliability will be essential due to the limited access during severe storm conditions, which can persist for extended periods.

While the onshore wind turbine industry is going strong, the USA is looking toward the possibility of adding offshore wind capacity in the future. Alstom recently signed a contract with Deepwater Wind to supply turbines for the 30-MW Block Island Wind Farm three miles off the coast of Rhode Island, and the company is also part of Dominion Virginia Power's Virginia Offshore Wind Technology Advancement Project, which was one of seven picked by the US DOE's advanced technology demonstration projects. The needs of offshore wind production require different solutions than onshore. The use of different technologies for onshore and offshore wind power projects is another change that has occurred over the past 10 years. While companies used to take the same wind turbine used on land and install it offshore, Alstom took a different approach with its current generation of offshore wind turbines. Alstom Haliade 150 has been designed from the very beginning to operate in offshore conditions. This has driven different aspects of the design of the wind turbine, with one of the main decisions being using direct drive technology. The Haliade 150 is a 6-MW turbine that uses a 150-m rotor. The company plans to continue developing and investing in the improvement of the direct drive solution for its offshore wind turbines, including improving its efficiency and weight [114].

The US DOE predicts that the country will have 404 GW of wind power capacity by 2050, enough to fulfill more than one-third of the nation's electricity demands. With growing consumer demand for clean, renewable energy, the wind industry is the second fastest renewable energy source to grow, according to the US Bureau of Labor. However, there are some drawbacks, such as high construction costs and poor durability associated with wind turbines. An important innovation made in wind turbines is DDGs, which have the ability to generate electricity at the speed of the rotor. However, expensive magnets are needed to substitute the original magnets in order to reach a specific frequency, but with lightweight DDGs and superconductor generators, no expensive material is required to reach optimal performance [118].

2.7.2 Pros and Cons of Onshore vs. Offshore Turbines

Onshore wind turbines are less expensive and one of the cheapest forms of renewable energy. It can boost local economies and less voltage drops between the windmill and the consumer. It creates fewer emissions associated with the transportation of wind structures. However, the onshore wind turbine has limited efficiency due to the unpredictable speed and direction of wind. It can endanger flying wildlife such as birds and bats. It can create a negative visual impact and create noise pollution. It is not able to create power year around due to unpredictable reliance on optimal wind conditions.

Offshore wind turbines are considered recent developments compared with conventional onshore wind turbines. Floating offshore wind turbines (FOWTs) are harder to get to, are more prone to getting damaged, and are more expensive to install and operate. They have increased O&M costs caused by increased wear from wind and waves. Longer wait times are required to correct any potential problems due to more limited access. Currently, they are limited in its ability to benefit local economies. They are, however, more efficient due to consistent wind speeds and direction. Fewer turbines are required to produce an equal amount of electricity. No risk of visual impact and interference with land usage is found They protect aquatic habitats by restricting access to certain waters. Finally, there are no physical restrictions in these locations due to backwind flow. Current technology advancements can strengthen the tower and provide more protection by handling the forces of waves or ice flows. In addition, upgrading nacelles on these turbines can prevent corrosion from the seawater from damaging the electrical internal components. As the direction and speeds of winds are actually getting more and more predictable offshore than onshore with advancements in technology, investing in offshore wind energy will grow at a faster-than-normal rate and may serve as the top renewable energy source [112,116,119,122].

2.8 FLOATING WIND TURBINE

Floating wind, which uses turbines located at sea but not attached to the ocean floor, has several key advantages over traditional onshore or fixed-bottom offshore wind farms. The floating wind has the potential to make a huge contribution to filling the installed capacity gap in renewables and represents a game changer for the energy sector globally. Current forecasts see 2035 as the moment when levelized costs for bottom-fixed and floating wind converge, but it is possible this moment could be reached much sooner. The concept of floating wind turbine is fairly simple: Floating wind uses turbines located at sea but not drilled into the ocean floor to generate clean electricity. The floating wind turbines are towed out to sea and then erected on floating platforms and can be towed back to shore for maintenance whenever required. The possibility to float turbines regardless of sea depth is a game changer in terms of capacity, and the potential for standardized manufacturing (compared to offshore fixed-bottom wind power) is a game changer in terms of cost. The floating wind has several key advantages compared with traditional onshore or fixed-bottom offshore wind farms. Statoil together with Siemens Wind Energy installed in 2009 the world's first full-scale floating spar-type wind turbine (Hywind) of 2.3 MW power output at the coast of Norway and at a depth of 200 m [6]. This wind turbine survived 11 m waves with seemingly no wear. In the summer of 2015, a 7-MW FOWT (semisubmersible concept) developed by MHI was moored at the Fukushima demonstration site, 20 km offshore [116].

Globally, the wind resource in deep water (depths >60 m) is very abundant. The potential of this wind resource in the USA is ranked second to China. The potential of wind resources within 50 nautical miles of the USA's coast can generate 900 GW of electricity in deep waters. This is greater than the total electricity generating capacity installed in the USA. Due to the abundance of potential at these depths, the wind turbines will need the design of a floating platform, because the wind turbines that are currently in operation are mostly fixed at the bottom and are dependent upon conventional concrete with a gravity base that is not feasible at these depths [138,139]. Offshore wind energy is of high importance among renewable energies; however, most of the sites with good wind resources are in deep water (>60 m). This makes up around 80% of European seas [140], 60% of oceans in the USA [138], and 80% of Japanese oceans [139–141]. With FOWT systems, deep water sites with high potential for wind energy utilization can be deployed, making offshore wind power no longer limited to water depths up to ~50 m. Park et al. [142] investigated the impacts of passive and semi-active structural control on a fixed bottom and a floating offshore wind turbine.

Floating wind turbine offers several advantages. First, a floating wind turbine has a far greater additional potential capacity than either onshore or bottom-fixed offshore wind turbine (i.e., wind turbines attached to the seabed). The onshore wind turbine is often seen by local communities as an eyesore, and planning permission for new sites is becoming increasingly difficult to obtain. Bottom-fixed offshore wind sites are limited by both water depth and existing maritime infrastructure, meaning turbines can only be installed in depths of up to 60 m and cannot interfere too heavily with local communities and wildlife. Floating wind, however, can be installed regardless of sea depth, opening up vast tracts of ocean, which could be used for energy generation. It is estimated that floating wind has roughly double the potential capacity of bottom-fixed offshore wind, and this is only taking into account areas located up to 200 km from shore. Second, wind levels are more consistent further from shore. This means that for similar installation costs, larger floating wind turbines with higher power outputs can be used, increasing energy and cost efficiency. Third, floating wind could facilitate the standardized design and production of turbines, which could bring costs down dramatically. Currently, different soil types and different sea currents mean each offshore wind farm needs to be designed separately, driving up capital expenditure costs significantly. Standardization would reduce these up-front costs and facilitate scale for production and maintenance, as well as the use of secondary materials, and recycling and reuse at the end of turbines' lives. This level of scale will be key to cost reduction, and CapEx costs are already sinking very quickly.

Along with generating electricity (which could potentially be fed into grids), the opportunity to leverage floating wind technology to produce green hydrogen (i.e., hydrogen produced using renewable energy) exists, potentially even transforming existing oil and gas assets (e.g., platforms in the North Sea) into decentralized hydrogen production units. Theoretically, if the shipping industry moves toward being fueled by hydrogen, decentralized fueling units for ships could also be created, lessening demand on ports. The outlook for jobs is also bright. The oil and gas offshore sector has traditionally been a large employer in many regions across the world (e.g., North Sea, Gulf of Mexico, and Persian Gulf). For years, workers trained in offshore oil and gas have developed skills and expertise in a dangerous environment: These same skills are directly transferrable to floating wind, e.g., equipment inspections and staffing supply boats. Floating wind will also bring local jobs: Ports will need to be upgraded, turbine maintenance will be done locally, and the production of turbines and support structures could also take place close to new wind farms. All of this will mean that local communities will benefit not only from clean energy, but also from the economic benefits in its wake [116,138–141,143–146].

The path ahead for floating wind is not obstacle-free, however. Floating wind turbine remains a new technology, and not all investors are willing to bet on it. Costs remain higher than fixed-bottom wind currently, with high costs for cables and connection to the grid required. Indeed, in some situations, connecting to local grids may not be feasible, requiring alternative measures to bring energy to the shore, such as green hydrogen. Huge and highly developed port facilities will also be required to service turbines effectively, infrastructure that cannot be built overnight. To jump-start investment, subsidies will initially be required (unless carbon externalities are adequately priced for), until standardization, learning, and scale bring floating wind costs below those of comparative methods.

Much of the knowledge about offshore wind energy generation is derived from the oil and gas industry, which, in the 1920s, designed and experimented with offshore structures. Nowadays, due to the existing technology of oil and gas, depths of thousands of meters can be reached. There is one significant difference between wind and oil platforms: The latter is designed with a focus on the actions of the wave and load of cargo, whereas the wind turbine platform additionally needs to focus on the speed

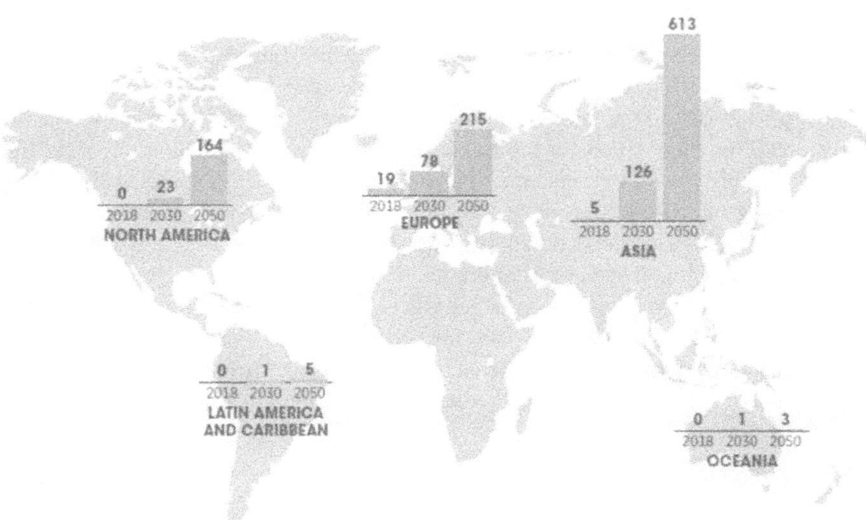

FIGURE 2.20 Installed capacities (GW) of offshore wind [116].

of the wind, which is not of much importance for the oil platforms due to their rigidity. Even though similar physical principles are involved in both cases, there exists a difference in primary actions, weight, and functions that evoke the need for increasing knowledge of offshore non-oil purpose lighter structure for building confidence to create and sustain floating platform structures [141]. The UK leads the world in offshore wind power installation with a capacity of 34% of total offshore wind installations, followed by Germany and China with 28% and 20% of total installations, respectively [144]. Offshore wind deployment is set to expand to Oceania and North America with the projects that will be built in the forthcoming years. Asia would become the prominent leader globally in offshore wind power commissioning in the next three decades, with a total capacity exceeding 100 GW by 2030 and 600 GW by 2050, as shown in Figure 2.20. The global offshore wind capacity increased by 5.2 GW in 2020, bringing the total capacity to 32.5 GW [145]. More than 30 FOWT concepts have been proposed [138,140]. However, this broad range of floater types being up to now investigated inhibits the fast achievement of high technology readiness levels (TRLs). Furthermore, less diversity in floating support structures would allow more focused research, development of required infrastructure, specification, and adaptation of suppliers and manufacturers, and realization of serial production [143]. This strategy will allow FOWTs to soon become cost-competitive with bottom-fixed offshore wind turbine systems.

2.8.1 Main Classification of Floaters

Floating support structures can be categorized based on the primary mechanism adopted to fulfill the static stability requirements. There are three main stabilizing

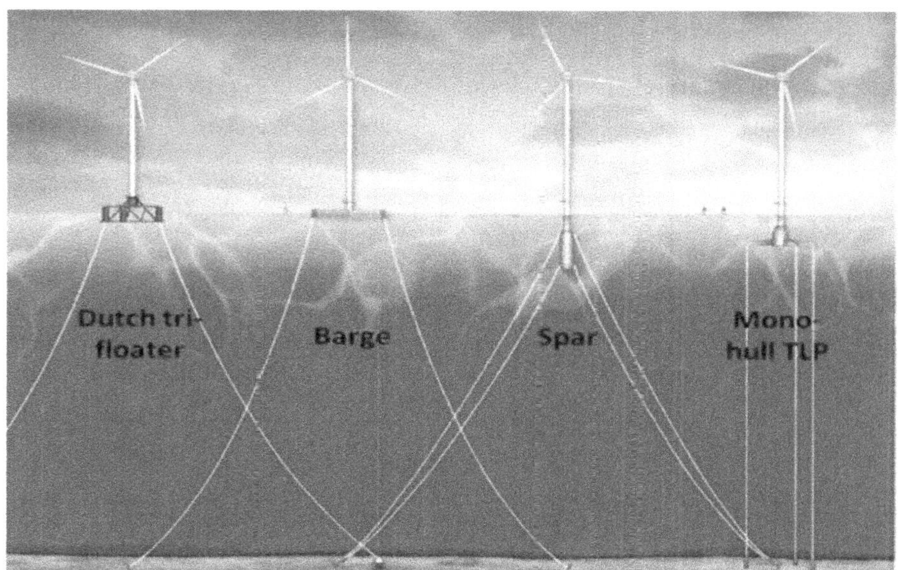

FIGURE 2.21 Types of floating wind turbine platforms [116].

mechanisms [143–145]: (1) *ballast stabilized (spar concept)*: Having large ballast deep at the bottom of the floating structure moves the center of gravity of the total system below the center of buoyancy. This leads, when tilting the platform, to a stabilizing righting moment, which counteracts rotational displacements. (2) *Waterplane (or buoyancy) stabilized (semisubmersible or barge concepts)*: The waterplane area is the main contributor to the restoring moment of the floater. Having a large second moment of area with respect to the rotational axis, either due to a large waterplane area or due to smaller cross-sectional areas at some distance from the system central axis, creates a stabilizing righting moment in case of rotational displacement, and (3) *Mooring stabilized (TLP concept):* High tensioned mooring lines generate the restoring moment when the structure is inclined. Spars, semisubmersibles or barges, and TLPs rely, respectively, on the abovementioned stabilizing mechanisms and thus make up the three cornerstones of floating support structures. Types of floating wind turbine platforms are illustrated in Figure 2.21.

Due to the different mooring systems (catenary mooring for spar, semisubmersible, and barge; tendons for TLP as shown in Figure 2.21), the floaters differ in their dynamics. For the catenary-moored floaters, the natural frequencies lie below the range of wave frequencies; however, for the TLP heave, roll, and pitch natural frequencies are above the first-order wave load frequencies. Table 2.3 describes the natural frequency of three types of floating platforms [117].

A balance among the two varying principles (i.e., the requirement for a stable foundation for the wind turbine's control and operation and the nature of the substructure being innate, to respond to environmental forces) is required for the design of the floating platform for wind energy [146]. As explained in Figure 2.22, the

TABLE 2.3
Representative Natural Frequencies of the Three Main Floater Types [117]

Degree of Freedom	Spar	Semisubmersible/Barge	TLP
Surge	0.02 Hz	0.02 Hz	0.04 Hz
Sway	0.02 Hz	0.02 Hz	0.04 Hz
Heave	0.07 Hz	0.07 Hz	0.44 Hz
Roll	0.05 Hz	0.05 Hz	0.43 Hz
Pitch	0.05 Hz	0.05 Hz	0.43 Hz
Yaw	0.02 Hz	0.02 Hz	0.04 Hz

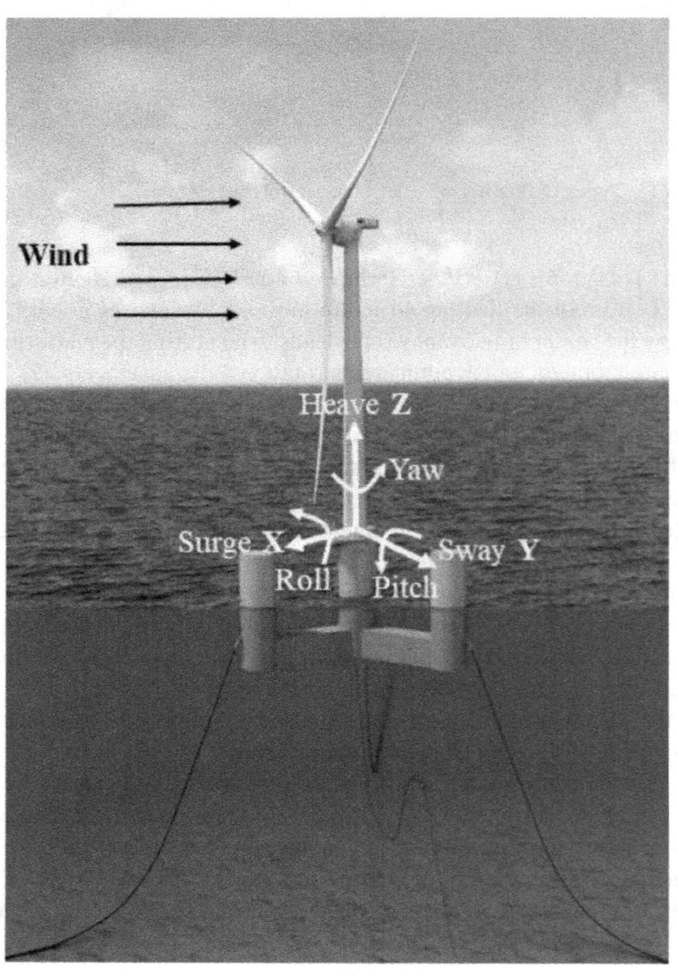

FIGURE 2.22 Six DOFs for a floating platform [116].

absence of rigid foundations results in an additional six degrees of freedom (DOFs) for the platform of floating turbines: three translational (surge X, sway Y, and heave Z) and three rotational (roll RotX, pitch RotY, and yaw RotZ). For the platforms of onshore wind turbines and bottom-mounted offshore wind turbines, the effect of soil–structure interaction (SSI) can be modeled with six DOFs: three translational (horizontal forces in X and Y and vertical force in Z) and three rotational (rocking moments in X and Y and a torsional moment in Z), respectively.

The SSIs have a positive effect on the structural vibrations of the system, as they add damping. The peak shear force and peak bending moment in the foundation are not affected by the SSI [116,117].

Most of the existing FOWT support structures can be assigned to the main categories presented earlier including platforms, such as catenary-moored semisubmersible platforms (CMSSP), TLPs, and the spar buoy. Some other designs are found to be a combination of different floater types, termed hybrid concepts. Finally, multipurpose floaters exist: a structure that carries more than just one wind turbine, so-called multi-turbine concepts, or a mixed-energy design, with which not only wind energy but also another energy source is captured. The market study reports about existing concepts and projects are presented in [147–150].

2.8.1.1 Spar Concepts

Spar buoys are commonly used platforms for wind turbines. The spar buoy is made from six sections all contributing to the stability of the wind turbine. The general principle of spar floaters includes a long cylindrical structure, ballasted at the bottom to obtain stability, and moored with three catenary lines. Some modifications for improving performance and floater characteristics could be a delta connection of the mooring lines to the floater, vacillation fins, or a reduced draft. Nowadays, the most well-known spar FOWT is the Norwegian Project Hywind by Statoil, which—after a single prototype—is already used in a prototype floating wind farm off the Scottish Coast [147,151–154]. This structure is currently very overdesigned [151], and further optimization is needed. Research is also conducted on the use of concrete: FLOAT (floating wind turbine) by GH-Tecnomare is a concrete buoy [151,155], the Hybrid Spar by Toda Construction uses steel at the upper and concrete at the lower section [147], the Universitat Politècnica de Catalunya designed a one-piece concrete structure for tower and floater [147], and within the Kabashima Island Project in Japan, a hybrid (concrete/steel) spar floater is developed [148,156]. Even some advanced spars, modified for improved performance, exist already. The delta connection, also called the crowfoot connection, of the mooring lines to the structure is often used, as well as redundant mooring lines, for example, for the double taut leg buoy by Massachusetts Institute of Technology (MIT) [25]. More advanced improvements, such as reduced draft or stabilizing fins for improving sway and heave response, are integrated into the advanced spar floater within the Fukushima-FORWARD Floating Project in Japan by Japan Marine United [147,156,157]. Finally, some quite different spar floaters are developed to support a VAWT. In these designs, such as the SeaTwirl by SeaTwirl Engineering in Sweden [147] or the DeepWind Spar by the DeepWind Consortium [147], the support structure is rotating together with the turbine.

2.8.1.2 Semisubmersible Concepts

The semisubmersible floater includes the catenary-moored three- or four-cylindrical structure, and heave plates are often attached to the bottom of the columns to reduce heave motion. Further improvements with respect to stability can be achieved by designing the geometry for wave cancelation or by using an active ballast system [158]. A braceless design would simplify manufacturing and inspection. The floating structure developed within the Fukushima-FORWARD Floating Project in Japan by Mitsui Engineering & Shipbuilding [147,156–158], as well as Winflo in France [151,158], VolturnUS by the DeepCwind Consortium [147], Drijfwind or FloatWind from the Netherlands [151,158,159], and Vertiwind in France by Technip and Nenuphar for a VAWT [4], represents the basic semisubmersible type with three or four columns, braces, and catenary moorings. Some simplified floaters without braces are the Dutch Tri-Floater by GustoMSC [146,160], Sea Reed by DCNS [147], OO-Star Wind Floater in Norway by Olav Olsen [161], SPINFLOAT (a type of vertical-axis wind turbine, this time on a braceless semi-submersible floating platform, being developed by French wind-power specialist Eolfi) by EOLFI (a name of the French company) for a VAWT [147], and TetraFloat by TetraFloat—a special lightweight design of the entire FOWT system [147]. Also, braceless, but more innovative, are the V-shaped semisubmersible of the Japanese Fukushima-FORWARD Floating Project by MHI [147,156–158] and the Nezzy super compact drive (SCD) by Aerodyn Engineering, which is a turret-moored Y-shaped structure but uses plastic–composite buoys instead of cylindrical columns [147]. The active ballast system is additionally used in the NAUTILUS (a name of the company NAUTILUS power) concept by NAUTILUS Floating Solutions [147,162] and the WindFloat by Principle Power in Portugal [147,151,154,158,163,164]. These platforms are composed of a series of connections between columns and steel braces. These steel braces are attached to mooring lines, which are embedded in the seabed.

As the depth of water increases, the manufacturing/repairing costs increase, but there are many benefits as well. Benefits include lower costs than fixed-bottom configurations in deep waters, easier installation, effortless part removal, and a broader range of installation sites. Additionally, these foundations have several advantages over other widely used foundations because they can be installed on a dock and transported to the sea, unlike TLPs or spar buoys. Other advantages include lower installation costs of the mooring system compared with other foundations and better hydrodynamic behaviors due to longer drafts and fewer wave exciting forces acting on it. Techniques have been developed to simulate wind fields, such as using EllipSys3D with FLEX5, which is a 3D flow solver that estimates the wind velocities in blade section coordinates [146]. The in-rotor flow of the turbines shows little effect of instabilities, proving the model accurate for the flow of wind on blades. The model is able to capture all important changes in the VAWTs with good quantitative evaluation of the flow field, acquire the radial expansion deformation of the wake, and rapidly calculate the aerodynamic performance of wind turbines under axial steady conditions [165].

2.8.1.2.1 Barge Concepts

Just like a semisubmersible, a barge floater is a waterplane area-stabilized structure. The main difference between these floaters, however, is that a semisubmersible has distributed buoyancy and consists of columns, while a barge is typically flat without interspaces. Only a few barge-type FOWT systems exist. ITI Energy Barge [152] is very standard. FLOATGEN (a name of the first 100% European floating wind turbine, built and assembled in the port of Saint-Nazaire. Powerful of 2 MW, once the wind turbine starts working, it will produce the equivalent of the annual power consumption of 5,000 people) by the French Ideol, however, is quite special with its concrete ring-shaped support structure utilizing a moon pool, also called a damping pool, a system for motion reduction [147,158,166.].

2.8.1.3 TLP Concepts

The TLP system is most reliant on the tendons and highly dependent on the soil conditions, and improvements can be achieved through redundant mooring lines and different, more soil-insensitive, anchors. An early design is the Eolomar ring-shaped TLP [151]. More contemporary and very basic is the TLP by MIT and NREL [151,152,160]. GICON in Germany with GICON-SOF (an innovation of GICON group; it is a floating offshore foundation for wind turbines, especially those that can be deployed in water depths between 30 and 500 meters) [147,167], the American Glosten Associates with PelaStar [147,154], Iberdrola with tension leg platform wind turbine (TLPWind) [147,168], and IDEAS with the tension leg wind turbine (TLWT) [163] have addressed the high-risk problem by equipping the floater with additional mooring lines, either via an increased number of arms or a supporting redundant mooring system. The strong soil dependence is solved by dielectric barrier discharge (DBD) Systems (Eco TLP) [147], Arcadis in Germany [151], and the Dutch Blue H Group (BlueH) [147,151,169] with (concrete) gravity anchors. TLPs are often used for deep water oil/gas developments due to their favorable motion characteristics, but they are more expensive than spar buoys and catenary-moored platforms. However, when TLPs are used as platforms for wind turbines, their displacement, tensile strength, and steel mass can be lowered. This is due to their overall weight being much lighter than when used for oil/gas platforms, allowing wave and wind forces to decrease when encountering the TLPs [170]. TLPs are experiencing greater popularity, as their conceptual design is an active area of research, and industrial workers are starting to get more interested in using these platforms for deep water wind turbines. Most TLP's structures consist of a single column and three to four pontoons, which allow for substantial buoyancy. The hull is made up of a steel cylinder and a protruding cylinder to support the tension legs. It is desirable for the tendons to have negligible weight.

The primary objective of the hull in TLP wind turbines is to decrease electricity costs through maximizing power conversion efficiency, while minimizing operational, manufacturing, and maintenance costs. To minimize manufacturing costs, steel mass, tensile strength, and overall size should be minimized. Limiting O&M fees requires a reduction in nacelle, tendon, tower, and blade loads. Optimizing the structure and design of the TLP is crucial to its effectiveness and lowering maintenance costs. In an experiment to determine which configuration would optimize the power conversion efficiency, lower the costs, and remain the most stable during harsh weather conditions, four different TLPs were designed for comparative analysis. The first design had the heaviest hull, most stability, and shortest installation time of these designs and had similar maintenance and operational costs as the others; however, it was the most expensive to manufacture. The second design had 60% of the first design's displacement but was unstable for towing out because only 30% of the displacement was on the three-legged pontoons [170]. The third and fourth designs were made after discovering the problems with the first and second designs. They both had 70% of the displacement on the pontoons, optimizing the distribution of the total weight of the TLP so that the other 30% from the hull could hold the wind turbines [170]. Although having less displacement than the other two designs made it more prone to damage, the costs were significantly less and the small column in the hull made the turbine more transparent to waves. The pros and cons of three major types of platforms are illustrated in Table 2.4.

Sophisticated and reliable tools are required for the manufacturing and design of cost-effective and optimized floating wind turbines that can generate the required response and dynamics properly and effectively [171]. Figure 2.23 shows the various operating conditions of a floating wind turbine [172,173].

Earlier work on single wind turbine floating platforms is well reviewed by NREL [114]. The NREL study concluded that in a TLP foundation, a mooring system is used to attach the structure to the seabed with a set of pre-tensed legs. The position and predetermined location of the foundation are recovered by adjusting the tension in the legs. The spar foundation, however, is stabilized using a large cylindrical buoy. The semisubmersible foundation design was formed by combining the designs of TLP and spar-type floating platforms. The semisubmersible platform provides adequate space at the base for carrying out O&M work [110]. The TLP platform can be lowered into the water to enable the maintenance of the wind turbine by setting it within a reasonable distance of the water [114]. The LCOE for the semisubmersible platform, spar platform, and TLP platform is 88.69–105.59 USD/MWh, 95.36–107.39 USD/MWh, and 113.34–135.11 USD/MWh, respectively [174]. The semisubmersible and spar platforms have a relative advantage over the TLP platform in economic terms. This economic analysis was based on the life cycle cost of a floating offshore wind farm that includes the costs of concept, design, construction, installation, operation, and decommissioning [175].

2.8.1.4 Hybrid Concepts

The combination of any of the three stability mechanisms, represented by spar, semisubmersible, or barge, and TLP in Figure 2.21, leads to so-called hybrid floating concepts. In this way, the advantages of different systems can be combined into one floating structure. Quite common is the tension leg buoy (TLB), which is a spar floater moored with tendons, such as the Floating Haliade by Alstom in France [150], the Ocean Breeze by Xanthus Energy in the UK [150], the TLB series by the Norwegian University of Life Science [163], and the SWAY (a name of company pioneer in floating wind turbine; the SWAY® system is a floating spar wind turbine for offshore locations in 60–300m + water depths) or Karmoy in Norway [147,151,163]. Nautica Windpower in the USA combined, in the single-point moored advanced floating turbine (AFT), a TLP with a semisubmersible to support a two-bladed wind turbine [147], while Concept Marine Associates added to a TLP a barge-shaped structure, which is ballasted offshore and, thus, functions as gravity-based anchor [146].

TABLE 2.4
Pros and Cons of Three Major Types of Floating Platforms [116,117]

Spar buoy

A cylinder with a low water plane area, ballasted to keep the center of gravity below the center of buoyancy. The foundation is kept in position by catenary or taut spread mooring lines with drag or suction anchors.

Pros:

- Tendency for lower critical wave-induced motions
- Simple design
- Lower installed mooring cost
- Higher stability
- Lower wave-induced motions

Cons:

- Offshore operations require heavy lift vessels and currently can be done only in relatively sheltered, deepwater
- Needs deeper water than other concepts (>100 m)
- Higher fatigue loads in the tower

Semisubmersible (or "spar-submersible")

Many large columns linked by connecting bracings/submerged pontoons. The columns provide hydrostatic stability, and the pontoons provide additional buoyancy. The foundation is kept in position by catenary or taut spread mooring lines and drag anchors.

Pros:

- Constructed onshore or in a dry dock
- Fully equipped platforms (including turbines) can float with drafts below 10 m during transport
- Transport to the site using conventional tugs, easy to tow
- Can be used in water depths of about 40 m
- Lower installed mooring cost
- Low draft requirements
- Low mooring costs
- Most viable for deepwaters

Cons:

- Tendency for higher critical wave-induced motions
- Tends to use more material and larger structures in comparison with other concepts
- Complex fabrication compared with other concepts, especially spar buoys and complex structure
- Higher wave induced

Barge Platform

Just like a semisubmersible, a barge floater is a waterplane area-stabilized structure. The main difference between these floaters, however, is that a semisubmersible has distributed buoyancy and consists of columns, while a barge is typically flat without interspaces.

Pros

Less complex

Low anchor costs

Easy decommissioning

Cons

Support structure is relatively expensive

Large wave-induced motions

(Continued)

TABLE 2.4 (*Continued*)
Pros and Cons of Three Major Types of Floating Platforms [116,117]

Tension leg platform

Highly buoyant, with a central column and arms connected to tensioned tendons that secure the foundation to the suction/piled anchors.

Pros:

- Tendency for lower critical wave-induced motions
- Low mass
- Can be assembled onshore or in a dry dock
- Can be used in water depths to 50–60 m, depending on met ocean conditions
- Lower fatigue loads
- Simple structure

Cons:

- Harder to keep stable during transport and installation
- Depending on the design, a special-purpose vessel may be required
- Some uncertainty about the impact of possible high-frequency dynamic effects on turbine
- Higher installed mooring cost

Besides these three, there are other types of floating platforms examined by the industry. These are briefly described below.

2.8.1.5 Multi-Turbine Concepts

Placing more than one wind turbine on top of one floater reduces the structural mass [147], as well as the mooring and anchoring costs per turbine, and increases stability [160]. However, the loads on the structure might increase, the overall size is enlarged, which complicates manufacturing and handling, and the turbines are likely to operate sometimes in the wake of the other turbine(s) [147,160]. This needs to be considered when designing a support structure for multi-turbine utilization. Two turbines are deployed on the Hakata Bay Scale Pilot Wind Lens by the Japanese Kyushu University [156], while the semisubmersible multiple unit floating offshore wind farm (MUFOW) [151,155] and the design by Lagerwey and Herema [151] support several turbines. Hexicon by Hexicon in Sweden carries three turbines in a row [147], and WindSea by FORCE Technology in Norway is a tri-floater with two upwind turbines and one downwind turbine [147,151].

2.8.1.6 Mixed-Energy Concepts

Another option for higher utilization of one floating support structure is to capture not only wind but also another energy source, such as wave, current, tidal, or solar energy. This way, the power density can be increased and the fluctuations in power production can be balanced to some extent. However, as for the multi-turbine floater, the complexity and overall dimension of the system, as well as the loads on the system, are increased [147]. Such multi-energy floaters are examined in the TROPOS (a name of project aimed at developing a floating modular multi-use platform system for use in deep waters), MERMAID (a name of offshore project (235 MW) 40–50 km

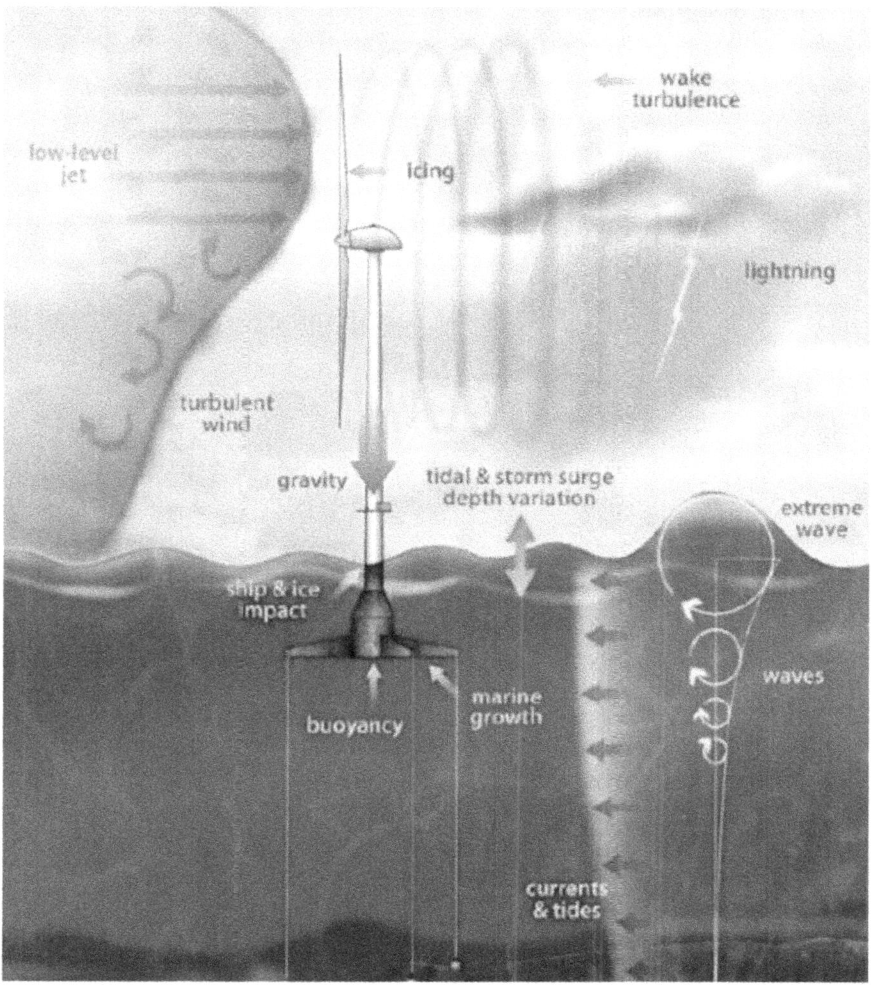

FIGURE 2.23 Offshore floating wind turbine loading sources [116].

off the coast of Ostend (Belgium)), H2OCEAN (a name of wind-wave power open-sea platform equipped for hydrogen generation), and MARINA (this Platform is a European project dedicated to bringing offshore renewable energy applications closer to the market by creating new infrastructures for both offshore wind and ocean energy converters) Projects [150,176]. A quite common combination is wind and wave energy, as realized by W2Power in Norway with the Pelagic Power floater [147] and by Floating Power Plant in Denmark with the Poseidon P80 semisubmersible [147]. Wind and ocean current turbines are combined in the Savonius Keel & Wind Turbine Darrieus (SKWID) by MODEC (a name of the company for offshore wind power generation) in Japan [147,156]. Finally, the multi-turbine floater Hakata Bay Scale Pilot Wind Lens also accommodates solar panels [147,156].

FIGURE 2.24 Blue H Technologies—world's first floating wind turbine (80 kW)—installed in waters 113 m (371 ft) deep—21.3 km (13.2 mi) off the coast of Apulia, Italy [177].

2.8.2 Typical Examples of Various Floating Platform Designs

Blue H Technologies of the Netherlands deployed the world's first floating wind turbine, 21.3 kilometers (13.2 mi) off the coast of Apulia, Italy, in December 2007. The 80 kW prototype was installed in waters 113 m (371 ft) deep to gather test data on wind and sea conditions and was decommissioned at the end of 2008. The turbine utilized a TLP design and a two-bladed turbine. Since then, numerous floating platform designs have emerged. We evaluate a few of them here (Figure 2.24).

2.8.2.1 Eolink

Eolink floating wind turbine is a single-point mooring system technology (see Figure 2.25). The patented structure of this French company based in Plouzané is a semisubmersible floating hull with a four-mast pyramidal structure. The structure supports the turbine with two upwind and two downwind masts. It gives more clearance for the blades and distributes stress. Unlike most of the floating wind turbines, the turbine rotates around its single mooring point to face the wind. The pivot point ensures the mechanical and electrical link between the turbine and the sea floor. Eolink grid connected its first 1/10th scale demonstrator in April 2018 [177].

FIGURE 2.25 Floating wind turbine single-point mooring Eolink [177].

2.8.2.2 Flowocean

Flowocean (a Swedish company) has developed a patented design for floating offshore wind power plants aiming to make floating offshore wind power cost-effective. FLOW (a flow around wind turbine; the presence of a wind turbine affects the airflow both upwind and downwind of the turbine) can be considered an assembly of three systems, the floater, the buoy, and the mooring system. FLOW is a semisubmersible FOWT technology with two wind turbine generators on one floating platform. The structure weather vanes passively so that the wind turbines always face the wind. Flow technology is a combination of TLP and semisubmersible, which gives the flow unit the benefits of both principles and allows the unit to be robust and light. The floater is all structure that is rotating. The buoy is of the turret type, is moored to the sea bed, and contains a bearing that allows the floater to rotate freely around it. The mooring system is the set of components that anchors the buoy to the sea bed, i.e., mooring lines/ropes/chains, chain stoppers, and anchors. The FLOW units are highly standardized with all subsystems well proven. Inter-array wind farm cabling and mooring systems are shared between the units [177].

2.8.2.3 GICON

The GICON-TLP is a floating substructure system based on a TLP developed by GICON GmbH. The system is deployable from 45 to 350 m water depth. It consists of six major components: four buoyancy bodies, horizontal pipes for the structural base, vertical pipes that pass through the water line, and angled piles for connection with the transition piece. Cast nodes are used to connect all components. The TLP can be equipped with an offshore wind turbine in a range of 6–10 MW.

The GICON-TLP is anchored to the seabed via four pretensioned mooring ropes with a buoyant gravity base anchor consisting of concrete. No pile driving or drilling for anchoring is necessary. All ropes are connected at the corners of the square-based system. The TLP for a 6-MW wind turbine is currently being developed. The main focus of the TLP design is on the modularity and the possibility of assembly in any dry dock near the installation site and without the use of construction vessels. After the offshore location is reached, joints of TLP and anchor will be decoupled and the gravity anchor will be lowered down using ballast water. Once the anchor has reached the bottom, it is filled with sand. One unique feature of the system is the sufficient floating stability during transport and operations [177].

2.8.2.4 Ideol

Ideol's engineers have developed and patented a ring-shaped floating foundation based on a central opening system (Damping Pool) used for optimizing foundation + wind turbine stability (see Figure 2.26). As such, the sloshing water contained

FIGURE 2.26 Ideol's 2-MW floating wind turbine installed off France [177].

in this central opening counteracts the swell-induced floater oscillations. Foundation-fastened mooring lines are simply attached to the seabed to hold the assembly in position. This floating foundation is compatible with all wind turbines without any modification and has reduced dimensions (from 36 to 55 m per side for a wind turbine between 2 and 8 MW). Manufacturable in concrete or steel, this floating foundation allows for local construction near project sites.

Ideol leads the FLOATGEN Project, a floating wind turbine demonstration project based on Ideol's technology, built by Bouygues Travaux Publics and operational off the coast of Le Croisic on the offshore experimentation site of Ecole Centrale de Nantes (SEM-REV – a fully fitted wave and wind energy test facility intended to test and improve the efficiency of wave energy converters and floating wind turbine at a prototype stage of development). The construction of this project, France's first offshore wind turbine with a capacity of 2 MW, was completed in April 2018, and the unit was installed on-site in August 2018. For February 2020, it had an availability of 95% and a capacity factor of 66% [177].

2.8.2.5 SeaTwirl

SeaTwirl develops a floating VAWT. The design intended to store energy in a flywheel; thus, energy could be produced even after the wind stopped blowing. The floater is based on a SPAR (a type of floating marine structure for wind energy) solution and is rotating along with the turbine. The concept limits the need for moving parts and bearings in the hub region. SeaTwirl (based in Gothenburg, Sweden) deployed its first floating grid-connected wind turbine off the coast of Sweden in August 2011. It was tested and decommissioned. In 2015, SeaTwirl launched a 30 kW prototype in the archipelago of Sweden, which is connected to the grid at Lysekil. The company aimed to scale the concept with a turbine of 1 MW size in 2020. The concept is scalable for sizes well over 10 MW [177].

2.8.2.6 Seawind Ocean Technology

Seawind Ocean Technology (Netherlands) develops two-bladed floating wind turbines (6.2 and 12.2 MW) suitable for installation in deepwaters with extreme wind conditions (see Figure 2.27). Seawind's technology stems from Glidden Doman's flexible two-bladed turbine system design that is *compliant* with the forces of nature rather than *resistant* to them. Seawind's robust design simplicity, which supports higher turbine rotation speeds, achieves lower torque, lower fatigue, a lighter drive train, and longer life due to its teetering hub technology. Seawind Ocean Technology's teetering hub technology works in conjunction with a yaw power control system that completely eliminates all blade pitch control mechanisms [177].

2.8.2.7 VolturnUS

The VolturnUS design utilizes a concrete semisubmersible floating hull and a composite material tower designed to reduce both capital and O&M costs and to allow local manufacturing (see Figure 2.28). VolturnUS is North America's first floating grid-connected wind turbine. It was lowered into the Penobscot River in Maine on 31 May 2013 by the University of Maine Advanced Structures and Composites Center and its partners. The VolturnUS floating concrete hull technology can support wind

FIGURE 2.27 Seawind's integrated turbine, tower, and substructure [177].

turbines in water depths of 45 m or more. With 12 independent cost estimates from around the USA and the world, it has been found to significantly reduce costs compared with existing floating systems [177].

2.8.2.8 WindFloat

WindFloat is a floating foundation for offshore wind turbines designed and patented by principle power (see Figure 2.29). A full-scale prototype was constructed in 2011 by Windplus, a joint venture between EDP (a name of company called EDP group), Repsol, Principle Power, A. Silva Matos, InovCapital, and flexibility area index (FAI). The complete system was assembled and commissioned onshore including the turbine. The entire structure was then wet-towed 400 km (250 mi) (from Southern to

FIGURE 2.28 VolturnUS floating wind turbine [177].

Northern Portugal) to its final installed location 5 km (3.1 mi) offshore of Aguçadoura, Portugal, previously the Aguçadoura Wave Farm. The WindFloat was equipped with a Vestas v80 2.0-MW turbine. This single wind turbine can produce energy to power 1,300 homes. It operated until 2016 and survived storms without damage.

The subsea metal structure is reported to improve dynamic stability, while still maintaining shallow draft, by dampening wave- and turbine-induced motion utilizing a tri-column triangular platform with the wind turbine positioned on one of the three columns. The triangular platform is then "moored" using a conventional catenary mooring consisting of four lines, two of which are connected to the column supporting the turbine, thus creating an "asymmetric mooring." As the wind shifts direction and changes the loads on the turbine and foundation, a secondary hull trim system shifts ballast water between each of the three columns. This permits the platform to maintain an even keel while producing the maximum amount of energy. This is in contrast to other floating concepts, which have implemented control strategies that de-power the turbine to compensate for changes in turbine thrust-induced overturning moment. This technology could allow wind turbines to be sited in offshore areas that were previously considered inaccessible, areas having water depth exceeding 40 m, and more powerful wind resources than shallow water offshore wind farms typically encounter. A 25 MW WindFloat Project received government permission in December 2016, with EU funding the €48 million transmission cable. A WindFloat with a 2 MW Vestas turbine installed near Scotland began delivering power in late 2018 [177].

FIGURE 2.29 A diagram of the WindFloat system [177].

Numerous other designs have been tested. A technological innovation developed and currently being upgraded is the spider float design. This floating substructure technology was developed to maximize power generation to reduce costs and improve the economic viability of offshore wind turbines [177].

2.8.3 Limitations and Problems for FOWT

In response to strong winds and sea conditions, floating wind turbines require strict reliability criteria. To achieve these criteria, a precise forecast of the movements and forces associated with environmental loads is required. The stability of the platform is affected by the thrust transmitted to the entire structure by the aerodynamic forces. The current model of collective control of the blades does not allow taking asymmetric loads into account. The same problem occurs in the case of hydrodynamic loads. The control methods then commonly used in the onshore wind industry are not appropriate. Additional parameters on the position of the platform, the environmental conditions, the tensions in the anchor lines, and the structural loads must appear in the control loops.

Just like for onshore and offshore fixed-bottom wind turbines, floating wind turbines can be exposed to different types of loads each affecting different rotors and wind turbines. The variability of the wind profile in magnitude and direction causes asymmetric aerodynamic loads that vary spatially through the rotor disk. For the current MW-scale WT, vertical wind variation increases causing the asymmetry of the aerodynamic loads. Additionally, the tower shadow implies an imbalance in the rotor loads. The airflow becomes more turbulent around the tower of the turbine, which creates abrupt movements on the blades. Cyclic loads (i.e., centrifugal force and gravitational force) damage the blades until they rupture. Gyroscopic forces can induce micro-cracking in composite blades. In the case of FOWT, these aerodynamic loads also create movements directly on the floating platform that further increase structural fatigue. An important dynamic of the floating system is the pitching motion caused by the wind. Jonkman [178] and Larsen and Hanson [179] showed that these types of controls can cause negative damping problems caused by blade angle control beyond rated speed. This negative damping may become greater than the radiation damping or viscous on the platform resulting in instability. The standard control strategy applied to the wind turbine operating in pitch control mode (Region III) is the angle change in the blades. This strategy reduces the thrust when the wind increases by modifying the attack edges of the blades. Jonkman [178] tested several control strategies involving torque control and blade control for a high-power turbine installed on a floating barge. Strategies based on proportional integral (PI) gain scheduling did not reduce pitch. The basic controllers show their limits in the case of floating systems.

In the offshore environment, the presence of additional forces caused by the dynamic behavior of waves (i.e., radiation and diffraction) must be considered. Additionally, currents induce stationary and oscillating forces. These hydrodynamic forces contribute to the dynamic motion of the floating platform. The motions of the floating wind platform are classically three translations (heave in the vertical, sway in the lateral, and surge in the axial) and three rotations (yaw about the vertical axis, pitch about the lateral, and roll about the axial). The x-axis is facing the main wind and wave direction. A floating wind turbine introduces additional control objectives aimed at reducing rotational motion and linear displacements. Both external loads and motions and their derivatives induce internal stresses in the structure. The bending moments created by these efforts can damage the support elements of the wind turbine (tower and nacelle). The value of these moments is closely related to the size of the wind turbine. While in an offshore environment, calculating hydrodynamic forces remains a complex task due to the stochastic behavior of the waves, two approaches make it possible to estimate the hydrodynamic forces. The first approach is a deterministic analysis and relies on three variables: wave height, period, and depth. The other approach, based on probabilistic theories, rests on the wave energy spectrum [24,180]. Two main formulations are available to calculate the hydrodynamic loads: the boundary element method (BEM) generally in a linearized context and the frequency domain, and the Morison formulation. Depending on the shape and size of the floater, one or both of these methods can be used. Regulating these movements requires more effective control strategies than the classic collective blade pitch and torque control. This is because, in the classical design approach, control of each element separately by a control loop is not adapted to the additional DOFs involved in FOWT.

2.8.3.1 Issues with Pitch Bearings

Besides platforms, there are new upgrades to pitch bearings, which can be installed in both onshore and offshore wind turbines. Pitch bearings are what connect the spinner to the blades and can adjust the blades to a certain angle to optimize capturing wind. The downsides to typical pitch bearings are that they cannot rotate more than 90°, have <5° of an oscillation angle, are held stationary for long periods, and are subject to constant vibrating when the turbine is running. This puts a lot of stress on the pitch-bearing components and can cause lubricant degradation and adhesive wear. Additionally, these bearings are observed either biannually or annually because of the isolated locations of most wind turbines. The primary causes of bearing failure are poor lubrication and grease degradation, which can cause corrosion, vibratory wear, and debris denting [20,181]. Due to this, proper lubricant/grease selection and continuous-feed lubrication machines are important to ensure maximum efficiency and minimal maintenance. The wind turbine components that experience friction and wear and require lubrication are the pitch bearings, main shaft bearings, gearbox, yaw drive, and generator bearing.

Another area of concern is overloading, which happens when bearings are not firmly supported, leading to a fraction of the raceway carrying most of the load. Overloading can lead to contact truncation (probability can increase because of decreasing external support), raceway core crushing, and component fractures. However, with bearing upgrades, all these problems can be minimized or resolved entirely. Upgrades can include strengthening the races, edge loading, addressing separator wear, and working directly with a manufacturer who can offer advanced bearing solutions that can help save time and money. Additionally, by inducing separator rings, the tensile and compressive load can be reduced. Alongside this, strict geometric dimensioning can create a near-perfect form, resulting in less friction, skidding, and tight spots, which minimizes internal degradation and upgrades the pitch system response and efficiency. Pitch-bearing seals play two crucial roles in preventing internal exposure and locking lubricants. Unfortunately, the old-fashioned seals are hydrogenated nitrile butadiene rubber, which degrades rapidly when exposed to UV, does not protect bearing internals, and has a slow response to changes in wave frequencies. However, an upgraded seal called the "H-profile" is made from thermoplastic polyurethane significantly improving seal effectiveness [20,181]. It is highly responsive, works efficiently even when deformed, reduces grease leakage, and has a significantly lower wear rate than a rubber seal. These collective improvements help to enhance robustness and reduce maintenance costs.

Finally, proper storage packaging and handling are crucial to maintaining a clean slate of the bearings. Bearings should be packaged in volatile corrosion inhibitor paper and corrosion preventive coatings for the prevention of contamination, degradation, and corrosion from hazards during transportation. It is also essential that before installing the bearings on the wind turbines, the bearing should still be wrapped because a little exposure to contaminants, especially dirt and water, can cause corrosion, hydrogen-induced fractures, static etching, and fretting [20,181].

2.8.3.2 Strategies Suitable for Problems Specific to FOWT

Because collective control of blade attack angles is not adapted to local wind variations and the "tower clearance" effect [182], individual pitch control is proposed in the literature for fixed OWT. Han et al. [183] suggested the use of FL while Kang and Kim [184] used NNC combined with individual pitch control. These methods allow for improving the management of power fluctuations, aerodynamic loads, local wind variations, rotor speed, and generator torque fluctuations. Although these kinds of control allow significant improvements in mechanical loads and disturbance reduction, they still do not reflect the dynamic behavior of FOWT. To increase platform stability, individual pitch control has also been proposed for FOWT. Bossanyi [185] and Namik and Stol [186,187] propose using individual control of the blades to reduce the loads. The conclusions of these studies show a decrease in the movements of the platform and consequently an improvement in stability. However, the individual control of the blades increases the number of DOFs, which complicates the control schemes. Pitch actuators' reliability is strongly decreased due to their constant and severe use. The blade pitch control, as explained in these papers, cannot reduce motion structure and tower loads at the same time.

Industry and academia propose a method to mitigate the motion of floating platform and consequently the structural loads. Inspired by Oil and Gas (O&G) and the shipbuilding industry, there are three main methods of structural controls: passive, semi-active, and active. Passive control is the simplest solution where the system uses tuned mass–spring–dampers (TMDs). No energy is needed for its operation. However, the use of constant parameters in the TMD limits the use of the system. The system is regulated to absorb specific excitation frequencies. A TMD system absorbs a portion of the energy along a given axis. Many DOFs increase the complexity (related to the number of axes to be controlled) of the TMD systems to be installed in the platform.

A control algorithm can be used to modify the TMD parameters to improve the stability as a function of the platform motion. Such a choice of system offers better performance than the passive method.

The last concept is the most effective in terms of control of the structure. The system is based on moving a mass through an actuator. The displacement of the mass counteracts the structure motion. The implementation of such a system requires providing energy for the actuator and very efficient control of the mass motion. Lackner [188] investigated a passive and active structural control system on a floating barge. The proposed control is called a hybrid mass–damper (HMD) and is a combination of both systems: It is a one-mass system moved by an actuator. Both systems are placed in the nacelle of the wind turbine. The results show a reduction in dynamic loads within the structure. The proposed system offers better performance than the passive system. Nevertheless, the energy balance between the consumption of the active system and the energy produced by the rotor becomes negative in Region I. For a spar buoy FOWT, Si et al. [189] simulated the dynamic response of a TMD model placed in the tower. Dinh et al. [190] simulated the dynamic behavior of a spar structure integrating a TMD in each blade, the nacelle, and the tower following a semi-active method. Li and Gao [191], however, used a robust controller called H_∞ to optimize the controller gain. The TMD is placed on the platform to reduce the oscillations induced by dynamic loads. All of these studies resulted in a reduction in dynamic loads but also observed degraded performance for winds below the rated wind speed.

Another alternative in active structural control is the method called tuned liquid column damper (TLCD). Here, the principle is to replace the spring–damper system with a liquid. Coudurier et al. [192] used a U-shaped tube with liquid inside placed on the orthogonal plane of the platform roll axis. The study used two methods: a semi-active control and a passive control for controlling the liquid transfer. In the first case, the head loss coefficient is constant, and in the second case, it is variable as a function of time and can be controlled. This coefficient corresponds to the dissipation, by friction, of the mechanical energy during the movement of the liquid. The results show a significant improvement in system stability with an advantage for the semi-active method.

Today's challenge for floating offshore wind is to optimize the production in different wind regions and reduce the cost of maintenance and operation. To achieve these objectives, it is possible to rely on many innovative control strategies developed for the industry. Based on new-generation algorithms and optimization techniques, these multi-objective strategies are used in industrial application to control complex nonlinear systems and can be used to improve the generated power and the reduction in structural loads. Most of the literature on FOWT is focused on platform design, dynamic loads, and stability analysis. It can be noticed that few papers propose innovative control strategies for floating turbines. However, dynamic system control has been present in the industry for many years. The linear quadratic Gaussian (LQG) method is used by Bossanyi [185] and Christiansen et al. [193] for control, which is based on a state representation of the system. The results show a significant reduction in bending moment without compromising power output. Model predictive control (MPC) is used when a PID controller becomes insufficient. This technique predicts future behavior thanks to a dynamic model implanted inside the controller. Raach et al. [194] tested the MPC method on an individual blade control strategy. The MPC method is compared against the work of Jonkman [178]. The results show significant improvements in electrical power and dynamic load mitigation. Nonlinear MPC with collective pitch angle has been adapted on a FOWT, and the results show good performance [195]. Some of these control methods have been simulated on offshore wind turbines. They are used to control the blades' pitch or the torque to improve performance and stability. FL can be used for onshore WT control. Tahani et al. [196] and Wenzhong et al. [197] used the FL control to reduce structural vibration. Salic et al. [24] and Qi and Meng [198] applied the method to the pitch control of the blades for an onshore wind turbine. Simulation results show improved response and better accuracy of the speed control. Other techniques such as proper mooring can be used to improve the control of FOWT. Control strategies for wind turbine in general were discussed earlier in Section 2.4.

2.8.4 Economics of Floating Wind Energy System

The technical feasibility of deep water floating wind turbines is not questioned, as the long-term survivability of floating structures has been successfully demonstrated by the marine and offshore oil industries over many decades. However, the economics that allowed the deployment of thousands of offshore oil rigs have yet to be demonstrated for floating wind turbine platforms. For deep water wind turbines, a floating structure will replace pile-driven monopoles or conventional concrete bases that are commonly used as foundations for shallow water and land-based turbines. The floating structure must provide enough buoyancy to support the weight of the turbine and to restrain pitch, roll, and heave motions within acceptable limits. The capital costs for the wind turbine itself will not be significantly higher than current marine-proofed turbine costs in shallow water. Therefore, the economics of deep water wind turbines will be determined primarily by the additional costs of the floating structure and power distribution system, which are offset by higher offshore winds and close proximity to large load centers (e.g., shorter transmission runs). With empirical data obtained from fixed-bottom installations in many countries since the late 1990s, representative costs and the economic feasibility of shallow water offshore wind power are well understood. In 2009, shallow water turbines cost US$2.4–3 million per MW to install, according to the World Energy Council, while the practical feasibility and per-unit economics of deepwater floating–turbine offshore wind were yet to be established. In 2021, a French auction closed below €120/MWh (US$141/MWh) of electricity for a 250 MW project, and the high cost, small project size, and lack of experience keep project developers and financial institutions from the risk of committing to the technology [199].

Initial deployment of single full-capacity turbines in deep water locations began only in 2009 [199]. The world's first commercial floating offshore wind farm, Hywind Scotland, was commissioned in 2017. Its capital cost was £264 million, or £8.8 m/MW, which is approximately three times the capital cost of fixed offshore wind farms and ten times the capital cost of gas-fired power stations. It uses five Siemens turbines of 6 MW each, has a capacity of 30 MW, and is sited 18 miles (29 km) off Peterhead. The project also incorporates a 1 MWh lithium-ion battery system (called batwind). Its operating costs, at approximately £150,000/MW, were also higher than for fixed offshore wind farms. A second UK project, the Kincardine Floating Offshore Wind Farm, has been reported as costing £500 million to build, or £10 m/MW. UK floating offshore wind could reach "subsidy-free" levels by the early 2030s, according to a study completed by the Offshore Renewable Energy (ORE)

Catapult's Floating Offshore Wind Centre of Excellence (FOW CoE). The UK's leading technology innovation and research center for offshore energy ORE Catapult has produced a report on Tugdock's technology: "Tugdock which could enable floating wind developments at sites without suitable port facilities nearby. It could also reduce substructure assembly costs by 10% when compared with conventional methods by reducing requirements for costly heavy lift vessels that are few and far between." As of October 2010, feasibility studies supported that floating turbines are becoming both technically and economically viable in the UK and global energy markets. The higher up-front costs associated with developing floating wind turbines would be offset by the fact that they would be able to access areas of deepwater off the coastline of the UK where winds are stronger and reliable. The offshore valuation study conducted in the UK has confirmed that using just one-third of the UK's wind, wave, and tidal resources could generate energy equivalent to 1 billion barrels of oil per year, the same as North Sea oil and gas production. A significant challenge when using this approach is the coordination needed to develop transmission lines. A 2015 report by the Carbon Trust recommends 11 ways to reduce cost. Also in 2015, researchers at the University of Stuttgart estimated the cost at €230/MWh. WindFloat Atlantic, sited 20 km off the coast of Viana do Castelo, Portugal, has a capacity of 25 MW and has operated since July 2020 [199,200]. In California, offshore wind coincides well with evening and winter consumption, when grid demand is high and solar power is low. One of the few ports large enough to prepare offshore wind equipment could be Humboldt Bay.

To achieve the lowest overall cost of the whole platform system for its entire life, it is necessary to optimize the floaters for wind turbines because as the depth increases, the construction cost of the wind farm increases due to the floating foundation [158]. The breakup of the total cost of the system for a typical TLP-type FOWT is shown in Figure 2.30, which includes the costs of maintenance, decommissioning, and operations. Although weighing the significance of every element differently, the design of wind platforms will include the same issues governing the oil and gas platforms. To control pitch, heave motions, and support the turbine's weight, the floating structure

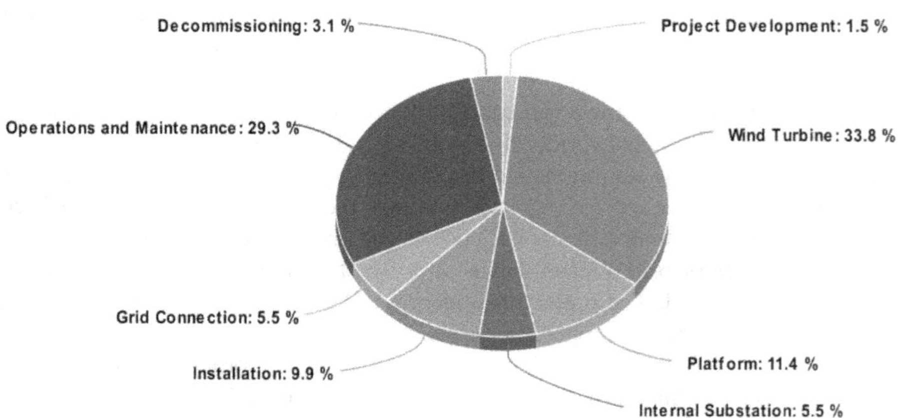

FIGURE 2.30 Cost Estimation for Offshore Wind Projects based on data from [116].

should provide enough buoyancy. Offset by high winds, the extra costs of the system of power distribution and the floating structure determine the economics of the wind turbines in deepwater. As proved by the integrated cost models of energy, if the platform cost is managed near to 10% of the overall capital cost of the system, then the USD 0.05/kWh goal of the DOE cost is possible to attain [146,201].

2.8.5 Single Wind Turbine versus Multi Wide Wind Turbine Floating Platforms

In 2003, various options for platform design were considered by Henderson et al. [202,203], which comprises multiple or single turbines per floating platform, and multiple or single rotor turbines. It was initially concluded, with the help of qualitative analysis, that the best economic and technical solution was a single turbine per floater including a spread mooring. In 2004, a confirmation was made by Musial et al. [160] on this discussion. The advantages and disadvantages of both concepts were laid out. The advantages of multi-wind turbine platforms over single-wind turbine platforms include wave stability, common anchors, fewer transmission costs, and the possibility of mass optimization. The disadvantage is that the yaw control for the multi-turbine platform is complex and the support structure is relatively expensive. However, the analysis carried out was not complete and many issues were to be considered including the following:

1. The requirement for the evaluation of barge platforms
2. Heave and roll were explained by simple equations of motion
3. The analysis did not explain the impact of waves and mooring lines

In 2006, a model based on the Fourier spectrum was applied by Zambrano et al. [204] to a 6 DOF moored-type floating platform with three turbines installed on it. The semisubmersible floating platform's dynamic response was studied by Ishihara et al. [205] and Shimada et al. [206], as shown in Figure 2.31. The development of the design of mobile, as well as a large structure with many wind turbines for offshore sailing type wind farms, was explained by Manabe et al. [207].

Resistance and towing tank tests were carried out by Kourogi [208,209] with the use of a scaled model of sailing-type floating wind turbine structure. In 2011, Lefranc and Torud [210] performed feasibility, design, and cost analysis for a semisubmersible floating platform with three turbines on it. It was shown that the wake interaction between upwind and downwind turbines was the most challenging issue and proved that the power loss due to the wake effect at low velocities was significant, while at high velocities it was almost unchanged. The economic analysis proved that the multi-wind turbine concept was comparable to the current single platforms in terms of cost-of-energy production.

NREL reviewed numerous other [211–219] theoretical analyses of multi-wind turbines in a single platform from different points of view. With the advent of high computing facilities, extensive numerical simulation of the multi-wind turbine floating platforms is possible, which can enable the development of better support structures

FIGURE 2.31 Semisubmersible offshore floating wind turbine [116].

for the platforms. The dynamics of the multi-wind turbine floating platform is an extension of the single platform dynamics with additional DOFs arising due to additional wind turbines. Based on this review, the advantages and disadvantages of single- and multi-wind turbine floating platforms narrated by NREL are summarized in Table 2.5.

2.9 AIRBORNE WIND ENERGY

In recent years, a completely new renewable energy sector, AWE, emerged in the scientific community. AWE aims at capturing wind energy at significantly increased altitudes. Machines that harvest this kind of energy can be referred to as AWE systems (AWESs). The high level and the persistence of the energy carried by high-altitude winds, which blow in the range of 200 m–10 km from the ground surface, have attracted the attention of several research communities. While the interest began in the 1980s, only during the last decade the interest has accelerated. The basic principle was introduced by the seminal work of Loyd [181] in which he analyzed the maximum energy that can be theoretically extracted with AWESs based on tethered wings. Several companies have entered the business of high-altitude wind energy, registering hundreds of patents and developing many prototypes and demonstrators.

TABLE 2.5
Comparison of Single and Multiple Wind Turbine Floating Platforms [116]

Floating Platform	Advantages	Disadvantages
Single wind turbine	Low requirements regarding the structure Standard control options for yaw	Individual cost for anchors Less stable Low possibility of mass optimization Relatively higher transmission costs
Multi-wind turbine	Stability due to wave loads Common anchors Possibility of mass optimization Fewer transmission costs	Support structure is relatively expensive Yaw control is complex

These are well reviewed in an excellent article by Cherubini et al. [220]. Several research teams all over the world are currently working on different aspects of the technology including control, electronics, and mechanical design. Some industrial innovations are briefly described later in this section.

The first work aimed at evaluating the potential of AWE as a renewable energy resource was presented by Archer and Caldeira [221]. Their paper introduced a study that assessed a huge worldwide availability of kinetic energy of wind at altitudes between 0.5 and 12 km above the ground, providing clear geographical distribution and persistency maps of wind power density at different ranges of altitude. Subsequently, Marvel et al. [222] estimated a maximum of 400 and 1,800 TW of kinetic power that could be extracted from winds that blow, respectively, near surface (harvested with traditional wind turbines) and through the whole atmospheric layer (harvested with both traditional turbines and high-altitude wind energy converters). Even if severe/undesirable changes could affect the global climate in the case of such a massive extraction, the authors show that the extraction of "only" 18 TW (i.e., a quantity comparable with the actual world power demand) does not produce significant effects at a global scale. This means that, from the geophysical point of view, a very large quantity of power can be extracted from wind at different altitudes. A more skeptical view on high-altitude winds was provided by Miller et al. [223] who evaluated in 7.5 TW the maximum sustainable global power extraction, but their analysis was solely focused on jet stream winds (i.e., only at very high altitudes between 6 and 15 km above the ground). Despite the large variability and the level of uncertainty of these results and forecasts, it is possible to conclude that an important share of the worldwide primary energy could be potentially extracted from high-altitude winds.

2.9.1 Classifications of Airborne Wind Energy Systems

The term AWESs is often used to identify the whole electromechanical machines that transform the kinetic energy of wind into electrical energy. AWESs are generally made of two main components, a ground system and at least one aircraft that are

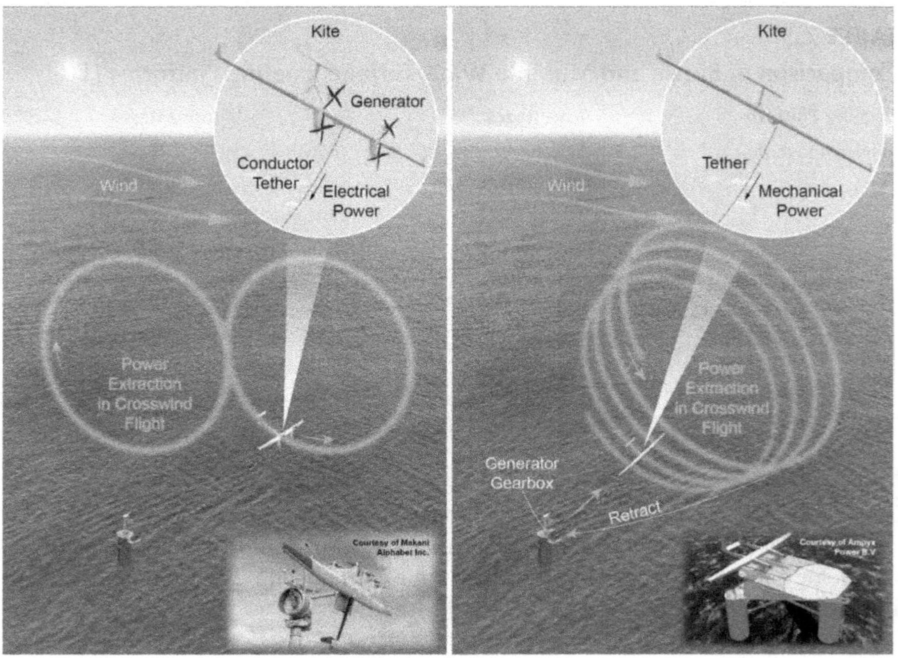

FIGURE 2.32 Operational principles of Fly-Gen (left) and Ground-Gen (right) crosswind AWE systems in an offshore setting [237].

mechanically connected (in some cases also electrically connected) by ropes (often referred to as tethers). Among the different AWES concepts, we can distinguish *Ground-Gen systems* in which the conversion of mechanical energy into electrical energy takes place on the ground and *Fly-Gen systems* in which such conversion is done on the aircraft [224] (Figure 2.32). In *Ground-Gen,* AWESs generate electricity by flying in figure-eight or circular crosswind trajectories while spiraling upward and reeling out the tether that connects the kite to a winch-generator unit on the ground station. Electricity is generated on the ground. Once the kite reaches its maximum operational altitude, it is retracted, and the tether is reeled in. The system is designed to achieve high lift on the wing for maximum mechanical power transmission through the tether. *In Fly-Gen,* AWESs fly onboard turbine–generator units that are connected to an airframe along figure-eight or circular crosswind trajectories defined by the tether length. Electricity is generated onboard the kite. The onboard-generated electrical power is transmitted to the ground station through a conductor in the tether. The system is designed to achieve high relative velocity on the flying wind turbines. The range of designs and approaches used to implement AWESs extends well beyond the simplified description of these two fundamental concepts. Figure 2.33 shows AWE prototypes in flight, a subset of the technologies that are currently under development.

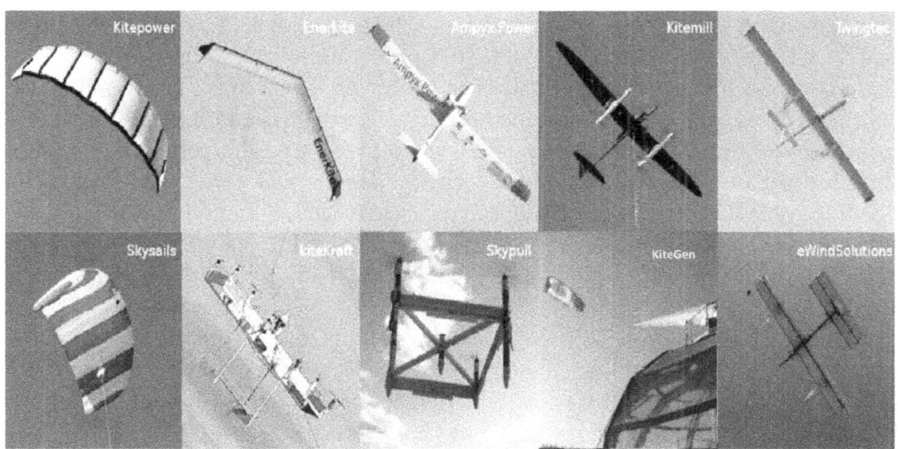

FIGURE 2.33 AWE system prototypes currently under development (in flight) [237].

In a Ground-Gen AWES (GG-AWES), electrical energy is produced by exploiting aerodynamic forces that are transmitted from the aircraft to the ground through one or more ropes, which produce the motion of an electrical generator. Among GG-AWESs, we can distinguish between fixed-ground-station devices, where the ground station is fixed to the ground, and moving-ground-station systems, where the ground station is a moving vehicle. As mentioned earlier, in a Fly-Gen AWES (FG-AWES), electrical energy is produced on the aircraft and it is transmitted to the ground via a special rope, which carries electrical cables. In this case, electrical energy conversion is generally achieved using wind turbines. FG-AWESs produce electric power continuously while in operation except during takeoff and landing maneuvers in which energy is consumed. Among FG-AWESs, it is possible to find crosswind systems and non-crosswind systems depending on how they generate energy.

Fixed-ground-station GG-AWESs (or pumping kite generators) are among the most exhaustively studied by private companies and academic research laboratories. Energy conversion is achieved with a two-phase cycle composed of a generation phase, in which electrical energy is produced, and a recovery phase, in which a smaller amount of energy is consumed. In these systems, the ropes, which are subjected to traction forces, are wound on winches that, in turn, are connected to motor–generator axes. During the generation phase, the aircraft is driven in a way to produce a lift force and consequently a traction (unwinding) force on the ropes that induce the rotation of the electrical generators. For the generation phase, the most used mode of flight is the crosswind flight with circular or so-called eight-shaped paths. As compared to a non-crosswind flight (with the aircraft in a static angular position in the sky), this mode induces a stronger apparent wind on the aircraft that increases the pulling force acting on the rope. In the recovery phase, motors rewind the ropes bringing the aircraft back to its original position from the ground. To have a positive balance, the net energy produced in the generation phase has to be larger than the energy spent in the recovery phase. This is guaranteed by a control system that adjusts the aerodynamic characteristics of the aircraft [225] and/or controls its flight path [226] in a way to maximize the energy produced in the generation phase and to minimize the energy consumed in the recovery phase. Pumping kite generators present a highly discontinuous power output, with long alternating time periods (in the order of tens of seconds) of energy generation and consumption. Such an unattractive feature makes it necessary to resort to electrical rectification means such as batteries or large capacitors. The deployment of multiple AWES in large high-altitude wind energy farms could significantly reduce the size of required electrical storage.

Moving-ground-station GG-AWESs are generally more complex systems that aim at providing an always positive power flow, which makes it possible to simplify their connection to the grid. There are different concepts of moving-ground-station GG-AWESs, but no working prototype has been developed up to date and only one prototype is currently under development. Differently from the pumping generator, for moving-ground-station systems, the rope winding and unwinding are not producing/consuming significant power but are eventually used only to control the aircraft trajectory. The generation takes place due to the traction force of ropes that induces the rotation (or linear motion) of a generator that exploits the ground station movement rather than the rope winding mechanism. Basically, there are two kinds of moving-ground-station GG-AWES: (1) "vertical-axis generator" (Figure 2.34a) where ground stations are fixed on the periphery of the rotor of a large electric generator with a vertical axis. In this case, the aircraft forces make the ground stations rotate together with the rotor, which in turn transmits torque to the generator, and (2) "rail generators" (closed loop rail (Figure 2.34b) or open loop rail (Figure 2.34c) where ground stations are integrated on a rail vehicle and electric energy is generated from vehicle motion. In these systems, energy generation looks like a reverse operation of an electric train.

As mentioned earlier, in GG systems the aircraft transmits mechanical power to the ground by converting wind aerodynamic forces into rope tensile forces. The different concepts that were prototyped are listed in Figure 2.35; examples of aircraft of GG systems that are currently under development are presented in Figure 2.33. They exploit aerodynamic lift forces generated by the wind on their surfaces/wings. The aircraft is connected to the ground by at least one power rope that is responsible for transmitting the lift force (and the harvested power) to the ground station. The flight trajectory can be controlled using onboard actuators, or with a control pod, or by regulating the tension of the same power ropes, or with thinner control ropes. There are also two GG concepts that are worth mentioning: One uses parachutes that exploit aerodynamic drag forces [227,228], and the other uses rotating aerostats that exploit the Magnus effect [229,230].

AWE is "the conversion of wind energy into electricity using tethered flying devices" [231]. Interest and investment in AWE have grown substantially in the last decade, with approximately 70 active research entities including over 20 technology developers globally. A wide variety of AWE technology concepts, operational principles, and designs have been under development primarily over the last two decades and in Europe and the USA. Developments target a diverse range of applications and

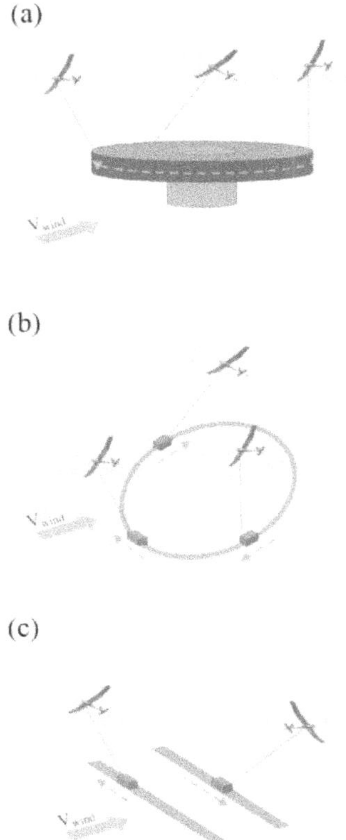

FIGURE 2.34 Scheme of three different concepts of moving-ground-station GG-AWES. (a) Vertical-axis generator: Ground stations are fixed on the periphery of the rotor of a vertical-axis generator. (b) Closed loop rail: Ground stations are fixed on trolleys that move along a closed loop rail. (c) Open loop rail: Ground stations are fixed on trolleys that move along an open loop rail [220].

markets and are currently at low-to-intermediate TRLs. Generally, they have not yet achieved market entry requirements, particularly, with respect to reliability and techno-economics, although developers have started delivering early-adopter market pilot projects [232] and are offering preorders [233] that are both in the 100-kilowatt (kW) unit capacity range. AWE technology developments targeting land-based and offshore competitive grid markets have achieved higher levels of technical sophistication. However, reliable, large-scale MW-class technology market rollout may require more than 10 years of dedicated research and development, depending on the level of support and effort provided. Overall, the technology development has not converged toward a single preferred archetype, and the design space has not been fully explored. AWE technology is fundamentally different from traditional wind technology, specifically wind energy produced by wind turbines that have towers, hereafter referred to as "traditional wind," in nearly every aspect and over the entire life cycle. AWE can thus be considered a separate renewable energy branch, and its

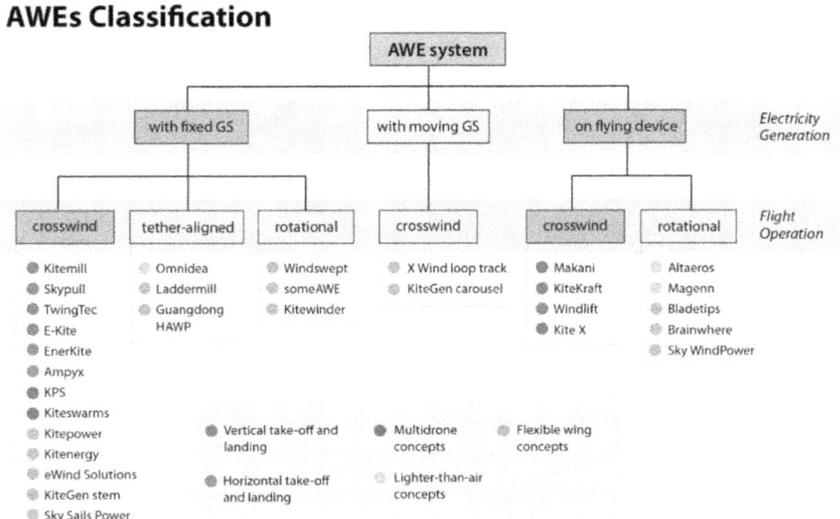

FIGURE 2.35 Classification of traditional AWE technology archetypes [237]. (GS, ground station).

success will hinge on the technology's ability to generate power in a cost-effective and reliable manner over the long term.

The ability of AWE to be cost-competitive with other generation sources at a commercial scale by 2030, the time frame of interest in the congressional request, depends primarily on the achievable power rating, maximum tether length, and capacity density. Airborne wind power plants across the USA comprising 5-MW devices may require capital expenditures of under $1,000/kW to produce the same levelized COE as traditional wind turbines that are anticipated to have capital costs of around $1,200/kW for commercial deployment in 2030. Airborne wind power plants with smaller 500-kW devices may have higher-capacity factors than the reference 5-MW traditional technology in broad regions of the USA. When coupled with potential capital expenditure reductions, AWE has the potential for low LCOE devices; however, commercial-scale deployment may be limited because of an estimated capacity density between 0.4 and 4 MW/square kilometers (km^2), which is typically lower than the average 3 MW/km^2 of traditional wind. The ability of AWE technology to achieve these performance metrics is currently unknown because of a lack of cost and operational data, which represents a significant source of uncertainty for the viability of the technology at a commercial scale.

The wind energy resource potential relevant to AWE (primarily targeting heights between 200 and 800 meters [m]) in the USA is significant. AWE's land-based technical potential for the conterminous USA varies drastically with AWES design between 420 and 34,573 GW and 1,615 and 92,469 TWh. For comparison in 2018, the total US consumption of electricity was 4,222.5 TWh. Positive wind shear above 200 m has not been consistently found, and selected sites and wind profiles appeared to be rather flat, with more evidence of positive shear on land than offshore. Using power curve projections of assumed mature, fully developed, and commercially operated AWE technology based on generic models and an overview of pilot experiment experience, NREL determined significant energy output at selected sites, showing annual energy production per installed capacity like the traditional wind. The exploitation of high-altitude (200 m and higher) winds is strongly dependent on and influenced by technology-archetype-driven design constraints, such as the experience or avoidance of high-tether drag and power-to-weight ratios.

The technical AWE potential in the USA considering social, environmental, and licensing constraints is substantial. The analysis is critically dependent on per-unit installed capacity and tether length, which is directly related to the setback assumptions employed. For example, the rigid wing, 500-kW AWE device with a 227-m tether length, could yield 9,029 GW of technical potential, which is comparable to the 7,827 GW of technical potential for traditional land-based wind technology. The absolute potential of airborne and traditional wind should not be interpreted as an additive, as there is significant spatial overlap between the two estimates. The available and targeted wind resources of both traditional wind and AWE technology are vast, many times greater than the US demand for electricity [234,235]. The technical potential for AWE and traditional wind occurs in similar areas spatially, and further investigation into where AWE may represent an increase in technical potential is recommended. The technical potential estimates are based on hypothetical, fully commercialized AWE technology, which is compared against traditional wind systems expected to be operational in 2030.

The prevailing technology development branches focus on fundamental working principles and design choices, such as Fly-Gen, with electricity generation in the air aboard the airborne subsystem, or Ground-Gen, with electricity generation on the ground, rigid-wing or soft-wing designs and variations thereof, and methods of take-off and landing. Overall, the AWE sector needs to advance significantly with respect to automation in all operational modes and design conditions, reliability and durability supporting system availability, and techno-economic performance at scale. These efforts will require a large dedicated additional effort and time and can be supported through valuable learning and maturation in early-adopter market operations.

Because AWE is aiming for deployment in a variety of markets and applications, many technology approaches, concepts of operation, and design are under consideration. Figure 2.35 classifies AWESs based on a sector overview from 2018 [236]. The first level of categories identifies where electrical generation takes place:

1. **Ground-Gen**. Generation occurs at a ground station that is either fixed or mobile.
2. **Fly-Gen**. Generation occurs onboard the airborne system. The second level of categories identifies groups with respect to flight operations.
3. **Crosswind**. A flight path in a crosswind with a large directional component is perpendicular to the wind direction to increase relative velocity on the kite and swept harvesting area. The crosswind flight path may be implemented through flying reciprocating patterns, such as figure eights, or returning flight patterns, such as circles.

4. **Tether-aligned**. A flight path aligns with the tether direction, which potentially reduces both the relevance to tether drag and the opportunity for area sweep.
5. **Rotational**. It is a rotor-type system with a circular flight path of aerodynamically active kite elements.

The different shades of black in Figure 2.35 indicate several additional design choices. Wings are categorized by their structural design as flexible (like parasail kites) or rigid (like traditional wind turbine blades or fixed-wing aircraft). Concepts are further classified with respect to takeoff and landing operations, distinguishing between vertical takeoff and landing and horizontal takeoff and landing. AWE concepts that are lighter than air make up an additional category.

Over the last two decades of global technology development, some AWE design approaches, such as lighter-than-air systems, have been identified as less promising and are no longer widely pursued. Instead, the prevailing technology development branches are focusing on fundamental working principles such as Fly-Gen or Ground-Gen, and fundamental flight dynamics, wing structure and wing designs, and takeoff and landing methods. There are complexities in the air, on the ground, and various methods of scaling through unit size, number of units, spatial distribution, combined multisystem arrangements, and cluster density. AWE R&D efforts have not converged to a single archetype and operational concept. The lack of convergence reflects the relative immaturity of the technology and may indicate uncertainty regarding the future success of the sector. At the same time, the diversity of design concepts provides ample opportunity for further improvement, as the sector is not locked into a confined design space.

Even though a variety of concepts and related design implementations are being considered, it is also evident that the design space has not been explored fully enough to draw decisive conclusions on the probability of success of individual concepts within different market segments. To illustrate this, the classification criteria employed in Figure 2.35 have been extended in Table 2.6 by increasing the number of variants, adding new concept classification attributes, and identifying related design variants. This preliminary selection of technology concept attributes and design options is incomplete yet illuminates the size of the potential design space.

As mentioned earlier, in the last decade, more than 50 companies and research institutions have been working to design a viable wind kite or energy kite, but to date, no models are broadly commercial. Industry watchers do not expect many commercial AWE products to be available before 2025. For example, Makani, based in Alameda, CA, began working on energy kites more than 10 years ago. Their first project was a 2 kW fabric kite. Today, the company is part of X, a research and development company owned by Alphabet, Google's parent company. Makani's current prototype in development is a 2016 model rated at 600 kW. Units have eight rotors, each connected to a PM generator. All are affixed to an 85-foot-long carbon fiber wing, making the M600 look like a small airplane. A tether approximately a third of a mile long connects the unit to the ground. Wind propels the unit in continuous loops and spins the rotors to generate electricity. Makani has not announced when this technology will be commercialized.

TABLE 2.6
Conceptual Design Attributes and Variants to Determine the Design Space and Possible Combinations of Concept and Design Options [237]

	Concept Attributes and Variants
Electrical generation location	In flight, ground-based fixed, ground-based mobile, other
Flight path	Stationary, in tether direction, crosswind with unidirectional rotation, crosswind with bidirectional rotation, other
Wing structure	Soft single layer, soft ram air multicell, soft with inflatable support, soft with highly nonlinear elastic reinforcement, semirigid, segmented rigid, rigid, multiwing, other
Takeoff	Vertical, horizontal, use of main aerodynamic system, use of ancillary embedded aerodynamic system, centrifugal, catapult, lift system, support system, fan, other
Landing	Vertical, horizontal, use of main aerodynamic system, use of auxiliary-embedded, dedicated aerodynamic system, other
Flight region	Low (up to 300 meters [m]), medium (300–600 m), high (above 600 m), combined, other
Control	Diverse active control surfaces: elevator, elevon, flap, wing deformation; tension control; part passive, other
Energy conversion stages	Direct generation, mechanical, hydraulics, other
Tether topology	Single, bridle, dual, multiple, other
Kite topology	Single, twin, rotor, staggered, combined (identical, varied), multiple, network architectures, other
Farm integration	Independent unity, staggered height levels, farm control, other
Reference	Fixed, moving, single, multiple, absolute, relative, other
Other attributes	Known, to be defined, other

Kiteenregy, an Italian start-up founded in 2010, is in the research and design phase of a small-scale prototype. Their technology uses kites flown at altitudes of 800–1,000 m to generate electricity “by converting the traction forces acting on the wing ropes into electrical power, using suitable rotating mechanisms and electric generators placed on the ground.” X-Wind (pronounced: crosswind), a German company, is piloting a system of wind energy kites connected to train bogies, also known as railroad trucks. As the trucks are pulled around a fixed track by fabric kites, electricity is generated through regenerative braking. The kites fly at an elevation of up to 500 m, enabling them to access wind speeds that are 40% higher than those of 100-m towers. X-Wind currently has a 2 MW pilot under construction. A joint pilot research project of four organizations in Germany, SkyPower100, is pursuing a 100 kW AWES that they hope will be the first to realize “autonomous long-term operation day and night, as well as automatic launching, landing, and stowing of the kite.” The project aims to eventually develop an industrial-scale system that lowers traditional wind energy development obstacles, such as noise, wildlife impacts, and permitting.

Lighter-weight material and reduced material use should help lower the LCOE. For example, the overall weight of Makani 600 is around 2 tons—about 98% less than a typical land-based wind turbine with a similar capacity (Vestas V44/600 kW). Fabric kites in development by X-Wind and others are significantly lighter. This also reduces visual impacts and noise. Furthermore, AWE technology is likely to be suitable for places that do not support conventional onshore and offshore development, such as islands, areas without crane access, land near deep water ports, and areas with transportation limitations. X-Wind points out that their kites can achieve a 60% capacity factor by accessing wind at 500 m (far higher than the 140-m towers in development) [237].

2.10 PERSPECTIVES ON TECHNOLOGICAL CHALLENGES AND FUTURE PROSPECTS

As outlined by DOE, the fundamental differences between traditional wind energy and AWE are outlined in Table 2.7. While these differences are striking, they also indicate that unlike traditional wind energy, air wind energy is at present strictly in the developmental stage with significant potential if it can be commercialized.

As shown earlier in Figure 2.33, a few AWES are under development to provide a sense of the variety of AWE architectures, but no claim is made or implied regarding the performance or feasibility of these concepts. Each architecture is distinguished by key design choices. The major design choices can be summarized as follows [238]:

- **Generator placement**. The first design choice is where to place the electrical generation equipment. If placed onboard (called *Fly-Gen*), the airborne device generates power continuously but must transmit the electricity to the ground over the tether connecting the airborne device to the ground. If the generator remains on the ground (called *Ground-Gen*), the airborne device can be simpler and lighter, but many designs only generate electricity in one phase of flight—called "reel out," as a tether is spooled from a drum—and consume energy during "reel in." Ground-Gen devices can have a stationary or moving generator, though the former is more commonly proposed.
- **Wing structure**. Second, the structure of the airborne device can be either *rigid* like most aircraft or *soft* and compliant like a parafoil kite. Rigid wings can have higher performance and durability but may be heavier. Soft wings sacrifice performance and may have to be regularly replaced due to wear on the fabric, but are lighter and fly slower, and thus are more likely to survive a crash and cause less damage. Soft wings may also inflate for strength or compress for easier transportation. Hybrid concepts such as intentionally flexible, jointed, or tailored composite designs are also emerging.
- **Flight operation**. The third key design choice is the method of flight operation. The device will (1) fly *crosswind* like a traditional wind turbine blade, leading to the highest relative wind speed and efficiency but challenging to control, or (2) be relatively *stationary*, generating power through subcomponent motion, autorotation, or by transmitting torque to the ground station (see "Windswept" concept above).

TABLE 2.7
Fundamental Differences between Traditional and Airborne Wind Energy [238]

	Traditional Wind	Airborne Wind
Concept	Spinning rotor comprised composite blades, a tower-mounted nacelle, and drivetrain	Self-supported airborne system tethered to a ground station, with an airborne or ground-mounted drivetrain
Response to a failure	Rotor blades pitch to stop rotation and the turbine waits for remote diagnostics or on-site technician	Airborne system must land safely and autonomously, while avoiding personnel/property
Installation/ maintenance	Crane lift and elevated assembly of major components. Inspection and maintenance also performed at height (80+ m)	All installation and maintenance performed near ground level
Market convergence	Upwind, three-bladed configuration dominates, developed over 40+ years with trusted international standards	Dozens of configurations, little market convergence, and no international design standards or requirements
Operating altitude	Typically below 250 m constant altitude	Typically 200–800 m variable altitude
Operational strategy	Annual OpEx ≈ 2%–3% of CapEx Designed for 25+ years of operational life	Annual OpEx ≈ 3%–20% of CapEx Major components may be replaced or upgraded often
Support structure	Tower and foundation must resist significant overturning moments	In minimal overturning moments, tether tension is dominant load on foundation
Overland transportation	Blades and towers are currently size constrained by the limits of highway and rail transportation	Kites may disassemble or compress for easier transportation. Larger rigid wings may become transportation size constrained in the future
Unit capacity	0–6 MW onshore 6–15+ MW offshore	0–2 MW onshore (notional) 2–5 MW offshore (notional)
Wind farm integration	2D placement (latitude, longitude)	3D placement (latitude, longitude, altitude) location depends on wind direction and speed

- **Takeoff and landing**. Fourth, how will the device become airborne? One could launch and land *horizontally* like a traditional aircraft, *vertically* like a helicopter or drone, use a combination of the two, or use a pilot kite or auxiliary system to loft the system.

According to the DOE report, there are at least 24 potential configurations, shown in Figure 2.36, which combine the above four design choices, and because of the potential for hybrid approaches, this represents a subset of the complete design space. It is not clear which of these concepts (or yet another) will emerge as dominant, and it is likely that multiple designs will survive to some extent, each best for a specific market [238].

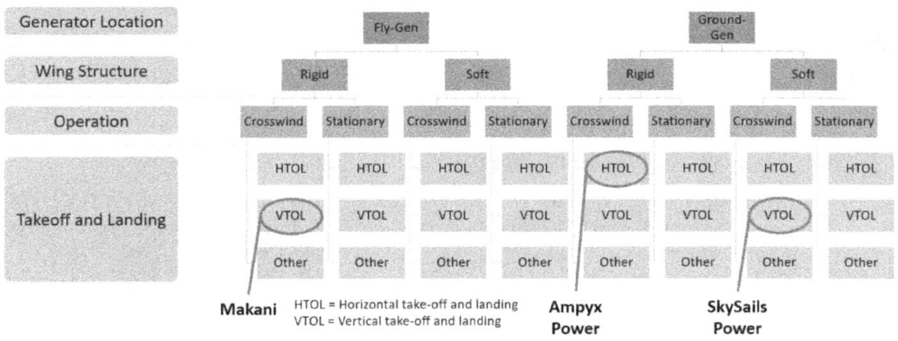

FIGURE 2.36 Some of the many possible AWES architectures and illustrative companies [238].

One of the critical questions around the technological advancement of AWE is how to scale up [239]. In this context, scaling ultimately relates to the installed capacity. There are inherently scientific and engineering challenges associated with upscaling AWE technology, and these challenges vary depending on the application: Fly-Gen or Ground-Gen systems, soft or rigid wings, and overall system integration. Relevant physics and engineering relate to aspects such as tether drag, weight, tension, and fatigue; conductivity; wing loads; flight path; speed; and accelerations [240]. One of the most important reasons why AWESs are so attractive is their theoretical capability of achieving the MW scale with a single plant. For example, in [224] a 34 MW plant is envisaged with a tethered Airbus A380, and many other publications present theoretical analyses with MW-scale AWES [181,241,242]. This scalability feature is rare in renewable energies and is the key to successful commercial development.

It is important to note that the requirements for upscaling installed capacity depend on the target market and application and that upscaling may be measured in different ways. Three relevant metrics are as follows: (1) installed capacity per unit, (2) installed capacity per infrastructure, and (3) installed capacity per land or sea area. Installed capacity per unit (or airborne device) is particularly relevant for AWE developers targeting utility-scale floating offshore wind farm deployment [243,244]. Maximizing the capacity in MWs per device is critical to achieving economic viability because of the high per-unit cost of the balance of plant—including the floating substructure, mooring system, and subsea electrical cable—as well as offshore O&M. For floating offshore AWE applications, engineering design scaling analysis based on LCOE results in optimal device scaling between 7.5 and 15 MW per unit [244]. Successful upscaling of the capacity per unit will depend on fundamental physical similarity considerations for the power-to-mass ratio and, most importantly, on determining achievable values through practical engineering design options [245].

Concepts for upscaling of installed capacity per critical infrastructure focus on maximizing capacity at a single ground station, floating platform, and other balance-of-plant subsystems. Options under study or development include multikite system configurations using a common tether, and parallel single-layer or two-layer multikite layouts [246–248]. Increasing the complexity of control in operation and transitional states with system architectures may also help reduce tether drag. Upscaling of installed capacity per occupied area of land or sea may be implemented by staggering operational heights, thereby using AWESs with different tether lengths and inclination angles, and developing plantwide control strategies [249,250]. Other concepts with higher levels of system integration have been proposed, such as rotary torsion-based, multiwing, multilayer AWESs [251–253] or branched tethers and large multi-tether kite networks [245]. Multikite system integration via a carousel-type system has also been proposed [254]. In summary, upscaling is critical for grid market applications and may be implemented by upscaling the unit capacity and through scaling by the number of units or degree of system integration.

The need to advance and demonstrate system reliability and continuous long-term autonomous operation and control has been identified across the AWE sector [255,256]. As with many functional requirements, autonomous operation is required across all markets, however, at different levels of sophistication. The requirement is particularly acute for offshore AWESs, which are highly capital and maintenance cost-intensive and demand the highest degree of reliability and automation to facilitate certification and long-term validation. While advancing through the precommercial development stages, developers concentrate on specific and narrow test objectives to demonstrate and validate equipment functionality and autonomous operation algorithms. Some systems require wind speeds of 3–5 meters per second (m/s) close to the surface for takeoff, whereas most AWESs use a variety of active launch mechanisms. Individual flight times vary between minutes and hours, and systems have performed several hundreds of operational flights [232,244,257]. Individual developers have reported continuous operation over several days [232,258]. Although some developers have demonstrated autonomous flight and other critical automated functions including takeoff and landing of reduced-scale prototypes [232,259], achieving reliable, fully automated, long-term operation represents the most significant challenge for the industry.

Critical subsystems for functional, reliability, and durability testing and validation include power conversion units [259], BTB tether spooling systems [244], and structural maximum load and fatigue testing [258]. SkySails Group developed, tested, validated, and commercially operated an airborne wind ship propulsion system with mechanical pulling power levels of up to 2 MW and reported a total of over 3,000 hours of operation since the first demonstration at sea over 10 years ago [232]. This system does not include energy conversion to electricity; thus, it does not represent an integrated AWES, but demonstrates the functionality of a key airborne subsystem in automated flight up to commercial readiness.

It is important to mention that fully autonomous operation is not a key requirement in many mini- and microgrid applications [260]. In specific applications in which part manual operation or system reset support is available—such as military operations, industrial microgrids, and remote installations—the requirement for autonomy may be somewhat softened, and criteria such as deployment transportation may dominate [233,261]. Such applications could serve as early-adopter markets, providing opportunities for the commercial operation of AWESs that initiate learning curves for improving system reliability, durability, and efficiency. The electrification of significant energy markets, such as agriculture, could potentially be supported in single or hybrid energy source microgrids with AWESs [258,262] soon, with the first commercial 100-kilowatt (kW) systems being sold to pilot projects [232].

Whether AWE technology can be regarded as (1) in direct competition with, (2) complementary to, or (3) independent of and thus additional to traditional wind, the following observations can be made with respect to AWES characteristics. The temporal correlation of the wind resource targeted by traditional wind and by airborne wind technology in each location is not shown to be anticorrelated in a consistent manner [263] and not of a complementary nature within the typical operational heights aboveground. Traditional wind and AWE technology largely target and access the wind resource through airflows at various heights and across different height ranges, with potential overlap depending on the technology concept. Most AWE technology developers target heights up to 800 m aboveground [232], with some developers targeting greater heights up to 2,000 m aboveground [254] and beyond [264]. Traditional wind and AWE technology may target and access different wind resource locations, which is driven by a variety of factors including minimal viable resource requirements [237,262], transportation and deployment requirements [237,261], site access, farm design, and terrain complexity.

Traditional wind and AWE technology partly require a different raw material resource at significantly different amounts per installed capacity, with AWE requiring significantly less raw material [265] estimates at around 20% of what is required for traditional wind [258]. Traditional wind and AWE technology requires access to different and largely independent supply chains, fabrication, and manufacturing processes. For example, foundations are not required to transfer significant torque load for AWESs [233]. Traditional wind and AWE technology has very different installation and deployment requirements. AWE offers faster deployment and installation with significantly lower installation equipment requirements [258]. Traditional wind and AWE technology has different workforce requirements. While both require manufacturing employees, the airborne wind is expected to require and provide a higher degree of O&M employment opportunities because of replacement requirements for key components, such as the tether and soft kite [233].

Because of the lack of operational experience, the environmental impact of AWE requires thorough investigation and is not well understood. The nature of traditional and AWESs in operation and their control needs and options may result in different environmental interactions and adaptation, including the interruption of operation of a traditional system and grounding of an AWES. Potentially low capital investment and high component and subsystem replacement and maintenance costs during the operational life span of AWE technology require diverse financing approaches, capital expenditure (CapEx) payback, and investment risk profiles. These circumstances may also allow shorter economically viable periods of operation than those of traditional wind technology in some specific markets such as hybrid microgrids [260].

Replacing components and subsystems (e.g., tether and soft kite) in AWE technology over the installation's life span is an operational cost burden. At the same time, it offers the possibility of upgrading the airborne subsystems as technology advances. It also allows adaptation to site conditions and to increase the capacity factor for a given ground station through kite adjustment and selection. The potentially low weight per installed capacity and modest deployment effort may allow the redeployment of AWE technology assets at different operational sites. Fast deployment, mobility, and swift relocation of assets are highly valued properties and satisfy imperative system requirements in some markets [261].

To date, several tens of millions of dollars have been spent for the development of AWESs, which is a relatively low amount of money, especially if one considers the scale of the potential market and the physical fundamentals of AWE technology. The major financial contributions came, so far, from big companies usually involved in the energy market [266]. The community is growing both in terms of patents and in terms of scientific research [267], but still, there is no product to sell and the majority of the companies that are trying to find a market fit are now focusing on off-grid markets and remote locations where satisfying a market need can be easier at first [268]. In all AWESs, increasing the flying mass decreases the tension of the cables. Since Ground-Gen systems rely on cable tension to generate electricity, a higher mass of the aircraft and/or cables decreases the energy production [269] and should not be neglected when modeling [270]. On the contrary, increasing the flying mass in Fly-Gen systems does not affect energy production even though it still reduces the tension of the cable. Indeed, as a first approximation, the basic equations of Fly-Gen power production do not change if the aircraft/cable mass is included and this is also supported by experimental data [271].

A question faced by many companies and research groups is whether rigid wings are better or worse than soft wings. On the plus side for soft wings, there are crash-free tests and lower weight (therefore higher power) because of the inherent tensile structure. Conversely, rigid wings have better aerodynamic efficiency (therefore higher power) and they do not share the durability issues of soft wings. It is unclear whether one of the two solutions will prove to be better than the other, but a trend is clearly visible in the AWE community: Even though a lot of academic research is being carried out on soft wings, more and more companies are switching from soft to rigid wings [268].

Starting and stopping energy production requires special takeoff and landing maneuvers. These are most difficult to automate and require a lot of research in private companies and academic laboratories [272–274]. Another interesting question is how much is the optimal flight altitude, i.e., how much are the optimal cable length and elevation angle that maximize the power output. Increasing the altitude allows to reach more powerful winds, but, at the same time, increasing the cable length or the elevation angle reduces the power output. Considering a standard wind shear profile, the optimal flight altitude is found to be the minimum that is practically achievable [271,275]. However, results greatly change depending on the hypotheses and, for example, a reduction in cable drag might lead to optimal flight altitudes around 1,000 m [275]. More detailed and location-specific analyses could therefore be useful to define an optimal flight altitude. Nowadays, many AWE companies are aiming at exploiting low-altitude winds with the minimum flight altitude set by safety concerns. Only a few companies and academic institutions are still trying to reach high altitudes.

The aerodynamics of the system is very important to the power production. It is easy to show that active control of the angle of attack is essential during the production phase of Ground-Gen or Fly-Gen devices. A variation of the angle of attack can be induced by a change in the tether sag or in the velocity triangle. As for the tether sag, the model would give [276] a numerical value between 7° and 11° for a large-scale AWES. With a simple velocity triangle, a variation of about 3.5° or 4° can be reasonably obtained. Such variations in the angle of attack can decrease the power output substantially or even make the flight impossible. For example, using the values of the aerodynamic coefficients for an airfoil specifically optimized for AWESs [277], a steady-state variation in the angle of attack of just 72° can lead to a decrease in power output between 5% and 42% with respect to the optimal angle of attack. Cables for AWESs are usually made of ultra-high-molecular-weight polyethylene (UHMWPE), a relatively low-cost material with excellent mechanical properties [278] even though many different materials are being used and studied [279]. The cables can be 1, 2, or 3, and in some concepts, they carry electricity for power generation or just for onboard actuation. Each of these choices has advantages and disadvantages, and at the present time, any prediction about the best tethering system would be highly speculative. The tethers also represent a known issue in the AWE community [276] because of wear, maintenance, and aerodynamic drag. In unsteady flight, the lower part of the cables could reasonably move less than in steady-state flight, thus dissipating less energy.

Some concerns have been raised regarding the behavior of tethers in the atmospheric environment [280]. An analysis performed on dry and wet polyethylene ropes without inner conductors [227] shows that nonconductive tethers will not trigger a flashover in typical static electric fields of thunder clouds; however, nonconductive tethers are very likely to trigger a flashover when subjected to impulsive electric fields produced by lightning. It is reasonable to say that AWESs will not work during thunderstorms and that lightning should not be an issue. A reduction in the cable drag coefficient would likely lead to an increase in power output by two or three times due to better aerodynamic efficiency, increased flight speed, and higher operational altitude [275]. As regards the aerodynamic cable drag, two patented concepts might provide an important improvement in the long term by setting zero the aerodynamic drag for the majority of length [259,281–283]. The first one described the so-called "dancing planes" concepts, and the second [259] describes a "multi-tether" AWES where three cables are deployed from three different ground stations and are eventually connected to each other in their top end.

In summary, high-altitude wind energy is currently a very promising resource for the sustainable production of electrical energy. The amount of power and the large availability of winds that blow between 300 and 10,000 m from the ground suggest that AWESs represent an important emerging renewable energy technology. In the last decade, several companies entered the business of AWESs, patenting diverse principles and technical solutions for their implementation. National Renewable Energy Laboratory (NREL) has developed a map of TRL versus technology performance level. This is illustrated in Figure 2.37. They followed DOE definitions for these levels. Overall, evaluating technology readiness and techno-economic performance potential is a rough estimation and is exclusively based on high-level, incomplete public domain information. It is important to note that within a specific AWES development, different degrees of readiness and techno-economic performance can

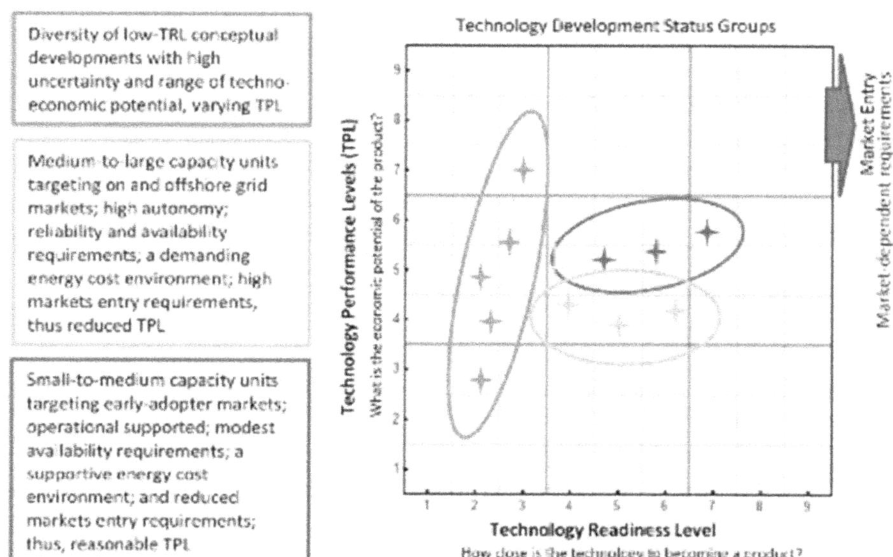

FIGURE 2.37 Technology readiness levels (TRLs) and technology performance levels (TPLs) of the main technology development status groups (left) across the global AWE sector [237].

be observed across the system levels of component, subsystem, and integrated system and may vary considerably across the various performance attributes. The visual representation in Figure 2.37 does not specify individual developers and technology developments may be identified at a lower or higher development status when subjected to more in-depth assessments. The figure does indicate three categories of developments: early stage (green), targeting large impact (yellow), and targeting early adoption (blue). Most technology developments first aim for market entry into an early-adopter market typically associated with an installed capacity per unit of well below 500 kW. A small number of developments have aimed [243], or are aiming [244], for market entry into competitive grid markets with a large installed capacity per unit and in offshore applications.

AWE technology is fundamentally different from traditional wind technology based on fundamental principles and through all life cycle stages, including engineering design, procurement, financing, manufacturing, deployment, installation, operations, maintenance, upgrading, decommissioning, circular economic, environmental, and societal integration, and could be regarded as a unique addition to the renewable energy sector. A wide variety of technology concepts, operational principles, and designs have been under development over the last two decades, originating from early theoretical modeling work in 1980 [181]. Current TRLs vary from low to medium and, in individual cases, medium to high (up to TRL 7 [232]). The potential for AWESs to achieve high techno-economic performance for market entry and to satisfy customer requirements varies significantly over technology concepts, depending on the requirements of their target markets and applications. Globally, a first developer is now entering pilot projects with sales of 100- to 200-kW units supporting remote island communities [232] and another is offering preorders of mobile 100-kW AWESs [233]. Both technologies are Ground-Gen devices with flexible wing designs.

Technology developments targeting land-based and offshore competitive grid markets may have achieved higher levels of technical sophistication yet are further away from fulfilling the demanding market requirements. AWE technologies may be able to offer competitive economic performance based on relatively low material requirements. However, achieving this performance depends on reliability, durability, survivability, and fully autonomous operations, some of which still require significant advancements to enter markets. Early-adopter and special application market entries are starting to emerge in the near term for up to 3 years, whereas reliable, large-scale multi-MW class technology for power to the grid may require 10 years for market rollout, depending on support and effort. Component, subsystem, and integrated system testing, validation, and operation will play a significant role in shaping technology development, increasing operational experience, and establishing reliability. The sector may benefit from the growth of enabling technology in related sectors such as sensors, controls, drone technologies, material science, engineering modeling, and high-performance computing.

Because of the uncertainty of technological success and the rather low global R&D investments to date and especially in the USA, AWE technology development requires significantly more support to increase knowledge, evidence, and validation to understand the impact potential that AWE has for the US electricity sector. In what form and to what degree the AWE sector will provide a significant contribution to the US renewable supply hinge on the rate of technology development and whether it can overcome the challenges associated with achieving commercial readiness, reliability, and techno-economic performance. The current initiation of pilot projects [232] in the 100-kW class may trigger the onset of the technological evolution of AWESs through commercial operational experience gathered in diverse early-adopter markets. Higher-capacity multi-MW AWESs for larger land-based and offshore grid markets may evolve from early-adopter markets, in parallel with or in addition to considerable precommercial research, development, and validation activities.

Both the risk of failure and the risk of lost opportunity need to be considered, and further research and analysis are needed to better understand and quantify these risks and the opportunities to overcome the challenges and realize the technical potential of AWE. The IRENA released an innovation report for offshore wind in 2016 [238], which called AWE a potential "game changer," and ranked the technology third most important, below next-generation offshore turbines and floating foundations, in which DOE's Wind Energy Technologies Office (WETO) is investing heavily. The projected commercialization timeline for airborne wind is vague, as seen in Figure 2.38, spanning nearly a decade from 2024 to 2033.

Despite its promise, AWE technology is still relatively immature and much remains to be done. The potential for reduced LCOE has not been validated given that the first commercial units were planned for deployment by SkySails Power in 2021 [4,5,7,238]. The reliability and availability of an AWES over many weeks, months, or

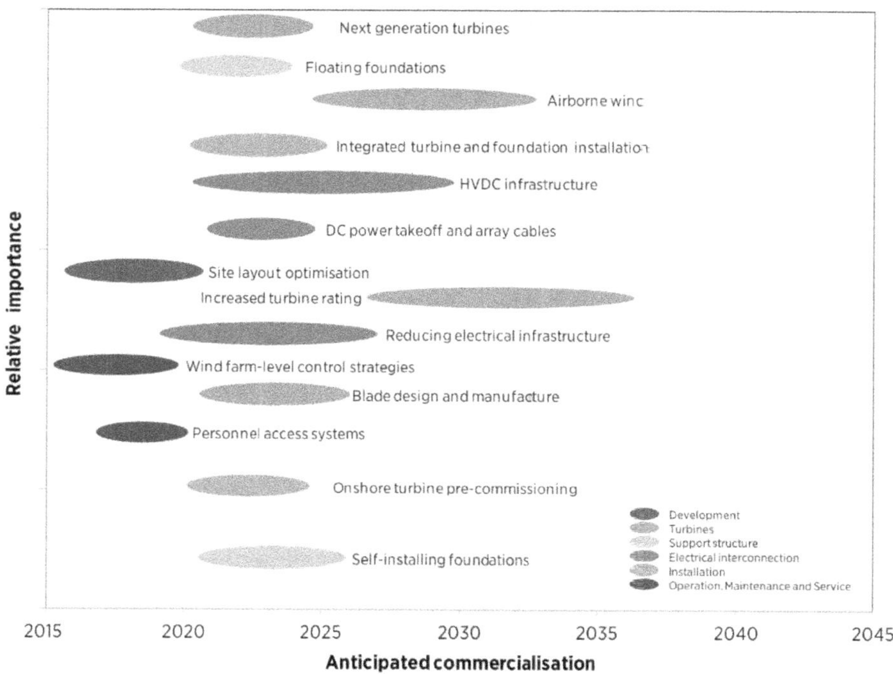

FIGURE 2.38 Timeline and relative importance of various wind power technologies [238].

years have not been demonstrated. Further, if any failure occurs, the device cannot stop midair and wait for help; it must be fail-safe, returning safely to the ground and avoiding any personnel or property nearby. Robust automatic launch and recovery and adequate protection against extreme weather (e.g., high winds, gusts, or lightning) must be developed for utility-scale deployment. There are regulatory and siting concerns related to noise, wildlife impacts, radar mitigation, airspace use, and grid compliance, which will need to be studied and addressed with the appropriate stakeholders. AWES must be more than simply cost-competitive; if the energy harvesting characteristics of mature AWES are not differentiated from traditional wind turbines in some way, the sector may not achieve significant market penetration [238]. None of these challenges are viewed as insurmountable, but they must be addressed for the successful application of the technology. In many cases, the challenges are unique to AWE and would not otherwise be addressed by traditional wind R&D. Overcoming these challenges will require sustained and coordinated R&D but may allow airborne wind to have an impact at scale.

There has been significant research carried out by NREL and others supported by the Department of Energy to evaluate techno-economic challenges for AWES commercial-scale applications. The key findings of this research [220,237,238,284] include the following: (1) If AWE can achieve ambitious performance metrics such as multi-MW devices requiring tether lengths on the order of 200–300 m, then it has the potential to produce lower LCOE in 2030 than land-based traditional wind systems;

(2) site-specific capacity factors and uncertainty in operational expenses lead to a wide range of breakeven CapEx values, clustered between \$500/kW and \$1,500/kW for a 5-MW rigid-wing AWE device and between \$1,000/kW and \$3,000/kW for a 500-kW rigid-wing device. NREL study also indicates that a 500-kW AWE device shows higher (i.e., more easily achievable) breakeven cost margins than a 5-MW AWE device. The broad range of values indicates that there are regions where site conditions are better suited for airborne wind than traditional wind; (3) the southeastern region of the USA is a good target region for AWE; (4) smaller devices may face challenges for commercial-scale deployment because of lower capacity densities than traditional wind turbines but may have a good opportunity for deployment in early-adopter markets (e.g., off-grid and remote locations) because of potential capital cost savings. Deployment in these remote markets would require the devices to achieve high reliability to lower O&M demands, and (5) the range of capacity densities for commercial-scale airborne wind power plants extends to higher-capacity densities than those in existing traditional wind power plants (2–19.6 MW/km^2 for a 5-MW rigid-wing kite and 0.6–4.0 MW/km^2 for a 500-kW rigid-wing kite); however, these densities do not account for wake losses or other siting constraints.

Both DOE reports and European analysis indicate that future research on AWES should focus on [238] (1) continuous operation and defining, achieving, and proving reliability targets; (2) substantiating the case for AWES and developing deep insight into resource potential and resource complementarity and better characterizing the wind resource from 100-meter (m) to 1,000-m altitude, (3) learning from past experiences and improving fundamental understanding, (4) creating a hub and concentrating testing activities in one geographic location; (5) utilizing technology crossovers and investing in enabling technologies such as composite materials and drone technology; and (6) building mutual trust between developers and public through strong communication and set realistic expectations and offer a conditional outlook of stable support that can help accelerate AWE technology development.

REFERENCES

1. Cohen, J., Schweizer, T., Laxson, A., Butterfield, S., Schreck, S., Fingersh, L., Veers, P., Ashwill, T., *Technology Improvement Opportunities for Low Wind Speed Turbines and Implications for Cost of Energy Reduction, July 9, 2005–July 8, 2006* (Technical report). NREL/TP-500-41036 February 2008. National Renewable Energy Laboratory (NREL), Golden, CO, 2006.
2. Carlin, P., Laxson, A., Muljadi, E., *The History and State of the Art of Variable-Speed Wind Turbine Technology* (Technical Report). NREL/TP-500-28607. National Renewable Energy Laboratory (NREL), Golden, CO, 2001.
3. IEA, *Wind Electricity*, IEA report, IEA, Paris, France, 2023. Available online: https://www.iea.org/reports/wind-electricity (accessed on June 1, 2023).
4. Thresher, R., Laxson, A., *Advanced Wind Technology: New Challenges for a New Century*. Conference Paper NREL/CP-500-39537. National Renewable Energy Laboratory (NREL), Golden CO, 2006.
5. Moorefield, L., *Advanced and Emerging Technologies for Wind Generation*. NRECA, Washington, DC, 2019.
6. Griffin, D. A., *WindPACT Turbine Design Scaling Studies Technical Area 1–Composite Blades for 80- to 120-Meter Rotor*. NREL/SR-500-29492. National Renewable Energy Laboratory (NREL), Golden, CO, 2001.

7. Thresher, R., Robinson, M., Veers, P., *The Future of Wind Energy Technology in the United States*. NREL/CP-500-43412. National Renewable Energy Laboratory (NREL), Golden, CO, 2008.
8. Wiser, R., Bolinger, M., *U.S. Wind Energy Update, Wind Exchange Webinar*. Energy Efficiency and Renewable Energy, DOE, Washington, DC, 2016.
9. *Next Generation of Wind Technology*, a website DOE Report by Wind Energy Technology Office, DOE, Washington, DC, 2017.
10. Elia, A., Taylor, M., Gallachóir, B. Ó., Rogan, F., Rogan Wind turbine cost reduction: A detailed bottom-up analysis of innovation drivers. *Energy Policy* 2020;147:111912.
11. Caduff, M., Huijbregts, M. A., Althaus, H. J., Koehler, A., Hellweg, S., Wind power electricity: The bigger the turbine, the greener the electricity? *Environ Sci Technol* 2012;46(9):4725–4733.
12. Lantz, E. J., Roberts, J. O., Nunemaker, J., DeMeo, E., Dykes, K. L., Scott, G. N., *Increasing Wind Turbine Tower Heights: Opportunities and Challenges* (Technical Report). NREL/TP-5000-73629 May 2019. National Renewable Energy Laboratory (NREL), Golden, CO, 2019.
13. IRE Future of Wind, *Deployment, Investment, Technology, Grid Integration and Socio-Economic Aspects (A Global Energy Transformation Paper)*. International Renewable Energy Agency, Abu Dhabi, IRENA, 2019.
14. Lantz, E., Hand, M., Larger turbines and the future cost of wind energy, a paper *Presented at the European Wind Energy Association Annual Event, March 14–17, 2011, Brussels, Belgium*. National Renewable Energy Laboratory (NREL), Golden, CO, 2011.
15. Malcolm, D. J., Hansen, A. C., *WindPACT Turbine Rotor Design Study: June 2000–June 2002 (Revised)*. NREL/SR-500-32495 April 2006. National Renewable Energy Laboratory (NREL), Golden, CO, 2002.
16. Tong, G., Li, Y., Tagawa, K., Feng, F., Effects of blade airfoil chord length and rotor diameter on aerodynamic performance of straight-bladed vertical axis wind turbines by numerical simulation, *Energy* 2023;265:126325.
17. Bortolotti, P., Berry, D. S., Murray, R., Gaertner, E., Jenne, D. S., Damiani, R. R., Barter, G. E., Dykes, K.L., *A Detailed Wind Turbine Blade Cost Model*. NREL/TP-5000-73585. National Renewable Energy Laboratory (NREL), Golden, CO, 2019. Available online: https://www.nrel.gov/docs/fy19osti/73585.pdf. (accessed on January 1, 2020).
18. Miller, L. M., Kleidon, A., Wind speed reductions by large-scale wind turbine deployments lower turbine efficiencies and set low generation limits. *Proc Natl Acad Sci U S A* 2016;113(48):13570–13575. DOI:10.1073/pnas.1602253113
19. Keith, D. W., DeCarolis, J. F., Denkenberger, D. C., Lenschow, D. H., Malyshev, S. L., Pacala, S., Rasch, P. J., The influence of large-scale wind power on global climate. *Proc Natl Acad Sci U S A*, 2004;101(46):16115–16120. DOI:10.1073/pnas.0406930101
20. Shah, R., Zhang, S., Kim, A., *Trending Developments in Wind Turbine Technology and the Future of Wind Energy*, a website report in AltEnergyMag, 2021.
21. Canet, H., Bortolotti, P., Bottasso, C. L., On the scaling of wind turbine rotors, *Wind Energ Sci* 2021;6:601–626. DOI:10.5194/wes-6-601-2021
22. Johnson, N., Bortolotti, P., Dykes, K. L., Barter, G. E., Moriarty, P. J., Carron, W. S., Wendt, F. F., Veers, P. S., Paquette, J., Kelly, C., Ennis, B., *Investigation of Innovative Rotor Concepts for the Big Adaptive Rotor Project* (Technical Report). NREL/TP-5000-73605 September 2019, Contract No. DE-AC36-08GO28308. National Renewable Energy Laboratory (NREL), Golden, CO, 2019.
23. Poore, R., Lettenmaier, T., *Alternative Design Study Report: WindPACT Advanced Wind Turbine Drive Train Designs Study*. NREL/SR-500-33196. National Renewable Energy Laboratory (NREL), Golden, CO, 2003.
24. Salic, T., Charpentier, J. F., Benbouzid, M., Le Boulluec, M., Control strategies for floating offshore wind turbine: Challenges and trends. *Electronics* 2019;8:1185. DOI: 10.3390/electronics8101185

25. Sawant, M., Thakare, S., Rao, A. P., Feijóo-Lorenzo, A. E., Bokde, N. D., A review on state-of-the-art reviews in wind-turbine- and wind-farm-related topics. *Energies* 2021;14:2041. DOI: 10.3390/en14082041
26. Kumar, A., Sadhu, P. K., Singh J., A technological review of wind power generation. In *IOP Conference Series: Materials Science and Engineering*, 2019, p. 012017. DOI: 10.1088/1757-899X/691/1/012017
27. Kumar, A., Singh, J., Sadhu, P. K., A review of wind power generation utilizing statcom technology. In *IOP Conference Series: Materials Science and Engineering*, 2019, p. 012016. DOI: 10.1088/1757-899X/691/1/012016
28. Biddulph, E., Identifying blade bearing issues, a website report by Greensolver. Available online: https://greensolver.net/blade-bearing-issues (accessed on 6 November 2021).
29. Krohn, S., Morthorst, P., Awerbuch, S., *The Economics of Wind Energy- A Report by the European Wind Energy Association*. European Wind Energy Association, 2009.
30. David, N., The doubly-fed induction generator in the rotor-tied configuration [Doctoral dissertation]. Iowa State University, 2014.
31. McInnis, D., Capezzali, M., Managing wind turbine generators with a profit maximized approach. *Sustainability* 2020;12(17):7139.
32. Gámez, M., Peñaloza, O., Nonlinear adaptive power tracking control of variable-speed wind turbines. *SAGE J* 2018;233(3):289–302.
33. Sahu, S., Behera, S., A review on modern control applications in wind energy conversion system. *Energy Environ* 2022;33(2):223–262.
34. Nguyen, H. M., Naidu, D. S., Advanced control strategies for wind energy systems: An overview. In *IEEE/PES Power Systems Conference and Exposition*, 2011, pp. 1–8.
35. Zhong, Q. C., Blaabjerg, F., Cecati, C., Power-electronics-enabled autonomous power systems. *IEEE Trans Ind Electron* 2017;64(7):5904–5906.
36. Sedky, J. S., Yassin, H. M., Hanafy, H. H., Ismail, F., Voltage and frequency control of standalone wind-driven self-excited reluctance generator using switching capacitors. *J Electr Syst Inf Technol* 2021;8(1):1–24.
37. Tanvir, A. A., Merabet, A., Beguenane, R., Real-time control of active and reactive power for doubly fed induction generator (DFIG)-Based wind energy conversion system. *Energies* 2015;8(9):10389–10408.
38. Jallad, J., Mekhilef, S., Mokhlis, H., Frequency regulation strategies in grid integrated offshore wind turbines via VSC-HVDC technology: A review. *Energies* 2017;10(9):1244.
39. Wu, Z., Gao, W., Gao, T., Yan, W., Zhang, H., Yan, S., Wang, X., State-of-the-art review on frequency response of wind power plants in power systems. *J Mod Power Syst Clean Energy* 2018;6(1):1–16.
40. Portillo, R. C., Prats, M. M., Leon, J. I., Modelling strategy for back-to-back level converters applied to high-power wind turbines. *IEEE Transaction Ind Electron* 2006;53(5):1483–1491.
41. Pena, R., Clare, J. C., Asher, G. M., A doubly fed induction generator using back-to-back PWM converter supplying an isolated load from a variable speed turbine. *IEE Proc - Electr Power Appl* 1996;143(3):231–241.
42. Rodriguez, P., Timbus, A. V., Teodorescu, R., Liserre, M., Blaabjerg, F., Flexible active power control of distributed power generation systems during grid faults. *IEEE Transaction Ind Electron* 2007;54(5):2583–2592.
43. Teodorescu, R., Blaabjerg, F., Flexible control of small wind Turbines with grid failure detection operating in stand-alone and grid-connected mode. *IEEE Trans Power Electron* 2004;19(5):1323–1332.
44. Muller, S., Deicke, M., & De Doncker, R. W., Doubly fed induction generator systems for wind turbine. *IEEE Ind Appl Mag* 2002;8(3):26–33.

45. Garcés, A., Molinas, M., A study of efficiency in a reduced matrix converter for offshore wind farms *IEEE Transaction Ind Electron* 2011;59(1):184–193.
46. Cárdenas, R., Peña, R., Wheeler, P., Clare, J. and Asher, G., Control of the reactive power supplied by a WECS based on an induction generator fed by a matrix converter. *IEEE Transaction Ind Electron* 2008;56(2):429–438.
47. Nikkhajoei, H., Iravani M., A matrix converter based micro-turbine distributed generation system. *IEEE Trans Power Del* 2005;20(3):2182–2192.
48. Cárdenas, R., Pena, R., Clare, J., Wheeler, P., Analytical and experimental evaluation of a WECS based on a cage induction generator fed by a matrix converter. *IEEE Trans Energy Convers* 2011;26(1):204–215.
49. Kumar, V., Joshi, R. R., Bansal, R. C., Optimal control of matrix-converter-based WECS for performance enhancement and efficiency optimization. *IEEE Trans Energy Convers* 2009;24(1):264–273.
50. Barakati, S. M., Kazerani, M., & Aplevich, J. D., Maximum power tracking control for a wind turbine system including a matrix converter. *IEEE Trans Energy Convers* 2009;24(3):705–713.
51. Gao, F., Iravani, M. R., Dynamic module of a space vector modulated matrix. *IEEE Trans Power Del* 2007;22(3):1696–1705.
52. Amin, M. M., Mohammed, O. A., DC-bus voltage control technique for parallel-integrated permanent magnet wind generation systems. *IEEE Trans Energy Convers* 2011;26(4):1140–1150.
53. Mohan, N., Undeland, T. M., Robbins, W. P., *Power Electronics: Converters, Applications, and Design*. John Wiley & Sons, New York, 2003.
54. Mwaniki, J., Lin, H., Dai, Z., A concise presentation of doubly fed induction generator wind energy conversion systems challenges and solutions. *J Eng* 2017;2017:4015102.
55. Yaramasu, V., Wu, B., Basics of wind energy conversion systems (Wecs). In *Model Predictive Control of Wind Energy Conversion Systems*, 2016, pp. 1–60.
56. Cheng, M., Zhu, Y., The state of the art of wind energy conversion systems and technologies: A review. *Energy Convers Manag* 2014;88:332–347.
57. Mwaniki, J., Lin, H., Dai, Z., A condensed introduction to the doubly fed induction generator wind energy conversion systems. *J Eng* 2017;2017:2918281.
58. Blaabjerg, F., Ma, K., Wind energy systems. *Proc IEEE* 2017;105(11):2116–2131.
59. Cho, C. D., Nam, S. R., Kang, S. H., Ahn, S. J., Modeling of DFIG wind turbines considering fault-ride-through grid code. In *APAP 2011 - Proceedings: 2011 International Conference on Advanced Power System Automation and Protection*, 2011, pp. 1024–1028.
60. Yaramasu, V., Dekka, A., Durán, M. J., Kouro, S., Wu, B., PMSG-based wind energy conversion systems: Survey on power converters and controls. *IET Electr Power Appl* 2017;11(6):956–968.
61. Tianyu, W, Guojie, L., Yu, Z., Chen, F., Damping for wind turbine electrically excited synchronous generators. *J Electr Eng Technol* 2016;11(4):801–809.
62. Abdelsalam, I., Adam, G. P., Williams, B. W., Current source back-to-back converter for wind energy conversion systems. *IET Renew Power Gener* 2016;10(10):1552–1561.
63. Wei, Q., Wu, B., Xu, D., Zargari, N.R., Overview of offshore wind farm configurations. In *IOP Conference Series: Earth and Environmental Science*, 2017, p. 012009.
64. Babu, B. P., Indragandhi, V., Analysis of back to back (BTB) converter control strategies in different power system Applications. In *IOP Conference Series: Materials Science and Engineering*, 2020, p. 012016.
65. Okedu, K. E., Barghash, H. F. A., Enhancing the performance of DFIG wind turbines considering excitation parameters of the insulated gate bipolar transistors and a New PLL scheme. *Front. Energy Res* 2021;8:373.

66. Ogidi, O. O., Khan, A., Dehnavifard, H., Deployment of onshore wind turbine generator topologies: Opportunities and challenges. *Int Trans Electr Energy Syst* 2020;30(5):e12308.
67. Yaramasu, V., Wu, B., *Model Predictive Control of Wind Energy Conversion Systems*. John Wiley & Sons, Hoboken, NJ, 2016.
68. Li, Z., Song, Q., An, F., Zhao, B., Yu, Z., Zeng, R., Review on DC transmission systems for integrating large-scale offshore wind farms. *Energy Convers Econ* 2021;2(1):1–14.
69. Meraj, S. T., Yahaya, N. Z., Hasan, K., Hossain Lipu, M. S., Masaoud, A., Ali, S. H. M., Hussain, A., Othman, M. M., Mumtaz, F., Three-phase six-level multilevel voltage source inverter: Modeling and experimental validation. *Micromachines* 2021;12(9):1133.
70. Diaz, M., Rojas, F., Donoso, F., Cardenas, R., Espinoza, M., Mora, A., Wheeler, P., Control of modular multilevel cascade converters for offshore wind energy generation and transmission. In *2018 13th International Conference On Ecological Vehicles And Renewable Energies*, 2018, pp. 1–10.
71. Moussa, M. F., Dessouky, Y. G., Design and control of a diode clamped multilevel wind energy system using a stand-alone AC-DC-AC converter. *Smart Innov Syst Technol* 2013;22:797–812.
72. Yaramasu, V., Milev, K., Dekka, A., Rodriguez, J., Modulated predictive current control of NPC converter-based PMSG wind energy system. In *IEEE Energy Conversion Congress and Exposition*, 2020, pp. 75–80.
73. Vahedi, H., Labbe, P. A., Al-Haddad, K., Balancing three-level neutral point clamped inverter DC bus using closed-loop space vector modulation: Real-time implementation and investigation. *IET Power Electron* 2016;9(10):2076–2084.
74. Pulikanti, S. R., Muttaqi, K., Suntanto, D., Control of five-level flying capacitor based active-neutral-point-clamped converter for grid connected wind energy applications. In *Conference Record - IAS Annual Meeting (IEEE Industry Applications Society)*, 2012, pp. 1–9.
75. Perez-Pinal, F. J., Editorial for the special issue on emerging power electronics technologies for sustainable energy conversion. *Micromachines* 2022;13(4):539.
76. Nguyen, H. M., Naidu, D. S., Advanced control strategies for wind energy systems: An overview. In *2011 IEEE/PES Power Systems Conference & Exposition (PSCE), Phoenix, AZ, USA*, 20–23 March 2011, pp. 1–8.
77. Desalegn, B., Gebeyehu, D., Tamirat, B., Wind energy conversion technologies and engineering approaches to enhancing wind power generation: A review. *Heliyon* 2022;8(11):e11263. DOI: 10.1016/j.heliyon.2022.e11263
78. Alsakati, A. A., Vaithilingam, C. A., Alnasseir, J., Naidu, K., Rajendran, G., Transient stability enhancement of grid integrated wind energy using particle swarm optimization based multi-band PSS4C. *IEEE Access* 2022;10:20860–20874.
79. Behabtu H. A., Messagie, M., Coosemans, T., Berecibar, M., Anlay Fante, K., Kebede, A. A., Mierlo, J. V., A review of energy storage technologies' application potentials in renewable energy sources grid integration. *Sustainability* 2020;12(24):10511.
80. Zhao, H., Wu, Q., Hu, S., Xu, H., Rasmussen, C. N., Review of energy storage system for wind power integration support. *Appl Energy* 2015;137:545–553.
81. Sitharthan, R., Sundarabalan, C. K., Devabalaji, K. R., Yuvaraj, T., Mohamed Imran, A., Automated power management strategy for wind power generation system using pitch angle controller. *Meas Control* 2019;52:169–182.
82. Lara, M., Garrido, J., Ruz, M. L., Vázquez, F., Adaptive pitch controller of a large-scale wind turbine using multi-objective optimization. *Appl Sci* 2021;11(6):2844.
83. Demirtaş, M., Şerefoȷlu, Ş., Design and implementation of a microcontroller-based wind energy conversion system. *Turk J Electr Eng Comput Sci* 2014;22(6):1582–1595.
84. Howlader, A. M., Urasaki, N., Yona, A., Senjyu, T., Saber, A. Y., A review of output power smoothing methods for wind energy conversion systems. *Renew Sustain Energy Rev* 2013;26:135–146.

85. Staino, A., Basu, B., Emerging trends in vibration control of wind turbines: A focus on a dual control strategy. *Phil Trans Math Phys Eng Sci* 2015;373(2035):20140069.
86. Xu, F., Zhang, J., Cheng, M., Analysis of double objectives control for wind power generation system with frequency separation. In *2011 4th International Conference on Electric Utility Deregulation and Restructuring and Power Technologies (DRPT)*, 2011, pp. 1366–1371.
87. Umemura, A., Takahashi, R., Tamura, J., Umemura, A., Takahashi, R., Tamura, J., The novel current control for virtual synchronous generator. *J Power Energy Eng* 2020;8(2):78–89.
88. Chu, J., Yuan, L., Chen, Z., Xu, W., Lin, Z., Du, Y., A torque coordination control strategy for wind turbine based on pole placement. In *2018 37th Chinese Control Conference (CCC)*, 2018, pp. 7387–7393.
89. Zhao, H., Wu, Q., Wang, J., Liu, Z., Shahidehpour, M., Xue, Y., Combined active and reactive power control of wind farms based on model predictive control. *IEEE Trans Energy Convers* 2017;32(3):1177–1187.
90. Frederik, J. A., Doekemeijer, B. M., Mulders, S. P., van Wingerden, J. W., The helix approach: Using dynamic individual pitch control to enhance wake mixing in wind farms. *Wind Energy* 2020;23(8):1739–1751.
91. Liu, C., Li, Q., Tian, X., Li, C., Optimal virtual inertial-based power system frequency regulation through multi-cluster wind turbines using BWOA. *Front Energy Res* 2022;10:848905.
92. Schuler, S., Adegas, D., Anta, A., Hybrid modelling of a wind turbine (benchmark proposal). In *3rd International Workshop on Applied Verification for Continuous and Hybrid Systems (ARCH)*, Vienna, Austria, 2017. pp. 18–26.
93. Kong, X., Ma, L., Liu, X., Abdelbaky, M. A., Wu, Q., Wind turbine control using nonlinear economic model predictive control over all operating regions. *Energies* 2020;13(1):184.
94. Nguyen, H. M., Subbaram, D., Thermal to mechanical energy conversion: Engines and requirements-evolution of wind turbine control systems-evolution of wind turbine control systems. Available online: https://www1.eere.energy.gov/wind/wind_how.html (accessed on January 1, 2013).
95. Shao, Y., Liu, J., Huang, J., Hu, L., Guo, L., Fang, Y., The implementation of fuzzy PSO-PID adaptive controller in pitch regulation for wind turbines suppressing multi-factor disturbances. *Front Energy Res* 2022;9:986.
96. el Kararoui, I., Maaroufi, M., Fuzzy sliding mode power control for wind power generation systems connected to the grid. *Int J Power Electron Drive Syst* 2022;13(1):606.
97. Maaoui-Ben Hassine, I., Naouar, M.W., Mrabet-Bellaaj, N., Model based predictive control strategies for wind turbine system based on PMSG. In *IREC2015, The Sixth International Renewable Energy Congress*, 2015, pp. 1–6.
98. Abdelrahem, M., Hackl, C., Kennel, R., Robust predictive control scheme for permanent-magnet synchronous generators based modern wind turbines. *Electronics* 2021;10(13):1596.
99. Heer, F., Esfahani, P.M., Kamgarpour, M., Lygeros, J., Model based power optimisation of wind farms. In *2014 European Control Conference (ECC)*, 2014, pp. 1145–1150.
100. Mirzaei, M., Soltani, M., Poulsen, N. K., Niemann, H. H., Model based active power control of a wind turbine. In *2014 American Control Conference*, 2014, pp. 5037–5042.
101. Stotsky, A., Egardt, B., Model based control of wind turbines: Look-ahead approach. *IFAC Proc Vol* 2012;45(13):639–646.
102. Jamila, E., Abdelmajid, S., Physical network approach applied to wind turbine modeling with simscape language. *Open J Model Simulat* 2014;2014(2):77–89.
103. Pöschke, F., Petrović, V., Berger, F., Neuhaus, L., Hölling, M., Kühn, M., Schulte, H., Model-based wind turbine control design with power tracking capability: A wind-tunnel validation. *Control Eng Pract* 2022;120:105014.

104. Evans, M. A., Lio, W. H., Computationally efficient model predictive control of complex wind turbine models. *Wind Energy* 2022;25(4):735–746.
105. Fontanella, A., Al, M., van Wingerden, J. W., Belloli, M., Model-based design of a wave-feedforward control strategy in floating wind turbines. *Wind Energy Science* 2021;6(3):885–901.
106. Costa, A., Orosa, J., Vergara, D., Fernandez-Arias, P., New tendencies in wind energy operation and maintenance. *Appl Sci* 2021;11(4):1386. DOI:10.3390/app11041386
107. Sheng, S., Fields, J., Cooperman, A., Shields, M., *Wind Plant Operations and Maintenance Challenges and Research Opportunities.* NREL/PR-5000-84457. National Renewable Energy Laboratory (NREL), Golden, CO, 2023.
108. McMorland, J., Flannigan, C., Carroll, J., Collu, M., McMillan, D., Leithead, W., Coraddu, A., A review of operations and maintenance modelling with considerations for novel wind turbine concepts. *Renew Sustain Energy Rev* 2022;165:112581.
109. Floating wind turbine, 2023. Wikipedia, The free encyclopedia, last edited 8 September 2023. https://en.wikipedia.org/wiki/Floating_wind_turbine.
110. Jiang, Z., Installation of offshore wind turbines: A technical review. *Renew Sustain Energy Rev* 2021;139:110576.
111. Möllerström, E., Wind turbines from the Swedish wind energy program and the subsequent commercialization attempts-A historical review, *Energies* 2019;12(4):690. DOI: 10.3390/en12040690
112. Chen, J., Kim, M. H., Review of recent offshore wind turbine research and optimization methodologies in their design. *J Mar Sci Eng* 2022;10:28. DOI:10.3390/jmse10010028
113. Barooni, M., Ashuri, T., Velioglu Sogut, D., Wood, S., Ghaderpour Taleghani, S., Floating offshore wind turbines: Current status and future prospects. *Energies* 2023;16:2. DOI: 10.3390/en16010002
114. Fulton, G. R., Malcolm, D. J., Elwany, H., Stewart, W., Moroz, E., Dempster, H., *Semi-Submersible Platform and Anchor Foundation Systems for Wind Turbine Support.* Subcontract Report NREL/SR-500-40282 December 2007. National Renewable Energy Laboratory (NREL), Golden, CO, 2007.
115. Coulling, A. J., Goupee, A. J., Robertson, A. N., Jonkman, J. M., Dagher, H. J., Validation of a FAST semi-submersible floating wind turbine numerical model with DeepCwind test data. *J Renew Sustain Energy* 2013;5:023116. DOI: 10.1063/1.4796197
116. Bashetty, S., Ozcelik, S., Review on dynamics of offshore floating wind turbine platforms. *Energies* 2021;14:6026. DOI: 10.3390/en14196026
117. Leimeister, M., Kolios, A., Collu, M., Critical review of floating support structures for offshore wind farm deployment. *J Phys Conf Ser* 2018;1104:012007. DOI: 10.1088/1742-6596/1104/1/012007
118. Map: Projected growth of the wind industry from now until 2050, a website report by DOE, Washington, DC, 2023.
119. Konstantinidis, E. I., Botsaris, P. N., Wind turbines: Current status, obstacles, trends and technologies. In *20th Innovative Manufacturing Engineering and Energy Conference (IManEE 2016)*, 2016, p. 012079. DOI: 10.1088/1757-899X/161/1/012079
120. Freeman, K., Hundleby, G., Nordstrom, C., Roberts, A., Valpy, B., Willow, C., Torato, P., Ayuso, M., Boshell, F., Floating foundations: A game changer for offshore wind power. In *Innovation Outlook: Offshore Wind (IRENA, 2016)*, 2016.
121. Wang, C. M., Utsunomiya, T., Wee, S. C., Choo, Y. S., Reserarch of floating turbines: A literature survey, research on floating wind turbines: A literature survey. *IES J Part A: Civ Struct Eng* 2010;3(4):267–277.
122. Wang, X., Zeng, X., Li, J., Yang, X., Wang, H., A review on recent advancements of substructures for offshore wind turbines. *Energy Convers Manag* 2018;158:103–119.
123. Wang, X., Zeng, X., Yang, X., Li, J., Feasibility study of offshore wind turbines with hybrid monopile foundation based on centrifuge modeling. *Appl Energy* 2018;209:127–139.

124. Kaiser, M. J., Snyder, B., *Offshore Wind Energy Installation and Decommissioning Cost Estimation in the US Outer Continental Shelf*. U S. Dept. of the Interior, Bureau of Ocean Energy Management, Regulation and Enforcement, Herndon, VA, 2010.
125. Zhixin, W., Chuanwen, J., Qian, A., Chengmin, W., The key technology of offshore wind farm and its new development in China. *Renew Sustain Energy Rev* 2009;13:216–222.
126. Fischer, T., Executive Summary-UpWind Project. WP4: Offshore Foundations and Support Structures. Available online: http://www.upwind.eu/pdf/WP4_Executive_Summary_Final.pdf (accessed on 12 December 2011).
127. Junginger, M., Agterbosch, S., Faaij, A., Turkenburg, W., Renewable electricity in the Netherlands. *Energy Policy* 2004;32:1053–1073.
128. Saleem, Z., *Alternatives and Modifications of Monopile Foundation or Its Installation Technique for Noise Mitigation*, TU Delft Report. TU Delft University, Delft, the Netherlands, 2011.
129. Pérez-Collazo, C., Greaves, D., Iglesias, G., A review of combined wave and offshore wind energy. *Renew Sustain Energy Rev* 2015;42:141–153.
130. Yang, H., Zhu, Y., Lu, Q., Zhang, J., Dynamic reliability based design optimization of the tripod sub-structure of offshore wind turbines. *Renew Energy* 2015;78:16–25.
131. Byrne, B., Houlsby, G., Foundations for offshore wind turbines. *Philos Trans R Soc Lond Ser A Math Phys Eng Sci* 2003;361:2909–2930.
132. Koh, J., Ng, E., Downwind offshore wind turbines: Opportunities, trends and technical challenges. *Renew Sustain Energy Rev* 2016;54:797–808.
133. Lozano-Minguez, E., Kolios, A. J., Brennan, F. P., Multi-criteria assessment of offshore wind turbine support structures. *Renew Energy* 2011;36:2831–2837.
134. Arshad, M., O'Kelly, B. C., Offshore wind-turbine structures: A review. *Proc Inst Civ Eng Energy* 2013:166;139–152.
135. Seidel, M., Jacket substructures for the REpower 5M wind turbine. In *Proceedings of the Conference Proceedings European Offshore Wind*, Berlin, Germany, 4–6 December 2007, pp. 1–8.
136. Gravity based structure, 2023. Wikipedia, the free encyclopedia, last edited 16 May 2023. https://en.wikipedia.org/wiki/Gravity-based_structure.
137. Suction Caisson, 2023. Wikipedia, The free encyclopedia, last edited 10 November 2022. https://en.wikipedia.org/wiki/Suction_caisson.
138. Jonkman, J. M., Matha, D., Dynamics of offshore floating wind turbines-analysis of three concepts. *Wind Energy* 2011;14:557–569.
139. Jonkman, J., Sclavounos, P., Development of fully coupled aeroelastic and hydrodynamic models for offshore wind turbines. In *Proceedings of the 44th AIAA Aerospace Sciences Meeting and Exibit*, Reno, NV, USA, 9–12 January 2006.
140. Wind Vision, Energy.gov. Available online: https://www.energy.gov/eere/wind/wind-vision (accessed on 27 October 2019).
141. Guillermo, O., Structural dynamic behaviour of a floating platform for offshore wind turbines [Minor Thesis]. UPC, Barcelona, 2013. Available online: https://upcommons.upc.edu/bitstream/handle/2099.1/24895/Structural%20dynamic%20behaviour%20of% 20a%20floating%20platform%20for%20offshore%20wind%20turbines.pdf?sequence=2&isAllowed=y (accessed on 10 October 2019).
142. Park, S., Lackner, M. A., Pourazarm, P., Rodríguez Tsouroukdissian, A., Cross-Whiter, J., An investigation on the impacts of passive and semiactive structural control on a fixed bottom and a floating offshore wind turbine. *Wind Energy* 2019;22(11):1451–1471. DOI: 10.1002/we.2381
143. Future of Wind. Available online: https://www.irena.org/publications/2019/Oct/Future-of-wind (accessed on 29 August 2021).

144. These 3 Countries Are Global Offshore Wind Powerhouses, World Economic Forum. Available online: https://www.weforum.org/agenda/2019/04/these-3-countries-are-global-offshore-wind-powerhouses/ (accessed on 29 August 2021).
145. Offshore Wind Energy Update and Outlook. Available online: https://www.orrick.com/en/Insights/2021/07/2021-Offshore- Wind-Energy-Update-and-Outlook (accessed on 29 August 2021).
146. Butterfield, S., Musial, W., Jonkman, J., Sclavounos, P., Engineering challenges for floating offshore wind turbines. In *Proceedings of the 2005 Copenhagen Offshore Wind Conference*, Copenhagen, Denmark, 26–28 October 2005, p. 13. Available online: https://www.nrel.gov/docs/fy07osti/38776.pdf (accessed on 18 May 2020).
147. James, R., Ros, M. C., *Floating Offshore Wind: Market and Technology Review.* The Carbon Trust, London, 2015.
148. Mast, E., Rawlinson, R., Sixtensson, C., *Market Study Floating Wind in the Netherlands: Potential of Floating Offshore Wind.* TKI Wind 0p Zee, Amsterdam, 2015.
149. Govindji, A. K., James, R., Carvallo, A., *Appraisal of the Offshore Wind Industry in Japan.* Carbon Trust, London, UK, 2014.
150. European Wind Energy Association, *Deep Water: The Next Step for Offshore Wind Energy* European Wind Energy Association (EWEA), Brussels, Belgium 2013.
151. Henderson, A. R., Witcher, D., Floating offshore wind energy - A review of the current status and an assessment of the prospects. *Wind Engineering* 2010;34:1–16.
152. Matha, D., *Model Development and Loads Analysis of an Offshore Wind Turbine on a Tension Leg Platform, with a Comparison to Other Floating Turbine Concepts.* Subcontract Report NREL/SR-500-45891. National Renewable Energy Laboratory (NREL), Golden, CO, 2009.
153. Rummelhoff, I., Bull, S., Building the world's first floating offshore wind farm, 2015. Available online: https://www.statoil.com/ (accessed on 16 December 2017).
154. ORE Catapult. 2015 Floating wind: Technology assessment. Available online: https://ore.catapult.org.uk/ (accessed on 17 December 2017).
155. Cruz, J., Atcheson, M. (Eds), *Floating Offshore Wind Energy: The Next Generation of Wind Energy*, Springer, New York, 2016.
156. Bossler, A., 2013 Japan's Floating Offshore Wind Projects: An Overview, a paper presented at Energy Ocean, June 3–5, 2014, Atlantic City, NJ, www.Energyocean.com #EO2014.
157. Fukushima Offshore Wind Consortium 2017 Fukushima Floating Offshore Wind Farm Demonstration Project. Available online: https://www.fukushima-forward.jp/english/ (accessed on 16 December 2017).
158. Liu, Y., Li, S., Yi, Q., Chen, D., Developments in semi-submersible floating foundations supporting wind turbines: A comprehensive review. *Renew Sustain Energy Rev* 2016;60:433–449.
159. Bulder, V. H., Henderson, H., Pierik, S., Wijnants, W., Studie naar haalbaarheid van en randvoorwaarden voor drijvende offshore windturbines, 2002. Available online: https://www.offshorewindenergy.org/ (accessed on 15 December 2017).
160. Musial, W., Butterfield, S., Boone, A., Feasibility of floating platform systems for wind turbines. In *Proceedings of the 42nd AIAA Aerospace Sciences Meeting and Exibit*, Reno, NV, USA, 5–8 January 2004.
161. Landbø, T., OO-Star Wind Floater: An innovative and robust semi-submersible for offshore floating wind. In *EU Research & Innovation Day*, Seoul, South Korea, November 23–24, 2017.
162. NAUTILUS 2017 Product. Available online: https://www.nautilusfs.com/en/ (accessed on 16 December 2017).
163. Myhr, A., Bjerkseter, C., Ågotnes, A., Nygaard, T. A., Levelised cost of energy for offshore floating wind turbines in a life cycle perspective. *Renew Energy* 2014;66:714–728.

164. Principle Power 2015 WindFloat. Available online: https://www.principlepowerinc.com/ (accessed on 16 December 2017).
165. Taboada, J. V., Comparative Analysis Review on Floating Offshore Wind Foundations (FOWF). In *Proceedings of the 54th Naval Engineering and Maritime Industry Congress*, Ferrol, Spain, 2015, pp. 14–16.
166. IDEOL, Ideol winning solutions for offshore wind, 2017. Available online: https://ideol-offshore.com/en (accessed on 16 December 2017).
167. GICON, The GICON SOF, 2016. Available online: https://www.gicon-sof.de/en/sof1.html (accessed on 16 December 2017).
168. ORE Catapult. 2016 Introducing TLPWIND UK. Available online: https://ore.catapult.org.uk/ (accessed on 16 December 2017).
169. Blue, H., Engineering 2017 Technology. Available online: https://www.bluehengineering.com/ (accessed on 16 December 2017).
170. Borg, M., Collu, M., A comparison between the dynamics of horizontal and vertical axis offshore floating wind turbines. *Philos Trans A Math Phys Eng Sci* 2015;373:20140076.
171. Cordle, A., Jonkman, J., State of the art in floating wind turbine design tools. In *Proceedings of the Twenty-First International Offshore and Polar Engineering Conference*, Maui, HI, USA, 19–24 June 2011; Available online: https://www.onepetro.org/ conference-paper/ISOPE-I-11-112 (accessed on 3 May 2020).
172. Dinh, V. -N., Basu, B., On the modeling of spar-type floating offshore wind turbines. *Key Eng Mater* 2013;569:636–643.
173. Jonkman, J., Dynamics modeling and loads analysis of an offshore floating windturbine [Ph.D. Thesis]. University of Colorado, Boulder, CO, 2007.
174. Castro-Santos, L., Vizoso, A. F., Couce, L. C., Formoso, J. F., Economic feasibility of floating offshore wind farms. *Energy* 2016;112:868–882.
175. Castro-Santos, L., Silva, D., Bento, A. R., Salvação, N., Soares, C. G., Economic feasibility of floating offshore wind farms in Portugal. *Ocean Eng* 2020;207:107393.
176. Koundouri, P., Giannouli, A., Souliotis, I., An integrated approach for sustainable environmental and socio-economic development using offshore infrastructure. In *Renewable and Alternative Energy*, 2017, pp. 1581–1601.
177. Floating wind turbine, 2023. Wikipedia, The free encyclopedia, last edited 8 September 2023. https://en.wikipedia.org/wiki/Floating_wind_turbine.
178. Jonkman, J. M., *Dynamics Modeling and Loads Analysis of an Offshore Floating Wind Turbine* (Technical Report). National Renewable Energy Laboratory (NREL), Golden, CO, 2007.
179. Larsen, T. J., Hanson, T. D., A method to avoid negative damped low frequent tower vibrations for a floating, pitch controlled wind turbine. *J Phys Conf Ser* 2007;75:012073.
180. Molin, B., *Hydrodynamique des Structures Offshore*. Editions Technip, Paris, France, 2002.
181. Loyd, M. L. Crosswind kite power (for large-scale wind power production). *J Energy* 1980;4(3):106–111.
182. Kim, H., Lee, S., Lee, S., Influence of blade-tower interaction in upwind-type horizontal axis wind turbines on aerodynamics. *J Mech Sci Technol* 2011;25:1351.
183. Han, B., Zhou, L., Yang, F., Xiang, Z., Individual pitch controller based on fuzzy logic control for wind turbine load mitigation. *IET Renew Power Gener* 2016;10:687–693.
184. Kang, M. J., Kim, H. C., Neural network based pitch controller. In *Proceedings of the World Congress on Electrical Engineering and Computer Systems and Science (EECSS 2015)*, Barcelona, Spain, 13–14 July 2015, pp. 1–6.
185. Bossanyi, E. A., Individual blade pitch control for load reduction. *Wind Energy Int J Prog Appl Wind Power Convers Technol* 2003;6:119–128.
186. Namik, H., Stol, K., Individual blade pitch control of floating offshore wind turbines. *Wind Energy Int J Prog Appl Wind Power Convers Technol* 2010;13:74–85.

187. Namik, H., Stol, K., Performance analysis of individual blade pitch control of offshore wind turbines on two floating platforms. *Mechatronics* 2011;21:691–703.
188. Lackner, M. A.; Rotea, M. A., Structural control of floating wind turbines. *Mechatronics* 2011;21:704–719.
189. Si, Y., Karimi, H. R., Gao, H., Modelling and optimization of a passive structural control design for a spar-type floating wind turbine. *Eng Struct* 2014;69:168–182.
190. Dinh, V. N., Basu, B., Nagarajaiah, S., Semi-active control of vibrations of spar type floating offshore wind turbines. *Smart Struct Syst* 2016;18:683–705.
191. Li, X., Gao, H., Load mitigation for a floating wind turbine via generalized structural control. *IEEE Trans Ind Electron* 2016;63:332–342.
192. Coudurier, C., Lepreux, O., Petit, N., Passive and semi-active control of an offshore floating wind turbine using a tuned liquid column damper. *IFAC-PapersOnLine* 2015;48:241–247.
193. Christiansen, S., Knudsen, T., Bak, T., Optimal control of a ballast-stabilized floating wind turbine. In *Proceedings of the 2011 IEEE International Symposium on Computer-Aided Control System Design (CACSD)*, Denver, CO, USA, 28–30 September 2011, pp. 1214–1219.
194. Raach, S., Schlipf, D., Sandner, F., Matha, D., Cheng, P. W., Nonlinear model predictive control of floating wind turbines with individual pitch control. In *Proceedings of the 2014 American Control Conference*, Portland, OR, USA, 4–6 June 2014, pp. 4434–4439.
195. Schlipf, D., Sandner, F., Raach, S., Matha, D., Cheng, P. W., Nonlinear model predictive control of floating wind turbines. In *Proceedings of the Twenty-Third (2013) International Offshore and Polar Engineering Conference*, Anchorage, AK, USA, 30 June–5 July 2013.
196. Tahani, M., Ziaee, E., Hajinezhad, A., Servati, P., Mirhosseini, M., Sedaghat, A., Vibrational simulation of offshore floating wind turbine and its directional movement control by fuzzy logic. In *Proceedings of the 2015 International Conference on Sustainable Mobility Applications, Renewables and Technology (SMART)*, Kuwait City, Kuwait, 23–25 November 2015, pp. 1–7.
197. Wenzhong, Q., Jincai, S., Yang, Q., Active control of vibration using a fuzzy control method. *J Sound Vib* 2004;275:917–930.
198. Qi, Y., Meng, Q., The application of fuzzy PID control in pitch wind turbine. *Energy Procedia* 2012;16:1635–1641.
199. Castro-Santos, L., deCastro, M., Costoya, X., Filgueira-Vizoso, A., Lamas-Galdo, I., Ribeiro, A., Dias, J. M., Gómez-Gesteira, M., Economic feasibility of floating offshore wind farms considering near future wind resources: Case study of Iberian Coast and Bay of Biscay. *Int J Environ Res Public Health* 2021;18(5):2553. DOI: 10.3390/ijerph18052553
200. Heidari, S., Economic modelling of floating offshore wind power calculation of levelized cost of energy [MS thesis]. School of Business, Malardalens Hogskola Eskilstuna Vasteras, Vasteras, 2017.
201. Kausche, M., Adam, F., Dahlhaus, F., Großmann, J., Floating offshore wind-Economic and ecological challenges of a TLP solution. *Renew Energy* 2018;126:270–280.
202. Henderson, A. R., Zaaijer, M. B., Bulder, B., Pierik, J., Huijsmans, R., van Hees, M., Snijders, E., Wijnants, G. H., Wolf, M. J., Floating windfarms for shallow offshore sites. In *Proceedings of the Fourteenth International Offshore and Polar Engineering Conference*, Toulon, France, 23–28 May 2004. Available online: https://www.onepetro.org/conference-paper/ISOPE-I-04-003?sort=&start=0&q=Floating+Windfarms+for+Shallow+Offshore+Sites&from_year=&peer_reviewed=&published_between=&fromSearchResults=true&to_year=&rows=10# (accessed on 4 May 2020).

203. Henderson, A. R., Bulder, B., Huijsmans, R., Peeringa, J., Pierik, J., Snijders, E., Van Hees, M., Wijnants, G. H., Wolf, M. J., Feasibility study of floating windfarms in shallow offshore sites. *Wind Eng* 2003;27:405–418.
204. Zambrano, T., MacCready, T., Kiceniuk, T., Roddier, D. G., Cermelli, C. A., Dynamic modeling of deepwater offshore wind turbine structures in Gulf of Mexico storm conditions. In *Proceedings of the 25th International Conference on Offshore Mechanics and Arctic Engineering*, Hamburg, Germany, 4–9 June 2006, pp. 629–634.
205. Ishihara, T., Phuc, P. V., Sukegawa, H., Shimada, K., A study on the dynamic response of a semi-submersible floating offshore wind turbine system Part 1: A water tank test. In *Proceedings of the 12th International Conference on Wind Engineering: ICWE 12*, Cairns, Australia, 1–6 July 2007, p. 4. Available online: https://citeseerx.ist.psu.edu/viewdoc/download?doi=10.1.1.505.278 8&rep=rep1&type=pdf (accessed on 26 October 2019).
206. Shimada, K., Ohyama, T., Miyakawa, M., Ishihara, T., Phuc, P. V., Sukegawa, H., A study on a semi-submersible floating offshore wind energy conversion system. In *Proceedings of the Seventeenth International Offshore and Polar Engineering Conference*, Lisbon, Portugal, 1–6 July 2006. Available online: https://www.onepetro.org/conference-paper/ISOPE-I-07-481 (accessed on 3 May 2020).
207. Manabe, H., Uehiro, T., Utiyama, M., Esaki, H., Kinoshita, T., Takagi, K., Okamura, H., Satou, M., Development of the floating structure for the sailing-type offshore wind farm. In *Proceedings of the OCEANS 2008-MTS/IEEE Kobe TechnoOcean*, Kobe, Japan, 8–11 April 2008, pp. 1–4.
208. Kourogi, Y., Takagi, K., Hotta, J., Experimental study on maneuverability coefficients for the navigation simulation of VLMOS. In *Proceedings of the OCEANS 2008-MTS/IEEE Kobe Techno-Ocean*, Kobe, Japan, 8–11 April 2008, pp. 1–6.
209. Korogi, Y., Hiramatsu, T., Takagi, K., Sailing performance of a very large mobile offshore structure for wind power plant. In *Proceedings of the ASME 2009 28th International Conference on Ocean, Offshore and Arctic Engineering*, Honolulu, HI, USA, 31 May–5 June 2009; pp. 1311–1318.
210. Lefranc, M., Torud, A., Three wind turbines on one floating unit, feasibility, design and cost. In *Proceedings of the Offshore Technology Conference*, Houston, TX, USA, 2–5 May 2011.
211. Hu, C., Sueyoshi, M., Liu, C., Liu, Y., Hydrodynamic analysis of a semi-submersible-type floating wind turbine. *J Ocean Wind Energy* 2014;1:7.
212. Bae, Y. H., Kim, M.-H., The dynamic coupling effects of a MUFOWT (multiple unit floating offshore wind turbine) with partially broken blade. *J Ocean Wind Energy* 2015;2:89–98.
213. Kim, K., Kim, H., Lee, J., Kim, S., Paek, I., Design and performance analysis of control algorithm for a floating wind turbine on a large semi-submersible platform. *J Phys Conf Ser* 2016;753:92017.
214. Kang, H., Kim, M.-H., Kim, K.-H., Hong, K., Hydroelastic analysis of multi-unit floating offshore wind turbine platform (MUFOWT). In *Proceedings of the 27th International Ocean and Polar Engineering Conference,* San Francisco, CA, USA, 25–30 June 2017. Available online: https://www.onepetro.org/conference-paper/ISOPE-I-17-637 (accessed on 4 May 2020).
215. Kim, K. H., Hong, J. P., Park, S., Lee, K., Hong, K., An experimental study on dynamic performance of large floating wave- offshore hybrid power generation platform in extreme conditions. *J Korean Soc Mar Environ Energy* 2016;19:7–17.
216. Bashetty, S., Ozcelik, S., Design and stability analysis of an offshore floating multi-turbine platform. In *Proceedings of the 2020 IEEE Green Technologies Conference (GreenTech)*, Oklahoma City, OK, USA, 1–3 April 2020, pp. 184–189.

217. Bashetty, S., Ozcelik, S., Effect of pitch control on the performance of an offshore floating multi-wind-turbine platform. *J Phys Conf Ser* 2021;1828:012055.
218. Bashetty, S., Ozcelik, S., Aero-hydrodynamic analysis of an offshore floating multi-wind-turbine platform-Part I. In *Proceedings of the 2020 IEEE 3rd International Conference on Renewable Energy and Power Engineering (REPE)*, Edmonton, AB, Canada, 9–11 October 2020, pp. 1–6.
219. Bashetty, S., Ozcelik, S., Aero-hydrodynamic analysis of an offshore floating multi-wind-turbine platform-Part II. In *Proceedings of the 2020 IEEE 3rd International Conference on Renewable Energy and Power Engineering (REPE)*, Edmonton, AB, Canada, 9–11 October 2020, pp. 7–11.
220. Cherubini, A., Papini, A., Vertechy, R., Fontana, M., Airborne wind energy systems: A review of the technologies. *Renew Sustain Energy Rev* 2015;51:1461–1476 DOI:10.1016/j.rser.2015.07.053
221. Archer, C. L., Caldeira, K., Global assessment of high-altitude wind power. *Energies* 2009;2:307–319.
222. Marvel, K., Kravitz, B., Caldeira, K., Geophysical limits to global wind power. *Nat Clim Change* 2012;3:118–121.
223. Miller, L. M., Gans, F., Kleidon, A., Jet stream wind power as a renewable energy resource: Little power, big impacts. *Earth Syst Dyn* 2011;2:201–212.
224. Diehl, M., Airborne wind energy: Basic concepts and physical foundations. In U. Ahrens, M. Diehl, R. Schmehl (Eds), *Airborne Wind Energy*. Springer, Berlin, 2013, pp. 3–22.
225. Williams, P., Lansdorp, B., Ockesl, W., Optimal crosswind towing and power generation with tethered kites. *J Guid Control Dyn* 2008;31:81–93.
226. Canale, M., Fagiano, L., Milanese, M., KiteGen: A revolution in wind energy generation. *Energy* 2009;34:355–361.
227. Ji, Y., He, J., Analysis on lightning triggering possibility along transmission tethers of high altitude wind energy exploitation system. *Electr Power Syst Res* 2013;94:16–23.
228. Zhang, J., Zou, N., Zhou, W. L., System and method for umbrella power generation. PCT Patent Application WO2010129124, 2010.
229. Perković, L., Silva, P., Ban, M., Kranjčević, N., Duić, N., Harvesting high altitude wind energy for power production: The concept based on Magnus' effect. *Appl Energy* 2012;101:151–160.
230. Magnus, G., Über die Abweichung der Geschosse, und über eine abfallende Erscheinung bei rotierenden Körpern. *Ann Phys* 1853;164(1):1–29.
231. Schmehl, R., "Airborne Wind Energy" PowerWeb webinar lecture, 2020. Available online: https://airbornewindeurope.org/wp-content/uploads/2020/06/20200520-Powerweb.pdf (accessed on January 1, 2021).
232. Brabeck, S., Private communication in stakeholder outreach with SkySails, 2021.
233. Peschel, J., Private communication in stakeholder outreach with KitePower, 2021.
234. Lopez, A., Mai, T., Lantz, E., Harrison-Atlas, D., Williams, T., Maclaurin, G., Land use and turbine technology influences on wind potential in the United States. *Energy* 2021;223:120044. DOI:10.1016/j.energy.2021.120044.
235. Musial, W., Heimiller, D., Beiter, P., Scott, G., Draxl, C., *2016 Offshore Wind Energy Resource Assessment for the United States* (Technical Report). NREL/TP-5000-66599. National Renewable Energy Laboratory (NREL), Golden, CO, 2016. Available online: https://www.nrel.gov/docs/fy16osti/66599.pdf (accessed on March 1, 2017).
236. Schmehl, R. (Ed), *Airborne Wind Energy: Advances in Technology Development and Research*. Springer, Singapore, 2018. DOI:10.1007/978-981-10-1947-0
237. Weber, J., Marquis, M., Cooperman, A., Draxl, C., Hammond, R., Jonkman, J., Lemke, A., Lopez, A., Mudafort, R., Optis, M., Roberts, O., *Owen Roberts, and Matt Shields Airborne Wind Energy* (Technical Report). NREL/TP-5000-79992, Contract No. DE-AC36-08GO28308. National Renewable Energy Laboratory (NREL), Golden, CO, 2021.

238. United States Department of Energy. *Challenges and Opportunities for Airborne Wind Energy in United States*, Report to Congress November 2021 United States Department of Energy, Washington, DC, 2021.
239. Anderson, M., Ready flyer one: Airborne wind energy simulations guide the leap to satisfying global energy demand. *IEEE Spectrum: Technology, Engineering, and Science News*, 2019. Available online: https://spectrum.ieee.org/energywise/energy/renewables/ready-flyer-one-airborne-wind-energy-simulations-guide-the-leap-to-satisfying-global-energy-demand (accessed on June 1, 2020).
240. Weber, J., AirborneMax - Scaling as the key issue for airborne wind. In *Airborne Wind Energy Conference 2019*, 2019. Available online: https://repository.tudelft.nl/islandora/object/uuid:ffd965bc-39ed-41cb-8d91-46504f9aef12 (accessed on March 1, 2020).
241. Fagiano, L., Milanese, M., Airborne wind energy: An overview. In *American Control Conference*, Fairmont Queen Elizabeth, Montreal, Canada, 2012 (accessed on September 1, 2012).
242. Argatov, I., Rautakorpi, P., Silvennoinen, R., Estimation of the mechanical power of a kite wind generator. In S. P. Lohani (Ed.), *Renewable Energy for Sustainable Future*, 2013, pp. 1–28.
243. Echeverri, P., Fricke, T., Homsy, G., Tucker, N., *The Energy Kite: Selected Results from the Design, Development, and Testing of Makani's Airborne Wind Turbines*. Makani Technologies, LLC, 2020. Available online: https://www.energykitesystems.net/AWEC2020Teleconference/PostersAWEC2020/AWEC2020POSTER023DocumentMakaniPowerLLC.pdf (accessed on January 1, 2021).
244. Kruijff, M., Private communication in stakeholder outreach with Ampyx Power, 2021.
245. Santos, D., Private communication in stakeholder outreach with kPower, 2021.
246. De Schutter, J., Leuthold, R., Bronnenmeyer, T., Paelinck, R., Diehl, M., Towards a modular upscaling strategy for utility-scale airborne wind energy. In *8th International Airborne Wind Energy Conference (AWEC 2019)*, Glasgow, UK, 2019a. Available online: https://repository.tudelft.nl/islandora/object/uuid%3Ac7970efe-b007-4d26-a00f-14c0cdba0ba1 (accessed on May 1, 2020).
247. De Schutter, J., Leuthold, R., Bronnenmeyer, T., Paelinck, R., Diehl, M., Optimal control of stacked multi-kite systems for utility-scale airborne wind energy. In *2019 IEEE 58th Conference on Decision and Control (CDC)*, 2019b. DOI: 10.1109/CDC40024.2019.9030026
248. Kiteswarms. Developing the future of airborne wind energy. Available online: https://www.youtube.com/watch?v=URyacqV5xI0 (accessed on May 1, 2020).
249. Leuthold, R., Gros, S., Diehl, M., Induction in optimal control of multiple-kite airborne wind energy systems. *IFAC-PapersOnLine* 2017;50(1):153–158. DOI: 10.1016/j.ifacol.2017.08.026.
250. Leuthold, R., De Schutter, J., Malz, E. C., Licitra, G., Gros, S., Diehl, M., Operational regions of a multi-kite AWE system. In *2018 European Control Conference (ECC). Presented at the 2018 European Control Conference (ECC)*, 2018, pp. 52–57. DOI: 10.23919/ECC.2018.8550199
251. Beaupoil, C., Practical experiences with a torsion based rigid blade rotary airborne wind energy system with ground based power generation. In *Airborne Wind Energy Conference*, 2019. Available online: https://repository.tudelft.nl/islandora/object/uuid:c50f7eb6-f116-4158-9e9b-aa1b05092ab2?collection=research(accessedonApril1, 2020).
252. Read, R., Kite networks for harvesting wind energy. In R. Schmehl (Ed), *Airborne Wind Energy, Advances in Technology Development and Research*. Springer, Singapore, 2018, pp. 515–537. DOI: 10.1007/978-981-10-1947-0_21

253. Tulloch, O., Kazemi Amiri, A., Yue, H., Feuchtwang, J., Read, R., Modeling Studies on Tensile Rotary Power Transmission for Airborne Wind Energy Systems, 2019. Available online: https://repository.tudelft.nl/islandora/object/uuid:fa2a20d6-51d7-4f85-9f64-65b6df1ca3e0?collection=research (accessed on January 1, 2020).
254. Ippolito, M., Private communication in stakeholder outreach with KiteGen, 2021.
255. Fagiano, L., Automation challenges in airborne wind energy systems and the role of academic research. In *Airborne Wind Energy Conference*, 2019. Available online: https://www.youtube.com/watch?v=ZptcYca_JBs ((accessed on July 1, 2020).
256. Vermillion, C., Cobb, M., Fagiano, L., Leuthold, R., Diehl, M., Smith, R., Wood, T., Rapp, S., Schmehl, R., Olinger, D., Demetriou, M., Electricity in the air: Insights from two decades of advanced control research and experimental flight testing of airborne wind energy systems. *Ann Rev Control* 2021;52:330–357. DOI:10.1016/j.arcontrol.2021.03.002
257. Harklau, T., Private communication in stakeholder outreach with Ampyx Power, 2021.
258. Bormann, A., Private communication in stakeholder outreach with Enerkite, 2021.
259. Tigner B., Multi-tether cross-wind kite power. US patent application US8066225, 2008.
260. Zywietz, D., What will it take for AWE to be successful in remote & mini-grid applications? In *Airborne Wind Energy Conference*, 2019. Available online: https://resolver.tudelft.nl/uuid:dbe7de7e-151d-483e-9658-0b4f9f26d26a (accessed on February 1, 2020).
261. Creighton, R., Private communication in stakeholder outreach with Windlift, 2021.
262. Schaefer, D., Private communication in stakeholder outreach witheWindSolutions, 2021.
263. Bechtle, P., Schelbergen, M., Schmehl, R., Zillmann, U., Watson, S., Airborne wind energy resource analysis. *Renew Energy* 2019;141:1103–1116. DOI:10.1016/j.renene.2019.03.118
264. Calverley, G., Private communication in stakeholder outreach with Abound, 2021.
265. Schmehl, R., Private communication in stakeholder outreach with the Delft University of Technolog, 2021.
266. Zillmann, U., Capital, D., The trillion dollar drone - a change of perspective. In R. Schmehl (Ed), *Book of Abstracts of the International Airborne Wind Energy Conference 2015*, Delft, the Netherlands, 15–16 June 2015.
267. Cherubini, A., AWE community is growing fast, 2014. Available online: https://www.antonellocherubini.com/awe-community-is-growing-fast.html (accessed on January 1, 2015).
268. Lütsch, G., Airborne wind energy network HWN500 - shouldering R&D in co- operations. In *Proceedings of the Airborne Wind Energy Conference*, Delft, the Netherlands, 15–16 June 2015.
269. Argatov, I., Rautakorpi, P., Silvennoinen, R., Estimation of the mechanical energy output of the kite wind generator. *Renew Energy* 2009;34:1525–1532.
270. Schmehl, R., van der Vlugt, R., Traction power generation with tethered wings - A quasi-steady model for the prediction of the power output. In *Book of Abstracts of the International Airborne Wind Energy Conference*, Delft, the Netherlands, 15–16 June 2015.
271. Lind, D. V., Analysis and flight test validation of high performance airborne wind turbines. In U. Ahrens, M. Diehl, R. Schmehl (Eds), *Airborne Wind Energy*, Springer, Berlin, 2013, pp. 473–490.
272. Haug, S., Design of a kite launch and retrieval system for a pumping high altitude wind power generator [M.Sc. thesis]. University of Stuttgart, 2012.
273. Enerkite website. Available online: https://www.enerkite.com (accessed on 30 December 2014).

274. Geebelen, K., Vukov, M., Zanon, M., Gros, S., Wagner, A., Diehl, M., Vandepitte, D., Swevers, J., Ahmad, H., An experimental test setup for advanced estimation and control of an airborne wind energy system. In U. Ahrens, M. Diehl, R Schmehl (Eds), *Airborne Wind Energy*, Springer, Berlin, 2013, pp. 459–471.
275. Cherubini, A., Kite dynamics and wind energy harvesting [M.Sc. thesis]. Politecnico di Milano, 2012.
276. Inman, M. and Davis, S., Near zero-Energy high in the sky: Expert perspectives on airborne wind energy systems. In M. Inman (Ed), 2012. http://nearzero.org/ (accessed on 9 September 2012).
277. Venturato, A., Analisi fluidodinamica del profilo alare Clark-Y ed ottimizza- zione multi-obbiettivo tramite algoritmo genetico [M.Sc. thesis]. Università degli studi di Padova, 2013.
278. Bosman, R., Reid, V., Vlasblom, M., Smeets, P., Airborne wind energy tethers with high-modulus polyethylene fibers. In U. Ahrens, M. Diehl, R. Schmehl (Eds), *Airborne Wind Energy*. Springer, Berlin, 2013, pp. 563–585.
279. Schneiderheinze, T., Heinze, T., Michael, M., High performance ropes and drums in airborne wind energy systems. In *6th Airborne Wind Energy Conference (AWEC 2015)*. Delft, the Netherlands, 2015.
280. Brandt, D., Busch, M., Bormann, A., Ranneberg, M., Adapting wind resource estimation for airborne wind energy converters. In *6th Airborne Wind Energy Conference (AWEC 2015)*. Delft, the Netherlands, 2015.
281. Payne, P. R. Self erecting windmill. US patent application US3987987, 1976.
282. Zanon, M., Gros, S., Andersson, J., Diehl, M., Airborne wind energy based on dual airfoils. *IEEE Trans Control Syst Technol* 2013;21(4):1215–1222.
283. Diehl, M., Horn, G., Zanon, M., Multiple wing systems - an alternative to upscaling? In R. Schmehl (Ed), *Book of Abstracts of the International Airborne Wind Energy Conference 2015*, Delft, The Netherlands, 15–16 June 2015.
284. Airborne wind energy, 2023. Wikipedia, The free encyclopedia, last edited 8 August 2023. https://en.wikipedia.org/wiki/Airborne_wind_energy.

3 Advances in Hydroelectricity

3.1 INTRODUCTION

Both in the USA and globally, hydropower is the largest renewable resource in the energy mix and certainly the largest source of renewably generated electricity. There are two fundamental ways power can be generated using hydro-energy: (1) hydropotential energy using dam for power generation and storage, and these dams can be very large or very small (or micro-level) run-of-river as described later; and (2) using hydrokinetic energy associated with ocean waves, currents, tidal waves, river flow, etc., to generate power. Here, kinetic energy is converted to power. As shown later, there are other ways to generate power as well from ocean such as ocean salinity gradient or ocean thermal gradient, but these are not as yet dominant and fully developed. While growth in the use of hydroelectricity (at least the traditional type—generated by very large dams) has slowed to near zero in the USA, many other countries in both the developed and developing world are pushing ahead with major projects to obtain energy from non-powered dams (NPDs), hydrokinetic energy from rivers, tidal waves, and ocean waves to generate electricity. As mentioned in Chapter 1, advances in hydropower have occurred in three different directions [1–9]. First, significant advances have been made in improving control, flexibility, and efficiency of large-scale hydropower. This is partly due to adjustments of dam-based hydropower in new reality of grid operation and its new mission. Second, significant progress has been made on small-scale river-based small hydropower (SHP) including the capture of hydrokinetic energy from flowing river water. Finally, in recent years, significant progress has been made to capture marine energy from ocean including energy from ocean waves, tidal waves, salinity gradients, ocean temperature gradients etc. This chapter covers advances made in all of these three directions.

There are three main types of dam-based conventional hydropower technologies: (1) impoundment (dam), (2) diversion, and (3) pumped storage. Impoundment is the most common type of hydroelectric power plant. An impoundment facility, typically a large hydropower system, uses a dam to store river water in a reservoir. Water released from the reservoir flows through a turbine, spinning it, which in turn activates a generator to produce electricity. Generation may be used fairly flexibly to meet baseload as well as peak load demands. The water may also be released either to meet changing electricity needs or to maintain a constant reservoir level. The layout of a typical impoundment hydropower facility is shown in Figure 3.1.

A diversion, sometimes called run-of-river, facility channels a portion of a river through a canal or penstock. It may not require the use of a dam but also has limited flexibility to follow peak variation in power demand. Thus, it will mainly be useful for baseload capacity. This scenario results in limited flooding and changes to river

DOI: 10.1201/9781003429906-3

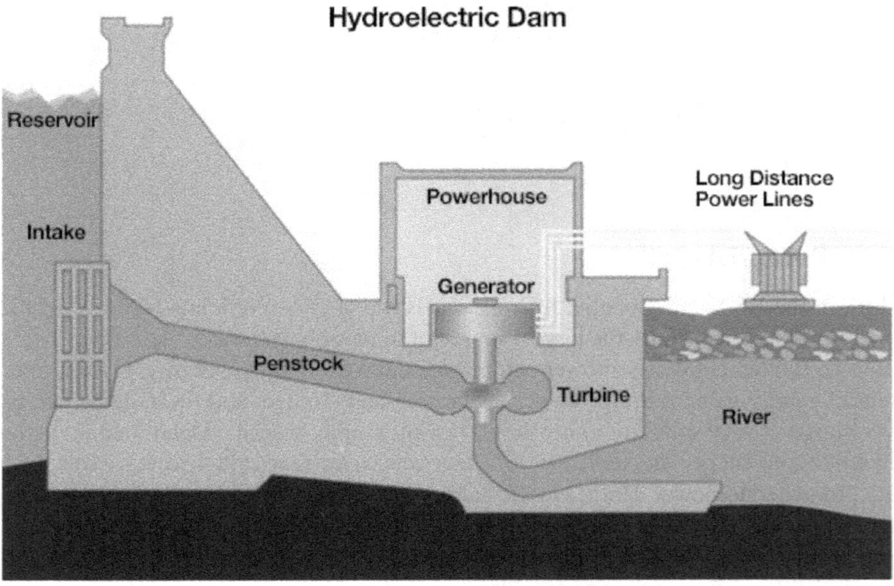

FIGURE 3.1 Conventional impoundment dam [2,5].

flow. In the USA, many of the dams in the Pacific Northwest (on the Columbia and Snake rivers) are diversion or run-of-river dams, with limited or no storage reservoir behind the dam. Figure 3.2 shows a schematic of run-of-river hydropower installation [10]. These facilities are generally considered as small, mini- or micro-hydropower stations.

A "pumped-storage" hydro-dam combines a small storage reservoir with a system for cycling water back into the reservoir after it has been released through the turbine, thus "re-using" the same water to generate electricity at a later time. When the demand for electricity is low (typically at night), a pumped-storage facility stores energy by pumping water from a lower reservoir to an upper reservoir. During periods of high electrical demand (typically during the day), the water is released back to the lower reservoir to generate electricity. Figure 3.3 shows a schematic of a pumped-storage hydro facility. Pumped-storage facilities are typically smaller in terms of generation capacity than their impoundment or diversion counterparts, but are sometimes combined with impoundment or diversion facilities to increase peak power output or flexibility. Globally, they are the major sources of energy storage at large scale. In these kinds of plants, reversible Francis devices are used for both pumping water and generating electricity. The energy conversion efficiency of recently pumped hydropower is over 80% [2,5]. Pumped-storage plants can be combined with intermittent renewable electricity sources. They can also serve as an optimal complement to nuclear-based electricity designed for baseload operation, but with only limited capability to adapt to daily and seasonal load fluctuations. Currently, pumped-storage capacity worldwide amounts to about 140 GW_e. In the European

FIGURE 3.2 A schematic representation of a run-of-river hydropower installation, highlighting the key components present for both high- and low-head diversion schemes [10].

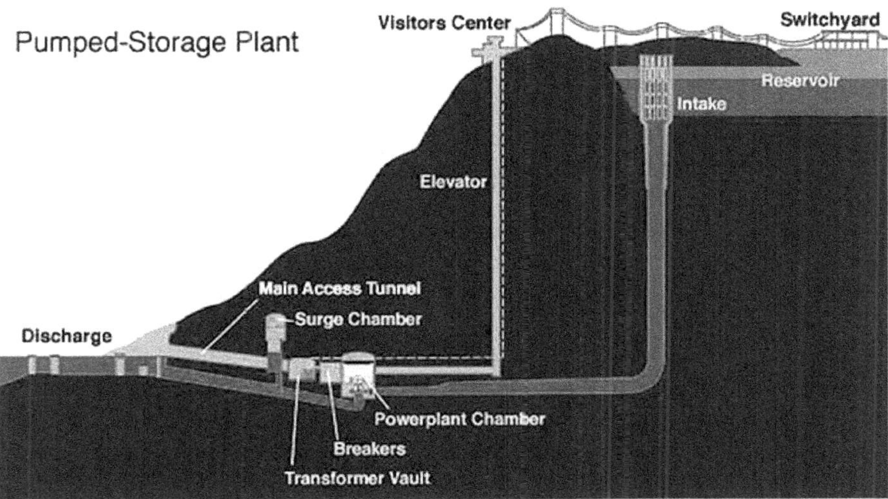

FIGURE 3.3 Schematic of a pumped-storage hydro-facility [2.5].

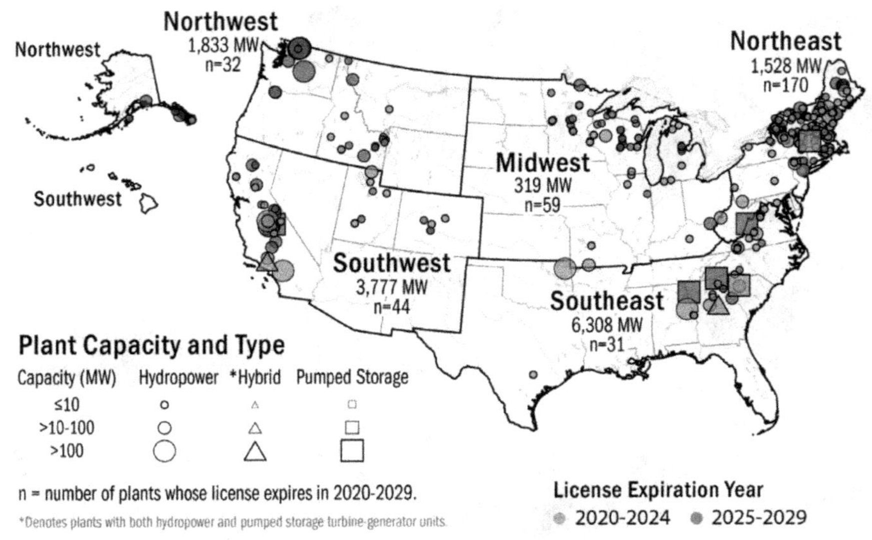

FIGURE 3.4 Hydropower and PSH plants with licenses expiring in 2020–2029 [2].

Union, there are 45 GW_e of pumped-storage capacity. In Asia, the leading pumped hydropower countries are Japan (30 GW_e) and China (24 GW_e). The USA also has a significant volume of the pumped-storage capacity (20 GW_e). In USA, hydropower and PSH plants with licenses expiring in 2020–2029 are illustrated in Figure 3.4.

Hydropower is a mature technology. Globally, hydroelectricity is a major electricity resource, accounting for more than 16% of all electricity produced on the planet. Energy production from hydro in 2018 was 4,300 terawatt-hours (TWh) which represented 17% of global electricity production [1–9,11]. In 2019, global installed hydroelectric power capacity reached 1,310 gigawatts (GW) [1–9,11]. More electricity is produced globally using hydropower than from plants fueled by nuclear fission or petroleum. Hydropower plants provide at least 50% of the total electricity supply in more than 35 countries. More than 150 countries produce some hydroelectricity, although around 50% of all hydropower is produced by just four countries: China, Brazil, Canada, and the USA. China is by far the largest hydropower producer on the planet, as shown in Table 3.1. Hydroelectricity production in China has tripled over the past decade, with the completion of some of the world's largest dam projects, in particular, the Three Gorges Dam (the world's largest), which could produce nearly enough electricity to power all of New England during a typical summer and left an area roughly the size of San Francisco flooded underwater.

Once hydroelectric dams are built, they run very cheaply and generally provide reliable supplies of electricity except during times of extreme drought. Developed countries, that have substantial hydro resources, have, by and large, already utilized those resources to produce electricity. In these countries, hydropower dominates the electricity supply system. Norway leads the pack here—the amount of hydropower

TABLE 3.1
Ten of the Largest Hydroelectric Producers as at 2020 [11]

Country	Annual Hydroelectric Production (TWh)	Installed Capacity (GW)	Capacity Factor	% of World's Production	% in Domestic Electricity Generation
China	1232	352	0.37	28.5%	17.2%
Brazil	389	105	0.56	9.0%	64.7%
Canada	386	81	0.59	8.9%	59.0%
USA	317	103	0.42	7.3%	7.1%
Russia	193	51	0.42	4.5%	17.3%
India	151	49	0.43	3.5%	9.6%
Norway	140	33	0.49	3.2%	95.0%
Japan	88	50	0.37	2.0%	8.4%
Vietnam	84	18	0.67	1.9%	34.9%
France	71	26	0.46	1.6%	12.1%

that it produces is not large in an absolute sense (it is the world's seventh-largest producer) but nearly all electricity generated in Norway comes from hydropower. Brazil and Canada are also highly dependent on hydropower. Other large hydropower producers, such as China and the USA, produce much less hydroelectricity relative to the size of their overall power sectors.

Hydropower plants also provide other key services such as flood control, irrigation, and water supply for emergency need. While its upfront investment cost is relatively high, hydropower is a cost-effective electricity source, and it offers high efficiency and low operating and generation costs. Hydroelectricity is valuable in an electricity system because it is easily capable of following variable load by varying the flow of water through the turbine. Hydropower is an extremely flexible electricity generation technology. The capacity factor of hydropower plants varies between 23% and 95%, depending on the targets and the service (i.e., baseload, peak load) of the specific power plant. Hydro reservoirs provide built-in energy storage that enables a quick response to electricity demand fluctuations across the grid, optimization of electricity production, and compensation for power losses from other sources. One of the advantages of hydropower is its operational flexibility. In other words, the capacity factor of hydropower varies depending on the specific plant and its services (i.e., baseload, peak load) between 23% and 95%.

Many countries including the USA will not likely pursue larger hydro-projects due to two main factors—first, we have already built hydroelectric dams in most of the best places, and second, there are concerns over the environmental and societal impacts of building more dams. These environmental impacts have even been used to justify dam removal in some cases, though weighing those environmental impacts against the societal benefits (such as irrigation, flood control, recreation, and so forth... not just electricity) has always been controversial. Some of these specific environmental and societal impacts include an impact on fish migration in which fish passages may be blocked or fish get killed going through turbines. The dam may also reduce water quality that can affect river ecosystems and aquatic plants. Reservoirs behind dams can flood large areas, and the dam project generally displaces many people living in the area surrounding the river. The decay of plant life in flooded areas can release significant amount of methane in the atmosphere. The decomposition of organic materials both upstream and downstream of dam can result in GHG emissions.

While in USA, there is little potential for growth in large-scale hydroelectric power generation and in some areas of the Pacific Northwest, the USA is removing some dams that produce electricity due to environmental concerns, and several hundred megawatts of hydroelectric generation in the Mid-Atlantic USA are planned for small pumped-storage facilities. It should also be noted that not all dams in the USA are equipped with turbines to generate electricity. There are, actually, quite a few that haven't potentially enough power generation to supply more than ten million homes. Many of these are located along major shipping routes, like the Ohio and Mississippi rivers. Others are located in areas where development might not make economic sense, because the power would need to be shipped across large and expensive transmission lines. Another unconventional technology—hydro-kinetics—could potentially supply enough electricity to power the state of Virginia, although these resources are highly concentrated in the lower Mississippi river and in more remote areas such as Alaska.

The market for large hydropower plants is dominated by a few manufacturers of large equipment and a number of suppliers of auxiliary components and systems. Over the past several decades, significant advances are made in the areas of digitalization of hydropower, variable-speed hydropower, advances in turbines and dynamic control of hydropower stations, innovative methods for pumped hydropower storage, and hybridization of hydropower with other renewable sources for power. These are examined in some details in the present chapter. Current hydropower growth options can be categorized into three types: *Large hydropower* (>10 MW_e); *small hydropower* ($\leq$10 MW_e) with *mini-hydro* (100 kW_e to 1 MW_e) as a subcategory; and *micro-hydropower* with capacity less than 100 kWe. *Upgrading potential also exists at hydropower plants* and dams. Small-scale and micro-hydropower are often used for distributed power applications as an alternative to, or in combination with, diesel generators or other small-scale power plants for rural applications. Some of the comparison between large and small hydropower systems is illustrated in Tables 3.2 and 3.3 as outlined by IRENA [8].

The capital cost of a river-based hydroelectric system is highly dependent upon local geology, geography, and hydrology. Such systems are frequently located in mountainous regions with difficult road access and lengthy distances for transmission.

The capital cost of an energy storage system has two components: an energy cost ($/GWh) and a power cost ($/GW). Sometimes, these components are conflated into a single number (e.g., $/GW) by using a fixed storage time such as 6 hours. This can sometimes be useful when comparing similar systems but is misleading when comparing different systems such as batteries and pumped hydro. A battery typically has a storage time of 1 hour; that is, it can operate at full power for 1 hour. Thus, a 1-hour battery with a power of 0.1 GW has an energy storage of 0.1 GWh. In contrast, a 1 GW off-river pumped hydro-system might have 20 hours of storage, equal to 20 GWh. Planning and approvals are generally easier, quicker, and lower cost for an off-river system compared with a river-based system.

The cost of storage energy ($/GWh) primarily relates to the cost of reservoir construction. The cost of constructing an off-river reservoir includes moving rock to form the walls, a small spillway and a water intake. Other significant costs could include road access, water access, lining the bottom of the reservoir to mitigate water leakage and placing evaporation suppressors on the water surface. Forming the walls is usually the dominant cost and can be approximated by the cost of moving rock ($/m^3). The amount of energy stored in a hydro-system is proportional to the head and to the usable water volume of the reservoirs. The important reservoir metrics are (1) the head and (2) the ratio of water impounded to the rock required to form the reservoir walls. Doubling the head or doubling the water/rock (W/R) ratio both approximately halve the effective cost of energy storage ($/GWh). The cost of storage power ($/GW) primarily relates to the cost of the water conveyance and the powerhouse. Additionally, transmission is sometimes a significant cost depending on distance to a high-voltage powerline. The expensive component of the water conveyance is the high-pressure pipe or tunnel that spans most of the altitude difference between the reservoirs.

In essence, the energy cost ($/GWh) is minimized by having large head and large ratio of usable water volume to volume of rock needed to form the upper and lower reservoirs [1–9,11]. The power cost ($/GW) is minimized by having large head and large average slope of the pressure pipe or tunnel. As with many engineering enterprises, systems with larger power and energy are cheaper per unit than smaller systems. Access to roads, water supply, and transmission lines is highly site-dependent. Since there is a wide range of sites to choose between, such costs can be minimized. Annual operation and maintenance costs plus major refurbishments after 20 and 40 years cost about 1% of the initial capital cost each year. This corresponds to about 20% of the annualized capital cost assuming 60 years of lifetime and 5% real discount rate. The cost of storage depends on the capital cost of the system, the annual cost of operations and maintenance, including periodic major refurbishments, the amount of energy sold by the storage each year and the price received relative to the price paid for the energy sent to the storage, losses in the energy storage cycle, the operational lifetime, and the real discount rate.

3.2 ADVANCES IN SMALL HYDROPOWER

Unlike large plants, small-scale hydropower installations comprise a huge variety of designs, layouts, equipment, and materials. Therefore, state-of-the-art technologies, knowledge, and design experience are key to fully exploiting local resources at competitive costs and without significant adverse environmental impact. *Upgrading* offers a way to maximize the energy produced from existing hydropower plants and may offer a less expensive opportunity to increase hydropower production. Gains of between 5% and 10% are realistic, cost-effective targets for most hydropower plants. Potential gains could also be higher at locations where non-generating dams are available. SHPs may be operated for around 50 years without substantial replacement costs. A recent study shows the total installed cost of large-scale hydropower facilities with storage range from as low as USD 1,050/kW to as high as USD 7,650/kW. The investment cost of small-scale hydropower plants of 1–10 MW ranges from less than USD 1,000/kW to about USD 4,000/kW. However, the cost of very-small-scale hydropower plants of less than 1 MW can range from USD 3,400/kW to USD 10,000/kW or more [12].

While the first hydropower plant produced 12.5 kW electricity, today, the US Fleet of hydropower plants consists of roughly 2,400 individual facilities spinning over 6,000 turbines, meeting roughly 7% of the annual electricity demand of the country with roughly 80 GW of installed capacity [13,14]. SHPs are generally classified as plants with less than 10 MW of nameplate capacity and provide 3.8 GW of installed capacity at over 1,700 individual facilities in 46 states (Figure 3.5). Over half of all SHP capacity is located in five states: California, New York, Idaho, Wisconsin, and Michigan. Though they represent only 4.7% of US hydroelectric installed capacity, over 73% of US hydropower plants are SHPs, and over 58% of US hydropower turbines are currently spinning at SHPs [13]. Hydropower systems usually require only minimal maintenance and have low operation costs. Annual O&M costs for typical hydropower plants range between 1% and 4% of investment costs per kW per year, while small-scale hydro can range from 1% to 6%. Table 3.2 provides R&D priorities for both large and small hydropower [12]. R&D and technical advances are also required for small hydropower, notably equipment design, materials, and control systems. One priority is the development of less expensive technologies for

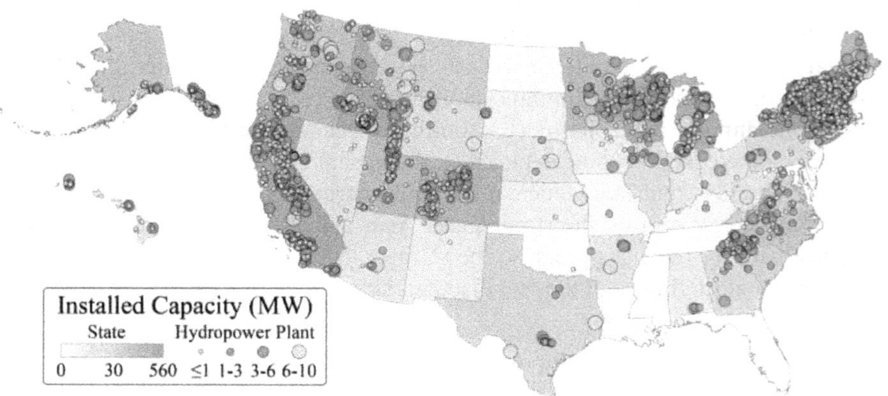

FIGURE 3.5 SHP Mode of Operation, SHP Year in Service, and SHP Turbine Type by Rated Head SHP Technical Specifications [12].

TABLE 3.2
Technology Advances for Hydropower [8]

	Large Hydro	Small Hydro
Equipment	Low-head technologies, including in-stream flow, advanced equipment, and materials	Low-impact turbines for fish populations Low-head technologies In-stream flow technologies
O&M	Maintenance-free and remote operations	Package plants with limited O&M
Storage and hybrid tech.		Wind-hydro and hydrogen-hydro systems.

TABLE 3.3
Summary Table: Key Data and Figures for Hydropower Technology [8]

Technical Performance	Typical Current International Values and Ranges		
Energy input		Hydropower	
Output		Electricity	
Technologies	Very small hydropower (VSHP, up to 1 MW_e)	Small hydropower (SHP, 1–10 MW_e)	Large hydropower (LHP, >10 MW_e)
Efficiency (turbine, Cp max), %	Up to 92	Up to 92	Up to 92
Construction time, months	6–10	10–18	18–96
Technical lifetime, yr.		Up to 100	
Load (capacity) factor, %	40–60 (50)	34–56 (45)	34–56 (45)
Max. (plant) availability, %	98	98	98
Typical (capacity) size, MW_e	0.5	5	50
(Existing) capacity, GW_e	75		925
Environmental Impact			
CO_2 and other GHG emissions, kg/MWh	Negligible		Under investigation
Costs (USD 2010)			
Investment cost, USD/kW	3400–10,000 or more	1000–4000	1050–7,650
O&M cost USD/kW/year	45–250 or more	40–50	45 (average)
Economic lifetime, yr.		30	
Interest rate, %		10	
Production cost, USD/MWh	270 or more	20–100	20–190

small-capacity and low-head applications to enable the exploitation of more modest resources. Key data for hydropower can also be found in Table 3.3 [12].

Small-scale hydro is mainly "run of river," so it involves construction of a quite small dam or barrage, usually just a weir, and generally little or no water is stored [15]. Civil works regulate the water level at the intake to the power plant [15]. In low (5 m of head) or medium (head 4–15 m) installations, a canal carries the water to

FIGURE 3.6 An example of a low-head in-weir scheme (Torrs Hydro Scheme, New Mills, Derbyshire, UK) [10].

the forebay/settling tank. Generally, for low-head installations (head of 5 m), water enters the turbine almost directly from the weir [15]. The size of a small hydropower scheme is about 10 MW or less, although most countries define the scheme differently. Run-of-river (see Figure 3.6) primarily uses the natural flow rate of water to generate power—instead of the power of water falling a large distance. For a run-of-river system to be possible in a given location, there needs to be two specific geographical features. The first is there must be a reasonably substantial flow rate, from either rainfall or a melting snowpack. In addition, there must be enough of a tilt to the river to speed the water up significantly. Therefore, run-of-river systems are best implemented in bodies of water with a fairly constant flow rate. In run-of-river systems, running water from a river is guided down a channel or penstock. There can be some change in altitude at this point (from a small dam or the natural landscape), so there may still be some contribution from "falling water."

There are several benefits that exist as a result of using run-of-river hydro instead of traditional, dam-based hydro. First, traditional hydro-dams are expensive and time-consuming to build. Comparatively speaking, run-of-river systems are less expensive to build and can be built over a shorter period of time. In addition, many areas where large hydro is used frequently—such as in Canada—have developed many new run-of-river hydropower sites. Run-of-river systems also avoid some of the environmental problems associated with the flooding, since the pondage (small storage) is much smaller than the lakes for traditional hydro. Small-scale hydropower systems are applicable to various sites such as dams, rivers, agricultural channels, city water and sewage water plants, and factories. River water is directly used for power generation without any storage facility. Since power generation is dependent on the flow rate of the river water, annual power yield always fluctuates throughout the year. Power generation is often dependent on agricultural water used for the irrigation of rice and vegetable fields. The generated power is fed to agricultural water irrigation facilities in order to reduce the maintenance cost. Power is also generated with the use of the residual pressure at the industrial water intake tank. Recent technical trends in small-scale hydropower generation include cost reduction, labor-saving maintenance, simplified digitalization of control units, and wider variety of control functions [1]. Modal et al. [16] compared the benefits of run-of-river hydroelectric energy production compared to reservoir-based hydroelectric generation. Their three-prong analysis indicates that smaller scale run-of-river hydropower will be most useful in remote settings by allowing people living therein to receive clean energy without the drastic environmental changes caused by damming.

Witt et al. [12] point out that if current trends are an indication of the future state of small hydropower in the USA, it is clear that few, if any, new greenfield SHPs will be supplying renewable energy to the grid in coming decades. In fact, the recent *Hydropower Vision* report [17] used advanced power system modeling to conclude no deployment of new hydropower projects will occur over the next 30 years under a business-as-usual modeling scenario. Witt et al. [12] point out the need for use of small modular hydropower research projects to alter this trajectory by incorporating modularity, standardization, and preservation of stream functionality into a new development paradigm. The new SMH project attempts to prove that standardization, modularity, and stream functionality are essential pathways for hydropower technology cost reduction. Deployment of new SHPs relies not only on lowering costs, however, but on demonstrating the value of hydropower as an energy resource, and of hydropower facilities as beneficial for the environment, for project owners, and for society. Consequently, SMH facilities may have opportunities to enhance the environmental, economic, and social benefits of new development compared to conventional approaches. Loots et al. [18] reviewed low head hydropower technologies and their applications in a south Africa context.

3.2.1 Advanced Turbines for Small-Scale Hydropower

As mentioned above, small hydropower can be characterized based on head height, discharge (flow rate), and capacity. A large flow rate and small head characterize large run-of-river plants equipped with *Kaplan* turbines, a propeller-type water turbine with adjustable blades. By contrast, low discharge and high-head features are typical of mountain-based dam installations driven by *Pelton* turbines, in which water passes through nozzles and strikes spoon-shaped buckets arranged on the periphery of a wheel. Intermediate flow rates and head heights are usually equipped with *Francis* turbines, in which the water comes to the turbine under immense pressure and the energy is extracted from the water by the turbine blades [12–15,19–24].

In remote areas where the connections to grid are not possible, small hydro plays an important role in mini-grid and rural electrification strategies. In developing countries, they support economic activities in remote areas [25,26]. Untapped hydro-potential lies in existing small dams developed in rural agricultural areas to meet various needs not related to energy production such as irrigation, drinking water supply [27], or flood mitigation [28]. The transformation of such dams to hydroelectric facilities, when possible, typically involves a fraction of the total cost and time. In USA, an analysis [29,30] indicates that NPDs could add up to 12 GW of hydropower capacity. In sub-Saharan Africa, similar estimate indicates 243.5 MW capacity [30]. Small-scale hydropower is generally more eco-friendly and can potentially offer an alternative clean energy solution in the variable electricity market. Novel designs of gravity hydraulic machines (waterwheels and Archimedes screws) and turbines (very low-head turbines and hydrokinetic turbines), and advanced designs and operation strategies for pumps as turbines, are being introduced. PATs are common pumps that can also operate in turbine mode. Additional designs in the PAT context under development include the Deriaz turbine.

The diversity of turbine types, technical specifications, and operational regimes makes SHPs a highly unique source of renewable energy. Of the reported modes of operation, approximately 1/3 of SHPs are operating in canals and conduits, nearly all of which were installed after 1980. The remaining SHPs are operated in streams as run-of-river, peaking, a combination of the two, or reregulating. The Francis turbine is the dominant turbine type in SHPs, with over 500 units installed at low-head (<30 ft) sites and roughly 700 units installed at not low-head (>30 ft) sites. Pelton, Kaplan, and fixed blade/propeller turbines make up the bulk of remaining turbines, with over 100 of each installed at SHPs in the USA. In total, there are over 21 different turbine types installed at US SHPs. While a handful of turbines are rated for greater than 1,000 ft of head, and even fewer for less than 10 ft of head, the majority of SHPs operate at a head between 10 and 60 ft, and with plant hydraulic capacity between 500 and 7,000 cfs.

3.2.2 Advances in Micro-Hydropower

A *micro*-hydropower plant has a capacity of up to 100 kW. A small or micro-hydroelectric power system can produce enough electricity for a single home, farm, ranch, or village. Micro-hydropower is almost always more cost-effective than any other form of renewable power. It is not difficult to operate and maintain [31]. Micro-hydropower is generally assuming distributed power function. If electric heating loads are excluded, 300–400 W of continuous output can power a typical North American house.

Power can be supplied by a micro-hydro-system in two ways. In a battery-based system, power is generated at a level equal to the average demand and stored in batteries. Batteries can supply power as needed at levels much higher than that generated, and during times of low demand, the excess can be stored. Most battery-based systems use an automotive alternator. Another type of generator used with micro-hydro-systems is the DC motor of permanent magnet types. Lead-acid deep-cycle batteries are usually used in hydro-systems. Hydro-systems with lead-acid batteries require protection from overcharge and overdischarge. Overcharge controllers redirect the power to an auxiliary or shunt load when the battery voltage reaches a certain level. This protects the generator from overspeed and overvoltage conditions [31]. If enough energy is available from the water, an AC-direct system can generate power as alternating current (AC). In a typical AC system, an electronic controller keeps voltage and frequency within certain limits.

There are two basic kinds of turbines for micro-hydropower: impulse and reaction. Impulse machines use a nozzle at the end of the pipeline that converts the water under pressure into a fast moving jet. This jet is then directed at the turbine wheel, which is designed to convert as much of the jet's kinetic energy as possible into shaft power. Common impulse turbines are Pelton, Turgo, and cross-flow. In reaction turbines, the energy of the water is converted from pressure to velocity within the guide vanes and the turbine wheel itself. Examples of reaction turbines are propeller and Francis turbines.

In the family of impulse machines, the Pelton is used for the lowest flows and highest heads. The cross-flow is used where flows are highest and heads are lowest. The Turgo is used for intermediate conditions. Propeller (reaction) turbines can operate on as little as 2 ft of head. A Turgo requires at least 4 ft and a Pelton needs at least 10 ft. These are only rough guidelines with overlap in applications. The cross-flow (impulse) turbine is the only machine that readily lends itself to user construction. They can be made in modular widths and variable nozzles can be used. Most developed sites now use impulse turbines. These turbines are very simple and relatively cheap. As the stream flow varies, water flow to the turbine can be easily controlled by changing nozzle sizes or by using adjustable nozzles. In contrast, most small reaction turbines cannot be adjusted to accommodate variable water flow. Those that are adjustable are very expensive because of the movable guide vanes and blades they require. If sufficient water is not available for lull operation of a reaction machine, performance suffers greatly.

An advantage of reaction machines is that they can use the full head available at a site. An impulse turbine must be mounted above the tailwater level, and the effective head is measured down to the nozzle level. For the reaction turbine, the full available head is measured between the two water levels while the turbine can be mounted well above the level of the exiting water. Centrifugal pumps are sometimes used as practical substitutes for reaction turbines with good results. They can have high efficiency and are readily available (both new and used) at prices much lower than actual reaction turbines. However, it may be difficult to select the correct pump because data on its performance as a turbine are usually not available or are not straightforward. One reason more reaction turbines are not in use is the lack of available machines in small sizes.

In the micro-hydropower field, there is an unexploited potential with low-head differences (few meters) available in rivers, irrigation canals, and at old mill sites [32,33]. Existing technologies are not always cost-effective at such scales [34,35], especially for power output below 50 kW. Recent advances in gravity hydropower converters (hydrodynamic screws and gravity waterwheels) have improved their cost-effectiveness; their environmental impacts, especially on fish populations, are minimal [36], and their efficiency is attractive. Gravity waterwheels have been extensively tested in terms of recent scientific projects [37,38]. They are distinguished into (1) overshot [39]; (2) breast shot [40] (3); and undershot waterwheels, depending on the head differences and the maximum flow rate per meter width [33]. Maximum hydraulic efficiency of gravity machines may exceed 80% [41], but typical values of global efficiency range at 50%–70%. Waterwheels cost is 33%–60% of that of Kaplan turbines, and lower than that of hydrodynamic screws [36]. Waterwheels are advantageous when it is possible to revamp existing civil structures such as old water mills, becoming attractive educational, tourist, and re-creational locations.

The rotational speed of waterwheels is generally low and requires a gearbox to match the generator frequency. This results in a very expensive PTO which needs to be improved. Preliminary works have been conducted to overcome this deficit, by testing a new transmission system [42], because permanent magnet generators could be used in these situations, but they require reasonably complex power electronics [43]. An additional approach tested the use of adjustable in flow structures that can be managed as a function of the flow rate. Some studies have already been carried out for gravity waterwheels [33,44]. Such approaches were tested in the laboratory [40] where a model of an existing breast shot waterwheel was tested at a 1:2 scale (Froude similarity). New prototype waterwheels include the hydrostatic pressure machine (HPM) and the turbine waterwheel (TWW). The HPM is a waterwheel that can be used in flowing water, without any canal drop, achieving a hydraulic efficiency of 60%–65%. The HPM generates an increase in the upstream water depth. This creates a hydrostatic force on the blades that compensates for the low levels of kinetic energy [45,46]. The TWW represents a waterwheel that can be used for the head differences of an overshot waterwheel (up to 6 m) and flow rates of an undershot waterwheel (few m^3/s^1). A 30-cm wheel diameter with scale <1:10 has shown TRL 4 in preliminary experiments [47].

Pump as turbines (PAT) is hydraulic pump operating in reverse mode as turbines, thus producing energy rather than consuming it by means of a connected induction motor working as generator [48]. Hydraulic pumps are mass-produced globally, and the main advantages of their application as turbines include compact dimensions, short delivery time, easy maintenance and availability of spare parts, and reduced installation cost [49–53]. Compared to conventional turbines, the PAT cost is 5–10 times lower [54]. This is particularly significant in the context of micro-hydropower schemes having installed power less than 100 kW where turbo-generator units typically cost 35% of the total scheme cost [55,56]. There are, however, some drawbacks of PAT. It has lower peak hydraulic efficiency compared to a conventional turbine. The lack of in-built regulation devices (e.g., wicket gates, movable pitch blades) commonly results in poor part-load performances. Its design uncertainties create risks for designers and users. Finally, very little performance data are provided by the manufacturers.

Agarwal [48] examined the role of pump as turbine (PAT) for micro-hydropower. His study concluded that PAT is outstanding solution to the micro-hydropower particularly in isolated areas. The initial cost of the project decreases substantially which makes it more viable. The limitations of PAT can be further reduced by selecting a proper PAT for a specific site. Conversion factors for PAT can be decided on the basis of theoretical and numerical studies, but its performance cannot be predicted accurately. Hence, there is still a need for further research to develop a general model for calculating the conversion factors. However, the efficiency can still be increased by using the various modifications such as redesign shape of blades, runner vane shape, and discharge head for vertical turbine pump, can modify the impeller by grinding the inlet ends of the impeller tips, etc. These and other modifications are discussed in details by the study of Agarwal [48].

Despite the benefits associated with the use of PATs, their share in the hydro turbine market until now has been negligible. This is partially attributable to the lack of knowledge or interest on the topic from pump manufacturers and hydropower consultants, and partially to the few technical challenges yet to be addressed regarding the PAT design and operation. Indeed, the current technology readiness level of PAT technology is estimated at TRL 4 due to limited knowledge on the design and operation characteristics of reversed pumps [57]. So far, existing PAT-based schemes typically feature a nominal power below 20 kW even though a few examples of larger installations exist [58]. The most outstanding field of application of PATs is powering off-grid rural electrification projects in remote areas, where local hydro turbine suppliers are not available [48,59,60] and energy recovery in pressurized water networks [61] is necessary.

According to Kougias et al. [9], the main research directions on the topic of PATs can be grouped into four topics. First, improved performance prediction of PATs and reduced design uncertainties should be pursued by reliably predicting the characteristic curves of any machine with respect to their known behavior as pumps. The main efforts should point toward the development of numerical methods based on empirical data from tested pumps/PATs or the refinement and validation of CFD models [62–66]. Second, improved control of PAT installations under variable flow and head conditions, typical in drinking and irrigation water networks, should be investigated. Such methods rely either on mechanical devices as automated valves and hydraulic bypass ducts (hydraulic regulation), or on the adoption of a variable-speed drive (electric regulation), or on a combination of both (hydraulic–electric regulation) [67–69]. Third, PAT geometry modification should be considered in order to improve the performance of a pump when used as the turbine. This includes measures such as inlet impeller rounding and suction eye enlargement [70]. Finally, coupling of PAT with innovative generator types and configurations other than induction alternators, such as permanent magnet or self-excited induction generator machines [71], should be investigated. The ultimate goal is to produce a design methodology which leads to a wider application of such units to tap a significant potential otherwise unused. Commercial R&D activities include solutions for PAT utilization in water networks [72] developed by EPFL/HES Wallis in terms of the DuoTurbo project [73], Tecnoturbines in the EU [74], Rentricity in the USA [75], and a modular PAT-based turnkey containerized powerhouse [76].

3.3 ADVANCES IN HYDROTURBINES AND ROTOR–TURBINE–GENERATOR ASSEMBLY

As the need for more power generations from intermittent wind and solar renewable energy grows, hydropower and its storage capacity have become more important for grid balancing and stability. This, however, resulted in new challenges associated with the great flexibility required in operation over an extended range of regimes far from the turbines' best efficiency point (BEP) [9]. When hydraulic turbines operate at off-design conditions, a moderate- or high-level residual swirl occurs in the draft tube due to a mismatch between the swirl generated by the wicket gates (guide vanes) and the angular momentum extracted by the turbine runner. These off-design and transient operating conditions (such as start-up, emergency shutdown, load rejection, and runaway) can result in an abrupt decrease in efficiency, severe pressure fluctuations that lead to vibrations, and unsteady phenomena [9]. These conditions can also result in damage of the mechanical components, failure of the runner blade and power swing, and general fatigue damages [9]. Furthermore, hydroturbines also damage water quality and fish mortality. In small hydropower operations, the efficiency of operation is significantly affected by the nature of hydro turbine. In recent years, significant advances are made in the designs of hydro turbine and rotor–turbine–generator assembly to counteract these harmful phenomena [9,11,77–86].

Among other changes, several active or passive techniques are adopted depending on the energy injected in the main flow [4]. A successful control technique that supports the flexible operation of hydropower plants within a wide range has the following features: (1) The control technique addresses the main cause of the self-induced instability rather than its effects; (2) the method has a minimal (no) effect on the efficiency; and (3) the control technique can be switched-off at operating points where it is not needed. The earlier attempts to analyze passive control techniques that address hydraulic instabilities in turbines' draft tubes were provided by Thicke [87] and others [88]. While passive control techniques do not require auxiliary power and control loop, the active ones require energy in different formats [9,89–91]. Although passive control techniques lead to significant improvements in turbine operation at far off-design regimes, they are not flexible and their components cannot be removed when their presence is no longer required which leads to unnecessary hydraulic losses and unexpected pressure fluctuations at different operating regimes. An alternative approach, however, can be the removal of a fraction of the discharge from downstream the runner at the discharge cone outlet [92].

The active flow control methods generally use either air or water injection, using an external energy source. The main active control techniques are described by Kougias et al. [9], Papillon et al. [93], and Zhong-dong et al. [94]. A recent study by Chirag et al. [95] considered magneto-rheological brake (MRB) to slow down the speed of the runner in order to control the swirling flow configuration downstream of it and associated self-induced instabilities. The swirling flow configurations and its associated unsteady effects were controlled by changing the speed of the MRB [96,97]. This active magneto-rheological technique diminishes the axial flux of the circumferential momentum by controlling the speed of the runner. Susan-Resiga et al. [98] used Francis turbine with tandem runners. With this approach, a downstream variable-speed runner named "low-pressure runner" operates in tandem with the Francis runner with constant speed.

In the operation of hydro turbines, the transient processes such as start-up, no-load, load rejection, and very low load are among the most damaging. In order to improve hydraulic and mechanical designs, machine stability, and reliability, to reduce stress, and to increase life span and identification of problematic regions for structural load or load changes, a better understanding of transient operating conditions of hydro turbines is important. An understanding of cavitation phenomena is also important because cavitation plays an important role during the start-up and runaway processes. Knowledge of the effects of a large number of frequent changes, weight optimization, and high-performance requirements on vibration behavior and fatigue along with the understanding of fatigue cracks created by pressure pulsations, swirling flow, and induced instability are important. The Francis project [99] performed experimental and numerical analyses of fluid–structure interaction focusing on the role of hydrodynamic damping, added mass effects on frequency, amplitudes of RSI, resonance, and corresponding mode shapes of high-head Francis runners. Evaluations of strain gauge measurements in Francis runners show that RSI-induced stresses are especially relevant for medium-to-high-head runners. Computational fluid dynamics (CFD) has helped the investigation of unsteady pressure pulsations during challenging operating conditions and has allowed a better understanding of transient operations. 3D-CFD simulations are also used to study the influence of the changes of internal flow to the external characteristics during transient processes. Simulation should also focus on the effect of cavitation on start-up and run-away processes [9].

3.3.1 Current-Controlled Segmented Generator Rotors for Better Dynamic Control

Novel power electronics with current-controlled power supplies can ensure better control of electrical machines during start and stop phases. Modern power electronics with current-controlled power supplies provide new opportunities for the control of electrical machines. The idea of a segmented rotor could be combined with novel ideas on excitation to open up possibilities for reduced investment cost and maintenance. Power electronic converters (PECs) also open up new possibilities for electrical machines which include the enhanced coupling of electrical machines with power electronics. Frequent start and stops required to provide secondary regulation result in additional wear on the energy conversion components. In order to reduce wear, condition-based maintenance is required. Nowadays, this is facilitated by dynamic measurements of vibration, temperature, voltage, and current by online sensors. The condition-based maintenance would provide some form of self-healing capability during emerging problems. Segmented rotors [100] and current-controlled rotor magnetization equipment have the potential of handling generator air gap unbalances.

Normally, an active system requires additional actuators, but in the case of the segmented—or split— rotor, the existing poles are used as the actuators. However, additional components are needed in the magnetization equipment in order to control the current. The idea is closely related to self-bearing machines where the radial force is controlled in the electrical machine from the stator side [101,102]. Here, the control is moved to the rotor instead and the circuit adapted to achieve the control. The system can also be thought of as a magnetic balancing system that evens out disturbances in the air gap flux density. This system reduces extra losses, voltage harmonics, vibration levels, and fatigue loads on the rotor and stator during operation, and extends lifetime of rotor–turbine–generator assembly. The idea of a segmented rotor could be combined with novel ideas on excitation [103,104] to open up possibilities for reduced investment cost and maintenance. The most obvious use of a segmented rotor system is to reduce unwanted forces as they occur inside the generator. However, since it is a controllable force, it can also be used to affect the rotor dynamics of the shaft. PECs thus open up new possibilities for electrical machines.

3.3.2 Advanced Fish-Friendly and High Water Quality Hydroelectric Turbines

Hydropower alters aquatic ecosystems with risks imposed on migrating fish and water quality. For example, the turbine may affect water quality (oil pollution and oxygen content) and disrupt the natural pathways for fish migration [105,106]. Ideal hydropower technologies should protect environment ecosystem and water quality. A dam that creates a reservoir or diverts water may obstruct fish migration. Hydropower turbines kill and injure some of the fish that pass through the turbine. The US Department of Energy has sponsored the research and development of turbines that could reduce fish deaths to lower than 2%. Many species of fish, such as salmon and shad, swim up rivers and streams from the sea to reproduce in their spawning grounds in the beds of rivers and streams. Dams can block their way. Different approaches to fixing this problem include the construction of *fish ladders* and elevators that help fish move around or over dams to the spawning grounds upstream. The vertical slot fish ladders [107] consist of a channel with typical bed slope between 5% and 10%, with pools separated by transversal baffles. Unfortunately, fish ladders are not suitable for downstream fish migration, because fish tend to follow the main river flow, that is, the flow running through the turbine. Despite the high TRL of fish passages, recent analyses show that most of them are not eco-efficient. In many cases, the river species are not able to use them [108,109]. The interaction between the fish populations and the passages is a complex phenomenon, and optimal designs should consider fish behavior and fish reaction to external stimuli coming from the turbulent and hydraulic flow field [110,111]. Fish passages adopted for the downstream migration are generally different from the fish ladder used for the upstream one. Accordingly, screens are placed before the turbine to prevent fish entrance and to divert them toward the passage. Such screens, however, induce head losses, that is, a reduction in the power output of the hydro-plant.

The research supported by DOE and EPRI is to develop environmentally enhanced turbines which can be further classified as fish-friendly turbine and ecological turbines. The turbines can also be classified as self-aerating turbines and self-lubricated turbines. Besides the development of fish-friendly turbines, the research is also carried out to develop fish-friendly hydropower operations and management. The main damage mechanisms that fish may undergo while passing through a turbine are as follows: mechanical injury (strike, grinding), decreased pressure and sudden pressure decrease, shear stresses and turbulence, and cavitation, which can cause external and internal injuries [112]. The pressure-related effects take precedence when a greater biodiversity is considered, since most migratory fish are prone to mortality due to barotrauma caused by the rapid decompression affecting the size of the swimming bladder almost instantly. Rapid change in barometric pressure, or barotrauma, is a potential cause of injury and mortality for juvenile salmonids passing through hydro turbines [113,114]. Barotrauma can lead to rupture of the swim bladder and exophthalmia (popped out eyes). In addition, gas bubbles can form in the blood and internal organs, leading to emboli [115]. Mechanical injuries are the main cause of fish mortality passing through a turbine [116]. Mechanical injury depends on fish dimensions, flow velocity and direction, and on turbine characteristics. The relation

TABLE 3.4
Comparison of Various Turbines for Fish Survivability (based on a 13.6 MW unit, Nielson et al., [121]) (also 20)

	Alden	Francis	MGR
HUB diameter (m)	3.9	2.5	2.7
Rot. speed (rpm)	120	190	277
Runner blades	3	13	5
Guide vanes	14	20	24
Survival rate for a fish of 200 m	98%	<50%	86%
Max. efficiency	93.6%	95%	95%

between these factors can be explained by the blade strike model [117], in particular by a dimensionless time T^*, that expresses the ratio between the time employed by the fish to pass through the turbine and the time of a blade passage: The higher the T^* is, and the higher is the injury risk. According to Deng et al. [117], mortality prediction can be carried out by stochastic model better than blade strike model.

Much of the ongoing research in the hydropower sector is concerned with the attainment of the ideal turbine—one that combines efficiency with environmental friendliness. A number of designs were proposed by a variety of vendors, and the DOE. Some examples are as follows: Ferguson et al. [118] for Francis and Kaplan turbines, Fu et al. [119] for Francis turbines, and Deng et al. [117] for Kaplan turbines. Pelton turbines are not generally considered, due to their intrinsically high-mortality behavior. In order to emphasize the novel strategies for fish protection via the redesign of turbines, a review work carried out by Hogan et al. [120] about the first generation of FFTs thoroughly described the conceptual development and implementation of two relevant technologies designed for better fish passage conditions, namely the Minimum Gap Runner (MGR) and the Alden turbines. Table 3.4 shows a comparison of the performances and dimensions of Alden, MGR, and Francis turbines. The Alden and MGR turbines can be currently used between 10 and 40 m head, and for flow rates above 17 m^3/s, but it is expected that they will be able to work up to 50 m head up to 325 m^3/s [121,122]. For hydropower applications below 10 m head, in order to ensure a good ecological behavior, as pointed out by Bozhinova et al. [123], the free surface hydraulic machines can be used (typically below 2.5 m). Due to their large dimensions, free surface operation, and low rotational speed (<100 rpm), they intrinsically exhibit a good environmental behavior in relation to fish passage. FFTs used in the very low-head context are, for example, waterwheels [124], the Vortex turbine [125], the Archimedes screw [126,127], hydrokinetic turbines [123], and the very-low-head (VLH) turbine [128]. More work to accommodate a larger biodiversity of fish and further improvement in the hydraulic performance of the machines is needed.

The MGR turbine is the optimization of the Kaplan–Bulb turbine. The gaps between the adjustable runner blade and the hub, and the gaps between the blades and the discharge ring, are minimized, reducing fish injury and mortality, and improving turbine efficiency [86]. The first field test of the MGR was done in Oregon at the Bonneville dam. The injury rate was 1.5%, smaller than the injury rate of 2.5% occurring at the adjacent Kaplan–Bulb turbine. The design and manufacture of the MGR is very similar to a standard Kaplan–Bulb turbine. Therefore, both turbines exhibit the same operating range in terms of power output, flow capacity, and hydraulic head, and the same dimensions. However, the higher efficiency and the survival rate of a MGR turbine are compensated by a higher cost [121]. The US Army Corps of Engineers installed an optimized MGR turbine at the Ice Harbor Lock and Dam, located in Washington State (27 m head, power slightly above 100 MW, depending on the unit), achieving a survival rate of more than 98% [129]. Another successful example is a rehabilitation project at the Eddersheim hydropower plant (km 15.55 in the River Main, Germany), which replaced the original Kaplan-type runner (4.1 m diameter, 2 MW nominal capacity) with a new 3-blade machine with thicker leading edges and minimized runner gaps.

Another approach developed by Alden Research Laboratory incorporates an integrated design, so that the runner blades are attached to a rotating shroud. The Alden turbine specifically works with head differences of up to 25 m. The Alden turbine (see Figure 3.7) is a relatively new design for a fish-friendly turbine. Since there is no gap between the blades' tips and wall, this eliminates the low-pressure vortices that occur near the blade tips, and also eradicates any chance of fish being caught between the blades and the turbine walls. In addition, in order to reduce the chance of blade strikes, this concept only uses three blades which are much longer than conventional blades and have nearly 180° of wrap. To test the principle, Alden constructed a 3:1 scale model of the turbine. These live fish tests found that American eels had a 100% survival rate, and that species such as sturgeon, trout, shad, and herring would have a better than 98% survival rate when passing through a full-sized turbine. The results also showed a maximum hydraulic efficiency of 93.6% and fish passage survival rates greater than 98% for fish less than 20 cm long [86].

The Alden turbine was initially conceptualized and studied using computational fluid dynamic (CFD) simulations and experimental tests at a pilot scale in the Alden Research Laboratory [130]. The current Alden turbine design rotates at a slower speed than conventional turbines in the same context. Table 3.4 shows a comparison between an Alden turbine and a 13-blade Francis turbine that rotates at 190 rpm. The cost of the electro-mechanical equipment of the Alden turbine is 39% more than the analogous cost of a Francis turbine, and 35% more than a MGR Kaplan–Bulb unit. Nevertheless, the global cost of a power plant equipped with the Alden turbine is lower than the cost of a Francis turbine hydropower plant [131], because of cost reduction in the balance of the system such as O&M, capital costs of downstream passage, costs of overall civil works, etc. [121]. In order to further improve the performance, the variable rotational speed allows to maintain the blades always in their maximum opening position reducing the strike probability with fish and avoiding dangerous gaps between the blades and other parts of the machine.

The DOE's *Marine and Hydrokinetic Technology Database* lists hundreds of other technologies at varying stages of development. Some of these are producing power for the grid, such as the Wavebob deployed in Ireland's Galway Bay in 2006, and the 0.45 MW Oceanlinx wave-energy converter (WEC) demonstration project

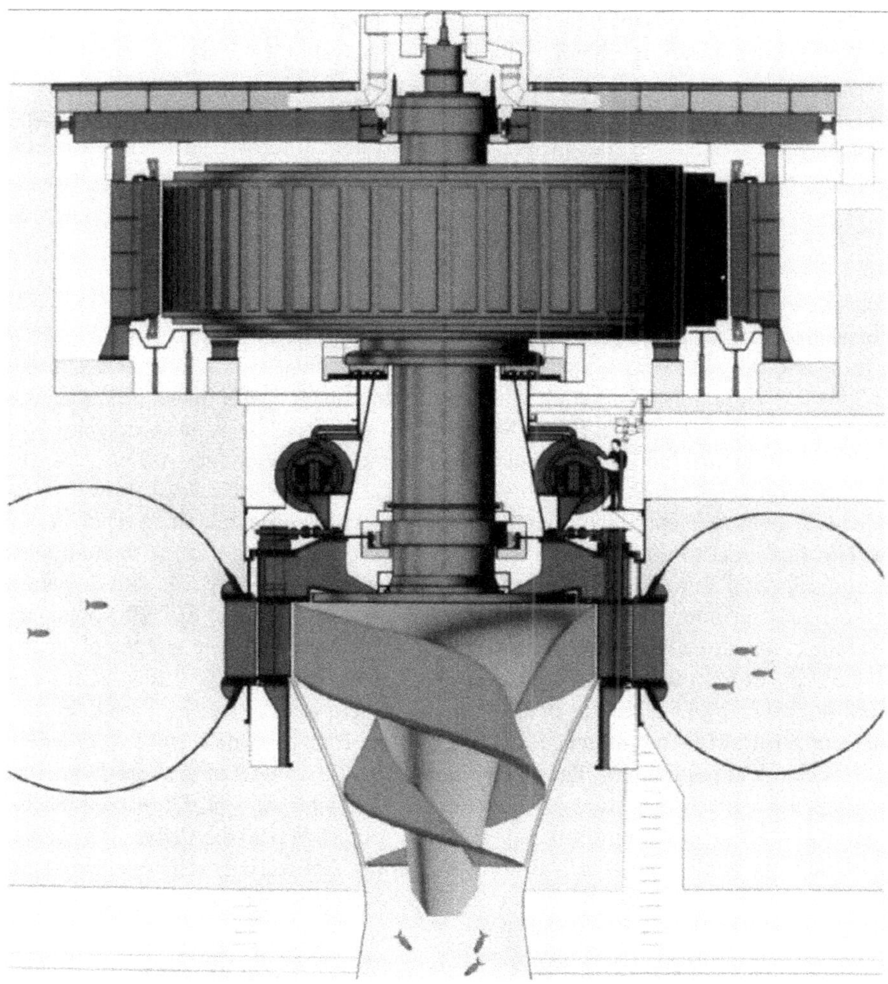

FIGURE 3.7 Schematic of the new ARL/NREC fish-friendly turbine (Courtesy of Alden Research Laboratory, Inc.) [80].

near Port Kembla in New South Wales, Australia, which operated for a few months in early 2010. EPRI study [130] indicated that for a reservoir depth greater than 15 m, installed power higher than 10 MW, reservoir volume greater than 61 Mm^3, densimetric Froude number below 7, and a retention time greater than 10 days, lower reservoir exhibits oxygen deficiency which can cause damage to the ecosystems downstream [132]. One method to resolve this issue is to utilize low-pressure regions below the runner to draw atmospheric air into the turbine during operation. Turbines where this system is implemented are called auto-venting turbines (AVTs). AVTs allow distributed, central, and peripheral aeration depending on the customer's requirements. This sustainable technology maximizes bubble distribution in draft tube and tailrace while minimizing the impact on the turbine. As shown by Kougias et al. [9], air injection is also an emerging technology for the flow control in the draft tube at off-design conditions. March and Jacobson [133] showed that in USA, the aerated turbines were 153 Francis turbines, 11 Deriaz turbines, 10 Bulb-Propeller, and 4 Kaplan turbines.

Literature has shown that proper aeration system improves efficiency for both Kaplan and Francis turbines. Foust and Coulson [134] described peripheral aeration system for a Kaplan turbine, where the bottom of the discharge ring was located up to 1.8 m below the tailwater elevation, due to cavitation constraints. Dorena Lake plant is equipped with a 4.4 MW vertical Kaplan turbine and a 1.2 MW horizontal Francis turbine, both turbines with peripheral aeration in the draft tubes, and additional diffusers in the tailrace [135]. March [136] reported an efficiency increase of 2%/1% without/with aeration with an aerating turbine with peripheral aeration, and of 3.8%/1% without/with aeration with an aerating turbine with distributed aeration, compared to the original turbine. March [136] reported the same efficiency with a new aerating turbine with central aeration, and a capacity increase of 21%/11% without/with aeration. Both March and Fisher [137] and March [136] showed some efficiency curves from which it is possible to assess the efficiency difference at part load with respect to the original turbine for different aerating turbine technologies. With peripheral aeration, the efficiency at part load increased by 5% with respect to the original turbine which had no aeration system. McIntosh et al. [138] presented the replacement of a Francis turbine with central aeration, showing that the plant efficiency improved by 3.7%, and the power output increase by 14%. Bunea et al. [139] examined the implementation of an innovative aeration device in a small Francis turbine to establish the influence of the aeration process on the turbines' energetic and mechanical parameters. The study showed that the impact of the aeration device implementation and operation on the energetic characteristics of the turbine did not exceed the efficiency measurements accuracy range. When the device was not in operation, its implementation had no effect on the turbine energetic performances. The study also showed that the influence of aeration device on the hydraulic performances of the turbine is admissible economically.

The aeration inside the casing of Pelton turbines is also very important in order to minimize oxygen deficit downstream [140], especially when they have to be operated with a backpressure. For Bieudron, the world's largest Pelton turbines, at full load, the natural aeration system delivers 6 m^3/s of air to the turbine casing [141]. For Kopswerk II pumped-storage hydropower plant in Austria, a 180 MW Pelton turbine is below the water level of the lower reservoir of 1–16 m. When the vertical distance between the Pelton runner and the water level is 16 m, the casing was pressurized at 3 bar [142].

Hydraulic turbines can also affect water quality through oil leakage. Adjustable blade runner hubs are typically filled with pressurized oil which lubricates runner blade trunnion bearing and sliding parts of the operating mechanism in the hub. St. Germain [143] points out that oil leakage from Kaplan–Bulb, Francis, and even Pelton turbines not only has a negative impact on the environment, but also causes some operational and maintenance problems. Oil leakage from Poatina hydropower plant was released to the river [144]. In recent years, several Kaplan–Bulb and Francis turbines have been upgraded so as to make them work free from oil [145–147], and new materials and lubricants are being developed [148]. Ingram and Ray [149] stated that water-lubricated guide bearings contribute to increase the overall plant efficiency and decrease the maintenance cost. Oguma et al. [150] described the performance of a water-lubricated guide bearing that was specifically designed for a multi-nozzle vertical Pelton unit. Oil-free hubs are a widely installed technology in Europe, particularly in the Scandinavian countries.

3.4 DIGITALIZATION OF HYDROPOWER OPERATION

In these days, digitalization of dynamic operations appears to be a more of norm then exception. In general, digitalization of processes has many positive attributes. Specifically for hydropower, it provides better utilization of the available hydraulic power, ensures safety, enables larger energy production on an annual level, and lowers the costs of maintenance. Digitalization will also provide the enhanced flexibility and stability of the electrical system, the prolongation of the lifetime of the hydropower equipment, and an increase in the overall efficiency [9,151–153]. It has been shown that using high-quality inflow forecasts, digitalization can provide 1% increase in efficiency and around 11% of energy generation of the existing HPPs [151]. When dispatchable hydropower is used along with intermittent solar and wind power generation, digitalization offers optimum conditions for balancing of the production and consumption of the grid.

An effective digitalization requires dynamic measurements of various process variables like temperature, pressure, flow rate, structural vibrations, etc., by a series of sensors in the dynamic and continuous manner which can be stored for process analysis, manipulation, and performance predictions. Signals-based data can be stored in a local database or on a cloud which allows access to the data from a distance. Both historical and present data can be used for various models and algorithms. A large number of data can be organized in a centralized platform. These data can be manipulated using Industrial IoT, Artificial Intelligence, Digital Twin, and other contemporary technologies. The digitally controlled hydropower operation along with other power generation by renewable energy technologies into the smart grid, electricity demand, and prices can be estimated to stimulate the end-user and to provide grid stability. The method can be applied to a single hydropower station or to a fleet of hydropower stations connected to the grid. Manipulation of both historical and present data thus allows identification of a game plan for optimization, generation, and maintenance predictions. Digital technologies are now used for design, construction, operation and maintenance, remote monitoring, and advanced grid-supporting services for hydropower without compromising system reliability and safety.

The increased use of digitalization of hydropower operation is facilitated by advances in computing capacity of the processors (which went from less than 1 MHz in the 1970s to 4 GHz in 2017) accompanied by the reduction of processor costs. The increase in the capacity of communication networks by digital systems both locally and remotely allowed an increase in the processing capacity and the development of new algorithms for the analysis and management of data (including "artificial intelligence"). These increasing computing and communication capabilities along with advances in artificial intelligence allowed the use of data-driven methods such as multivariate regressive methods and artificial neural network to the data collected during the hydropower plants operation. This in turn allowed the predictions of phenomena of oil leakage, cavitation erosion, fatigue, performance degradation, even hydraulic turbine failure, etc., during hydropower operation. More research is needed to use digital technology for predictive maintenance and optimization. Digital technologies are now used for design, construction, operation and maintenance, remote monitoring, and advanced grid-supporting services for hydropower without compromising system reliability and safety. It is estimated that a total of 42 TWh could be added to present hydropower energy production by implementing hydropower digitalization. Such an increase could lead to annual operational savings of US$5 billion and a significant reduction of greenhouse gas emissions.

While hydrodynamic phenomena occurring within hydropower limit its flexibility, a key challenge for modern hydropower plants is to enhance drastically their flexibility by providing storage capacity and advanced system services. Advanced monitoring facilitated by digitalization would also provide advanced levels of safety that are currently not available. Future hydropower plants need to fulfill the future EPS requirements, by enabling fast frequency containment reserve (FCR), frequency restoration reserve, and black start in emergencies. Such a technological advance would build on the so-called digitalization of hydropower, which will transform the way projects are designed, developed/upgraded, operated, and maintained. Apart from the prolongation of the lifetime and addressing cyber security risks, rehabilitation and digitalization involve increasing the overall efficiency and, thus, the produced energy. The enhanced services to the grid can be achieved by increasing the operating range of turbines to enhance the operational flexibility of hydropower plants. Digitalization has a number of other benefits: it (1) will reduce the response time of generating units or reversible pump turbines, (2) will allow assessment of economic impact of additional reserve flexibility, and (3) will provide high-level safety and reliability standards for the hydropower plants which are required to extend operating range and operate under fast dynamics and unexpected outage. A concept of a digital turbine has received increased attention [154]. A recent study [155] has shown that with digitalization of a reversible pump–turbine unit, the time to change from the pumping operation mode to generating mode can be drastically reduced to make this technology complying with the new grid specifications. Digitalization also allows the production of dispatchable hydropower in a changing environment which in turn allows an operation that provides inertia to the EPS and facilitates a higher penetration of RES in the grid.

In recent years, digitalization has taken hydropower operation to a new level of flexibility, predictability, and optimization through intelligence maintenance, more efficient operation, and digital twins [153]. In an intelligence maintenance, a model supported by continuous data can become increasingly accurate in detecting deviation and faults in the plants. Through more accurate measurement of the input and output parameters (flow, pressure, power), and with the "intelligent" control systems that allow to achieve more precise adjustments in the regulators, high (more than 90%) efficiency can be achieved. Extra energy generated by an additional percentage point of efficiency can easily justify additional investment in existing plants. The "digital twin" allows replication of the operation of the plant in a virtual world in which different modes of operation can be simulated. Being an intelligent model, that "learns" the behavior of the plant with input data, its accuracy improves over time. These technologies can also be used for the verification of failure modes for risk analysis (e.g., of dams), the simulation of future operation scenarios with different levels of generation of other renewables, and among others.

A digital twin is a virtual representation that serves as the real-time digital counterpart of a physical object or process. In general, DTs consist of three components: physical product, virtual product, and connected data that link physical and virtual product via various data communication schemes. DT makes full use of all the data and integrates multidimensional and multifaceted simulation processes to real-time to reproduce the dynamics of a physical system in the virtual space. With the help of IoT, AI and cloud computing DT can provide information of hydropower in the entire life cycle of a product, from its design and production to its operation and maintenance.

The virtual simulation provided by DT can allow the HPP to perform more reliably with higher efficiency and to increase its generation Simulation can also be used for further development and optimization of hydropower operation. Along with numerous sensors, data for both upstream and downstream operations, weather data, and grid data can also be taken into account for optimization. For a pumped-storage station in Ireland, the Turlough Hill Power Station, a DT was formed for fatigue assessment and predictive maintenance in order to provide the plant's life extension [156]. Dreyer et al. [157] implemented a DT called the hydroclone for monitoring and fatigue assessment of the penstock of the 200 MW La Bâtiaz hydropower plant. The use of DT also allows to assess and identify the risk of structural failure and simulate future events. This is important as the plant may need to operate in more than one cycle a day because of the dynamics introduced by other renewable energy sources. The digital tool, the Smart Power Plant Supervisor (SPPS), was developed within the framework of the EU-H2020 project XFLEX HYDRO. The goal of the SPPS was to improve the flexibility and lifespan of the hydropower plant by optimizing its operation and maintenance due to a model-based advanced control.

Hydropower operations are becoming more complex and demanding because they play a major role in grid balancing, reliability, and resiliency in the face of increased contributions from variable solar and wind energy. To meet this new reality, hydropower technology will require the integration and full benefit of the best available and future advancements in sensors, data and control systems, analytics, simulation, optimization, and computing capabilities to remain competitive. In order to address this challenge, the development of digital twins for hydropower systems is necessary. Funded by the Water Power Technologies Office of the US Department of Energy, ORNL, and PNNL launched a project in 2020 to develop a Digital Twin for Hydropower Systems (DTHS) open-platform framework (OPF) as a key initiating research activity and outcome. The goal was to optimize the plant operation for managing electric power demand, and to perform fault diagnostics, condition and health monitoring, and management of hydropower systems operation using real-time data. The overall structure of the DTHS-OPF (DTHS open platform framework) is shown in Figure 3.8a, and component structure is shown in Figure 3.8b [151], which shows

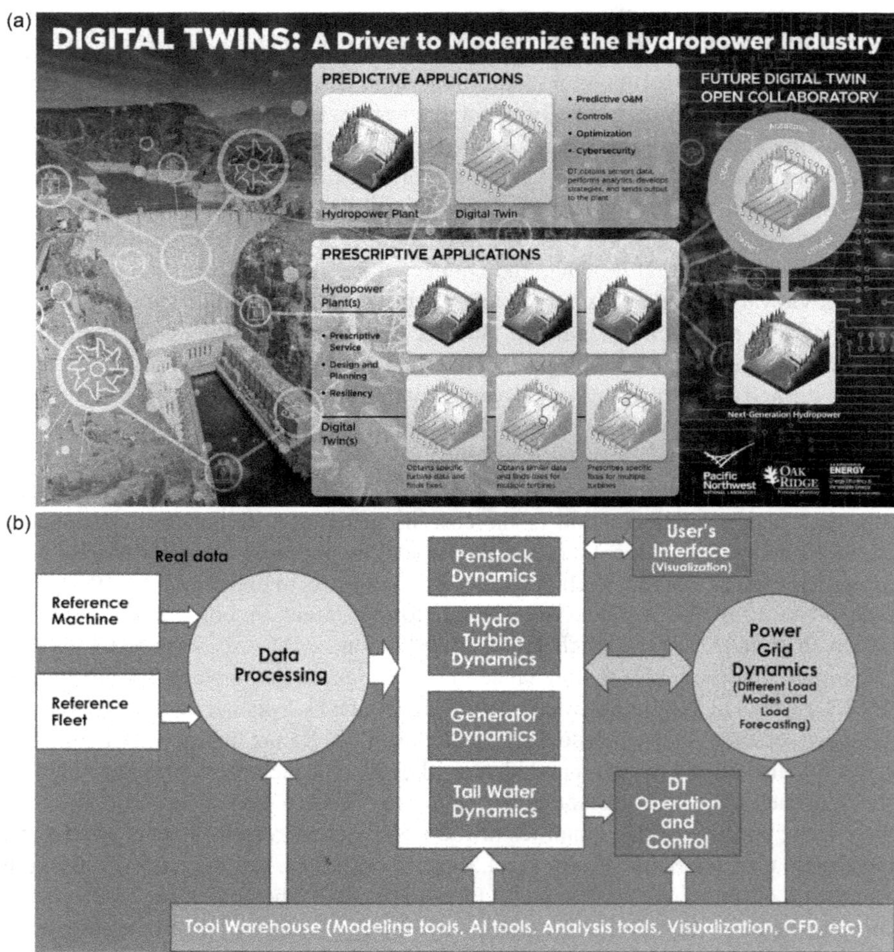

FIGURE 3.8 (a) DTHS-OPF overview. (b) The overall component structure of DTHS-OPF [151].

that DTHS-OPF will collect data from real hydropower systems and continuously update its dynamic models for various components of hydropower systems. This enables the DT to comprehensively represent the actual plant operation in a digital form. It also has a powerful user interface including visualization, augmented reality, etc., to allow user-friendly functionalities for the hydropower industry, hydropower system equipment manufacturers, and academia.

DT can perform in parallel operation on actual hydropower systems as it expands from a single generation unit to a fleet, and any operational dynamics and status can therefore be reflected and mimicked by DT. Therefore, plant operators, manufacturers, and researchers can use DTHS to (1) optimize the actual plant operation in response to changes and requirements of the overall power grid for managing electrical power flows and demand and (2) initiate testing on operational optimization before fully implementing these operational strategies. This will allow economic and in-time repair of damaged equipment used in hydropower operation. ORNL indicates that new algorithms on modeling and control can be developed using models and interface capabilities of DTHS open-platform framework [1]. Thus, DTHS-OPF can be effectively used to modernize and optimize the hydropower industry as it moves toward full-scale digitalization, and it will help in integrating hydropower industry in the country's power grid. Manufacturers in the hydropower industry agree that digitalization is a key to enabling growth in the hydropower sector by capturing the value of data. Manufacturers also believe that by implementing diagnostic services, acoustic monitoring, and remote expert support, hydropower maintenance can not only be improved but can be made more cost-effective. Industry researchers expect the DT market to grow from $3.8 billion in 2019 to $35.8 billion by 2025, according to the latest report from https://www.marketsandmarkets.com/Market-Reports/digital-twin-market-225269522.html [1].

The progress in digitalization can be affected by human barriers which include management attitudes toward its cost-effectiveness. The cyber security is described as being one of the top digitalization issues for the hydropower sector. The advent of "big data" and more complex IT systems to support digitalization can create the need for greater security measures. Digitalization can also be an enabler for enhanced cyber security. The advances and breakthroughs in the IT and advanced automation and control capabilities will shape hydropower's future. In future, hydropower operation needs to consider different markets (i.e., spot, balancing, frequency reserve) in its scheduling. This will, however, require significant computational power and big data analytics which includes real-time simulations and modeling to respond to specific ancillary service requests with the use of sophisticated algorithms, simulation, and optimization techniques. With digitalization, hydropower systems will be more aligned with the climate policies and the electricity markets.

Looking to the digital future, International Hydropower Association (IHA) discussions have also sought to address various concerns of the hydropower industry [151–156]. Human resources can be a barrier, and it can be alleviated by continuous learning, proper management strategy, and facilitation of digital process implementation. Economics can also be a barrier, and in order to alleviate this barrier, digitalization must be cost-effective and demonstrated by a proven track record. Finally, cyber security can be the top digitalization issues. IHA says the advent of "big data" and more complex IT systems to support digitalization can create the need for greater security measures. Digitalization can also be an enabler for enhanced cyber security.

3.5 VARIABLE-SPEED HYDROPOWER GENERATION

Since hydropower plant can last more than a hundred years, a periodic refurbishing of existing hydropower plant is needed. This refurbishing can involve retrofitting to improve plant performance by employing new technologies for control scheme, fault protection, measurement of important variables, automation of some auxiliary equipment, and even changing some parts of important equipment [158]. These changes generally increase the power plant's efficiency, while maintaining its capacity. Refurbishing can also involve uprating the turbine or the generator, increasing the height of a dam, as well as other entities that aim to improve the overall power plant capacity.

Due to new adjustments needed in the grid by intermittent renewable technologies, in recent years, refurbishing also requires changing a constant-speed hydropower plant to a variable-speed hydropower plant. Generally, variable operating conditions in a hydropower plant occur due to flow instabilities which can reduce plant efficiency and lifetime. By varying the rotational speed of the hydropower plant's units with respect to their synchronous speed, the plant can better adapt to the hydrological regime of the river, thereby increasing the plant's global efficiency and the units' lifetime, and can also increase its contribution to the EPS ancillary services. Variable-speed hydropower also allows better active and reactive power control and larger spinning reserve capacity. Variable-speed pumped-storage units are able to control both active and reactive power in pump mode, as well as start-up in pump mode and change the operation mode in a time shorter than conventional pumped-storage units.

Hydropower plant is generally designed to operate at a designed head. Unfortunately, both upstream and downstream heads can be changed for a number of reasons such as a variation in inflow, use of stored water due to increased electrical demand, etc. The operation of a turbine at a speed lower than the design reduces its efficiency, and the lower frequency than the system frequency generated by lower speed makes it impossible to establish a synchronous, direct connection to the power grid [159–162]. Compared to their fixed-speed counterparts, variable-speed hydraulic machines are able to operate over wider head ranges. For a given head, the variable speed allows power variation in pumping, while a unique power is possible at fixed speed, but leaves the power range unchanged in turbine mode. Variable-speed PSH is thus able to provide FCRs in both pump and turbine modes. Variable hydropower system allows support to the ancillary services such as frequency control and reactive supply, which offer new economic benefits to the system operation [163,164].

During the last decades, developments in power electronics have enabled the supply of electrical machines with variable-frequency voltages, resulting in the possibility to vary the speed of PSH plants. Large machines, of more than 50 MW, use a doubly-fed induction machine (DFIM) with a power converter rated to only a few percent of the nominal power while smaller machines use a synchronous machine with a full-size power converter [158–161,165]. This variable-speed possibility can be used to always operate the hydraulic machine at its BEPs, as these are related to different speeds in pump and turbine modes, thereby increasing the revenues from price arbitrage on the energy markets. Sometimes, the variable speed becomes a necessity in order for the pump mode to support high-head variations and be able to operate between its stability and cavitation limits [158–161,165]. Another option is to use the variable speed to provide transmission system operators (TSOs) with ancillary services, in particular primary and secondary frequency control, in both pump and turbine modes. The task to convert a constant-speed power plant into a variable speed may require either a deep change in the electric machine or large physical space in the power plant both of which can be done while refurbishing the power plant [166]. A majority of the solutions that obtain variable speed make use of power electronics and rectifier commutation. This comes with numerous technical problems and concerns related to harmonics and related filters, degradation of components, air gap torque pulsations, insulation stress, increased losses, and issues related to bearings and operating temperature, which leads to more auxiliary systems [167,168].

There are four methods to obtain variable speed from constant speed in a power plant [158,160,161,165]. The first one is the use of back-to-back frequency converter. In this case, the electrical machine, that is, a synchronous generator (SG), remains the same. If the speed is variable, the generation frequency will also be variable, and this generated variable frequency is converted to a direct current (DC) power system before being inverted to alternate current (AC) power in a constant and desired frequency that is synchronous to the system. The main limitations of this solution are the capacity of the power electronics, which must be equal to the total capacity of the connected synchronous machine, the space required to accommodate a new device, and the necessary auxiliaries for each of the units of a power plant. In this arrangement, the electrical machine can be a synchronous machine or an asynchronous one with a squirrel cage rotor. The first option has the advantage of generating reactive power under excitation control. The second one offers reduced maintenance costs and requires no excitation system but adds cost.

The second method uses double-fed asynchronous machine. In this process, the conventional SG is changed to a brushed asynchronous generator called a double-fed induction generator (DFIG). The armature of the machine does not need to be modified. The rotor should be a three-phase wound rotor of an asynchronous generator and excited by a back-to-back converter with variable frequency three-phase AC. The power necessary to feed the excitation, which is a function of the expected speed variation, needs to be converted. The DFIG rotor must be excited with a complement frequency to achieve the rated frequency in the stator generated voltage [169,170]. The only demanded power is that which is necessary to feed the losses, which is proportional to the efficiency of the excitation systems. The rotor replacement is best done during retrofitting, and additional space is required in order to install the new excitation.

The third method is to use variable frequency transformer. If the frequency of input and output is different, the machine must rotate at a given speed that is proportional to this difference in order to match primary and secondary frequencies [171,172].The VFT is driven by a DC motor and the rotation speed of the VFT is proportional to the difference in frequencies. All of the slip rings and brushes must be conveniently sized to afford all of the power generated [173,174]. This alternative has the advantage of maintaining the used SG along with its regular excitation system. Both active and reactive powers are normally controlled. In this method, only the installation of the VFT is required which demands a large amount of space. The VFT can be installed for one machine or for the whole power plant. In general, it is manufactured at the rated power of 100 MW.

The fourth method is direct current transmission. In this method, the generator is connected through an HVDC to the power system. The generated power is rectified and connected to the grid. This is a practical solution for a retrofit scenario, mainly

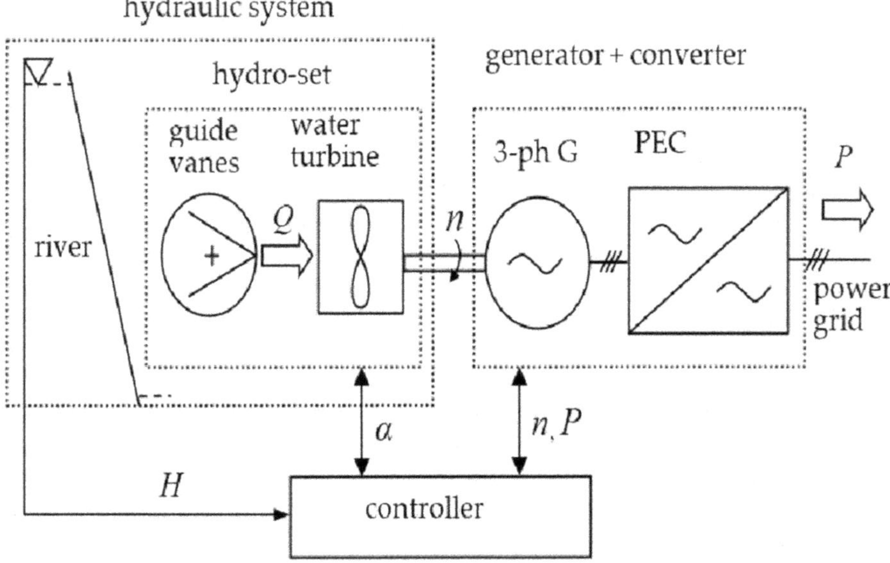

FIGURE 3.9 Schematic diagram of the energy conversion system of a variable-speed SHP [158].

because the generation system remains almost the same [175,176]. Space must be provided only to accommodate the converter, which inherently consumes some reactive power with trivial synchronization. Power to feed auxiliary systems must also be provided.

In recent years, conversion of SHPs (under 10 MW) to a variable-speed mode has gained significant attention because variable hydropower plants with reservoirs can play an important role in hybrid power systems with VRE sources [177] for grid stability. The modern solutions for water turbines decrease the production costs of water turbines with a guide vane governor; furthermore, PECs, which are widely available, can increase the productivity of hydropower plants by using the principle of variable-speed operation [178] (Figure 3.9). Control over the speed allows adjustments in the operation parameters which is especially important in low-power systems, where investment and exploitation costs can be reduced by using simple propeller turbines with fixed rotor blades. The possibility of adjusting speed to mimic actual hydrological conditions widens the operating range of such hydro-sets and increases their efficiency at operating points far from the nominal values [161] and enhances the production of electrical energy. Such solutions require, however, the use of a PEC to match the parameters of the generated energy to the power system requirements. For this purpose, the use of permanent magnet synchronous generators (PMSGs), which increase the total efficiency of the system, may be employed. This solution offers the possibility of integrating a water turbine with an electrical generator by inserting turbine blades into the generator rotor [179].

Recently, a number of control techniques have been developed for low-power hydropower systems with variable-speed operation. Most of the literature studies analyzed energy conversion systems with a specific water turbine modeled in a simplified manner or by empirical formulas. The study by Borkowski and Majdak [158] shows that control over SHPs with variable-speed operation is troublesome, especially in the low-head scheme, where the dynamics of the hydraulic system influences system control significantly and the measurement of turbine discharge is difficult. Moreover, turbine type has a significant effect on the SHP steady state. The study led to a recommendation that indirect methods should be used to define the initial operation curve due to their simplicity and high dynamics. Further, the operation curve should be periodically adapted to the actual system features using direct control techniques. This approach, which is termed the hybrid method, seems to be the most promising for controlling SHPs working at variable speed. The main disadvantages of hybrid methods lie in their algorithm complexity and algorithm parameter adjustment for a given SHP. In order to simplify the determination of the optimal operating curve, the study proposes the use of an analytical formula based on the geometrical dimensions of the turbine. The verification of this formula with actual measurements showed that it can be used to identify the initial performance curve in hybrid methods.

The study by Borkowski and Majdak [158] also compares the constant and variable rotational speed operation for a SHP. The results of the analysis of efficiency and annual energy production showed that in the case of a propeller turbine, the annual energy production increased by almost 16%. The energy gain using this technique was much less with a Francis turbine, and its profitability was questionable. In addition, the presence of PECs in the energy conversion chain reduces overall system efficiency. On the other hand, PEC minimizes the negative impact of the power plant on the power system by controlling the power factor, improves system flexibility, and facilitates integration with photovoltaic, wind, and battery systems, creating a hybrid power microgrid.

The literature studies [158,160,161,165] also outline details of various methods used for hybrid control and conclude that hybrid method is the best control method. The studies assert that a hydropower plant model should include the static and dynamic features of the turbine and hydraulic system (in case of long penstocks that provide water to the turbine or surge tank presence). An analytical model of the hydro-set (used in indirect methods) should be based on polynomial equations related to turbine discharge and efficiency (e.g., approximated by neural networks). The turbine type (slow, normal, and fast) and SHP function (run-of-river and reservoir) determine the control objective (control curve). The control algorithm should include the influence of the generator and PEC efficiency on the optimal operating curve (the total efficiency or output power needs to be analyzed). Maximization of the SHP output power (pure MPPT) is justified only for nonregulated slow turbines in the reservoir scheme, where the water level can be controlled by switching the SHP on and off. The control algorithm must include hydraulic parameter regulation (e.g., water level), which may influence the control technique significantly. Adjustment of the system parameters that influence turbine discharge (usually the guide vane angle) needs to be minimized in order to prevent hydraulic system disturbance. Finally, measurement of turbine discharge (necessary in the MEPT algorithm) in the low-head scheme needs to be reduced (e.g., by using a discharge estimator).

Gish et al. [180] studied the possibility of using DFIMs in hydropower plants. The first variable-speed pumped-storage unit was commissioned in Japan in 1987 by Kansai Electric Power Company (KEPCO) [181,182]. A few more variable-speed pumped-storage units have been commissioned in Japan since then [183–189], namely two 400-MW units at Okhawachi PSPP, one 100-MW unit at Takami PSPP, one 300-MW unit at Shiobara PSPP, one 300-MW unit at Okukiyotsu, one 30-MW unit at Yanbaru PSPP; four 340-MW units at Omarugawa PSPP, and one 400-MW unit at Kazunogawa PSPP [190]. In China, the first variable-speed pumped-storage unit was commissioned in 1989 as part of the Panjiakou PSPP [191]. In the 1990s and the early 2000s, several projects were undertaken to demonstrate the feasibility of variable-speed operation in SHPs, some of which include the 378-kW variable-speed Kaplan tubular turbine with fixed guide vanes, installed in Ingel gen [192], the 60-kW variable-speed PAT installed in the drinking water network of Sion, Switzerland [192], the replacement of the salient pole rotor, a 10-MW unit of computer to hydroplant with a three-phase wound rotor fed by a cyclo-converter [193], the addition of a 21.3-MW frequency converter to Forbach PSPP [194], and the 50-kW variable-speed hydropower unit with permanent magnet excitation installed in the River Tirva6, Finland [195].

However, it was not until 2004 that the first large variable-speed hydropower units were commissioned in Europe when Goldisthal PSPP began operation. The plant was equipped with four 265 MW pumped-storage units, for which a DFIM is used to connect two units to the grid. Since then, three new variable-speed PSPPs have been commissioned in Europe, all of them equipped with Francis pump–turbines and DFIMs, namely Avče, in Slovenia [196,197], Linthal, in Switzerland [198,199], and Frades II, in Portugal [196,200]. In addition, the refurbishment project of Grimsel II PSPP, in Switzerland, involved variable-speed operation. Three more variable-speed PSPPs in Switzerland, China, and India [201–203] began operation. Two of which will be coupled to a DFIM [202], and Tehri PSPP, in India, is equipped with 4,250-MW pump–turbines, each coupled to a DFIM [203]. In addition, two old 303-MW pumped-storage units of Okutataragi PSPP, in Japan, are upgraded for variable-speed operation [181,204].

Variable-speed hydropower generation has already reached the highest possible technology readiness level. There are, however, only a few variable-speed hydropower units (most of which are pumped-storage units) in operation all over the world. There are four reasons for slow pace of installation: (1) slow pace for project commissioning, (2) environmental concerns, and (3) the extra revenue a variable-speed hydropower plant can gain in the electricity and ancillary services markets is not always worth the extra cost necessary for the plant to operate at variable speed, and (4) regulatory uncertainty in hydropower investments. In future, the advantages of variable-speed hydropower generation can be further expanded by enlarging the stable operating range of hydraulic machines in order to take full advantage of variable-speed operation and enhance the insulation system of converter-fed machines [158–161,165].

3.6 ADVANCES IN HYDROELECTRIC POWER STORAGE

Hydroelectricity production entails diversion of water from a river through a turbine to generate electricity. As mentioned earlier, some hydroelectric systems include dams that store large volumes of water. Others are run-of-river which include small or nearly zero storage, with energy production rising and falling according to day-to-day rainfall in the river catchment. A run-of-river hydroelectric power station that is downstream of a large dam takes advantage of storage in that dam to reduce dependence on day-to-day rainfall. Water is conveyed from the water intake to the turbine and returned to the river through use of tunnels or pipes (penstocks), sometimes augmented with aqueducts. The pipe/tunnel must withstand large pressures. Depending on local geology, geography, and the head, tunnels are sometimes partially or fully lined with concrete or steel. The turbine spins in response to flow of high-pressure water, and it is attached to the generator which spins to produce electricity.

The largest hydroelectric storage systems involve pumped hydroelectric storage (PHS). In pumped hydroelectricity storage systems, the turbine can become a pump: Instead of the generator producing electricity, electricity can be supplied to the generator which causes the generator and turbine to spin in the reverse direction and pump water from a lower to an upper reservoir. Sometimes, the pump and the turbine are separate items of equipment, but more commonly they are combined. The head refers to the altitude difference between the water intake and the water egress. Since the cost of most components is largely independent of the head, a larger head will generally allow cheaper electricity generation and storage on a per-unit basis. While typical heads are in the range 100–800 m, larger and smaller heads are sometimes used. The efficiency of generation is about 90%. Hydroelectric systems that include large reservoirs can offer seasonal storage. Sufficient water is harvested from a river during the wet season and stored to allow significant electricity generation for many months. Large reservoirs can store thousands of GWh of energy. PHES comprises about 96% of global storage power capacity and 99% of global storage energy volume [2,3]. Some countries have substantial PHES capacity to help balance supply and demand. For example, Japan's PHES capacity was constructed to help follow varying power demand, allowing its nuclear and fossil fuel fleet to operate at nearly constant power output at the highest efficiency.

In order to compensate the increase of variable RES in power systems, energy storage such as that provided by pumped hydropower storage (PHS) is needed for stable grid operation. PHS plants operate both in turbine and in pump mode for peaking regulation. Across the USA, 43 pumped-storage hydropower (PSH) facilities have the capacity to generate and store 21 GW of renewable energy. The long-term storage that PSH provides is becoming more important to balance the grid against more and more use of intermittent renewable energy. PSH can, however, impact environment. Recently, significant efforts have been made to innovate PSH using different mechanisms for upper and lower reservoirs.

The grid balance for inverter-based large variable solar and wind systems can be generally achieved using PHES with large lower reservoir and smaller upper reservoir. Storage and electricity generation in PHES can be adjusted according to supply and demand needs. PHES has rapid response (from idle to full output in a time span of 20 seconds to a few minutes). PHES has rotational inertia if the generator is spinning, to replace the loss of the rotational inertia associated with conventional thermal generators when they retire. PHES has black-start capability, meaning that an electricity system can be restarted after complete collapse of supply without the need for electricity supply to start the generators. Together, batteries and PHES can completely replace the ancillary services hitherto provided by fossil and nuclear generators. Nearly all existing pumped hydro-systems are river-based. In many places, there is substantial environmental and social opposition to damming or modifying more rivers. However, there are alternative methods of constructing PHES that do not require significant modification to river systems. One method is to connect closely spaced existing reservoirs using underground tunnels and powerhouses. With care, there is low disturbance at the surface. This method is used for the 2 GW, 350 GWh, Snowy 2.0 system currently under construction underground in the World Heritage Kosciuszko National Park in Australia.

3.6.1 Closed-Loop Systems versus Open-Loop Systems

PNNL recently conducted a comparative study to evaluate the environmental effects of building and operating two types of PSH: the traditional open-loop system and a relative newcomer, closed-loop. The PNNL team found that environmental risks of closed-loop PSH systems are generally lower than those of open-loop PSH [2,3,205] (see Figure 3.10).

All of the country's currently operating PSH projects are considered open-loop, which involves connection to a natural water source to create a lower reservoir. In contrast, closed-loop PSH is not continuously connected to a natural water source. Through detailed investigations, PNNL study showed that closed-loop PSH systems provide more opportunities to minimize environmental effects to aquatic resources such as surface water quality and quantity, groundwater quality and quantity, and aquatic ecology and to terrestrial resources such as geology and soils, terrestrial ecology, land use, recreation, visual resources, and cultural resources. The study also showed that while off-stream located closed-loop projects have greater siting flexibility, these closed-loop systems using groundwater to fill and replenish their reservoirs could potentially have greater impacts on geology, soils, and groundwater quantity and quality than open-loop systems, which use surface water. Even though pumped storage is a mature storage technology, it continues to evolve to respond to the faster and more frequent transition requirements from pump to turbine and vice versa.

Closed-loop PHES (off-river) systems are particularly useful when they are located away from any significant river. An off-river PHES system comprises a pair of artificial reservoirs spaced several kilometers apart, located at different altitudes, and connected with a combination of aqueducts, pipes, and tunnels. The reservoirs can be specially constructed (greenfield) or can utilize old mining sites or existing reservoirs (brownfield). Off-river PHES utilizes conventional hydroelectric technology for construction of reservoirs, tunnels, pipes, powerhouse, electromechanical equipment, control systems, switchyard, and transmission, but in a novel configuration.

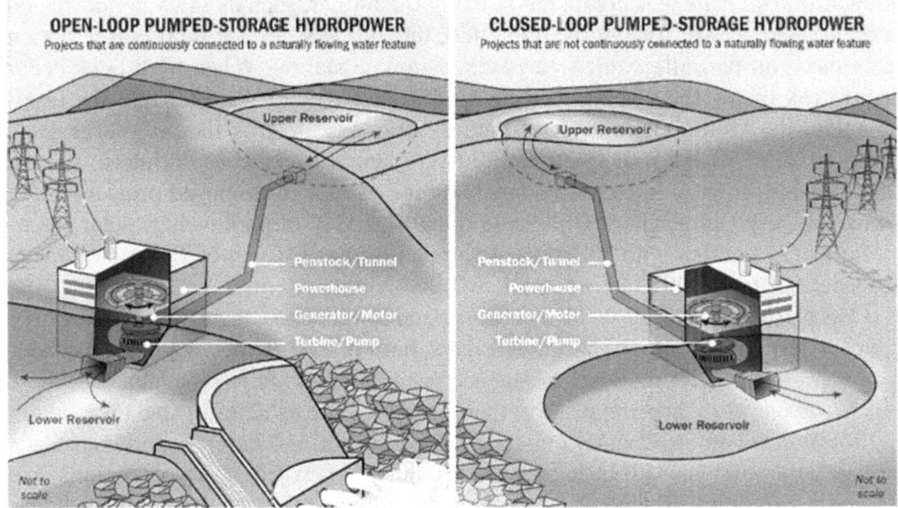

FIGURE 3.10 Closed-loop PSH offers more opportunities to minimize environmental effects to aquatic and terrestrial habitats than open-loop PSH, according to a study by PNNL researchers (Illustration/NREL [205]).

The energy that is stored in an off-river PHES system is usually lower than in a major river-based hydroelectric dam with similar power rating. An off-river PHES system has the advantage that flood mitigation costs are minimal compared with a river-based PHES system. Environmental costs of damming rivers are avoided with off-river PHES, which helps with social acceptance. The much greater number of off-river sites compared with on-river sites allows much wider site choice from environmental, social, geological, hydrological, logistical, and other points of view. Another advantage is that construction of off-river pumped hydro can be much faster than other storage methods. Bespoke engineering in mountainous river valleys is unnecessary. Work can proceed in parallel on the two reservoirs, the water conveyance, the powerhouse, and the transmission.

3.6.2 Innovative Approaches for PSH

Conventional hydroelectric dams may also make use of pumped storage in a hybrid system that both generates power from water naturally flowing into the reservoir and stores water pumped back to the reservoir from below the dam. The Grand Coulee Dam in the USA was expanded with a pump-back system in 1973. Existing dams may be repowered with reversing turbines, thereby extending the length of time the plant can operate at capacity. Optionally, a pump-back powerhouse such as the Russell Dammay should be added to a dam for increased generating capacity. Making use of an existing dam's upper reservoir and transmission system can expedite projects and reduce costs. Pumped-storage plants can operate with seawater, although there are additional challenges compared to using freshwater, such as saltwater corrosion and barnacle growth. Inaugurated in 1966, the 240 MW Rance tidal power station in France can partially work as a pumped-storage station. When high tides occur at off-peak hours, the turbines can be used to pump more seawater into the reservoir than the high tide would have naturally brought in. It is the only large-scale power plant of its kind. In 1999, the 30 MW Yanbaru project in Okinawa was the first demonstration of seawater pumped storage. A pair of proposed projects in the Atacama Desert in northern Chile would use 600 MW of photovoltaic solar (Skies of Tarapacá) together with 300 MW of pumped storage (Mirror of Tarapacá) raising seawater 600 m (2,000 ft) up a coastal cliff.

PHS projects generally involve an upper and lower reservoir. Another interesting concept being considered is to locate one or both reservoirs below ground (subsurface). While a project utilizing subsurface reservoirs has yet to be completed, these types of projects are attractive due to their perceived site availability and their potential for reduced environmental impacts. Abandoned mines, caverns, and man-made storage reservoirs have all been proposed as potential project reservoir options, and there are examples of several projects under initial phases of development [206]. The underground excavation or materials-handling costs, construction risk, and time required for underground excavation and construction could make the economics of such a project difficult, so most developers are looking to utilize existing subsurface structures or minimize/offset underground costs through the sale of excavated materials (ore, aggregate, etc.). Recent examples include the proposed Summit project in Norton, Ohio, the proposed Maysville project in Kentucky (underground limestone mine), and the Mount Hope project in New Jersey, which was to have used a former iron mine as the lower reservoir. Kitsikoudis et al. [207] described the use of Martelange slate mine in Belgium for the underground pumped-storage hydropower. The Callio site in Pyhäjärvi (Finland) utilizes the deepest base metal mine in Europe, with 1,450 m (4,760 ft) elevation difference. Several new underground pumped-storage projects have been proposed. Cost-per-kilowatt estimates for these projects can be lower than for surface projects if they use existing underground mine space. In Bendigo, Victoria, Australia, the Bendigo Sustainability Group has proposed the use of the old gold mines under Bendigo for pumped hydro-energy storage.

Several factors are often used to make the case for the potential use of abandoned mines to develop PSH plants. One is the topography: For example, in Indiana, the land is mostly flat, and there are no opportunities to construct conventional PSH plants. In addition, there are the economic reasons: Using an existing underground mine as lower reservoir is thought to reduce the overall project costs. Environmental reasons are also considered: Developing a closed-loop PSH project on a brownfield site would allow its re-use and revitalization, which may be much more acceptable to surrounding communities than greenfield development.

The underground mine PSH concept envisions using the tunnels and galleries of an existing abandoned mine as a lower reservoir and constructing or using an existing surface reservoir to serve as an upper reservoir for the PSH plant. An underground powerhouse needs to be constructed to contain the electromechanical equipment. A penstock between the upper reservoir and the powerhouse also needs to be constructed, as well as access shafts and tunnels, ventilation shafts, a surge tank, and other features necessary for PSH operation. It is also highly likely that some existing mine tunnels and galleries will need to be structurally reinforced to ensure their stability during rapid changes of water level during PSH operations. Finally, there may be a need to excavate additional underground tunnels to increase the volume of the lower reservoir.

The Elmhurst Quarry Pumped-Storage Project (EQPS) is a conceptual underground pumped-storage project that would utilize an abandoned mine and quarry for both the upper and lower reservoirs. The project would be located in the City of Elmhurst, Illinois, within 20 miles of downtown Chicago. The project would divert and gravity-feed water from an aboveground source into an underground powerhouse, where it would travel through the pump–turbine. The EQPS has an initial design capacity of between 50 and 250 MW with an estimated 708.5 GWh of energy storage potential. *Riverbank Wisacasset Energy Center* (RWEC) is a proposed 1,000-MW PHS facility located 2,200 ft underground in Wisacasset, Maine. The RWEC project would divert water into its underground shaft down 2,000 vertical feet to drop into a powerhouse containing four 250-MW pump–turbines. Similar to the EQPS, the water passing through the pump–turbines would be stored in large underground reservoirs (caverns) before being pumped up using low-cost, off-peak power. Abandoned coal mines in Germany (Prosper-Hanel in Bottrop, Porta Westfalica, or Hartz mine) and Australia (Centennial Fassifern coal mine) are also used for PSH. In the USA, ORNL performed a study in 2015 on the feasibility of using abandoned coal mines for small modular PSH plants [208]. Carbon Solutions LLC in collaboration with Indiana University–Purdue University Indianapolis (IUPUI) are investigating potential use of this technology in the state of Indiana. Figure 3.11 is a conceptual layout of an underground powerhouse in a coal mine [208].

Another concept similar to the one described above proposes to use the infrastructure of decommissioned open-pit mines for the development of PSH projects. This is not a new idea, and one PSH project of this type has already been constructed—the 1,728-MW Dinorwig PSH plant in the UK, which was commissioned in 1984. Dinorwig is a closed-loop PSH project that utilizes an abandoned slate quarry as lower reservoir. In Australia, construction has been approved for the 250-MW Kidston PSH project, which will be located on the site of a decommissioned open-pit gold mine [209]. Two existing pits of the gold mine, which are at different elevations, will be converted into the upper and lower reservoirs of a closed-loop PSH project. In the USA, the 1,300-MW Eagle Mountain PSH project is planned on the site of a decommissioned iron ore mine. Two open mine pits will serve as upper and lower reservoirs of the proposed closed-loop PSH project [4].

Gravity power, LLC, is proposing to develop a grid-scale electricity storage system. The company's gravity power module (GPM) uses the established principles of pumped storage combined with a large piston that is suspended in a deep, water-filled shaft. Once the shaft is initially filled with water, no additional water is required. As the piston drops, it forces water down the storage shaft, up the return pipe, and through the turbine, and spins a pump–turbine motor/generator to produce electricity. To store energy, power purchased off-peak drives the pump–turbine in reverse, spinning the pump to force water down the return pipe and into the shaft, lifting the piston.

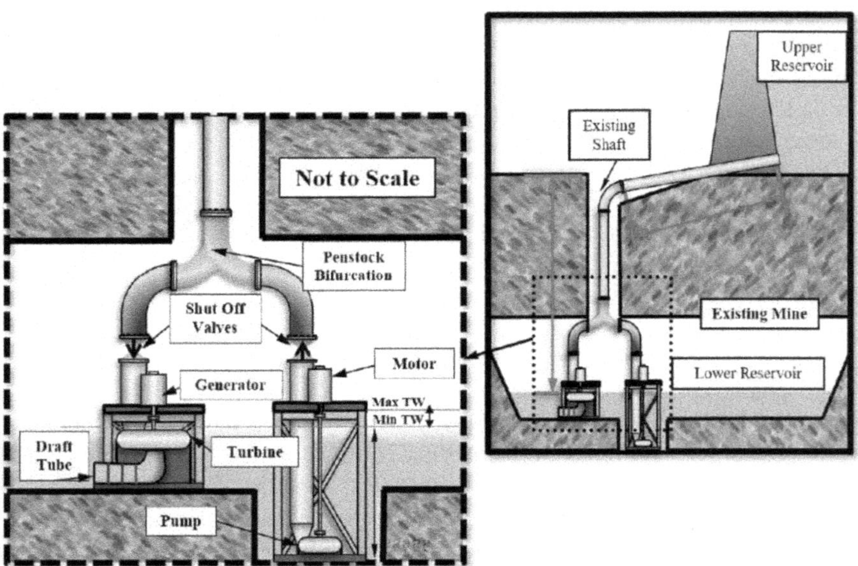

FIGURE 3.11 Conceptual configuration of a closed-loop PSH plant in an underground coal mine [4].

Obermeyer Hydro, Inc. is developing PSH technology using submersible pump–turbines and motor–generators in the USA. While conventional PSH plants typically use reversible pump–turbines that are submerged below water level and non-submerged motor–generators above them in the powerhouse, this technology proposes that both pump–turbine and motor–generator can be submerged in a vertical shaft (or "well"), thus avoiding the need for the construction of powerhouse. Obermeyer estimates that the diameter of the well would be approximately 2–3 m [210]. Figure 3.12 is a cross section of a power plant using this technology. For this technology to work, Obermeyer is developing a special type of pump–turbine with a flow inverter, which redirects the flow of water by 180° because the water enters and exits the well from above in both the generating and pumping modes of operation.

An important limitation of PHES is that it can only be developed in geographically suitable locations. The underwater pumped hydro-energy storage (UPHES) is a novel pumped-storage concept in which the upper reservoir is the sea itself and the lower reservoir is a hollow deposit (or a set of) located at the seabed. The seawater entering the deposit drives a turbine and generates electricity. The deposit is emptied by pumping the water back to the sea, thus storing part of the electricity consumed in the form of potential energy. The concept was devised with the aim to enlarge the number of potential locations for PSPP. The technical viability of UPHES was analyzed for the first time by a research team from the Massachusetts Institute of Technology (MIT) between 2008 and 2011 [211]. For this purpose, the research team built a concrete spherical deposit with an inner diameter of

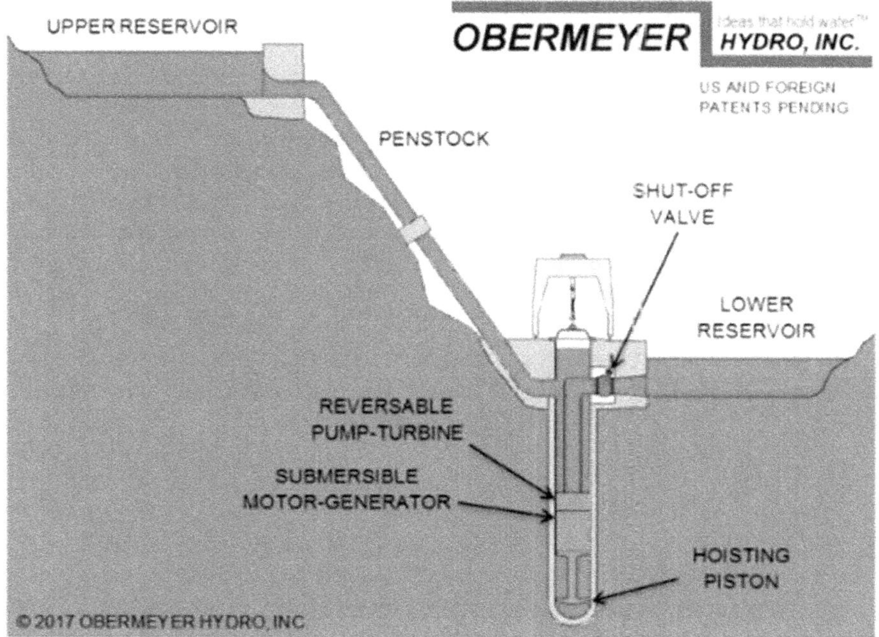

FIGURE 3.12 Cross section of PSH plant using submerged pump–turbine and motor–generator [4].

75 cm, equipped with a micro-pump and a micro-turbine of approximately 125 and 400 W, respectively. The test unit was successfully tested in both turbine and pump modes with and without a vent line (i.e., a pipe from inside to outside the sphere aimed to keep the pressure inside the sphere near the atmospheric pressure). The tests showed that without the vent line, the energy stored per volume capacity and the power output of both the turbine and the pump increase. However, the power output significantly varies with the internal pressure. Slocum et al. [211] identified Spain as the location with the largest suitable area for installing UPHES. An alternative idea included a closed-loop underwater system that would consist of two large polymeric bags at different depths under the sea. One bag would be anchored deep under the sea and serve as high-pressure reservoir, while the other would be a low-pressure reservoir floating close to the sea surface. The difference in the hydrostatic pressure between the two bags could be used to store energy and produce power when the water from the high-pressure bag is released up to the floating surface bag [4].

The UPHES concept has recently been brought to a TRL 5/6 under the framework of the Storing Energy at the Sea (StEnSea) project [212]. The techno-economic viability of UPHES was addressed in the first phase of the StEnSea project. Main design parameters of the system were material, location, shape, size, and installed power capacity of each deposit. The conditions for potential sites included a water depth ranging from 600 to 800 m, a slope of the seabed not greater than 1°, and a distance to the electrical grid/maintenance bases/installation bases lower than 100/100/500 km. A smaller feed pump was also proposed to be used instead of the vent line to avoid cavitation in pump mode [213]. The results of the phase 1 study indicated that in Europe, Spain and Italy have the highest potential for such storage. In the second phase of StEnSea, a 1:10 scale model was installed at 100 m depth in the Constance Lake. While StEnSea's test took place at a depth of 100 m in the freshwater lake, Constance Lake, the technology is foreseen to be used in saltwater at greater depths. Since the submerged reservoir needs only a connecting electrical cable, the depth at which it can be employed is limited only by the depth at which the turbine can function, currently limited to 700 m.

The saltwater pumped storage in an underwater configuration offers a range of advantages. No land area is required, and no mechanical structure other than the electrical cable needs to span the distance of the potential energy difference. In the presence of sufficient seabed area, multiple reservoirs can scale the storage capacity without limits and should a reservoir collapse, and the consequences would be limited apart from the loss of the reservoir itself. Evaporation from the upper reservoir has no effect on the energy conversion efficiency, and the transmission of electricity between the reservoir and the grid can be established from a nearby offshore wind farm limiting transmission loss and obviating the need for onshore cabling permits. The StEnSea project plans to use concrete spheres of outer diameter 34 m and wall thickness 2.7 m, each with a mass of about 12,000 tons, installed at depth $H = 750$ m at the bottom of the ocean, which must be flat to better than one degree. With a round-trip efficiency of 73%, the storage capacity then is $E = 18$ MWh per sphere [4,9]. In a full-scale offshore PHS system, farms of large numbers of spheres are planned. The StEnSea proposal was recently extended to the case of concrete spheres submerged not offshore, but in a deep artificial lake created by further excavating and flooding an exhausted open-pit coal mine named "Hambacher Loch" in North Rhine-Westphalia, Germany [4,9]. In order to avoid excessive transmission cost and loss, in general, the reservoirs should be placed off deep water coasts of densely populated areas, such as Norway, Spain, USA, and Japan.

Koritarov et al. [4] examined several other novel approaches to improve PSH technologies. These include small PSH with reservoirs of corrugated steel and floating membranes, geomechanical PSH, integrated PSH and desalination plant, pressurized vessel PSH, thermal underground PSH (TUPH), and high-density fluid PSH. The concept of PSH with corrugated steel and floating membranes was originally proposed by Shell Energy North America (SENA), in which a small modular PSH technology can use reservoirs made of corrugated steel or floating membranes. In principle, this technology could be designed and configured as either an open- or a closed-loop PSH system. SENA proposed a concept that uses corrugated steel for the upper reservoir and a floating membrane reservoir (floating in a larger body of water, e.g., a lake) for the lower reservoir. The water in this closed-loop PSH system, which was referred to by SENA as a "hydro-battery," would circulate between the upper and lower reservoirs and would not mix with the water in the lake in which the lower floating membrane reservoir is situated, except for occasional small amounts of water additions to make up for losses and evaporation.

The geomechanical PSH technology is an innovative energy storage concept that is currently being developed by Quidnet Energy, Inc. (Quidnet). The main idea is to pump water down into the ground, between rock layers where the water would be kept under pressure. The natural elasticity of certain rock formations will act like a spring and keep the water under pressure, until the valve is opened and the water is released through a hydroelectric turbine to generate electricity. IPHROCES combines a seawater PSH plant with a reverse-osmosis desalination plant to provide an integrated energy storage and freshwater supply system. By combining the PSH plant and the desalination plant, the hybrid facility may achieve synergistic benefits that would not be available if each plant were operating separately. The addition of a desalination process to the PSH plant brings an additional value stream and revenue from providing large amounts of fresh, potable water. The desalination plant also benefits from synergy with the PSH plant, which provides the energy needed for desalination, thus making desalination more cost-effective.

Researchers from ORNL in the USA proposed a new concept to store energy by using water to pressurize air in a vessel. The concept is called ground-level integrated diverse energy storage (GLIDES), and it uses water as a liquid piston to pressurize air in a high-pressure reservoir [214,215]. The energy is stored when the water is pumped into the vessel to pressurize the air, while energy is generated when the water is released from the vessel and pushed by the compressed air to turn the turbine and generate electricity. Figure 3.13 illustrates the pumping and generating cycles for GLIDES.

The TUPH concept was proposed by Professors Georg Pikl, Wolfgang Richter, and Gerald Zenz of Graz University of Technology in Austria [216,217]. This is an interesting concept that envisions a closed-loop underground PSH system that uses hot water for PSH operations. Both the upper and lower reservoirs of the PSH plant are built underground, so that geothermal energy can be used to heat the water in the reservoirs up to 95°C. The heat energy of the water can be used to provide heating and cooling functions for the local community. For this purpose, heat exchangers would be placed in the upper and lower reservoirs of the PSH plant and connected to the district heat network. An illustration of the TUPH concept is shown in Figure 3.14.

High-density fluid PSH concept is similar to traditional PSH technologies, except that instead of water, it uses a high-density fluid for PSH operations. A representative example of this technology is being developed by RheEnergise, Ltd., a start-up

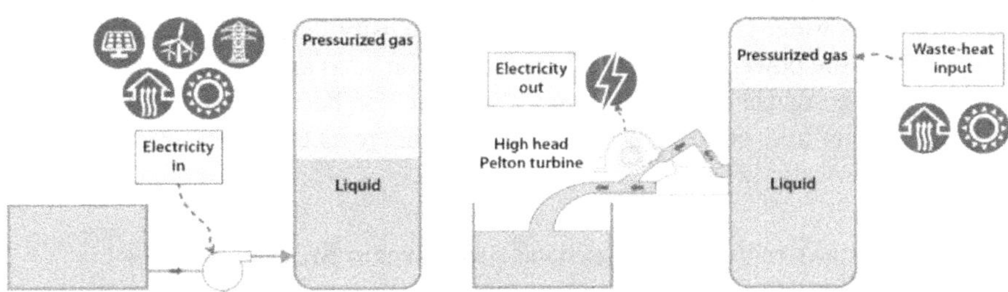

FIGURE 3.13 GLIDES pumping and generating cycles [4].

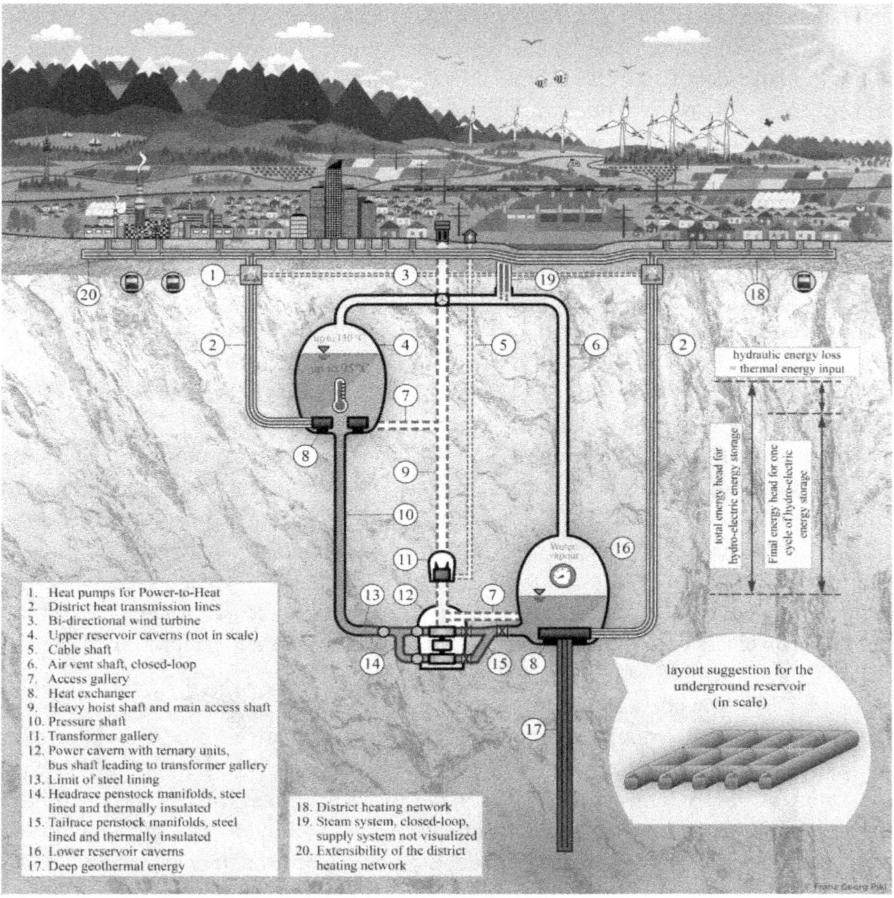

FIGURE 3.14 Schematic of TUPH system [4].

company based in the UK and Canada. Because it uses a high-density fluid instead of water for PSH operations, this technology can only operate as closed-loop PSH system. The main reason is that high-density fluid is used to obtain more power from a smaller PSH project, thus achieving cost savings. Compared to a traditional PSH plant that uses water as working fluid, the high-density fluid PSH should be able to achieve the same power output for a lower hydraulic head and turbine flow. The footprint of the high-density fluid PSH plant is also smaller, because smaller of upper and lower reservoirs may contain the same energy storage as those of a traditional PSH plant that uses water [218].

3.6.3 Integrated Virtual Reservoirs and Hybrid Storage Systems

Small (or micro) applications for pumped-storage could be built on streams and within infrastructures, such as drinking water networks and artificial snow-making

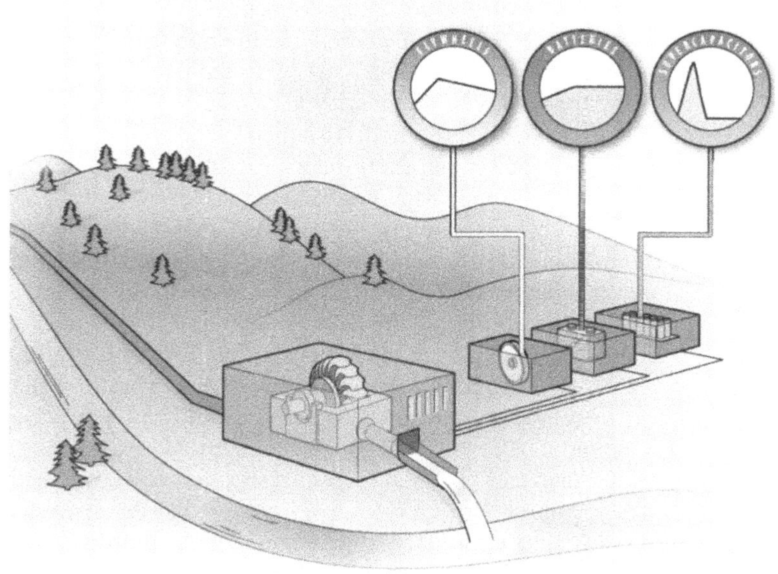

FIGURE 3.15 Concept design of INL's integrated project (*Source*: Courtesy of INL [219].)

infrastructures. In this regard, a stormwater basin has been concretely implemented as a cost-effective solution for a water reservoir in a micro-pumped hydro-energy storage. Such plants provide distributed energy storage and distributed flexible electricity production, and can contribute to the decentralized integration of intermittent renewable energy technologies, such as wind power and solar power. Reservoirs that can be used for small pumped-storage hydropower plants could include natural or artificial lakes, reservoirs within other structures such as irrigation, or unused portions of mines or underground military installations.

During 2019–2020, Idaho National Laboratory (INL) worked closely with Argonne and NREL to demonstrate the technical potential and economic benefit of co-locating and coordinating multiple run-of-river hydropower plants with different types of energy storage devices, creating "virtual reservoirs" with potential to function similar to conventional reservoir-based hydropower plants (see Figure 3.15). Partnering with Siemens, the project team developed a centralized control scheme, the Smart Energy Box, to coordinate operation of different energy storage devices at one or multiple hydropower plants. The project also succeeded in applying market participation modeling to assess the ability of hybrid energy storage systems to perform similar to conventional hydropower plants in responding to grid needs, opening the door for potential new markets for run-of-river owner/operators to participate in.

Smart Energy Box, which coordinates operation of multiple units/plants and energy storage devices, can be optimized for providing services and performance desired by the owner/operator. The Smart Energy Box monitors grid operator signals and determines how to meet programmed objectives using all connected devices most effectively. It can be used to enable a broad range of benefits, including provision of ancillary services, or reduction of ramping speed and the frequency of hydropower generators, which can have positive outcomes on wear-and-tear rates. The project also accomplished the application of Argonne's Conventional Hydropower Energy and Environmental Systems (CHEERS) model to simulate and optimize market participation of integrated hydropower and energy storage systems. CHEERS optimize day-ahead scheduling and real-time operations for hydropower by considering multiple objectives (e.g., cost, power, environmental considerations) to support decision-making on unit commitment and turbine-level operating points. Using Argonne's CHEERS model and real-world market, water, and plant capability data, the team found integrating energy storage with Idaho Falls Power's hydropower facilities can boost revenue by 12%–16%. Black-start capabilities at the distribution level would enhance system reliability and allow hydropower owners/operators to restore services in islanded mode, reducing downtime for customers if an event occurs and improving grid resilience.

Hydropower turbines currently provide inertial response and are well suited to provide load frequency control. However, when the system inertia is low, fast-acting frequency-responsive units are needed to maintain the system frequency within the standard frequency range [9]. The coordinated operation of fast energy storage systems (i.e., in converter-coupled) and hydropower would allow a better frequency control in the electrical power system, with lower wear and tear of the hydropower units. For this purpose, flywheels and super capacitors appear to be the best existing inverter-coupled energy storage technologies. They are best suited for frequency control applications, both are able to withstand a large number of continuous charge–discharge cycles, and have a small energy storage capacity which will allow them to control their active power output/input for a small amount of time (from a few seconds to a few minutes) [9]. Hydropower generation has, in turn, lower frequency response but can usually control its active power output for a longer time (from a few hours to a few days, or even longer). Flywheels and supercapacitors can be easily integrated into existing hydropower plants and fully operational in a very short time (few months). Their integration in existing hydropower plants might provide additional benefits to the voltage control in the vicinity of the power plant.

It would be also possible to coordinate the operation of existing hydropower plants with a fast energy storage system connected to the transmission power system in another network node, or even with a set of them geographically distributed. The coordination of fast energy storage systems and hydropower has not yet been demonstrated in an operational environment. The technology was partially brought to a TRL 4 by a research team of the PNNL, as a result of a research project titled "Wide-area energy storage and management system" to balance intermittent resources in the Bonneville Power Administration and California ISO control project. The results of this study have shown that frequency quality in EPS can be improved by providing an inertial response with fast-acting energy storage systems and frequency regulation with medium-fast generation units. The system rate of change of frequency can be reduced by the virtual inertial response provided by the fast-acting energy storage systems. Thus, medium-fast generation units are able to contain the frequency deviation within a narrower band.

3.7 HYBRID HYDROPOWER

While solar power and wind power continue to dominate the renewable energy expansion, jointly accounting for more than 90% of the new capacity installed in 2019, hydropower still represents nearly half of the global renewable capacity already in operation. As shown earlier, hydropower can be large scale or small scale, cost competitive, flexible, and easy to store—in contrast to so-called variable renewables like solar and wind. As indicated earlier, in recent years, significant efforts are made to modernize existing hydropower plants along with building new ones. Modernization is not only about maintaining or increasing safety, performance, and reliability, but also to extend lifespan for its new role in the renewable infrastructure. Future power systems will have higher needs for so-called dispatchable generation, that is, plants that can adjust their power output on demand at the request of power grid operators, like opening the water tap. This flexibility will be necessary when a bigger share of the power will be coming from variable, non-dispatchable, renewable energy sources (solar and wind) relying on the unpredictable weather in order to generate power. One method is to combine solar and/or wind power with hydropower. This type of hybrid power can provide dispatchable solution for the grid.

One type of hybrid power that is getting worldwide attention is the combination of floating solar photovoltaics and hydropower. Floating solar photovoltaics (FPV) is emerging as a complement to conventional solar power on buildings and ground-mounted and as an interesting solution to expand energy generation by hydropower. The concept is quite simple using conventional solar panels installed on floating structures such as floats, pontoons, or membranes, while the whole system is firmly anchored and connected to the electrical connection onshore. When FPV is co-located with hydropower, the benefits also include increased PV efficiency due to the temperature regulating effect of water, reduced algae growth, reduced evaporation, reduced shading effects on modules, and reduced capital cost. Previous studies estimate the global potential of FPV is around 400–1,000 GW and the global potential of FPV paired with hydropower is 4,400–5,700 GW. Unlike FPV, hydropower requires steeper terrain so that water can flow, making it in theory quite complicated for both energy sources to be located in the same spot. In addition, hydropower operators are struggling with more and more stringent constraints: providing flexibility to power systems, but also coping with more demanding water management. Hydropower projects are often multipurpose: managing drinking water, water supply to the industry, irrigation, and recreational activities. Dealing with multiple stakeholders adds complexity, and conflicts can emerge in relation to the operation of the reservoir. Even so, floating solar power opens wide the possibility to deploy large-scale solar power close to existing hydropower projects. More energy, meaning more revenues, can be produced from the same location, sharing some of the existing assets like electrical infrastructure and existing transmission lines to capture the power.

A hybrid power plant, operating simultaneously the solar and hydro-parts, can answer to the challenges of both energy sources. Hydropower compensates for the unstable solar power production by its rapidly adjustable output, whereas solar power contributes to saving water on mid- to long-term scheduling, providing seasonal and daily flexibility. With an even more firm and dispatchable power output, the operator can then increase electricity sales and seek higher prices. The best hybrid example (which is not floating) is currently located in China: the Longyangxia power plant, in operation since 2014. Hybrid in this setting means that the hydropower and the solar power are connected in the same system. Murphy et al. [220] define hybrid systems as those in which net economic benefits are anticipated from the coupling of multiple generation technologies, relative to the cost and/or value associated with comparable, independent, stand-alone technologies [221–226]. Murphy et al. [220] propose three possible hybridization configurations, each offering different cost and performance-added values:

1. **Co-location hybrid systems (cost improvements)**. Two or more technologies sited together to achieve cost savings, but operations are separately optimized.
2. **Virtual hybrid systems (performance improvements)**. Two or more technologies are sited separately, with operations linked through bilateral agreements and some co-optimized operation.
3. **Full hybrid systems (cost and performance improvements)**. Both cost and performance improvements are achieved through co-optimized planning and operation. These often consist of at least one dispatchable technology paired with one or more variable renewable energy technologies, which, when paired, offer operational benefits.

Alternatively, the World Bank [221] proposes three different hybridization levels for PV and reservoir-based hydropower differentiated by the peak PV power relative to the hydropower capacity. The highest level includes solar peak power greater than the capacity of hydropower and includes a pumped-storage plant beyond the reservoir-based hydropower where the FPV is sited [222–226]. As an example, the study by Gadzanku et al. [227] considered full hybrid FPV-hydropower systems with FPV and hydropower coupled at a common substation—allowing for their operations to be co-optimized and dispatched in concert, as depicted in Figure 3.16. Hydropower refers specifically to reservoir-based impoundment and pumped-storage systems. These reservoir-based systems are often highly controllable—allowing for the complementary operational benefits discussed in this work. The study excludes diversion (run-of-river) hydropower systems as the lack of reservoir area and the presence of water currents may be problematic for FPV deployment [221–224].

The establishment of a synergy between hydroelectric dams and floating solar allows more generation of energy. The hybridization allows panels to produce solar energy during the day while saving water for hydroelectricity to complete during intermittent times when the sun goes down. When water storage is possible, it also allows high-value hydropower to be produced at peak demand time. Another great advantage when installing floating solar power on a dam is the benefit of using existing electrical infrastructure, including high-voltage grid access and transformation devices. This drastically lowers the overall capex costs and makes projects happen quicker. Since solar and hydropower are smartly hybridized, exporting either solar or hydroelectricity according to the hour of the day, it is not necessary to augment

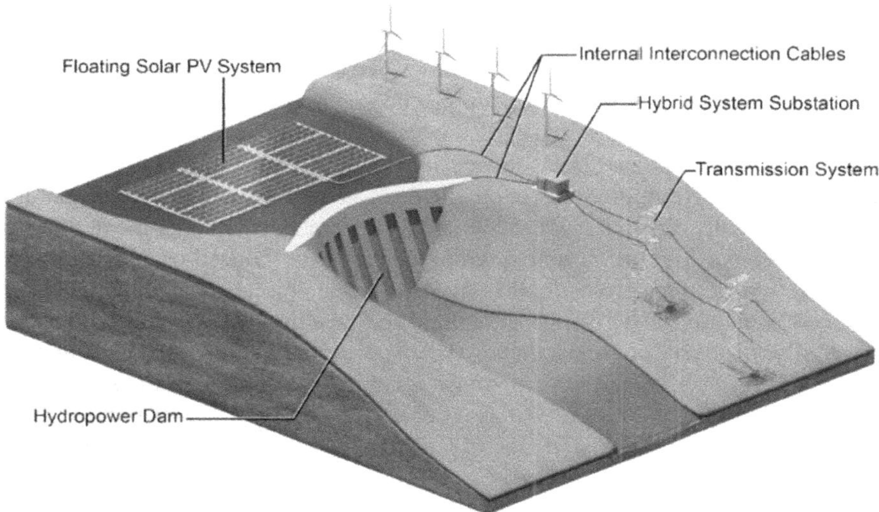

FIGURE 3.16 Schematic of a hybrid FPV-hydropower system [222].

transformation or transport capacity if the maximum peak output of the solar array does not exceed the maximum hydro peak capacity.

Dams seldom reach a full power production ratio over 4,000 hours/year, and often have a much lower ratio, leaving a large opportunity for energy generation to complete the grid output with solar. Due to its intermittency, solar plants are required to have a storage device. The battery storage systems, even if prices are decreasing, remain very expensive. Storing only 1 hour of peak power raises the capex price of a solar plant by 50% (with consideration to the cost of the solar plant equal to $800 MWp, and cost of a 1 MWh battery system equal to $400). For a hybrid dam FPV plant, the reservoir is the battery, and so the extra cost of storage is saved. In hybrid systems, several benefits are achieved with respect to the independent operation of the solar and hydro-plants. In general, hydro-plants are easy to access and already grid-connected; thus, the installation of PV panels requires less work and infrastructure. Other benefits include the following [221–226]: (1) PV panels protect the dam surface from direct solar radiation that may negatively affect the stability of the dam itself, reducing thermal excursion of the dam surface and increasing dam durability; (2) PV panels are mounted on an inclined area, minimizing the distance required between two panels with respect to an analogous installation on a flat area, thus increasing solar energy generation; (3) the intermittency of solar PV generation due to weather conditions, spatial resource qualities, and daily patterns can be counterbalanced by highly controlled hydropower at different timescales on a seasonal, monthly, daily, hourly, and even sub-hourly level, and also the cooling effect provided by water below the panels increases panel efficiency; (4) FPV-hydropower hybrids could provide multiple options for energy storage, these include coupling FPV with pumped-storage hydropower to use excess solar generation to pump water into an upper reservoir to store for later use, or the full hybrid in which water resources can be conserved during peak solar production hours—utilizing the reservoir as storage for the non-dispatchable solar PV, and energy generated by PV can be used for pumping in pumped-storage hydropower plants; (5) connecting FPV to existing transmission infrastructure may increase the utilization rates of transmission lines where additional transfer capacity exists; (6) since system operator views the hybrid FPV-hydropower system as a single generator able to provide predictable, controllable, dispatchable generation (not separate generators), the hybrid system may receive a higher capacity credit—allowing for reduced or no curtailment. Also, the water surface provides areas free of shading objects along with higher sunlight reflection, improving PV generation; (7) co-locating FPV systems with hydropower allows for unified connection to existing transmission infrastructure, and this may reduce the additional costs of transmission extensions, substations, and other infrastructure requirements as well as reducing time and siting constraints (such as land acquisition); and (8) installation of FPV on hydropower reservoirs may help to decrease water evaporation through decreased air flow and absorption of solar radiance—increasing resources for hydropower generation. PV panels are installed on an existing structure (the dam surface), reducing land use. The shading provided by PV panels on the water reduces algae growth and water evaporation, improving hydroenergy generation and water quality.

Floating panels can increase the capacity factor of a hydropower plant by 50%–100%, where the capacity factor of the hydro-plant is the ratio of total generated energy to the maximum energy than can be generated if the hydro-plant would always work at its maximum installed power capacity. Floating panels can gain 7%–14% more energy than a land installation due to the reduction of temperature. However, floating PV has an important limit: It cannot resist strong wind gusts, necessitating a very large number of mooring points in order for it to remain intact. The solution devised by the company Upsolar Floating is based on a much more robust concept where rafts are built with polyethylene pipes and steel beams supported by 20–24 panels. They have been shown to resist damage by winding up to 140 km/hour. Despite potential benefits, challenges exist that could hinder further investment and deployment [221–226].

1. The capacity (MW) of the FPV systems in hybrid applications may be constrained by factors beyond the available reservoir surface area. Fang et al. [223] suggested establishing limitations on the peak FPV power to guarantee that hydropower can compensate for power deficiencies caused by variableness of FPV output. This may be done by setting a conservative limit of peak FPV capacity equal to the installed hydropower capacity [223]. A second, related approach is to constrain FPV development to ensure existing transmission infrastructure is sufficient for the additional power supply from FPV when paired with hydropower [222–226].
2. The hydropower operation range may limit the full dispatchability of the hybrid system, as reservoir levels are often regulated to avoid erosion on reservoir banks, conserve water for other uses, and maintain dam safety (may require water curtailment in rainy seasons). Hydropower turbine

ramping rates may also limit hybrid operation as variations in FPV may occur too quickly for some turbines without high degrees of generation control [221–226]. These challenges may be overcome in new systems designed for hybrid operation or retrofitted systems; however, it is important to point out that benefits may not be available for all existing systems.

3. Solar modules, inverters, and floating structures are not novel systems; however, their installation as hybrid systems may present multiple material-related challenges [222]. Previous stand-alone FPV research indicated that insulation failures can lead to frequent unplanned downtime, and long-term exposure to humidity may expedite PV modules degradation, corrosion of floating structures, water body contamination, and/or material fatigue of joints, which may result in debris that interferes with the local ecosystem or hydropower systems [222,223].
4. Capital and operation and maintenance costs may also present challenges, as the required electrical and anchoring systems and mooring costs vary dramatically from vendor to vendor and could be higher than land-based systems [222–226].
5. Other potential challenges include concerns related to aesthetics, intrusion on recreational water bodies, and environmental impacts, such as temperature fluctuations that could affect local ecosystems. Hydropower development has historically faced significant attention due to social and environmental impacts, and it is reasonable to assume further development of FPV on reservoir surfaces could also require significant dialogue and analysis to ensure alignment with the broader values of societies [222–226].

Lee et al. [222] state that some of the above challenges can be overcome in new systems designed for hybrid operation or retrofitted systems. Since early phases of the project development, different research institutions have been investigating the project's complementary operations, scheduling and, optimization. Unfortunately, and despite evidence of the benefits provided by such hybrid plants, it appears that the industry has failed to repeat the implementation of projects of this type so far. Most of the commercial projects on hydropower reservoirs currently in development are so-called collocated solutions. This means that, even if floating solar is implemented on the reservoir, both energy sources each have their own power purchase agreement, operating independently. The full potential of joint operations is not exploited. This is for example the case of the Norwegian consortium Statkraft/Ocean Sun and their pilot project in Albania, which got some media attention in recent years. Only a handful of small projects are currently planned with hybrid operations worldwide which include EDP's pilot project on Alto Rabagão reservoir, Portugal.

The reasons for the slow development so far are directly linked to the intricacy of local and national regulations and policy frameworks, but also to the perceived technical complexity of hybrid operations. In addition, solar power and hydropower plants have up until recently been developed, built, and operated by very different industries. Floating solar power technology provides new occasions as solar companies are now looking for opportunities on water, driven by the need to identify large areas with low conflicting interests and short distance to the grid.

3.7.1 Global Potential of Hybrid FPV-Hydropower

In spite of various benefits and challenges, the global potential of FPV-hydropower hybrid is enormous. A study from Lappeenranta University of Technology stated that approximately 6,270 TWh of extra electricity could be produced per year from floating solar power on reservoirs worldwide (with only 25% of surface coverage). This corresponds to approximately 40 times the annual electricity generation from hydropower in Norway. It does not only provide additional capacity; it also adds flexibility allowing further integration of renewables. In simple words, the output of these hybrid plants is higher than the plain sum of both energy sources: one + one does not equal two, but three. Figure 3.17 presents the median scenario capacity (MW) results for hybrid FPV systems by global region. North America has the largest estimated technical potential (1,785 GW); however, other regions, such as South America (739 GW), Eastern (473 GW) and Southern Asia (362 GW), and Northern Europe (404 GW), also have significant potential (see Figure 3.17). According to the authors, their estimated annual generation results range from 4,251 to 10,616 TWh/year for the scenarios considered [221–226].

The above-described potentials do not consider the project-siting constraints that developers will eventually face, and that a deep knowledge of local ground-verified data and development regulations is needed to assess the feasibility of a project. Additionally, technical potential does not capture the economic or market potential for floating solar PV or potential future technology improvements. Future, studies at the regional, national, or subnational scale are required.

Of all possible hybrid hydropower systems, hybrid FPV-hydropower is most advanced. Hybrid FPV-hydropower system was just perfected in Portugal at the Alto Rabagão Dam. Located in Montalegre, Portugal, it is the world's first hybrid FPV

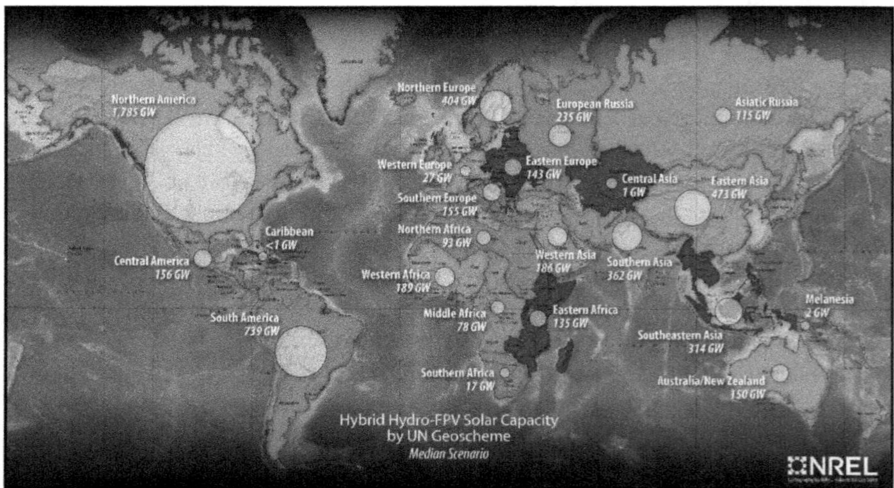

FIGURE 3.17 The median scenario capacity (GW) results for hybrid FPV systems by global region [222].

and hydroelectric dam power plant system. With a total capacity of 68MWp, the dam adds an additional 220 kWp through the floating PV installation. The installed 840 floating PV panels are expected to generate 332 megawatts per hour in its first year—equivalent to the annual consumption of around 100 homes. This array will be extended as soon as the first results are computed. The location of the Alto Rabagão dam, built in the 60s, presented new technical challenges due to the water variation and depth. The operator EDP, that operates the pumped-storage hydroelectric power plant on the 94 meters tall wall of the dam, wanted confidence that the floating system would not create any possible conflict of use with their asset even in case they have to have an emergency drawdown of the water. Following those conditions, the floating solar array has been designed and was moored at more than 60m in depth, while dealing with a water level variation of 30m. After success at the Alto Rabagão Dam, another experiment will start in Brazil, at the Balbina Dam, in Amazonia state, with an initial 5 MW peak capacity. The Balbina Dam currently suffers from drought issues, sedimentation, and high greenhouse gas emissions because of the drowned forest. By adding solar energy, it could easily double the total power of the dam to 250 MWp, by covering less than 1/1,000th of the lake surface. One other example is the 200-kWp floating solar project in Suvereto, Italy [221–226].

A Norwegian consortium led by Scatec is planning to build a hybrid hydropower-floating PV plant at an unspecified location in West Africa. Building both facilities simultaneously will help its developers define a series of parameters for proper sizing, optimization, and design, and set a benchmark for future projects of this kind. The Norwegian consortium includes floating PV specialist Ocean Sun, independent research organization SINTEF, software provider Prediktor, the Institute for Energy Technology (IFE), and Norwegian solar industry body the Solenergiklyngen. This project will be developed specifically as a greenfield combo plant with overall low LCOE. PV and hydropower are complementary on a seasonal basis, and hydropower can convert intermittent PV into higher-value steady power. The project includes the development of a platform for planning, scaling, and optimizing the operation of hybrid power plants. The optimization of the two power-generating elements gives benefits for water remediation technology and the dimensioning of the reservoir and the basin, as well as for defining early production levels from the solar plant during the construction phase. The project will also explore adaptations of the Ocean Sun system for amphibious use during the gradual filling of the reservoir. The reduced water evaporation and PV cooling effects will be of key importance in the project. The consortium will develop novel control mechanisms for the optimization and fine-tuning of the hybrid power [plant's] production [221–227].

Today, a floating solar project costs 10% more than a solar plant on the ground, but this higher cost is overcome by the increased efficiency. The final kWh cost is 20% lower than a ground-based project. The cost for large projects is about ~$763 per kWp, all included (mooring, cable, inverters, electric cabinet), while the final kWh price is in the range of $33–$54 per MWh, depending on the local radiation conditions. Floating solar panels installed on a dam surface can be applied to several dams across the globe. For example, a recent study carried out by the European Commission's Joint Research Center revealed that the application of such hybrid systems to ten selected dams in South Africa can generate an annual electricity amount of 72 GWh from PV from an installed peak power of 42 MWp. An example of a real project can be found in Japan's Kutani Dam, with an installed PV capacity of 4.99 MWp and a 20-year revenue of ~$5.4 million. By assuming coverage of 25% of the 265.7 thousand sq. km that represent all existing hydropower reservoir surfaces with floating PV, 4,400 GW could be generated (6,270 TWh) that can reach 5,700 GW (8,000 TWh) using all existing reservoirs, both for hydropower and for other purposes. Floating solar could prevent about 74 billion cubic meters of water evaporation, increasing water availability by 6.3% and hydropower generation by 142.5 TWh. This application to water reservoirs worldwide has already been considered, for a total installed capacity of floating PV of 376 MW in China, 22.7 MW in Japan, 9.3 MW in the UK, 6 MW in South Korea, 4 MW in Australia, 0.77 MW in Italy, and 0.67 MW in the USA. Other examples, but at a minor scale with respect to the previously mentioned countries, can be found in Spain, Portugal, France, and India [221–227].

Due to the advantages of this type of combined hybrid plant, the potential and market opportunities are expected to grow in the future, especially the floating solution, thus reducing the combined (hydro + solar) GHG emissions per kWh produced, particularly in tropical regions. Dams have several purposes: providing electricity of course, regulating the river flow, and supplying water for irrigation. Water for irrigation is extremely valuable, because uses are endless, mainly for agriculture, the largest water user at global scale accounting for around 70%. Therefore, future trends are heading toward a higher need of water for this purpose. If power capacity is increased with the addition of a solar plant, then the dam managers will be able to allow more water for irrigation. Also, any dams equipped with a pumped storage with a double reservoir (PHES) are placed in an even better situation for solar hybrid systems. During the sunny hours, solar electricity can be, without export limitation, used to raise water in the upper reservoir to provide power capacity in the evening or during cloudy periods. Reservoirs are usually much smaller here in those facilities; however, they are still large enough to generate a significant part of the required pumping power [221–226]. Borkowski et al. [228] examined run-of-river hydro–PV–battery hybrid system to supply local electrical loads.

3.7.2 Hybrid PSH and Wind Plant

While there are many ways to design a hybrid wind and PSH plant, here we will highlight one interesting design that was proposed by the Max Boegl Wind AG (Max Boegl) company in Germany [4] that includes the construction of small concrete reservoirs around the foundations of wind turbines located on a hill. The combination of these small reservoirs then serves as a multipart upper reservoir, and they are connected via water conduits with a lower reservoir at the bottom of the hill. The pilot project includes four wind turbines with towers made of concrete and water reservoirs around their bases, which are also made of concrete. Both the wind towers and the water reservoirs can be built either from prefabricated elements that are transported to the project site or using mobile fabrication facilities at the project site.

FIGURE 3.18 Water reservoir around the foundation of wind tower [4].

Figure 3.18 illustrates the water reservoir being constructed around the base of one of the wind towers. The four wind turbines at the Gaildorf pilot project have an installed capacity of 3.4 MW each, for a total of 13.6 MW. The installed capacity of the PSH plant is 16 MW. This is a closed-loop PSH system where the upper reservoirs at the bases of wind turbines are connected to each other and then to the lower reservoir using an underground penstock. The hydraulic head between upper and lower reservoirs is 200 m, or about 656 ft.

Some advantages of this technology include lower investment costs due to the standardization of construction and prefabrication of components, short construction period, long PSH plant lifetime (Max Boegl estimates about 50 years), and possible hybrid operations with other renewable resources (e.g., solar) in addition to wind. Koritarov et al. [4] have evaluated estimated project cost, LCOS, construction time, project development risk, scalability and applicability, potential market size, environmental impacts, physical siting limitations, and TRL (which is estimated as 7 to 8). Their evaluation indicates that this technology provides an interesting option for hybrid PSH and wind plants that has potential to reduce the cost, time, and risk for PSH project development. The key innovation is an upper PSH reservoir consisting of multiple small reservoirs built around the foundations of wind turbine towers. Max Boegl has developed a technology that uses prefabricated modular elements made of concrete for fast construction of both water reservoirs and wind turbine towers.

3.7.3 Theoretical Analysis of Other Hybrid Hydropower Systems

Besides FPV-hydropower hybrid system, several other hybrid systems are theoretically analyzed in the literature. Hybrid hydro-energy systems are usually analyzed with pumped hydro-storage systems, which can facilitate energy accumulation from other sources. Despite the lack of water storage, run-of-river hydropower plants are also attractive for hybrid systems owing to their low investment cost, short construction time, and small environmental impact. In the study by Borkowski et al. [228], a hybrid system that contains run-of-river SHPs, PV systems, and batteries to serve local loads was examined. Low-power and low-head schemes that used variable-speed operation were considered. The novelty of this study was the proposal of a dedicated steady-state model of the run-of-river hydropower plant that is suitable for energy production analysis under different hydrological conditions. The presented calculations based on a real SHP of 150 kW capacity showed that a simplified method can result in a 43% over estimation of the produced energy. Moreover, a 1-year analysis of a hybrid system operation using real river flow data showed that the flow averaging period has a significant influence on the energy balance results. The system energy deficiency and surplus can be underestimated by approximately 25% by increasing the averaging time from day to month.

Wei and Liu [229] examined the integration of wind–solar–hydropower generation in enabling economic robust dispatch. Currently, the renewable energies including wind power and photovoltaic power have been increasingly deployed in power system to achieve contamination free and environmental-friendly power production. However, due to the natural characteristics of wind and solar, both wind power and photovoltaic power contain uncertainty and randomness which may significantly impact the stability, security, and economic efficiency of the conventional power system mainly consisted of hydropower and thermal power. To deal with the issue, this paper presented a two-stage robust model which was able to achieve the optimal day-ahead dispatch strategy in the worst-case scenario of wind and photovoltaic outputs. Because of the strong interactions between the two stages, the original optimization was decomposed into the day-ahead dispatch master problem and the additional adjustment subproblem considering the uncertainty and randomness of the wind and the photovoltaic outputs. Also, the piecewise linearization technique was employed to convert the original problem into a MILP problem. Afterward, the dualization of the additional adjustment subproblem can be obtained by using linear programming strong duality theory. Additionally, the Big-*M* method enabled the linearization of the dual model. The interacted iterations between the master problem and the subproblem were successfully implemented which can ultimately figure out the optimal day-ahead dispatch strategy of the power system with conventional and renewable energies. Based on the model decomposition and linearization, the original model can be converted into MP and SP, which are finally handled by C&CG approach.

Based on the experimental results, the contributions of the presented model were as follows: (1) The two-stage robust model was able to achieve the optimal day-ahead dispatch scheme under the worst wind and photovoltaic output scenarios, and the scheme proved to be economical and had great potential to improve the capability of the power system to accommodate wind and photovoltaic power; (2) hydropower can effectively handle the uncertainties caused by wind power and photovoltaic power, by considering the adjusting ability of the hydropower, the adjustment demand of the thermal unit can be significantly reduced by the model, and additionally, the wind power curtailment and PV power curtailment can also be potentially avoided; (3) The adjustable robust parameter enabled the performance adjustment of the optimized dispatch scheme by tuning their values, and as a result, the compromise between economy and robust could be achieved under different dispatching requirements.

Simão and Ramos [230] examined hybrid pumped hydro-storage energy solutions towards wind and PV integration for the improvement on flexibility, reliability, and energy costs. This study presented a technique based on a multi-criteria evaluation, for a sustainable technical solution based on renewable sources integration. It explored the combined production of hydro, solar, and wind, for the best challenge of energy storage flexibility, reliability, and sustainability. Mathematical simulations of hybrid solutions were developed together with different operating principles and restrictions. An electrical generating system composed primarily of wind and solar technologies, with pumped-storage hydropower schemes, was defined, predicting how much renewable power and storage capacity should be installed to satisfy renewables-only generation solutions. The study demonstrated that technically the pumped-storage hydropower system integrating other renewable sources is an attractive energy solution. The dynamic contribution of individual sources follows different patterns, due to the stability of hydro by pumping and random variability of other energy sources and the energy demand. Employing the three technologies in a complementary and balanced manner, the hybrid system could generate and store electricity at low cost, facing climate changes and reducing the footprint of electricity in a self-sufficient solution. Thus, a consistent multi-criteria framework was developed to optimize the availability and storage of renewable energy, selecting the best combination of peak factors to achieve the optimum solution in terms of efficiency, energy use, costs, and footprint. The optimization showed that in a hybrid solution, turbines and pumps can be used at the same time depending on the intermittency, availability, and optimized variables, which include different renewable sources, the storage capacity, and the load demand. The pumping system can be supplied by intermittent renewable sources when available and at the same time can be guaranteed a constant power production by hydraulic turbines. The three sources were combined considering different pump/turbine (P/T) capacities, PV solar power, and different reservoir volume capacities. The results demonstrate that technically the pumped hydro-storage with wind and PV is an ideal solution to achieve energy autonomy and to increase its flexibility and reliability. After selecting the best installation power for P/Ts, four scenarios were tested, changing the wind/solar powers and the water storage capacity. The results obtained show the process of selecting the best scenario is not straightforward, and it depends on the final goal.

Agarkar and Barve [231] examined hybrid wind and hydropower and hybrid wind–solar and hydropower system. The study found that hybrid (wind/hydro/solar) system was more economical and environmental friendly. The power-generation capacity of hybrid system was more than the power-generation capacity of the individual system. The results showing the power output of individual system and hybridized system obtained from metrological station for the period of 5 years 2010–2015 were used for comparison between the different hybrid systems.

3.8 POWER FROM HYDROKINETIC ENERGY FROM RIVERS

In Section 3.2, we examined power from small hydropower created by off-river hydro-dam. Here we examine hydrokinetic energy created in various ways by kinetic energy from river [232,233]. A hydrokinetic system is an electromechanical device that converts the kinetic energy of water flow into electrical energy through a generator and power electronics converter. Even though the output capacity is small, capacity can be increased by an array or modular installation [234,235]. In addition, a hydrokinetic system is based on free-flowing water without the construction of a reservoir or impoundment. The system is easy to transport and relocate due to the small size of the plant. Moreover, the system can be installed along the riverside either mooring to a fixed structure or on a floating pontoon [236]. Water currents have been used as sources of energy for over a century. One of the technologies using water flow is the watermill. The system consists of a waterwheel or water turbine to drive a mechanical process such as grinding, rolling, and hammering. These technologies have been installed at fast-flowing rivers for food, textile, and paper production, among other applications [232]. Electricity can also be generated using the flow of water.

3.8.1 Assessment of Technologies and Mechanisms

Based on the literature, energy harnessing from free stream rivers is attributed to Peter Garman, who developed the water current turbine (WCT) [232]. The WCT is used for water pumping and electricity generation in remote areas. In 1978, the Intermediate Technology Development Group (ITDG) developed the Garman turbine for water pumping and irrigation. During the early 1980s, a free rotor with 15 kW output power at 3.87 ms^{-1} water velocity was installed by the US Department of Energy for an ultra-low-head hydro-energy program as reported in R and H [237]. In 1986, the in-stream turbine with the straight blade Darrieus turbine was designed by Nova Energy Systems and ITDG. The system was able to harness 0.5 kW output power at a flow speed of 1 ms^{-1}. Experiments on the use of WCT for electricity generation and irrigation have been carried out in several countries, such as Canada [232], Zaire, and Australia [238]. The straight blade Darrieus turbine has been used in Canada and Africa with 5 and 15 kW output power, respectively. Australia developed the horizontal-axis Tyson turbine with the generator submerged under water. In 1990, the idea to manipulate WCT technology for large scales emerged [232].

Many of the technologies used to extract energy from the tides (or similar technologies) could be deployed in freshwater river systems rather than the saltwater ocean, effectively acting as very small run-of-river facilities. These "hydrokinetic" power-generation systems are individually small (each generating about 100 kW or less of power) and could be situated in two ways. First, a propeller-like or turnstile-like turbine could be deployed directly into the riverway, operating much like a small-scale tidal power system. Second, a "micro-hydro" type of system could be employed, where river water is channeled to a turbine housing via a channel or pipeline, as shown in Figure 3.19.

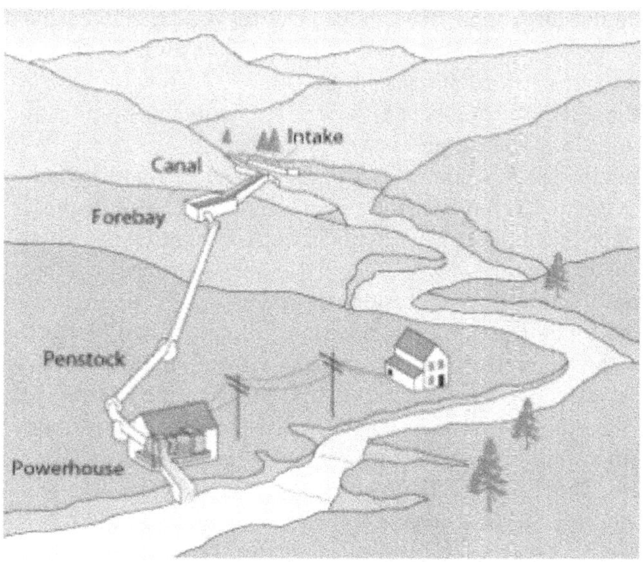

FIGURE 3.19 Schematic for a micro-hydro-system [239].

As one of the promising renewable energy technologies, the hydrokinetic energy conversion system (HECS) offers an economical and reliable option for remote and off-grid areas, compared to conventional hydropower. HECS does not require a head, large dam, or reservoir to operate, and a free stream velocity as low as 0.3 m/s is acceptable in order to rotate the small turbine [232]. Even though the capacity for power generation of hydrokinetic systems is small compared to conventional hydropower, using an array system or hydrokinetic farm, the capacity of HECS can be increased up to 100 MW [232]. Several studies have reported on hydrokinetic array systems. For example, Vennell et al. [240] proposed a design layout for macro–micro array turbines in HECS. The controller and details design for a modular hydrokinetic system connected to a smart grid was presented in Alvarez Alvarez et al. [234]. Most researchers, such as Behrouzi et al. [241], Kumar and Chatterjee [242], and Vermaak et al. [243], reported that a hydrokinetic system is similar to the wind turbine system in terms of concept, operation, and electrical hardware. Bahaj and Myers [244] indicated that because the density of water is 800 times that of air [232], with the water velocity between 2 and 3 ms^{-1}, a hydrokinetic system is able to generate four times the output power compared to a similarly rated wind turbine. In other words, the size of the hydrokinetic turbine could be much smaller than that of a WECS with the same output power.

The significant difference between HECS and WECS is the range of tip speed ratio (TSR). Ginter and Pieper [245] reported that HECS has a lower TSR than WECS. The optimal TSR for WECS is typically between 5 and 6. In contrast, the TSR value for HECS is less than 2.5 to avoid cavitation [232]. Romero-Gomez and Richmond [246] reported that HECS is less dependent on weather conditions compared to

WECS. While the direction and velocity of water are practically fixed and can be predicted reliably, wind speed and direction depend on air pressure and temperature, air turbulence, and earth rotation, and are difficult to predict [232]. Muljadi et al. [247] found that the level of mechanical stresses, inducing significant fatigue of the physical turbulence in the air and water, is similar for HECS and WECS. The turbine design and the use of a control strategy, such as maximum power point tracking (MPPT), pitch control, and robust control, are important for reducing the mechanical stress and fatigue due to turbulent effects in harsh marine environments.

A hydrokinetic system consists of a hydrokinetic turbine, a generator (PMSG), power electronics converter, and battery or grid-tie connection system. The flowing water is able to rotate the turbine at a certain velocity. The PMSG rotor is coupled to the turbine shaft directly without a gearing system, and the movement automatically turns the generator rotor. The output power from the PMSG is controlled and converted by the power electronics conversion system. In the stand-alone system, the variable AC (three-phase) system converts to the variable DC voltage through three-phase rectifiers. Then, the DC–DC converter converts the variable DC voltage into a constant DC bus voltage. In contrast, in the grid-tie connection system, an inverter is used to convert the constant DC bus voltage into AC power prior transporting it to the grid system. As pointed out by Khan et al. [248] and Lago et al. [249], a hydrokinetic system can be classified based on the energy conversion scheme and the working principle of the system. Generally, the system is classified into two classes based on conversion scheme as a turbine and a non-turbine system.

Conversion schemes using turbines, such as the horizontal axis, vertical axis, and cross-flow, are widely used in HECS [232]. According to Magagna and Uihlein [250], the horizontal-axis turbine has dominated almost 76% of the research and development into turbine design worldwide. In the horizontal-axis turbine, the rotational axis is parallel or inclined toward the direction of the flowing water. The advantage of a horizontal-axis turbine is that the turbine has a self-starting capability for slow water currents [232]. Nevertheless, the turbine clogs easily with debris in the river, and the cost of manufacturing is higher than that of the vertical-axis turbine. The vertical-axis turbine is also commonly used to extract the kinetic energy in the rivers [232]. The vertical-axis turbines have the rotor's axis of rotation at a right angle to the surface of the water [232]. This property means that vertical-axis turbines can do without a yawing device since it can handle incoming flows from any direction. Besides, the turbines are quieter in operation, and the mechanical complexity has been reduced. Furthermore, this type of turbine requires no gearing coupling, and the costs will decrease because of placement above water [232].

Gorlov helical turbine (GHT) is based on Darrieus turbine, and it is used for ocean and hydrokinetic energy. The standard Darrieus design, with foils parallel to the axis, produces a sinusoidal power cycle depending on the angle of the foils to the direction of the air or water. The rotor speed must therefore be controlled so that it doesn't remain at the resonant frequency of the foils. In addition, the flow must be at a high enough speed to start the turbine spinning. The GHT uses helical rather than strait foils. These foils then present a constant angle of attack, no matter what direction the flow is coming from. This eliminates the pulsing problem as well as allowing the turbine to spin at a lower flow velocity. Two GHTs which are installed in the free tidal flow of Uldolmok Strait (South Korea) began producing a combined 1 MW of electricity. A spherical form of the GHT was developed by Lucid Energy Technologies in 2009, for use inside water transmission pipelines. This is now being marketed as the *Northwest PowerPipe* by the Northwest Pipe Company and Lucid. These units take the energy from the change in elevation along the pipeline route and convert it to electricity. A 60'' pipe with a 12 ft head drop and 7 ft/sec. velocity can reportedly produce 60 kW, more than 500 MWH/year [232].

Turbines with a diffuser and augmentation channel are still the focus for research in the field [251–253]. Augmentation channels can increase the velocity of the water; this can in turn provide greater energy extraction [99]. The increase in pressure within the confined area in the augmentation channel leads to an increase in the velocity of the flow. If the turbine has been placed on a channel, the velocity around the rotor will be higher than that of the free rotor. Different terms are used widely to represent the augmentation channels, including ducts, shrouds, wind lenses, nozzles, concentrators, or diffusers, all used synonymously [254]. The Betz limit does not apply to turbines with augmentation channels. Nevertheless, this limit is dependent on the inlet–outlet pressure gradient as well as the volume of flow through the duct. This factor is dependent upon the duct's shape and the duct-turbine area ratio [255].

The design of channels can be categorized into two types, namely hybrid and diffuser types.

Hybrid-type augmentation is suitable for vertical-type turbines while the diffuser type is more suitable for horizontal-axis turbines. Several groups have produced systematic reviews and analyses regarding diffuser-augmented turbines. For example, Nunes et al. [256] found that the TSR has an approximately 90% narrow operational interval by using the diffuser augmentation on a horizontal-axis turbine. Wong et al. [257] showed that the maximum output power increases dramatically using the vertical-axis turbine augmentation system. Bontempo and Manna [258] proposed diffuser augmentation based on a divided duct surface, into an internal and external part, and developed axial momentum theory approach which is an extended version of the free-wake ring-vortex actuator-disk model. Wake studies also constitute a research trend in hydrokinetic systems. For example, Lust et al. [259] presented a survey of the near wake of the horizontal axis in the presence of surface gravity waves. Dou et al. [260] proposed a wake model to predict the turbine wake in a yawed condition. The wake measurement data from Guerra and Thomson [261] can be used for numerical validation and hydrokinetic turbine array design. The interaction between the turbine wakes and sediment has also been investigated by Musa et al. [262].

The cross-flow turbine has an orthogonal rotor axis with respect to the flow of water but parallel with reference to the surface of the water [263]. It is economical in terms of space, and the rectangular swept area increases the output power. It can be operated at a lower speed which reduces the possibility of cavitation, produces less noise, and is safer for marine animals [232]. The cross-flow turbine can operate without the yawing mechanism, similar to vertical-axis turbine [232]. As noted in Saini and Saini [264], the configuration of a cross-flow turbine can be classified into three groups based on lift force, drag force, and combination of lift and drag forces. ORPC successfully installed the RivGen in a remote Alaskan village. The company also installed the first grid-connected hydrokinetic system to harness tidal energy, using the TidGen at Eastern Maine in 2012 [232]. The venturi turbine which is based on funnel-like devices can be applied at low water velocity with shallow water depth [232]. It will increase the water velocity and decrease the pressure subsequently driving a turbine. On the other hand, the vortex turbine requires a round basin with central drain, and it is able to generate power at low-head and low flow rate using gravitational vortices [232].

A non-turbine system can also be used to extract power from marine, river, or open channel flows. This system mimics the motion of animals. The flapping foil consists of a series of sails that are connected and rotate in a rectangular motion. As the water flows through the device, the sails produce a lift force perpendicular to the water flow that is able to power the generator [265]. Wang et al. [266] classified a non-turbine system based on flow-induced vibration (FIV) energy harvesting into four categories such as vortex-induced vibration (VIV), buffeting, galloping, and flutter. The VIVACE converter utilizes VIV, galloping, and flow-induced motions (FIM). The early model of the VIVACE converter was a combination of a physical spring, damper, and generator [232]. The latest VIVACE is more complicated, with a cylinder, a belt, pulley transmission, a generator, and a controller to control the damping and spring forces. The flutter flag has a two-layer piezoelectric polymer PVDF with an electrode sandwiched in between [232]. The differential pressure around the flag results in bending and will activate the charge separation inside the piezoelectric materials to produce the energy.

Yuce and Muratoglu [99] classified existing hydrokinetic technology according to the principle of operation. Hydrokinetic systems are divided into current energy converter (CEC) systems and WEC systems. River current energy conversion systems (RECSs), tidal-in stream energy converters, and marine current turbines (MCTs) are placed under CEC. Oscillating water columns (OWC), overtopping devices, and wave-activated bodies fall under WEC. According to Niebuhr et al. [267], new developments will expand these classifications. The RECSs employ the same principle as the tidal systems, but with lower output power, and are suitable for remote communities. The systems are based on floating structures and are placed at river channels. Smart hydropower has developed two types of river turbines: the Smart Monofloat and the Smart Free Stream, both with debris protection and a 5 kW underwater generator. Smart Monofloat has a diffuser system to increase the velocity of the water. The Smart Free Stream is very reliable and requires almost no maintenance. The turbine is installed on a river-bed or canal with the slightly curved blades to reduce debris effects.

The Waterotor can produce a high-energy output while operating in shallow waters with low flow speeds [232]. The system can extract the energy at as slow as 2 mph flow consistently when submerged in rivers or canals. This device has no blades, it is safe for aquatic life [268], and it can either be suspended from buoys or anchor it to the seafloor. The Idénergie is a novel form of subwater electricity generator which has high efficiency even at low water velocity, and with the fully sealed housing, the generator is able to produce more than 500 W continuously [232].

3.8.2 Assessment of Potentials and Commercialization

New York University conducted a US DOE-funded in-stream resource assessment for rivers systems in the USA, with the final report published in August 1986 [269].

This resource assessment only considered rivers with volumetric flow rates greater than 113.3 m³/s (4,000 ft3/s) and mean velocities greater than 1.31 m/s (4.3 ft/s). For river sections that met these criteria, this study assumed that 1/16 of the rivers' cross sectional area was occupied by turbine arrays, with each device having an efficiency of 40%. Within these arrays, this study assumed that each device had a diameter equal to 80% of the water depth and that devices were separated by a very small margin across the flow and 5 diameters downstream. Using these assumptions, the average estimated power potential from all US rivers is 12,600 MW. This study also estimated the average power potential for 16 distinct regions within the USA with an estimated 4,500 MW for Alaska; 3,200 MW for the Northwest USA including northern Washington State, northern Idaho, and western Montana; and up to 1,500 MW in each of the other regions.

Assessment of potential river flow contributions to electrical power generation was also obtained from the USGS National Hydrography Dataset provided with the National Map enhancements [270] and various sources on Alaskan rivers. For rivers with mean flows of over 93 m³/s (1,000 ft³/s) and water depths greater than 2 m (95% of the time), the theoretical power general potential was 1,381 TWh/year for the continental USA. The technically recoverable fraction of this potential was estimated at 120 TWh/year (an average power of 13,700 MW, 9% greater than the New York University study) by using the Corps of Engineers' Hydrologic Engineering Canter's River Analysis System (HEC-RAS) numerical hydrodynamic model [271] to account for flow statistics, maximum number of devices, device efficiency, etc., and an empirical recovery coefficient [272].

The first in-stream hydrokinetic device deployed on a river system in the USA was deployed in Ruby, Alaska, for 1 month in 2008, and briefly redeployed in 2009 and 2010 [273]. This 5 kW in-stream hydrokinetic device was developed by New Energy Corporation to harness electricity from the free-flowing Yukon River and to test the viability of using a hydrokinetic generator to offset the high-cost diesel fuel used to power the community's electrical grid [273]. This vertical-axis turbine was deployed from a moored catamaran barge, and the power was transmitted to shore using a cable laid on the river bed. While this test was in large part successful, it did experience significant problems caused by in-water debris.

A 25 kW New Energy Corporation turbine was also briefly installed in the Yukon River in 2010, this time near Eagle, Alaska [273]. Similar to the installation in Ruby, Alaska, this vertical-axis turbine was deployed from a moored catamaran barge and the power was transmitted to shore using a cable laid on the river bed. While this system was successful, the turbine and the generator power cable were damaged in two successive operations due to very heavy debris. The City of Hastings, Minnesota, in conjunction with Hydro Green Energy, began operating an in-stream hydrokinetic project in December 2008 that can generate up to 250 kW of power [274]. This was the first commercially operated, FERC-licensed in-stream hydrokinetic power facility in the USA [274]. This in-stream hydrokinetic project utilized a shrouded horizontal-axis turbine that is rigidly attached to a fixed structure. This project operated until 2012 [275].

Free Flow Power Corporation operated its first full-scale hydrokinetic turbine generator in the Mississippi River starting 20 June 2011. The 3m-diameter shrouded horizontal-axis turbine, rated at 40 kW, was installed on a floating research platform in Plaquemine, Louisiana [276]. Verdant Power installed and operated a tidal energy project in the East River in New York City that used six turbines and delivered 70 MWh of energy to end-users during the 2-year period 2006–2008. These horizontal-axis turbines were attached to vertical monopiles and were completely submerged during operation. To optimize performance, these systems rotated with tidal change to align themselves with the direction of the flow. On January 03, 2012, FERC issued Verdant Power a pilot project license for their 1 MW, 30-turbine RITE Project, which was the first commercial license for a tidal project in the USA [277].

3.9 POWER FROM OCEAN—MARINE ENERGY

Marine energy or *marine power* refers to the energy carried by ocean waves, tides, salinity, and ocean temperature differences. The movement of water in the world's oceans creates a vast store of kinetic energy, or energy in motion. Some of this energy can be harnessed to generate electricity to power homes, transport, and industries. The term marine energy encompasses both wave power, that is, power from surface waves, and tidal power, that is, obtained from the kinetic energy of large bodies of moving water. Offshore wind power is not a form of marine energy, as wind power is derived from the wind, even if the wind turbines are placed over water. The oceans have a tremendous amount of energy and are close to many if not most concentrated populations. Ocean energy has the potential of providing a substantial amount of new renewable energy around the world.

There is the potential to develop 20,000–80,000 terawatt-hours per year (TWh/y) of electricity generated by changes in ocean temperatures, salt content, movements of tides, currents, waves, and swells (see Table 3.5). Indonesia, as an archipelagic country that is three-quarters ocean, has 49 GW recognized potential ocean energy and has 727 GW theoretical potential ocean energy.

TABLE 3.5
Global Potential of Marine Energy [278]

	Global Potential
Form	**Annual Generation**
Tidal energy	>300 TWh
Marine current power	>800 TWh
Osmotic power salinity gradient	2,000 TWh
Ocean thermal energy thermal gradient	10,000 TWh
Wave energy	8,000–80,000 TWh

Source: IEA-OES, Annual Report 2007 [3].

3.9.1 Power from Salinity Gradient

In many situations within an ocean such as areas where the ocean and river meet, two bodies of water have salinity differences and the resulting salinity gradient induces energy. It is estimated that there are approximately 3.1 TW of salinity gradient energy throughout the globe [278,279]. Salinity gradient energy can be extracted through three different methods. One method is the pressure-retarded osmosis which uses semi-permeable membranes to transport water from the river to the sea which will cause an increase in static energy that is used to power the turbine. This method is not very reliable because currently membranes needed for pressure-retarded osmosis on a commercial scale [280,281] are not available and need more research. Some types of membranes used in this process are cellulose acetate, polybenzimidazole, and poly(amide-imide) due to their increase in water flux, strength, and resistance against wear [278–282].

Another method of extraction is reverse electro-dialysis which pumps the saltwater and freshwater into membranes filled with anions and cations. This produces an electrochemical potential which produces a current. Like the pressure-retarded osmosis, reverse electro-dialysis also depends on the properties of the membranes. Recently, it was discovered that by improving the properties of the membrane, scientists were able to improve the efficiency of reverse electro-dialysis [278–282]. The last method is an electric double-layer capacitor which stores the charges in the saltwater, transports them into the freshwater, and allows the charged ions to diffuse which creates electrostatic energy [278–282]. The introduction of ions from the saltwater into the freshwater may harm the marine life in freshwater as they are not used to the charged ions from the saltwater.

Although these approaches have been attracting considerable attention for some decades at a research level, their relatively high costs and low efficiency still inhibit their commercial development. However, in the past few years, new nanomembranes have been found to be capable of generating voltage effectively when an electrolyte solution flows through narrow channels driven by a pressure gradient. For example, Siria et al. [280,282] reported membranes made of boron nitride nanotubes that can produce power with a density of several kilowatts per square meter; Feng et al. [279] reported a power density as high as 10^3 kW/m^2 when using MoS_2 nanomembranes. Surface charges play a key role in achieving such osmotic energy conversion, and this can be explained by the fact that charged solid surfaces induce the formation of an electric double layer at the interface with aqueous solutions. The effect can be modulated by other surface properties, including surface structures of the materials, surface friction, and substrate hydrophobicity. Because water in these cases is flowing, playing with fluid dynamics by designing specific surface morphologies and chemistries can help to maximize the osmotic energy harvesting, as explained by Siria et al. [282,280] in their review.

3.9.2 Ocean Thermal Gradient Conversion to Power

The upper layer of the ocean absorbs the energy from the sun and is used in the ocean thermal energy conversion (OTEC) cycles. The OTEC cycles contain the warm

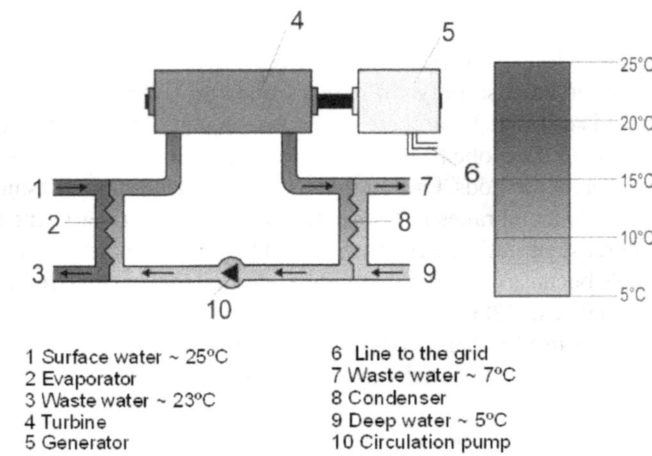

FIGURE 3.20 Diagram of a closed cycle OTEC plant [283].

seawater from the upper layers of the ocean and the cold seawater from the bottom layers of the ocean. The energy can be extracted from these cycles using open-cycle, closed-cycle, and hybrid-cycle techniques [283].

Closed-cycle systems (see Figure 3.20) use fluid with a low boiling point, such as ammonia (having a boiling point around –33°C at atmospheric pressure), to power a turbine to generate electricity. Warm surface seawater is pumped through a heat exchanger to vaporize the fluid. The expanding vapor turns the turbo-generator. Cold water, pumped through a second heat exchanger, condenses the vapor into a liquid, which is then recycled through the system. In 1979, the Natural Energy Laboratory and several private-sector partners developed the "mini OTEC" experiment, off the Hawaiian coast and produced enough net electricity to illuminate the ship's light bulbs and run its computers and television.

Open-cycle OTEC (see Figure 3.21) uses warm surface water directly to make electricity. The warm seawater is first pumped into a low-pressure container, which causes it to boil. In some schemes, the expanding vapor drives a low-pressure turbine attached to an electrical generator. The vapor, which has left its salt and other contaminants in the low-pressure container, is pure freshwater. It is condensed into a liquid by exposure to cold temperatures from deep-ocean water. This method produces desalinized freshwater, suitable for drinking water, irrigation, or aquaculture. In other schemes, the rising vapor is used in a gas-lift technique of lifting water to significant heights. Depending on the embodiment, such vapor-lift pump techniques generate power from a hydroelectric turbine either before or after the pump is used. In May 1993, an open-cycle OTEC plant at Keahole Point, Hawaii, produced close to 80 kW of electricity during a net power-producing experiment.

A hybrid cycle combines the features of the closed- and open-cycle systems. In a hybrid, warm seawater enters a vacuum chamber and is flash-evaporated, similar to the open-cycle evaporation process. The steam vaporizes the ammonia working

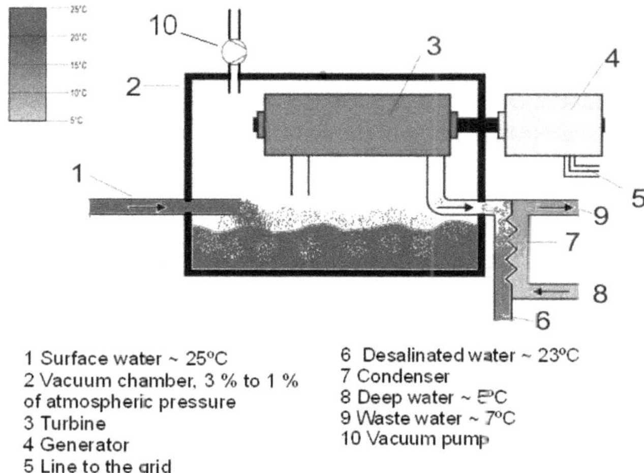

FIGURE 3.21 Diagram of an open cycle OTEC plant.

fluid of a closed-cycle loop on the other side of an ammonia vaporizer. The vaporized fluid then drives a turbine to produce electricity. The steam condenses within the heat exchanger and provides desalinated water [283].

Major issues with this concept are efficiency, cost, and long-term sustainability. More demonstration of the concept is needed.

3.9.3 Power from Tidal Energy

Water in the oceans is constantly in motion due to waves and tides, and energy can be harvested from these kinds of motions. Waves, driven by the winds, make the water oscillate in roughly circular orbits extending to a depth of one-half of the wavelength of the wave (distance between peaks). Tides, related to the gravitational pull of the Moon and Sun on the oceans, are like very long-wavelength waves that can produce very strong currents in some coastal areas due to the geometry of the shoreline. In terms of power-generation technologies, wave power and tidal power have both similarities and differences. Both refer to the extraction of kinetic energy from the ocean to generate electricity (again, by spinning a turbine just as hydroelectric dams or wind farms do), but the locations of each and the mechanisms that they use for generating power are slightly different. Furthermore, tide and current are not the same. Tide is the vertical rise and fall of the water, and tidal current is the horizontal flow. In simple words, the tide rises and falls, and the tidal current floods and ebbs. As mentioned earlier, the principle of tidal forces is generated by the Moon and Sun. The Moon is the main tide-generating body. Due to its greater distance, the Sun's effect is only 46% of the Moon's. To extract energy from these tides, hydrokinetic turbines such as vertical-axis turbines, horizontal-axis turbines, the venturi effect turbines, the oscillating hydrofoil, and cross-flow devices are used. These tidal current turbines use the kinetic energy of the tides. In the vertical-axis turbines, the turbine blades rotate at an angle perpendicular to the flow of water, whereas in the horizontal-axis turbine, the turbine blades rotate at an angle parallel to the flow of water. In the tidal current turbines, the flow of the water causes the blades to move which turns the generator on through the gearbox. Horizontal-axis turbines are more efficient, easier to maintain as it is self-starting, and cheaper than vertical-axis turbines [6].

There is a difference between ocean currents and tidal currents. Ocean currents are caused by the disturbances of winds on the ocean and the circulation of the ocean. Tidal currents are a result of the movements of the tides. The energy from the tidal currents is usually found between the seabed and the sea surface and can be extracted using tidal CECs. Ocean currents are found to be more consistent than tidal currents because ocean currents move in one direction whereas tidal currents can change directions because of the changes in flood and ebb cycles. Since the tidal currents change directions, it requires more energy which means tidal currents contain more energy than ocean currents. This might explain why ocean currents are generally slower than tidal currents. It was discovered that when tidal currents flow in the same direction as the wave, the amplitude is decreased which decreases the tidal energy. Additionally, when the tidal current flows in the opposite direction of the wave, the amplitude of the wave increases which increases tidal energy [8]. Therefore, to increase the yield of electricity, the tidal energy from the tidal current that flows in the opposite direction of the waves should be used in the conversion of tidal energy into electricity.

There are also three types of tides: diurnal, semidiurnal, and mixed [284]. Tidal energy is one of the new and evolving technologies, which is as yet commercially not viable and still to a large degree in Research & Development (R&D) stage. Tidal energy is inexhaustible and can be considered as a renewable energy source [285]. Unlike solar and wind energy, it is less vulnerable to climate change [286]. Most of the existing technology used for tidal energy conversion is derived from the wind power industry [287–289]. Extreme tides are found in many locations across the globe. Tidal energy projects typically work by forcing water through a turbine or a "tidal fence" that looks like a set of subway turnstiles. The systems depend on regular tidal activity to generate power. Because this tidal activity is predictable (each coast sees at least one tidal cycle per day—high tide and low tide—and some areas actually see two tidal cycles on a daily basis), tidal energy projects have the advantage of being able to provide a fairly predictable source of electricity. The use of tidal power, globally, has been quite limited because there are only a few sites in the world that see sufficiently large variations in tides to produce enough power, as shown in Table 3.6.

Tidal power offers numerous advantages. It produces no liquid or solid pollution. It has little visual impact. Tidal power exists on a worldwide scale from deep-ocean waters. Tidally driven coastal currents provide an energy density four times greater than air, which means that a 15-m-diameter turbine will generate as much energy generated by a 60-m-diameter windmill. Finally, tidal currents are both predictable and reliable, a feature which gives them an advantage over both wind and solar systems. Power outputs can be accurately calculated far in advance, allowing for easy integration with existing electricity grids. There are, however, several drawbacks

TABLE 3.6
Tidal Ranges in the Different Areas of the World [290]

Tidal ranges in different areas of the world

Country	Site	Tidal Range (m)
Canada	Bay of Fundy	16.2
England	Severn Estuary	14.5
France	Port of Granville	14.7
France	La Rance	13.5
Argentina	Puerto Rio Gallegos	13.3
Russia	Bay of Mezen	10.0
Russia	Penzhinskaya Guba	13.4
USA (Alaska)	Turnagain Arm	9.2
USA (Alaska)	Cook Inlet	7.6

of tidal power. It involves high cost of construction, installation, and generation. Barrages can disrupt natural migratory routes for marine animals and normal boating pathways. Turbines can kill up to 15% of fish in area; although technology has advanced, the turbines have to move slow enough not to kill many. It can cause flooding and ecological changes. Finally, the research on tidal power is still in the initial stages. The most important variables generally considered for tidal power are as follows [290–295]:

1. **The local water depth**. Existing device technology concepts are generally limited to operational water depths of 25–45 m.
2. **The location of the nearest exploitable grid connection**. For an immature industry, the economics of tidal energy extraction require easy access to a nearby grid connection with spare capacity; otherwise, the capital cost cannot be viably recouped across the life of the project.
3. **An energetic and persistent resource**. Large mean spring and neap tide velocities are highly desirable. Some sites have the added advantage of minimizing the low velocity periods of the tidal cycle as the local dynamics ensure that the tidal flow reverses through the slack period at an accelerated rate. The sites that the developers are interested to extract energy tend to have peak spring tidal velocities of 3+ m/s.

If these three primary criteria are met, a site is considered to have solid potential for future development. The majority of coastal locations can be rejected out of hand by consideration of just these three variables [296]. In order for the tidal power to be a fully viable commercial enterprise: (1) The tidal energy industry has to develop a new generation of efficient, low-cost, and environmentally friendly apparatus for power extraction from free or ultra-low-head water flow; (2) since the negative environmental impacts of tidal barrages are not well understood at this time, it is important to consider the influence of energy extraction while estimating the available energy from a potential tidal energy site; (3) the method for extracting energy from tidal streams needs to approach the design convergence, horizontal-axis turbines have shown to be the most employed technologies, and alternative designs include vertical-axis turbines and tidal kites; and finally (4) the technology needs to be tested with the deployment of full-scale devices and first arrays in real sea conditions.

In order to realize the benefits of tidal energy on a commercial scale, it will be important for researchers to identify new technologies and methods that significantly lower installation and maintenance costs, reduce environmental effects, and increase the suitability of more locations. There are a few tidal projects in operation; however, the industry is growing slowly due to barriers to entry and lack of supply chain.

Tidal turbines can be installed in places with strong tidal activity, either floating or on the seafloor, individually or in arrays. They look and operate much like wind turbines, using blades to turn a rotor that powers a generator, but must be significantly more robust given their operating environment and, as tidal turbines are much smaller than large wind turbines, more turbines are required to produce the same amount of energy. Multiple tidal demonstration projects are under way in the USA. Turbines placed in tidal streams capture energy from the current, and underwater cables transmit it to the grid. Tidal stream systems can capture energy at sites with high tidal velocities created by land constrictions, such as in straits or inlets. When fully operational, the MeyGen project in Scotland will be the largest tidal stream generating station in the world, with up to 398 MW generation capacity.

Another way to convert tidal energy into electricity is through tidal barrages, which is also known as a dam and deals with the tide's potential energy. Single-basin tidal barrages use three methods to convert tidal energy. One method is the ebb generation which traps the water in the basin during high tide and the water flows through low-head turbines which are then used to generate electricity. Another method is the flood generation method which uses a hydrostatic head that controls when the water can be allowed to flow through the turbine and into the basin. The last method is the two-way generation which combines the ebb generation and the flood generation methods. In a two-way generation, the water flows through the turbines when it is high tide or when the flood cycle is ending. Double-basin tidal barrages use the ebb generation method whereby only the water travels through the first basin at high tides and is stored at the second basin. In 2009, there were four tidal plants that were reliable in generating energy. Tidal barrages are like dams built across tidal rivers, bays, and estuaries to form a tidal basin. Turbines inside the barrage enable the basin to fill during incoming tides and release through the system during outgoing tides, generating electricity in both directions. It operates much like a river dam in capturing the power in surrounding water. Two of the world's largest tidal power stations are barrages in South Korea and France, with 254 MW and 240 MW electricity generation capacities, respectively. The next largest in Canada has much lower generation capacity at 20 MW.

Tidal lagoons are like barrages in using man-made retaining walls to partially contain a large volume of incoming tidal water, with embedded turbines to capture its energy. They also rely on a large tidal range to generate power. Unlike barrages, tidal lagoons could be placed along natural coastline for continuous power generation as the tide changes and designed to minimize their environmental footprint.

Though the energy output from tidal lagoons is unproven, with no current examples in operation, a few are under development in China, North Korea, and the UK. Due to the environmental challenges they pose, tidal barrages and lagoons are not the focus of tidal energy development efforts in most areas of the world. The predominant application for tidal energy has been the generation of electricity for use on shore via the national power grid. There is also potential value in tidal energy to serve the needs of other existing or emerging ocean industries (e.g., aquaculture, ocean mineral mining, oceanographic research, or military missions), as captured in DOE's Powering the Blue Economy initiative. The "blue economy" is defined as the sustainable use of ocean resources for economic growth, improved livelihoods, and jobs, while preserving the health of ocean ecosystems.

As technology advances, the production of energy from tidal waves will continue to increase. One of the main challenges due to these tidal plants is that the electromagnetic fields that are generated from these tidal plants negatively affect the growth of marine animals [232,290–295]. Tidal power is created using a head difference between two bodies of water. To create this difference, a wall separates the two water bodies. As the tide flows in or out, the wall blocks the flow of the tide and generates this head difference. When the head difference has reached the optimum design level, the water is forced to pass through holes in the wall, where turbines are placed to generate power. Since two tidal cycles occur per day, this head difference develops four times each day (in one cycle, the tide comes in and out).

3.9.3.1 Assessment of Commercial Potentials and Operations

The first tidal plant was built in 1967 in Brittany's Rance Estuary with a capacity of 240 megawatts [297]. In 2011, the biggest tidal plant was built in South Korea and has a capacity of 254 MW. Turbines convert the water energy into mechanical energy, and then into electricity by means of an electric generator. Since tidal head differences generally measure a few meters, typical turbines include waterwheels, Archimedes screws, and bulb turbines. One other way to generate power from tides uses turbines on the seabed, driven by the kinetic energy of the moving tidal flow, similar to wind turbines in airflows. Tidal power projects exist in France, South Korea, Russia, the UK, and China, among other places. Researchers have predicted that UK has the capability to produce over 20% of its electrical needs from its tidal resources [298]. Studies have shown that the European territorial waters have 106 locations for extracting tidal energy that would provide electricity of 48 TW/year. It is estimated around 50,000 MW of installed capacity being achievable along the coasts of British Columbia alone. There are greater predictions of extracting energy of about 90,000 MW off the northwest coast of Russia and about 20,000 MW at the inlet or Mezen river and White Sea. There are also estimations along the west coast of India having potential to generate 8,000 MW. There are estimates that the energy that can be globally extracted is around 1,800 TWh/year [286]. The main characteristics of four large-scale tidal power plants that were constructed after World War II and currently exist are given in Table 3.7.

Hydrokinetic technology in tidal and marine settings has been rapidly emerging since the early 1990s. In early development, the underwater electric kite was developed by UEK Corporation in the USA with the diffuser-augmented solid pontoon

TABLE 3.7
Existing Large Tidal Power Plants [290,299]

Site/Country	Bay Area (km^2)	Avg. Tide (m)	Installed Power (MW)
La Rance, France	22	8.55	240
Kislaya Guba, Russia	1.1	2.3	0.4
Annapolis, Canada	15	6.4	18
Jiangxia, China	1.4	5.08	3.9

[300]. The most significant success story of tidal energy comes from MCT Ltd. In the late 1990s, MCT started the Seaflow Project that was financed by the UK DTI, the European Commission, and the German government. In 2003, the Seaflow was installed and rotated for the first time with 300 kW output power. By November 2005, Seagen launched a twin-rotor turbine with capacity of more than 1,000 kW output power [301].

Verdant Power (see Figure 3.22) was the first company to acquire a commercial license for a tidal power project in the USA. From 2006 to 2009, the company tested six full-scale prototypes at the East River in New York City. Verdant Power has advanced the kinetic hydropower system to the 5th generation (Gen5) based on operational experience gained from the Roosevelt Island Tidal Energy (RITE) project [302]. Scot renewables Tidal Power Limited launched SR250, a 250 kW prototype of a large floating tidal turbine in 2011. In 2016, the company successfully launched the 2 MW SR2000, which is the world's largest tidal energy converter [232]. The company claims that the floating tidal turbine can sustain 20 years of operation in a harsh marine environment, in contrast to a bed-mounted system.

Open Hydro Canada was established in 2014 to commercialize tidal technology. Several projects have been carried out successfully, such as 4 MW tidal array at the Bay of Fundy, Nova Scotia, Canada, and 100 MW tidal farm at Antrim Coast, Northern Ireland in 2012. Open Hydro's design philosophy is to keep the turbine as simple as possible to reduce build and maintenance costs [303]. The RER Hydro TREK (Kinetic Energy Recovery Turbine) [304] is a ducted, multistage turbine. There are three rows of blades, in which the first and last rows acting as stators [305]. The TREK has been in full-scale testing since 2010. In 2012, RER Hydro-partnered with Boeing, giving Boeing the rights to sell and market the RER hydro-technology. The Nautricity Cormat consists of two rows of contra-rotating blades and is moored by a single point at the front of the floating turbine. In this design, the turbine can align with the flow stream passively.

Tidal energy as an industry remains limited by a few significant barriers, cost being its most challenging. Developing tidal arrays and connecting them to the power grid requires extensive and costly engineering and manufacturing work. While there are numerous tidal technologies being tested that may improve affordability, none have emerged as a market leader that could help establish supply chains and begin reducing installation and maintenance costs. Tidal energy technologies have been slow to develop, and some industry participants have exited the market. Suitable

FIGURE 3.22 Verdant Power Company is a US tidal energy developer with turbines in the East River of New York. Pictured here is one of their tidal devices. (Photo: Tethys Engineering. Public Domain [295].)

locations for tidal energy facilities are inherently limited, given that not all coastal bays and tidal channels experience the conditions required for effective power generation. Also, among those limited locations, some are not near the grid, requiring further investment to install lengthy undersea cables for transmitting generated electricity.

In addition to cost and geographic limitations, there is also significant concern about environmental effects. Constructing and operating tidal energy arrays based on massive underwater structures may change the ambient flow field and water quality, as well as negatively affect sea life and their habitats, potentially threatening collisions by marine animals and fish with rotating turbine blades and affecting marine animal navigation and communication with underwater noise. This may cause some sensitive species to shy away from electromagnetic fields from power cables or changes to their habitats. Achieving cost reductions, developing devices that can endure ocean forces, and minimizing environmental effects to improve tidal energy's commercial viability are and must be the primary focus of research investments in this area.

Some tidal plants in development are in Russia, South Korea, India, and the Philippines. To support these projects, tidal power engineers are refining existing technologies, with a special focus on the improvement of turbines. Modified bulb turbines with an additional set of guide vanes allow better management and control of the flow through the turbine. Bulb turbines can reach very high power output. Two examples include the 7-m-diameter bulb turbines in the Swansea Bay tidal lagoon that can produce 16 MW and the innovative Straflo turbines can reach values of 20 MW, as in the Annapolis plant.

Archimedes screws could be employed as a fully submerged tidal stream device, or they can be enclosed in a pipe system. Archimedes screws also have the additional advantage of "fish friendliness." Finally, waterwheels under testing use inflow hydraulic structures to better control flow and power. Tidal power arrays of varying sizes are being developed or have been deployed recently around the world, with much focus on energy generation from tidal streams or currents. A tidal stream array located in the Pentland Firth in Scotland—the body of water between the Scottish mainland and the northern islands—is the newest to begin operating and is the first of its kind. The MeyGen tidal energy project began phased operations in 2018, and its first four turbines had generated and delivered more than 35 GWh of power to the grid by the end of 2020. At full deployment, 61 turbines submerged on the seabed will generate up to 400 MW of energy from high-speed currents in the area.

There are multiple projects under way in Wales, an emerging hot spot for the industry. There are also other test sites and technology deployments at various stages in countries including Scotland, France, Japan, Korea, China, Canada, and the USA as developers bring forward new and improved tidal current technologies that show promise for clearing key hurdles to commercial viability. The ability to assess the performance and environmental effects of new technologies in real sea conditions is critical to sustainable industry advancement. Engineers are working to improve tidal energy generation technologies to increase their energy production efficiency, reduce biofouling, decrease their environmental effects, and find a path to commercial profitability. Nova Innovation's MECmate is a wet-mate connection solution specifically designed for marine energy converters (MECs). The project builds on its proven Nova Can technology currently in operation in Nova's Shetland tidal array. This program will develop technology that will reduce the cost of wave and tidal energy and ultimately help marine energy play a part in Scotland's net-zero future. The overall aim of the quick connection systems program is to reduce the duration, cost, and risk of offshore operation for wave-energy convertors by supporting projects to design and develop quick connection and disconnection systems between devices and moorings and/or electrical systems. The Korea Institute of Ocean Science and Technology (KIOST) has contracted EMEC to support the development of its tidal energy projects test site at Jang-Juk Strait, near Jindo Island, Korea. KIOST is developing a grid-connected tidal energy test site development on the Jang-Juk Strait offshore southwestern Korea. The Korea Tidal Current Energy Centre (K-TEC) site has about 4.5 MW grid capacity.

In the USA, the tidal energy locations with the largest power potential can be found in northern locations such as Maine, Puget Sound in Washington State, and Alaska [232,290–295]. Detailed information on US tidal resources is available from the Georgia Tech [306] with the predictions in this report made using the ocean circulation numerical model (ROMS). The hotspots in this study are defined as areas where the annually averaged power density is at minimum 500 W/m^2, the water depth is at least 5 m, and the surface area where these two criteria are met is over 0.5 km^2. Nationally, 151 hotspots are identified in this study, and this number would have been more than triple if the size criterion were relaxed [306]. The study also calculated total theoretical available power estimates. Using this approach, the national available energy from tidal streams is calculated at 50 GW, with 47 GW of this power potential located in Alaska [306]. In Alaska, Cook Inlet has the maximum calculated available power (18 GW), and Chatham Strait has 12 GW [306]. Following Alaska, other US states have the following estimated power potential: Washington (683 MW), Maine (675 MW), South Carolina (388 MW), New York (280 MW), Georgia (219 MW), California (204 MW), New Jersey (192 MW), Florida (166 MW), Delaware(165 MW), Virginia (133 MW), Massachusetts (66 MW), North Carolina (66 MW), Oregon (48 MW), Maryland (35 MW), Rhode Island (16 MW), and Texas (6 MW) [306]. The report also identifies the specific locations where these resources are located within each state.

The Ocean Renewable Power Company became the first company to generate electricity from Bay of Fundy tidal currents without the use of dams in 2008, during a yearlong program of in-water testing. In 2010, they operated a precommercial version of their device, which was the largest in-stream hydrokinetic device to ever be deployed in the USA—a power-generation capacity of 150 kW. These systems, which were deployed from custom barges, were both horizontal-axis turbines designed to operate with the flow perpendicular to the rotor axis. In September 2012, an Ocean Renewable Power Company turbine deployed in Cobscook Bay became the first tidal energy system to deliver power to the US electrical grid [307].

3.9.4 Power from Ocean—Wave Energy

Wind waves are mechanical waves that propagate along the interface between water and air; the restoring force is provided by gravity, and so they are often referred to as surface gravity waves. As the wind blows, pressure and friction forces perturb the equilibrium of the water surface. These forces transfer energy from the air to the water, forming waves. In the case of monochromatic linear plane waves in deep water, particles near the surface move in circular paths, making wind waves a combination of longitudinal (back and forth) and transverse (up and down) wave motions. When several wave trains are present, as is always the case in nature, the waves form groups. In deep water, the groups travel at a group velocity which is half of the phase speed. Following a single wave in a group, one can see the wave appearing at the back of the group, growing, and finally disappearing at the front of the group.

As the water depth decreases toward the coast, this will have an effect: Wave height changes due to wave shoaling and refraction. As the wave height increases, the wave may become unstable when the crest of the wave moves faster than the trough. This causes surf a breaking of the waves. The movement of wind waves can be captured by wave-energy devices. The energy density (per unit area) of regular sinusoidal waves depends on the water density Ò, gravity acceleration *g*, and the wave height *H* (which is equal to twice the amplitude, *a*): The velocity of propagation of this energy is the group velocity. Waves are generated by wind passing over the sea surface. As long as the waves propagate slower than the wind speed just above the waves, there is an energy transfer from the wind to the waves. Both air pressure differences between the upwind and the lee side of a wave crest and friction on the water surface by the wind shear stress cause the growth of the waves. Oscillatory motion is highest at the surface and diminishes exponentially with depth. However, for standing waves (clapotis) near a reflecting coast, wave energy is also present as pressure oscillations at great depth, producing microseisms. These pressure fluctuations at greater depth are too small to be interesting from the point of view of wave power. The waves propagate on the ocean surface, and the wave energy is also transported horizontally with the group velocity. The mean transport rate of the wave energy through a vertical plane of unit width, parallel to a wave crest, is called the wave-energy flux (or wave power, which must not be confused with the actual power generated by a wave power device). Waves continue to roll, even after the wind stops blowing, which leads to a higher degree of utilization than for wind power. The circumstances, with moderate wave conditions, indicate a degree of utilization of 60% to 80% (depending on the beach). However, in bigger seas and large oceans, this can go up to 90%. Furthermore, the energy density is a lot higher than for wind or solar power. The physical conditions for wave power are therefore very good, and the relatively high degree of utilization makes waves a predictable source. Energy generation from wave power should thus have a considerable potential to contribute to electrical energy production. This is especially the case along the coastlines of the big oceans, provided that suitable technologies can be developed. About 70% of the earth's surface is covered by water. Various estimations show that the world's potential for wave energy is 10,000–15,000 TWh/year, which is about the same as the economic potential of hydropower in the world. Excellent reviews on wave energy and converters are published in the literature [308–312].

Vast and reliable wave power has long been considered as one of the most promising renewable energy sources. In a special IPCC 2011 report, several estimates are presented for the world total potential of ocean wave-energy resources. The theoretical maximum has been estimated at about 30,000 TWh/year (3.10^{13} kWh/year), which is about 20% of the 2019 world energy consumption. However, due to technological and economic constraints, the exploitable resource is almost a factor 10 less. Regions with the highest wave power are the Southern Ocean and the North Atlantic. Focusing on the annual wave resource near to the shore, in the northern hemisphere, the highest levels are the ones registered in the west coast of the British Isles, Iceland, and Greenland. In the Southern Hemisphere, the highest energy levels are found in Southern Chile, South Africa, and the entire south and southwest coasts of Australia and New Zealand. Medium levels are located in equatorial waters, with the highest coastal resources of Northern Peru and Ecuador, although El-Nino may induce significant inter-annual variability in this area.

The wave climate strongly fluctuates with high extremes in most regions with a high wave power, especially in the northern hemisphere. This places severe requirements on the robustness of the design and the service lifespan of WEC devices, which translate into high production and maintenance costs. Areas with a more consistent wave climate and narrow-band wave spectrum may therefore be preferred for wave-energy recovery even if the maximum energy yield is somewhat lower. Waves that are practical for conversion are long swells with amplitudes of 2–3 m or less. These types of waves also offer the advantage of producing minimal impact forces, aiding in the survivability of the device [5].

3.9.4.1 Types of WECs

WECs convert wave power into electricity. Although attempts to utilize this resource date back to at least 1890, wave power is currently not widely employed. The operational wave power installed worldwide in 2020 totaled approximately 16 MW [2], which is about 5 orders of magnitude less than the 2–3 TW required to exploit the global wave-energy potential. An important reason is the production costs per kWh, which were in 2020 about a factor 10 higher compared to offshore wind farms.

A plethora of innovational methods for wave power conversion have been invented in the last three decades, resulting in thousands of patents over recent years. At present, a number of different wave-energy concepts are being investigated by companies and academic research groups around the world. This research effort is mainly directed to technical optimization of the WEC performance of the three energy transfer stages: (1) the conversion of wave power to mechanical power by the wave–structure interaction between ocean waves and device structures (geometric optimization), (2) the transfer of mechanical power into electrical power (through direct-drive generator, or indirect via, e.g., rotational motion by air turbines, hydraulic rams, gearboxes, or mechanical motion rectifiers) using optimal control strategies to tune the system dynamics to maximize power output, and (3) power electronics to improve power quality to transfer the nonstandard AC power into direct current (DC) power for energy storage or standard AC power for grid integration.

Although many working designs have been developed and tested through modeling and wave tank tests, only a few concepts have progressed to sea testing. Strong cost reductions, which are only possible with a sharp increase in global application, might enable wave plants to compete favorably with conventional power plants in the future.

3.9.4.1.1 OWCs

OWCs consist of an air turbine and a chamber partially submerged below the water surface. The air in the chamber is compressed by the oscillating motion of the wave surface and flows through the air turbine to generate electric power. Falcao et al. [8] discussed the theoretical, numerical, and experimental modeling techniques associated with OWC converters. Lopez et al. [9] optimized the turbine–chamber coupling for the OWC via a 2D numerical model based on the RANS equations and the VOF surface capturing scheme. Gomes et al. [11] proposed to improve the wave-energy extraction by the OWC by optimizing the dimensions of the floater and the tube under certain geometric constraints.

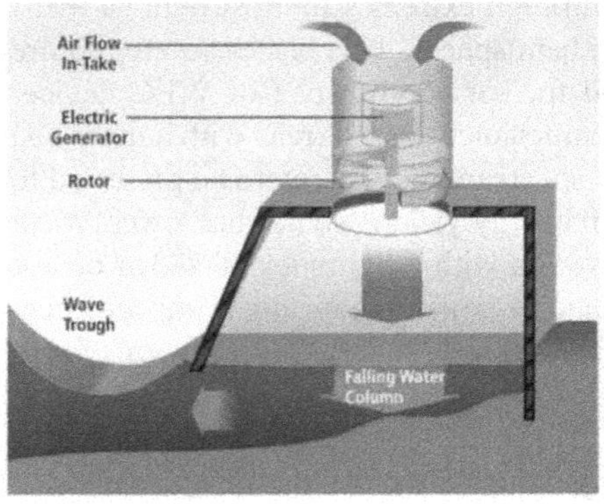

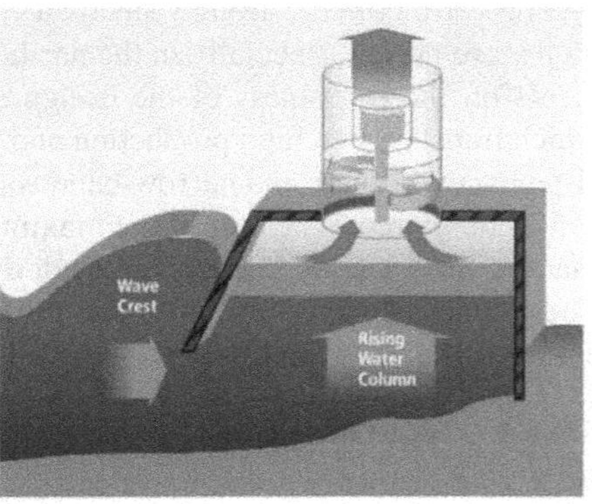

FIGURE 3.23 Wells turbine (From [313]). Wave energy conversion using principles of oscillating water column from IPCC (2011).

The principle of the OWC is illustrated in Figure 3.23, for use in conjunction with a fixed structure (e.g., breakwater). The functioning of the OWCs is somewhat similar to that of a wind turbine, being based on the principle of wave-induced air pressurization. The device is set upon a closed air chamber, which is placed above the water while the base is open to the ocean. The passage of waves changes the water level within the closed housing, and the rising and falling water level increases and decreases the air pressure within the housing introducing a bidirectional air flow. By placing a turbine on top of this chamber, air will pass in and out of it with the changing air pressure levels.

There are two options to separate the bi-directional flow: a Wells turbine to create suction or alternatively, pressure generating valves. The Wells turbine is constructed such that it rotates always in the same direction regardless of airflow direction (see Figure 3.24). The efficiency is lower (50%–60%) than with conventional turbines, but higher than achievable with conventional turbines in alternating mode [313].

OWC devices can be moored offshore, but they can also be placed near the shore where waves break. This results in significant cost savings. The disadvantage is the shallow water depth along the shore, which dampens the largest waves. Special buoys have been developed for the application of OWC converters in deep water, according to the principle of Figure 3.25. The length of the shaft determines the resonance frequency, allowing optimum energy efficiency to be achieved. An example of an offshore OWC is the Spar Buoy, Figure 3.26. The original concept was invented by Yoshio Masuda (1925–2009), who developed navigation buoys powered by wave energy, equipped with an air turbine. Due to the cylindric shape, it is invariant to wave direction (Figure 3.26). The size varies according to the sea conditions at the deployment site, but maximum dimensions are estimated at 30 m diameter, 50 m height, and 35 m draft, which could deliver up to 450 kW [313].

3.9.4.1.2 Overtopping Devices

Another type of WEC is the overtopping device, which operates somewhat similar to a hydroelectric dam. The "Wave Dragon" created by Wave Dragon ApS is the

FIGURE 3.24 Wave-energy conversion using the principle of the OWC (From IPCC 2011 [313]). Wells turbine from Falcao and Henriques (2016).

best known example of an offshore overtopping device (Figure 3.27). Its floating arms focus waves onto a slope from which the wave overtops into a reservoir. The resulting difference in water elevation between the reservoir and the mean sea level then drives low-head hydro turbines. It was estimated that an optimal size design of 260 m width and 150 m length can produce up to 4 MW. In wave climates above 33 kW/m, this technology was expected to be economically competitive with offshore wind power in the near future. After a combined cost-saving and power efficiency increase, the power price could eventually be in line with costs of fossil fuel generation. However, a feasibility study for a Wave Dragon deployment at the northern Spanish coast published in 2020 reported costs which were still a factor 10 higher.

Near the shore, overtopping converters can be installed in front of or as part of caisson breakwaters. An example is the SeaWave Slot-Cone Generator, which collects seawater by wave overtopping over several reservoirs placed above each other, resulting in high hydraulic efficiency (see WECs in coastal structures) [313].

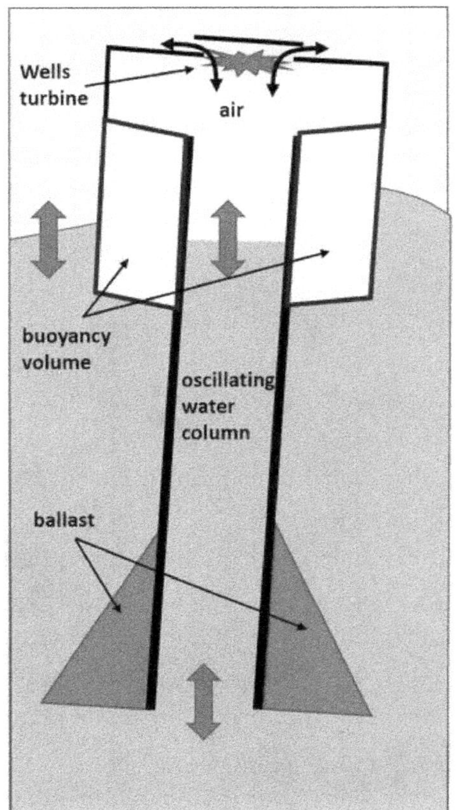

FIGURE 3.25 Wave-energy buoy based on the OWC principle [313].

3.9.4.1.3 TAPCHAN Wave-Energy Conversion

The tapered channel, or TAPCHAN method of wave-energy conversion, is a very simple device.

Tapchans, or tapered channel systems, consist of a tapered channel that feeds into a reservoir constructed on cliffs above sea level. The narrowing of the channel causes the waves to increase in height as they move toward the cliff face. According to conservation of energy, as the wave width decreases, the amplitude increases, enabling the wave travel up a ramp and pour into the reservoir as shown. Once the water is in the reservoir, it flows through a traditional hydroelectric turbine back to the sea.

The principle of operation can be divided into the following four subsystems: Firstly, a collector which is designed to concentrate on the water energy and optimize collection efficiency for a range of frequencies and directions; secondly, the energy converter, in which the energy of the collected waves is transformed into potential energy in an onshore water reservoir, this is the unique part of the power plant, it consists of a gradually narrowing channel with wall heights equal to the filling level

FIGURE 3.26 Spar buoy WEC [313].

of the reservoir (typical heights 3–7 m), and the waves enter the wide end of the channel, and as they propagate down the narrowing channel, the wave height is amplified until the wave crests spill over the walls; thirdly, a reservoir which provides a stable water supply for the turbines; and finally, the hydroelectric power plant, where well-established techniques are used for the generation of electric power. The water turbine driving the electric generator is of a low-head type, such as a Kaplan or a tubular turbine. It must be designed for saltwater operation and should have good regulation capabilities [313].

3.9.4.1.4 Wave-Absorbing Devices

A great number of different devices have been developed to harness wave energy directly by using wave-induced water motion. The most popular types are displayed schematically in Figure 3.28.

FIGURE 3.27 Wave Dragon overtopping WEC [313].

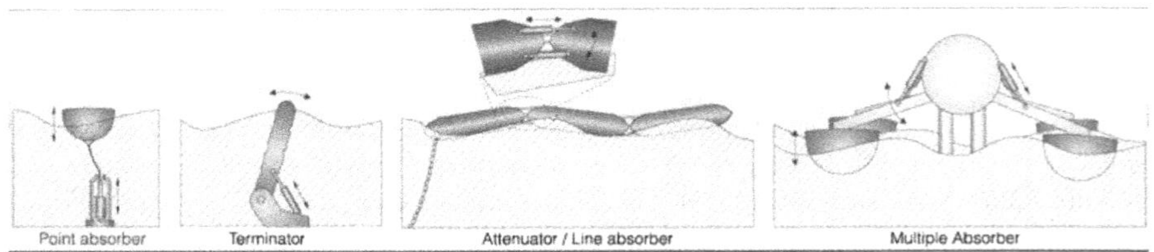

FIGURE 3.28 Different energy conversion principles based on absorption of wave energy. (Image from https://www.mdpi.com/journal/energiesCreative Commons license [313].)

3.9.4.1.4.1 Point Absorbers Point-absorber WECs employ buoys excited by heavy wave motion to drive hydraulic pumps or linear generators to produce electric power. Zurkinden et al. [314] developed a nonlinear numerical model to analyze the most significant nonlinear effects of a point-absorber WEC; the dynamical properties of the semi-submerged hemisphere buoys, oscillated by waves, were investigated. Bozzi et al. [315] presented a numerical model of the coupled buoy-generator system to simulate the behavior of the WEC under different wave heights and periods. This numerical model based on linear potential wave theory simulated the influence of the hydrostatic forces upon a point-absorber WEC, and also considered the radiation impedance and excitation force. Nevertheless, those WECs had heavy, bulky

FIGURE 3.29 The FO3 point absorber (top) and the Wave Star attenuator (bottom) [313].

generator, secular wave-power device that has linear generators, rotary generators, or hydraulic pumps.

Point absorbers are buoy-type WECs that harvest incoming wave energy from all directions. They are placed offshore at the ocean surface or just below. A vertically submerged floater absorbs wave energy which is converted by a piston or linear generator into electricity. One such a point-absorber WEC is the FO3 concept developed by Norwegian entrepreneur Fred Olsen. It consists of several (12 or 21) heaving floaters attached to a 36-by-36 m rig (Figure 3.29 top). By means of a hydraulic system, the vertical motion is converted into a rotational movement that drives the hydraulic motor. This motor in turn powers the generator that can produce up to 2.52 MW [313].

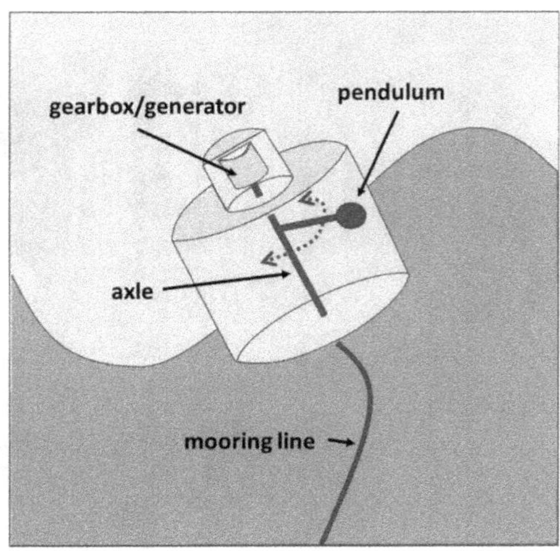

FIGURE 3.30 VAPWEC [313].

The multiple-point-absorber-type WEC "Wave Star," developed by Wave Star ApS, has a number of floaters on movable arms (Figure 3.29 bottom). The energy of the motion of the arms is again captured in a common hydraulic line and converted into electric current. Most noticeably, being able to raise the entire installation along its pillars, this system has a high endurance for rough storm conditions. So far, this method has not been deployed at full scale. A 1:2 scaled installation has been built at Hanstholm which turns out 600 kW. However, production is thought to be scale-able up to 6 MW. A major benefit of these types of exploitation is the minimal contact with water, placing any delicate machinery and electrics out of reach of any corrosion or physical forcing of the waves. The Wave Star development was abandoned in 2016.

3.9.4.1.4.2 Vertical-Axis Pendulum Pendular wave-power devices consist of a rectangular box that is open to the sea at one end. A flap is hinged over the opening, and the action of the waves causes the flap to swing back and forth. The motion powers a hydraulic pump and a generator. A vertical-axis pendulum WEC (VAPWEC) is a point-absorber-type device whose power takeoff (PTO) system is based on the motion of a pendulum that is connected to a generator, which is all within the device's hull (see Figure 3.30). As the device floats on top of the surface, the pitching and rolling moments due to the incoming waves cause the VAPWEC pendulum to swing around a vertical axis. Advantages of this device type are as follows: (1) a robust design since all major components are sealed within the protective hull and (2) the simplistic energy conversion design, which has minimal moving parts and PTO system that has few hard stops.

3.9.4.1.4.3 Terminators Terminators consist of flaps that rotate with the wave orbital motion around an axis parallel to the wave front. Examples are the

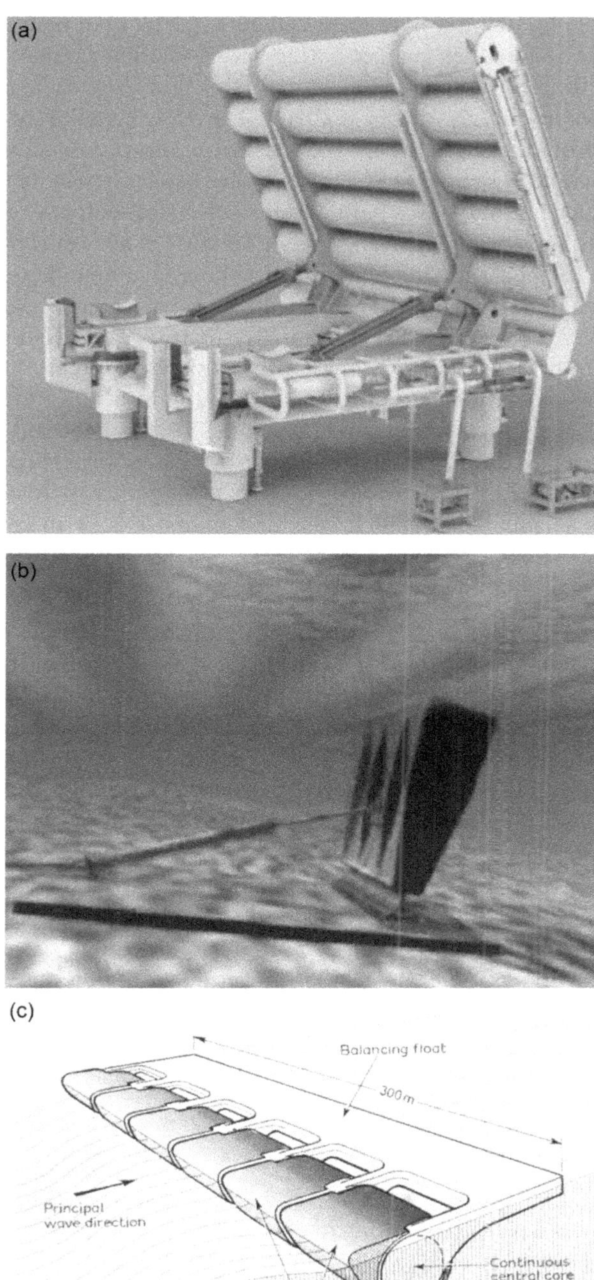

FIGURE 3.31 (a) Bottom-hinged WEC Oyster. (b) Bottom-hinged WEC WaveRoller. (c) Top-hinged WEC Salter Duck.

Oyster (Figure 3.31a) and the WaveRoller (Figure 3.31b), both bottom-hinged terminators and the Salter Duck (Figure 3.31c), with a rotation axis near the surface. Although these designs capture wave energy with a high efficiency, no large-scale operational deployments have been realized. They are installed in intermediate depths, not far from the shore, where the wave front direction is most of the time close to shore-parallel. Oyster and WaveRoller have been tested in nature; for Salter Duck, no full-scale prototype was built. Oyster's electrical output is generated by a hydroelectric turbine onshore, driven by high-pressure water via subsea pipelines. WaveRoller is equipped with an onboard hydraulic system that drives an electricity generator, which is connected to the electric grid via a subsea cable. Hinged wave surge converters generally use hydraulic systems for PTO. Hydraulic systems are well suited to harvest energy from high-force, slow-oscillatory motions which have to be converted to rotary motion and drive a generator. In order to rectify the fluctuating wave power, which would result in variable electrical power output unsuited to the electrical grid, some sort of energy storage system (or other means of compensation, such as an array of devices) is usually incorporated in the PTO system, such as accumulators, which can function as short-term energy storage, helping the system handle the fluctuations [313].

3.9.4.1.4.4 Wave Attenuators These devices lie parallel with the incident wave direction. The "DEXA," developed and patented by DEXA wave-energy ApS, is an illustrative example of a wave attenuator. The device consists of two hinged catamarans that pivot relative to the other (Figure 3.32). The resulting oscillatory flux at the hinge is harnessed by means of a water-based low-pressure power transmission that restrains angular oscillations. Flux generation is optimized by placing the floaters of each catamaran half a wavelength apart. A scaled prototype (dimensions 44 × 16.2 m) placed in the Danish part of the North Sea should generate 160 kW. Full-scale models are thought to be able to generate up to 250 kW. However, the DEXA development was terminated in 2012. Another example of this type of device is the 750 kW Pelamis, consisting of five tube sections linked by universal joints which allows for flexing in two directions as the device "rides" over the incoming waves. The flexing movement is converted into electricity via hydraulic PTO systems [313].

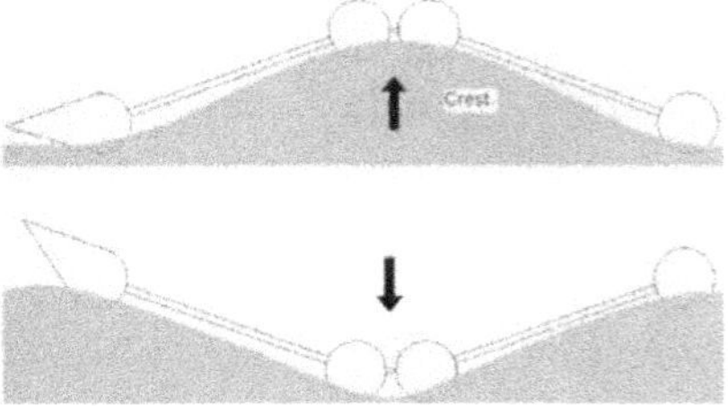

FIGURE 3.32 DEXA WEC and working principle [313].

3.9.4.1.5 PTO Systems

The wave-activated bodies use a PTO system driven by wave-induced motions of masses or large bodies to generate electricity. Numerous devices have been developed based on this concept; one well-known example is the Pelamis, developed by the Scottish company Pelamis Wave Power (formerly Ocean Power Delivery) in 2004 [316]. Henderson et al. [317] would further study the Pelamis WEC to enhance control of the PTO system. Ruellan et al. [318] attempted to develop a SEA REV WEC based on this operational concept.

The objective of WECs is to convert the kinetic energy from incident waves to compatible electrical energy that can be integrated into the utility grid. Extracting the maximum energy from a WEC system is a key challenge. A WEC device generally contains two controller systems: the primary controller for the PTO and the secondary controller for the grid power converter. The primary controller aims to maximize the harvesting of electrical energy from the wave-generated mechanical device motions. The secondary controller regulates the voltage, current, and frequency to make sure it can be integrated into the power grid. Another function control systems' offer is their ability to limit the mechanical forces by damping and tuning the device motion, providing improved device survivability and potentially lower maintenance costs. For highest power extraction, damping must be adjusted to achieve maximum energy conversion efficiency. If the damping is too high, then the motions are limited and little power is produced. If the damping is too light, then the damper absorbs little power and little power is taken off. With any PTO system, the correct damping is vital for an efficient system. Active device tuning can range from adjusting parameters of the primary converter for a particular sea state to wave-by-wave adaptation (also known as fast tuning).

The PTO system of a WEC has a direct impact on the capital cost of a project by usually accounting for between 20% and 30% of the overall investment. The economic viability, efficiency, and complexity of a WEC depend largely on its PTO system. Maintenance at sea is a demanding and expensive task; high reliability and durability of all components of the PTO system are therefore required. This is technically challenging for systems exposed to the harsh marine environment, especially systems that consist of many moving parts that are exposed to corrosion and fouling. Several popular PTO systems have been reviewed by Ahamed et al. [319], from which the summary below is extracted (see Figure 3.33).

OWC wave converters generally use the previously described Wells turbine as PTO system. These turbines are vulnerable due to the relatively large number of moving parts. Overtopping devices are generally equipped with hydroturbines for PTO. Conventional hydroturbines require for high efficiency more head and flow than provided by overtopping ocean waves. Wave-absorbing devices use hydraulic motor systems or direct mechanical or electrical drive PTO systems. Hydraulic motor-based PTO systems are suited for converting the low-speed oscillating motion into energy. However, the hydraulic motor-based PTO system consists of many mechanical moving parts, and due to compression and decompression of the fluid, there is a risk of hydraulic oil leakage. Direct mechanical drive uses linear-to-rotary conversion systems without pneumatic or hydraulic systems. The efficiency is high,

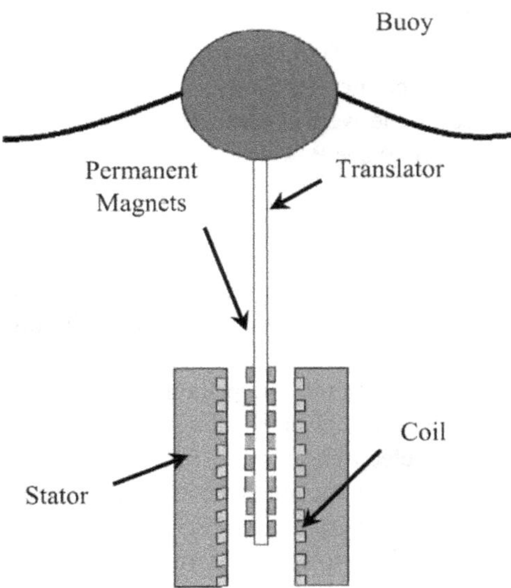

FIGURE 3.33 Schematic of a linear electrical generator based on a permanent magnet generator [313].

but the lifetime is relatively short and maintenance costs are substantial. Direct electrical drive transmits the wave energy into electrical energy directly by coupling the mechanical energy to the moving part of a linear generator (Figure 3.33). Issues are the need for a heavy structure due to the attractive forces between the stator and the translator and the complicated power transmission system due to the unequal generated voltage created by the irregular wave motion. To smooth out the power fluctuations of such a direct-drive WEC, energy storage components such as batteries, supercapacitors, or a hybrid combination of the two can be employed to stabilize the generated power.

The PTO system is a crucial part of WECs, which largely determines the cost, efficiency, and reliability of the WEC. Many studies focus on optimization of the PTOs; several challenges related to PTOs are mentioned in the previous section.

There are three main types of PTO methods that extract energy from waves. The direct mechanical drive system uses an electric generator to convert the wave energy into electricity. Although it is found to have a 97% efficiency rate, the direct mechanical drive system is associated with high maintenance costs and is found to have a short life cycle. To make the direct mechanical drive system a reliable method to convert wave energy into electricity, there must be more research done on the electric generator [316]. The triboelectric nanogenerator (TENG) is a recent invention that uses triboelectrification and electrostatic induction to convert wave energy into electricity. Unlike many generators that can only be used on low-frequency waves, the TENG can be used on waves with any frequency and it is not as costly. The TENG may be a great contender for converting wave energy into electricity; however, it is

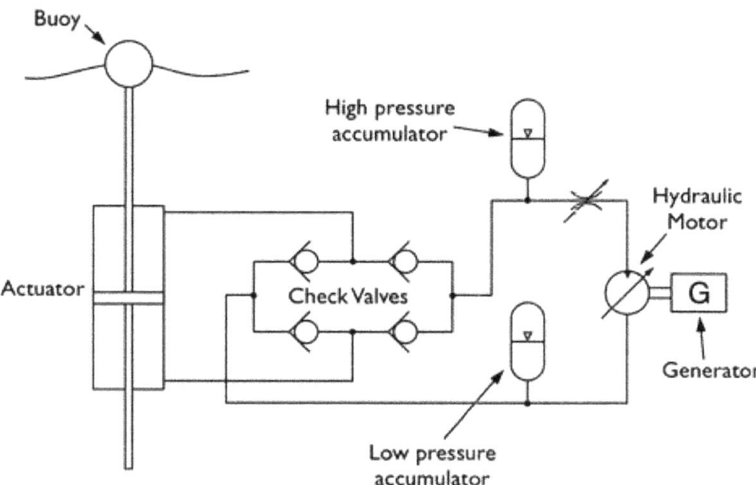

FIGURE 3.34 Example of a hydraulic PTO system for wave-energy conversion [316].

unknown what the conditions of the ocean may do to the nanogenerator. More discussion on the use of TENG (TENG) for power generation from wave energy is given in the section 3.9.4.4.

Turbine transfer describes the use of liquid to power a turbine that is connected to a generator. In the air turbine transfer system, a WEC system is first used to convert the wave energy into pressurized air which travels through the turbine and into the generator which allows for the production of electricity. The air turbine transfer system is used when the waves are weak and slow. One benefit of air turbines is that they do not have to be located in the middle of the ocean like other devices, which means they will not be easily corroded and can be easily maintained. However, there is a high cost associated with the use of air turbines [316]. In the hydroturbine transfer system, the water travels through the turbine and drives the generator to produce electricity. It was discovered that the hydroturbine transfer has a 90% efficiency rate whereas the air turbine system has approximately 62.5% efficiency rate. Like other devices used to extract energy from the ocean, the hydroturbine can be easily damaged by the ocean water as it may harm the seals and the valves of the turbine which will decrease its efficiency [316].

Hydraulic systems use pressure to force the water through valves and into the actuator which converts the wave energy into electricity. The hydraulic system consists of a buoy, ram, hydraulic motor, accumulator, and generator (see Figure 3.34). The energy from the waves is absorbed from the ram and generates pressure which runs through the motor. The motor then allows the generator to convert the wave energy into electricity. The hydraulic system can produce a high yield of energy from low-frequency waves. Although they are successful, it is difficult to contain the fluid in the system, the system also requires constant maintenance, and it may be costly and difficult to store energy. In addition to these challenges, the hydraulic system is found to have a lower efficiency when used in the real world when compared to the efficiency when used on the laboratory scale because when used on the laboratory scale, it doesn't account for other factors like the effect of the ocean on the system. The fluid in the system may also lead to oil leakage which will harm marine life [3]. The hydraulic system may be a promising way to retrieve energy from waves, but there are many obstacles that need to be overcome before they can be applied on the commercial level [316].

Portugal presents an ideal environment for HiWave-5 wave-energy project due to its natural assets and environmental consciousness. Grid operator REN recently installed a new offshore cable servicing floating wind, and there is significant commercial interest from utilities and project developers for next-generation wave-project development. The HiWave-5 program is recognized as one of the most ambitious efforts in ocean energy [320]. The HiWave-5 demonstration project aims to convert CorPower's wave technology into a bankable product by 2024, by proving the survivability, performance, and economics of a grid-connected array of WECs in northern Portugal. In Wales, a first-of-its-kind, full-scale "cell module" made its way to the assembly workshop. This marked a milestone for Bombora's 1.5 megawatt (MW) mWave Pembrokeshire Demonstration Project, which was installed off the coast of Pembrokeshire in the first half of 2021.This cell is the first of four structures that form Bombora's 75 m-long, subsea mWave. The patented cell module presents a step change from previous approaches to wave-energy capture. Each cell module will be covered in a robust rubber membrane. As waves pass over mWave, underwater pressure increases, causing the rubber membranes to compress in sequence, forcing air inside the membranes along a duct and through a turbine, and spinning a generator converting this rotation into electricity.

The project is made up of four elements: (1) the Marine Energy Test Area within the Milford Haven Waterway led by Marine Energy Wales, enabling technology developers to test their marine energy devices close to their base of operation, (2) a 90 km^2 Pembrokeshire Demonstration Zone delivered by Wave Hub Limited that will enable the deployment of future energy-generating technologies, including floating wind, (3) a technology, innovation, and research center delivered by the Offshore Renewable Energy (ORE) Catapult, and (4) redevelopment of land at Pembroke Dock, led by the Port of Milford Haven, to deliver the infrastructure needed by the industry as it continues to mature. The other two successful systems in the Wave Energy Scotland call are being designed by Blackfish Engineering Design and Nova Innovation. Blackfish Engineering Design's C-DART provides a remote installation system for a WEC or other floating system. The novel system allows quick connection and disconnection of a WEC to an offshore buoy, providing both a mechanical mooring and an electrical connection [320].

3.9.4.2 WECs Combined with Offshore Wind

The high cost of stand-alone wave-energy conversion schemes is a major obstacle for its large-scale application. However, costs can be significantly reduced by combining WECs with structures offshore or along the coast that are being built for other applications. A good example is the integration of WECs with breakwaters in the coastal zone. Another possible combination is the integration of wind and wave-energy production. This is especially interesting in areas where the conditions for optimal wind-energy production do not systematically coincide with conditions for optimal wave-energy production. It is also a way to make optimal use of marine space. The main advantage of integrated wind power generation is shared infrastructure costs, especially foundations and grid connections. Hybrid power-generation architectures that integrate WEC with offshore wind turbine generators or energy storage systems can be a promising solution for power quality improvement and sustainable electric power production. However, with the existing WEC techniques, the costs per kWh produced are still higher with a combined wind-wave application than with wind energy alone. Synergy benefits can also be sought through improved stability of the structure, for example in the case of an OWC-WEC integrated into an offshore wind turbine monopile. Stability improvement can be a major benefit for designs in which the interaction between the wind and wave substructures is strong, as in the case of a WEC combined with a floating wind turbine. WECs can also reduce wave heights inside a wind farm, increasing in this way the weather windows to access the wind turbines [313].

3.9.4.3 Issues with Conventional WECs

In order to be able to produce energy at competitive prices, several obstacles have to be overcome. The following factors in particular determine the high costs of wind energy. A very robust construction made of high-quality materials is required that remains intact under severe storms and withstands the demanding conditions at sea leading to corrosion, fouling, and fatigue. The classical protection measure against fouling and corrosion of steel structures is regular maintenance and repainting. But, this is time-consuming and costly because of the difficult access to offshore installations, especially under harsh conditions when damage occurs. In addition, the use of antifouling paints may be detrimental to the marine environment (e.g., tributyltin paints). The performance of the installation may suffer not only from colonization by microorganisms, but also from attracting fish and other marine mammals and seabirds that feed on fish. The cost of repairs is a major component of the wave farm costs. The reliability of the components, especially due to the cost of eventual repair actions, is pivotal for the economic viability of a project. Deployments should be capable of lasting 30 years or more. This is a tall order as demonstrated by the mooring of the Wave Dragon that failed after 2-year deployment during a severe storm on the 8 January 2004.

The power generated by ocean waves fluctuates strongly due to the irregular wave climate, which makes connection to the electricity grid difficult. In addition, connection points to the electricity grid may not be available in areas where the conditions for wave-energy generation are most favorable.

Crucial for any design is the mooring which ensures a maintained position under both normal operating loads and extreme storm load conditions. It should not exert excess tension loads on the electrical transmission cables and ensure the suitable safety distances between devices in multiple installations. Most commonly, a free-hanging catenary configuration is used for mooring but multi-catenary systems and flexible risers are also utilized. The mooring configuration should be sufficiently compliant to accommodate tidal variations and environmental loading while remaining sufficiently stiff to allow berthing for inspections and maintenance. Variations in wave directionality in deep waters also pose a challenge for device power generation. It is essential that wave devices (also arrays of wave devices and non-axisymmetric devices) are able to align themselves accordingly on compliant moorings that allow them to absorb energy from any direction.

Due to the great variability of the wave climate with incidental extremes, it is important to be able to test prototype designs over a long period before they are deployed on a large scale. Mathematical simulation models can be of great value here, because this is a quick and relatively inexpensive way to test a design for its effectiveness and efficiency at the long term, both in terms of capital and operational expenses, in order to achieve the lowest overall cost of electricity. For this purpose, the modeling tool supports the decision-maker in the pursuit of the most reliable and easy-to-maintain device design and also informs about the trade-off between energetic yield and operational and maintenance efforts [313].

3.9.4.4 Ocean Wave-Energy Harvesting Using TENG

The "TENG" and its extension, the "Triboelectric–Electromagnetic Hybrid Nanogenerator" (TENG-EMG), are new developments that can efficiently harvest energy in any frequency range, are low-cost lightweight, are easy to fabricate, and are easy to be scaled (size of individual units being below the micrometer scale). These nanogenerators use a polymer–metal pair to create contact electrification (triboelectric effect) between two materials sliding against each other and to induce charge transfer between their electrodes due to electrostatic induction, either in a layer structure or in a spherical-shell structure (see Figure 3.35). Challenges for application in a prototype lie in the use of TENGs for power transfer to the shore, cost to scale, lifetime of the TENG materials in the ocean environment, and connection methods of the thousands of TENG units that are needed [308–312,321–323].

Ocean wave energy can also be harvested using TENG concept with a variety of structure. The structures that are most effective are spherical-shell structure, wavy structure, spring-assisted structure, and bionic structure. Here, we briefly examine the effectiveness of these structures.

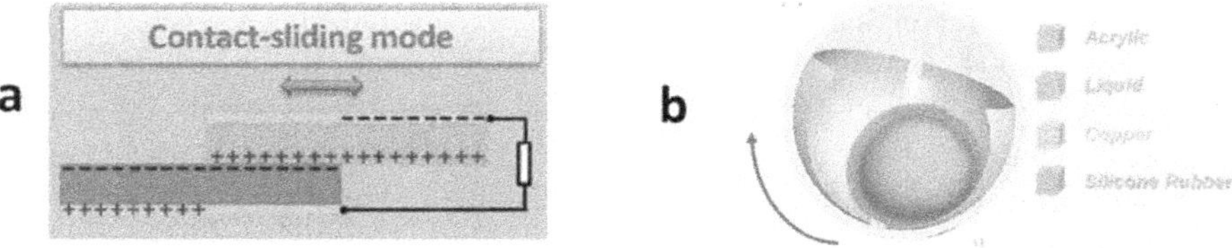

FIGURE 3.35 Principle of triboelectric nanogenerators. (a) Triboelectric charges are generated on the surface of two laterally sliding dielectric films, due to frictional effects. Polarization occurring in the sliding plane drives an electron flow between the metal electrodes that generate an AC. (b) Soft contact rolling-mode spherical TENG (SS-TENG). When receiving external vibration from the ocean waves, the ball will roll back and forth between the electrodes to provide AC power to the external load. The soft ball maximizes the contact area. [313].

The spherical-shell structure is very promising in TENG for ocean wave-energy harvesting. It has the characteristics of lightweight and simple structure, which can harvest ocean wave energy in any direction. The spherical-shell structure is easy to connect to a network. This structure is also named as freestanding triboelectric-layer-based nanogenerator (RF-TENG). In this design, under the water wave of 1.43 Hz, this RF-TENG can directly light up several ten commercial LEDs and has excellent power-generation performance. RF-TENG uses a rolling rigid nylon ball to contact the Kapton film in a closed spherical shell. The Kapton film is connected by two curved electrodes on the back to form an independent triboelectric-layer structure. When receiving external vibration from the ocean waves, the ball will roll back and forth between the two electrodes to provide AC power to the external load. Experiments show that RF-TENG has a stable output performance between 1.05 and 2.35 Hz in water-wave frequency. The experimental results show that the output power density of the nylon/Kapton device is better than that of the PTFE/Al device. As they are all hard balls, the obvious disadvantage of these shell structures is that the contact area is too small, resulting in low-power generation. And hard balls can greatly reduce the durability of the TENG due to heavy wear.

In order to solve the small contact area and improve durability, Xu et al. [324] have manufactured a TENG using silicone rubber instead of hard nylon balls (S-TENG); the softness of silicone rubber can increase the actual contact area and help improve the durability of the device. To enhance the contact electrification of silicone rubber, the silicone rubber is UV-treated and mixed with polyoxymethylene particles in the dielectric layer to produce microstructures on the surface. To further increase the contact area of the ball, Cheng et al. [325] further optimized the inner ball. By using rolling soft water/silicone as the inner rolling ball, a better TENG (SS-TENG) was manufactured. This design maximizes the contact area, which can generate more charge and adjust the output performance by changing the thickness of the silicone shell. Compared with the S-TENG, the output of the design is ten times higher at 5 Hz (Figure 3e and f) and two times higher at 2 Hz. Xia et al. [326] also designed a similar internal soft ball structure to harvest ocean wave energy, named multiple-frequency TENG based on the water balloon (WB-TENG).

Shi et al. [327] created a highly symmetrical 3D spherical water-based TENG (SW-TENG). In this device, the rolling ball was directly replaced with a liquid, because in a complex water-wave environment, a liquid-based 3D symmetrical structure is more desirable and more efficient and not easy to leak. Similarly, Lee et al. [328] made a spherical hybrid TENG (SH-TENG) based on solid–solid and liquid–solid power generation; in this work, the liquid was replaced with a hardball inside the spherical-shell structure, and solid–solid TENG and solid–liquid TENG were coupled together using a single electrode. In contrast, Yang et al. [329] obtained greater output performance, the inside of the shell is layered, and multiple TENGs are integrated into each layer to achieve the maximum utilization of space. Liu et al. [330] designed the spherical shell into an oblate shape and integrated two kinds of TENG inside, which can perform well in rough seas and relatively calm seas. Spherical-shell structure is a very promising structure, because using soft ball can increase the contact area and improve durability, and using multilayer structure can provide more power density, which will be the first choice in the future practical applications.

The wavy structure is one of the earliest structures used to harvest wave energy, and it is also a popular structure in the initial research. The wavy structure was first proposed by Wen et al. [331], and it is based on a wavy structure of Cu/Kapton/Cu film, which is sandwiched between two flat nanostructured PTFE membranes. When subjected to external mechanical vibration/shock/compression, the membrane will extend the vertical impact transition laterally, which causes the sliding charge between the electrode and the PTFE membrane. After the impact, the TENG will automatically restore the initial position due to the elasticity of the membrane. This repeated pressing and release will allow the charge to flow between the planar electrode and the electrode, thereby generating AC.

Wavy structure TENG (WS-TENG) can be used as part of an integrated device, which can more efficiently collect wave energy. Jiang et al. [332] manufactured a boxed TENG consisted of a wavy-shaped TENG wall and a closed ball. Each WS-TENG consists of a wavy Cu-Kapton-Cu film and two fluorinated ethylene propylene (FEP) films, and the metal electrodes are sputtered into a sandwich structure. The size and mass of the box-shaped TENG spheres are optimized to obtain an optimal sphere size and mass, which is proved that the maximum output power and electrical energy can be obtained at this time by experiments. The work of Yao et al. [333] indicate that under direct water-wave impact, the energy storage and maximum energy storage efficiency are determined by the depth of deformation, which can be optimized by the size of the sphere under the impact of a closed ball. In order to make full use of the inner wall area, Zhang et al. [334] changed the previous box shape to a dodecahedron shape. Each surface is fixed with a wavy-shaped TENG (WS-TENG). Twelve WS-TENGs are integrated to form a dodecahedron device. Research on wavy-shaped TENG shows that the structure has a good ability to harvest wave energy and will have better performance output in the future. However, the intermediate wave structure requires a great deal of elasticity, and when the elasticity decreases, the performance of the generator will decrease. Therefore, the wave structure faces a great durability challenge which can be addressed using some composites [335].

TENG without assistant components is more effective for harvesting transient mechanical energy. However, the waves provided by the ocean have a lower trigger frequency, causing most of the impact potential energy to dissipate, making themselves face a problem of generating very limited electrical energy in a short time. In some TENGs, using springs to store unstable and discontinuous kinetic energy and then apply these kinetic energies to the TENG unit, which can convert low-frequency water waves into high-frequency oscillations and can act for a longer period, ultimately enhance energy conversion efficiency. Xu et al. [336] first proposed a novel design, which was based on an elastic suspension oscillator structure and a mechanism for transmitting and distributing the collected water-wave energy using air pressure. A spring as an energy storage element can effectively drive a series of integrated TENG. This design has been proven to store unstable and discontinuous kinetic energy under the impact of water waves and then further act on the moving components, and then convert low-frequency ocean wave motion into high-frequency oscillations, resulting in higher average output power.

Jiang et al. [337] designed a spring-assisted TENG and connected a TENG unit based on the vertical contacts separation mode at each end of spring. In this way, the potential energy generated during the mechanical triggering process was spring store and then releases it, causing the TENGs separated with a certain frequent contact, thereby generating AC. Xiao et al. [338] used the same spring-assisted structure, but using a silicone rubber/carbon black composite electrode. To improve space utilization, they designed a TENG array composed of spherical TENG elements based on elastically assisted multilayer structures [339], Due to the use of springs and increased space utilization, the output current of a spherical TENG unit can reach up to 120 μA, which is two orders of magnitude larger than the output current of the previous rolling spherical TENG, and the TENG can achieve a maximum output power of 7.96 mW. Although the effect of frequency on TENG has been solved, the most significant challenge to this structure is the durability issue, which is associated with friction and wear. This is a mechanism associated with TENG, so the solution to this problem must be to improve the durability of materials.

Inspired by nature, researchers have also invented many TENG devices with bionic structures, which have superior performance. Ducks have an efficient hydrodynamic structure, and they have been shown to extract 80% of mechanical energy in waves under laboratory conditions [340]. Therefore, a duck-shaped TENG based on free-standing triboelectric-layer mode is designed. The working principle of this design is that the structure repeats pitching motion caused by water waves, causing the inner nylon ball to roll back and forth on the Kapton membrane, thereby generating AC. In order to further explain the excellent performance of duck TENG, Saadatnia et al. [341] performed a comparative analysis between a TENG device and an equivalent EMG to harvest wave energy. They obtained the electrical output characteristics of the two technologies under various mechanical and electrical conditions. The analysis shows that at low operating frequencies of 2.5 Hz, the peak power densities of TENG and EMG reach 213.1 and 144.4 W/m^3, respectively. Harvesting wave energy at low frequencies, the structure has superior performance output than EMG.

Inspired by the jellyfish movement, Chen et al. [342] demonstrated a biomimetic jellyfish TENG (bi-TENG) based on the contact-separation mode. This structure is considered as a priority technology, due to its high sensitivity, portability, and adaptability for continuous detection of water levels. Inspired by the movement of sea snakes in the water, Zhang et al. [343] produced a sea snake-based TENG (SS-TENG). The interior of the sea snake consists of polytetrafluoroethylene (PTFE) balls, nylon membranes, sputtered copper layers, and soft anti-loosening springs. This spring is an enlarged schematic diagram of the SS cross section. To increase the frequency of the balls rolling over the electrodes, the copper wires of TENG are also arranged in an interdigitated manner, which will increase the total output current of the device. The sea snake repeatedly shakes under the impact of the waves, causing the nylon ball to roll back and forth, thereby generating AC. This structure also solves the problem that the electrolyte affects the output of the generator. Wang et al. [344] inspired by seaweed designed a bio-inspired TENG (BI-TENG) to mimic the movement of kelp. Kelp will sway gently with the waves and use energy in the process. Lei et al. [345], inspired by a butterfly, manufactured a butterfly-type TENG (B-TENG) with a spring-assisted four-link mechanism. The bionic structure is of great uncertainty, because the imitation can be changeable without uniformity, and the structure is relatively complex, which is unfavorable to the future commercialization.

3.9.4.4.1 Liquid–Solid Contact TENG to Harvest Ocean Wave Energy

Harvesting wave energy based on solid–solid contact TENGs requires quite tight sealing, and the influence of the electrolyte in the water wave on the output performance of the generator needs to be considered. TENG based on liquid–solid contact power generation has many advantages, such as the ability to collect wave energy from different directions, a larger contact area, and easy sealing. Li et al. [346] reported a liquid–solid TENG based on nanowires by etching on the FEP film, increasing the contact area between the FEP film and the water, thereby generating more charge. The TENG can produce a maximum output current and voltage of 10 μA and 200 V, respectively, and the performance output was much higher than previously reported [346,347], which has an important guiding role in wave-energy harvesting.

To study the effect of liquid properties on TENG performance, Pan et al. [348] designed and manufactured a U-shaped tube TENG based on the liquid–solid contact mode. To study the effect of liquid characteristics on the output performance of TENG, they used 11 liquids as an experimental medium. The experimental results showed that the output performance of TENG depended on the polarity of the liquid, the dielectric constant, and the FEP affinity. The pure water U-shaped TENG had the highest output among these 11 liquids. Zhao et al. [349] demonstrated the device of water-wave energy harvesting using a sliding freestanding water-TENG device. In this device, multiple pairs of electrodes were mounted on top of a flexible substrate. The electrode material was made of conductive fabric, and the PTFE membrane is used as a hydrophobic coating after dry etching. Liquid–solid contact TENG is an effective method of collecting wave energy because of its simple structure, large contact area, and excellent output performance. It is an effective device for collecting wave energy in the future.

3.9.4.5 Hybrid Generator to Harvest Ocean Wave Energy

Although traditional EMG is not enough to collect wave energy due to its bulkiness and other characteristics, in recent years, EMG and TENG hybrid generators to collect wave energy appeared to have significant potential. In general, these hybrid nanogenerators can be divided into two ways to cut magnetic induction lines. One is to rotate the magnetic induction lines to generate electricity like traditional electromagnetic power generation [350], and the other is to take into account the characteristics of the ocean waves themselves and cut the magnetic induction lines in wave mode to generate electricity [321,351].

For rotary cutting, Shao et al. [350] reported the design of a hybrid generator based on the contact-separation mode TENG (CS-TENG) and the rotating self-supporting EMG (RF-EMG). The water wave makes the EMG rotor rotate to cut the magnetic induction line, thereby generating AC. During the rotation process, due to the magnetic force, TENG periodically contacts and separates and also generates an AC. The EMG of the existing hybrid nanogenerators is mostly based on the wave mode, and TENG is mainly based on contact-separation and freestanding triboelectric-layer mode. Of course, there are also two modes that integrate power generation [321,352]. Zia et al. [353,354] designed a typical TENG hybrid nanogenerator based on sliding mode which used the traditional point-absorptive structure for collecting wave energy. In this device, TENG is based on the grating structure mechanism and the dielectric–dielectric independent TENG operation. Under the action of high and low waves, the slider moves up and down in the barrel, cutting the magnetic induction wire while sliding across the dielectric layers at different positions so that TENG generates an AC. For the TENG hybrid generator based on the contact-separation mode, its working principle is the same as that of the EMG hybrid generator based on the rotary cutting mode [321,355,356]. The TENG friction layer is periodically contacted and separated by magnetic force, thereby generating AC. The structure of the hybrid nanogenerator is based on a single TENG integrated with an EMG, but it can obtain more than double the output performance, make full use of the limited space, and be an important way to collect wave energy in the future. EMG is, however, relatively bulky when compared with other types of TENG due to added magnets and coils.

3.9.4.6 TENG Network to Harvest Ocean Wave Energy

The concept of using the TENG network to harvest ocean wave energy was proposed by Wang et al. [321,357]; the TENG network consists of thousands of TENG units through a certain connection method, which can output high-power electrical energy. The connection method of the TENG network will affect the performance output of the entire network. Xu et al. [324] studied the impact of three connection modes of rigid, flexible, and wire connection on the performance output of the TENG network. Research showed that for a group of TENG if the units were not connected to each other, the efficiency will be very low. Therefore, the mechanical connection between the various units plays a vital role in the TENG network, because it can provide a high output by coupling the relevant mechanical motion between them. The experimental research also showed that under actual water-wave conditions, rigid connections imposed too many internal constraints between the units. Compared with rigid connections, flexible connections are a better network strategy.

The ocean environment is harsh, and the connections between TENG networks may be broken due to natural disasters such as storms and large waves. Therefore, Yang et al. [329] invented a self-assembling structure that can automatically reorganize the scattered TENG units. A plurality of self-adaptive magnetic joints (SAM-joint) is installed on the closed TENG shell to realize self-assembly. When two nodes of two TENG units are close, the spherical magnet will rotate rapidly and self-adjust to the state of the opposite external magnetic pole and attach due to magnetic interaction. To maintain the flexibility of the energy collection performance while maintaining the assembly structure of the network, a limit block is also designed at the node to limit the anisotropy of the degree of freedom of the unit. The shape of the network can be adjusted by installing a different number of SAM joints on each unit. This design greatly enhances the autonomy and mechanical robustness of the network, which is conducive to large-scale manufacturing and maintenance.

In the past few years, the TENG device has achieved great success in collecting ocean wave energy. TENG with different structures has been applied, and the output voltage of the TENG device has been greatly improved. The output voltage of TENG harvesting wave energy is now up to 1,780 V [358]. TENG method to harvest wave energy, however, requires additional research. The durability of TENG has always been a problem, and it is necessary to choose durable materials and reasonable designs to solve this problem. The connection mode is between TENG networks, the sea environment is harsh, and the network connection method is the key link for the final transmission of the collected energy. The network connection method needs to be comprehensively considered. Finally, management and distribution of TENG output power needs to be examined.

3.9.4.7 Power from Wave Energy Using Piezoelectricity

Piezoelectric materials, which have the advantages of a small size and large energy density, have been used in applications such as wind and vibration energy [359–362]. Kan et al. [359] proposed a piezoelectric windmill that could harvest wind energy at low speeds and over a wide speed range. The operational frequency of the piezoelectric component was increased by excitation with a rotating magnet. A prototype of this piezoelectric windmill was fabricated and tested to prove the analytical result. Zhao et al. [360] optimized the performance of the galloping piezoelectric energy harvester using an effective analytical mode to incorporate both electromechanical coupling and the aerodynamic force. Dai et al. [361] investigated a piezoelectric energy harvester consisting of a multilayered piezoelectric cantilever beam, which was applied to energy harvesting from base excitations and VIVs. Mutsuda et al. [362] proposed a wave-energy harvester using painted flexible piezoelectric device (FPED) that had an elastic material deformed by the wave and a piezoelectric paint to generate electric power. The FPED was tested in various wave conditions to prove its function, and the experimental results had good agreement with the proposed theoretical model. Fan et al. [363] proposed a nonlinear harvester capable of collecting energy from various vibration directions. The low-frequency vibration in the environment was converted into high-frequency motion by improving performance through the action of magnetic coupling.

Chen et al. [364] developed a WEC with a piezoelectric generator. The size of the WEC equipped with a piezoelectric generator might be reduced, potentially enhancing the electrical output by parallel or series circuits. However, the use of piezoelectric materials in wave environments requires the challenge of lower driving frequencies that are far from the natural frequency of the piezoelectric component to be overcome. Li et al. [365] discussed the application and operation of piezoelectric energy harvesters in low-frequency environments (0–100 Hz). There are three methods for applying piezoelectric materials to low-frequency environments, including frequency tuning, broadband response, and frequency up-conversion techniques. Fan et al. [366] designed and developed a piezoelectric energy harvester using a beam roller configuration to convert low-frequency sway and vibration into high-frequency vibration of the piezoelectric beam. Another approach was to use the flexible substrate to obtain larger deformation of the piezoelectric material to enhance the piezoelectric effect.

Lin et al. [367] proposed a wave-energy harvester with a mechanical impact-driven frequency up-converted device. A mathematical model of this device was established and compared with experimental results. The results showed that the varying frequency of the waves could be converted into a higher preset frequency of the vibration of the beams. Renzi [368] derived a fully coupled model for investigating the hydro-electromechanical-coupled dynamics of a piezoelectric WEC. The relationship between the plate motion and the power extraction was also determined by the mathematical model. Yang et al. [369] developed a prototypical vibration-energy-harvesting system using a large-fiber composite material distinct from lead zirconate titanate; this new material has the property of being flexible under large deformation. Orrego et al. [370] developed a wind-energy harvester to collect energy through self-sustained oscillations of a flexible piezoelectric membrane. They evaluated the flapping behavior and resultant energy output by studying the influence of the geometrical parameters. In a wave environment, increasing the drive frequency through the design of a frequency up-conversion mechanism and increasing the deformation of the piezoelectric material can greatly improve the material's power generation efficiency.

Existing WECs utilize linear generators, rotary generators driven by mechanical linear-to-rotary converters, or hydraulic pumps. However, these generators are bulky and heavy. Chen et al. [364] developed a novel WEC using a piezoelectric power generation component, including a flexible piezoelectric composite film (piezoelectric film) (see Figure 3.36). A piezoelectric film offers a small size and simple structure. To achieve larger electrical power, higher operating frequencies and larger deformation in the deflection range of the piezoelectric film are preferable. A frequency up-conversion mechanism based on a geared-linkage mechanism is developed to convert low-frequency wave motion into higher-frequency mechanical motion. Mechanical deformation with limited amplitude is used to drive the piezoelectric generator. In this work, the piezoelectric performance of the film is tested via scanning electron microscopy, X-ray diffraction, and capacitance testing. The kinematic performance of the frequency up-conversion mechanism is analyzed by computer and tested in a

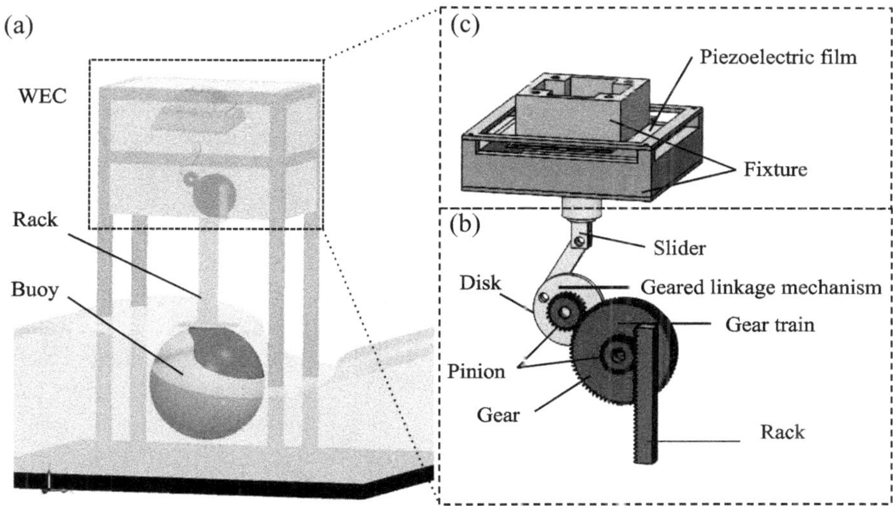

FIGURE 3.36 Schematic diagrams of the piezoelectric WEC. (a) Piezoelectric WEC; (b) frequency up-conversion mechanism; and (c) piezoelectric power generation component [364].

wave flume. The analytical and experimental results are compared. Finally, the electrical output of the PWEC is measured via an actual wave test in the wave flume. The output voltage, RMS voltage, and average electric power are discussed.

In the study by Chen et al. [364], a PWEC consisting of a buoy, a frequency up-conversion mechanism, and a piezoelectric power-generator component was developed. The operational frequency of the piezoelectric power generation component driven by the frequency up-conversion mechanism was six times that of the wave motion. The developed, flexible piezoelectric composite films of the generator component were used to produce electrical voltage under a low driving frequency and larger deformation. The deformation range was 30 mm, limited by the range of the slider of the geared-linkage mechanism. This limitation allowed fracturing of the flexible piezoelectric composite film to be avoided.

The developed PWEC was tested in the wave flume at the Tainan Hydraulics Laboratory, National Cheng Kung University, Taiwan (THL, NCKU). The results led to the conclusion that low-frequency wave motion was converted into mechanical motion with a six-times-higher frequency. The results indicated that the PWEC functioned well and outputted electrical energy. The authors expect that the PWEC will be able to generate electricity when the wave amplitude is greater than 15 mm, which can drive the slider to have a 30-mm displacement. The period of output voltage will depend on the frequency of the waves. In another study, Kim et al. [371] investigated wave power generation by piezoelectric sensor attached to a coastal structure. A piezoelectric sensor was installed to the seaward position of an existing coastal structure. The efficiency and power of the generated energy was evaluated by measuring the wave pressure and voltage according to the wave activity using the piezoelectric element and the harvesting system which are mechanical energy conversion devices.

Carlos [372] suggested a method to capture wave energy using nanotechnology. He suggested a rubbery structure moving with waves with thousands of piezoelectric crystals inside converting wave motion into electricity. Furthermore, one can connect wires crystal to crystal and take out electric energy to store in ultra capacitors for a later use. In order to trap the energy from the waves, a floating surface over the water moves parallel to the wave front, which results in maximizing the energy conversion. By trapping wave surface under the floating piezoelectric sponge surface (FPSS), all kind of wave motion can be harnessed including secondary currents through deflectors. The floating sponge can by compressed and expanded and shear stress recovered from every movement maximize the energy conversion. The converters are compact, more efficient, lighter, cheaper, easy to install, and can be taken out from the water.

Offshore, in the case of ships for example, the floating sponge can be carried without problems, because it does not represent a big load or drag force to the ship. The power from FPSS can produce the hydrogen for the engine of the ship, and hydrogen can be easily mixed with air in the intake manifold and then with fossil fuel inside the engine. In shore, over artificial or natural breakwater, a combination of FPSS and Brazilian wave power plant can be installed to recover the energy at interface, where the energy effect is more powerful, because sea waves are travel waves that deliver energy at the interface conformed by breakwaters. The random motions of the FPSS produce movements of the concave or convex form through the surface in all directions. Random compression will convert into random voltage, if random electric signals are added, and strong power will be obtained.

The random wave signals can be converted to electrical signals and voltage using diodes and condenser. The power generated from FPSS, which is located about 100 or 300 m near beach, can be taken out as DC voltage using air wires in towers. Electronic devices are needed to collect individual electric voltages inside FPSS closed to nanocrystals. High voltages are taken out by air wires. The beach inverters take DC voltage and convert into AC voltage and synchronized with the national grid system. The impedance coupling of the FPSS is soft, and this phenomenon permits high-energy conversion at all sea wave frequency spectrum. Other wave converters can have a lot mechanic element that does not resist corrosion and movement and need significant maintenance. On the other hand, in FPSS, the piezoelectric device with electrical elements is imbedded in a polymer cover that resists corrosion and abrasive behavior. The advantages of FPSS power converter over other technologies are that converters are more efficient, lighter, more compact, easy to install, functional, and have low environmental impact and low cost, and other important aspects are easy assembly, disassembly, and repair can be towed by boats to self-production of synthetic fuels at sea.

The most important concept here is that the piezoelectric crystals convert compression into electricity. The inverse effect can convert electrical signal into structural deformation such as in muscle of animals. This effect can permit in the future the movement of ships and submarines like fish. Actually, it is possible to recover energy of the vehicle using the well-known device called KERS. This device converts kinetic energy loss from brakes into electric energy in sport cars. The power obtained of the waves can also be used to water desalinization, hydrogen production, ocean mining, synthesis of liquid and solid fuels, ice production, power barge rental, etc. The power of plants depends only on the length of beach covert by FPSS amplitude and wave frequency. For big plants, it is possible to synchronize more and more FPSS units to obtained more powerful plants of about 1–10 and even 100 MW.

3.9.4.8 Assessment of Wave Energy Potentials

Ocean currents can be found throughout the world's oceans with the greatest electricity-producing potential primarily found in western boundary currents. These currents can contain significant amounts of extractable energy, with the entire Gulf Stream current system having an estimated 44 GW of theoretically extractable energy [373]. Of this resource, it is estimated that about 5.1 GW can be extracted from the Florida current portion of the Gulf Stream alone extracted [233]. If the entire portion of the Gulf Stream within 200 miles of the US coast from Florida to North Carolina is considered, then 18.6 GW can be extracted [233].

To help estimate the average amount of power that a single device, or small array of devices, could produce, data from the hybrid coordinate ocean model (HYCOM) ocean circulation model are utilized [233]. These data are taken from water velocity snapshots calculated each day for 00h GMT, and do not include tidal components. HYCOM is a three-dimensional, real-time, ocean prediction simulator with a global 1/12° resolution [374]. The HYCOM versions (90.6, 90.8, and 90.9) used to calculate the values represented pictorially assimilate observed data from both historical and near real-time measurements collected by satellites and in situ instruments (ocean currents, sea surface temperature, and sea surface height), conductivity temperature and depth sensor, and moorings (temperature and salinity profiles), and the special sensor microwave imager [375]. The four vertical coordinate schemes utilized by HYCOM in different regions are one of its notable features, and these include an isopycnal structure utilized for deep-ocean modeling, constant depth or pressure schemes utilized for mixed layer modeling, terrain-following coordinate approach utilized for coastal regions, and level coordinates utilized for very shallow regions. HYCOM data utilized for this analysis were taken for a depth of 50 m.

Globally, at a depth of 50 m, 844,500 km^2 is calculated to have a kinetic energy flux over 0.5 kw/m^2. 87,600 km^2 has a kinetic energy flux over 1.0 kw/m^2, and 14,300 km2 has a kinetic energy flux over 1.5 kw/m^2 [376]. For each of the selected eight regions, the maximum temporally averaged kinetic energy flux is as follows: 1.93 kW/m^2 off the SE US mainland, 1.78 kW/m^2 off Japan, 1.66 kW/m^2 off SE Africa, 1.57 kW/m^2 off the Philippines, 1.34 kW/m^2 off NE Africa, 1.08 kW/m^2 off Northern Brazil, 0.86 kW/m^2 off Eastern Madagascar, and 0.73 kW/m^2 off eastern Australia [376]. It is worth noting that the HYCOM model has been shown to slightly underestimate resources at several locations, and therefore, these can be viewed as conservative estimates.

The portion of the Gulf Stream off the coast of southeast Florida is located between Florida and the Bahamas near a major load base. Water velocity measurements taken over a 19-month period during 2001 and 2002 show that approximately 21 km from the coast of Southeast Florida, there is a mean water speed of 1.6 m/s near the surface, with this near-surface speed ranging between 0.2 and 2.4 m/s [377]. A second set of measurements taken at a nearby location over a 13-month period during 2009 and 2010 also show a mean water speed of 1.6 m/s near the surface, with a similar surface speed range of between 0.4 and 2.5 m/s [378]. While it is neither feasible nor desirable to extract all of the energy from an ocean current, the kinetic energy flux does give an idea of the magnitude of these flows. The entire average kinetic energy flux in the portion of the Gulf Stream, between southeast Florida and the Bahamas, has been estimated by Duerr et al. [379] at 19 GW. The Kuroshio Current off the coast of Taiwan has lesser kinetic energy flux, with an estimated value of 5.5 GW at several representative cross sections [380]. While an estimate of the kinetic energy flux of the Agulhas current could not be found, this current is suggested to have the largest volumetric transport of any western boundary current and a mean water velocity near the surface of 1.2 m/s [381]. Pictures of some of the precommercial wave converters are illustrated in Figure 3.37.

3.10 FAR-REACHING INNOVATIONS IN HYDROPOWER

Numerous new inventions for hydropower are constantly occurring. In this section, we examine some far-reaching ideas people are examining to generate power from different types of water-wave energy. Some of these are at the conceptual stage while others are further developed. For example, Taylor, owner of Verdant Power, tested a special three-bladed electric turbine, made of plastic and layered fiberglass, into the waterway that runs along Manhattan's east side. Taylor designed the turbine, which

FIGURE 3.37 Pictures of some precommercial wave converters being examined [382].

looks like a modern fan attached to a torpedo-shaped body, to generate electricity from the push and pull of the river's rushing currents. Five years later, 30 of those turbines were placed in the river, with each unit generating 35 kW of electricity. A new way to boost the power of existing hydroelectric stations by nearly 10% is to apply the spiral-shaped design of human blood vessels to create a similarly shaped *penstock pipe*. The helicoid penstock is similar to a rifle barrel, which has spiral grooves etched inside. Rushing water flows through the helicoid penstock is like a bullet passed through a rifled barrel, begins to spin. The pipes focus the flow of the water directly on the electric turbine, improving the turbine's performance.

In another study, Dickson used a hydroelectric generator that takes advantage of the immense pressure differentials in the deep water of lakes or oceans. He calls it a *hydrosphere* and believes it can generate up to 500 MW of continuous renewable energy. Although he has yet to build a prototype or secure a patent, Dickson's hydrosphere is a type of enclosed, cylindrical hydroelectric dam that works off the varying pressure of ocean or lake water. The hydrosphere led Dickson to another invention, the *air–water–gravity (AWG) generator*, which he believes is the hydropower plant of the future. The AWG is a large, hollow cylinder filled with air and anchored to the seafloor at varying depths. An electrical generator sits inside the cylinder. To generate power, a valve lets water into the device under great pressure. The flowing water enters a vacuum chamber and forces a piston to climb a *stator*, the stationary part of the generator on which a rotor spins. As the piston moves up the stator, it generates electricity. When the piston reaches a metal stop at the top of the stator, it releases a valve connected to a hollow snorkel pipe at the base of the cylinder. The pipe opens, allowing air to decompress. That forces the rotor down the stator, once again generating electricity. Water is also pushed out of the cylinder at great force and out the snorkel pipe to the surface of the ocean. The water shoots out of the top of the pipe like a geyser. The release valve then closes, the water intake reopens, and the cycle repeats itself. Depending on its size and the depth on which it is placed in the ocean, the AWG can produce up to a half a GW of continuous power. The device is as yet not prototyped or patented.

In 2012, one of the first tidal power projects in the USA began delivering power to the electrical grid. The project, an underwater turbine off the coast of Maine, was built by Ocean Renewable Power Co. The turbine resembles an old-fashioned lawn mower, but in essence is a type of undersea windmill. The turbine's foils rotate when the tide rushes in and out of Cobscook Bay near Eastport. The tides in the area are some of the highest in Maine, reaching 20 ft (6 m). The $21 million generator can power 25–30 homes. The company's RiverStar system harvests kinetic energy all along a river rather than in one spot, as dams do. In this system, engineers place a number of "modules" across a river. Each module is made up of a turbine, a stabilizer, a mooring system, and an energy conversion system. High-tension steel cables hold each unit in place and connect one to another in an array. Flowing water passes through the turbines, and as they spin, they collect the river's energy, which drives a generator. Bourne officials say RiverStar can generate 50 kW in a river with a water speed of 4 knots. The company adds that RiverStar does not affect the migration patterns of fish or impede river traffic.

In 2007, Michael Bernitsas, a professor at the University of Michigan, found a way to harness the kinetic energy of a flowing river by looking at how fish move through water. He created a device he calls *VIVACE*, which is short for VIV for aquatic clean energy. As any trout fisherman can tell you, when a school of fish moves through water, the fish curve their bodies and create tiny swirling vortices. The fish push their bodies off the vortices to propel themselves forward. VIVACE works the same way. Engineers place a series of cylinders on the river or ocean bottom. The passing current flows over the cylinders creating vortices, which move the cylinders up and down. Inside each cylinder is a magnet that moves over a metal coil, generating an electrical DC. VIVACE then takes the DC and converts it into AC. Unlike other hydrokinetic technologies, Bernitsas' invention can harness energy from slow-moving rivers. An Israeli company called Leviathan has created a water turbine that can be enclosed in a pipe. When water rushes pass the turbine, it generates electricity. The device, called the Benkatina turbine, works off the water flowing through enclosed water pipes, sewer pipes, canals, and pipes that remove wastewater from factories.

Fulton Innovation, a Michigan-based company, has created Lilliputian hydroelectric technologies that can power electric radios, speakers, clocks, and TVs, using the water coming out of a bathroom faucet. At the heart of the Fulton's splash power technology is a device called a miniature hydroelectric generator. Such generators produce electricity by using the water flow in a house or building. Every time you run the water with a splash power mini-generator, you can power sprinkler systems, emergency lighting, a water softener, and even outdoor camp lighting. The mini-hydroelectric devices can also be used to charge batteries.

Aside from osmotic energy and thermal energy conversion mentioned earlier, it is also possible to convert thermal and chemical energy into electricity by the evaporation-induced and moisture-induced electricity generation processes, respectively. Zhang et al. [383] explore a range of phenomena, which they refer to as "emerging hydro-voltaic effects," in which the electricity is generated by the direct interaction between the materials and the water. On properly designed nanostructures, water can be forced to flow, or create waves or drops, or evaporate naturally. The most interesting aspect of the evaporation and moisture-induced approaches is that no mechanical inputs are needed in the processes. The electric voltage generated in a centimeter-sized carbon black sheet under ambient evaporation condition can be up to 1 V. To improve these technologies and make them ultimately viable, however, greater effort should be devoted to clarifying the electricity generation mechanism.

Recently, PolyU's Department of Building Services Engineering and the Water Supplies Department (WSD) of the Hong Kong Special Administrative Region Government have been working together to turn water mains into an alternative source of power. Hong Kong has a network of water mains travelling over 7,800 km, which is comprehensively monitored by WSD using monitoring devices to make sure that water supply remains clean and well-maintained. Water mains need power for those devices. Conventionally, they are putting small turbines into Hong Kong pipes to produce electricity from drinking water. Hong Kong water main pipes present a real challenge. They are just 1 m across and hold far less water volume and potential energy compared to giant water dams. The resulting turbine is small enough to fit into a pipe, and uses just a fraction of hydro-energy to generate about 80 V, enough to power four compact fluorescent light bulbs. The novel device consists of an external hydroelectric generator and highly efficient spherical water turbine which dips into flowing water and reclaims residual pressure. When water passes through, the turbine drives a central rotating shaft and a micro-generator to produce electricity. The key lies in a number of intelligent designs to extract more energy from flowing water. The eight-blade turbine would only take away a fraction of kinetic energy because it strikes an accurate balance between water volume, water pressure, and consumption of hydrokinetic energy, which boosts efficiency without reducing the momentum of running water to guarantee a reliable water supply. Turbine blades are carefully sized to intersect the largest possible area of water flow and minimize water bypassing. To achieve maximum power output, a revolutionary design made the central rotating shaft hollow on the inside to minimize energy losses when driving the generator and utilize the harvested energy in full. The team also made the water more energetic and produced a strong current with a special metal block placed at the center of the pipe to compress and accelerate the water flow. To further protect the drinking water, the turbine does not have moving parts and does not need any lubricant to eliminate the slightest change of contamination. This technology also points to viable turbine solutions for waters that have little potential for energy generation.

REFERENCES

1. Blakers, A., Stocks, M., Lu, B., Cheng, C., A review of pumped hydro energy storage. *Prog Energy* 2021;3:022003. DOI: 10.1088/2516-1083/abeb5b
2. Uria-Martinez, R., Johnson, M., Shan, R., *U.S. Hydropower Market Report*, a report by Office of Energy Efficiency and Renewable Energy. U.S. Department of Energy, Washington, DC, 2021.
3. Moran, E. F., Lopez, M. C., Moore, N., Müller, N., Hyndman, D. W., Sustainable hydropower in the 21st century. *Proc Natl Acad Sci U S A* 2018;115(47):11891–11898. DOI:10.1073/pnas.1809426115
4. Koritarov, V., Ploussard, Q., Kwon, J., Balducci, P., *A Review of Technology Innovations for Pumped Storage Hydropower, Hydrowires*. ANL-22/08 U.S. Department of Energy, Washington, DC, 2022.
5. Ardiansyah, H., Hydropower technology: Potential, challengers and future. In H. Ardiansyah, P. Ekadewi (Eds), *Indonesia Post-Pandemic Outlook: Strategy towards Net-Zero Emissions by 2060 from the renewables and Carbon-Neutral Energy Perspectives*, 2022, pp. 89–107. BRIN Publishing. DOI: 10.55981/brin.562.c6

6. Zhao, J. F., Oh, U. J., Park, J. C., Park, E. S., Im, H. B., Lee, K. Y., Choi, J. S., A review of world-wide advanced pumped storage hydropower technologies. *IFAC PapersOnLine* 2022;55(9):170–174.
7. Goldman, P., Ahgrimm, J., *Hydropower-Setting a Course for Our Energy Future*, a report by Energy Efficiency and Renewable Energy, DOE/GO-102004-1981. U.S. Department of Energy, Washington, DC, 2004.
8. Simbolotti, G., Tosato, G., Koyama, M., *Hydropower-Technology Brief*, a report by IRENA. IEA-ETSAP, Paris, France, 2015.
9. Kougias, I., Aggidis, G., Avellan, F., Deniz, S., Lundin, U., Moro, A., Muntean, S., Novara, D., Pérez-Díaz, J.I., Quaranta, E., Schild, P., Analysis of emerging technologies in the hydropower sector. *Renewable Sustainable Energy Rev* 2019;113:109257.
10. Anderson, D., Moggridge, H., Warren, P., Shucksmith, J., The impacts of 'run-of-river' hydropower on the physical and ecological condition of rivers. *Water Environ J* 2015;29:268–276. DOI: 10.1111/wej.12101
11. "Hydroelectricity", 2023. Wikipedia, The free encyclopedia, last edited 26 August 2023. https://en.wikipedia.org/wiki/Hydroelectricity.
12. Witt, A. M., Fernandez, A., Mobley, M. H., DeNeale, S. T., Bevelhimer, M. S., Smith, B. T., *How Standard Modular Hydropower Can Enhance the Environmental, Economic, and Social Benefits of New Small Hydropower Development*. This manuscript has been authored by UT-Battelle, LLC under Contract No. DE-AC05-00OR22725 with the U.S. Department of Energy, Washington, DC, 2016.
13. Samu, N., Kao, S.-C., O'Connor, P., *National Hydropower Plant Dataset, Version 1 (Update FY18Q2)*. No. eha-3224. Oak Ridge National Laboratory (ORNL), Oak Ridge, TN, 2016.
14. Uria-Martinez, R., Johnson, M., O'Connor, P., *2014 Hydropower Market Report*. DOE/EE-1195. Wind and Water Power Technologies Office, U.S. Department of Energy, Washington, DC, 2015.
15. Okot, D., Review of small hydropower technology. *Renew Sustain Energy Rev* 2013;26:515–520.
16. Modal, C., Solomon, M., Tew, B., Gerhman, B., Lehner, C., *Analysis of Reservoir-Based Hydroelectric versus Run-of-River Hydroelectric Energy Production*. Land Resources and Environmental Sciences Montana State University, Bozeman, MT, 2014
17. Department of Energy (DOE). Hydropower Vision Report, 2016. Available online: https://energy.gov/eere/water/new-vision-united-states-hydropower (accessed on Feb 15, 2017).
18. Loots, I., Van Dijk, M., Barta, B., Van Vuuren, S. J., & Bhagwan, J. N., A review of low head hydropower technologies and applications in a South African context. *Renew Sustain Energy Rev* 2015;50:1254–1268. DOI: 10.1016/j.rser.2015.05.064
19. Fujimori, H., *Small-Scale Hydropower System*, MEIDEN REVIEW Series No.169 2017 No. 1, p. 30.
20. *Hydro-Small and Mini Hydropower Solutions*, a report by Andritz, Vienna, Austria, 2020, pp. 1–20.
21. Kosnik, L., The potential for small scale hydropower development in the US. *Energy Policy* 2010;38:5512–5519.
22. Nachman-Hunt, N., *Small Hydropower Systems: Energy Efficiency and Renewable Energy Clearinghouse*. No. DOE/GO-102001-1173; NREL/BR-810-29065. National Renewable Energy Laboratory (NREL), Golden CO, 2001.
23. Korkovelos, A., Mentis, D., Siyal, S.H., Arderne, C., Rogner, H., Bazilian, M., Howells, M., Beck, H., De Roo, A., Geospatial assessment of small-scale hydropower potential in Sub-Saharan Africa. *Energies* 2018;11:3100. DOI: 10.3390/en11113100
24. Walczak, N., Operational evaluation of a small hydropower plant in the context of sustainable development. *Water* 2018;10:1114. DOI: 10.3390/w10091114

25. Kougias, I., Karakatsanis, D., Malatras, A., Monforti-Ferrario, F., Theodossiou, N., Renewable energy production management with a new harmony search optimization toolkit. *Clean Technol Environ Policy* 2016;18(8):2603–2612.
26. Kougias, I., Szabo, S., Monforti-Ferrario, F., Huld, T., Bódis, K., A methodology for optimization of the complementarity between small- hydropower plants and solar PV systems. *Renew Energy* 2016;87:1023–1030.
27. Kougias, I., Patsialis, T., Zafirakou, A., Theodossiou, N., Exploring the potential of energy recovery using micro hydropower systems in water supply systems. *Water Utility J* 2014;7:25–33.
28. Patsialis, T., Kougias, I., Kazakis, N., Theodossiou, N., Droege, P., Supporting renewables penetration in remote areas through the transformation of non-powered dams. *Energies* 2016;9(12):1054.
29. Hadjerioua, B., Wei, Y., Kao, S. C., *An Assessment of Energy Potential at Non-Powered Dams in the United States.* GPO DOE/EE-0711, Wind and Water Power Program, Department of Energy, Washington, DC, 2012.
30. Szabó, S., Moner-Girona, M., Kougias, I., Bailis, R., Bódis, K., Identification of advantageous electricity generation options in Sub-Saharan Africa integrating existing resources. *Nat Energy* 2016;1(10):16140.
31. Cunningham, P., Atkinson, B., *Micro Hydro Power Systems: Overview.* A website report by altE store, 2015.
32. European Small Hydropower Association (ESHA), *Small and Micro Hydropower Restoration Handbook.* National Technical University of Athens, Athens, 2014.
33. Quaranta, E., Müller, G., Sagebien and Zuppinger water wheels for very low head hydropower applications. *J Hydraul Res* 2018;56(4):526–536.
34. Bozhinova, S., Hecht, V., Kisliakov, D., Müller, G., Schneider, S., Hydropower converters with head differences below 2.5 m. *Proc Inst Civ Eng Energy* 2013;166(3):107–119.
35. Müller, G., Kauppert, K., Performance characteristics of water wheels. *J Hydraul Res* 2004;42(5):451–460.
36. Studies on the effectiveness alternative engine techniques and protection concepts for migrating fish in the operation of small hydropower plants. Zusammenarbeit, 2008. [In German language].
37. Quaranta, E., Revelli, R., Gravity water wheels as a micro hydropower energy source: A review based on historic data, design methods, efficiencies and modern optimizations. *Renew Sustain Energy Rev* 2018;97:414–427.
38. Quaranta, E., Stream water wheels as renewable energy supply in flowing water: Theoretical considerations, performance assessment and design recommendations. *Energy Sustain Develop* 2018;45:96–109.
39. Quaranta, E., Revelli, R., Output power and power losses estimation for an overshot water wheel. *Renew Energy* 2015;83:979–987.
40. Quaranta, E., Revelli, R., Performance characteristics, power losses and mechanical power estimation for a breastshot water wheel. *Energy* 2015;87:315–325.
41. Senior, J., Wiemann, P., Muller, G. U., The rotary hydraulic pressure machine for very low head hydropower sites. In *International Conference on Small Hydropower. Hidroenergia 2008: "On the Crossroads", Bled, Slovenia, 11th–13th June 2008, 10–12 June 2008*, 2008.
42. Quaranta, E., Müller, G., Butera, I., Capecchi, L., Franco, W., Preliminary investigation of an innovative power take off for low speed water wheels. In *Proceedings of the International IAHR Conference, New Challenges in Hydraulic Research and Engineering*, Trento, Italy, 2018.
43. Dietz, A., Groeger, A., Klingler, C., Efficiency improvement of small hydroelectric power stations with a permanent-magnet synchronous generator. In *2011 1st International Electric Drives Production Conference*, 2011, pp. 93–100.

44. Quaranta, E., Revelli, R., Optimization of breastshot water wheels performance using different inflow configurations. *Renew Energy* 2016;97:243–251.
45. Senior, J., Saenger, N., Müller, G., New hydropower converters for very low-head differences. *J Hydraul Res* 2010;48(6):703–714.
46. Butera, I., Fontan, S., Poggi, D., Quaranta, E., Revelli, R., Laboratory results on the effect of channel geometry on a rotary hydraulic pressure machine. In *EGU General Assembly Conference Abstracts*, 2018, vol. 20, p. 7705.
47. Helmizar, H., Turbine wheel-a hydropower converter for head differences between 2.5 and 5 m [Ph.D. Thesis]. University of Southampton, 2016.
48. Agarwal, T., Review of pump as turbine (PAT) for micro-hydropower. *Int J Emerg Technol Adv Eng* 2012;2(11):163–169.
49. Williams, A. A., Pumps as turbines for low cost micro hydro power. *Renew Energy* 1996;9(1–4):1227–1234.
50. Alatorre-Frenk, C., Cost minimisation in micro-hydro systems using pumps- as-turbines [Ph.D. Thesis]. University of Warwick, 1994.
51. Chapallaz, J. M., Eichenberger, P., Fischer, G., Manual on Pumps Used as Turbines. Vieweg, Braunschweig, Germany, 1992.
52. Garay, P. N., Using pumps as hydro-turbines. *Hydro Rev* 1990;9(5):52–61.
53. Jain, S. V., Patel, R. N., Investigations on pump running in turbine mode: A review of the state-of-the-art. *Renew Sustain Energy Rev* 2014;30:841–868.
54. Novara, D., Derakhshan, S., McNabola, A., Ramos, H. M., Estimation of unit cost and maximum efficiency for pumps as turbines. In *Proceedings of the 9th Eastern European IWA Young Water Professionals*, Budapest, Hungary, 24–27 May 2017.
55. Ogayar, B., Vidal, P. G., Cost determination of the electro-mechanical equipment of a small hydro-power plant. *Renew Energy* 2009;34(1):6–13.
56. Gallagher, J., Harris, I. M., Packwood, A. J., McNabola, A., Williams, A. P., A strategic assessment of micro-hydropower in the UK and Irish water industry: Identifying technical and economic constraints. *Renew Energy* 2015;81:808–815.
57. Kougias, I., Moro, A. (Eds), Emerging technologies in the hydropower sector. In *JRC Conference and Workshop Reports*, 2018, pp. 1–47. JRC111048. Low Carbon Energy Observatory Deliverable 360.
58. KSB SE & Co. KGa A. KSB pumps used as turbines: Trend-setters in energy generation and recovery. KSB Magazine, 2018.
59. Arriaga, M., Pump as turbine-a pico-hydro alternative in Lao People's Democratic Republic. *Renew Energy* 2010;35(5):1109–1115.
60. Motwani, K. H., Jain, S. V., Patel, R. N., Cost analysis of pump as turbine for pico hydropower plants-a case study. *Procedia Eng* 2013;51:721–726.
61. Choulot, A., Energy recovery in existing infrastructures with small hydropower plants. FP6 Project Shapes (Work Package 5WP5), 2010.
62. Fecarotta, O., Carravetta, A., Ramos, H. M., Martino, R., An improved affinity model to enhance variable operating strategy for pumps used as turbines. *J Hydraul Res* 2016;54(3):332–341.
63. Yang, S. S., Derakhshan, S., Kong, F. Y., Theoretical, numerical and experimental prediction of pump as turbine performance. *Renew Energy* 2012;48:507–513.
64. Sharma, K., Small Hydroelectric Project-Use of Centrifugal Pumps as Turbines. Kirloskar Electric Co., Bangalore, India, 1985.
65. Rawal, S., Kshirsagar, J. T., Numerical simulation on a pump operating in a turbine mode. In *23rd International Pump Users Symposium*, 2007.
66. Lima, G. M., Luvizotto Júnior, E., Method to estimate complete curves of hydraulic pumps through the polymorphism of existing curves. *J Hydraul Eng* 2017;143(8):04017017.
67. Carravetta, A., Del Giudice, G., Fecarotta, O., Ramos, H. M., Energy production in water distribution networks: A pat design strategy. *Water Resour Manag* 2012;26(13):3947–3959.

68. Carravetta, A., Del Giudice, G., Fecarotta, O., Ramos, H. M., Pat design strategy for energy recovery in water distribution networks by electrical regulation. *Energies* 2013;6(1):411–424.
69. Carravetta, A., Del Giudice, G., Fecarotta, O., Ramos, H. M., Pump as turbine (pat) design in water distribution network by system effectiveness. *Water* 2013;5(3):1211–1225.
70. Singh, P., Optimization of internal hydraulics and of system design for pumps as turbines with field implementation [PhD. Thesis]. Universit'at Karlsruhe, 2005.
71. Capelo, B., Pérez-Sánchez, M., Fernandes, J. F., Ramos, H. M., López-Jiménez, P. A., Branco, P. C., Electrical behaviour of the pump working as turbine in off grid operation. *Appl Energy* 2017;208:302–311.
72. Biner, D., Hasmatuchi, V., Violante, D., Richard, S., Chevailler, S., Andolfatto, L., Avellan, F., Münch, C., Engineering & performance of DuoTurbo: Microturbine with counter-rotating runners. *IOP Conf Ser Earth Environ Sci* 2016;49:102013.
73. Gomes Borga Delgado, J. N., Pumps running as turbines for energy recovery in water supply systems. EPFL, 2018.
74. Tecnoturbines Paterna, *Study Case: Turbine for A Drinking Water Tank in La Coma* (Technical report). Tecnoturbines Powering Water, Alicante, Spain, 2016.
75. Rentricity Inc., *12-kW Flow-to-Wire System Installed in Distribution System*, 2014.
76. KSB SE & Co. KGa A. KSB's Powerhouse: Small turnkey hydropower system. KSB Magazine, 2018.
77. Tănasă, C., Bosioc, A., Muntean, S., Susan-Resiga, R., A novel passive method to control the swirling flow with vortex rope from the conical diffuser of hydraulic turbines with fixed blades. *Appl Sci* 2019;9:4910. DOI: 10.3390/app9224910
78. Bunea, F., Ciocan, G. D., Bucur, D. M., Dunca, G., Nedelcu, A. Hydraulic turbine performance assessment with implementation of an innovative aeration system. *Water* 2021;13:2459. DOI: 10.3390/w13182459
79. *Types of Hydropower Turbines*, a website report by Office of Energy Efficiency and Renewable Energy. Department of Energy, Washington, DC, 2017.
80. Odeh, M., *A Summary of Environmentally Friendly Turbine Design Concepts*, DOE/ID/13741. U.S. Department of Energy, Idaho Operations Office, Idaho, 1999.
81. Yang, C., Zheng, Y., Zhang, Y., Luo, H. A review of research on the design of fish- friendly hydraulic turbines, Strategic Study CAE 2018;20(3): 96–101. DOI: 10.15302/J-SSCAE-2018.03.014
82. Postacchini, M., Darvini, G., Finizio, F., Pelagalli, L., Soldini, L., Di Giuseppe, E., Hydropower generation through pump as turbine: Experimental study and potential application to small-scale WDN. *Water* 2020;12:958. DOI: 10.3390/w12040958
83. AV, D., Bade, M. H., Pump as turbine: Review of simple modifications for performance improvement. In *International Conference & Expo on "Advances in Power Generation from Renewable Energy Sources (APGRES 2017)" December 22–23, 2017 at GEC Banswara.* www.apgres.in
84. Susan-Resiga, R., Vu, T. C., Muntean, S., Ciocan, G. D., Nennemann, B., Jet control of the draft tube vortex rope in Francis turbines at partial discharge. In *23rd IAHR Symposium Conference*, October 2006.
85. Wang, L., Cui, J., Shu, L., Jiang, D., Xiang, C., Li, L., Zhou, P., Research on the vortex rope control techniques in draft tube of Francis turbines. *Energies* 2022;15:9280. DOI:10.3390/en15249280
86. Quaranta, E., Pérez-Díaz, J. I., Romero–Gomez, P., Pistocchi, A., Environmentally enhanced turbines for hydropower plants: Current technology and future perspective. *Front Energy Res* 2021;9:703106. DOI: 10.3389/fenrg.2021.703106
87. Thicke, R. H., Practical solutions for draft tube instability. *Water Power Dam Constr* 1981;33(2):31–37.

88. Nishi, M., Liu, S., An outlook on the draft-tube-surge study. *Int J Fluid Mach Syst* 2013;6(1):33–48.
89. Kurokawa, J., Kajigaya, A., Matsui, J., Imamura, H., Suppression of swirl in a conical diffuser by use of J-groove. In *Proceedings of 20th IAHR Symposium on Hydraulic Machinery and Systems*, August 6th–9th 2000, Charlotte, USA, 2000, pp. 1–10.
90. Kurokawa, J., Imamura, H., Choi, Y. D., Effect of J-groove on the suppression of swirl flow in a conical diffuser. *J Fluids Eng* 2010;132(7):071101.
91. Nishi, M., Wang, X. M., Yoshida, K., Takahashi, T., Tsukamoto, T., An experimental study on fins, their role in control of the draft tube surging. In *Hydraulic Machinery and Cavitation: Proceedings of the XVIII IAHR Symposium on Hydraulic Machinery and Cavitation*, 1996, pp. 905–914.
92. Tănasă, C., Susan-Resiga, R., Muntean, S., Bosioc, A. I., Flow-feedback method for mitigating the vortex rope in decelerated swirling flows. *J Fluids Eng* 2013;135(6):061304.
93. Papillon, B., Sabourin, M., Couston, M., Deschenes, C., Methods for air admission in hydro turbines. In *Proceedings of the 21st IAHR Symposium on Hydraulic Machinery and Systems*, Lausanne, Switzerland, 2002, pp. 1–6.
94. Qian, Z. D., Yang, J. D., Huai, W. X., Numerical simulation and analysis of pressure pulsation in Francis hydraulic turbine with air admission. *J Hydrodyn* 2007;19(4):467–472.
95. Trivedi, C., Cervantes, M. J., Dahlhaug, O. G., Gandhi, B. K., Experimental investigation of a high head Francis turbine during spin-no-load operation. *J Fluids Eng* 2015;137:061106.
96. Javadi, A., Bosioc, A., Nilsson, H., Muntean, S., Susan-Resiga, R., Experimental and numerical investigation of the processing helical vortex in a conical diffuser, with rotor–stator interaction. *J Fluids Eng* 2016;138(8):081106.
97. Muntean, S., Bosioc, A. I., Szakal, R. A., Vékás, L., Susan-Resiga, R. F., Hydrodynamic investigations in a swirl generator using a magneto-rheological brake. In *Materials Design and Applications*. Springer, 2017, pp. 209–218.
98. Susan-Resiga, R. F., Stuparu, A., Muntean, S., Francis turbine with tandem runners: A proof of concept. In *Proceedings of the 29th IAHR Symposium on Hydraulic Machinery and Systems*, Kyoto, Japan, 2018, pp. 1–8.
99. Yuce, M. I., Muratoglu, A., Hydrokinetic energy conversion systems: A technology status review. *Renew Sustain Energy Rev* 2015;43:72–82.
100. Pérez-Loya, J. J., Abrahamsson, C. J., Lundin, U., Demonstration of active compensation of unbalanced magnetic pull in synchronous machines. *CIGRE Sci Eng* 2017;8:98–107.
101. Chiba, A., Deido, T., Fukao, T., Rahman, M. A., An analysis of bearingless AC motors. *IEEE Trans Energy Convers* 1994;9(1):61–68.
102. Chiba, A., Fukao, T., Ichikawa, O., Oshima, M., Takemoto, M., Dorrell, D. G., *Magnetic Bearings and Bearingless Drives*. Elsevier, 2005.
103. Yao, F., An, Q., Sun, L., Lipo, T. A., Performance investigation of a brushless synchronous machine with additional harmonic field windings. *IEEE Trans Ind Electron* 2016;63(11):6756–6766.
104. Dai, J., Hagen, S., Ludois, D. C., Brown, I. P., Synchronous generator brushless field excitation and voltage regulation via capacitive coupling through journal bearings. *IEEE Trans Ind Appl* 2017;53(4):3317–3326.
105. Nguyen, T., Everaert, G., Boets, P., Forio, M., Bennetsen, E., Volk, M., Hoang, T. H. T., Goethals, P. L., Modelling tools to analyze and assess the ecological impact of hydropower dams. *Water* 2018;10(3):259. DOI: 10.3390/w10030259
106. Geist, J., Editorial: Green or red: Challenges for fish and freshwater biodiversity conservation related to hydropower. *Aquat Conserv: Mar Freshw Ecosyst* 2021;31:1551–1558. DOI: 10.1002/aqc.3597
107. Quaranta, E., Katopodis, C., Revelli, R., Comoglio, C., Turbulent flow field comparison and related suitability for fish passage of a standard and a simplified low-gradient vertical slot fishway. *River Res Appl* 2017;33(8):1295–1305.

108. Fjeldstad, H. P., Pulg, U., Forseth, T., Safe two-way migration for salmonids and eel past hydropower structures in Europe: A review and recommendations for best-practice solutions. *Mar Freshw Res* 2018;69:1834–1847.
109. Wilkes, M. A., Mckenzie, M., Webb, J. A., Fish passage design for sustainable hydropower in the temperate southern hemisphere: An evidence review. *Rev Fish Biol Fish* 2018;28:117–135.
110. Silva, A. T., Katopodis, C., Santos, J. M., Ferreira, M. T., Pinheiro, A. N., Cyprinid swimming behaviour in response to turbulent flow. *Ecol Eng* 2012;44:314–328.
111. Thorstad, E. B., Økland, F., Kroglund, F., Jepsen, N., Upstream migration of Atlantic salmon at a power station on the River Nidelva, Southern Norway. *Fish Manag Ecol* 2013;10(3):139–146.
112. Mueller, M., Sternecker, K., Milz, S., Geist, J., Assessing turbine passage effects on internal fish injury and delayed mortality using X-ray imaging. *PeerJ* 2020;8:e9977. DOI: 10.7717/peerj.9977
113. Brown, R. S., Carlson, T. J., Gingerich, A. J., Stephenson, J. R., Pflugrath, B. D., Welch, A. E., Langeslay, M. J., Ahmann, M. L., Johnson, R. L., Skalski, J. R., Seaburg, A. G., Quantifying mortal injury of juvenile Chinook salmon exposed to simulated hydro-turbine passage. *Trans Am Fish Soc* 2012;141(1):147–157. DOI: 10.1080/00028487.2011.650274
114. Richmond, M. C., Serkowski, J. A., Ebner, L. L., Sick, M., Brown, R. S., Carlson, T. J., Quantifying barotrauma risk to juvenile fish during hydro-turbine passage. *Fish Res* 2014;154:152–164. DOI: 10.1016/j.fishres.2014.01.007
115. Brown, R. S., Colotelo, A. H., Pflugrath, B. D., Boys, C. A., Baumgartner, L. J., Deng, Z. D., Silva, L. G., Brauner, C. J., Mallen-Cooper, M., Phonekhampeng, O., Thorncraft, G., Understanding barotrauma in fish passing hydro structures: A global strategy for sustainable development of water resources. *Fisheries* 2014;39:108–122. DOI: 10.1080/03632415.2014.883570
116. Amaral, S. V., Hecker, G. E., Dixon, D. A., Designing leading edges of turbine blades to increase fish survival from blade strike. In *EPRI-DOE, Conference on Environmentally-Enhanced Hydropower Turbines*, EPRI, California, 19–20 May 2011.
117. Deng, Z., Carlson, T. J., Ploskey, G. R., Richmond, M. C., Dauble, D. D., Evaluation of blade-strike models for estimating the biological performance of Kaplan turbines. *Ecol Model* 2007;208(2–4):165–176. DOI: 10.1016/j.ecolmodel.2007.05.019
118. Ferguson, J. W., Ploskey, G. R., Leonardsson, K., Zabel, R. W., Lundqvist, H., Combining turbine blade-strike and life cycle models to assess mitigation strategies for fish passing dams. *Can J Fish Aquat Sci* 2008;65(8):1568–1585. DOI: 10.1139/f08–078
119. Fu, T., Deng, Z. D., Duncan, J. P., Zhou, D., Carlson, T. J., Johnson, G. E., Hou, H., Assessing hydraulic conditions through Francis turbines using an autonomous sensor device. *Renew Energy* 2016;99:1244–1252. DOI: 10.1016/j.renene.2016.08.029
120. Hogan, T. W., Cada, G. F., Amaral, S. V., The status of environmentally enhanced hydropower turbines. *Fisheries* 2014;39(4):164–172. doi:10.1080/03632415.2014.897195
121. Nielson, N. M., Brown, R. S., Deng, Z. D., Review of Existing Knowledge of the Effectiveness and Economics of Fish-Friendly Turbines (Technical Paper No. 57). Phnom Penh, Mekong River Commission, Vientiane, Lao PDR, 2015
122. Twaróg, B., An analysis of the application of the Alden turbine–a case study of the Dobczyce hydroelectric power plant. *Czasopismo Techniczne* 2015:147–156. DOI:10.4467/2353737XCT.15.193.4398
123. Hammar, L., Andersson, S., Eggertsen, L., Haglund, J., Gullström, M., Ehnberg, J., Molander, S., Hydrokinetic turbine effects on fish swimming behaviour. *PLoS One* 2013;8(12):e84141. DOI: 10.1371/journal.pone.0084141

124. Quaranta, E., Wolter, C., Sustainability assessment of hydropower water wheels with downstream migrating fish and blade strike modelling. *Sustainable Energ Tech Assessments* 2021;43:100943. DOI:10.1016/j.seta.2020.100943
125. Müller, S., Cleynen, O., Hoerner, S., Lichtenberg, N., Thévenin, D., Numerical analysis of the compromise between power output and fish-friendliness in a vortex power plant. *J Ecohydraulics* 2018;3(2):86–98. DOI: 10.1080/24705357.2018.1521709
126. Brackley, R., Interactions between migrating salmonids and low-head hydropower schemes [PhD thesis]. University of Glasgow, Glasgow, 2016.
127. Piper, A. T., Rosewarne, P. J., Wright, R. M., and Kemp, P. S., The impact of an Archimedes screw hydropower turbine on fish migration in a lowland river. *Ecol Eng* 2018;118:31–42. DOI: 10.1016/j.ecoleng.2018.04.009
128. Bozhinova, S., Hecht, V., Kisliakov, D., Müller, G., Schneider, S., Hydropower converters with head differences below 2·5 M. *Proc Inst Civil Eng Energ* 2013;166(3):107–119. DOI: 10.1680/ener.11.00037
129. Quaranta, E., Bonjean, M., Cuvato, D., Nicolet, C., Dreyer, M., Gaspoz, A., Rey-Mermet, S., Boulicaut, B., Pratalata, L., Pinelli, M., Tomaselli, G., Hydropower case study collection: Innovative low head and ecologically improved turbines, hydropower in existing infrastructures, hydropeaking reduction, digitalization and governing systems. *Sustainability* 2020;12(21):8873. DOI: 10.3390/su12218873
130. EPRI and DOE, Fish Friendly Hydropower Turbine Development and Deployment. Alden Turbine Preliminary Engineering and Model Testing. EPRI and U.S. Department of Energy, Palo Alto, CA, Washington, DC, 2011.
131. Dixon, D., Hogan, T., Session B3: Alden fish-friendly hydropower turbine: History and development status. In *International Conference on Engineering and Ecohydrology for Fish Passage*, Groningen, Netherlands, 22–24 June, 2015.
132. Pleizier, N. K., Nelson, C., Cooke, S. J., Brauner, C. J., Understanding gas bubble trauma in an era of hydropower expansion: How do fish compensate at depth. *Can J Fish Aquat Sci* 2020;77(3):556–563. DOI: 10.1139/cjfas-2019-0243
133. March, P., Jacobson, P., Industry Experience with Aerating Turbines. HydroVision International, Portland, OR, 2015.
134. Foust, J. M., Coulson, S., Using dissolved oxygen prediction methodologies in the selection of turbine aeration equipment. In *EPRI-DOE Conference on Environmentally Enhanced Hydropower Turbines: Technical Papers*, EPRI. Report No. 1024609. Electric Power Research Institute, Palo Alto, California, 2011.
135. Douglas, J., Tong, K., Building 7.5-MW Dorena Lake at an Existing Dam. Renewable Energy World. Hydro Review, 2012, pp. 16–19. Available online: https://www.renewableenergyworld.com/storage/new-development-building-7-5-mw-dorena-lake-at-an-existing-dam/#gref (accessed on June 10, 2013).
136. March, P., Hydraulic and environmental performance of aerating turbine technologies. In *EPRI Conference on Environmentally Enhanced Hydropower Turbines*, Washington DC, 19–20 May 2011, pp. 1–32.
137. March, P. A., Fisher, R. K., IT'SNOTEASYBEINGGREEN: Environmental technologies enhance conventional hydropower's role in sustainable development. *Annu Rev Energ Environ* 1999;24(1):173–188. DOI: 10.1146/annurev.energy.24.1.173
138. McIntosh, D., Cherwinski, D., Kahl, T., Hardy Unit No. 3 Turbine Rehabilitation. In Proceedings of HydroVision *2010*. PennWell Corporation, Tulsa, Oklahoma, 2010.
139. Bunea, F., Ciocan, G. D., Bucur, D. M., Dunca, G., Nedelcu, A., Hydraulic turbine performance assessment with implementation of an innovative aeration system. *Water* 2021;13:2459. DOI: 10.3390/w13182459
140. Kramer, M., Wieprecht, S., Terheiden, K., Minimising the air demand of micro-hydro impulse turbines in counter pressure operation. *Energy* 2017;133:1027–1034. DOI: 10.1016/j.energy.2017.05.043

141. Keck, H., Vullioud, G., Joye, P., Commissioning and operation experience with the world's largest Pelton turbines Bieudron. In *Proceedings of HydroVision 2000 Conference*, Charlotte, USA, 2000, pp. 1–12.
142. Hirtenlehner, K., *The Initial Experience of the Start-Up and the First Operation of Pelton Turbine in Back Pressure Operation*. ZR Hirtenlehner, 2008.
143. St Germain, F., *Addressing Pressure Loss and Oil Leakage in Kaplan-Bulb Turbines and the Impact on Efficiency, BBA, Mont-Saint-Hilaire*. Québec, Canada, 2018. Available online: https://www.bba.ca/publication/addressing-pressure-loss-issues-for-the-kaplan-turbine-runner-blade-and-impact-on-efficiency/ (accessed on June 1, 2019).
144. Zulović, E., *Preventing Problems through Past Experience*. Renewable Energy World, 2014. Available online: https://www.renewableenergyworld.com/baseload/preventing-problems-through-past-experience-2/#gref (accessed on March 15, 2015).
145. Värlind, K. -E., Oil-free operation. *Int Water Power Dam Construction* 2002;54(3):32–34
146. Auger, G., Ren, G., Reducing Maintenance with Water Lubricated Turbine Guide Bearings-Design Principles & Case Studies, Tribodays, 28–29 November 2017. Hydropower and Dams, Älvkarleby, Sweden, 2017.
147. Falkenhem, R., Nakagawa, N., Havard, D., *Technologies for Eliminating Oil in Kaplan-Bulb Turbines Canals, Tunnels and Penstocks, Rehabilitation and Repair, Technology and Equipment*. HydroReview, 2011, p. 6. Available online: https://www.hydroreview.com/world-regions/technologies-for-eliminating-oil-in-kaplan/#gref (accessed on Nov 15, 2011).
148. Quaranta, E., Davies, P., *Emerging and Innovative Materials for Hydropower Engineering Applications: Turbines, Bearings, Sealing, Dams and Waterways, and Ocean Power Engineering*. Elsevier, 2021.
149. Ingram, E., Ray, R., *Bearings & Seals: Examples of Innovations and Good Ideas*. Renewable Energy Worlds, 2010. Available online: https://www.renewableenergyworld.com/storage/bearings-amp-seals/#gref (accessed on Jan 15, 2011).
150. Oguma, T., Nakagawa, N., Mikami, M., Thantrong, L., Kizaki, Y., Takimoto, F., Water lubricated guide bearing with self-aligning segments. *Int J Fluid Machinery Syst* 2013;6(2):49–55. doi:10.5293/IJFMS.2013.6.2.049
151. Wang, H., Ahmed, O., Smith, B. T., Bellgraph, B., Developing a digital twin for hydropower systems – An open platform framework. *Int Water Power Dam Constr* 2021;81(3):24–25. Available online: http://energy.gov/downloads/doe-public-access-plan (accessed on Jan 23, 2021).
152. Vagnoni, E., Gerini, F., Cherkaoui, R., Paolone, M., Digitalization in hydropower generation: Development and numerical validation of a model-based smart power plant supervisor. In 30th IAHR Symposium on Hydraulic Machinery and Systems IOP Publishing IOP *Conference Series*: Earth and Environmental Science, 2021, vol. 774, p. 012107 DOI: 10.1088/1755-1315/774/1/012107
153. *Hydropower Embraces Digitalization*, a website report by Water Power & Dam Construction, 27 March 2020.
154. Jeff St. John. Behind New York power authority's "digital avatar" project with GE. October 2016.
155. Nicolet, C., Braun, O., Ruchonnet, N., Hell, J., Béguin, A., Avellan, F. Simulation of pump-turbine prototype fast mode transition for grid stability support. *J Phys Conf Ser* 2017;813:012040.
156. Inside the world's first digital twin of a hydroelectric power station, a website report by Power technology, April 20, 2020.
157. Dreyer, M., Nicolet, C., Gaspoz, A., Gonçalves, N., Rey-Mermet, S., Boulicaut, B., Boulicaut monitoring 4.0 of penstocks: Digital twin for fatigue assessment. In *30th IAHR Symposium on Hydraulic Machinery and Systems IOP Publishing IOP Conference* Series: Earth and Environmental Science, 2021, vol. 774, p. 012009. DOI: 10.1088/1755-1315/774/1/012009

158. Borkowski, D., Majdak, M., Small hydropower plants with variable speed operation—an optimal operation curve determination. *Energies* 2020;13(23):6230. DOI: 10.3390/en13236230
159. Singh, K., Singal, S. K., Operation of hydro power plants-a review. *Renew Sustain Energy Rev* 2017;69:610–619.
160. Bortoni, E., de Souza, Z., Viana, A., Villa-Nova, H., Rezek, A., Pinto, L., Siniscalchi, R., Bragança, R., Bernardes Jr, J., The benefits of variable speed operation in hydropower plants driven by Francis turbines. *Energies* 2019;12(19):3719. DOI: 10.3390/en12193719
161. Valavi, M., Nysveen, A., Variable-speed operation of hydropower plants: a look at the past, present, and future. *IEEE Ind Appl Mag* 2018;24(5):18–27.
162. Borkowski, D., Analytical model of small hydropower plant working at variable speed. *IEEE Trans Energy Convers* 2018;33:1886–1894.
163. Inage, S., The role of large-scale energy storage under high shares of renewable energy. *Wires Energy Environ* 2015;4:115–132.
164. Bessa, R., Moreira, C., Silva, B., Filipe, J., Fulgêncio, N., Role of pump hydro in electric power systems. *J Phys Conf Ser* 2017;813:012002.
165. Reigstad, T. I., Uhlen, K., Optimized control of variable speed hydropower for provision of fast frequency reserves. *Electr Power Syst Res* 2020;189:106668.
166. Kouro, S., Rodriguez, J., Wu, B., Bernet, S., Perez, M., Powering the future of industry: High-power adjustable speed drive topologies. *IEEE Ind Appl Mag* 2012;18:26–39.
167. Pannatier, Y., Kawkabani, B., Nicolet, C., Simond, J. J., Schwery, A., Allenbach, P., Investigation of control strategies for variable-speed pump-turbine units by using a simplified model of the converters. *IEEE Trans Ind Electron* 2010;57:3039–3049.
168. Padoan, A. C., Kawkabani, B., Schwery, A., Ramirez, C., Nicolet, C., Simond, J. J., Avellan, F., Dynamical behavior comparison between variable speed and synchronous machines with PSS. *IEEE Trans Power Syst* 2010;25:1555–1565
169. Lung, J. K., Lu, Y., Hung, W. L., Kao, W. S., Modeling and dynamic simulations of doubly fed adjustable-speed pumped storage units. *IEEE Trans Energy Convers* 2007;22:250–258.
170. Bocquel, A., Janning, J., Analysis of a 300 MW variable speed drive for pump-storage plant applications. In Proceedings of the European Conference on Power Electronics and Applications, Dresden, Germany, 11–14 September 2005.
171. Ambati, B. B., Khadkikar, V., Variable frequency transformer configuration for decoupled active-reactive powers transfer control. *IEEE Trans Energy Convers* 2016;31:906–914.
172. Ambati, B. B., Kanjiya, P., Khadkikar, V., El Moursi, M. S., Kirtley, J. L., A hierarchical control strategy with fault ride-through capability for variable frequency transformer. *IEEE Trans Energy Convers* 2015;30:132–141.
173. Merkhouf, A., Doyon, P., Upadhyay, S., Variable frequency transformer—Concept and electromagnetic design evaluation. *IEEE Trans Energy Convers* 2008;23:989–996.
174. Bakhsh, F. I., Khatod, D. K., A new synchronous generator based wind energy conversion system feeding an isolated load through variable frequency transformer. *Renew Energy* 2016;86:106–116.
175. Naidu, M., Mathur, R. M., Evaluation of unit connected, variable speed, hydropower station for HVDC power transmission. *IEEE Trans Power Syst* 1989;4:668–676.
176. Arrillaga, J., Sankar, S., Arnold, C. P., Watson, N. R., Characteristics of unit-connected HVDC generator-convertors operating at variable speeds. *IEE Proc C Gener Trans Distrib* 1992;139:295–299.
177. Yang, W., Yang, J., Advantage of variable-speed pumped storage plants for mitigating wind power variations: Integrated modelling and performance assessment. *Appl Energy* 2019;237:720–732.

178. Nababan, S., Muljadi, E., Blaabjerg, F., An overview of power topologies for micro-hydro turbines. In Proceedings of the 3rd International Symposium on Power Electronics for Distributed Generation Systems, Aalborg, Denmark, 25–26 June 2012, pp. 737–744.
179. Borkowski, D., Węgiel, T., Small hydropower plant with integrated turbine-generators working at variable speed. *IEEE Trans Energy Convers* 2013;28:452–459.
180. Gish, W. B., Schurz, J. R., Milano, B., Schleif, F. R., An adjustable speed synchronous machine for hydroelectric power applications. *IEEE Trans Power Apparatus Syst* 1981(5):2171–2176.
181. KEPCO (Kansai Electric Power Company). Environmental report, 2018.
182. Desingu, K., Selvaraj, R., Chelliah, T. R., Khare, D., Effective utilization of parallel connected megawatt three-level back-to-back power converters in variable speed pumped storage units. *IEEE Trans Ind Appl* 2018;55(6):6414–6426.
183. Kuwabara, T., Shibuya, A., Furuta H., Design and dynamic response characteristics of 400 MW adjustable speed pumped storage unit for Ohkawachi power station. *IEEE Trans Energy Convers* 1996;11(2):376–384.
184. Fujihara, T., Imano, H., Oshima, K., Development of pump turbine for seawater storage power plant. *Hitachi Rev* 1998;47(5):199–202.
185. Nagura, O, Higuchi, M, Tani, K, Oyake, T., Hitachi's adjustable-speed pumped-storage system contributing to prevention of global warming. *Hitachi Rev* 2010;59(3):99–105.
186. JICA (Japan International Energy Agency). Final report on feasibility study on adjustable speed pumped storage generation technology 2012. Available online: http://open_jicareport.jica.go.jp/pdf/12044822.pdf (accessed on April 7, 2013).
187. Koritarov, V., Veselka, T.D., Gasper, J., Bethke, B.M., Botterud, A., Wang, J., Mahalik, M., Zhou, Z., Milostan, C., Feltes, J., Kazachkov, Y., Guo, T., Liu, G., Trouille, B., Donalek, P., King, K., Ela, E., Kirby, B., Krad, I., Gevorgian, V., *Modelling and Analysis of Value of Advanced Pumped Storage Hydropower in the United States*. ANL/DIS-14/7. Argonne National Laboratory, Argonne, IL, 2014.
188. Kubo, T., Tojo, H., Mori, J., Shiozaki, T., Watnabe, T., *Large-Capacity Adjustable-Speed Pumped-Storage Power System*. The Japan Society of Mechanical Engineers Medal for New Technology, 2015. Available online: https://www.jsme.or.jp/award/jsme2015/mnt2015-2.pdf (accessed on August 10, 2016).
189. Suganuma, S., Operation of pumped storage (PSHP) hydropower in TEPCO, workshop on pumped storage and variable renewables integration. Mexico City, Mexico, July 28, 2018.
190. Iliev, I., Trivedi, C., Dahlhaug, O. G., Variable-speed operation of Francis turbines: A review of the perspectives and challenges. *Renew Sustain Energy Rev* 2019;103:109–121.
191. Wang, D., Zhang, L., Yang, B., Li, G., Tao, Y., Fu, J., Li, J., Ji, L., Developing and simulation research of the control model and control strategy of static frequency converter. In Proc. 2nd International Conference on Intelligent System Design and Engineering Application, Sanya, Hainan, China, 6–7 January, 2012.
192. KWI Architects Engineers Consultants. *Status Report on Variable Speed Operation in Small Hydropower*. European Commission, Directorate-General for Energy and Transport, Energie, New Solutions in Energy, St. Pölten, Austria, 2000.
193. Merino, J. M., López, Á. ABB Varspeed generator boosts efficiency and operating flexibility of hydropower plant. *ABB Rev* 1996(3):33–38.
194. AEG. *Control of the Line by a Pump-Storage Power Generator in the Water Power Station Forbach*. AEG Technik Magazin, 1993, p. 4. Available online: http://www.aeg-ie.com/englisch/download.htm (accessed on Feb 27, 1994)
195. Bard, J., Pirttiniemi, H., Goede, E., Mueller, A., Upadhyay, D., Rothert, M., VASOCOMPACT – a European project for the development of a commercial concept for variable speed operation of submersible compact turbines. Presented at HIDROENERGIA International Conference, Crieff, Scotland, UK, June 7–9, 2006.

196. Basic, M., de Oliveira e Silva, P. C., Dujic, D., High power electronics innovation perspectives for pumped storage power plants. In Proc. HYDRO Conference, Gdansk, Poland, October 15–17, 2018.
197. Aubert, S., Power on tap. A pumped storage solution to meet energy and tariff demands. *ABB Rev* 2011;3:26–31.
198. Münch, C., *Innovative Technologies for Hydropower.* HES·SO Valais-Wallis PhD School, 2016. Available online: https://www.hevs.ch/media/document/2/munch_sccer_soe_phdschool2016.pdf (accessed on Jan 1, 2017).
199. eStorage. New rotor design guidelines for both doubly and full-fed solutions, Deliverable 1.4. Publishable Summary, eStorage project, 2017. Available online: http://www.estorage-project.eu/document-library/ (accessed on March 7, 2018).
200. Hildinger, T., Ködding, L., Eilebrecht, P., Kunz, A., Henning, H., Frades II – europe¡s largest and most powerful doubly-fed induction machine. A step ahead in variable speed machines. In *Proc. HYDROVISION International*, Charlotte, North Carolina, June 26–28, 2018.
201. Seingre G., Nant de Drance 900 MW pumped storage power plant, presented at International Tunneling and Underground Space Association. (ITA) Awards, 2014.
202. Ingram E., New Chinese pumped-storage hydro plant to be the "world's largest" when completed in 2021. *Hydro Rev* 2017;9.
203. Joseph, A., Desingu, K., Semwal, R. R., Chelliah, T. R., Khare, D., Dynamic performance of pumping mode of 250 MW variable speed hydro-generating unit subjected to power and control circuit faults. *IEEE Trans Power Syst* 2018;33(1):430–441.
204. IEA (International Energy Agency). *Renewal and Upgrading of Hydropower Plants.* Volume 2: Case Histories Report. IEA Technical Report, 2016.
205. Blankenship, G., *Open or Closed: Pumped Storage Hydropower is on the Rise*, a website report by Pacific Northwest National Laboratory, August 11, 2020.
206. Madlener, R., Specht, J. M., An exploratory economic analysis of underground pumped-storage hydro power plants in abandoned deep coal mines. *Energies* 2020;13(21):5634. DOI: 10.3390/en13215634
207. Kitsikoudis, V., Archambeau, P., Dewals, B., Pujades, E., Orban, P., Dassargues, A., Pirotton, M., Erpicum, S., Underground pumped-storage hydropower (UPSH) at the Martelange Mine (Belgium): Underground reservoir hydraulics. *Energies* 2020;13(14):3512. DOI: 10.3390/en13143512
208. Witt, A., Hadjerioua, B., Uría-Martínez, R., Bishop Jr, N. A., *Evaluation of the Feasibility and Viability of Modular Pumped Storage Hydro (m-PSH) in the United States.* ORNL/TM-2015/559. Oak Ridge National Laboratory, Oak Ridge, TN, 2015.
209. IFPSH (International Forum on Pumped Storage Hydropower), *Innovative Pumped Storage Hydropower Configurations and Uses.* Capabilities, Costs & Innovation Working Group, September, 2021. Available online: https://www.hydropower.org/publications/innovative-pumped-storage-hydropower-configurations-and-uses (accessed on 24 October, 2021).
210. Obermeyer, H., George, L., Wells, J., Robichaud, R., *Submersible Pump-Turbine Configuration to Reduce the Civil Costs of Pumped Storage Hydropower*, HydroVision International, Portland, OR, 2019.
211. Slocum, A. H., Fennell, G. E., Dundar, G., Hodder, B. G., Meredith, J. D., Sager, M. A., Ocean renewable energy storage (ORES) system: Analysis of an undersea energy storage concept. *Proc IEEE* 2013;101(4):906–924.
212. Garg, A., Lay, C., Füllmann, R., The feasibility of an underwater pumped hydro energy storage system. In 7th International Renewable Energy Storage Conference, 2012.
213. Puchta, M., Bard, J., Dick, C., Hau, D., Krautkremer, B., Thalemann, F., Hahn, H., Development and testing of a novel offshore pumped storage concept for storing energy at sea – StEnSea. *J Energy Storage* 2017;14:271–275.

214. Chen, Y., Odukomaiya, A., Kassaee, S., O'Connor, P., Momen, A. M., Liu, X., Smith, B. T., Preliminary analysis of market potential for a hydropneumatics ground-level integrated diverse energy storage system. *Appl Energy* 2019;242:1237–1247. DOI: 10.1016/j.apenergy.2019.03.076
215. Kassaee, S., Abu-Heiba, A., Ally, M. R., Mench, M. M., Liu, X., Odukomaiya, A., Chen, Y., King Jr, T.J., Smith, B.T., Momen, A.M., PART 1 - Techno-economic analysis of a grid scale Ground- Level Integrated Diverse Energy Storage (GLIDES) technology. *J Energy Storage* 2019;25:100792. DOI: 10.1016/j.est.2019.100792
216. Pikl, F. G., Richter, W., Zenz, G., Pumped-storage technology combined with thermal energy storage - Power station and pressure tunnel concept. *Geomechan Tunn* 2017;10(5):611–619.
217. Pikl, F. G., Richter, W., Zenz, G., Large-scale, economic and efficient underground energy storage. *Geomech Tunn* 2019;12(3):251–269.
218. Crosher, S., Personal communication from Crosher (RheEnergise, Ltd.) to Argonne National Laboratory, May 14, 2021.
219. Menser, P., *Virtual Reservoirs can Boost Flexibility of Small-Scale Hydropower*, a website report by INL Laboratory, IDAHO, 2019.
220. Murphy, C., Brown, P., Carag, V., *The Roles and Impacts of PV-Battery Hybrids in a Decarbonized U.S. Electricity Supply.* No. DE-AC36-08GO28308, NREL/TP-6A40-82046. National Renewable Energy Laboratory (NREL), Golden, CO, 2022.
221. World Bank Group; ESMAP; SERIS. Where Sun Meets Water: Floating Solar Market Report-Executive Summary. The World Bank. pp. 1–24. Available online: https://documents1.worldbank.org/curated/en/579941540407455831/pdf/Floating-Solar-Market-Report-Executive-Summary.pdf (accessed on 31 January 2019).
222. Lee, N., Grunwald, U., Rosenlieb, E., Mirletz, H., Aznar, A., Spencer, R., Cox, S., Hybrid floating solar photovoltaics-hydropower systems: Benefits and global assessment of technical potential. *Renew Energy*, 2020. DOI: 10.1016/j.renene.2020.08.080
223. Fang, W., Huang, Q., Huang, S., Yang, J., Meng, E., Li, Y., Optimal sizing of utility-scale photovoltaic power generation complementarily operating with hydropower: A case study of the world's largest hydro-photovoltaic plant. *Energy Conv Manag* 2017;136:161–172. DOI: 10.1016/j.enconman.2017.01.012
224. Kakoulaki, G., Sanchez, R. G., Amillo, A. G., Szabo, S., De Felice, M., Farinosi, F., De Felice, L., Bisselink, B., Seliger, R., Kougias, I., Jaeger-Waldau, A., Benefits of pairing floating solar photovoltaics with hydropower reservoirs in Europe. *Renew Sustain Energy Rev* 2023;171:112989.
225. Solomin, E., Sirotkin, E., Cuce, E., Selvanathan, S. P., Kumarasamy, S., Hybrid floating solar plant designs: A review. *Energies* 2021;14(10):2751. DOI: 10.3390/en14102751
226. Dursun, M., Saltuk, F, Reservoir effect on the hybrid solar-hydroelectric (SHE) system. *Politeknik Dergisi* 2023:1–1. DOI: 10.2339/politeknik.1074180
227. Gadzanku, S., Mirietz, H., Lee, N., Daw, J., Warren, A., Benefits and critical knowledge gaps in determining the role of floating photovoltaics in the energy-water-food nexus. *Sustainability* 2021;13(8):4317. DOI: 10.3390/su13084317
228. Borkowski, D., Cholewa, D., Korzeń, A., Run-of-the-river hydro-PV battery hybrid system as an energy supplier for local loads. *Energies* 2021;14:5160. DOI: 10.3390/en14165160
229. Wei, P., Liu, Y., The integration of wind-solar-hydropower generation in enabling economic robust dispatch", *Mathematical Problems in Engineering* 2019;2019:4634131. DOI: 10.1155/2019/4634131
230. Simão, M., Ramos, H. M., Hybrid pumped hydro storage energy solutions towards wind and PV integration: Improvement on flexibility, reliability and energy costs. *Water* 2020;12:2457. DOI: 10.3390/w12092457

231. Agarkar, B. D., Barve, S. B., A review on hybrid solar/wind/hydro power generation system. *Int J Current Eng Technol* 2016;4(4):188–191.
232. Ibrahim, W. I., Mohamed, M. R., Ismail, R. M. T. R., Leung, P. K., Xing, W. W., Shah, A. A., Hydrokinetic energy harnessing technologies: A review. Energy Rep 2021;7:2021–2042.
233. VanZwieten, J., McAnally, W., Ahmad, J., Davis, T., Martin, J., Bevelhimer, M., Cribbs, A., Lippert, R., Hudon, T. and Trudeau, M., In-stream hydrokinetic power: Review and appraisal. *J Energy Eng* 2015;141(3):04014024.
234. Alvarez, E. A., Rico-Secades, M., Corominas, E. L., Huerta-Medina, N., Guitart, J. S., Design and control strategies for a modular hydrokinetic smart grid. *Int J Electr Power Energy Syst* 2018;95:137–145.
235. Shafei, M. A. R., Ibrahim, D. K., Ali, A. M., Younes, M. A. A., Abou El-Zahab, E. E. D., Novel approach for hydrokinetic turbine applications. *Energy Sustain Dev* 2015;27:120–126
236. Anyi, M., Kirke, B., Evaluation of small axial flow hydrokinetic turbines for remote communities. *Energy Sustain Dev* 2010;14:110–116.
237. Radkey, R. L., Hibbs, B. D., Definition of Cost Effective River Turbine Designs. U.S. Department of Energy, 1981. Available online: https://www.osti.gov/biblio/5358098 (accessed on 27 November, 2019).
238. Levy, D., Power from natural flow at zero static head. *Int Power Gener* 1995;18:19.
239. Microhydro power systems, a website DOE report, Washington, DC, 2017.
240. Vennell, R., Funke, S. W., Draper, S., Stevens, C., Divett, T., Designing large arrays of tidal turbines: A synthesis and review. *Renew Sustain Energy Rev* 2015;41:454–472.
241. Behrouzi, F., Nakisa, M., Maimun, A., Ahmed, Y. M., Global renewable energy and its potential in Malaysia: A review of hydrokinetic turbine technology. *Renew Sustain Energy Rev* 2016;62:1270–1281
242. Kumar, D., Chatterjee, K., A review of conventional and advanced MPPT algorithms for wind energy systems. *Renew. Sustain Energy Rev* 2016;55:957–970
243. Vermaak, H. J., Kusakana, K., Koko, S. P., Status of micro-hydrokinetic river technology in rural applications: A review of literature. *Renew Sustain Energy Rev* 2014;29:625–633.
244. Bahaj, A. S., Myers, L. E., Fundamentals applicable to the utilisation of marine current turbines for energy production. *Renew Energy* 2003;28:2205–2211.
245. Ginter, V. J., Pieper, J. K., Robust gain scheduled control of a hydrokinetic turbine. *IEEE Trans Control Syst Technol* 2011;19:805–817.
246. Romero-Gomez, P., Richmond, M. C., Simulating blade-strike on fish passing through marine hydrokinetic turbines. *Renew Energy* 2014;71:401–413.
247. Muljadi, E., Wright, A., Donegan, J., Marnagh, C., Muljadi, E., Wright, A., *Power Generation for River and Tidal Generators Power Generation for River and Tidal Generators*. No. NREL/TP-5D00-66097. National Renewable Energy Laboratory (NREL), Golden, CO, 2016, pp. 1–55.
248. Khan, M. J., Bhuyan, G., Iqbal, M. T., Quaicoe, J. E., Hydrokinetic energy conversion systems and assessment of horizontal and vertical axis turbines for river and tidal applications: A technology status review. *Appl Energy* 2009;86:1823–1835.
249. Lago, L. I., Ponta, F. L., Chen, L., Advances and trends in hydrokinetic turbine systems. *Energy Sustain Dev* 2010;14:287–296.
250. Magagna, D., Uihlein, A., Ocean energy development in Europe: Current status and future perspectives. *Int J Mar Energy* 2015;11:84–104. DOI: 10.1016/j.ijome.2015.05.001
251. Song, K., Wang, W. Q., Yan, Y., Numerical and experimental analysis of a diffuser-augmented micro-hydro turbine. *Ocean Eng* 2019;171:590–602.
252. Vaz, J. R. P., Mesquita, A. L. A., Amarante Mesquita, A. L., de Oliveira, T. F, Junior, A. C. P. B. Powertrain assessment of wind and hydrokinetic turbines with diffusers. *Energy Convers Manage* 2019;195:1012–1021.

253. Nunes, M. M., Mendes, R. C., Oliveira, T. F., Junior, A. C. B., An experimental study on the diffuser-enhanced propeller hydrokinetic turbines. *Renew Energy* 2019;133:840–848.
254. Khan, M. J., Iqbal, M. T., Quaicoe, J. E., A technology review and simulation based performance analysis of river current turbine systems. In *2006 Canadian Conference on Electrical and Computer Engineering*, 2006, pp. 2288–2293.
255. García, E., Pizá, R., Benavides, X., Quiles, E., Correcher, A., Morant, F., Mechanical augmentation channel design for turbine current generators. *Adv Mech Eng* 2014;6:650131.
256. Nunes, M. M., Brasil Junior, A. C. P., Oliveira, T. F., 2020. Systematic review of diffuser-augmented horizontal-axis turbines. *Renew Sustain Energy Rev* 2020;133:110075. DOI: 10.1016/j.rser.2020
257. Wong, K. H., Chong, W. T., Sukiman, N. L., Poh, S. C., Shiah, Y. C., Wang, C. T., Performance enhancements on vertical axis wind turbines using flow augmentation systems: A review. *Renew Sustain Energy Rev* 2017;73:904–921. DOI: 10.1016/j.rser.2017.01.160
258. Bontempo, R., Manna, M., On the potential of the ideal diffuser augmented wind turbine: An investigation by means of a momentum theory approach and of a free-wake ring-vortex actuator disk model. *Energy Convers Manage* 2020;213:112794. DOI: 10.1016/j.enconman.2020.112794
259. Lust, E. E., Flack, K. A., Luznik, L., Survey of the near wake of an axial-flow hydrokinetic turbine in the presence of waves. *Renew Energy* 2020;129:92–101.
260. Dou, B., Guala, M., Lei, L., Zeng, P., Wake model for horizontal-axis wind and hydrokinetic turbines in yawed conditions. *Appl Energy* 2019;242:1383–1395.
261. Guerra, M., Thomson, J., Wake measurements from a hydrokinetic river turbine. *Renew Energy* 2019;139:483–495.
262. Musa, M., Hill, C., Guala, M., Interaction between hydrokinetic turbine wakes and sediment dynamics: Array performance and geomorphic effects under different siting strategies and sediment transport conditions. *Renew Energy* 2019;138:738–753.
263. Laws, N. D., Epps B. P., Hydrokinetic energy conversion: Technology, research, and outlook. *Renew Sustain Energy Rev* 2016;57:1245–1259.
264. Saini, G., Saini R. P., Study of installations of hydrokinetic turbines and their environmental effects. In *AIP Conference Proceedings*, 2020, p. 2273. DOI: 10.1063/5.0024338
265. Van Arkel, R., Owen, L., Allison, S., Tryfonas, T., Winter, A., Entwistle, R., Keane, E., Parr, J., Design and preliminary testing of a novel concept low depth hydropower device. In *OCEANS'11 MTS/IEEE KONA*, 2011, pp. 1–10.
266. Wang, J., Geng, L., Ding, L., Zhu, H., Yurchenko, D., The state-of-the-art review on energy harvesting from flow-induced vibrations. *Appl Energy* 2020;267:114902. DOI: 10.1016/j.apenergy.2020.114902
267. Niebuhr, C. M., Van Dijk, M., Neary, V. S., Bhagwan, J. N., A review of hydrokinetic turbines and enhancement techniques for canal installations: Technology, applicability and potential. *Renew Sustain Energy Rev* 2019;113:109240.
268. Neil, S., Could sea-current generator be used here? 22 July 2015. Available online: https://www.royalgazette.com/article/20150722/BUSINESS/150729918 (accessed on 21 March, 2017).
269. Miller, G., Francheschi, J., Lese, W., and Rico, J., *The Allocation of Kinetic Hydro Energy Conversion Systems (KHECS) in USA Drainage Basins: Regional Resource and Potential Power* (Technical Report). NYU/DAS 86-151. Department of Applied Science, U.S. Department of Energy, New York University, New York, 1986.
270. USGS, *National Hydrographic Dataset Applications.* United States Geological Survey, Reston, VA, 2013. Available online: https://nhd.usgs.gov/applications.html (accessed March, 2013).

271. United States Army Corps of Engineers (USACE), Hydrologic Engineering Centers River Analysis System HEC- RAS. U.S. Army Corps of Engineers Hydrologic Engineering Center, Davis, CA, 2013. Available online: https://www.hec.usace.army.mil/software/hec-ras/ (accessed March, 2013).
272. Electrical Power Research Institute (EPRI), *Assessment and Mapping of the Riverine Hydrokinetic Energy Resource in the Continental United States.* Electric Power Research Institute, Palo Alto, CA, 2012.
273. Johnson, J. B., Pride, D. J., *River, Tidal, and Ocean Current Hydrokinetic Energy Technologies: Status and Future Opportunities in Alaska* (Technical Report). Prepared for the Alaska Energy Authority, Alaska Center for Energy and Power, 2010.
274. Hydro Green Energy, 2011. Available online: https://hgenergy.com/index.php/projects/hastings-project/ (accessed on July, 2011).
275. Hydro Green Energy, 2013. Available online: https://hgenergy.com/index.php/projects/hastings-project/ (accessed on April, 2013).
276. Free Flow Power, 2013. Available online: https://free-flow-power.com (accessed on July, 2013).
277. Verdant Power, 2014. Available online: https://verdantpower.com/ (accessed on March, 2014).
278. Marine energy, 2023. Wikipedia, The free encyclopedia, last edited 28 June 2023. https://en.wikipedia.org/wiki/Marine_energy.
279. Feng, J. D., Graf, M., Liu, K., Ovchinnikov, D., Dumcenco, D., Heiranian, M., Nandigana, V., Aluru, N. R., Kis, A., Radenovic, A., Single-Layer MoS_2 nanopores as nanopower generators. *Nature* 2016;536:197–200. DOI: 10.1038/nature18593
280. Siria, A., Bocquet, M. L., Bocquet, L., New avenues for the large-scale harvesting of blue energy. *Nat Rev Chem* 2017;1:0091. DOI: 10.1038/s41570-017-0091
281. Wang, L., Wang, Z., Patel, S. K., Lin, S., Elimelech, M., Nanopore-based power generation from salinity gradient: Why it is not viable, *ACS Nano* 2021;15(3):4093–4107.
282. Siria, A., Poncharal, P., Biance, A. L., Fulcrand, R., Blase, X., Purcell, S. T., Bocquet, L., Giant osmotic energy conversion measured in a single transmembrane boron nitride nanotube. *Nature* 2013;494:455–458. DOI: 10.1038/nature11876
283. Ocean thermal energy conversion, 2013. Wikipedia, The free encyclopedia, last edited 30 August 2023. https://en.wikipedia.org/wiki/Ocean_thermal_energy_conversion.
284. Hagerman, G., Polagye, B., Bedard, R., Previsic, M., *Methodology for Estimating Tidal Current Energy Resources and Power Production by Tidal in-Stream Energy Conversion (TISEC) Devices*; Technical Report for Electric Power Research Institute (EPRI), Palo Alto, CA, 2006.
285. Tousif, S. M. R., Taslim, S. M. B., 2011, Tidal power: An effective method of generating power. *Int J Sci Eng Res* 2011;2(5):1–5.
286. Nicholls-Lee, R. F., Turnock, S. R., Tidal energy extraction: Renewable, sustainable and predictable. *Sci Prog* 2008;91(1):81–111.
287. Bahaj, A. S., Myers, L. E., Thompson, G., Characterising the wake of horizontal axis marine current turbines. In *Proceedings of the Seventh European Wave and Tidal Energy Conference*, 2007.
288. Batten, W. M. J., Bahaj, A. S., Molland, A. F., Chaplin, J. R., Experimentally validated numerical method for the hydrodynamic design of horizontal axis tidal turbines. *Ocean Eng* 2007;34(7):1013–1020.
289. Fraenkel, P. L., Power from marine currents. *Proc Inst Mech Eng Part A* 2002;216(1):1–14.
290. Mendi, V., Rao, S., Seelam, J. K., Tidal energy: A review. In *Proceedings of International Conference on Hydraulics, Water Resources and Coastal Engineering (Hydro2016)*, CWPRS, Pune, India, 8th–10th December 2016.
291. Qin, Z., Tang, X., Wu, Y. T., Lyu, S. K., Advancement of tidal current generation technology in recent years: A review. *Energies* 2022;15:8042. DOI: 10.3390/en15218042

292. Neill, S. P., Haas, K. A., Thiébot, J., & Yang, Z., A review of tidal energy-Resource, feedbacks, and environmental interactions. *J Renew Sustain Energy* 2021;13:062702. DOI: 10.1063/5.0069452
293. Chowdhury, M.S., Rahman, K.S., Selvanathan, V., Nuthammachot, N., Suklueng, M., Mostafaeipour, A., Habib, A., Akhtaruzzaman, M., Amin, N., Techato, K., Current trends and prospects of tidal energy technology. *Environ Dev Sustain* 2021;23:8179–8194. DOI: 10.1007/s10668-020-01013-4
294. Tidal power, 2023. Wikipedia, The free encyclopedia, last edited 12 August 2023. https://en.wikipedia.org/wiki/Tidal_power.
295. Tidal energy, a website report by Pacific Northwest national laboratory, INL, Idaho, 2017.
296. Couch, S. J., Bryden, I. G., Tidal current energy extraction: Hydrodynamic resource characteristics. *Proc Inst Mech Eng M: J Eng Marit Environ* 2006;220:185–194.
297. Charlier, R. H., Finkl, C. W., *Ocean Energy: Tide and Tidal Power.* Springer, Berlin, Heidelberg, 2009.
298. Callaghan, J., Boud, R., *Future Marine Energy: Results of the Marine Energy Challenge: Cost Competitiveness and Growth of Wave and Tidal Stream Energy* (Technical report). Carbon Trust, 2006.
299. Gorlov, A. M., *Tidal Energy.* Northeastern University, Boston, MA, 2001, pp. 2955–2960.
300. Vauthier, P., The underwater electric kite east river deployment. In *OCEANS'88. 'A Partnership of Marine Interests'. Proceedings*, Baltimore, MA, 1988, pp. 1029–1033.
301. Fraenkel, P., Marine current turbines: An emerging technology. In *Paper for Scottish Hydraulics Study Group Seminar in Glasgow*, 2004, pp. 1–10.
302. Verdant Power. Kinetic hydropower system (KHPS) verdant power 2017. Available online: http://www.verdantpower.com/kinetic-hydropower-system.html (accessed on 11 November 2017).
303. OpenHydro. Openhydro naval energies 2017. Available online: https://www.openhydro.com/Company/Overview (accessed November 11, 2017).
304. Laws, N. D., Epps, B. P., Hydrokinetic energy conversion: Technology, research, and outlook. *Renew Sustain Energy Rev* 2016;57:1245–1259.
305. Hanson, K., RER Hydro. Shop Metalwork Technology, 2014. Available online: https://shopmetaltech. com/fabricating-technology/fabricating/river-trek.html (accessed on 23 November, 2019).
306. Haas, K., Fritz, H., French, S., Smith, B., Neary, V., *Assessment of Energy Production Potential from Tidal Streams in the United States.* Final Project Report June 29, 2011 Georgia Tech Research Corporation Award Number: DE-FG36-08GO18174, Atlanta, Georgia, 2011.
307. ORPC, *Utility Grid.* ORPC, Inc., 2019. Available online: https://www.orpc.co/markets/utility-grids (accessed on 13 August, 2019).
308. Qiao, D., Haider, R., Yan, J., Ning, D., Li, B., Review of wave energy converter and design of mooring system. *Sustainability* 2020;12:8251. DOI: 10.3390/su12198251
309. Guo, B., Wang, T., Jin, S., Duan, S., Yang, K., Zhao, Y., A review of point absorber wave energy converters. *J Mar Sci Eng* 2022;10:1534. DOI: 10.3390/jmse10101534
310. Drew, B., Plummer, A. R., Sahinkaya, M. N., A review of wave energy converter technology. *Proc Inst Mech Eng A: J Power Energy* 2009;223:887. DOI: 10.1243/09576509JPE782
311. Rusu, E., Onea, F., A review of the technologies for wave energy extraction. *Clean Energy* 2018;2(1):10–19. DOI: 10.1093/ce/zky003
312. Bruno, M., Maccanti, M., Pulselli, R. M., Sabbetta, A., Neri, E., Patrizi, N., Bastianoni, S., Benchmarking marine renewable energy technologies through LCA: Wave energy converters in the Mediterranean. *Front Energy Res* 2022;10:980557. DOI: 10.3389/fenrg.2022.980557

313. Dronkers, J., Wave energy converters, Coastal Wiki, 2023. The free encyclopedia, last visited 25 February 2023. https://www.coastalwiki.org/wiki/Wave_energy_converters.
314. Zurkinden, A. S., Ferri, F., Beatty, S., Kofoed, J. P., Kramer, M. M., Non-linear numerical modeling and experimental testing of a point absorber wave energy converter. *Ocean Eng* 2013;78:11–21. DOI: 10.1016/j.oceaneng.2013.12.009
315. Bozzi, S., Besio, G., Passoni, G., Wave power technologies for the Mediterranean offshore: Scaling and performance analysis. *Coastal Eng* 2018;136:130–146. DOI: 10.1016/j.coastaleng.2018.03.001
316. Têtu, A., Power take-off systems for WECs. In A. Pecher, J. Kofoed (Eds), *Handbook of Ocean Wave Energy. Ocean Engineering & Oceanography, vol 7.* Springer, Cham, 2017. DOI: 10.1007/978-3-319-39889-1_8
317. Yemm, R., Pizer, D., Retzler, C., Henderson, R., Pelamis: Experience from concept to connection. *Philos Trans A Math Phys Eng Sci* 2012;370(1959):365–368. DOI: 10.1098/rsta.2011.0312
318. Ruellan, M., BenAhmed, H., Multon, B., Josset, C., Babarit, A., Clement, A., Design methodology for a SEAREV wave energy converter. *IEEE Trans Energy Convers* 2010;25(3):760–767. DOI: 10.1109/TEC.2010.2046808
319. Ahamed, R., McKee, K., Howard, I., Advancements of wave energy converters based on power take off (PTO) systems: A review. *Ocean Engineering* 2020;204:107248.
320. CorPower Ocean HiWave-5 project, a website report by TETHYS, Pacific Northwest National Laboratory (.gov). Available online: https://tethys.pnnl.gov/project-sites/corpower-ocean-hiwave-5-project (accessed on 1 December, 2022).
321. Huang, B., Wang, P., Wang, L., Yang, S., Wu, D., Recent advances in ocean wave energy harvesting by triboelectric nanogenerator: An overview. *Nanotechnol Rev* 2020;9(1):716–735. DOI: 10.1515/ntrev-2020-0055
322. Liang, X., Liu, S., Yang, H., Jiang, T., Triboelectric nanogenerators for ocean wave energy harvesting: Unit integration and network construction. *Electronics* 2023;12:225. DOI: 10.3390/electronics12010225
323. Song, C., Zhu, X., Wang, M., Yang, P., Chen, L., Hong, L., Cui, W., Recent advances in ocean energy harvesting based on triboelectric nanogenerators. *Sustain Energy Technol Assess* 2022;53:102767.
324. Xu, L., Jiang, T., Lin, P., Shao, J. J., He, C., Zhong, W., Chen, X. Y., Wang, Z. L., Coupled triboelectric nanogenerator networks for efficient water wave energy harvesting. *ACS Nano* 2018;12(2):1849–1858. DOI: 10.1021/acsnano.7b08674
325. Cheng, P., Guo, H., Wen, Z., Zhang, C., Yin, X., Li, X., Liu, D., Song, W., Sun, X., Wang, J., Wang, Z. L., Largely enhanced triboelectric nanogenerator for efficient harvesting of water wave energy by soft contacted structure. *Nano Energy* 2019;57:432–439. DOI: 10.1016/j.nanoen.2018.12.054
326. Xia, K., Fu, J., Xu, Z., Multiple-frequency high-output triboelectric nanogenerator based on a water balloon for all-weather water wave energy harvesting. *Adv Energy Mater* 2020;10:2000426. DOI: 10.1002/aenm.202000426
327. Shi, Q., Wang, H., Wu, H., Lee, C., Self-powered triboelectric nanogenerator buoy ball for applications ranging from environment monitoring to water wave energy farm. *Nano Energy* 2017;40:203–213. DOI: 10.1016/j.nanoen.2017.08.018
328. Lee, K., Lee, J. W., Kim, K., Yoo, D., Kim, D. S., Hwang, W., Song, I., Sim, J. Y., A spherical hybrid triboelectric nanogenerator for enhanced water wave energy harvesting. *Micromachines* 2018;9:598. DOI: 10.3390/mi9110598
329. Yang, X., Xu, L., Lin, P., Zhong, W., Bai, Y., Luo, J., Chen, J., Wang, Z. L., Macroscopic self-assembly network of encapsulated high-performance triboelectric nanogenerators for water wave energy harvesting. *Nano Energy* 2019;60:404–412. DOI: 10.1016/j.nanoen.2019.03.054
330. Liu, G., Guo, H., Xu, S., Hu, C., Wang, Z. L., Oblate spheroidal triboelectric nanogenerator for all-weather blue energy harvesting. *Adv Energy Mater* 2019;9:1900801. DOI: 10.1002/aenm.201900801

331. Wen, X., Yang, W., Jing, Q., Wang, Z. L., Harvesting broadband kinetic impact energy from mechanical triggering/vibration and water waves. *ACS Nano* 2014;8:7405–7412. DOI:10.1021/nn502618f
332. Jiang, T., Zhang, L. M., Chen, X., Han, C. B., Tang, W., Zhang, C., Xu, L., Wang, Z. L., Structural optimization of triboelectric nanogenerator for harvesting water wave energy. *ACS Nano* 2015;9:12562–12572. DOI: 10.1021/acsnano.5b06372
333. Yao, Y., Jiang, T., Zhang, L., Chen, X., Gao, Z., Wang, Z., Charging system optimization of triboelectric nanogenerator for water wave energy harvesting and storage. *ACS Appl Mater Interfaces* 2016;8:21398–21406. DOI: 10.1021/acsami.6b07697
334. Zhang, L. M., Han, C. B., Jiang, T., Zhou, T., Li, X. H., Zhang, C., Wang, Z. L., Multilayer wavy-structured robust triboelectric nanogenerator for harvesting water wave energy. *Nano Energy* 2016;22:87–94. DOI: 10.1016/j.nanoen.2016.01.009
335. Lapčík, L., Vašina, M., Lapčíková, B., Hui, D., Otyepková, E., Greenwood, R. W., Waters, K. E., Vlček, J., Materials characterization of advanced fillers for composites engineering applications. *Nanotechnol Rev* 2019;8(1):503–512. DOI: 10.1515/ntrev-2019-0045
336. Xu, L., Pang, Y., Zhang, C., Jiang, T., Chen, X., Luo, J., Tang, W., Cao, X., Wang, Z. L., Integrated triboelectric nanogenerator array based on air-driven membrane structures for water wave energy harvesting. *Nano Energy* 2017;31:351–358. DOI: 10.1016/j.nanoen.2016.11.037
337. Jiang, T., Yao, Y., Xu, L., Zhang, L., Xiao, T., Wang, Z. L., Spring-assisted triboelectric nanogenerator for efficiently harvesting water wave energy. *Nano Energy* 2017;31:560–567. DOI: 10.1016/j.nanoen.2016.12.004
338. Xiao, T. X., Jiang, T., Zhu, J. X., Liang, X., Xu, L., Shao, J. J., Zhang, C. L., Wang, J., Wang, Z.L., Silicone-based triboelectric nanogenerator for water wave energy harvesting. ACS Appl Mater Interfaces. 2018;10:3616-23. DOI: 10.1021/acsami.7b17239
339. Xiao, T. X., Liang, X., Jiang, T., Xu, L., Shao, J. J., Nie, J. H., Bai, Y., Zhong, W., Wang, Z. L., Spherical triboelectric nanogenerators based on spring-assisted multilayered structure for efficient water wave energy harvesting. *Adv Funct Mater* 2018;28:1802634. DOI: 10.1002/adfm.201802634
340. Salter, S., Recent progress on ducks. *IEE Proc Part A: Phys Sci Meas Instrum Manage Educ Rev* 1980;127:308. DOI: 10.1049/ip-a-1.1980.0049
341. Saadatnia, Z., Asadi, E., Askari, H., Zu, J., Esmailzadeh, E., Modeling and performance analysis of duck-shaped triboelectric and electromagnetic generators for water wave energy harvesting. *Int J Energy Res* 2017;41:2392–2404. DOI: 10.1002/er.3811
342. Chen, B. D., Tang, W., He, C., Deng, C. R., Yang, L. J., Zhu, L. P., Chen, J., Shao, J. J., Liu, L., Wang, Z. L., Water wave energy harvesting and self-powered liquid-surface fluctuation sensing based on bionic-jellyfish triboelectric nanogenerator. *Mater Today* 2018;21:88–97. DOI: 10.1016/j.mattod.2017.10.006
343. Zhang, S. L., Xu, M., Zhang, C., Wang, Y. C., Zou, H., He, X., Wang, Z., Wang, Z. L., Rationally designed sea snake structure based triboelectric nanogenerators for effectively and efficiently harvesting ocean wave energy with minimized water screening effect. *Nano Energy*. 2018;48:421–429. DOI: 10.1016/j.nanoen.2018.03.062
344. Wang, N., Zou, J., Yang, Y., Li, X., Guo, Y., Jiang, C., Jia, X., Cao, X., Kelp-inspired biomimetic triboelectric nanogenerator boosts wave energy harvesting. *Nano Energy* 2019;55:541–547. DOI: 10.1016/j.nanoen.2018.11.006
345. Lei, R., Zhai, H., Nie, J., Zhong, W., Bai, Y., Liang, X., Xu, L., Jiang, T., Chen, X., Wang, Z. L., Butterfly-inspired triboelectric nanogenerators with spring-assisted linkage structure for water wave energy harvesting. *Adv Mater Technol* 2019;4:1800514. DOI: 10.1002/admt.201800514
346. Chen, M., Li, X., Lin, L., Du, W., Han, X., Zhu, J., Pan, C., Wang, Z.L., Triboelectric nanogenerators as a self-powered motion tracking system. *Adv Funct Mater* 2014;24:5059–5066. DOI: 10.1002/adfm.201400431

347. Liu, J, Fei, P, Zhou, J, Tummala, R, Wang, Z. L., Toward high output-power nanogenerator. Appl Phys Lett. 2008;92:173105. DOI: 10.1063/1.2918840
348. Pan, L., Wang, J., Wang, P., Gao, R., Wang, Y. C., Zhang, X., Zou, J. J., Wang, Z. L., Liquid-FEP-based U-tube triboelectric nanogenerator for harvesting water-wave energy. *Nano Res* 2018;11:4062–4073. DOI: 10.1007/s12274-018-1989-9
349. Zhao, X. J., Kuang, S. Y., Wang, Z. L., Zhu, G., Highly adaptive solid-liquid interfacing triboelectric nanogenerator for harvesting diverse water wave energy. *ACS Nano* 2018;12:4280–4285. DOI: 10.1021/acsnano.7b08716
350. Shao, H., Cheng, P., Chen, R., Xie, L., Sun, N., Shen, Q., Chen, X., Zhu, Q., Zhang, Y., Liu, Y., Wen, Z., Triboelectric-electromagnetic hybrid generator for harvesting blue energy. *Nano-Micro Lett* 2018;10:54. DOI: 10.1007/s40820-018-0207-3
351. Hao, C., He, J., Zhai, C., Jia, W., Song, L., Cho, J., Chou, X., Xue, C., Two-dimensional triboelectric-electromagnetic hybrid nanogenerator for wave energy harvesting. *Nano Energy* 2019;58:147–157. DOI: 10.1016/j.nanoen.2019.01.033
352. Wen, Z., Guo, H., Zi, Y., Yeh, M. H., Wang, X., Deng, J., Wang, J., Li, S., Hu, C., Zhu, L., Wang, Z. L., Harvesting broad frequency band blue energy by a triboelectric-electromagnetic hybrid nanogenerator. *ACS Nano* 2016;10:6526–6534. DOI: 10.1021/acsnano.6b03293
353. Saadatnia, Z., Asadi, E., Askari, H., Esmailzadeh, E., Naguib, H. E., A heaving point absorber-based triboelectric-electromagnetic wave energy harvester: an efficient approach toward blue energy. *Int J Energy Res* 2018;42:2431–2447. DOI: 10.1002/er.4024
354. Saadatnia, Z., Esmailzadeh, E., Naguib, H. E., Design, simulation, and experimental characterization of a heaving triboelectric-electromagnetic wave energy harvester. *Nano Energy* 2018;50:281–290. DOI: 10.1016/j.nanoen.2018.05.059
355. Shao, H., Wen, Z., Cheng, P., Sun, N., Shen, Q., Zhou, C., Peng, M., Yang, Y., Xie, X., Sun, X., Multifunctional power unit by hybridizing contact-separate triboelectric nanogenerator, electromagnetic generator and solar cell for harvesting blue energy. *Nano Energy* 2017;39:608–615. DOI: 10.1016/j.nanoen.2017.07.045
356. Wang, J., Pan, L., Guo, H., Zhang, B., Zhang, R., Wu, Z., Wu, C., Yang, L., Liao, R., Wang, Z. L., Rational structure optimized hybrid nanogenerator for highly efficient water wave energy harvesting. *Adv Energy Mater* 2019;9:1802892. DOI: 10.1002/aenm.201802892
357. Wang, Z. L., Catch wave power in floating nets. *Nature* 2017;542:159–160. DOI: 10.1038/542159a
358. Xie, Y., Wang, S., Niu, S., Lin, L., Jing, Q., Su, Y., Wu, Z., Wang, Z. L., Multi-layered disk triboelectric nanogenerator for harvesting hydropower. *Nano Energy* 2014;6:129–136. DOI: 10.1016/j.nanoen.2014.03.015
359. Kan, J., Fan, C., Wang, S., Zhang, Z., Study of piezo-windmill for energy harvesting. *Renew Energy* 2016; 97:210–217. DOI: 10.1016/j.renene.2016.05.055
360. Zhao, D., Zhou, J., Tan, T., Yan, Z., Sun, W., Yin, J., Zhang, W., Hydrokinetic piezoelectric energy harvesting by wake induced vibration. *Energy* 2021;220:119722.
361. Dai, H. L., Abdelkefi, A., Wang, L., Piezoelectric energy harvesting from concurrent vortex-induced vibrations and base excitations. *Nonlinear Dyn* 2014;77:967–981.
362. Mutsuda, H., Tanaka, Y., Doi, Y., Moriyama, Y., Application of a flexible device coating with piezoelectric paint for harvesting wave energy. *Ocean Eng* 2019;172:170–182. DOI: 10.1016/j.oceaneng.2018.11.014
363. Fan, K., Chang, J., Pedrycz, W., Liu, Z., & Zhu, Y., A nonlinear piezoelectric energy harvester for various mechanical motions. *Appl Phys Lett* 2015;106:223902. DOI: 10.1063/1.4922212
364. Chen, S. -E., Yang, R. -Y., Wu, G. -K., Wu, C. -C., A piezoelectric wave-energy converter equipped with a geared-linkage-based frequency up-conversion mechanism. *Sensors* 2021;21:204. DOI: 10.3390/s21010204

365. Li, H., Tian, C., Deng, Z. D., Energy harvesting from low frequency applications using piezoelectric materials. *Appl Phys Rev* 2014;1:041301.
366. Fan, K., Chang, J., Chao, F., Pedrycz, W., Design and development of a multipurpose piezoelectric energy harvester. *Energy Convers Manag* 2015;96:430–439.
367. Lin, Z., Zhang, Y., Dynamics of a mechanical frequency up-converted device for wave energy harvesting. *J Sound Vib* 2016;367:170–184.
368. Renzi, E., Hydroelectromechanical modelling of a piezoelectric wave energy converter. *Proc R Soc Lond* 2016;472:20160715.
369. Yang, Y., Tang, L., Li, H., Vibration energy harvesting using macro-fiber composites. *Smart Mater Struct* 2009;18:115025.
370. Orrego, S., Shoele, K., Ruas, A., Doran, K., Caggiano, B., Mittal, R., Kang, S. H., Harvesting ambient wind energy with an inverted piezoelectric flag. *Appl Energy* 2017;194:212–222.
371. Kim, K. H., Cho, S. B., Kim, H. D., Shim, K. T., Wave power generation by piezoelectric sensor attached to a coastal structure. *J Sensors* 2018;2018:7986438. DOI: 10.1155/2018/7986438
372. Carlos, D., *Sea Wave Energy Based on Nano Technology*. OSTI. GOV (gov), 2010. Available online: https://www.osti.gov/etdeweb/servlets/purl (accessed on March 9, 2011).
373. Yang, X., Haas, K. H., Fritz, H. M., Theoretical assessment of ocean current energy potential for the gulf stream system. *Marine Technol Soc J* 2013;47(4):101–112.
374. Chassignet, E. P., Hurlburt, H. E., Metzger, E. J., Smedstad, O. M., Cummings, J. A., Halliwell, G. R., Bleck, R., Baraille, R., Wallcraft, A. J., Lozano, C., Tolman, H. L., US GODAE: Global ocean prediction with the hybrid coordinate ocean model (HYCOM). *Oceanography* 2009;22(2):64–75.
375. Chassignet, E. P., Hurlburt, H. E., Smedstad, O. M., Halliwell, G. R., Hogan, P. J., Wallcraft, A. J., Baraille, R., Bleck, R., The HYCOM (hybrid coordinate ocean model) data assimilative system. *J Marine Syst*2007;65(1–4):60–83.
376. VanZwieten, J. H., Jr., Duerr, A. E. S., Alsenas, G. M., and Hanson, H. P., Global ocean current energy assessment: An initial look. In *Proceedings of the 1st Marine Energy Technology Symposium (METS13) hosted by the 6th Annual Global Marine Renewable Energy Conference*. Foundation for Ocean Renewables, Darnestown, MD, 2013. Available online: https://www.foroceanenergy.org/mets/2013-peer-reviewed-mets-papers/ (accessed on 3 April, 2014).
377. Raye, R. E., Characterization study of the Florida current at 26.11 north latitude, 79.50 west longitude for ocean current power generation [M.S. thesis]. Ocean Engineering Department, Florida Atlantic University, Boca Raton, FL, 2000.
378 VanZwieten, J. H., Jr., Oster, C. M., Duerr, A. E. S., Design and analysis of a rotor blade optimized for extracting energy from the Florida current. In *Proceedings of the ASME 2011 International Conference on Ocean, Offshore, and Arctic Engineering, American Society of Mechanical Engineers (ASME)*, New York, 2011.
379. Duerr, A. E. S., Dhanak, M. R., Van Zwieten, J. H., Utilizing the hybrid coordinate ocean model data for assessment of Florida current hydrokinetic renewable energy resource. *Marine Technol Soc J* 2012;46(5):24–33.
380. Chen, F., Kuroshio power plant development plan. *J Renew Sustain Energy Rev* 2010;14(9):2655–2668.
381. Bryden, H. L., Beal, L. M., Duncan, L. M., Structure and trans- port of the Agulhas current and its temporal variability. *J Oceanogr* 2005;61(3):479–492.
382. Guo, B., Ringwood, J. V., A review of wave energy technology from a research and commercial perspective. *IET Renew Power Gener* 2021;15:3065–3090. DOI: 10.1049/rpg2.12302
383. Zhang, Z., Li, X., Yin, J., Xu, Y., Fei, W., Xue, M., Wang, Q., Zhou, J., Guo, W., Emerging hydrovoltaic technology. *Nat Nanotechnol* 2018;13(12):1109–1119. DOI: 10.1038/s41565-018-0228-6

4 Advances in Fuel Cells for Power Generation

4.1 INTRODUCTION

Fuel cells (FCs) are direct electrochemical fuel-to-electrical energy conversion devices and offer higher efficiency (50%–70%) compared with conventional technologies such as internal combustion engines (~35% efficiency) and others (2%). If the waste heat of the FC is also used, fuel efficiencies of 90% are possible. FCs consist of an electropositive anode, where oxidation occurs to fuels (hydrogen, methanol, ethanol, methane, etc.), an electronegative cathode, where reduction occurs (to oxygen, air, etc.), and an electrolyte (e.g., a proton-conducting polymer membrane and ionic conductivity-doped ceria), where ions carry the current between the electrodes. The scheme of reactions and processes that occur in the various FC systems is depicted in Figure 4.1 as shown by Du and Pollet [1]. They also described the characteristics of the various power-generating electrochemical FC systems as shown in Table 4.1 [1].

FCs can be roughly divided into low-temperature (LT) (ca. <200°C), medium-temperature (around 250°C), and high-temperature (ca. >450°C) FCs. Alkaline FC (AFC), polymer electrolyte membrane (PEM) FC (PEMFC, also known as proton-exchange membrane FC under the same acronym), and direct methanol FC (DMFC) are typical LT FCs. Phosphoric acid FC (PAFC) is a medium-temperature FC. Molten carbonate FC (MCFC), direct carbon FC (DCFC), and solid oxide FC (SOFC) belong to the high-temperature FC class. Besides these, microbial FC is a biological or enzymatic FC, which is used for waste treatment along with power generation. In general, LT FCs (AFC, DMFC, and PEMFC) feature a quicker start-up, which makes them more suitable for portable applications, especially PEMFCs, which have recently gained momentum for applications in transportation and as small portable power sources. However, AFC, DMFC, and PEMFC require relatively pure hydrogen (minimum 99.999%) as fuel, and consequently an external fuel processor, which increases the complexity and cost and decreases the overall efficiency. They also require a higher loading of precious metal catalysts [3]. In LT-PEMFCs and DMFCs, the high cost originates from the expensive Nafion® membrane and the use of notable platinum (Pt) catalyst; their poor durability is caused by the instability of the catalyst support and the degradation of membrane and catalyst performance. LT MFC is a biological or enzymatic FC, which provides multiple functions of power generation and waste valorization. It is at the early stages of development, and it suffers from short active lifetimes (typically 8 hours to 7 days) [4] and limited power generation [5]. It is extensively discussed in Chapter 5, and its recent developments are also discussed in the recent reviews by Barton et al. [6] and Harnisch and Schröder [7]. The applications of LT FC vary. AFC is useful for military and space applications. PEMFC is useful for automobile and stationary applications, and DMFC is useful for portable electronics.

DOI: 10.1201/9781003429906-4

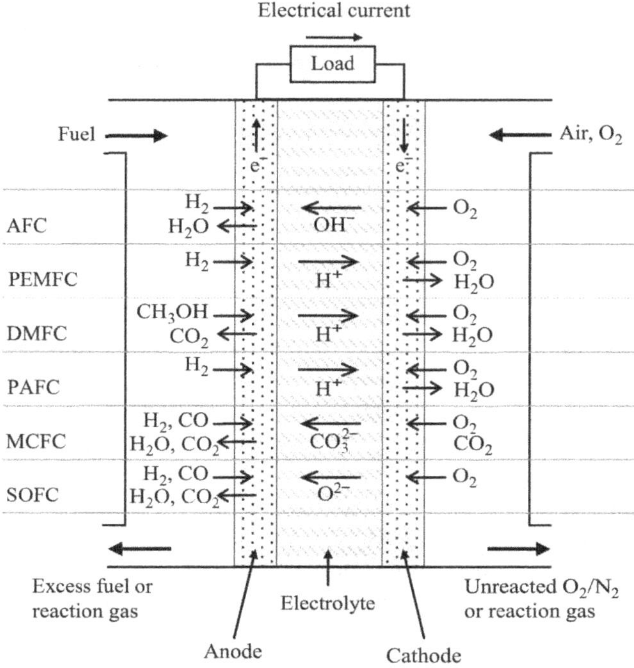

FIGURE 4.1 Operation scheme of typical fuel cells [1].

TABLE 4.1
Typical Characteristics of Various Fuel Cell Systems [1]

Type	Anode Composition	Cathode Composition	Electrolyte	Operating Temperature (°C)
AFC	Carbon/platinum catalyst	Carbon/platinum catalyst	Aqueous KOH	Ambient–100
PEMFC	Carbon/platinum catalyst	Carbon/platinum catalyst	Acidic polymer	Ambient–90
DMFC	Carbon/platinum catalyst	Carbon/platinum catalyst	Acidic polymer	60–90
PAFC	Carbon/platinum catalyst	Carbon/platinum catalyst	Phosphoric acid in SiC matrix	150–220
MCFC	Porous Ni	Porous NiO	Molten Li_2CO_3 in LiAlO–	550–700
SOFC	Ni-YSZ	Strontia-doped lanthanum manganite perovskite	YSZ	600–1,000

Source: Winter and Brodd [2].

In contrast to LT FCs, PAFC, DCFC, MCFC, and SOFC are more flexible regarding fuel because they can reform various fuels (methanol, ethanol, natural gas, gasoline, etc.) inside the cells to produce hydrogen, still offering advantages for stationary applications, and especially for cogeneration. They are also less prone to catalyst *poisoning* by carbon monoxide (CO) and carbon dioxide (CO_2) [8]. However, their slower start-up limits them to more stationary applications [9]. All high-temperature FCs are actively under research and development and since they employ solid electrolytes their operation and maintenance are easier. However, their commercialization is still hampered by high cost, poor durability issues, and operability problems that are directly linked to severe material challenges and system issues. High temperature in SOFC leads to high cost, use of expensive interconnect and sealing materials, microstructure decline, carbon deposition, and sulfur poisoning of the catalyst. These factors also affect SOFC durability and commercial viability. The applications of different high-temperature FCs also vary. PAFC is useful for distributed power generation. MCFC is useful for electric utility and large distributed generation. SOFC is useful for auxiliary power, electric utility, and large distributed generation.

Several of the above issues for both LT and high-temperature FCs can be addressed with the use of high-performance materials, with novel design and preparing technologies, in which nanomaterials have played a critical role. For example, in LT FCs, nearly half of the cost of the FC is linked to the electrocatalyst cost. To reduce the cost, in PEMFCs and DMFCs, Pt catalyst with novel nanostructure and high performance has been developed to reduce the loading amount of Pt, or directly by using less expensive alternative nanostructured electrocatalysts, such as N-doped carbon nanotube (CNT) [10] or iron-based catalysts [11]. Regarding high-temperature SOFCs, to decrease the operating temperature to the intermediate range (450°C–600°C) for a low system cost and improved durability, nanocomposite electrodes and electrolyte are developed, such as Samaria-doped ceria (SDC) nanowires (NWs)/Na_2CO_3 nanocomposite electrolyte [12] or gadolinium-doped ceria (GDC) anode with ionic NW nanoarchitectures [13].

We discuss the advances in FCs in two parts. In this chapter, we strictly focus on LT, medium-, and high-temperature FCs, which are largely used for power generation. In Chapter 5, we examine multifunctional FCs such as MCFC, regenerative FCs, MFC and hybrid FCs, which enhance the applications of FCs for energy and environment industries. FCs are very versatile and provide power in a wide range. Power ratings for different types of FCs are described in Figure 4.2. Different types of FCs are at different stages of development. The study by Akinyele et al. [14] points out the

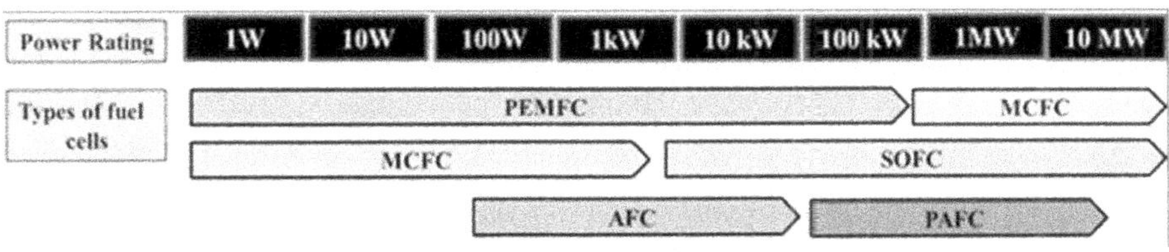

FIGURE 4.2 Different groups of fuel cells based on their power ratings and advantages [15].

advantages of the various FC technologies and the technical barriers associated with the mode of operation of each system. The study discussed the developmental trends of different technologies. The DMFCs are in the early phase of technological development, and as such, they are not as developed as other systems. PAFCs are technically matured systems that are ready for commercialization and they are regarded as the first-generation FCs, while the proton-exchange methanol FCs are likely to be in the commercial phase in the next 5–10 years. MCFCs are referred to as second-generation systems that have reached the early demonstration phase, with their developments and possibility of attaining commercial status after the PAFCs within the next 5 to less than 10 years. SOFCs are regarded as third-generation systems as their attainment of commercial status is expected to be after PAFCs and MCFCs, and they are in the developmental phase and are also likely to be commercialized within the next 5–10 years. Microbial FC can be considered fourth-generation FC system, and it is at very early stages of development and may take more than 10 years to achieve a respectable level of power generation to be commercialized. Based on the mentioned factors, there is the possibility that electrical power systems will experience a wider application of FCs within the next 5–10 years as they have the technical characteristics and advantages to compete with other energy resources.

4.2 ALKALINE FUEL CELL TECHNOLOGY

AFCs employ the "alkaline" electrolyte, namely potassium hydroxide (KOH), diluted in water, and operate with pure hydrogen fuel, while pure oxygen serves as the oxidant [16,17]. In their operation, hydrogen fuel is supplied to the anode, but oxygen is transported to the cathode. The exchange of ions is permitted between the cathode and the anode in the liquid KOH, which leads to the generation of direct current [17]. Materials such as potassium titanate, ceria, and zirconium phosphate gel have been used as "microporous" separators in AFCs [18]. While early AFCs were operated in the temperature range of 100°C and 250°C, recent cells can be operated at 70°C and LTs [16,19]. This implies that they can operate within a wider range of temperatures [17]. Generally, anode and cathode are made of Ni and Ag materials, respectively. Equations (4.1) and (4.2) describe the anode and cathode reactions in AFCs [20].

$$H_2 + 2OH^- \rightarrow 2H_2O + 2e^- \tag{4.1}$$

$$0.5O_2 + H_2O + 2e^- \rightarrow 2OH^- \tag{4.2}$$

AFC offers high efficiency of up to 0.6 in some applications. According to Alhassan et al. [21], depending on the system design, a single AFC can achieve a voltage output of 0.5–0.9 V with an efficiency as high as 0.65. Furthermore, AFCs can deliver electrical output in the range of 5–150 kW [17]. KOH electrolyte is, however, corrosive, and because of its liquid state, the sealing of the anode and cathode gases becomes a very difficult task compared with the use of a solid electrolyte [16]. Furthermore, KOH adsorbs the CO_2, thus reducing the conduction power of the electrolyte [17,22].

The design of AFCs is classified into three categories, viz. mobile electrolyte, static electrolyte, and the dissolved fuel type, and is used in military, space, and submarine applications [17] where low-cost KOH electrolyte is desirable. It is mechanically rechargeable and has limited activated life. In an AFC, humidified hydrogen gas is supplied to the anode, which reacts with the hydroxide ions in the electrolyte to produce water and electrons after penetrating the gas diffusion layer (GDL) and reaching the catalyst layer (CL). A humidified oxygen source, typically purified air or oxygen, is supplied to the cathode together with water. Oxygen gas, solvated in water, is reduced at the cathode CL to form hydroxide ions, which diffuse through the electrolyte to participate in the hydrogen oxidation reaction (HOR) that takes place on the anode.

The red–ox reactions mentioned here can be combined to form the overall mechanism as follows:

$$O_2 + 2H_2 \rightarrow 2H_2O \qquad E_0 = 1:229\ V_{RHE} \tag{4.3}$$

AFC suffers from various inefficiencies and losses, which are categorized as charge transfer, ohmic, or mass transport overpotential. These overpotentials variably affect a cell at specific current densities. Charge transfer overpotential is mostly linked with activation losses and is most visible at low current densities. Ohmic overpotential originates in the electrical resistance of the cell components, determines the slope of the current–voltage graph, and is most visible at intermediate current densities. At high current densities, the FC will be limited by the speed at which reactants can reach the electrode, inducing mass transportation losses. The HOR and its complement, the hydrogen evolution reaction (HER), are two important reactions in FCs. There is a notable asymmetry between the anodic (HOR) and cathodic (oxygen reduction reaction (ORR)) FC reactions.

While AFC anode can use a variety of fuels such as borohydride, hydrazine, ammonia, methanol, ethanol, and ethylene glycol, the choice of fuel determines the choice of the useable materials for the catalyst. Historically, the first real, "useful" AFCs were based on liquid alkaline electrolytes and metal electrodes using a form of "zero gap technology" in which the metal electrodes were placed back to back against a thin electrically insulating diaphragm immersed in the liquid electrolyte. The idea of the diaphragm was to prevent the crossover of hydrogen and air and allow the chemical reactions to take place without interference. To supply fuel and air, it was necessary to develop a gas-permeable electrode that allowed the reactants to reach the active sites of the catalyst but at the same time prevent the flow of liquid electrolyte into gas channels. This led to the invention of the gas diffusion electrode (GDE) whose character changed over subsequent years. An ideal GDL should be hydrophobic and electronically conductive, where the latter may be achieved by adding carbon black (CB) to the mixture featuring 25–60 vol.% polytetrafluoroethylene (PTFE) [23], enabling usage in a bipolar design. The catalyst and its support material bonded with PTFE, where high surface area support materials such as CB are frequently used to increase power density (PD). CB is active due to its high surface area. Other carbon structures, such as CNTs [24,25], carbon spheres [26,27], and other carbon architectures [28,29] can substitute CB as support with equal effectiveness.

However, stability issues with carbon supports led to a new direction utilizing transition metals (TM)-based support materials for catalysts.

Complete utilization of CB's high surface area and the affiliated electrochemical activity is coupled to the catalyst's malleability toward uniform distribution over a substrate. Areas not covered are electrically conductive; however, they do not participate in the electrochemical reaction. Furthermore, the triple-phase boundary (TPB) between gas, electrolyte, and catalyst should cover the entire surface, as catalyst agglomeration causes uneven wear leading to decreased performance and endurance. The PTFE creates a TPB on the substrate and works primarily as a binder, where the loading (5–25 vol.%) depends on the required hydrophobicity. A metal sheet/grid fixed to the GDL facilitates current collection [30,31]. The traditional design has been challenged by integrated anodes, where the GDL and CL are combined through a PTFE-adhered palladium-plated nickel foam [32]. While this design increased the PD, the usage of palladium is questionable. PTFE can be substituted by the amorphous fluoropolymer material Teflon-alkaline fuel cell (AF) [33] and Gor-Tex (expanded PTFE) [34]. Both of these have displayed improved AFC performance when measured against PTFE GDEs.

AFC system designs are split between mono- and bipolar systems. Bipolar designs have the benefit of the interconnection between cells (bipolar plates (BPs)), which requires electrically conductive GDL (CB-PTFE composites). Monopolar designs benefit from lower costs due to a lack of BPs, lower stack thickness due to the singular gas chamber between the electrodes, and the ability to disconnect a cell with the remainder of the stack still operating. A significant disadvantage includes ohmic losses from inefficient current collection [35]. Bipolar designs feature a uniform current density and a higher terminal voltage, rendering it the preferred commercial design [36,37]. Complete AFC systems are branched into either spatial or terrestrial, the former being pressurized, without cost restraints, and impelled by pure H_2/O_2, and the latter with low-cost components and powered by H2/air at ambient pressure and LT.

Over the years, the commercial interests in AFC have declined in favor of PEMFCs, anion-exchange membrane (AEM) FCs (AEMFCs), or producing underlying components such as GDEs, membranes, and ionomer [36,37]. Today, existing companies supplying AFC solutions include the UK-based company AFC Energy and GenCell Energy from Israel. While the former still supplies AFCs at the >10 kW to >MW scale, the shift toward a solid-state electrolyte (AEMFCs) is pronounced. GenCell announced in 2018 a commercial system including a 4 kW AFC employing cracked ammonia (99.5%) as a hydrogen source (Project Alkammonia) as a stationary off-grid power supply [38]. The future of AFCs appears to predominantly lie in transferring relevant experiences over to AEMFC technology to accelerate its progress. While there are some companies left that still supply complete AFC systems, the potential for further development is deemed greater in PEMFCs and in the near future AEMFCs.

4.2.1 Anionic Exchange Membrane Fuel Cells (AEMFC)

Generally, AFC contains liquid electrolytes. If these electrolytes, however, are replaced by solid electrolyte membranes, they can have higher power densities, simplified operations, and easier maintenance [39]. FCs with solid PEM can be further classified as PEMFCs and alkaline AEMFCs. AEMs are gaining popularity for use in alkaline electrolysis and FC technologies to replace traditional AFCs. This type of FC, however, poses a number of challenges such as its performance, stability, durability, mechanical strength, and low-cost production methods that need to be addressed before it can become a mainstream commercial product.

AEM electrolysis offers better CO_2 tolerance and decreased gas crossover compared with liquid electrolyte alkaline electrolysis, which results in purer hydrogen and for FCs a higher open-circuit voltage (OCV) and increased efficiency [36,37]. Moreover, the problem of flooding and "weeping" is also largely avoided in AEMFC due to the solid-state electrolyte. One of the greatest AEM challenges is creating a thin membrane (~10 mm), which can reach a high enough conductivity (>100 mS/cm) to allow it to deliver sufficient current density (>0.5 A/cm^2) while simultaneously achieving the target of 5,000 hours of stable operation at 80°C [36,37]. The research has shown that anion conductive block copolymers display promising properties for electrochemical device application, where the unique block architecture induces phase separation leading to the dramatic improvement of ionic conductivities. As a commercial block copolymer, polystyrene-b-polybutadiene-b-polystyrene (SBS) has been wildly investigated, and its block structure renders it a promising material for the preparation of AEMs with microphase separation structure. Recently, a series of side chain-type AEMs based on quaternized SBS have been reported by Liu et al. [40]; these AEMs showed exciting alkaline stability and high conductivity [41].

Average hydroxide conductivities in AEMs have reached proton conductivities in PEMs [36,37] (~ 100 mS/cm at 60°C [36,37]), and several exemplary papers report values in the range of 150–300 mS/cm [36,37]. Just like for traditional PEMFCs, the conductivity of the membrane is adversely affected by low humidity, because the mobility of the OH-group in the polymer body of the membrane depends on the level of dissociation of the cation–anion groups, which require a higher level of solvating water molecules than would be present in the dry membrane. Additionally, AEMs can suffer from embrittlement if stored in dry conditions after activation in 1.0 M KOH. For this reason, AEMs are always kept humidified and may be refreshed using a dilute alkali, e.g., 0.1 M KOH, thereby removing any buildup of carbonate species.

The durability and mechanical strength are of great importance for AEM since they can easily rupture through stresses caused by attaching them to electrode frames and especially in the process of creating membrane electrode assemblies (MEAs), via the application of catalysts and current collectors (CCs). Generally, AEMs are high tensile strength thin (10–30 mm) membranes, which can operate at high peak power densities (PPDs) due to lower mass transport resistance and improved water management [36,37]. This, however, places extensive demands on the mechanical/chemical stability of the AEM [36,37] during its creation. This is often achieved using bonding methods other than the hot-pressing method. Often AEMs are also supplied on a backing material providing additional mechanical strength, or supplied with fiber reinforcement for similar reasons. Commercial AEM membranes include Ionomr (Canada—AFN-HNN8–50-X, AFN-HNN8–25-X, etc.), FuMA-Tech (Germany—FAA, FAB, FAD, etc.), and the US companies such as Xergy (Xion Durion™, Pention™), Orion (Orion™), and Dioxide Materials (Sustainion®). Other companies producing AEM-related products include Tianwei (China), MEGA (Czech Republic), Asahi Chemical Industry Co. (Japan) [42–44], and Pention AEMs (3,370 mW/cm^2) from Xergy Inc. [45,46]. Many of the commercially available membranes for FC and electrolyzer applications have evolved from the chloralkali industry or desalination plants.

The catalysts developed for the HOR using pure hydrogen gas as fuel for AEMFC are reviewed by Ferriday and Middleton [36]. Their review indicates that recent research has focused on developing non-Pt group metal (PGM) HOR/ORR catalyst materials with high performance and longevity. Compared with that of PEMFCs with Nafion® membranes (electrolyte), AEMFCs that operated under high pH conditions enable the use of non-precious metal catalysts (such as cobalt, nickel, or silver) instead of Pt-based catalysts [39]. As a key component of AEMFCs, an ideal AEM should possess high hydroxide conductivity, excellent mechanical property, good thermal stability, and robust alkaline stability to play an important role in separating fuels and transporting OH^- from the anode to the cathode of AEMFCs [39]. Typically, AEMs are composed of polymer backbone and cationic groups and are connected by covalent bond [39]. In recent years, various AEMs based on aliphatic or aromatic polymers [such as poly(sulfone)s, poly(arylene ether)s, poly(phenylene) s, poly(styrene)s, polypropylene, poly(phenylene oxide)s, poly(olefin)s, poly(arylene piperidinium), and poly(biphenyl alkylene)s] with different cationic groups (such as quaternary ammonium (QA), guanidinium, imidazolium, pyridinium, tertiary sulfonium, spirocyclic QA, phosphonium, phosphatranium, phosphazenium, metal cation, benzimidazolium, and pyrrolidinium) have been synthesized to prepare AEMs with high conductivity and excellent alkaline stability [39]. Although the performance of AEMs has been greatly enhanced during the past few years, the foundational properties of AEMs are not comparable to those of PEMs (such as Nafion) due to the intrinsic low mobility of OH^- and the well-known base-induced decomposition of organic cations and polymer backbones [39]. In recent years, inspired by Nafion, various AEMs with hydrophilic/hydrophobic microphase separation structure have been developed and they showed improved conductivity and alkaline stability. Typical alkaline AEMFC is illustrated in Figure 4.3.

4.2.2 Microphase Separation

The development of an appropriate hydrophilic/hydrophobic microphase separation structure is an effective way to solve the conductivity and stability limitation of AEMs. Microphase separation is the term used to describe the type of chain segregation, which is able to occur in block and graft copolymers in the bulk state and in concentrated solution, and the structures generated by the packing of domains on the macro-lattice are termed mesomorphic structures. During the past few years, a series of AEMs with microphase separation structure have been reported, and most of them possess high conductivity, relatively low swelling ratio, and good alkaline stability. Since each of the AEMs could not develop its microphase separation structure, it is necessary to design the chemical structure of polymer backbone and its side cationic groups. In recent years, several approaches, including synthesis of block, graft, clustered, and comb-shaped [39] polymers with tethered organic cations, have

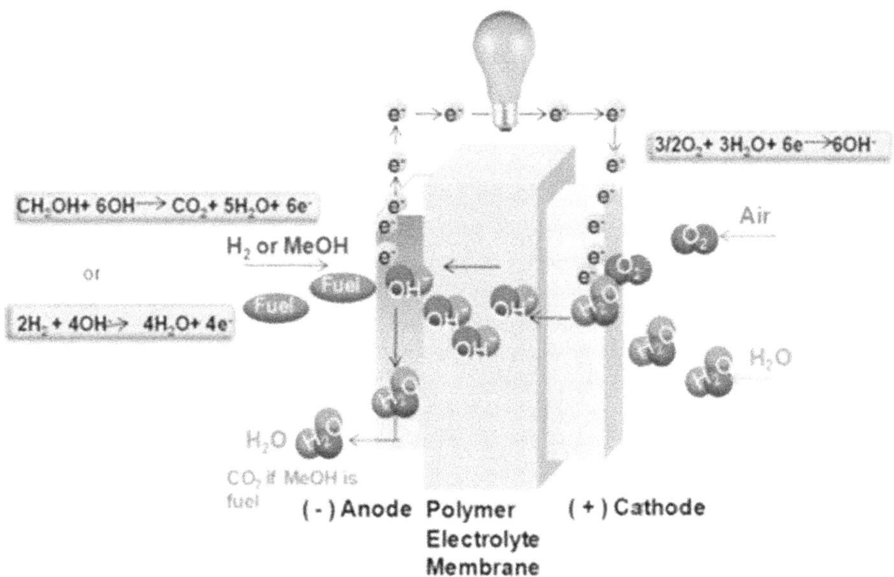

FIGURE 4.3 Typical alkaline anion-exchange fuel cell [47].

been pursued to obtain AEMs with microphase separation structure (Figure 4.4 a–d). Among various polymers, comb-shaped polymers are the most popular ones that are used to prepare AEMs with microphase separation due to their good designability and can be obtained from both pre- and post-modification methods. Although the AEMs with microphase separation were obtained by different methods, most of them showed high conductivity and robust alkaline stability, and they showed obviously higher conductivity than those membranes with similar ion-exchange capacity (IEC) and chemical structure but with no microphase separation [39].

AEM ionic conductivity depends on the number of functionalized cationic groups; however, there is a trade-off with mechanical stability because AEMs tend to swell up in volume as more groups are functionalized. This leads to the main challenge with AEMs keeping a high anionic conductivity at the same time as high mechanical strength. Another challenge lies in the paradox that the presence of OH– in the AEM causes long-term degradation, because the OH-group also acts as a nucleophile that can attack the cation group, thus splitting it into methanol and an amine. The effect of nucleophilic degradation has been found to be mitigated by placing the cation group at the end of the pendant alkyl chain [48]. A second degradation process is where OH^- attacks via an a-b Hoffman elimination reaction form a carbon–carbon double bond and the elimination of water. Many different strategies have been devised in recent years to overcome these issues, such as microphase separation, cross-linking, and organic–inorganic composites [36]. Microphase separation is utilized to increase the performance of anionic membranes by augmenting their ionic conductivity and adjusting their dimensional stability. Cross-linking and organic–inorganic composites have been utilized with great success to alter the mechanical stability of an AEM.

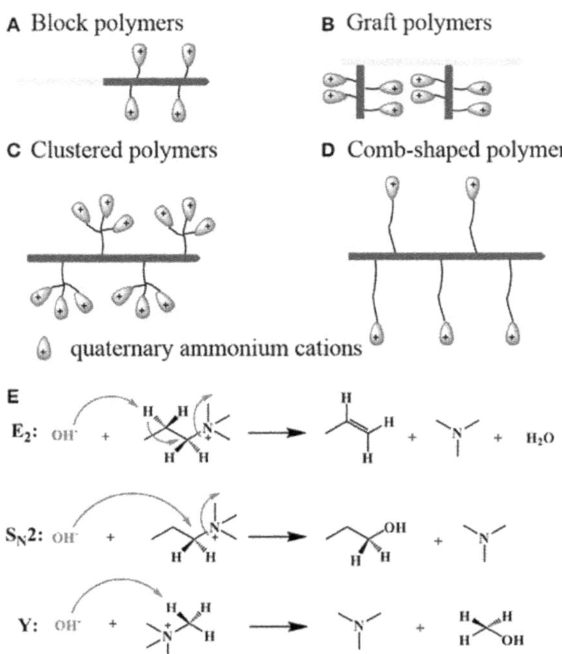

FIGURE 4.4 Illustrations of several polymer architectures of AEMs with well-defined microphase separation morphology: block (a), graft (b), clustered (c), and comb-shaped polymers (d) and the possible degradation mechanisms of QA cations in alkaline solutions: Hofmann elimination (E_2), nucleophilic substitution (S_N2), and ylide formation (Y) (e) [39].

The concept of microphase separation implies that the backbone of the polymer is hydrophobic and the cation-containing polymer is hydrophilic [39,49]. In Nafion, this is achieved by replacing the hydrogen atom with fluorine atoms in the main polymer chain, and non-fluorinated hydrocarbon-based polymers are used in AEMs. The microphase separation creates a region where water can be absorbed locally to improve the hydroxide ion conductivity while simultaneously controlling the amount of swelling that would otherwise degrade the mechanical strength. With this type of architecture, AEMs can be fabricated with a higher anionic conductivity, while still possessing the same IEC. There are four main types of AEM architectures that are based on microphase separation. These have contributed to great advancement in AEM performance in recent years.

In one approach, a long alkyl chain was grafted onto a poly phenylene oxide (PPO) backbone, forming a comb-like structure with microphase separation resulting in excellent stability maintaining 91% of its conductivity despite the PPO backbone [36]. The comb-shaped morphology was considered responsible for high hydroxide conductivity (92.6/46.6 mS/cm at 80°C/30°C). Similarly, long-side chains are employed in another approach to induce microphase separation where using spacers that are chemically incorporated into polymer backbones allows more room for the water of solvation to decrease swelling and mechanical degradation [36]. With respect to using functional cationic end groups, there are many approaches to making AEMs, including QA groups, quaternary phosphonium, imidazolium, guanidinium, and cobaltocenium [36]. Historically, QAs have received more attention; however, initial instability and manufacturing difficulties caused attention to swing toward the other options. Today, the initial stability of AEM issues has been ameliorated with techniques such as cross-linking [36,50]. Creating an organic–inorganic composite by adding TiO_2, SnO_2, or layered double hydroxides as filler in AEMs represents another method for enhancing an AEM mechanical properties [36,51]. Due to inadequate FC performance associated with this method, it has not garnered notable attention.

Recently, aryl ether-free proprietary acidic polymer (PAP) AEMs have amassed attention as stable, high-performance membranes [36]. To decrease the adverse effects associated with phenyl groups, an aliphatic chain was integrated into a PAP AEM/ionomer resulting in an aliphatic chain-containing poly(diphenyl-terphenyl piperidinium) (PDTP) copolymers [36]. Another aryl–aryl type polymer backbone that has shown great alkaline stability is polyphenylene [36,51]. Polyphenylene was functionalized with hexyl trimethyl ammonium (HTMA) and benzyltrimethylammonium (BTMA) resulting in two AEMs [52]. Membranes based on materials such as chitosan have attracted some attention in regard to sustainable AEM production. Chitosan is a nontoxic, biodegradable cycloaliphatic polymer, which contains hydroxide and active amino groups (- (NH2) on the polymer backbone, which provide hydrophilicity, high mechanical strength, and chemical adaptability into the resulting membrane [36]. Several FC systems incorporate both an AEM and liquid electrolytes in an attempt to maintain long-term performance and at the same time reduce gas crossover [53]. However, this approach caused decreased CO_2 tolerance and the adverse formation of carbonate.

4.2.2.1 Performance of AEMs with Microphase Separation

It is a huge challenge to develop AEMs that possess high conductivity, low swelling ratio, robust alkaline stability, and excellent mechanical properties at the same time. However, the formation of hydrophilic/hydrophobic microphase separation structure in the AEMs is probably a promising approach to solve these issues. Recently, it has been found that the formation of microphase separation structure in membranes could improve the conductivity of AEMs [39]. The microphase separation and the aggregation of ionic channels were driven by the hydrophilic/hydrophobic segment of the polymer, which constructs the ionic highway for OH^- conduction, shortens the pathway, and enhances the OH^- conducting efficiency in membranes [39]. In addition, hydrophobic segments control dimensional swelling in water and enhance the mechanical properties that make it possible for AEMs to possess high conductivity and good mechanical strength with relatively low IEC and dimensional swelling. Recently, Gao et al. [54] prepared AEMs with rigid-side-chain symmetric piperazinium structures possessing high conductivity. Jin et al. [55] reported poly(arylene piperidine)-based AEMs with microphase separation that resulted by introducing long-side heterocyclic ammonium cations onto the backbone, which showed a much higher conductivity (25 mS/cm at 20°C).

The alkaline stability of AEMs also limited the application of AEMFCs. Generally, the chemical structure of cationic groups determined the alkaline stability of AEMs. The most commonly used cationic groups for AEMs are QA cations. However, as is shown in Figure 4.4e, QA cations are unstable under high pH conditions, especially at elevated temperatures, due to the degradation via Hofmann elimination (E2), nucleophilic substitution (SN2), and (or) ylide formation (Y) [39]. The alkaline stability of cations could be enhanced by introducing proper groups due to their steric hindrance effect and electron donor effect. However, all organic groups, more or less, will be degraded under alkaline condition at high temperature because the nucleophilic attack forms OH^- to organic cations [36,39]. More recently, some papers reported that the chemical structure of the backbone also influences the alkaline stability of AEMs, and the polymer backbones with ether bonds degrade quickly under alkaline conditions [36,39].

Ferriday and Middleton [36] and Xu et al. [39] reported a series of AEMs with comb-shaped side chain in which the cation groups are separated from the polymer backbone by the long flexible alkyl spacers, and the comb-shaped AEMs showed much higher alkaline stability than the traditional AEMs with BTMA groups due to the low electro-withdrawing and steric effects on QA groups. Improved alkaline stability was observed for other AEMs with microphase separation structure based on various polymer backbones and cationic groups [36,39]. To obtain AEMs possessing high conductivity, low-dimensional swelling, and excellent alkaline stability at the same time, the fabrication of hydrophobic/hydrophilic microphase separation structure in AEMs is a promising method. The PPDs of AEMFCs have advanced from mW/cm^2 to W/cm^2 and durability surpassing >1,000 hours in the last decade. The current PPD and stability records paint a promising picture of AEMFC development; however, there is a wide variation in PPDs. In an FC with a PAP with aliphatic chain, AEM displayed a PPD of 2.58 W/cm^2 [36]. Radiation-grafted AEMFCs comprising low- and high-density polyethylene AEMs were compared yielding PPDs of 2.01 and 2.55 W/cm^2 [36]. This increase in performance was believed to be a result of increased water transport due to the change in polyethylene. More importantly, the stability was greatly enhanced as shown by a 440-h durability test. The AEMFC with low-density polyethylene (LDPE) and high-density polyethylene (HDPE) AEM yielded degradation rates of 790 and 68 mV/h, respectively.

AEMFC degradation also depends on temperature and relative humidity. Low humidity and high temperature accelerate AEMFC degradation. Water management is a critical challenge on the system level, as optimizing this aspect can make a great difference in the degradation rate. PPD also depended on the materials used for cathode and anode. Wang et al. [56] described the radiation-grafted LDPE AEM, which performed admirably at 110°C, yielding a PPD of 2.1 W/cm^2 [36]. Mechanical stability is of high importance in this regard, as producing very thin membranes enables fast water transport through the membrane and decreases the influence of the voltage drop across the membrane. Moreover, losses associated with poor water management are recoverable by increasing the dew points of the cell [36].

Significant studies have been carried out for the development of anionic ionomer. While ionomers are often created by dissolving the AEM using a mixture of solvent and water (thus yielding an ionomer with a noncompeting chemistry to the membrane), ionomers should ideally be developed separately from the AEM [36]. This is of great importance if the MEA preparation method requires a large difference in the solubility of the AEM and the ionomer. Due to the intrinsic characteristics of the anode/cathode, it is likely that optimized ionomers will be created separately from the AEM. Li et al. [57] showed that the adsorption of ionomer components (phenyl groups) notably impacted AEMFC anode performance. The design of ionomers less reliant on phenyl groups is a topic that requires further research. AEMFC performance degrades due to the slow diffusion of hydrogen gas by phenyl adsorption along with co-adsorption of cation–hydroxide–water. AEMFC performance increases as the adsorption of phenyl groups weakens. The adsorption energies of the ionomers were determined as follows: p-terphenyl-m-terphenyl > biphenyl > diphenyl ether > benzene-o-terphenyl > 9,9-dimethylfluorene. Generally speaking, Pt-Ru catalysts are less affected by phenyl groups, which is why this catalyst is generally preferred over Pt for the anode. Varcoe et al. [44] showed that ORR catalysts can also be adversely affected by anionic ionomers. Ex situ studies on the interaction between fully dissolved cations (free from a polymer) and Pt/C as an ORR catalyst exhibited that imidazolium cations caused a terrible degradation of performance, causing an increase in the production of peroxide species.

Ul Hassain et al. [58] optimized the ionomer, thereby doubling the PD (1.6/3.2 W/ cm^2). This study compared the tetrablock poly(norbornene) copolymers (GT32, GT64, and GT78), where the excellent aforementioned performance increment was brought about using a combination of these ionomers. By utilizing the hydrophilic GT78 in the anode and the hydrophobic GT32 in the cathode, the characteristics of the ionomers harmonize with the intrinsic water requirements in the anode and cathode. Leonard et al. [59] synthesized ionomers with different ammonia concentrations to optimize anodic/cathodic water management, thus enabling the AEMFC to operate under low rice husk (RH). The research also shows [36] that Pt catalysts produced high activities with all ionomers. Generally, all catalyst–Nafion combinations resulted in greater diffusion-limited current densities than any other ionomer. This was attributed to a lower H_2/O_2 permeability of the PPO-based ionomers and/or an incomplete ion exchange (converting the PPO ionomers from bromide to hydroxide form).

In summary, according to Ferriday and Middleton [36] and Xu et al. [39], AEMFCs have attracted enormous attention as clean and high-efficient conversion devices. As the key component of AEMFCs, AEMs act both as a barrier to separate the fuel and an electrolyte to transport OH^- from the anode and the cathode. To meet AEMFCs' practical application and commercialization, AEMs should possess high conductivity and excellent alkaline stability under alkaline conditions. The micromorphology and chemical structure of backbone and cations have a great impact on the properties of AEMs. Although various AEMs with different chemical structures and micromorphology have been prepared to improve their conductivity and alkaline stability, AEMs with excellent performance are highly desirable. Various polymer architectures, including block, graft, clustered, and comb-shaped polymers with tethered organic cations, have been synthesized for the preparation of AEMs with microphase separation structure. The formation of hydrophilic/hydrophobic microphase separation structure in the AEMs has three benefits. Firstly, the OH^- conductivity could be greatly improved due to the formation of ion transport channels in the membranes.

Secondly, the hydrophobic segments restrict the membranes' dimensional swelling in water and make it possible that AEMs possess good mechanical strength. Thirdly, the alkaline stability of AEMs could be enhanced because the hydrophobic phase weakens the nucleophilic attack from OH^- to the backbone. After recent development, the conductivity and alkaline stability of AEMs have been improved greatly. In fact, the fabrication of AEMFCs is a complex procedure; besides the properties of AEMs, there are so many parameters such as gas pressure, catalysts, and work temperature that affect the performance of AEMFCs.

Ferriday and Middleton [36] concluded that great advancements in polymer science have resulted in highly conductive and mechanically stable AEM classes such as poly(norbornene) (PNB), PAP, and HDPE. Advancements such as these have engendered great AEMFC performances with notable durability. Both in and ex situ experiments have been effective in mapping out issues and determining solutions such as characterizing the ionomer–catalyst relationship, though additional advancements on these topics are necessary. The average rate of degradation for AEMFCs is notably higher (~500 mV/h), and accounting for these rates renders AEMFCs more typical of technology at the beginning of its life. While significant progress has been made during the last five years, numerous challenges remain on all levels, from the chemical/mechanical stability of the AEM and the interaction between the ionomer and the catalyst on the component level, to issues such as water management and carbonate buildup on the system level. While the realization of AEMFCs with PPDs >1.0 W/cm^2 operating at >1,000 hours with low degradation rates (~5–10 mV/h) paints a promising picture, these targets will initially have to be met again with ultra-low-PGM loadings (<100 mg/cm^2) to capture and keep commercial interest and prove the viability of AEM technology compared with PEMFCs. Further on, these targets must be met a third time with a non-PGM AEMFC. The advancement in AFC indicates that alkaline-based discoveries mostly relate to AEMs. Further improvement in the stability of abundant and low-cost catalyst materials is required, although notable progress has been made in creating AEMs with good stability and conductivity. Ionomer–catalyst compatibility is essential during MEA creation, and a transition from laboratory scale to application level is required to advance AEMs. Future research should include the creation of efficient and stable HOR/ORR non-PGM catalysts, which are able to match or surpass the performance of the PGM benchmark. More work is needed to establish long-term stability (>1,000 hours). AEM needs to be more stable, and its performance, which satisfies the rigorous demands, needs to be demonstrated. Overall system optimization is required with particular attention to large-scale (>100 cm^2) AEMFC units. Finally, more design strategies are required for catalysts, ionomer, electrolyte, and miscellaneous electrode components to avoid a mismatch of electrochemical properties.

4.2.3 Low-Cost Anion-Exchange Membrane Electrolysis for Large-Scale Hydrogen Production

AEM electrolysis is a promising solution for large-scale hydrogen production from renewable energy resources. However, the performance of AEM electrolysis is still lower than what can be achieved with conventional technologies. The performance

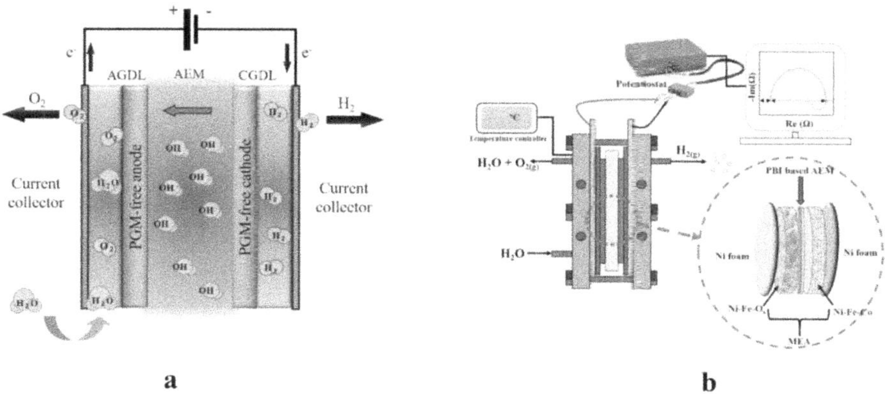

FIGURE 4.5 (a) Schematic of the anion-exchange membrane (AEM) electrolysis. *AEM:* anion-exchange membrane, *AGDL:* anode gas diffusion layer, *CGDL:* cathode gas diffusion layer. (b) Schematic diagram of AEM water electrolysis with EIS experimental setup.

of AEM electrolysis is limited by integral components of the MEA and the reaction kinetics, which can be measured by ohmic and charge transfer resistances. Vincent et al. [60] investigated and then quantify the contributions of the ohmic and charge transfer resistances, and the rate-determining steps, involved in AEM electrolysis using electrochemical impedance spectroscopy analysis. The factors that have an effect on the performance, such as voltage, flow rate, temperature, and concentration, were studied at 1.5 and 1.9 V. The study by Vincent et al. [60] observed that increased voltage, flow rate, temperature, and concentration of the electrolyte strongly enhanced the anodic activity, and here, the anodic reaction offered a greater contribution to the overpotential than that offered by the cathode. A schematic of AEM electrolysis for large-scale hydrogen production is illustrated in Figure 4.5.

The goal of the study by Vincent et al. [60] was to investigate the various resistances involved in AEM electrolysis, both qualitative and quantitatively, using electrochemical impedance spectroscopy (EIS). In particular, the study determined the contributions of the individual resistances toward the overall resistance and the rate-determining step of AEM electrolysis. The study successfully diagnosed and quantified the ohmic and charge transfer resistances from the overall resistances. Furthermore, the study demonstrated, for the first time, how the liquid electrolyte flow rate also affects performance. At higher flow rates, the removal of OH^- ions was very rapid. This reduced the available reaction time for oxidation and reduction, subsequently leading to lower availability of OH^- ions for the catalysts, resulting in a significant increase in the overpotentials (especially for the oxygen evolution reaction [OER]). At higher temperature, activation of the CL expedites the OER reaction, which reduces both the ohmic and charge transfer resistances. Furthermore, an increase in electrolyte concentration increases the availability of OH^- ions and the mobility of ions, directly reducing the charge transfer resistance. However, at lower flow rate, the catalyst active sites may be blocked by oxygen bubbles. In all the cases considered in the study, the cathode resistance was not significantly improved;

hence, the rate-determining step is the anodic OER process. The optimum working conditions for AEM electrolysis, offering high performance with this MEA, were the following: 1 M KOH liquid electrolyte, temperature 60°C, and flow rate 40 mL/min. Under these conditions, the following was achieved: best performance of 500 mA/cm^2, with cell potential 1.85 V, and cell resistance as low as 20 mΩ/cm^2. Following on from results by Vincent et al. [60], suggestions for advancing the design of AEM electrolyzers include the following: maintain the optimum flow rate and reduction of anodic ohmic resistance using plasma-sprayed electrodes.

4.3 DIRECT METHANOL FUEL CELL

In LT FCs, hydrogen can be obtained from numerous different types of fuels. This has led to the development of numerous types of FCs such as direct methanol, direct ethanol, direct ethylene glycol, direct borohydride, and direct formic acid FCs. As a representative of these FCs, here we examine in some detail DMFC. However, one should point out that direct ethanol FCs (DEFCs) are gaining attraction, as this fuel, compared with methanol, offers higher energy density and a low price, in addition to being less corrosive and nontoxic. Ethanol (C_2H_5OH or EtOH) can be obtained in large amounts through the fermentation of agricultural renewable crops and waste, such as corn and sugarcane.

DMFC has a design and internal configuration similar to PEMFC with operating temperatures between 60°C and 130°C [16,17]. The generation of CO_2 as a waste product in DMFC differentiates it from PEMFC. The polymer membrane perfluorosulfonic acid (Nafion membrane) used in DMFC is a variant of one used in PEMFC, and the catalyst on their negative electrode (i.e., anode) extracts hydrogen from the liquid methanol [16] through a reforming reaction between methanol and steam fed to the anode. The reaction produces CO_2 and hydrogen, and the electrons then flow through the external circuit to generate the current, i.e., before flowing back to the cathode, and the protons are transported to the cathode through the electrolyte. Water is formed at the cathode by the reaction between the protons and electrons with oxygen. The anode and cathode reactions in DMFCs can be described by the following Eqs. (4.4) and (4.5) [17]:

$$CH_3OH + H_2O \rightarrow CO_2 + 6H^+ + 6e^- \tag{4.4}$$

$$1.5O_2 + 6H^+ + 6e^- \rightarrow 3H_2O \tag{4.5}$$

At a temperature of 120°C, DMFC can develop an efficiency of around 0.4 [16], the lowest of all FC technologies [17]. This low operating efficiency is due to "methanol crossover," which is the cause of unproductive methanol consumption in the process [37]. LT reforming reaction to generate hydrogen requires noble metal catalysts at the anode and the cathode [17]; the anode and cathode materials are Pt or Pt–ruthenium (Pt-Ru) and Pt, respectively. Methanol is cost-effective, easy to produce, safer to store, and can be used directly in the cell. This allows a simple cell structure and design with relatively low weights, which makes them a low-weight alternative to battery technologies for military and other applications and very suitable for portable power for laptops and mobile appliances, including small plants that are less than 5 kW [17,37]. For DMFC to be competitive with PEMFC for applications requiring around 50–200 kW of power, it must be able to achieve 250 mW/cm^2 above 0.5 V with less than 50 ma/cnr' of methanol crossover (MCO). Current state-of-the-art Pt-Ru catalysts can provide around 300 mW/cm^2 at high temperature with pressurized oxygen on at least the cathode at 0.5 V. Whether the goal of MCO is being met with Nafion membranes is a current subject of debate in the literature. Several researchers have shown that CO_2 crossover results in an overestimation of the amount of MCO in DMFCs.

While the desired performance goals for DMFCs have been obtained for ideal conditions and with large Pt loadings, the cost of Pt is an issue. High Pt cost does not allow the desirable system cost of $300/kW for 5–100 kW stationary power system. The need for a pressurized cathode also requires a compressor that both increases system cost and significantly reduces system efficiency. To reduce Pt loadings and cathode pressure, high-temperature operation is recommended. This, however, is unachievable without higher pressure on the cathode. MCO also increases due to higher methanol diffusion coefficient in the membrane at higher temperatures. More research is needed to generate stable and conductive membrane under these conditions. Functionalizing acid groups in the membrane with ZrO, SiO_2, heteropoly acids, and palladium led to reduced crossover, but also reduced membrane conductivity. Researchers have not been able to show good performance by novel polymers at temperatures above the boiling point of water at any cathode pressure.

Making MEAs from novel membrane materials is also complicated by needing to get a good interface between the ionomer in the membrane and electrode. A suggestion to laminate a layer of methanol impermeable polymer between two Nafion membranes so that the electrodes will make good contact with the membrane requires hydration to perform at high temperatures. The issue is also analyzed using complex multistep reaction kinetic models for anode where the CO oxidation limits the reaction at high potentials and adsorption of methanol limits the reaction rate at low potential. The solution, however, generally requires numerical methods. Furthermore, modeling mass transfer in the anode is complex because of the two-phase flow in all regions. Wang and Wang [61,62] have used computational fluid dynamics (CFD) to include the two-phase effects in the backing layer. The cathode kinetics are usually included using a Tafel expression. Most cathode models consider oxygen concentration to be constant due to a large stoichiometric excess of pure oxygen at high pressures. While two-dimensional models have been explored, very little is done on transient models.

While DMFCs are generally operated at low concentration of methanol due to MCO, the use of high-concentration methanol is highly demanded to improve the energy density of a DMFC system. Feng et al. [63] examined a selective electrocatalyst-based DMFC operated at a high concentration of methanol. They showed that at an operating temperature of 80°C, the as-fabricated DMFC with core–shell–shell Au@Ag_2S@Pt nanocomposites at the anode and core–shell Au@Pd nanoparticles

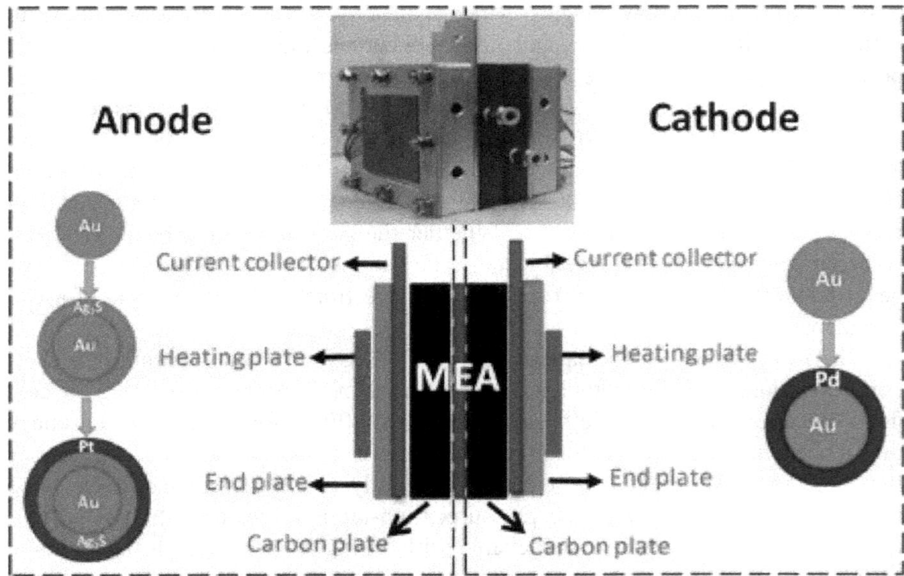

FIGURE 4.6 Schematic showing the DMFC fabricated with selective electrocatalysts at the anode and cathode chamber. Inset: Photograph of a practical cell [63].

(NPs) at the cathode produced a maximum PD of 89.7 mW/cm^2 at a methanol feed concentration of 10 M and maintained good performance at a methanol concentration of up to 15 M. The high selectivity of the electrocatalysts achieved through structural construction may have accounted for the successful operation of the DMFC at high concentrations of methanol. Figure 4.6 illustrates the basic DMFC structure. Furthermore, the use of Au nanoclusters with fine diameters as starting materials or a design of a semiconductor–metal system with energy-level alignment more favorable for the electronic coupling effect and further enhancement in activity or selectivity for DMFC reactions might be possible.

DMFC technology has the potential to prevail as a leader in the booming market for portable electronic devices because of its advantages of high-energy density and quick refueling, which are crucial characteristics of portable power systems [64–67]. In general, dilute methanol solutions, for example, 1–2 M for active DMFCs or about 3 M for passive DMFCs, are often used as fuel in DMFCs to inhibit the crossover of methanol from the anode to the cathode to achieve high performance [68,69]. However, to compete with lithium-based rechargeable batteries that currently dominate the portable power market, the use of high-concentration methanol as a fuel is highly demanded to capitalize on the high-energy density of DMFCs. It has been reported that the specific energy of a DMFC system can be comparable with that of conventional Li-ion batteries only when a methanol solution with a concentration of 9 M or higher is fed as fuel (assuming that the overall efficiency of a DMFC system is 20%) [70].

4.3.1 Passive Small Direct Alcohol Fuel Cells for Low-Power Portable Applications

Another advance in the DMFC is the development of passive small direct alcohol FCs (PS-DAFCs). PS-DAFCs are compact, standalone devices capable of electrochemically converting the chemical energy in the fuel/alcohol into electricity, with low pollutant emissions and high-energy density. Thus, PS-DAFCs are extremely attractive as sustainable/green off-grid low-power sources (milliwatts to watts), considered alternatives to batteries for small/portable electric and electronic devices. PS-DAFCs benefit from long-life operation and low cost, assuring an efficient and stable supply of inherent nonpolluting electricity. In recent years, this technology is significantly advanced.

DAFCs are attractive because they generate electric power directly from the electrooxidation of liquid fuels (alcohols), with water and CO_2 as the main products, along with heat. Moreover, alcohols have a higher volumetric energy density, in addition to being easier to handle and store [71–74]. Among the different possible alcohols, methanol (CH_3OH or MeOH) is the simplest and the most used one. The technology fundamentals of a DAFC device can be roughly divided into its core components: the MEA and the general system/cell structuring and packaging (e.g., the CCs and the end plates). Figure 4.7 presents an exploded-view schematic of the DAFC main components.

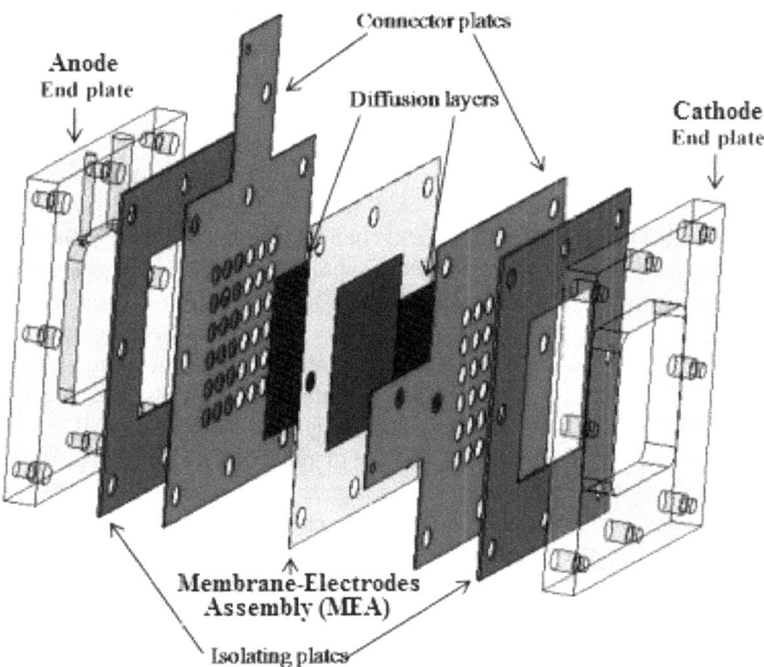

FIGURE 4.7 DAFC main components: exploded-view schematic [71].

MEA can be identified as the anode diffusion layer (ADL), anode CL (ACL), PEM, cathode CL (CCL), and cathode diffusion layer (CDL). Both electrodes are supported on macroporous carbon paper or cloth called the "diffusion layer," often named the "gas diffusion layer" (GDL), and sometimes smoothed out by a thin microporous layer (MPL), made of CB and PTFE. The GDLs provide the transport channels for the reactants (fuel and oxidant) and reaction products. The PEM most commonly used in DAFCs is Nafion®, a perfluorinated sulfonic acid ion-exchange membrane, developed by DuPont (Wilmington, DE, USA). This polymer/proton conductive membrane, in addition to being the electrolyte, serves as the barrier between the two electrodes and a barrier to fuel crossover/permeation. The electrons released in the alcohol oxidation reaction (AOR) at the anode are not conducted by the membrane, but by the anode CC and returned to the cathode by the cathode CC, via an external electric circuit. On the cathode side, the ORR occurs in air, promoted by the CCL, to form water. In this manner, the redox pair (AOR and ORR) provides the path to directly, in an efficient and clean way, convert the fuel's chemical energy into electrical energy, and these power generator devices can be used in the different applications (as presented in Table 4.2) [71,75,76].

The performance of these devices can be characterized by their OCV or voltage at which no load is applied, peak/maximum PD obtained from polarization (current–voltage) curves, and operating temperature (T) [71,77]. These devices are generally operated at LT (typically below 95°C) [71,77]. Their low operation temperature allows an uncomplicated start-up, in addition to being a quick and easy response to changes in load and/or operating conditions, even at room temperature. In practical applications, an FC stack (a series of connected single cells) is used instead of a single cell. Partitioning the adjoining cells stacked on one another and serving as the transmitter/connector of an electric current to the outside, like the CCs of the single cell, there are conductive separators called BPs. Thus, the CCs or BPs are also key components in these devices, since they account for about 80% of the total weight of the cell and are essential to ensure the uniform compression of the MEA and minimum contact resistance. They connect electrically and supply reactants/remove reaction products through different flow field/channel designs, to/from both electrodes. Depending on how the fuel and oxidant are supplied, DAFCs can be categorized into two major operation modes: active or passive (Figure 4.8) [78].

For small/portable power applications, there is a strong need to minimize the size of the components [71]. Therefore, passive DAFCs seem more suitable with an attractive cost-to-power ratio compared with active DAFCs. The passive feeding system does not require auxiliary supplying devices; thus, a fuel pump along with an air blower is not present. The flow of reactants and products is based on diffusion, as well as natural convection. Hence, the fuel is provided to the ADL and ACL from a reservoir/tank build on the anode end plate, whereas the oxygen to the CDL and CCL comes from the surrounding air through the open cathode end plate (often called "air-breathing" mode). In this way, no additional power consumption is needed, and the parasitic power phenomenon due to the external components is minimized. Thus, a significant system volume reduction is achieved [78]. Moreover, with the rapid development of micromachining technologies, miniature (millimeter or micrometer scale) passive DAFCs (PS-DAFCs) represent a promising application

TABLE 4.2
DAFC Portable Commercial Applications [71]

Field	Fuel	Products	Manufacturer	Characteristics
Laptop	Methanol	CD-ROM-sized fuel cell pack	Antig Technology	Electric power of 45 W for 8 hours of normal laptop use
	Methanol	A laptop docking station	Samsung	Maximum output of 20 W
	Methanol	Battery charger	Panasonic	Provided between 10 W and 20 W of power with 200 mL of methanol
	Methanol	Fuel cell notebook PCs	Antig Technology & Toshiba	Produced power of 10 W and voltage of 7.2 V
Military/industrial	Methanol	Off-grid power generators	SFC Energy	Power output of 500 W
	Methanol	Army field power pack	SFC Energy	Power output of 250 W for the larger unit at 12 or 24 V and 100 W at 28 V for the smaller one
	Methanol	Portable generators, chargers, and batteries	SFC Energy	Capacity of 50 W
Medical	Methanol	Hearing aid	Danish Technological Institute (DTI)/ WIDEX®	Supply 2.5 mW continuously for 24 hours at a voltage above 350 mV with less than 200 μL of methanol
Telecommunication	Methanol	Battery charger	Antig Technology	Power output 3 W and voltage of 5.5 V
	Methanol	Dynario™ (Battery charger)	Toshiba	Power output of 2 W with a single injection of 14 mL of concentrated methanol solution
	Methanol	Mobile and remote power source	Neah Power System	Power density level exceeding 80 mW/cm^2
Others	Ethanol	Power pack	NDC Power	Power output of 3–250 W and operational time of 3,700 hours
	Methanol	Bioenergy discovery kit	Horizon Fuel Cell Technology	Generated power of 10 mW
	Methanol	Mobile audio player	Toshiba and Hitachi	Capacity of 100 mW for 35 hours player usage with single 3.5 mL or 300 mW for 60 hours with single 10 mL of concentrated methanol solution

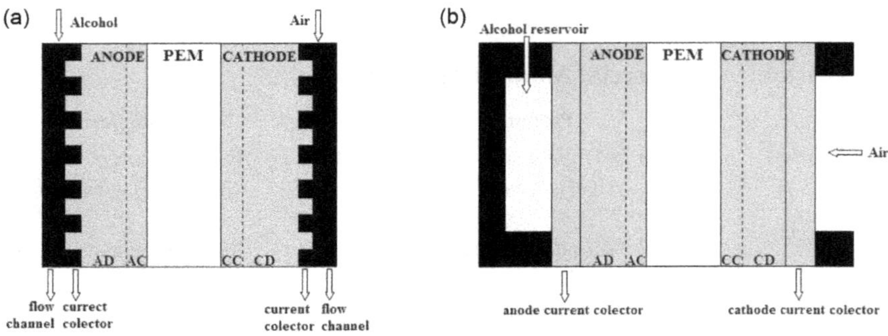

FIGURE 4.8 General main operating modes of DAFCs: (a) active and (b) passive [71].

as micro-energy power sources aimed at a niche market. These can be used as a substitute for batteries in small portable devices, such as mobile phones (0.1–3 W) or laptops (5–50 W) and digital cameras (5–20 W), as well as other cordless systems (3–50 W), including next-generation portable consumer electronics, such as wearable electronics and medical devices, very sensitive to the weight, and volume of micro-power sources [71,79,80]. The decrease in size, however, results in internal two-phase gas–liquid transport by CO_2 bubbles and water droplets. In the absence of external flowing force to remove them, they are constantly produced by the anode and cathode reactions. This hinders further reactions by blockage of mass transfer and the electrode surface. Moreover, PS-DAFCs can also suffer from fuel crossover from the anode to the cathode, since the Nafion® membrane is not completely impermeable to alcohol molecules.

While the use of high methanol concentration leads to MCO, which reduces its performance, low alcohol concentrations could lead to an inadequate fuel supply, reducing the cell performance and rendering the power output of the passive DAFC unacceptable for real applications. The water flux through the Nafion® membrane is due to the coupled effect of diffusion and electroosmotic drag (EOD). In a proper balanced operation, the water present on the cathode side should be removed to avoid oxygen transfer limitation and water should be provided on the anode side to compensate for the water crossover through the membrane. All these factors are substantially critical for PS-DAFCs, compared with their active counterparts.

The major setback of PS-DAFC commercialization is the lack of low-cost materials (electrocatalysts, electrodes, and electrolyte/membrane) providing stable and satisfactory performance [71,74,78]. Up to now, the most effective catalysts in PS-DAFCs have been precious metal group (PMG)-based NPs including Pt, Ru, Pd, and Au, supported on high-surface-area carbon. Pt-based alloys such as Pt-Ru show high tolerance to the poisoning effect of intermediary CO, providing stability and improved kinetics toward the AOR, particularly for methanol. Even so, the high loadings of PMG NPs required for the reasonable power output of PS-DAFCs raise the costs and impact of the technology. This aspect, together with the catalyst durability (improving the performance of the catalysts and their supports), is driving recent research in the field [71,20,80–82]. For FCs in general, two material-based strategies have been tackled. One is to decrease PGM usage by increasing its activity (so-called low-PGM catalysts) [71,83,84], while the other one is to develop alternative PGM-free catalysts to completely replace PGM with earth-abundant materials [71,85–88]. Due to the significant efforts in recent years on alternative catalysts for active DMFC systems in acidic medium, PGM-free catalysts, such as Fe–N–C catalysts (commercially available from Pajarito Powder Manufacturing Company), have demonstrated an improved ORR activity and methanol-tolerant behavior, when used as a cathode electrode, with reasonable power outputs [71,89,90]. Overall, these results demonstrated the promising prospect for high-performance and inexpensive DAFCs using PGM-free or low-PGM-based catalysts for ORR and AOR, with tuned selectivity and tolerance to a range of fuels and chemicals [91].

With PGM-free catalysts, however, the acidic environment of conventional PEM-DAFCs still imposes stability challenges, mainly due to Fe dissolution and carbon corrosion under operational conditions, involving mechanisms such as dissolution, agglomeration, and/or detachment of the carbon support from the catalytic NPs [71]. The use of solid AEMs as an alternative electrolyte has led to increasing interest in alkaline DAFCs (AEM-DAFCs), since non-precious metal catalysts can be employed to overcome problems associated with PEM-DAFCs [71,92,93]. AEMs usually contain a main polymer backbone covalently bound to cationic functionalities (such as simple QA groups or other more complex and stable groups) that confer anion selectivity. The most common anion species transported in AEMFCs is represented by hydroxyl ions (OH^-) produced by ORR at the cathode, and the MEAs can be fabricated using commercial AEMs such as Tokuyama A201 (Tokuyama Corporation, Tokyo, Japan) and Fumasep® FAA3 (FUMATECH BWT GmbH, Bietigheim-Bissingen, Germany). However, despite a flurry of recent research activity and the improved performance of AEMFCs, their durability is still lower than that of PEMFCs [71,94,95]. Thus, further advances in AEM-DAFCs depend on improving the performance of electrocatalysts along with AEM stability and durability, required to achieve highly effective devices [71,96,97]. Finally, the devices (PEM-DAFC or AEM-DAFC) generally require oxygen as the electron acceptor (oxidant), typically from the ambient air in the passive mode.

The application of DAFC in outer space and underwater requires an additional oxygen tank in the system, lowering the energy density of the whole FC system. An alternative approach of using the hydrogen peroxide (H_2O_2) reduction reaction (HPRR) instead of ORR leads to low activation loss and the possibility of using non-Pt catalysts. HPRR is also a simpler and easier (two-electron transfer) process. In addition to improving cell performance, water flooding problems are avoided due to their intrinsically liquid nature. Yan et al. [98] demonstrated the concept for an acid DMFC, as well as for an alkaline DEFC [99], also in active mode with non-Pt catalysts. For FC-based power systems for unmanned aerial vehicles (UAVs), some advances and challenges are described in [100].

While DAFC has been investigated in detail, more efforts are needed to increase the durability and reliability of the cells [101,102] for its commercialization. Moreover, a recent cost analysis study of DMFC stacks for mass production pointed out that the introduction of innovative approaches can result in further cost savings [103]. PS-DAFCs are considered promising alternatives to batteries as small power sources (milliwatts to watts), as they benefit from a high PD, long lifetime, and low cost. Thus, they are capable of ensuring an efficient and stable supply of inherent nonpolluting electricity for next-generation portable applications. Further developments should consider other alcohols and other oxidants, as well as AEM applications for both single cells and stacks. New effective catalysts and membranes are also needed to further improve performance. Researchers must also control the content and location of reactants and products when operating the devices. The ultimate goal should be to support and facilitate the deployment of different configurations of PS-DAFCs into broad commercial markets.

4.4 PROTON-EXCHANGE MEMBRANE FUEL CELL (PEMFC)

PEMFC is the most advanced LT FC. The distinctive advantages of rapid start-up time, wild range operating temperature (−40°C–90°C), and high-specific energy have made PEMFC stand out from all types of FC and wildly be used in FC vehicles and stationary applications. While the work on its cost and durability continues, it has significant potential for large-scale commercialization. Here, we examine advances made in various components of the PEMFC.

4.4.1 Membrane–Electrode Assembly

A membrane–electrode assembly for PEMFC is illustrated in Figure 4.9. A single PEMFC composed of an MEA and two fluid flow patterns (FFPs) produces less than 1 V, which is very low for most applications. Therefore, to increase the potential to a practical level, the individual cells are connected in series; i.e., the cathode of one cell is electrically connected to the adjoining cell's anode to form an FC stack [104]. PEMFCs utilize redox reactions to generate electricity. Hydrogen gas is pumped into

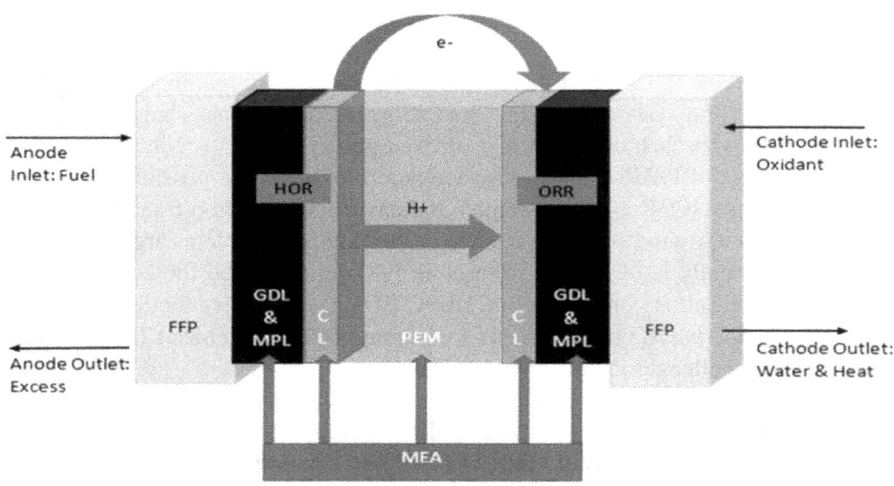

FIGURE 4.9 Membrane–electrode assembly [104].

the anode, where a HOR occurs. The hydrogen ions travel through the PEM from the anode to the cathode, while the electrons travel externally to the cathode. Air or oxygen gas is pumped into the cathode and combined with the other components to go through an ORR.

Two predominant mechanisms are possible for the ORR process in the cathode of the PEMFC (i.e., in the acidic environment), and they involve multielectron reactions. In general, electrocatalysts that boost a direct $4e^-$ ORR are preferred for high-performance H_2/O_2 FC development over $2e^-$. The $4e^-$ and $2e^-$ ORR are as follows [104]:

i. Direct four-electron ($4e^-$) oxygen reduction

$$O_2 + 4H^+ + 4e^- \rightarrow 2H_2O \quad (4.6)$$

ii. The two-electron ($2e^-$) oxygen reduction

$$O_2 + 2H^+ + 2e^- \rightarrow H_2O_2 \quad (4.7)$$

$$H_2O_2 + 2H^+ + 2e^- \rightarrow 2H_2O \quad (4.8)$$

The overall HOR in the acidic medium, the so-called anode half-cell reaction, involves the oxidation of an H_2 molecule on the electrode surface (Eq. 4.9). Electron and proton transfer from the electrode surface and the electrolyte, respectively [104], is given as follows:

$$H_2 \rightarrow 2H^+ + 2e^- \quad (4.9)$$

Thus, the overall reaction of the PEMFC is as follows:

$$2H_2 + O_2 \rightarrow 2H_2O \quad (4.10)$$

While PEMFCs are actively fabricated and commercialized, work continues to optimize the cost and durability of cells and their components. In recent years, nature with fully hierarchical scaling structures has been a source of design inspiration for different parts of the FC. Bioinspired designs and materials for flow fields, catalysts, and membranes have lived up to the performance expectations (see Figure 4.10). Bioinspiration can be used to optimize PEMFCs and have an environmental impact by repurposing materials previously determined to be waste as viable materials [105].

MEA includes three layers, namely membrane, GDL, and CL, to provide micro-channels for mass transport and electrochemical reactions, and influence the performance, durability, and cost of PEMFC [106,107]. GDL consists of carbon, water, alcohol, PTFE, or another hydrophobic substance and provides channels for gas and electrons, supports the CL, conducts electrons, and discharges water generated from the reactions of the PEMFC [106]. PTFE is used to promote the transport of gas and water during operation under flooding conditions [106]. The CL is where the

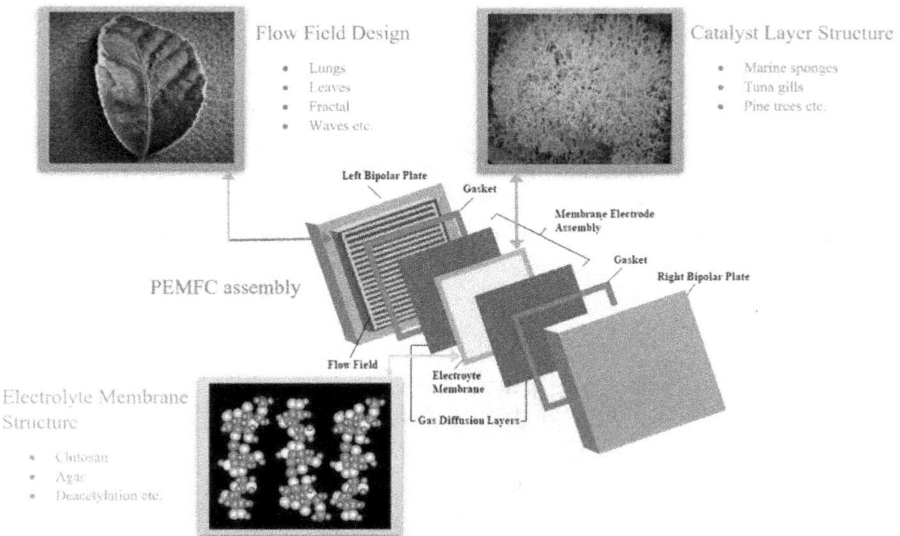

FIGURE 4.10 Bioinspired approach for various components of PEMFCs.

electrochemical reaction occurs that converts hydrogen gas and oxygen (in air) into water and electricity. The CL thickness typically varies in the range of 5–100 μm, with a porosity of 40%–70%, and it contains a well-dispersed catalyst with a particle size of 1–10 nm [108].

Pedram et al. [104] point out that three generations of MEA, GDE, catalyst-coated membrane (CCM), and oriented MEA, have developed with the increasing demand for large-scale commercialization of PEMFCs, significantly improving their performance and lifespan and reducing the cost. While the preparation of GDE [109] in the first generation is easy and simple, involving the coating of a CL on a GDL to form a PEM, and then forming an MEA by hot pressing, the amount of catalyst is difficult to control accurately, resulting in low utilization of the catalyst (lower than 20%) and high cost. The interface between GDL and PEM can also be partially isolated resulting in increased internal contact resistance. This led to the second generation of preparation of CCM by pressing a GDL with two catalyst-coated sides and a PEM into an MEA with evenly dispersing catalyst particles. In this way, the prepared MEA demonstrates a better performance with low resistance, good proton transfer ability, high Pt utilization, and low cost in comparison with the first generation, resulting in its wide application in large-scale commercialization. The third and the latest generation is oriented MEA, which was first proposed in 2002 by Middelman et al. by coating uniform Pt particles (2 μm) and then coating a proton conductor (10 μm) on the surface of oriented carbon [110]. 3M developed a commercial nanostructured thin film (NSTF) by growing whiskers onto a microstructured substrate and then coating Pt on the whiskers with controlled loading through roll-to-roll vacuum sputtering [106]. NSTF exhibits higher catalyst utilization, activity, and stability with a thickness of 0.25–0.4 μm (5% of the thickness of traditional Pt/C electrodes) and almost 100% utilization of Pt.

Considerable research has been conducted and reviewed on MEA to improve its performance and lifespan [106]. The active area and the thickness of CL are influenced by the Pt loading, which in turn affects the mass transfer and water management and in turn cell performance. Qu et al. [111] concluded that the performance degradation of MEAs with Pt/C catalyst could be mitigated by increasing the Pt loading. However, a higher Pt loading would lead to a thicker CL and hence poorer mass transfer and possibly carbon material corrosion. Van Deo et al. [108] fabricated a Pt/C double-layer catalyst to enhance the electrocatalytic properties of electrodes with more Pt active sites than obtained by the electrophoresis deposition method.

4.4.2 Proton-Exchange Membrane (PEM)

Most FCs use Nafion membranes manufactured by DuPont as PEM. A PEM has the main functions of conducting protons, separating fuel oxidizer, and insulating protons, and its performance directly affects the performance of PEMFCs. An ideal PEM should exhibit a high proton conductivity rate, proper water content and gas molecular permeability, good electrochemical stability, and mechanical stability, with ideal characteristics of a decomposition temperature of 250°C–500°C, water absorption rate of 2.5–27.5 H_2O/SO_3H, and conductivity in the range of 10^{-5} to 10^{-2} S/cm. The current production process for Nafion membranes is expensive, and reducing production costs and improving chemical and mechanical stabilities are important goals for PEMs. Ballard developed a *perfluorosulfonic acid* (PFSA) membrane with properties comparable to those of Nafion membranes with low cost and simpler production process. Apart from the PFSA membrane, other membranes such as polybenzimidazole (PBI)-based membranes, sulfonated aromatic (such as polyphenylsulfone and SPEEK) membranes, phosphonic-based membranes, polyphosphazene-based membranes (SPE), and polystyrene sulfonic acid (PSSA) membranes [106] are also developed. Among them, PBI exhibits better chemical and physical properties, with an operating temperature range from 100°C to 200°C, while the temperature range of polyphenylsulfone is up to 230°C and SPE membranes are applied in high-temperature PEMFCs. Su et al. [112] developed five types of GDEs, namely polyvinylidene fluoride (PVDF), sulfonated polymer (Nafion), PBI, fluorinated ethylene propylene (FEP), and PBI/PVDF. In a single cell, the PTFE- and PVDF-based GDEs are considered to be very promising for commercial applications.

There is a promising trend toward developing hybrid membranes based on existing backbones with known functions, to achieve a particular function efficiently. Sutradhar et al. [113] synthesized sulfonated polyphenylene benzophenone membrane (SPPBP), through carbon–carbon coupling polymerization for PEMFCs with the properties of high thermal and chemical stabilities and high proton conductivity. Neethu et al. [114] investigated a new PEM incorporated with activated carbon extracted from coconut shell (ACGS), which could improve the proton exchange with high porosity and superior specific surface area, and natural clay, which could lower the cost to approximately 45 $/m². Haragirimana et al. [115] fabricated a series of four sulfonated poly aryl ether sulfones (SPAES) copolymer blend PEMs with different sulfonation degrees, leading to better thermal and mechanical stabilities, better hydrolytic and oxidative stabilities, and lower methanol permeability than the control membrane.

Advances are also made in the preparation method, which includes thin-film coating methods, vacuum deposition, and electrodeposition. Teixeira et al. [116] used impregnation and casting methods to incorporate a derivative of aryl monoacid or bisphosphonic acid into the Nafion membrane, demonstrating that the proton conductivity of the membrane created by casting was 1.55 times higher than that of the regular membrane at 30°C, 40°C, and 50°C. Wang et al. [117] invented a porous nanofiber composite membrane (PNFCM) by impregnating a nanofiber mat, prepared by a hybrid electrospinning and soft template method, with chitosan resulting in a membrane with superior properties. Generally, two strategies are adopted to develop PEM. The first one is to use the fabrication method to reduce the thickness of membrane from 50–150 nm to 5–25 μm resulting in a significant drop in the cell ohmic impedance and intensity at a high current density. The second one is to follow the success of PFSA in reducing cost and ensuring safety by exploring in more detail hydrocarbon polymers. Finally, following bioinspired solution strategy, He et al. [118] reported an artificial acid–base pair inspired by mussel bio-adhesion. Polydopamine-modified graphene oxide (DGO) sheets are fused into a sulfonated poly (ether ketone) (SPEEK) matrix to form the nanocomposite membrane. The interfacial electrostatic attractions between the sulfonic acid groups in SPEEK chains and $-NH_2$/–NH– groups in DGO sheets generate acid–base pairs that provide a long-range low-energy barrier pathway for proton hopping. This membrane gave very positive results for OCV and maximum current density. In the future, more work is needed to further improve electrolyte conductivity and thermal and mechanical stabilities of the PEMs. Also, nature-inspired designs and bio-based membrane materials show great potential to benefit PEMFC development and improve its efficiency and performance.

4.4.3 Gas Diffusion Layer

GDL is responsible for underpinning the homogeneous distribution of the reaction gas on the catalyst surface, as well as removing water from the catalyst surface. Drainage, gas permeability, and electrical conductivity are three compulsory conditions for GDL. Thickness is an important factor that has a significant effect on the above three elements. The thickness of the GDL is the sum of the thicknesses of the macrolayer and microlayer. There are four types of GDL configurations, namely two layers of (CC), a layer of the mixture of carbon–polymer (CP) and carbon cloth (CPCC), two layers of CP and CPCC, and four layers in the order of CPCPCC-CC-CPCC; an X–Y robotic spraying technique can be used to fabricate a GDL by spraying carbon ink on a heated CC and then drying [106]. The carbon powder can influence the pore structure of the prepared GDL significantly owing to the difference in physical properties, such as the specific surface area, pore distribution, particle size, and electrical conductivity. At present, Vulcan XC-72R and acetylene black (AB) are commonly used in PEMFCs. In Ballard mark V cells, carbon cloth provides an outstanding advantage at high current densities and has been shown to be more efficient for the transport of liquid water owing to the hydrophobicity of the cloth surface. Lin and Chang [119] examined an MPL composed of composite carbon material with multiwall CNTs and AB and compared the results with the cases of MPLs made with pure AB and CNT. Their experiments demonstrated that the cathode MPL with a mixing ratio of 1:4 of AB and CNT by mass demonstrates the best performance in all cases.

4.4.4 Catalyst Layer

Catalysts are the key materials in FCs, but the cost of Pt material used for the catalyst is very high. Many researchers are focusing on finding cheaper and durable substitutes for Pt, although they may not exhibit comparable performance to that of Pt. These catalysts fall into three categories: (1) Pt-based catalysts; (2) modified Pt-based catalysts containing other metals, such as Cr, Cu, or Co; and (3) non-Pt-based catalysts, such as non-noble metal or organometallic catalysts [120]. Pt-based catalysts are also being examined for improving both morphology and synthetic protocols [106].

The type of carbon substrate, CL and catalyst properties, different surface structures, placement of CL, and other factors are also studied to explore their influences. Additionally, current studies are focused on developing modified carbon-based or non-carbon-based materials as catalyst supports to promote the catalytic performance of Pt [106]. The research has shown that while the utilization of Pt alloys has a positive effect on promoting catalytic performance by increasing the active catalyst size, its durability is yet to be proven. The oxygen reduction process (ORR) in FCs takes place on the cathode. ORR can take two paths. The first, commonly referred to as partial reduction, involves a process with the formation of two electrons resulting in the production of adsorbed hydrogen peroxide. The full reduction follows the more efficient pathway of the four electrons, which does not involve the production of H_2O_2 [120,121]. Due to the improved full reduction efficiency and the relatively high reactivity of hydrogen peroxide compared with water stability, the full reduction is the path sought when choosing a catalyst for ORR. Meenakshi found that the Pt3Sc/photo electrochemical carbon nanotube (PECNT) cathode catalyst exhibited a high ORR activity and higher PD (760 mW/cm^2 at 60°C) in single-cell measurements [106]. The use of NPs can reduce the utilization of noble metals and thereby reduce cost [122]. Pillai et al. [123] examined continuous flow synthesis of nano-structured bimetallic Pt-Mo/C catalysts in milli-channel reactor for PEM fuel cell application.

While the use of NPs and several composites of Pt such as Pt-Cr, Pt-Pd, and Pt-Fe has been investigated, the most promising group of compounds is found to be polyoxometalates: $H_3SiW_{12}O_{40}$, $H_3SiMo_{12}O_{40}$, $H_3PW_{12}O_{40}$, and $H_3PMo_{12}O_{40}$. The heteropoly compound group shows a relationship between size and physicochemical properties, these compounds have optical properties (right and left torsion structures), and aqueous solutions of HTPA are strong completely dissociated acids. The undoubted advantage of polyoxometalates is the simplicity of making a multilayer and, at the same time, thin film, a layer-by-layer method, and consists in applying successive layers to the carbon paper substrate [121,124]. Recent studies have also shown that, depending on the type of crystallographic Pt: Pt (111), Pt (100), or Pt

TABLE 4.3
Materials Commonly Used as Catalysts of ORR in Fuel Cells [124]

Nanoparticles	Solid Materials	Organometallic Composites
TiO2 deposited onto Au	Pt, Ru, Cu-Ru, Au, Pd-Co Cu, Ni, TiO2, Ti, V2O5	Transition metal complexes with porphyrin ligands
Pd deposited onto Au		
Pd deposited onto CNTs		Metal complexes

(110) ORR kinetics follow different mechanisms [125,126]. Pt, palladium, and Pt metal alloys have excellent catalytic properties on the negative electrode (anode). According to the literature data, depending on the electrode's shape, these metals are used in an amount of 1–20 mg/cm^2 [127] and lower [124]. The classification of nanocatalysts (NCs), which accelerate the ORR process in FCs, is presented in Table 4.3. It is divided into three groups: precious metals, base metal electrodes, and organometallic compounds. Pt is the most commonly used NC for oxygen reduction in an acidic environment. A change in the diameter of the NP from 12 to 2 nm increases the development of the Pt surface from 25 to 150 m^2/g [128,129], thereby improving the use of catalyst.

The high cost and short-term durability of PEMFCs imposed by PGM require the development of PGM-free and iron-free electrocatalysts. High-temperature pyrolyzed FeN(x)/C catalysts have been recognized as effective non-precious metal electrocatalysts for ORR and have attracted considerable attention from researchers. Wang et al. [122] explored an FeN(x)/C catalyst derived from poly-m-phenylenediamine (PmPDA-FeN(x)/C) with higher ORR activity and a lower H_2O_2 production rate (<1%). Liu et al. [130] conducted computational and experimental research to study the activity, mechanism, and durability of Mn and N co-doped carbon (denoted as Mn–N–C), and the results suggest that Mn–N–C would exhibit high catalytic performance for ORR in an acidic medium. N-doped iron-based carbon materials (FeNx/C) can be fabricated by employing peptone as a precursor and molten salt, NaCl, as a template, providing a more environmental-friendly and low-cost method of catalyst preparation that is available. While non-Pt-based catalysts and modified Pt-based catalysts are promising substitutes, more work is needed for practical manufacturing and industrial applications. For example, more examination is needed for rich carbon and metal content in sludge biochar-based catalysts (SBCs), which are suitable to serve in PEMFC, owing to both the adsorptive and catalytic properties [106].

When hydrogen is fed to the cell along with CO and CO_2, the chemisorption of CO on Pt NPs reduces the active metal surface area [71]. There are two methods to reduce the scale of catalyst poisoning on the electrode [131,132]. The first method is to carry out selective oxidation using a catalyst before the anode. This method can reduce CO level below 10 ppm. The second method is to capture CO by Pt or Al in front of the cell. This will reduce CO concentration in the cell below 100 ppm.

Alloys that bind adsorbed CO, poisoning catalyst (reducing the surface of the active catalyst), are listed by Wlodarczyk [124]. Currently, catalysts based on Pt NPs are still the most active materials for LT hydrogen FCs [133]. Pt alloying with transition metals increases the electro-catalyzation of O_2 reduction. In LT FCs, it was observed that Pt-Fe, Pt-Cr, and Pt-Cr-Co electrocatalysts have high-specific activity for oxygen reduction compared with Pt electrocatalysts [124,134]. Pt alloys with Mn, Co, Fe, and Ni exhibit higher catalytic activity than their spherical counterparts in the field of oxygen reduction, oxidation of formic acid. and methanol oxidation. Zhang et al. [135] developed a method for the synthesis of a solution in a single vessel to produce a Pt–iridium catalyst with nanodendrometric morphology and homogeneous dispersion of particles with a diameter of 15 nm. Shao in al. [136] experimented with depositing the Pt monolayer on gold NPs. The resulting catalysts were monodispersed (~3 nm) particles. Pt catalysts doped with precious metals or transition metals are the most effective alloys in terms of ORR activity. High hopes are related to the future use of Pt-free catalysts, which are cheaper and exhibit catalytic properties of reduction processes. Research on the use of ORR catalysts has been classified into generations. The third-generation research is to develop precious metal catalysts, and the fourth-generation research is to search for nonmetal ORR catalysts.

The study of Pedram et al. [104] has shown that following bioinspiration mimicking the structure of cytochrome c oxidases and multicopper oxidase enzymes for the cathode and hydrogenase enzymes for the anode of the PEMFC is useful. These approaches offer promise for reducing the use of costly Pt catalysts. While nature-inspired PGM-free materials can reduce cost, these biostructured designs can also increase the performance of the CL. Furthermore, the membrane can use bioinspiration in the form of biopolymers (chitosan and agar) and proton transport channels in living organisms such as Halobacterium to replace the current membrane layer used in PEMFCs. Another approach is to reduce Pt NP usage in the CLs using unique catalyst support materials with a high surface area. Bioinspired CLs and electrolyte membranes require further research before they can adequately compete with the current materials and replace them commercially. These biocomponents can improve performance while decreasing cost and toxicity. A detailed review of this subject is given by Morozan et al. [137]. Laccase is a group of multicopper oxidase enzymes that can catalyze O_2 reduction, similar to the process that occurs with cytochrome oxidase [138]. In this regard, numerous copper complexes as the ORR electrocatalyst layer have been investigated based on this biological system [104,139,140]. This research field presents appealing approaches for synthesizing multinuclear copper-based complexes to imitate the natural rates of ORR for reasonable performance and stability at PEMFC operating conditions.

Hydrogenase enzymes are among the most efficient biocatalysts for the oxidation of molecular hydrogen (H_2) into protons, an electron, and vice versa. These enzymes containing iron and/or nickel metal centers as the active sites are potent sources of inspiration to design and fabricate catalysts as an alternative to Pt-based catalysts [104,141]. Goff et al. [141] consider the structure of bis-diphosphine nickel complexes to synthesize CLs. In this process, the electrode is fabricated by covalently bonding the nickel complexes to the MWCNTs to take advantage of their high surface areas and facilitate higher catalyst loading. A comparison of catalytic properties between

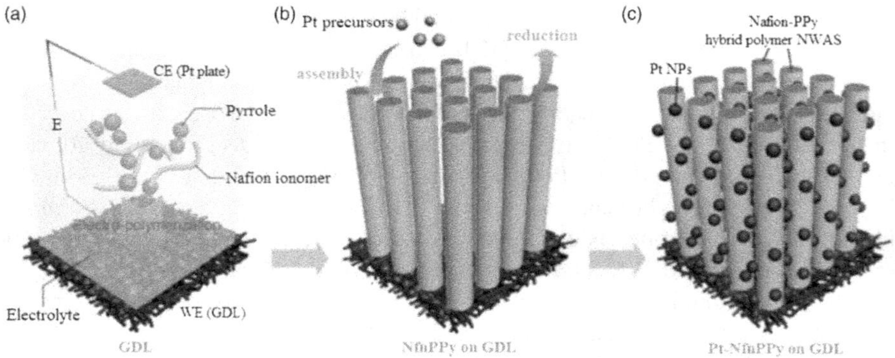

FIGURE 4.11 Schematic of the fabrication of the electrode arrays. (a) Synthetic scheme of the electrode arrays via electrochemical polymerization, (b) Pt loading, and (c) MEA fabrication. (Reproduced under the terms of the CC [104].)

the bioinspired molecular catalyst (synthetic ligand molecules bonded to a metal) of [Ni $(P^{Cy}{}_2N^{Gly}{}_2)_2]^{2+}$ complex and [NiFe]-hydrogenases is conducted by Macia et al. [142] for the HOR. The results demonstrate superior overpotential and catalytic performance for Ni catalyst at lower pH, whereas the hydrogenase enzyme outperforms the synthetic Ni catalyst at neutral pH. Moreover, the immobilized Ni complex shows appropriate stability in the presence of CO, which empowers the utilization of lower-purity H_2 as a fuel.

A hierarchical nanoarray electrode structure is proposed by Xia et al. [143] inspired by the metalloenzymes molecular structure, an active metallic center surrounded by dual-function charge transport media. Vertically aligned nanowire arrays are developed directly on a GDL of FCs through electrochemical polymerization of Nafion and polypyrrole mixture. Pt NPs are deposited on these NWs by wet chemistry method followed by hydrogen reduction. In this copolymer configuration, the sulfonate group of Nafion and polypyrrole functions as the protonic and electronic conductor, respectively. At the same time, interspaces between the NWs facilitate the continuous transport of reactant and product. A schematic of the fabrication of this electrode is shown in Figure 4.11. This electrode architecture with simultaneous electron/proton transfer shows comparable PEMFC performance with a conventional Pt–C catalyst with three times higher Pt loading. The durability test also shows about 37% degradation of the cathode for the novel electrode, whereas 64% for the conventional cathode with Pt–C for 5,000 cyclic voltammetry cycles.

4.4.5 Catalyst Support

The structure and properties of support material also play an important role in the performance of Pt–polymer composite catalysts. The high surface area of the support provides more sites for catalyst NPs. To this end, Wang et al. [144] examined a hierarchical structure design of a PEMFC CL, mimicking the configuration of a pine tree. A carbon nanofiber mat with Pt NWs was fabricated through annealing electrospun

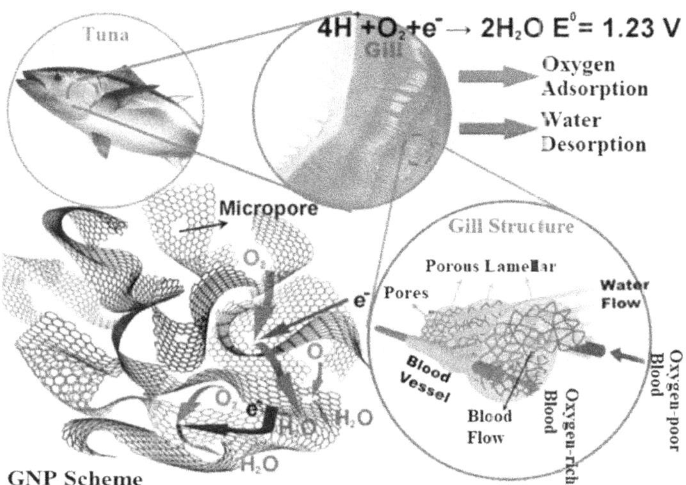

FIGURE 4.12 Schematic illustration of the sorption behavior of oxygen for GNP catalysts analogous to the oxygen exchange in "tunas" gills [104].

polyvinylpyrrolidone titanium dioxide (PVP-TiO_2) film. These Pt NWs were further embellished by porous bismuth coating to modify the surface atomic structure and catalytic activity. The study showed that this support structure provided a higher specific surface area, easier reactant diffusion, and enhanced catalytic activity toward formic acid electrooxidation compared with a pure Pt surface.

Kong et al. [145] study the mimicry of the marine sponge lattice architectures as carbon–carbon composite support for Pt-based FC catalyst. This mesoscale design configuration was synthesized using two different carbon morphologies consisting of CB and MWCNTs at different compositional ratios. The size of the deposited Pt nanocrystal fabricated with wet chemistry was controlled by the carbon matrix pore size and porosity. The study showed that the optimum Pt NP size for electrochemical performance was 5–6 nm. As shown in Figure 4.12, Yao et al. [146] investigate microporous graphene nanoplatelet (GNP)-based Fe/N/C electrocatalysts for ORR activity, inspired by the porous lamellar structure of tunas' gills. The GNP-based Fe/N/C catalysts were fabricated through pyrolysis of highly dispersed polyimide and molten salt mixtures. The study showed increased ORR kinetics resulting from improved microporosity of GNP structure and good catalytic performance and stability due to increased O_2 adsorption and desorption (O_2 diffusion).

4.4.6 Bipolar Plate

BPs of PEMFCs connect a series of single cells to form a stack and collect and transport the generated current from one cell to the next cell. This requires good electrical and thermal conductivity and durability. BP should be designed such that even distribution of fuel gas at the anode and oxygen at the cathode occurs such that maximum chemical reaction rate and higher current density are obtained without creating localized hot spots. Uniform distribution and removal of unused gas and water should also occur to avoid water flooding in the CL and GDL at temperatures below 100°C. BPs are exposed to water and gas phases under relatively high temperature and humidity, and thus, good thermal conductivity and mechanical strength to remove heat and support the stack are required to ensure a long lifespan. The requirements of evenly distributing fuel gases, maximizing the reaction area, minimizing the total resistance, transporting the proper amount of water, and simple manufacturing on large scales have made it a long and difficult path to optimize the flow field design in BPs. In recent years, the Department of Energy (DOE) indicates that the focus of BP design should be switching from reducing resistance to enhancing water removal and now to enhancing mass transfer.

BP materials widely used in PEMFCs include nonporous graphite plates, surface-modified metal plates, and composite BPs coated with anticorrosion paints. Composite plates can increase the volumetric power and mass ratio power of a cell and combine the advantages of graphite and metal plates using a thin metal plate or other high-strength conductive plates as a partition plate and a graphite plate as a flow field plate. In general, nonporous graphite plates can adjust the electrical conductivity and mechanical intensity of the BP according to the ratio of the conductive filler and the resin and can also be mass-produced using molding or injection molding processes, thereby reducing the manufacturing cost of the BP. Selecting suitable conductive fillers and resins, improving the uniform mixing of carbon materials and resins, and optimizing the preparation process enable the realization of composite BPs with high electrical conductivity and mechanical strength. Composite BPs combine the advantages of graphite BPs and metal BPs and are suitable for applications in which high corrosion resistance, high electrical conductivity, and high intensity are required. Additionally, appropriate coatings have the potential to prevent corrosion and improve performance.

4.4.7 Flow Channel

Flow channels are utilized for removing water and delivering and distributing reactants to the MEA, whereas the region between the flow channels is used for collecting and transporting current from one cell to another and transferring the heat generated in the MEA. The flow channel is the only part that exchanges mass and removes moisture from the system. If the moisture in the flow channel cannot be removed in time, it will not only affect the transmission of the reaction gas but also increase the system pressure drop, make the current distribution uneven, lower the performance, and, even more critically, adversely affect the operation safety. Kahraman and Orhan [147] experimentally studied three main flow field channels and indicated that water would accumulate and cause flooding in the center of a parallel flow field, downstream channels of an interdigitated flow field, and corners of serpentine flow field channels. The structure of a flow channel determines the flow state of the reactants and products in the flow fields. A well-designed flow field plate should be able to (1) evenly distribute the fuel and oxidant required for FC discharge to ensure uniform current density distribution and avoid local overheating and (2) make the FC generate water smoothly under the reaction exhaust purge and entrainment exclusion, which requires the fluid to have a certain linear velocity in the flow channel.

Flow fields in PEMFC also control mass transport. Following the design of biological mass transport systems in humans, several bioinspired designs following Murray's law such as various fractal structures, leaves, lungs, porous media, and wave-like designs have been explored. These designs offer multiple advantages over conventional designs [124]. The use of CFD to simulate the environment and operating conditions and optimize performance by minimization of material requirements, uniform distribution of gases, effective water transport, and manufacturability [104] has also been attempted. Bioinspired design of the flow field is also used to enhance thermal and water management capabilities, which is vital for the optimal performance of a PEMFC. To understand two-phase flow in FC flow channels, 2D and 3D modeling is also required. Such analysis indicated that both capillary and viscous forces govern the two-phase flow. The liquid water tends to interact with the gas flow inside the flow channel, forming distinct droplets that attach themselves to the surface of the GDL, resulting in the clogging of pores and inhibition of reactant flow. Both theoretical analysis and examination of nature's living systems design indicate that shorter, consecutive, and increased branching channels provide lesser pressure drop, better water management, and more favorable conditions for two-phase flow in PAMFC.

To better manage heat flow and cooling of the GDL layer, a branching network provided by bioinspired design of leaf vein and lung-based structures or in general porous structures provides large surface areas functioning as heat sinks to avoid localized hotspot formations [104]. These designs and their analysis by CFD simulations provide a better thermal management system. Another important bioinspired design, the fractal flow field, was studied by Tuber et al. [148] in comparison with conventional parallel and serpentine structures. The results reveal that a steady voltage of 0.7 V is achievable even under a constant load of 128 mA/cm^2 for serpentine design, whereas the voltage drops immediately in the case of fractal and parallel flow field designs. The study attributes this voltage drop to the accumulation of water, clogging fractal and parallel channels, and reducing the active area of the cell. However, the study suggests that an improved performance is achievable for a fractal design at higher temperatures, which do not involve removing water in the liquid phase. Lorenzini et al. [149] reveal that an increase in the level of bifurcation in fractal tree-like flow fields substantially improves the PPD, exhibiting a 25.25% increase for three-level versus one-level bifurcation fractal design. Wang et al. [124,150] pointed out that in fractal flow fields channel-to-rib ratio and size of the flow channel are also important. The study points out that under similar operating conditions, the PD of the FC increases with a decrease in the cross-sectional area of the channels and channel-to-rib ratio due to an increase in flow velocity facilitating efficient water removal. CFD analysis has also shown that most bioinspired FFP designs demonstrate superior FC performance over conventional designs by reducing pressure drop and developing more uniform reactant gas distribution,

While PEMFC has been commercialized for portable power sources, novel designs inspired by veins in leaves, lungs, and fractal structures showed improvements between 5% and 40% depending upon configuration and other cell properties. Bioinspiration is useful in FC CLs and membrane design. Focusing on PGM-free catalysts, several options are shown to be promising, for example, mimicking the structure of cytochrome c oxidases and multicopper oxidase enzymes for the cathode and hydrogenase enzymes for the anode of the PEMFC. These approaches offer promise for reducing the use of costly Pt catalysts. While nature-inspired PGM-free materials can reduce cost, these biostructured designs can also increase the performance of the CL. Similar to the CL, the membrane can use bioinspiration in the form of biopolymers (chitosan and agar) and proton transport channels in living organisms such as Halobacterium to replace the current membrane layer used in PEMFCs.

The study by Marappan et al. [151] indicates that the performance of PEMFC for cars, buses, and portable electronics can be significantly improved by bioinspired fluid flow channel design in electrodes, membranes, and other accessories. The study indicates that a trap-shaped flow channel is simple, inexpensive, and has high performance, and it will be one of the potential designs for the coming generations. Furthermore, the vertical orientations of AB-PEMFC are the best due to the easy removal of water at the cathode side. The wave, bioinspired wave, and M-shaped blockages and the arrangement of pin-fins and cylinders in the flow channel have improved the performance. The porous flow field (PFF) design is simple and easy to manufacture and can be improved by (1) the staggered arrangement of rectangular and trapezoidal blockages along the channel, (2) the use of metal foams with high porosity, (3) the provision of sub-channels, transition areas, inclinations, micro-grooves in rib walls, and micro-distributor, (4) provision of bends and curves in channels, number of funneled in/out, small holes in landing/rib areas of the flow field, and two sets of parallel channels operating at different pressures, and (5) generation of circumferential flow of reactant near GDL by the insertion of wire coils in channels, and Marappan et al. [151] indicate that due to the high cell performance, reliability, and durability, the serpentine flow field (SFF) is widely used for many applications and it is considered the "industry standard." Marappan et al. [151] also describe important points obtained from the advanced integrated flow field (IFF) design. While the spiral, concentric spiral, radial, cylindrical, and tubular flow fields showed higher performance than planar flow fields, their design and fabrication are complex. The cylindrical SFF has better performance than cylindrical PFF and cylindrical IFF. The cylindrical PEMFC is having high gravimetric and volumetric PD compared with the planar PEMFC. The innovative radial flow field with control rings showed superior flow distribution and water removal, utilization of active area, and lower pressure drop than parallel and serpentine types.

Marappan et al. [151] outline the advantages of various bioinspired flow fields. The intersectant FF with optimum porosity gave superior performance than the SFF. The honeycomb FF having the hexagonal pins increased the diffusion rate of oxygen through GDL by ten times. Compared to the uniform type, the zigzag pin-type FF showed a higher performance due to the even distribution of reactant. The fractal designs of flow fields have lower performance than serpentine. For portable applications, the tubular cylindrical configurations are deemed as a possible choice over conventional designs. The length and number of channels, width and height of landing and channel, and number of inlets have an influence on the cell performance. The bending curvature, slope, and width have an influence on FC performance. In automotive applications, a wide channel with narrow width and the co-current flow gives higher performance, whereas, for stationary applications, a narrow channel with wide width and the co-current flow gives higher performance.

The rectangular cross section of the channel with a sharp curve resulted in better performance when compared to other cross sections such as triangular, trapezoid, and stepped channel. For commercial applications of FC, the flow field with an active area of 100, 200, and 300 cm^2 and above is more suitable. For increasing the output voltage and current, the cells can be connected in series and parallel, respectively, to form a stack. The converging–diverging flow channels, the convergent serpentine channels, and the taper in width and height of channels have improved the reactant flow and cell performance. The performance of the taper channels is better than the conventional ones. More details on some of these conclusions are outlined by Marappan et al. [151].

4.4.8 Thermal and Water Management

Thermal management of PEMFC includes heat generation and transmission, temperature distribution, and the methods of cooling. The cooling method is categorized into three types: liquid coolant cooling, phase change cooling, and air cooling. Air cooling methods include cathode air cooling, reactive air, and cooling air separation, which are suitable for small FC systems of less than 100 W and 100–1,000 W, respectively. Water cooling reactors are widely used in practical appliances, owing to their smaller size, more compact structure, and higher specific heat capacity. The main source of heat is from ohmic heat and reaction heat. There are three main ways in which the residual heat of the FC can be discharged from the stack: internal water vaporization, heat radiation, and external circulating cooling water to take away heat. The balance between heat generation and removal determined the operating temperature and thermal distribution. Electrolyte dehydration and cathode flooding caused by the nonuniform temperature distribution, which exists in the vertical and horizontal directions, become the most critical challenges for PEMFCs.

Water is the main product of PEMFC, and it exists in all parts of its key components. An efficient and stable operation of the FC requires good management of water transport. Water is transported in a PEMFC by many different mechanisms, such as EOD, thermal–osmotic drag (TOD), hydraulic permeation (HP), and back diffusion (BD) [106]. TOD appears because of the temperature difference in a membrane, where water flows from a cold area to a hot area. Richard et al. indicated that the heat pipe effect in the CL of PEMFC is one of the most important factors of TOD. A temperature gradient in the cell causes evaporated water to condense along the path of the cell. The process of water transport across the membrane due to dragging by protons is called EOD, which is relevant to the relationship between proton conductivity and humidity. The EOD coefficient is defined as the moles of water transferred across the membrane by 1 mole of protons. Park and Caton [152] showed that EOD has a negative correlation with current density. Excess water generated at the cathode compared with the anode causes BD [106]. A pressure gradient exists between the anode and the cathode, resulting from a capillary pressure difference or gas-phase pressure difference. As a result, HP occurs and can be measured by the methods of liquid–liquid and liquid–vapor permeation [153].

While water movement in the cell can be measured by numerous techniques such as nuclear magnetic resonance (NMR) imaging, beam interrogation, X-ray, neutron imaging, and high-speed photography, it is difficult for a transparent cell to observe water movements accurately. Nikiforow et al. [154] derived a water content distribution model along the flow channel in the cathode and anode with flow fully developing at a steady state. They concluded that using dry air and saturating the hydrogen feed would prevent flooding and the formation of a water droplet. EOD and BD are recognized as the dominant mechanisms for water transport within the membrane. Wu et al. [155] proposed an unsteady non-isothermal 3D model demonstrating that the cell current output response time is influenced by the finite rates of sorption/desorption of water, and the relative permeability exhibited greater effects on the liquid–water transport than the capillary pressure. By measuring adsorption, desorption, and diffusion coefficients, Srinivasan et al. [156] showed that the desorption coefficient of water differed from the adsorption coefficient by an order of magnitude, indicating the imbalance of water absorption and water loss of the membrane.

The production and condensation of water from the electrochemical reaction cause water flooding, resulting in the degradation of cell performance. Poor mass transport and carbon corrosion also occur in the stack because of water flooding. According to the principle of bionics, a three-step water removal was realized by Chen et al. [157], indicating a three-step water removal process in which the gravity of the droplet itself overcomes the viscous force and causes the self-removal of the droplet. Fabian et al. [138] proposed a new type of BP, which includes a layer of conductive hydrophilic wick laid on the bottom of the flow channel and an electroosmotic pump. When liquid water appears on the membrane electrode, it is absorbed by the hydrophilic water collector and discharged along the wick structure under the action of the electroosmotic pump [106].

Maintaining sufficient moisture in the PEM is a prerequisite for proper FC operation. In the past, this was provided by internal and external humidification approaches. In modern systems, the water generated inside the stack is sufficient to wet the membrane electrode with a substantial increase in the PD of FC stacks. In the future, Toyota will use a hydrogen circulation pump to humidify the PEM with a humidified hydrogen cycle, thereby eliminating the humidifier and greatly reducing the complexity of the stack, whose principle is the same as that of the self-humidification method. It is clear that self-humidification technology will become the mainstream direction of humidification methods in the future. More work is needed to develop a complete water transport model.

4.4.9 High-Temperature Proton-Exchange Fuel Cell (HT-PEMFC)

HT-PEMFC was developed in 1995 for operation at higher cell temperatures aiming at lower sensitivity of PEMFC regarding impurities. Thus, HT-PEMFC technology is one of the youngest FC types, and HT-PEMFC systems are produced since the 21st century by several companies such as Advent Technologies (USA), Blue World Technologies (Denmark), and Siqens (Germany).

By changing the electrolyte from being water-based to a mineral acid-based system, HT-PEMFCs can operate at around 150°C–180°C and as high as 200°C. This overcomes some of the current limitations with regard to fuel purity for LT-PEMFC. The CO tolerance for LT-PEMFC is less than 50 ppm, while HT-PEMFC can tolerate

1%–5% by volume of CO. Other impurities' tolerance is also higher for HT-PEMFC. While LT-PEMFC must usually be operated with hydrogen with high purity of more than 99.9%, HT-PEMFC can be operated with reformate gas with a hydrogen concentration of about 50%–75%. The low sensitivity to impurities allows the use of fuels such as methanol, ethanol, natural gas, liquid petroleum gas (LPG), and dimethyl ether (DME), which are reformed in a reformer to hydrogen-rich reformate gas. Often methanol is used as a source of fuel for HT-PEMFC. Because of the low sensitivity to impurities and the proton conductivity of the membrane, the MEA, which is used for HT-PEMFC, can also be used for hydrogen separation to separate ultrapure hydrogen efficiently from diluted or impure hydrogen-containing gases. While LT-PEMFC can have a cold start, HT-PEMFC requires high temperature. Finally, LT-PEMFC has complex water management, while HT-PEMFC does not require water management or a humidifier. The balance-of-plant (BOP) system efficiency for methanol-fueled HT-PEMFC systems is typically between 35% and 45% and can reach up to about 55% depending on system design and operating conditions. Cell efficiency of up to 63% can be reached.

Other benefits and weaknesses of HT-PEMFC include the following:

Benefits:

1. Waste heat of the stack (130°C–180°C) can be used making combined heat and power (CHP) possible for further usage of the heat in contrast to LT-PEMFC, which has too low waste heat temperature below 80°C.
2. Simple cooling of stack is possible because of the higher stack temperature compared with LT-PEM FC.
3. Use of plastic components and elastomer seals in the stack is possible in contrast to SOFC FCs.
4. There is higher system efficiency of methanol-fueled HT-PEM FC systems (35%–45%) compared with DMFC (20%–30%). There is low methanol fuel consumption. There is no need for high methanol fuel purity for methanol-fueled HT-PEMFC system compared with DMFC. A higher lifetime for the methanol-fueled HT-PEMFC system than for the DMFC system is possible. Pure fuels or water–fuel mixtures are applicable (depending on the FC system design). The use of renewable fuels is possible. Hydrogen with low purity can be used as fuel. Hydrogen with low purity is cheaper than high-purity hydrogen, which has to be usually used for LT-PEMFC. The use of fuels such as methanol makes cheaper fuel costs per kWh possible compared with hydrogen (e.g., LT-PEMFC) or diesel (e.g., gensets) as fuel.
5. Cold storage temperatures below 0°C are no problem for the FC membrane in contrast to DMFC and LT-PEMFC.

Weaknesses

1. It has longer start-up time compared with LT-PEMFC (time for heating of stack and reformer). So, hybridization with a larger battery than for LT-PEMFC systems is sometimes necessary. A system component for stack heating during start-up is necessary for contrast to LT-PEMFC and DMFC.

2. PD for LT-PEMFC is higher compared with HT-PEMFC. More cells are needed compared with LT-PEMFC for reaching high-power output or the same efficiency as for hydrogen LT-PEMFC because of the inferior characteristic curve of HT-PEMFC: Higher stack costs, stack volume, and stack weight compared with LT-PEMFC. Technologies for reaching better characteristic curve properties are in the basic research state.
3. Pt loading in LT-PEMFC is 0.2–0.8 mg/cm^2, while for HT-PEMFC it is 1–2 mg/cm^2. Pt recycling should be considered. The development of Pt-free electrodes for HT-PEM FCs is in the early research state.
4. When organic fuels are used CO_2, perhaps traces of CO are emitted (concentration depending on system design, typically CO concentration by far lower than emitted from combustion engines).
5. Some system components must be able to resist higher temperatures than in LT-PEMFC and DMFC, which limits the choice of applicable materials (e.g., polymers with resistance up to 120°C–180°C).

HT-PEMFC systems are used for stationary and portable applications. For example, methanol-fueled HT-PEMFCs are used as replacement for generators (e.g., off-grid applications, backup power, emergency power supply, and auxiliary power unit) and for range extension of electric vehicles (e.g., sports car Gumpert Nathalie). Typically, the HT-PEMFC system is used in hybrid operation with a battery. HT-PEMFC systems fueled with natural gas are also used for CHP applications in buildings. More details on HT-PEMFC are given in references [158–161].

One of the essential components of an HT-PEMFC is the PEM, which has to possess good proton conductivity and stability and durability at the required operating temperatures. The membrane consists of an acid and temperature-resistant polymer, which has the ability to uptake phosphoric acid, which acts as an electrolyte. The membrane used in HT-PEMFC makes portable applications possible for HT-PEMFC. Among the various membrane candidates, phosphoric acid-impregnated PBI-type polymer membranes (PBI/PA) are considered the most mature and some of the most promising, providing the necessary characteristics for good performance in HT-PEMFCs (see Figure 4.13). The review by Subianto [158] aims to examine the recent advances made in the understanding and fabrication of PBI/PA membranes and offers a perspective on the future and prospects of deployment of this technology in the FC market.

The review by Guo et al. [162] focuses on recent advances in the limitations of acid-based PEM (acid leaching, oxidative degradation, and mechanical degradation) and the approaches mitigating membrane degradation. Preparing multilayer or polymers with continuous network, adding hygroscopic inorganic materials, and introducing phosphoric acid (PA) doping sites or covalent interactions with PA can effectively reduce acid leaching. Membrane oxidative degradation can be alleviated by synthesizing cross-linked or branched polymers and introducing antioxidative groups or highly oxidative stable materials. Cross-linking to get a compact structure, blending with stable polymers and inorganic materials, preparing polymer with high molecular weight, and fabricating the polymer with PA doping sites away from backbones are recommended to improve the membrane mechanical strength. Also,

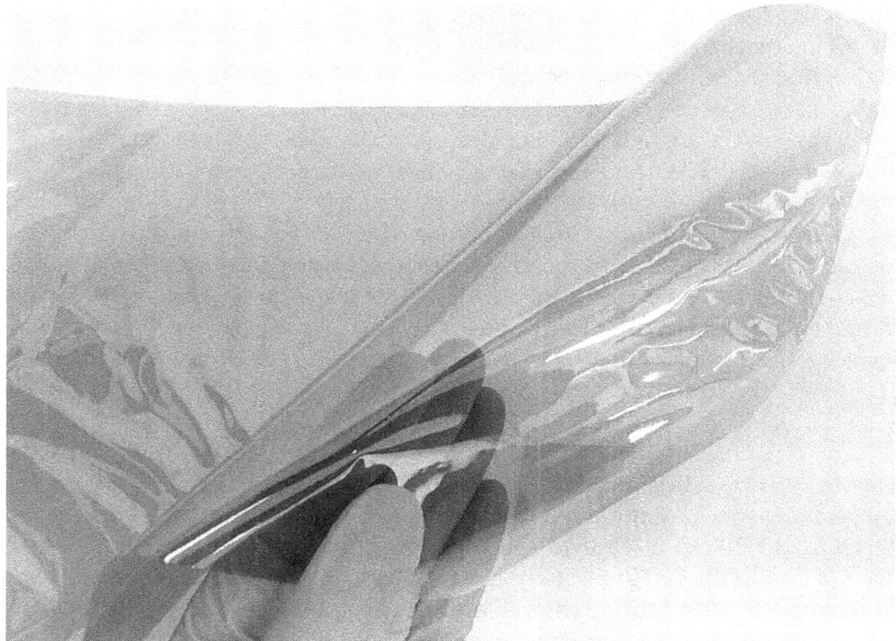

FIGURE 4.13 Phosphoric acid-doped PBI membrane for HT-PEM fuel cell [177].

by comparing the running hours and decay rate, three current approaches, (1) cross-linking via thermally curing or polymeric cross-linker, (2) incorporating hygroscopic inorganic materials, and (3) increasing membrane layers or introducing strong basic groups and electron-withdrawing groups, have been concluded to be promising approaches to improve the durability of HT-PEMFCs.

4.4.10 Future Perspectives

More work is being done on hydrogen generation and storage, reduction in generator production costs, and their miniaturization. An obstacle in the commercialization of these solutions is the high cost of materials, high weight of instrumentation, and hydrogen storage. FCs are still too expensive, and their service life (about 10,000 hours) is too short. The operation of FCs depends to a large extent on the purity of the fuel and the condition of the catalyst. Contaminants cause catalyst "poisoning." At the high cost of the catalyst, which is Pt, taking care of its stability and durability during the work of the cell is extremely important. Also, the water management in the cell must be optimized: Too high a level of moisture can cause the membrane to "flood"; too low a level will cause the diaphragm to dry out and its irreversible destruction. The extremely important role of catalysts, in the form of nanomaterials, requires continuous research on the materials from which cell elements are built. There are two ways to obtain perfect materials: modification of already known materials or reaching for new materials and use of nanotechnology. As can be seen from the above considerations, the transition from bulk materials to the nanoscale can significantly change the electrode and durability of FCs. NPs have catalytic capabilities for the processes taking place in the cell, but also nanostructures with other morphologies (nanotubes, NWs, nanofoams, nanospheres, nanodendrites, nanorods, and nanooctahedra) are also effective. However, nanomaterials also have disadvantages, such as low thermodynamic stability, high surface reaction, and possible toxicity problems. The solution to challenges related to the design of optimized hybrid nanostructures and surface modification, the implementation of systematic testing of nanomaterials properties and control of their operation on the environment, and determination of procedures related to the detection of nanostructures will be a major breakthrough in the development of nanoscience in hydrogen technology and FC development.

4.5 PHOSPHORIC ACID FUEL CELLS

The PAFCs may be classified as a technology that falls between the LT and high-temperature systems (medium-temperature), and they operate with liquid phosphoric acid (H_3PO_4) electrolyte and operating temperatures between 150°C and 220°C [16–18,37,163]. The electrolyte is essentially an acid in a Teflon-bonded SiC structure [22]. The positively charged hydrogen ions are transported to the cathode via the electrolyte. The electrons produced at the anode flow to the cathode through the external circuit, thus giving rise to direct current. Water is also formed through the reaction of electrons and hydrogen ions with oxygen at the cathode. The construction details of PAFC are illustrated in Figure 4.14 [164,165].

The electrode materials in PAFCs are Pt or Pt–Ru and Pt for anode and cathode, respectively, and the fuel-to-electrical efficiencies between 0.35 and 0.40 could be achieved in this technology [16,17]. However, it is possible to realize an efficiency of about 0.85 when the technology is engaged for CHP applications. The PAFCs have

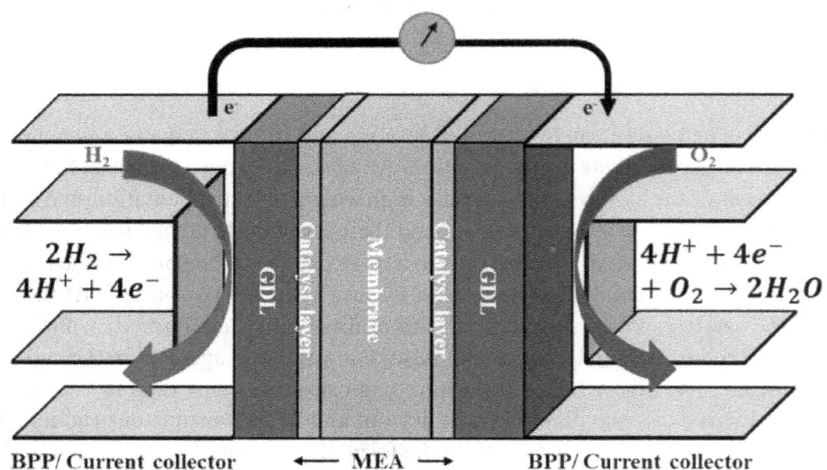

FIGURE 4.14 Constructional details and working of a PAFC [164].

a simple structure design and are less prone to CO poisoning and electrolyte volatility and are generally used for small- to medium-sized plants between 50 kW and 11 MW [17]. They are, however, weighty [16,22], have low efficiency, and require an integration of corrosion-resistant components to mitigate the effect of acid on the electrolyte. PAFCs can generate hydrogen in situ from methanol or similar liquid fuels. This allows PAFCs to be employed using available refueling and distribution systems [166].

A PAFC typically houses phosphoric acid (H_3PO_4, PA) trapped in a matrix (conventionally of silicon carbide, SiC) as the electrolyte. Pt with different loadings on carbon-based GDL forms the electrodes, i.e., anode and cathode. The fuel source and the oxidant are circulated via BPPs on opposite sides of the electrolyte (Figure 4.14). While doing so, hydrogen is oxidized to protons and electrons at the anode. The protons pass through the electrolyte, whereas electrons pass through the external circuit resulting in a current. At the cathode, protons and electrons combine with oxygen to form the byproduct of water. The heat released during the reaction is normally utilized for heating water, space heating, etc. [167]. Reactions that occur at the respective electrodes are summarized as follows:

Anode:

$$2H_2 \rightarrow 4H^+ + 4e^- \tag{4.11}$$

Cathode:

$$4H^+ + 4e^- + O_2 \rightarrow 2H_2O \tag{4.12}$$

Overall reaction:

$$2H_2 + O_2 \rightarrow 2H_2O \tag{4.13}$$

The oxidation of hydrogen is a fast process; however, ORR occurring at the cathode is slow and determines the overall rate of reaction, i.e., current density. The reaction of hydrogen with oxygen at standard temperature and pressure (STP) conditions (1 atm, 298.15 K) produces 286 kJ per mol of hydrogen gas [168]. This is the maximum amount of energy that can be withdrawn from the cell. However, due to inevitable losses incurred due to thermodynamic barrier (entropy loss), operational deviation from STP conditions, fuel crossover and overpotentials, and the potential of the cell settle to lower values.

The concentration and distribution of PA in the membrane are very important in the case of PAFCs. Leakage of acid from a membrane leading to a loss in ionic conductivity has been the primary cause of degradation in membranes. The conduction of protons through an acid-doped membrane follows the Grotthus mechanism [169]. Since the partial pressure of water in the membrane is very low due to the low evaporative nature of acid species, the higher humidification of gases is not required. However, during the cell operation, water generated from reactions dilutes the acid, thereby changing its concentration. It is further exacerbated when membrane swelling is arrested by the cell. This promotes acid leaching (migration and evaporation) from the membrane causing a decrease in ionic conductivity leading to performance loss of the cell over time. Nevertheless, the leaked acid affects other cell components such as GDL and BPPs, leading to their collective degradation.

To enhance the reaction rate, the appropriate choice of catalyst becomes very important. Since thermodynamics often represents the amount of energy that can be harnessed from the reaction, the electrons and their transfer rates (i.e., current) between the species are limited and determined by the electrode design and choice of the materials used. Such materials catalyze the process such that the reaction proceeds with minimum activation energy invested (η_{act}). A good catalyst, therefore, should have (1) high mechanical stability, (2) high electrical conductivity, (3) low corrosion, (4) high porosity, (5) ease of manufacturing, and (6) high catalytic activity, i.e., high exchange current density.

HOR being most easy process is elaborated to understand the importance of catalyst and its support material. ORR, however, is a more complex reaction and therefore requires a higher quantity of catalyst. The materials for both the anode and the cathode remain similar to the difference in the catalyst loading (e.g., 0.1 and 0.5 mg/cm^2 of PtC loading on the anode and the cathode, respectively). Since the PAFC is supplied with reformated fuel, which contains some amount of CO, the catalysts used for HOR should be tolerant to CO. Pt-Ru/C catalysts were found to be more tolerant to CO compared with Pt/C catalysts [170]. Furthermore, researchers have given attention to ORR being a more complex process compared with that of HOR.

The manner in which the CL is placed in the MEAs is important. The common types of MEA available are CCM and catalyst-coated electrode (CCE) also called GDE. CCM is prepared by coating the catalyst directly on the membrane surface using a suitable coating method (spray coating), whereas the GDE is prepared by coating the catalyst on GDLs. Although both CCM and GDE are used widely, CCM offers few advantages over GDE such as reduced contact resistance, higher ECSA (electrochemical surface area), and lower charge transfer resistance [171]. PA is stored in GDL in the case of CCM, and the acid is stored in the membrane in the case of GDE. During the performance, the acid from the membrane reaches CL that is placed on GDL (in GDE), thereby developing TPB. In CCM, the acid has to travel from GDL to the membrane through the entire CL. This in fact generates more TPB, thereby enhancing the performance.

A catalyst helps in increasing the rate of reaction when its surface area is enhanced. It is normally achieved by employing support that offers better crystallinity, high surface area, and high electrical conductivity. Carbon 60 or C_{60}, single/multiwalled CNTs/fibers, graphite, graphene, silicon carbide (SiC), titanium dioxide (TiO_2), tin oxide (SnO_2), etc., have been widely used as catalyst supports in various FC applications. Carbon supports, specifically activated CBs, have been primarily used as catalyst supports in both LT and high-temperature FCs [172]. Their physical and chemical stability and availability and cost-effectiveness have made them researchers' choice as catalyst support. Nanotubes (NTs) have also been found as a better option as support material [164]. Galbiati et al. [173] fabricated an array of self-standing NTs by evaporating Pt on porous alumina substrates. Better catalyst utilization was observed for NTs compared with conventional Pt/C due to the absence of porous carbon as support. Such preparation of catalyst without support was observed to yield enhanced performance in an FC despite offering lesser electrochemical active area (ECSA) and lower relative humidity of gases. Bhosale et al. [164] described the deposition of Pt NPs over CNTs as GDLs using dry platinization and wet chemical platinization methods.

The BPPs are an important component in the FCs, wherein they function as gas and air distribution channels and conduct electricity from one cell to the other in the stack. They also help in removing heat from the active area and prevent leakage of gases and coolant. An ideal BPP [174] should have (1) high electrical conductivity (>100 S/cm), (2) low gas permeability ($<2\times10^{-6}cm^3\ (cm^2s)^{-1}$), (3) better corrosion resistance (<16 $mAcm^2$), (4) good tensile strength (>41 MPa), (5) better flexural strength (>59 MPa), (6) high thermal conductivity ($>10W\ (m\ K)^{-1}$), (7) high chemical and electrochemical stability, (8) low thermal expansion, and (9) efficient processability. The conventional choice for BPPs by the researchers and industries has been graphite satisfying almost all the properties listed above except machinability and gas permeability. However, the bulkiness of graphite lowers the volumetric power of the cell. Attention should be focused on metallic BPPs due to their better electrical conductivity, machinability, and strength. S.S., Ti, Al, etc., have been favorites among metal BPPs [164]. These, however, possess higher contact resistance due to the formation of an oxide layer on the surface, if not masked with anticorrosive coatings such as Au, TiN, and CrN [164,175].

The situation in the case of PAFCs compared with LT PEMFCs is different in the sense that the BPP of PAFC also needs to face high temperature in combination with the acidic environment present in the reaction chambers. The additional burden on the operating system complicates the durability issue of BPPs. Yan et al. [176] underlined the effect of temperature, time of operation, and PA on Au-coated SS304 compared with that of uncoated one. It was evident that the presence of Au protected the bare metal, which followed a similar degradation slope to that of graphite. Furthermore, the formation of ferric oxide and iron phosphate on an uncoated steel plate was the major cause of the multifold increase in the ohmic resistance of the cell, which was prevented when replaced with Au-coated steel as the BPP. Al-based BPP-coated with Ni-P coating showed promise compared with graphite plates. Another study involved a coating of polylaminate TaN/Ta on 430 SS using a magnetron sputtering method. Ta and TaN are the innermost and outermost layers. A summary of the different materials used in PAFC is illustrated in Table 4.4.

Bhosale et al. [164] addressed many designed issues such as nonuniformity in the planarity of electrodes, crushing of electrodes by peripheral ring gasket, and nonuniformity in the clamp pressure distribution. The first issue can be resolved by the use of specially designed springs on one end of the stack. The second issue can be resolved by use of a rectangular gasket made of Viton rubber. The third issue can be resolved by the use of porous rib-like structures for anode and cathode flow field, the use of 30–50 μm soft hydrophobized/perforated carbon paper, and the optimization of electrode fabrication time contact pressure distribution. Bhosale et al. [164] also addressed various operational time issues such as diagnosing faults in an active stack, failure due to thermal expansion, and variation of clamp load under operation. They suggested that the first issue is resolved by the measurement of internal

TABLE 4.4
Summary of Different Materials Used in PAFC [164]

No.	PAFC Component	Materials Used	Comments
1.	Catalyst	Pt, Pt-V, Pt-Cr, Pt-Co, Pt-Ni, Pt-Ru, Pt-Sn, etc.	Alloying of Pt with other materials reduces Pt poisoning and enhances its durability, close to maturity
		Fe-Nx, Fe-N-C	Non-precious catalysts, lower catalytic activity and durability than Pt-based catalysts, under development
2.	Catalyst support	Activated carbon black, CNTs, CNFs, graphene, NCNTs, etc.	Higher electronic conductivity, better stability, higher surface area, close to maturity
		SiC, SiC/C, TiO_2, IrO_2, WO_3, SnO_2, etc.	Lower electronic conductivity, low stability, under development
3.	Membrane	Nafion/ZrSPP composite, Nafion/sulfonate d, Nafion/silica (SBA-15)	Modifications in Nafion membrane for higher water uptake, difficult to maintain the membrane hydrated at elevated temperature, thereby losing the proton conductivity
		Matrix for holding PA: SiC, PBI, polyacrylamide (PAM), polyolefin, composite membrane of glass microfiber (GMF), and polytetrafluoroethylene (PTFE)	SiC to be stable in acid environment, however, has thickness of 0.1–0.2 mm leading to higher ohmic resistance, PBI-based structure employed for its excellent chemical, thermal, and mechanical stability in the acid
3.	Gas diffusion layer (GDL)	Carbon paper/cloth coated with a microporous layer and a hydrophobic agent	180–400 μm thick, offers mechanical strength to catalyst layer, holds PA if made by CCE method, mature technology
4.	Bipolar plate (BPP)	Substrate — Coating material Al — Ni-P Steel — Au Cu — Graphene Steel — ZrN Titanium — TiN	Anticorrosive coatings on substrate materials enhance their corrosion resistance and durability. Steel preferred choice among substrate materials due to its better machinability, high strength, low cost, etc.

resistance of the stack using the current interruption method; the second issue can be resolved by applying nominal load during assembly, and the clamp load is increased once the stack is heated up; and the third issue can be resolved by online measurement of deformation and tightening of bolts if required.

There are also some challenges regarding the stack component and its quality. These include maintaining uniformity in the thickness of electrode/MEA and maintaining the quality of microporous membrane. Bhosale et al. [164] point out that the first challenge can be addressed by thickness measurement of a micro-area using technical association of pulp and paper industry (TAPPI) methodology and the second challenge can be addressed by performing a matrix adhesion test to check the adhesion between the electrolyte matrix and the CL and acid migration speed test is performed to understand the pore size and acid retention under compression.

The review by Guo et al. [162] focuses on recent advances in the limitations of acid-based PEM (acid leaching, oxidative degradation, and mechanical degradation) and the approaches mitigating membrane degradation. Preparing multilayer or polymers with continuous network, adding hygroscopic inorganic materials, and introducing PA doping sites or covalent interactions with PA can effectively reduce acid leaching. Membrane oxidative degradation can be alleviated by synthesizing cross-linked or branched polymers and introducing antioxidative groups or highly oxidative stable materials. Cross-linking to get a compact structure, blending with stable polymers and inorganic materials, preparing polymer with high molecular weight, and fabricating the polymer with PA doping sites away from backbones are recommended to improve the membrane mechanical strength. Also, by comparing the running hours and decay rate, three current approaches, (1) cross-linking via thermally curing or polymeric cross-linker, (2) incorporating hygroscopic inorganic materials, and (3) increasing membrane layers or introducing strong basic groups and electron-withdrawing groups, have been concluded to be promising approaches to improve the durability of HT-PEMFCs.

4.5.1 Applications of PAFC

PAFCs are considered the first-generation FCs to have been commercialized successfully. It was predominantly due to the poorer performance of AFCs and their intolerance to CO_2. PAFCs' tolerance rates of CO are much higher than the PEMFCs with acceptable limits up to 1% of CO at 200°C, thereby widening the scope of choice of fuels used, i.e., methanol, ethanol, and natural gas. Furthermore, considering a high heat dissipation and inferior performance compared with PEMFCs, PAFCs are normally preferred in stationary applications. Best efficiencies are offered when the heat generated by the cell/stack is utilized along with the electrical output (CHP). In such cases, a thermoelectric heater and a thermoelectric cooler can be employed to use the waste heat produced by PAFCs. Such a hybrid system ensures an enhancement in efficiency by ~3% compared with a conventional PAFC system [177].

PAFCs can also be coupled with heat-driven refrigerators such that the heat produced by the cell/stack at temperature T is supplied to the refrigerator that maintains the cold reservoir at temperature T_c ($T_c < T$). Also, the reactants to be supplied to PAFCs can be heated using the high-temperature outlet gases of the cell/stack using a heat exchanger/regenerator. Such a hybrid system is expected to increase the overall efficiency by ~6% [164]. An absorption refrigerator, proposed by Yang et al. [178], that is coupled to a PAFC system along with an auxiliary regenerator, is expected to enhance the system efficiency by 3% compared with that of a PAFC system alone. It uses the waste heat generated by the PAFC system to run the refrigerator such that the electrical demand from the grid is reduced. However, the losses in terms of the thermodynamics of the refrigerator and the operating current density along with the temperature and pressure of the PAFC system should also be accounted for while determining their effect on the hybrid system. The auxiliary regenerator is used to exchange the heat between hot exit gases and the room temperature inlet gases supplied to the PAFC stack, thereby enhancing the effectiveness of the system. Generating electricity from the heat produced by PAFCs is possible using a thermally regenerative electrochemical system (TREC). It primarily employs two cells, viz. hot cell that remains in contact with the hot source carrying waste energy from the PAFC system and a cold cell to handle the cold source. This allows a dual function as a heat exchanger and an electrochemical reactor to generate electricity [179].

Apart from such small- and medium-scale stationary applications, PAFCs are also used in powering residential apartments. The heat generated by the stack can also be utilized in heating the water and the space in the apartment, which reduced the load on fuel required to run the boiler. The amount of CO_2 emitted is observed to be greatly dependent on the direct supply of electricity from the grid as it manipulates the stack operation in CHP mode [164]. It becomes very important to check the economic viability of a PAFC system despite its attractive features. One such plant (200 kW) installed and operated by ONSI corporation in Korea revealed that the PAFC system was able to generate electricity at a cost of 0.068 $/kWh, which was 125% higher compared with that of 1,000 kW gas turbine power plant. Also, this FC system could be made a viable option if the investment cost is reduced to 1500 $/kWh [164].

Other applications of the PAFC system include its operation for more than 40,000 hours along with a boiler to serve as a heating system to a local residential area in Germany. The application was also extended to heating the public baths in Saale as such baths were used throughout the year demanding both heat and power. This PAFC system set a European record of continuous operation at such places for more than 46,500 hours. Combining the FC system with a chemical industry having a hydrogen-rich mixture as a byproduct was a good idea as the system used the mixture to power some of the industry operations. Furthermore, hospitals also benefitted from the usage of PAFC system in Germany. During summer, the heat released by the system was used to operate the air conditioning equipment of the hospital. Such an FC system was shown to run for more than 30,000 hours, thereby generating 6 million kWh by 2002 [164]. Maintaining a PAFC stack intact during shifting has been quite challenging as it sometimes may result in damage to the sealants, thermal insulations, etc., thus requiring overhauling of the entire PAFC system. Periodic checking of any leakages for the coolant, integrity of thermal insulation, cleaning the interior, etc., are also required. Since systems are normally used outside, cleaning the entire system becomes a necessity followed by a check for leakages. In recent years, overhauling has become a liability instead of a necessity and several old PAFC plants have therefore been terminated.

The wide range of fuels that can be used for PAFC applications ranging from natural gas, naphtha, and LPG to biogas makes PAFCs a desirable candidate in the energy industry. PAFCs were traditionally made for stationary applications with output power ranging from 100 to 400 kW but are now finding use in mobile applications such as buses, trains, and even submarines. PAFC can be scaled up for power plants ranging from 20 to 500 kW. Future development in PAFC systems should be targeted toward the development of cheaper catalysts (other than Pt) and expanded use of other fuel types. These changes can make them more viable for large-scale commercial applications.

The major players commercializing the PAFC technology across the globe are Fuji Electric Company Limited and Doosan Fuel Cell America, Inc. Doosan Fuel Cell America has supplied their product, PureCell© 400, to commercial organizations, industries, educational institutions, data centers, and telecom centers and hospitals throughout the world (Doosan Fuel Cell America, 2020). PureCell© 400 can produce 440 W of electric power and 500 W of heat. FP-100i can be used in hospitals, apartments, and commercial buildings, FP-100iB can be used in sewage and water treatment plants, and FP-100iH can be used in chemical manufacturing plants. Fuji Electric has delivered 84 units of their PAFC package throughout the world, and 67 units are still in operation as of July 2019 [164]. Both Doosan Fuel Cell America and Fuji Electric Company report that the efficiency of their respective products is 90%, when operated in CHP mode. Several educational and research establishments have also been working on PAFC systems or adopting PAFC systems for different applications. The PAFC stacks developed by Naval Materials Research Laboratory (NMRL) are being adopted by the Defence Research and Development Organization (DRDO) for its Air Independent Propulsion (AIP) technology. PAFC-based AIP systems increase the underwater endurance of the submarines, and a land-based prototype of the same has been demonstrated at NMRL. DRDO is planning to integrate its AIP technology into Scorpene submarines when the submarines are due for refit [164,180].

4.6 DIRECT CARBON FUEL CELL

At present, the utilization of carbon fuels is still a preferred option for the development of FCs, due to the abundant resources, easy access, and relatively high-energy density. DCFCs, as an energy conversion technology, generate electricity from solid carbon fuels through electrochemical reactions [181]. The device has a higher energy conversion efficiency rate than traditional power generation devices of heat engines, which minimizes system complexity and thus ensures a lower cost [181]. Edison [182] reported the first DCFC in the middle of the 19th century, which used a carbon rod suspended in a KNO_3 solution. In 1896, William Jacques [183] developed a large-scale DCFC using 100 single cells and baked coal as the anode. He used a mixture of KOH and/or NaOH as an electrolyte and the iron pots as a cathode. Subsequently, researchers from Stanford Research Institute (SRI, International) investigated DCFC technology thoroughly and invented the actual devices for electricity generation by the electrochemical oxidation of carbon [181].

In DCFC, carbon is oxidized to CO_2 in two steps: first to CO and subsequently to CO_2. DCFC is an electrochemical device that is static, clean, and highly efficient that can use a variety of carbonaceous fuels. It finds its application in small electronic devices such as laptops and radios to huge power plants. The high efficiency of DCFC makes it an ideal fit for power generation applications. Using recycling, the device can allow up to 100% fuel consumption. The performance of the device depends on the properties of fuel. The device emits pure CO_2, which facilitates subsequent CO_2 capture and separation [181]. DCFC power station has a modular structure design, which can be adjusted regarding the cost [181]. Some solid carbon fuels, such as graphite, CB [181,184,185], and different types of coal fuel [181,186,187] and renewable fuels such as biomass and organic waste have been used in various DCFC configurations. Biochar is suitable in DCFC for electric energy generation because the biochar obtained by pyrolysis is amorphous, which is conducive to the exposure of carbon fuel surface active sites. The high porosity of biochar also promotes gas transportation in the reaction process [181]. Moreover, natural metal ions dispersed uniformly in biochar catalyze the carbon gasification reaction [181]. It is worth mentioning that the efficiency of biomass-fueled DCFC is generally 50%–60%, while it is up to 80% if heat and electricity cogeneration is applied [181].

The earliest reviews of DCFC were carried out by Howard [188] and Liebhafsky [189]. Later, Cao et al. [190,191] and Giddey et al. [192] summarized the fundamental electrochemical performance and the development of DCFC technology. The reaction mechanism of direct carbon oxidation and the conversion was explained by Cooper et al. [193] and Gür [192,194]. Rady et al. [195] reviewed the performance of various fuels used in MCFCs and SOFCs, and Zhou et al. [196] published a review paper that discussed the anode used in DCFCs. Recently, Jiang et al. [197] presented an overview of the impact of different parameters on the resistance and power output and the electrochemical behavior of DCFCs and also summarized the challenges associated with developing DCFCs. Glenn et al. [198] reviewed the carbon electrocatalysis mechanism of alkali metal molten carbonates in DCFCs.

There are three types of DCFC; these are categorized according to the electrolyte material, i.e., hydroxide, carbonate, and solid oxide. Of these, the molten carbonate DCFC (MC-DCFC) was one of the earliest cells, and it has been widely studied [45]. MCFC has been considered superior because of its comparative high efficiency and low operating temperature. The molten carbonate also remains stable compared with hydroxide-based cells, even in an environment rich in CO_2, which helps to prevent electrolyte damage [181,199]. Recently, an investigation was done on hybrid DCFCs (HDCFCs), using mixed solid carbon and carbonate in the anode [181]. This overcomes corrosion issues associated with the MCFC and yields slightly faster kinetics compared with the SOFC system [181]. DCFC produces high-quality waste heat that can be used to improve fuel conversion efficiency. At present, DCFCs are at the early stage of development with only a few authors or industrial organizations reporting performance parameters on single cells or small stacks, which are somewhat similar to those obtained for gas-fed FC systems. Although a wide range of designs and concepts have been tested, there is currently no clear leader within the field with respect to optimal design or operating parameters including operating temperature. An EPRI report on DCFC concludes that DCFC technology has large benefits, which include low-cost fuel, high efficiency, simple system to operate, and ease of CO_2 sequestration [181,196].

DCFCs convert the chemical energy in a carbon fuel directly into electricity without the need for gasification. Fine (submicron) carbon particles in an electrochemical cell are electrochemically oxidized at high temperatures (600°C–900°C) with the overall FC reaction being $C + O_2 = CO_2$. The reactions in DCFC produce almost pure CO_2, provided high-purity carbon is used as the fuel. Besides differentiating by electrolytes, the DCFCs are also differentiated from each other via the materials used within the anode, the design of the anode chamber, and the method of fuel delivery to the electrode/electrolyte interface. For example, systems with ceramic electrolytes

TABLE 4.5
Main Types of Direct Carbon Fuel Cells and Fuel Cell Reactions [192,200]

Fuel/Anode	Electrolyte	Cathode	T (°C)
Solid graphite rod as fuel and anode $C+4OH^-=2H_2O+CO_2+4e^-$	Molten hydroxides OH^-	Air as oxidant $O_2+2H_2O+4e^-=4OH^-$	~ 600
Carbon particle as fuel in molten carbonate and anode $C+2CO_3^{-2}=3CO_2+4e^-$	Molten carbonates CO_3^{2-}	Air as oxidant $O_2+2CO_2+4e^-=2CO_3^{2-}$	~ 800
Carbon particles in fluidized bed $C+2O^{-2}=CO_2+4e^-$	Oxygen-ion conducting ceramic electrolyte O^{2-}	Air as oxidant $O_2+4e^-=2O^{2-}$	800–950
Fuel in contact with molten tin $Sn+2O^{-2}=SnO_2+4e^-$			
Carbon particles as fuel in molten carbonate and anode $C+2O^{-2}=CO_2+4e^-$			

may adopt a range of strategies to deliver fuel to the electrode/electrolyte interface including a fluidized bed, carbon mixed with a molten metal, or carbon mixed with a molten salt.

The theoretical or thermodynamic efficiency of DCFC is almost twice those of current-generation coal-fired plants and significantly higher than other FC types. Thus, compared with conventional coal-fired power plants, there is a potential for 50% reduction in greenhouse gas emissions and significantly less CO_2 to be sequestered. The projected cost, including BOP of around US$1500/kW, is lower than most other FC types with substantially lower operating costs due to the availability of cheap fuel sources [181,196]. Some major challenges that need to be resolved before commercialization include the mode of solid fuel delivery to electrode–electrolyte interface (fluidized bed, molten salt, or molten metal); fuel processing and fuel quality requirements (effect of ash and other contaminants in coal on DCFC performance); understanding the electrochemical reaction kinetics and mechanism for carbon oxidation; corrosion of cell components, especially where molten salts are used either as the electrolyte or fuel carrier; lifetime (currently far too short even for reasonable demonstration); degradation rates and causes; improvement in materials performance and power densities; overall systems design; and technology upscaling.

As mentioned earlier, the DCFCs are classified into three main classes based on electrolyte used as described in Table 4.5 [200].

I. Molten salt (KOH, NaOH)—operating at 500°C–600°C.
II. Molten carbonate (Li, Na, K)—operating at 750°C–800°C.
III. Oxygen-ion conducting ceramic (doped zirconia, ceria)—operating at 800°C–1,000°C.
   a. Fluidized bed (direct contact of carbon particles with anode).
   b. Molten metal anode (carbon in contact with molten metal anode).
   c. Molten salt (carbon particles suspended in a slurry).

Apart from electrolytes, DCFCs can further be subcategorized based on the material and design of anode and the method by which fuel is delivered to electrodes within the cell [201].

- Solid carbon—fluidized bed
- Carbon mixed with a molten metal
- Carbon mixed with a molten salt

Along with carbon FCs having solid electrolyte, there are DCFCs in which carbon is oxidized internally or externally to produce CO. Although the reaction of carbon to produce CO is not electrochemical and thus will not produce any voltage, the reaction of CO to produce CO_2 will be electrochemical and produces voltage for the process. In this type of cell system, fuel is CO instead of carbon. Such FC can be referred to as in DCFC. This type of system has an advantage, since CO is produced externally; therefore, impurities, if any, can be removed that can disturb the cell. Moreover, this type of system produces a pure form of CO_2, which can be separated easily. However, the utilization of CO reduces the theoretical efficiency of the system [200]. There is also another but less studied type of direct FC based on an aqueous alkaline electrolyte, which uses an aqueous hydroxide solution of potassium, lithium, sodium, cesium, and magnesium and operates at temperature less than 250°C [200]. However, the performance of such cells is not satisfactory. A tree diagram for different DCFCs is described in Figure 4.15 to clarify many technologies under development.

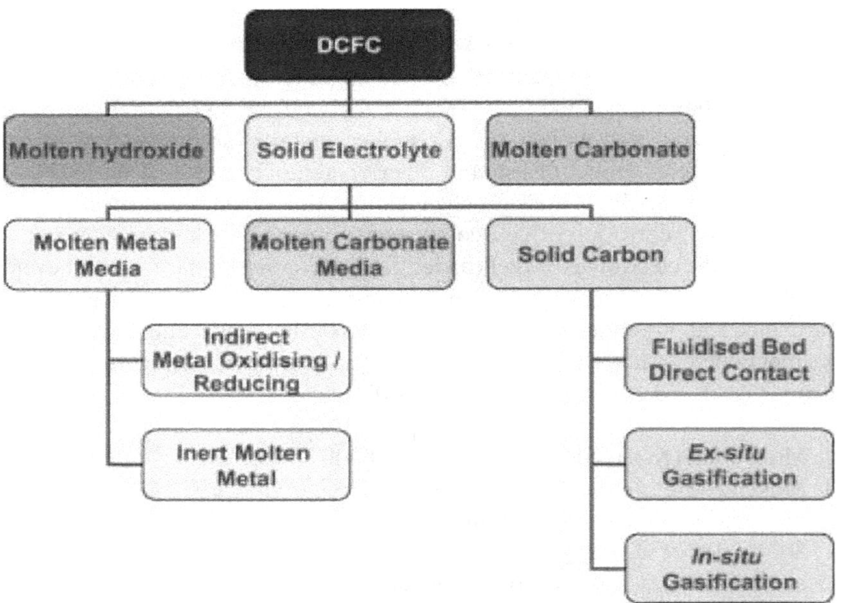

FIGURE 4.15 A tree diagram of various DCFC technologies under development [192,200].

4.6.1 Molten Salt DCFC

In this type of DCFC, molten hydroxide such as NaOH or KOH is used as an electrolyte The electrolyte is kept in a metallic pot, which also acts as a cathode [200]. The anode makes up of graphite or coal-derived carbon is submerged in the electrolyte to act as fuel. So, this carbon rod acts as fuel and anode at the same time. Oxygen is supplied at the cathode by bubbling the air from the bottom of molten electrolyte [200]. Typically, this type of DCFC operates at temperature of 500°C–600°C [202,203]. Electrochemical reactions include the following:

At anode:

$$C + 4OH^- \rightarrow 2H_2O + CO_2 + 4e^- \tag{4.14}$$

At cathode:

$$2H_2O + O_2 + 4e^- \rightarrow 4OH^- \tag{4.15}$$

Although molten hydroxide has a many advantages with the main being relatively lower operating temperature, high ionic conductivity, and higher electrochemical activity of carbon, it has a disadvantage of the formation of carbonates [202]. These carbonates are formed by the combination of CO_2 and carbon with hydroxide ions of electrolyte [200,202].

$$2OH^- + CO_2 \rightarrow CO_3^{2-} + H_2O \tag{4.16}$$

$$C + 6OH^- \rightarrow CO_3^{2-} + 3H_2O + 4e^- \tag{4.17}$$

DCFCs with molten hydroxide electrolyte can be operated at LTs allowing the fabrication of cell with comparatively less expensive material, thus reducing the cost of cell. Goret and Tremillon [204] showed that the rate of carbonate formation is dependent on the concentration of oxygen ions and water concentration. As a result an increase in water content in the cell, CO_3^{2-} can be considerably reduced [204]. Zecevic et al. [205] examined DCFC with hydroxide electrolyte. Despite its compelling cost and performance advantages, the use of molten metal hydroxide electrolytes has been ignored due to the reaction between hydroxide and CO_2. The study describes the cell performance during the initial stage of a long-term operation and discusses the causes of the initial cell performance degradation. To date, five successive generations of medium-temperature DCFC prototypes have been built and tested at SARA Inc. to demonstrate the technology, all using graphite rods as their fuel source.

The main challenge of DCFC with hydroxide electrolyte is that the cell performance decreases over time mainly due to oxygen cathode polarization. There are three possible causes for this performance decay: carbonate formation, electrolyte evaporation due to air bubbling, and corrosion product build-up. The detailed study examining various operating parameters such as water content in the melt, current density, and carbonate content of the melt indicated that the performance of DCFC with hydroxide electrolyte during the initial 200 hours is governed by the oxygen cathode performance that is mainly affected by corrosion products. The corrosion products catalyze the decomposition of peroxide ions, which are reacting species at the cathode resulting in an increase in cathode polarization over time. The effect of carbonate ions on the initial cell performance decay is insignificant as compared to the effect of corrosion product.

4.6.2 Molten Carbonate DCFC

The electrolyte in this type of cell is molten carbonate, and fine particles of carbon are dispersed in the electrolyte that acts as fuel. Ibrahim and Ayub [200] showed that carbonate ions CO_3^{2-} act as the carrier to carry charge. Because of high carbonate conductivity and good stability molten carbonates of potassium, lithium, and sodium are used, the operating temperature for the cell ranges from 750°C to 800°C [206,207]. The reactions in the cell are as follows:

At anode:

$$C + 2CO_3^{2-} \rightarrow 3CO_2 + 4e^- \tag{4.18}$$

At cathode:

$$O_2 + 2CO_2 + 4e^- \rightarrow 2CO_3^{2-} \tag{4.19}$$

An eutectic mixture of lithium and potassium carbonates is present in the cell. Anode current collector (CC) is made up of nickel, and at the cathode, the current is collected by sintered frit of fine nickel particles. Electrodes are separated by zirconia felt [200,202]. Corrosion of metal-clad BPs, high cathode polarization losses, and upscaling are the main disadvantages of this type of cell. Moreover, poor understanding of relationship between carbon structure and its electrochemical activity, lack of a suitable fuel delivery system, and electrolyte tolerance to high percentages of contaminants are some other issues being investigated in the literature.

There are numerous studies using molten carbonate reported in the literature. An MC-DCFC built by Ohio State University suspended graphite, activated carbon, and coal particles in a bath of $(Li/Na/K)_2$ CO_3 slurry. This cell could be operated under different conditions. With the increase in overpotential, particle loading rate, temperature of cell, and rate of agitation, the current density was found to increase [39]. In other studies, Rao et al. [208] and Kapteijn et al. [209] compared CO formation rates for Li_2CO_3-added CB and K_2CO_3-added CB. The results showed that K_2CO_3 is a slightly more effective catalyst than Li_2CO_3. Chen et al. [210] used ternary mixtures of Li_2CO_3, K_2CO_3, and Al_2O_3 with different anode combinations in direct electrochemical oxidation with feedstocks mainly vapor-grown carbon fiber, CB, flake graphite, and green needle coke. Ibrahim and Ayub [200] investigated the performance of DCFC, which employed graphite and activated carbon derived from bamboo or oakwood. MCFC is considered superior because they offer high electric energy conversion (>45%) and relatively low operating temperature [200,206]. This leads to considerably reduced fuel consumption along with low environmental impact (reduced CO_2 emissions). These properties of MCFC make it a fit for medium- to large-scale power generation. Moreover, MCFC operates at a lower temperature of around 600°C, which omits the need for expensive metal catalysts [200]. This reduces the production cost of the cell that can be used for commercial purposes [211].

The study by Cui et al. [181] examines three types of carbonate-based DCFC: MC-DCFC, CO_3^{2-} + mixed ionic–electronic conductors (MIEC), and CO_3^{2-} + SOFC. The configurations of three cell systems and some possible reactions related to carbonate are shown in Figure 4.16. The review focused on the functions of carbonate for (1) transport ions as an electrolyte; (2) catalyzed carbon oxidation as a catalyst; and (3) enlarged reaction area, that is, the anode/electrolyte reaction interface zone.

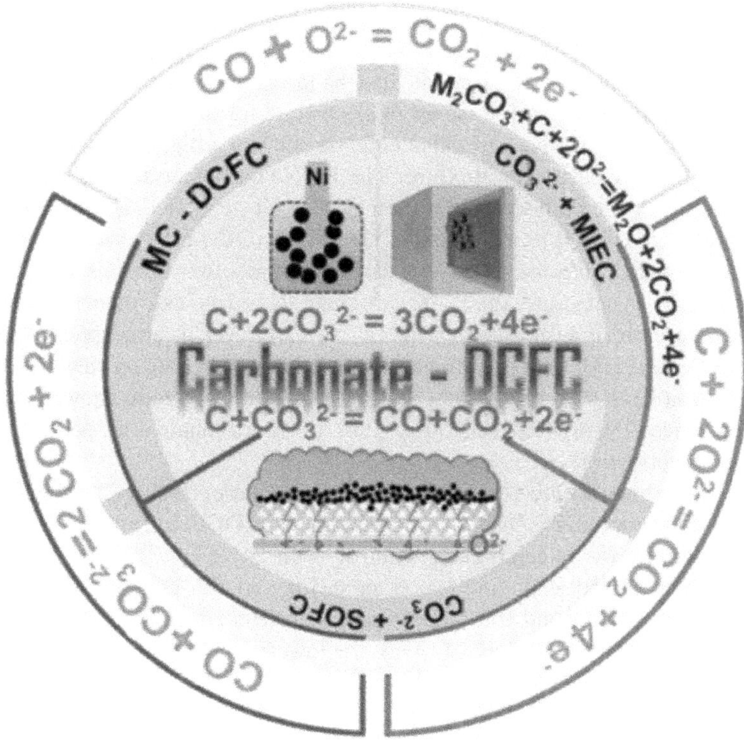

FIGURE 4.16 Schematic of DCFC in the presence of carbonate using three cell system (MC-DCFC, CO_3^{2-} + MIEC, and CO_3^{2-} + SOFC) and some possible reactions: The primary reaction between carbonate and carbon; the electro-catalytic reaction between carbonate and carbon and possible side reaction [181].

Molten carbonate is commonly used in MCFC and has received wide attention in DCFC [181,212]. It shows good compatibility with CO_2 when used as an electrolyte [181] and provides excellent ionic conductivity at a relatively LT [213]. Furthermore, the molten carbonate in the anode chamber can significantly enlarge TPBs, which favor ion diffusion to the electrochemical reaction sites [181]. Many studies have shown that doping alkali metal ions in molten carbonate can effectively accelerate the reaction rate of carbon gasification in a CO_2 gas stream [181,214]. In particular, potassium salt delivers high catalytic activity for the carbon gasification reaction [181]. The operating temperature of DCFCs is reduced by changing the carbonate composition while maintaining cell performance, which is another positive factor when using molten carbonate as the electrolyte [181,215]. A suitable amount of other carbonates or oxides can also further reduce the melting temperature of the carbonate [181]. For example, adding an appropriate amount of Ru_2CO_3 and $SrCO_3$ to a binary carbonate reduces the surface tension of molten carbonate, which, in turn, increases the solubility of gas and reduces the melting temperature [216]. Therefore, using carbonate with different components in DCFC is very useful.

Carbonate can be used as a binary or tertiary mixture. It is well known that a lower melting point is obtained with a binary carbonate eutectic (Li_2CO_3–Na_2CO_3, Li_2CO_3–K_2CO_3, and K_2CO_3–Na_2CO_3) than with a single carbonate [217,218]. Of the three binary carbonates, Li–K carbonate has the lowest melting point (below 550°C). Therefore, it is more suitable for use as an electrolyte for the cell. However, an early study suggested that a Li–Na carbonate eutectic showed a higher ionic conductivity compared with Li-K carbonate mixture. The Li–Na carbonate eutectic has a lower dissolution rate in the cathode (NiO) than that of Ni generated by NiO reduction, and it is quickly oxidized to nickel oxide [181]. Due to the difference in cell life of different carbonate components, Li–Na carbonate has often been used as an electrolyte in MCFC to substitute the Li–K carbonate. Nevertheless, there may be a risk of a rapid decrease in cell voltage with Li–Na carbonate at atmospheric pressure and LT ($\leq$600°C) [181]. One investigation showed that, when using sawdust biofuel, HDCFC with a Li–K carbonate eutectic as the medium in anode showed excellent cell performance (789 mW/cm^2) at 750°C [190]. Li–K carbonate is a good catalyst for gasification reactions [181].

A ternary carbonate eutectic of lithium, sodium, and potassium carbonate as the carbon oxidation medium is also a potential choice in DCFCs. Initially, the ternary carbonate eutectic was selected as an electrolyte in DCFC due to its good compatibility with unexpected pollutants, such as sulfur and ash contents in coal [219]. Furthermore, compared with the binary carbonate eutectic mixed DCFC, the ternary system (Li_2CO_3–Na_2CO_3–K_2CO_3) has the lowest melting point [220]. Vutetakis et al. [221] showed that a ternary carbonate of Li_2CO_3–K_2CO_3–Na_2CO_3 was capable of reducing the cell operating temperature to 500°C. Jiang et al. [222] showed that the same tertiary mixture in coal-fueled HDCFC generated a maximum PD of 50.02 mW/cm^2 with an OCV of 1.1 V at 700°C. In DCFC, carbonate can contact carbon particles in the electrolyte or anode compartment directly, to catalyze the reaction of carbon oxidation. It can also be used as an electrolyte for ion transportation in MC-DCFC and CO_3^{2-} + MIEC. Here, carbonate combines with MIEC materials in a solid electrolyte form. By introducing carbonate in the HDCFC anode, cell performance is improved, possibly due to the catalytic effect on carbon oxidation and the enlargement of the reaction interface.

MCFC is a commercial FC. Its scale is gradually expanded from kilowatt to megawatt [181]. It can be operated by both gaseous and solid fuels. Some progress has been made in the development of different cell structures, materials, and the electrode reaction mechanisms for MC-DCFC. Baur et al. [223] were the first ones to use carbonate electrolyte followed by SRI International who first dispersed a carbon particle in a molten Pb using a molten alkali metal carbonate as the electrolyte [181,224], and some researchers used coal as the anode with Li–Na–K ternary carbonate electrolyte [219]. Some alkali metal salts (such as Li_2CO_3) were regarded as a catalytic agent for accelerating carbon gasification [181,225]. Significant progress was also reported for materials related to cathode catalysts, aerogels, and xerogel carbon anodes [181,223] for MC-DCFC by Lawrence Livermore National Laboratory.

Cui et al. [181] point out that numerous MC-DCFC designs have been developed over the years. As shown in Figure 4.17a, an unconventional tilted design was invented by Cooper et al. [226] using molten carbonate as the electrolyte. The cell cathode was composed of a lithiated NiO, while the anode was made of the carbon–carbonate slurry mixture and the Ni CC. The key aspect of this device was the electrolyte bed of ZrO_2 fabric filled with molten salt that conducted ions while preventing a short circuit of the electrodes. The cell surface area was expanded from the conventional 2–60 cm^2 with no significant polarization loss [181]. Subsequently, as shown in Figure 4.17b, "planar MC-DCFC" was designed in which the mixture of carbon and carbonate is located above the anode where the redox reactions occur. The composite electrode composed of foam nickel and stainless steel (SS) made a great contribution to the stability of the cell structure [227]. This design gave improved performance due to low cell resistance. Subsequently, as shown in Figure 4.17c, Li et al. [228] and Vutetakis et al. [229] designed an MC-DCFC, which consists of three electrodes: a working electrode (WE); a gold counter electrode (CE); and a 12 mm diameter alumina sheath that serves as the reference electrode (RE). An Inconel stirring bar was also introduced into the molten carbonate electrolyte to ensure the uniform distribution of carbon particles and improve the mass transfer process inside the cell. However, the PD in this configuration was not good enough because the reaction area (fuel/electrode/electrolyte contact area) is severely limited.

In an effort to extend the formation zone of TPBs, a fluidized bed cell with a three-dimensional (3D) electrode was adopted by Gür and Huggins [230]. As shown in Figure 4.17d, Zhang et al. [231,232] showed another one with a self-designed fluidized bed electrode anode. In this design, the bubbling gas was applied in DCFC to ensure mass and heat transfer. Recently, a tubular DCFC with a closed-end structure was conceived by Ido et al. [233] (see Figure 4.17e). This design can effectively protect the short circuit between the electrodes using carbon powder. Additionally, the continuous solid fuel supply can be realized by calcinating the anode nickel particles outside the DCFC.

In the MC-DCFC system, the solid carbon fuel can be oxidized directly to CO_2. Thereafter, it is circulated to the cathode compartment through the molten carbonate electrolyte to achieve mass balance. The possible oxidation reactions of carbon

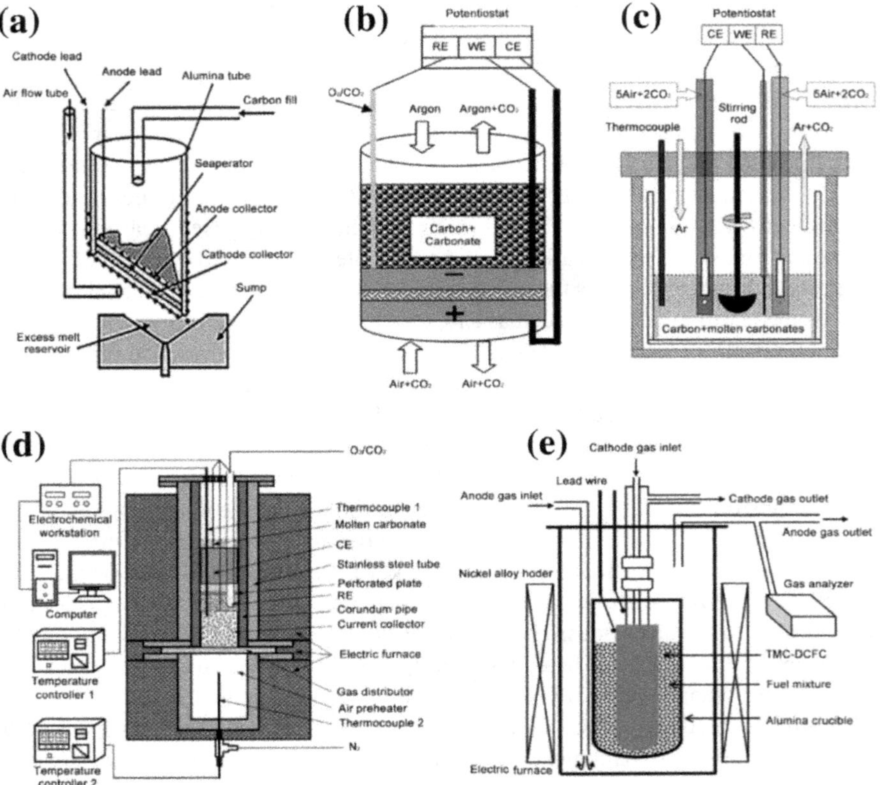

FIGURE 4.17 Different configurations of MC-DCFC: (a) unconventional tilted MC-DCFC favoring the continuous supply of fuel that ensures there is no corrosion of the cathode; (b) planar MC-DCFC with the mixture of carbon and carbonate located above the anode to ensure low cell resistance; (c) MC-DCFC with a stirring rod in the molten carbonate to ensure uniform distribution of carbon particles and improve the mass transfer process inside the cell; (d) FBEDCFC using the bubbling gas to agitate a mixture of carbon and molten carbonate to accelerate mass and heat transfer; (e) TMC-DCFC with a closed-end structure that ensures there is no short circuit and that there is a continuous solid fuel supply. (Reproduced from refs. [181].)

in MC-DCFC and the gas produced show that (Figure 4.18) the decomposition of carbonate can also produce CO_2; it has an influence on the DCFC performance [181]. One investigation showed a noticeable increase in overpotential at a higher current density, owing to the mass transfer process being prevented, but it was easy for the released CO_2 gas to make contact with the carbon and ions at the anode again, with a long-term discharge recorded [234].

Generally, it requires a high operating temperature to enhance the anode reaction rate. Early experimental results show that the predominant product with a carbon anode is CO above 700°C, and this is dependent on the reversal of the Boudouard

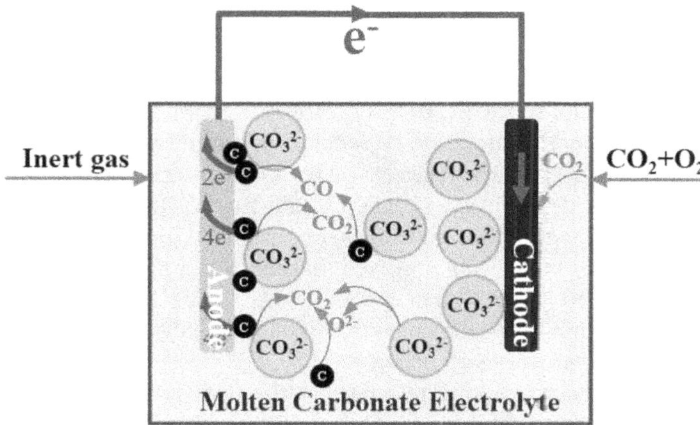

FIGURE 4.18 Schematic of the MC-DCFC system: CO_2 and O_2 gain electrons to produce carbonate ion at the cathode; carbonate ion and carbon generate CO or CO_2 and electrons at the anode. At the same time, carbonate ion is likely to decompose into O^{2-} and CO_2 [181].

reaction. The expected energy from reaction (4.12) is halved, because only two equivalent charges are obtained from one mole of carbon without a significant voltage change at 750°C. The electrochemical formation of CO is not an insurmountable problem in DCFC, as this reaction in the anode compartment occurs only in the process of cell operation [229]. Cooper et al. [193,227] proposed that the mechanisms of carbon oxidation in MCFC are the same as in the Hall process, while MCFC also forms oxygen ions. The oxygen ions trigger the subsequent oxidation reactions of carbon.

Recently, more attention has been paid to tailoring the cell structure to improve the performance of MC-DCFC. Lee et al. [235] proposed that the addition of Gd_2O_3 to a Ni anode improved cell performance due to the enlarged TPBs and the reduced charge transfer resistance. Lee et al. [236] also reported that the addition of lanthanum strontium cobalt ferrite (LSCF) and MIEC to the Ni anode, at a molar ratio of 1:1, showed a better PD of 111 mW/cm^2 at 700°C compared with the single Ni anode. Bie et al. [237] designed a novel syringe-type anode, which ensured extended region TPBs by pressing the carbon powder into the molten electrolyte and preventing carbon oxidation.

Cui et al. [181] point out that the performance of DCMCFC also depends on the extent of corrosion, ash content of solid carbon fuels, and wettability of carbonates. Most molten carbonates are corrosive and reactive, which raises widespread concern about the thermal corrosion and chemical stability of cell materials and the chemical composition of solid fuels. Besides the ash of solid carbon fuels, the wettability of carbonates also affects cell performance. These problems ultimately lead to a significant reduction in conversion efficiency and long-term stability. From 1996 to 2006, the cell lifetime greatly increased from just a few months to 2 years [181]. The corrosion issue gives rise to severe challenges in improving the chemical and physicochemical stability of electrode materials for MCFC.

At present, nickel is one of the most widely used electrode materials because of its low price and good performance in conductivity and electrocatalytic ability. However, Ni and NiO are easily dissolved into the molten carbonate, which causes an internal short circuit in MCFC or a short lifespan [238]. It was reported that the Ni/NiO solubility in carbonate could be reduced by adding SrO or MgO or $CaCO_3$, $BaCO_3$, or $SrCO_3$ to alkali carbonates [181,238]. Doping a rare earth metal [239] or a rare earth metal oxide (such as La_2O_3, Y_2O_3, and Yb_2O_3) [181,239] is another practical approach to protect the nickel or nickel oxide electrode against dissolution in molten carbonate. Liu et al. [240] observed that adding the rare earth metal Dy to a mixture of Li_2CO_3 and K_2CO_3 in a mole ratio of 62:38 could reduce the NiO/Ni passive anodic current, inhibiting the outward diffusion of Ni at 650°C, and eventually increase the resistant ability to corrosion.

When the solid oxide electrolyte separates the two electrodes, it prevents the cathode from carbonate corrosion. Nevertheless, the chemical compatibility between the electrolyte and the carbonate is essential because of the contact attack. Suski et al. [241] reported a double cell with yttrium-stabilized zirconia (YSZ) and an electrolyte of 53% Li_2CO_3–47% Na_2CO_3 gave long operation hours. Jiang et al. [14] indicated that the formation of oxides and superoxides was more likely to occur in an oxygen environment, which was the primary reason for YSZ corrosion [181]. It is necessary to investigate the stability of YSZ in both the reducing atmosphere and the oxidizing atmosphere to simulate both operating conditions [181,242]. According to Xu et al. [243], YSZ is a good choice of electrolyte for HDCFC, due to the reducing atmosphere of the anode chamber. The doped cerium oxide-based material is a typical solid electrolyte in DCFC that shows good stability below 650°C [181]. Xu et al. [243] explored the stability of SDC in 62 mol% Li_2CO_3–38 mol% K_2CO_3 in air and found that SDC is more resistant to corrosion than YSZ, when using air. The study also indicated that crystal boundaries might be the most preferred site for corrosion development [181]. Since corrosion in molten carbonate cannot be avoided, it is vital to develop nascent corrosion-resistant materials, to maximize the advantages of carbonate for DCFC devices.

Solid carbon fuels come from a wide range of sources and have a unique constituency. The molten carbonate is likely to change chemically due to interaction with some ash components, including heavy metals and their oxides, sulfides, and chlorides, which ultimately lead to cell failure [181,229]. However, some metals in ash have a positive effect on cell performance. For example, Cao, MgO, and Fe_2O_3 have a catalytic effect on the oxidation reaction of solid carbon, which increases the current density [181]. Therefore, studying the effect of impurities in the ash on cell materials is key to improving the performance of DCFC. It has been reported that coal with a low ash content shows better cell performance than high ash [181,242,243]. Ju et al. [244] found that ash-free coal (low ash content) exhibited the most prolonged stability The high ash content in raw coal is considered detrimental to the short-term stability with a Ni-YSZ anode, as the ash blocks the anode and causes contact between the reaction interface and the carbon interface during cell operation [181]. Tulloch et al. [245] reported that coal with about 70 wt% SiO_2 ash showed a significant decrease in current density for MC-DCFC.

Recently, different pretreatment methods have been proposed to remove unwanted impurities and enhance cell performance [171–176,178–180]. Eom et al. [246] reported that using HCl to pretreat coal could ensure maximum reduction in ash content and the sensitivity of cell performance to surface silicon content may be reduced when the temperature is higher than 733°C. Xie et al. [247] modified bituminous coal with acetic acid and effectively removed ash of about 84 wt% Si and 64 wt% Al, to optimize cell performance. As mentioned earlier, some impurities in ash are beneficial due to their catalytic effects. Cai et al. [248] employed orchid leaf char in SO-DCFC with YSZ electrolyte and an Ag-GDC electrode. The researchers concluded that the natural Ca in biochar exerted a catalytic effect on the reverse Boudouard reaction and enhanced the performance of DCFC. Hao et al. [249] showed the positive effect of calcite and magnesium calcite in magazine waste paper, which indicated that more amorphous carbon exhibited a higher degree of carbon oxidation and also catalyzed the gasification reaction of carbon fuel and delivered a higher PD. The enhanced cell performance was also observed because the KCl in raw reed ash led to the high oxidation activity of reed char [181].

Cui et al. [181] reported that complete wetting of the carbon particles in eutectic carbonate is critical for the charge transfer at the TPB zone, as this is a dominant factor in ensuring uninterrupted round-the-clock power generation by the DCFC. Hong et al. [250] found that the wetting of carbon in carbonate is driven by capillary force, as well as being largely determined by CO bubbles produced from the reverse Boudouard reaction. Therefore, no matter what carbonate is employed in DCFC, as long as the solid carbon is completely soaked in the molten carbonate to form the interface where the carbon oxidation reaction occurs, the electrons can be continuously generated and transferred. The wettability of solid carbon fuel in molten carbonate is enhanced by pretreatment to increase the surface area and the pore volume of the carbon particles [251]. The effect of base or acid pretreatment on the electrochemical activity of carbon in lithium and potassium carbonate has also been examined [251] with HF, HNO_3, and NaOH. Of all the samples, activated carbon pretreated with HF exhibited the highest electrooxidation activity, with an increased current of approximately 50 mA/cm^2. It has also been shown that carbon particle wettability relies on carbonate composition. Watanabe et al. [252] indicated that the surface tension of solid carbon could be reduced by lowering the amount of sodium carbonate and thus improving its wettability.

Cui et al. [181] also reported that for MC-DCFC, the degree of stirring is also an important variable that affects carbon particle wetting. Li et al. [228] observed that mass transfer has been significantly improved at 400 rpm and that current density can be further increased in the whole electrode potential range when the stirring rate is increased up to 600 rpm. Vutetakis et al. [229] also reported improved current density when increasing the stirring rate. However, the performance of DCFC is greatly reduced if the stirring rate exceeds a specific value, because the fuel splash phenomenon leads to a fuel shortage [181]. Cui et al. [181] point out that the improvement of cell performance by stirring also depends on cell operating temperature and electrolyte viscosity.

MC-DCFC is also attractive for direct biomass fuel because of their excellent wetting property for solid fuels and their high tolerance to impurities [253]. Ahn

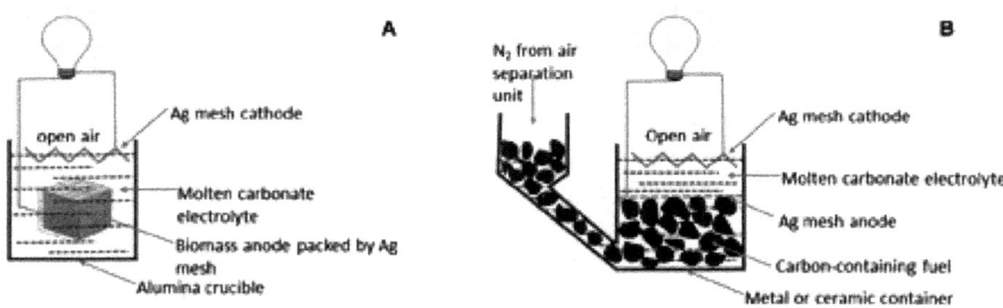

FIGURE 4.19 Schematic diagram of a matrix-free MCFC without flowing CO_2 at the cathode. (a) Caged anode. (b) Design for continuous operation [253].

et al. [212] reported an MC-DCFC, where carbon fuel was premixed with a molten carbonate electrolyte, with air and CO_2 bubbled through a porous cathode tube. In another report, a porous anode tube was used for an MC-DCFC [254]. In both cells, flowing CO_2 at the cathode was essential [212,254]. In a recent report where a porous nickel rod was used as the anode, mixed graphite and (Li, K)$_2$CO$_3$ were used as slurry fuel, and mixed CO_2 and O_2 were flowed at the cathode [255] and good performance was achieved. Another approach is to use matrix-free MCFC as illustrated in Figure 4.19a. In this design, solid fuel is put in a cage made of conducting materials such as silver mesh, eliminating the need to flow CO_2 at the cathode. This greatly simplifies the cell manufacturing process and reduces the operating cost. Continuous FC operation can be achieved through a cell design shown in Figure 4.19b. Solid fuels such as carbon or biomass are ideal for this new type of MCFC.

In the design shown in Figure 4.19, the only possible direct chemical reaction between the fuel and oxygen was through dissolved molecular oxygen (O_2). It has been reported that, at a temperature of 600°C, the dissolved oxygen in binary mixtures of Li-K and Li-Na carbonates is in charged forms, such as superoxide (O^{-2}) or peroxy monocarbonate (CO_2^{4-}), and the dissolved molecular oxygen (O_2) at high temperatures is negligible [256]. In the reaction between charged oxygen species and the anode, there must be a loss of electrons to generate electricity. Under this circumstance, the charged dissolved oxygen is used as the charge carrier. Thus, an MCFC can work without flowing CO_2. In this new cell design, at high working temperatures, the efficiency loss due to the direct reaction between molecular oxygen and fuel will be minimized because of the low solubility of molecular O_2 in molten carbonates [256]. Moreover, the possible stability issues associated with the decomposition of carbonates can also be avoided because of the in situ formation of CO_2 at the anode, which suppresses carbonate decomposition. CO_2 solubility in molten carbonates is much higher than that of O_2 [254–257]. They also formed CO_2^{4-} and CO_2^{3-} ions, which diffuse to the anode to react with the fuel. Under this circumstance, flowing CO_2 at the cathode is not required in the matrix-free MCFCs. If the CO_2 generated at the anode is beyond the solubility limit in molten carbonates, the "extra" CO_2 may also pass through the electrolyte to the cathode and then react with O_2 in the air, forming carbonate ions to take part in the FC reaction.

4.6.3 Oxygen-Ion Conducting Ceramic Fuel Cell

Oxygen-ion conducting ceramic is used as an electrolyte in this type of cell. Stabilized zirconia (8–10 mol % Y_2O_3, balance ZrO_2) is the most common electrolyte used in DCFCs. The cell typically operates at temperature of 800°C–1,000°C [253]. There are three subcategories that use oxygen-ion conducting solid electrolyte. They differ in the anode design and the mechanism of fuel delivery to the anode/electrolyte interface.

4.6.3.1 Solid Carbon or Fluidized Bed as Fuel

In this technology, a direct electrochemical reaction occurs between the oxygen ions and carbon fuel at the anode. The oxygen ions are transported through ceramic electrolyte from the cathode to the anode. The reaction is as follows:

$$C + 2O_2^- \rightarrow CO_2 + 4e^- \quad (4.20)$$

The carbon particles are in direct contact with the anode [258]. Most of the work on this type of cell is focused on button cells comprised of ceramic electrolyte disk with nickel-based anode and lanthanum strontium manganite (LSM)-based cathode [253]. Apart from the technical issues related to SOFC, solid fuel delivery to the anode/electrolyte interface and lack of understanding of carbon oxidation reaction mechanisms at the interface are major issues in its development.

4.6.3.2 Solid Carbon in Molten Metal as Fuel

This type of technology employs a molten metal such as tin that acts as the carbon fuel carrier and is used in anode. The reaction of tin with oxygen ions is given as follows:

$$Sn_{(liq)} + 2O^{-2} \rightarrow SnO_2 + 4e^- \quad (4.21)$$

The cell uses the above reaction in the presence of oxygen at the cathode with OCV of 0.78 V. Thus, the tin oxide can be converted back to tin by a reaction with carbon fuel. This type of cell is operated at 1,000°C with an exothermic reaction. Electricity is produced by direct oxidation in a FC that has molten tin as an anode and carbon fuel. Porous ceramic separators are used to maintain the contact of molten tin with carbon fuel while avoiding formation of amalgam. However, the use of porous ceramic separator causes excessive anodic polarization losses [259].

4.6.3.3 Solid Carbon in Molten Carbonate

In this technology, molten carbonates containing carbon fuel act as anode and oxygen-ion conducting ceramics as the electrolyte. Carbon fuel mixed with molten carbonates is supplied to anode. Various types of fuel such as biomass, coal, and tar can be used in this type of cell since it is a hybrid between molten carbonate and SOFCs. The carbon is first oxidized inside the cell or externally to CO as per the following reaction:

$$C + CO_2 \rightarrow 2CO \quad (4.22)$$

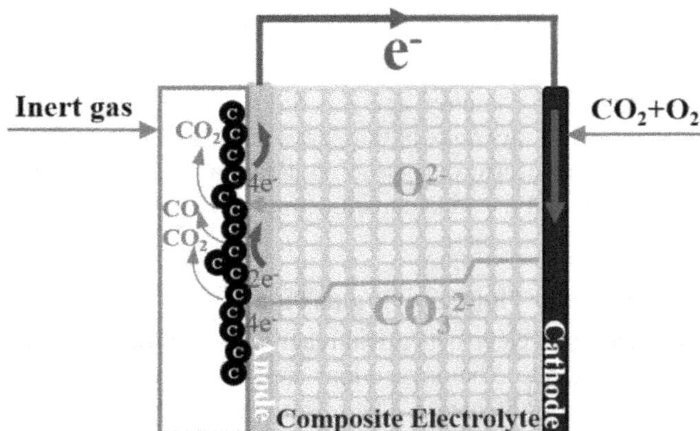

FIGURE 4.20 Schematic of DCFC with a composite electrolyte: CO_2 and O_2 receive electrons and produce carbonate ion; only O_2 receives electrons and produces O^{2-} at the cathode. Then, the O^{2-} and carbonate ions are transferred to the anode through a composite electrolyte and react with carbon to produce CO, CO_2, and electrons [181].

This CO reacts with the oxygen ions at the electrode to form CO_2. In the cell, the carbon particles that come in contact with oxygen ions are converted to CO_2, while the other ones first form CO, which is further oxidized to CO_2. For this type of cell, the SDC–carbonate composite electrolyte shows excellent conductivity of 10^{-2} to 1.0 S/cm in the range of 400°C–700°C and this composite electrolyte is not subject to corrosion issues in the normal MCFC. The cell performance was improved with this composite electrolyte because molten carbonate with mobility can expand the TPB at the anode, which in turn reduced electrode polarization resistance. It was also found that the redox reaction of carbon fuel was enhanced when using doped ceria materials. A diagram of the electrochemical process between the electrolyte and the electrode is presented in Figure 4.20. It shows that two electrochemical mechanisms might be responsible for forming carbonate ions and oxygen ions in the cathode chamber filled with O_2 and CO_2.

In the composite electrolyte of SDC carbonate, the charged species are carbonate ions and oxygen ions and carbonate ions are transferred in the molten electrolyte, while oxygen ions are the conducting species within the SDC [181]. The carbon particles in the anode cavity combine with carbonate ions, while others are in contact with the electrolyte of SDC directly and react with oxygen ions releasing CO or CO_2 and generating electrons. CO can also be produced from the Boudouard reaction, with further oxidation by oxygen ions [260].

The oxidation reactions within MC-DCFC are still not well understood [181]. Elleuch et al. [261] used low-cost solid carbon for DCFC with an SDC-NiO anode, an SDC electrolyte, and $Li_xNi_{1-x}O$-SDC as the cathode. The results showed that the carbonized almond shell with more oxygen-containing functional groups has good cell performance at 700°C, with a power output of 127 mW/cm^2. The researchers further explored the electrochemical oxidation of carbon using graphite in the same cell [181]. Recently, a dual 3D ceramic textile electrode was integrated into a GDC–carbonate composite electrolyte-supported DCFC, and thus, the TPB region was expanded [181]. Bian et al. [262] developed a unique electrolyte-supported DCFC consisting of a GDC electrolyte, a NiO-GDC anode, and a $Sm_{0.5}Sr_{0.5}CoO_3$-GDC cathode. It exhibited unprecedented cell performance at 600°C, with a power output of 392 mW/cm^2 when using graphitic fuel, due to the enhanced charge and mass transfer on the electrode below 600°C [181]. HDCFC was also combined with SOFC technology and MCFC technology to provide a new way for carbon fuel to reach the reaction region [263]. This method ensures that the molten carbonate is incorporated into the anode cavity, which significantly expands the reaction region from a two-dimensional region to a three-dimensional region, and this accelerates the mass transfer to the solid anode/electrolyte.

HDCFC is based on two typical FCs—SOFC and MCFC—in which the solid oxide electrolyte (which includes YSZ, GDC, or SDC) separates the electrode chambers; the molten carbonate electrolyte has fluidity at a high temperature and expands the oxidation reaction area [263]. The oxygen ions reduced from the oxygen molecule are transmitted from the cathode to the anode compartment through the solid oxide electrolyte. In the anode compartment, carbon particles may be completely oxidized to CO_2 or partially oxidized to CO [181,264]. An early tubular HDCFC using a Pt cathode and YSZ electrolyte was designed by the University of St. Andrews, with the anode nickel mesh placed in a mixture of carbonate and carbon [181]. The researchers at the University of St. Andrews also developed a planar button cell with better sealing for improved gas purification [181]. The electrochemical reaction mechanism in the anode was also investigated in detail [263,265].

Figure 4.21 shows that some reactions and some gas production occur in the anode compartment of HDCFC, which is full of CO_2 and CO at the same time. In the anode

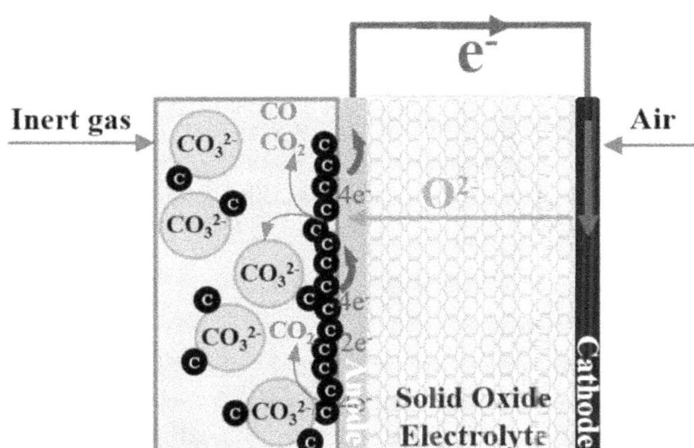

FIGURE 4.21 Schematic diagram of the SOFC+MCFC system: O_2 gains electrons and then produces O^{2-} at the cathode; O^{2-} then passes through a solid oxide electrolyte and arrives at the electrolyte/anode interface, where it reacts with carbon to produce CO, CO_2, and electrons [181].

chamber filled with nitrogen, the reaction processes of carbon (both electrochemical and chemical) are complicated. In the anode, both oxygen ions and carbonate ions are active species of electrochemical oxidation in the slurry of carbon/carbonate [263]. When the number of oxygen ions is sufficiently high, the CO_2 could be converted to carbonate ions, which would result in a slow decline of CO_2 activity in the molten carbonate. With continued consumption of the oxygen ions, the carbonate ions oxidize carbon to CO_2 or CO. Carbonate ions then regenerate from oxygen ions and CO_2, which maintains the electric charge balance in the molten carbonate solution. The presence of molten carbonate, which facilitates the flow of carbon particles to the anode chamber and expands the TPBs, is expected to act as an electrochemical mediator and accelerate the oxidation reaction kinetics of the carbon particles [181,227,230].

In addition, the non-electrochemical reaction of the Boudouard reaction that occurs at 750°C has a strong influence on the entire anode reaction, which consumes carbon through a chemical reaction and causes a sharp decrease in current density [181]. Deleebeeck et al. [266] compared the effects of different anode gases on cell performance and found that introducing pure CO_2 could reduce the mass transfer limitation by facilitating the Boudouard reaction or preventing carbonate decomposition. Recently, Li et al. [267] found that introducing CO_2 into the anode chamber could improve cell performance through the electrochemical reaction at 700°C–800°C. Lee et al. [125] examined the effects of different contacts of carbon fuels on the anode. Jiang et al. [268] designed an experiment to investigate the possible reaction active sites at the anode chamber in HDCFC. Apart from the temperature, the carbon oxidation process at the anode also depends on the electrode and electrolyte. Jiang et al. [265] further explored the reaction mechanism of the anode by exchanging the carbonate content in the HDCFC system with a configuration of NiO-YSZ/YSZ/LSM, while using 62 mol% Li_2CO_3–38 mol% K_2CO_3 in the anode chamber. They observed reduced polarization resistance of the cell with a carbonate content of 20 mol% or 50 mol%. They also observed that carbon particles are prevented from reaching the electrode by a high concentration of molten carbonate, which has a limited effect on various carbon oxidation reactions. Some corrosion and sealing issues are significant concerns when using high carbonate concentrations. Cantero-Tubilla et al. [269] proposed that the rate of movement of carbon fuel slowed in the presence of high carbonate concentrations. The details of all of these investigations are given by Cui et al. [181].

4.6.4 Fuels for DCFC

DCFC used solid carbon as fuel. The performance of cell is dependent on the structure and chemical characteristics of the solid carbon fuel. In the case of coal, the properties and characteristics of fuel vary greatly depending upon the source of coal and pretreatment methods used. DCFCs can be used in wide variety of fuels including coal, liquid hydrocarbon fuels, biomass, and organic waste [200]. Some of these fuels are used directly with a little pretreatment, while others require heavy pretreatment and purification to be used in DCFCs [200]. Researchers believed that the performance of DCFC depends on the properties of fuel such as crystal structure

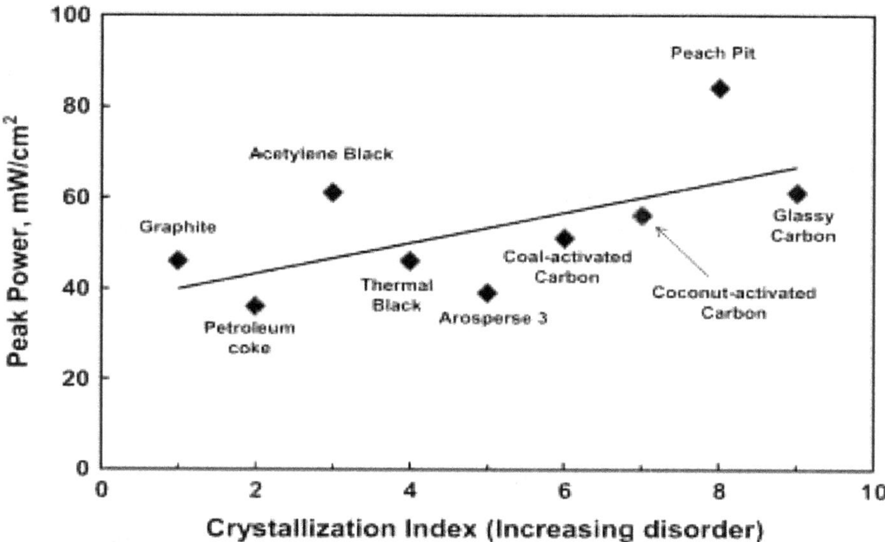

FIGURE 4.22 Performance of different carbon fuels in DCFCs [200].

(degree of graphitization), particle size, and impurity, since the mechanism of carbon electrochemical oxidation is not fully understood. The oxidation mechanism varies with the type of DCFC. The chemical and physical properties of carbon vary with its allotropic form. Studies are conducted to understand the correlation between crystal structure and oxidation reaction mechanism. It was found that the oxygen reactivity in graphitic or partially graphitic carbons is highly anisotropic [270,271]. Kintoshita et al. [272] found that the rate of carbon oxidation is faster at crystal edges. Moreover, the rate of carbon oxidation is also affected by structural defects, dislocations, lattice vacancies, and grain boundaries, and as expected, highly disordered structures, such as amorphous carbon, have much higher oxidation rates than highly ordered structures [270,271]. For electrode characterization, crystallite size and lattice spacing are considered significant parameters. Studies of the reactivity of carbon showed that these factors also play a significant role in DCFCs. This has been shown by the number of studies that correlate the crystal structure of carbon to its electrochemical reactivity. In general, structures with a greater degree of crystal disorder have higher electrochemical reactivity due to the greater availability of reaction sites. The crystallization index is the factor on which the properties of fuel depend. Figure 4.22 shows the crystallization index of different fuels.

Palniandy et al. [273] examined the use of biochar as a fuel source for DCFC. The performance of a DCFC using fuel sources derived from woody and nonwoody biomass was compared in this study. The effect of pyrolysis temperature ranging from 550°C to 850°C on the preparation of biochar from rubber wood (RW) and RH was evaluated for power generation from DCFCs. In addition, the effect of applying chemical pretreatment and posttreatment on biochar was further investigated for DCFC performance. In general, the PD derived from RW biochar is significantly higher (2.21 mW/cm²) compared with RH biochar (0.07 mW/cm²). This might be due to the presence of an oxygen functional group, higher fixed carbon content, and lower ash content in RW biochar. The study concludes that woody biochar is more suitable for DCFC application, and alkaline pretreatment in the preparation of biochar enhances the electrochemical activity of DCFC. Because charcoal is normally produced through the pyrolysis of biomass such as wood, bamboo, starch, and sawdust, during this pyrolysis, tar, bio-oil, and biogas are formed along with the charcoal. These are also potential fuels for MCFCs for power generation. The combustion of wood, etc., produced biogas in the air can provide heat to maintain the high operating temperature. Therefore, directly using biomass such as wood as fuel in an MCFC can maximize energy utilization. The composition of wood is also well suited for MC-DCFC with molten carbonate electrolyte.

4.6.5 Benefits and Drawbacks of Direct Carbon Fuel Cell

DCFC has many properties or features that make it different from the rest of power generation technologies. It is an emerging technology that can be used for commercial power generation through rigorous research and development. DCFC produces pure CO_2, which can easily be separated and dumped without being released into the environment. No other pollutants such as sulfur oxide (SO_x) and nitrous oxide (NO_X) are produced, since the cell allows only carbon to convert electrochemically. DCFCs have high conversion efficiencies when compared to other power generation technologies as shown in Figure 4.23.

DCFC has higher efficiency because of a lower number of energy transformations in the cell as compared to any combustion-based device [200]. This is because with

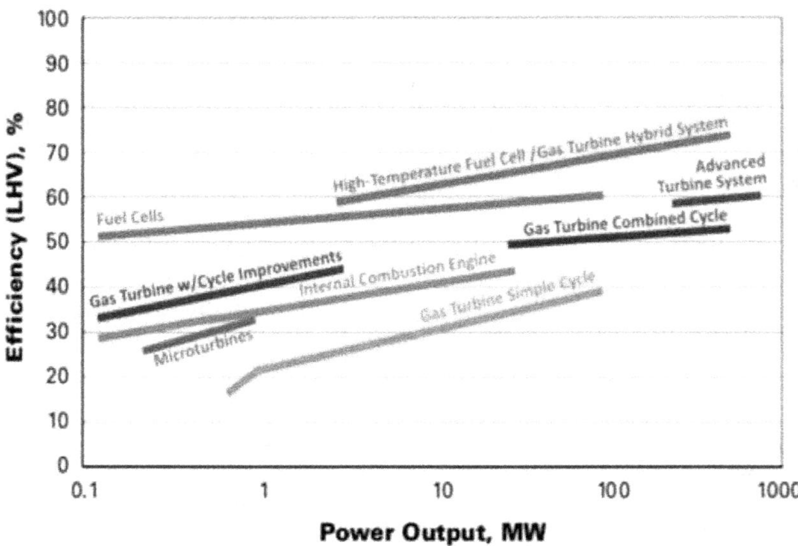

FIGURE 4.23 Efficiency comparison of power generation technologies [200].

each conversion (energy transformation) some of the energy is lost, resulting in the overall reduction in system's efficiency [274]. DCFCs have outstanding modularity. The output power of any FC system can be controlled by varying the number of cells-per-stack and/or stacks-per-system. The efficiency of a DCFC is not dependent on the size of system, which allows it to be used at both small and large scale [275]. DCFC can use a wide variety of fuels including coal, liquid hydrocarbon fuels, biomass, and organic waste [200]. Some of these fuels are used directly with a little pretreatment, while others require heavy pretreatment and purification to be used in DCFCs [200]. Along with this flexibility, annually in the USA billions of kilograms of CB are produced by pyrolysis. Moreover, carbon releases much more energy per unit volume with oxygen (20.0 kWh/L) as compared to hydrogen (2.4 kWh/L), methane (4.0 kWh/L), gasoline (9.0 kWh/L), and diesel (9.8 kWh/L). Literature research indicates that the performance of DCFC depends on the properties of fuel such as crystal structure (degree of graphitization), particle size, and impurity. The oxidation mechanism varies with the type of DCFC. DCFC is an inherently silent device with simpler manufacturing, design, assembly, operation, and analysis of the cell than that of heat engines. Such characteristic of the cell promotes its use in auxiliary power and distributed generation applications in addition to portable applications [276]. DCFC has many unique features. It has significant thermodynamic advantages. DCFC can allow a full conversion of carbon fuel in a single pass making the fuel efficiency of 100%. This makes DCFC the most efficient power generation technology. DCFCs have 50% lower emissions with ten times reduced off-gas volume compared with coal-burning power plants. Furthermore, released pure CO_2 can be used for tertiary oil recovery among other utilization possibilities. There is no emission of sulfur dioxide and nitrogen oxide. Finally, the absence of reformers and heat engines makes DCFCs mechanically simple. It can be built near coal mine, thus eliminating the need for coal transportation, saving cost, and energy.

DCFC also has some drawbacks. It has usually low-power densities. Cells are exposed to high degradation rates that cause corrosion of various cell components. Methods of fuel delivery to the anode within the cell are complex. The solid fuel must be delivered to the anode in such a way that there is maximum interaction of fuel with an electrode at the electrode and electrolyte interface. Also, the electrochemical oxidation of carbon fuel inside FC is a complex phenomenon and requires more research. The performance of cell largely depends on the properties of fuel. Therefore, fuel processing is crucial in direct FC technology and low-cost fuel processing technology is required. These factors greatly affect the upscaling of DCFC technology and more work is required for its commercialization.

4.6.6 Applications of DCFC

DCFCs have the potential to be used in a broad range of applications. As a result of their high efficiency, static nature, modularity, and fuel flexibility, they have applications ranging from vehicles to power plants. Portable applications mainly include portable power generators and electronic devices [277,278]. Portable power generators are designed for light outdoor personal uses such as camping and climbing, light commercial applications such as portable signage and surveillance, and power required for emergency relief efforts. In electronic devices such as laptops and cell phones, FCs can replace battery. The power of portable FCs typically ranges between 5 and 500 W [277]. Residential, commercial, and industrial stationary power generation sectors can also employ DCFCs. Emergency backup power supply (EPS) [279], remote area power supply (RAPS) [280,281], and distributed power generation (CHP) [282] are the main stationary applications of DCFCs. Around 70% of the FCs are utilized in stationary applications [200]. DCFC is also a very viable use in transportation industry with near-zero harmful emissions without having to compromise the efficiency of the vehicle. Considering the advantages such as static operation, fuel flexibility, modularity, and low maintenance requirements, DCFC can become an ideal future alternative for current combustion engines.

4.6.7 Future Perspectives

While DCFC may provide a clean electrochemical device for generating electricity from solid carbon, further research on the basic theory of carbon oxidation is still needed. In DCFC, the carbonate accelerates ion transfer as a medium or is a catalyst for carbon oxidation and gasification reaction. However, the specific reaction mechanism with carbonate involved in the whole-cell system still lacks sufficient proof of experiments due to the complexity of reactions inside the cell. Further theoretical model and experimental tests should be designed to determine what substances affect the electrode reaction process.

Three major issues that need further investigations to commercialize DCFC are material corrosion, role of ash properties, and wetting of coal particles. The issue of material corrosion in molten carbonate has not been solved satisfactorily. More research on the corrosion of carbonate to the solid-state electrolyte for emerging HDCFCs is needed. It is unknown whether adding some carbonates, oxides, or rare earth elements has a positive or negative effect on cell performance. It has been confirmed that the addition of carbonate to the anode favors expanding TPBs, because the fluidity of molten carbonate could promote the transfer of carbon fuel to the anode, resulting in a dramatic improvement of electrochemical reactions. However, carbonate attacks the solid electrolyte. While SDC has better corrosion resistance than YSZ, the corrosion issue still cannot be avoided. Corrosion thus affects the long-term viability of DCFC. Hence, more attention needs to be focused on the carbonate corrosion to other cell components (including solid-state electrolytes), instead of only common electrode materials.

While fuel flexibility is one of the major advantages of DCFC, the role of ash content in coal and biomass fuels is not well known. While some ash components have a catalytic effect on carbon oxidation, in general high ash content results in poor cell performance. For example, a large amount of Fe, Mg, and Ca promotes the gasification of carbon, while Si and Al have a negative effect on the cell performance of DCFC with carbonate. It is difficult to identify the effect of each inorganic salt on cell performance because there is often more than one salt in the fuel. For these reasons, removing ash from non-pure carbon fuels such as biomass should also be considered seriously. Pretreatment should be attempted as an essential treatment method applied to DCFC (including heat treatment and acid or base washing and air plasma). The effect of pretreatment on the properties of the fuel itself and the reactivity of the pretreated fuel with carbonates also need to be determined. The parameters that affect the proper wetting of carbon by the electrolyte also need to be further investigated. While many cell designs have been developed to provide better continuous power generation, the fluidized bed cell can provide a more convenient feeding mode. Thus, designing different cell configurations is still key to the commercialization of DCFC.

4.7 SOLID OXIDE FUEL CELL

SOFC is gaining more attention because of its high-power generation with enough electrical efficiency for household devices and automobiles [283–286]. SOFCs do not have problems of leakage, lubrication, and heat loss of traditional heat engines, and they offer diversified benefits such as fuel flexibility and desirable chemical-to-electrical conversion efficiency, which is not limited by Carnot cycle, chemically non-pollutant, lower emission of gases, and generation of heat and electricity [283,284,287]. A single SOFC consists of a cathode and an anode separated by a solid oxide electrolyte as shown in Figure 4.24 [283,284,288–290]. The fuel (hydrogen, methane, etc.) is continuously provided to the anode side, and an oxidant is continuously provided to the cathode side. The fuel is decomposed into negative and positive ions at the anode terminal. The intermediate electrolyte acts as an insulator for negative ions (electrons) and allows only positive ions (protons) to flow from the anode to the cathode. An external circuit flows electrons from the anode to the cathode to generate power, and these electrons are accepted by the cathode for oxygen

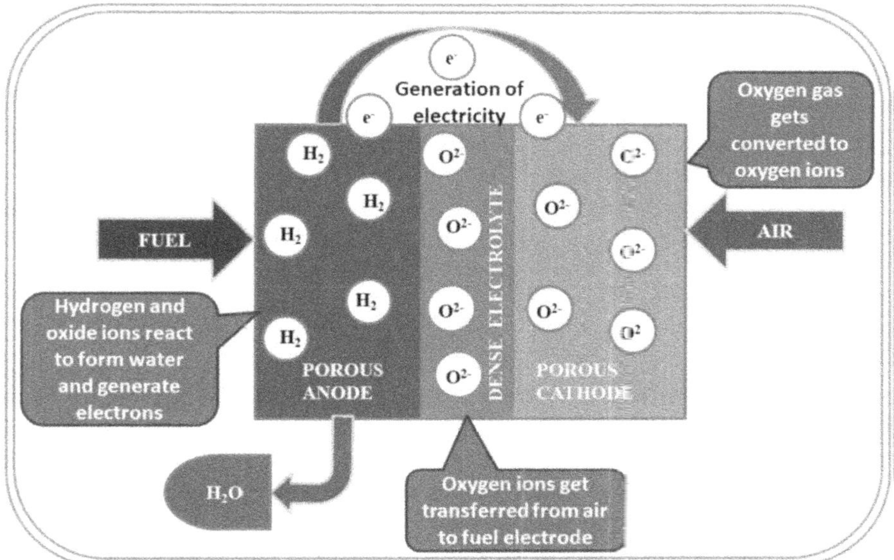

FIGURE 4.24 Schematic diagram of solid oxide fuel cell [283].

reduction to complete the external circuit. For its practical use, several SOFCs are combined in a stack for higher power output [283–285].

Electrolytes are mostly ceramic materials and conduct ions between anode and cathode. It may be a proton (H^+) or an oxide-ion (O^{2-}) conducting medium. The electrolyte materials should have higher conductivity (0.01–0.1 S/cm for 1–100 μm thickness of electrolyte) of oxide, ion, or proton. A low oxide-carrying property of solid electrolyte due to ohmic losses results in nonlinear conducting properties. The electronic conductivity of the solid electrolyte should be low since the high degree of electronic conductivity results in leakage of O_2 and a larger amount of voltage loss without the generation of sufficient electricity. The solid electrolyte must also have chemical, thermal, phase, dimensional, and morphological stability and high mechanical strength to endure stress. Finally, SOFC should have a low cost and easy cell fabrication technology [283,284].

4.7.1 Materials for Electrolytes

In SOFC, the property of the oxide electrolyte material defines the whole structure of the FC [283,284]. The main characteristics of electrolytes in SOFC include the following: (1) Electrolyte should have high ionic and negligible electronic conductivity, (2) chemically stable at elevated temperature, (3) gas-tight/free of porosity, (4) low ohmic losses, and (5) matching thermal expansion coefficient (TEC) with adjacent cell components (anode/cathode). In SOFCs, the oxygen ions are transferred from cathode to anode through oxide-ion conducting electrolyte. The role of point defect is most important for oxygen diffusion, and the defect–defect interactions are significant in this motion [291]. In the case of oxygen-ion conductors, current transports due to the movement of oxide ions through the crystal lattice. The ionic conductivity is temperature-dependent, and at high temperatures, it can approach values close to 1 S/cm.

During the last decade, YSZ is extensively used as solid electrolyte material in SOFC [283,292]. The main drawback of YSZ is that ionic conductivity decreases when the temperature of SOFC reduces below 1,000 K due to the increasing resistance of the cell [283]. The nature and concentration of dopants are important to enhance ionic activity. Doping with 8 mol% Y_2O_3 stabilizes cubic zirconia and gives the highest ionic conductivity (~0.18 S/cm at 1,273 K), while doping with 9–10 mol% Y_2O_3-ZrO_2 conductivity decreases slightly. Scandia-stabilized zirconia (ScSZ) has shown higher conductivity and better stability compared with YSZ [283,292], but it is costly and not easily available [283]. The most effective material that can be used as an electrolyte in SOFC does depend on its operating temperature. While cerium gadolinium oxide (CGO) or GDC shows higher conductivity at LTs as compared to YSZ or ScSZ [292], its mechanical stability, mixed ionic and electronic conduction behavior at low oxygen partial pressure, and cost and availability of gadolinium are the major drawbacks. Some other useful oxides are GDC, SDC, cerium oxide doped with yttrium (YDC), and cerium doped with calcium (CDC). While perovskite-based electrolyte materials such as lanthanum gallate ($LaGaO_3$) doped with Mg on the Ga site and Sr on the La site $(La_{1-x}Sr_x)$ $(Ga_{1-y}Mg_y)$ O_3[283,292] have higher ionic conductivity at LT, its phase stability, evaporating property of Ga at low oxygen partial

TABLE 4.6
Comparison of Common Electrolyte on their Working Temperature [283]

Temperature Operational	High (800°C–1,000°C)	Intermediate (600°C–800°C)	Low (< 600°C)
Electrolyte	YSZ	SDC, GDC	SDCC
Descriptions	High ionic conduction (~0.1S/cm) within the operating regime The thin electrolyte is preferred to offset the ohmic resistance below ~800°C	High ionic conduction (0.1 S/cm) within the operating regime Drastic performance loss at < 600°C due to chemical and mechanical instabilities, i.e., reduction of $Ce^{4+} \rightarrow Ce^{3+}$	Superionic conductions of H^{+}/O^{2-} < 600°C Resolve issues faced by ceria-based electrolyte The lower activation energy for charge transfer due to the existence of proton conduction H^{+}

pressure, and mechanical stability are questionable. It is also incompatible with NiO, a material normally used for the anodes. In general, the most effective material that can be used as an electrolyte in SOFC depends on its operating temperature. SOFC is often separated into LT-SOFC , IT-SOFC, and HT-SOFC. Table 4.6 compares common electrolyte to their working temperatures. In IT-SOFC, the contribution of grain boundary increases [293]. This is often achieved with the fabrication of nano-structured materials. Fergus et al. [294] examined the problem of the low operating temperature of the zirconia–ceria- and lanthanum gallate-based electrolyte materials. Ceria is another doping material of fluorite structure with good ionic conductivity at low operating temperatures and low polarization resistance in comparison with zirconia [283,294].

An alternating layer of nanostructure on ceria and zirconia has excellent ionic conductivities and ionic mobility [294]. Ceria-based electrolytes have higher ionic conductivity than YSZ at lower operating temperature, chemical inertness, and thermal expansion match with high-performing cathode materials such as LSCF and LSM. It also shows enhanced performance when used in composite electrodes. Ramesh et al. [295] showed that the ceramic composition $Ce_{0.84}(Gd_{0.5}Pr_{0.5})_{0.16}O_2$ gave the highest ionic conductivity (1.059×10^{-2} S/cm) at operating temperature of 500°C. Huang et al. [296] synthesized $Ce_{0.8}Sm_{0.2}O_{1.9}$ (SDC) carbonate electrolyte, and this electrolyte showed high ionic conductivity at LTs of 400–600°C and had the ability to conduct both oxygen and proton ions at similar temperatures and chemically stable, which is a more favorable feature for LT-SOFC. Zhu et al. [297] prepared mixed RE carbonates for LT-SOFC. Leng et al. [298] prepared $La_{0.6}Sr_{0.4}Co_{0.2}Fe_{0.8}O_3$ (LSCF) powder by glycine nitrate combustion method. Ferreira et al. [299] showed that at temperatures above 700°C and at low oxygen partial pressures, ceria-based electrolytes resulted in a decreased OCV and internal short circuiting. Thus, pure ceria has a serious problem with degradation in performance with time at elevated temperature [283]. The working mechanism of the SDC carbonate superionic transport due to proton and ionic conductive (coexistence) phases is as shown in Figure 4.25.

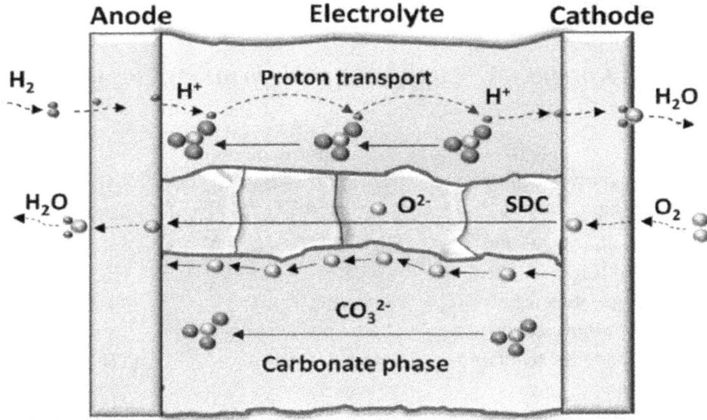

FIGURE 4.25 Diagram of ionic and proton conductions [283].

4.7.2 Materials for Anode Systems

In SOFC, hydrogen fuel reacts with the oxygen ion in a reducing environment and varying temperature (973–1,273 K) at the anode–electrolyte interface. The main characteristics of SOFC anodes should be (1) high electronic conductivity, (2) sufficient electro-catalytic activity for fuel oxidation reactions, (3) chemically stable and thermally compatible with adjacent cell components (electrolyte/interconnector) and sufficient porosity for efficient gas transportation, (4) operating in reducing atmosphere, and (5) matching TEC with adjacent cell components.

The best material for the anode depends on the SOFC operating temperature. Some of the metals such as Ni, Co, Cu, Ru, and Pt are those that have better catalytic activity for hydrogen–oxygen recombination reaction, but they have too high TEC as compared to the electrolyte material. To reduce TEC, anode materials are usually mixed with ceramic (electrolyte) material. Among all metals, Ni is widely used as the anode material in SOFC due to its low cost and high catalytic activity. The minimum amount of Ni for percolation in the YSZ matrix is 30%. The most extensively used anode materials for SOFCs are Ni-YSZ, Ni-GDC, and Cu-GDC (GDC). The metallic phase in an anode serves a dual purpose, namely catalyst and electrical conduction path and ceramic (electrolyte) component, and serves the purpose of TEC matching with that of electrolyte material. Due to this mixed phase, the anode material is also referred to as "cermet" (ceramic+metal). The TPB length directly affects the electrode performance. At high temperatures, an anode with high nickel content degrades fast due to coarsening of nickel particles.

Ni-YSZ cermet material is used for anode because of its unique properties such as high order of porous in nature, high electrical conductivity, structure stability, and compatibility of thermal expansion with solid-state material [283,287]. Ni-YSZ-based anodes contacted with hydrocarbon fuels have contamination issues and result in carbon deposition. Mahato et al. [287], Jacobson [300], and da Silva and de Souza [301] found that the addition of alumina (Al_2O_3), silver (Ag), and niobium oxide

(Nb_2O_5) can successfully modify the Ni-YSZ cermet properties. Song et al. [302] prepared a three-layer structured anode of Al_2O_3-YSZ, Al_2O_3-NiO, and NiO-YSZ layers. The results indicated a significant improvement in electrical conductivity and 39% increase in stability compared with the samples without alumina. Wang et al. [303] reported that adding alumina to Ni-YSZ cermet enhanced coking resistance and FC efficiency. Using a higher amount of alumina reduced the electrical conductivity by 17.3% and the device operated in a stable manner at 750°C for 130 hours. Tian et al. [304] doped Ni-YSZ cermet with silver (Ag) and observed excellent electrochemical performance and achieved high PD. The contamination problems were controlled by the reduction in carbon fuel. Gd or Sm-doped ceria (SDC) (CGO or CSO) and Mg-doped lanthanum gallate (lanthanum strontium gallium magnesium oxide [LSGM]) are also extensively used for the applications of FC systems [283]. The minimum operating temperature of SOFCs based on CGO or CSO and LSGM was ~ 550°C, the desired thickness is 10 μm, and ionic conductivity is 1×10^{-2} S/cm [300]. Perovskite oxide electrode materials containing lanthanum have the reacting property at high operating temperature, which aims for the formation of layers ($La_2Zr_2O_7$) with high resistivity [283,300]. Perovskite-based cathodes have a high order of compatibility with LSGM and higher ionic conductivity with lanthanum transition oxide [283,300]. Ceria-doped-based rare earth (0.1–0.2 Sm_2O_3 or Gd_2O_3) composite has higher ionic conductivity at low operating temperature.

With natural gas as a fuel, carbon deposition on nickel catalyst reduces its catalytic activity. An alternate anode, tidoped YSZ, used with nickel showed improved thermal stability and better electrical conductivity and a lower degradation at 1,000°C [283]. The doped ceria shows good catalytic activity for carbon oxidation than YSZ. Hence, fuel cell with nickel–ceria anode operating on hydrocarbon fuels have resulted in decreased carbon deposition at the anode. Ceria has some electronic conductivity contribution along with ionic conductivity, which helps in increasing electronic conductivity contribution in anode performance. $La_{1-x}Sr_xCr_xMn_{1-x}O_3$ (LSCM) material has also been established as the anode for SOFC [283]. The advantage of this material is that it has good electrochemical activity in both cathode and anode environment and compatibility with many solid electrolytes. This material has very low ionic transport and low electrical conductivity, which can be improved by YSZ/ceria addition. Due to dual catalytic activity (anodic and cathodic), LSCM can be used as dual electrode in a single-chamber SCFC.

The electrochemical performance of SOFC also depends on the microstructure and the method of fabrication of anode. To maximize the performance, a large surface area of TPB is needed. The electrochemical reaction takes place at the TPB, which is the point of contact between electronic and oxygen-ion conductor and gas as shown in Figure 4.26a. A prominent porous microstructure facilitates quick gas transportation and reaction by-product [283,305]. Besides, anodic materials must be good in stability, high order of electronic conductivity, be efficient thermally with other components of the cell, and high electrocatalytic activity. However, utilizing noble metals has shown the extended lengths of the TPB electrodes, and the charge transfer accelerates significantly to oppress polarization resistance as shown in Figure 4.26b. All of these factors combined to form a high-performance anode by minimizing the polarization losses.

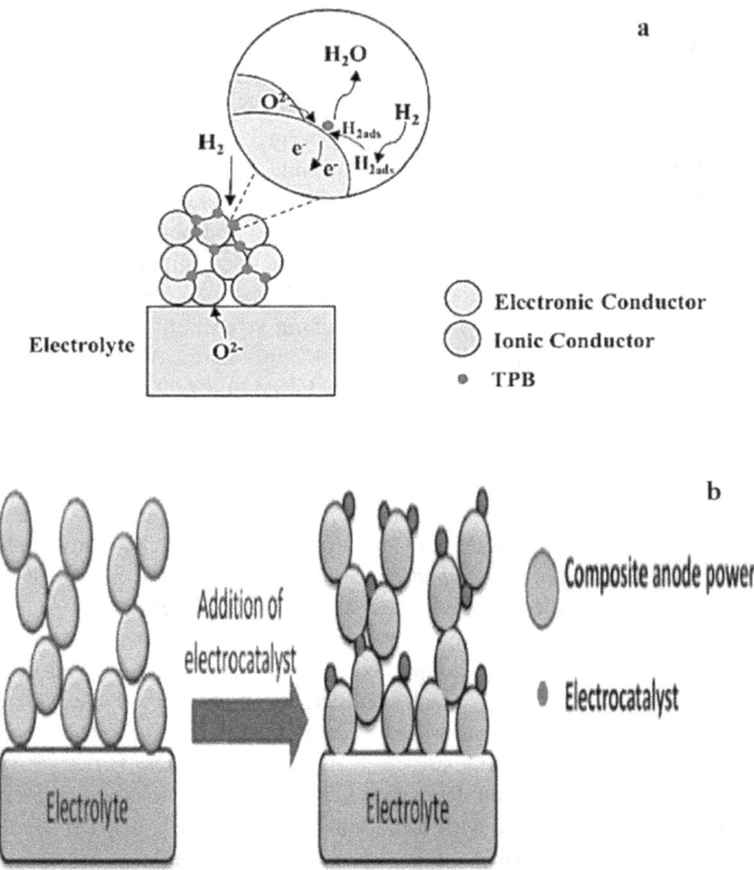

FIGURE 4.26 (a) TPB anode material. (b) Schematic diagram of metallic electrocatalyst on anode composite substrate [283].

LSCM composite material is an effective redox stable for both an anode and a cathode. LSCM is conductive and stable in both reducing and oxidizing atmospheres with excellent performance. Bastidas et al. [306] reported that the usage of LSCM composites for both electrodes is possible to construct the symmetrical SOFC with this specific property. Liu et al. [307] prepared Ni exsolved Ba $(Ce_{0.9}Y_{0.1})_{0.8}Ni_{0.2}O_{3-\delta}$/GDC novel composite anode by solution impregnation. The prepared composite shows long-term stability and good electrochemical performance in the presence of CH_4 fuel. The anode composition of $Sm_{0.2}Ce_{0.8}O_{1.9}/Co_{0.5}Fe_{0.5}$-$Sm_{0.2}Ce_{0.8}O_{1.9}$ achieved 1,200 mW/cm^2 PD at T = 800°C in the presence of fuel 3% H_2O–H_2/O_2 with the cathodic material of SCF and electrolytic material of LSGM. Lu et al. [308] and Wei et al. [309] showed that for perovskite anode materials Sr_2CoMoO_6 (SCMO), the highest PD was 1,017 mW/cm^2 for LDC. Cowin et al. [310] showed that the anode material with 70 vol% Ni-SDC had the highest electronic conductivity (~4,000 S/cm) at $T = 800°C$. Mumtaz et al. [311] showed that $Al_{0.1}Mn_{0.1}Zn_{0.8}O$ (AMZ) and $Al_{0.1}Mn_{0.1}Ni_{0.1}Zn_{0.7}O$ (AMNZ) had ionic and electronic conduction suitable for anode materials in LT-SOFCs. Benamira et al. [312] showed that $Ba_{0.5}La_{0.5}In_{0.3}Ti_{0.1}Mn_{0.6}O_3$ (BLITIM) and $Ba_{0.5}La_{0.5}Ti_{0.3}Mn_{0.7}O_3$ (BLTM) had the electronic conductivity of 11.3 S/cm and 13.4 S/cm at $T = 700°C$ under the oxidative atmosphere of air, respectively. Moura et al. [313] indicated that porous metals such as Fe, Ag, Pt, Co, Ni, Ru, and Mn group of electrode materials are beneficial for electronic conductivity and fuel permeability for the utilization as anode materials.

Sinha et al. [314] confirmed that titanium oxy-carbide is stable in reducing conditions and compatible with $Ce_{0.9}Gd_{0.1}O_{3-\delta}$ at intermediate operating temperature electrolytic material. Rossmeisl et al. [315] indicated that Fe, Ni, Ru, Mn, Co, Cu, Au, Ag, Rh, Pt, and Pd may be suitable for SOFC anodes. Yang et al. [316] examined $Sr_{2-x}La_xFeMoO_{6-\delta}$ ($0 \leq\leq x \leq\leq 1$) (SLFM), for SOFC anode materials. Verbraeken et al. [317] synthesized $La_{0.20}Sr_{0.25}Ca_{0.45}TiO_3$ ($LSCT_{A-}$) perovskite-type structure with co-doping of Ca and La at A-site by the conventional solid-state route. Steiger et al. [318] synthesized metal oxides of $La_{0.3}Sr_{0.55}TiO_{3-\delta}$ (LST) and LSTN mixed perovskite-type structure for SOFC anode application by the citrate gel method. Zha et al. [319] prepared a single-phase complex anodic material pyrochlore $Gd_2Ti_{1.4}Mo_{0.6}O_7$ by solid-state reaction method. Li et al. [320] successfully synthesized NiO-YSZ ceramic anode material by dry-pressing method with different concentrations of Bi_2O_3, increasing the concentration of Bi_2O_3, reduced sintering temperature, and increased relative density, weight loss, and bending strength. Dong et al. [321] successfully synthesized Sn-doped double perovskite $PrBaFe_{(2-x)}Sn_xO_{5+\delta}$ ($x = 0 - 0.3$) anodic material by a combustion method. Shaheen et al. [322] synthesized $Cu_{0.5}Sr_{0.5}$ (CS) and $La_{0.2}Cu_{0.4}Sr_{0.4}$ (LCS) mixed metal oxide nanocomposites by the Pechini method. The electrical conductivity of various types of fluorite oxides can be arranged in an order such as $Bi_2O_3 > CeO_2 > ZrO_2 > ThO_2 > HfO_2$ [18]. Bi_2O_3 is predominantly an electronic conductor at room temperature. However, at high temperature δ-Bi_2O_3 phase with fluorite-related structure has the maximum known oxide-ion conductivity $\sigma_0 \approx 2.3$ S/cm at 1,063 K [323]. In terms of conventional anode material, there are generally Ni-based, including Ni-YSZ, Ni-GDC, and Ni-SDC, and perovskite-based materials, such as $Sr_2FeMoO_{6-\delta}$ (SFM) and $La_{0.7}Sr_{0.3}Cr_{0.5}Mn_{0.5}O_{3-\delta}$ (LSCM) [324].

4.7.3 Materials for Cathode Systems

At the cathode and electrolyte interface, oxygen/air is reduced to oxygen ions with the help of a cathode catalyst and two electrons arriving from the external circuit. The cathode operates at 1,273 K in an oxidizing environment (air/oxygen) and also contributes to the ORR. The main characteristics of SOFC cathodes should be as follows: (1) They should have sufficient electronic conductivity, (2) thermal and chemical stability during cell operation and cell fabrication, (3) TEC should matches with cell components, (4) compatibility and minimum reactivity with adjacent cell components, and (5) low cost.

ABO_3 perovskite structures have unique properties and are integrated with A- and B-cations. The electronic properties of perovskite material can be altered by changing the order of distortion and by a partial replacement in the sub-lattices of both A- and B-cations. The conductivity of the perovskite depends on the structure distortion. The selection of perovskite-type cathodic material based on transition metal oxide depends on the variation of oxidation state. $LaMO_3$ perovskite-type transition metal oxide where M = Mn, Cr, Co, Ni, and Fe of p-type conductivity is stable in the oxidation atmosphere in a certain range of temperature [283]. The conductivity can be enhanced by substituting heterovalent La^{3+} with Sr^{2+}, Ca^{2+}, or Ba^{2+}.

TEC is another important parameter for cathodic material in SOFC technology, which should be matched with other components of FC, and it depends on the electronic or ferroelectric magnetic properties and chemical composition of the perovskite-type oxide and its increase in temperature. TEC also depends on thermally activated transitions of different states of spin such as $LaCoO_3$, which has higher TEC in contrast to $LaFeO_3$ or $LaNiO_3$. Cobalt (Co)-perovskite has high TEC, and it is highly reactive with the electrolytes of YSZ-based. To avoid reactivity, a protective layer of GDC can be deposited between the YSZ electrolyte and cathode material, and the deposited GDC layer is inert with respect to cobalt (Co)-oxide at the temperature of deposition. To avoid other complexities caused by diffusion via the GDC layer, cobalt can be partially replaced by another B-site cation such as Mn, Ni, Fe, or Cu [283].

Doped lanthanum manganites and doped lanthanum cobaltites are widely used as cathode materials. $La_{1-x}Sr_xCoO_{3-\delta}$ composite has high conductivity and TEC at 800°C and high catalytic activity and high O_2 self-diffusion coefficient [283]. Manganites ($Ln_{1-x}A_xMnO_3$) (Ln-lanthanides, A = Ca, Sr, Ba, Pb) ($LaMnO_3$) work well at an operating temperature above 800°C due to its high electronic conductivity and equivalent TEC with many electrolyte materials (e.g., YSZ and GDC). It also has good catalytic activity for ORR at the cathode. LSM is generally used as a cathode material for FCs operating at 800–1,000°C. For LT operation, lanthanum strontium ferrite (LSF), $(LaSr)(Fe)O_3$, lanthanum strontium cobaltite (LSC), $(LaSr)CoO_3$ and samarium strontium cobaltite (SSC), and $(SmSr)CoO_3$ can be good electrode materials [283].

Sun et al. [324] showed that $NdB_{0.96}CO$ oxide appeared to have enough porosity and appropriate grain size to be established as a potential cathodic material for IT-SOFCs technology. Choi et al. [325] showed $LnBa_{0.5}Sr_{0.5}Co_{2-x}Fe_xO_{5+\delta}$ (LnBSCF), where Ln = Pr and Nd; x = 0, 0.25, 0.5, 0.75, and 1.0, or co-doped PBSCFO ($PrBa_{0.5}Sr_{0.5}Co_{2-x}Fe_xO_{5+\delta}$) has fast ionic diffusions through pore channels, high degree of catalytic activity at LTs, good stability, and splendid compatibility with electrolyte materials under the FC operating conditions. Ma et al. [326] showed that for ceramic oxides of $La_{0.5-x}Pr_xSr_{0.5}FeO_{3-\delta}$ (x = 0, 0.25, 0.5) hydration energy was decreased with the increasing concentration of Pr.

SDC ($Sm_{0.2}Ce_{0.8}O_{1.8}$) as an electrolyte and $Sm_{0.5}Sr_{0.5}CoO_3$ (SSC) as porous cathodes were also successfully used in SOFC. The thickness of the SDC electrolyte was about ~ 25 μm, and NiO-SDC was used as an anode by co-pressing technique to form a bilayer structure. The study showed that 950°C firing temperature was more suitable for the application of SDC cathodic systems. An appropriate selection of

SDC: SSC ratios gave an enhanced catalytic performance. Dai et al. [327] added Co in SDC and successfully prepared $Sm_{0.5}Sr_{0.5}CoO_{3-\delta}$ (SSC) powder for the cathodic application. Baumann et al. [328] synthesized two materials, $Ba_{0.5}Sr_{0.5}Co_{0.8}Fe_{0.2}O_{3-\delta}$ (BSCF) and $(La_{0.6}Sr_{0.4})_{0.9}Co_{0.8}Fe_{0.2}O_{3-\delta}$ (LS09CF), and showed that the substitution at A-site of La by Ba improved surface exchange kinetics. A change in the ratio of Co/Fe between 0 and 1 at around 750°C did not affect surface exchange kinetics. Kim et al. [329] prepared $NdBa_0Sr_{0.5}Co_{1.5}Fe_{0.5}O_{5+\delta}$ (NBSCF) and showed that the triple-conducting oxide ($H^+/O^{2-}/e^-$) cathode NBSCF is an effective candidate for IT-SOFC applications. Wang et al. [330] showed that SCN ($SrCo_{0.9}Nb_{0.1}O_{3-\delta}$) cathodes are suitable materials for proton-conducting applications. When $BaCe_{0.4}Zr_{0.4}Y_{0.2}O_{3-\delta}$ (BCZY442) electrolyte was employed with these SCN cathodes to form proton-conducting SOFC, it resulted in the maximum PD of 348 mW/cm^2 at the operating temperature of 700°C.

While Co-containing cathodes are a well-known ability to operate at high temperature in SOFC technology, Co-free cathodes were the best alternative way for the higher electrochemical efficiency cells in the range of IT to LT (IT-LT) [283]. Ding et al. [331] prepared $GdBaFe_2O_{5+\delta}$ (GBF), a promising perovskite Co-free cathode for IT-SOFC, and showed good catalytic activity. Lee et al. [332] showed that the mixing of GDC and $Ca_2Fe_2O_5$ particles reduced the TEC values. Zhou et al. [333] showed that Co-free $SrFe_{0.9}Nb_{0.1}O_{3-\delta}$ (SFN) cubic perovskite oxide gave the maximum PD of 407 mW/cm^2 at 800°C. Jiang et al. [334] also showed enhanced performance by Co-free SNF cathode material. Yu et al. [335] prepared $SrFe_{1-x}Ti_xO_{3-\delta}$ (SFT, $x=0.00$–0.15) oxides by solid-state reaction. Ling et al. [336] synthesized Co-free cubic perovskite oxide $Sm_{0.5}Sr_{0.5}Fe_{0.8}Cu_{0.2}O_{3-\delta}$ (SSFCu) by combustion method. Zhu et al. [337] enhanced the mismatch of TPB and TEC by mixing $Sm_{0.6}Sr_{0.4}FeO_{3-\delta}$ (SSF) with $Ce_{0.8}Cm_{0.2}O_{2-\delta}$ (SDC) electrolyte. In perovskite structure, materials, such as Mo, Ti, and Cu-substituted cobalt at B-cations, showed an excellent performance compared with cobalt-enhanced cathodic material.

Duan et al. [338] applied Y and Zr co-doped $BaCo_{0.4}Fe_{0.4}Zr_{0.1}Y_{0.1}O_{3-\delta}$ (BCFZY0.1) perovskite structure for LT-SOFCs. It showed high O_2 reduction reaction activity, long-term stability, and lower order of activation energy, LT response, large lattice parameter, and excellent compatibility with Ce-based SOFC electrolytes. Zhang et al. [339] showed an improved electrochemical performance by $SrFeO_{3-\sigma-\delta}F_\sigma$ (SFF_σ, $\sigma=0$, 0.05, and 0.10) and $SrFe_{0.9}Ti_{0.1}O_{3-\sigma-\delta}F_\sigma$ ($SFTF_\sigma$, $\sigma=0$, 0.05, and 0.10) perovskite oxy-fluorides. The prepared perovskite oxy-fluorides also showed high catalytic activity. Li et al. [340] used Ta and Nb co-doped $SrCo_{0.8}Nb_{0.1}Ta_{0.1}O_{3-\delta}$ (SCNT) perovskite material for the cathodic application. Doping of Nb and Ta produced desirable ionic mobility and O_2 vacancies at $T \leq\leq 500$°C. Hussain et al. [341] showed that $La_{0.54}Sr_{0.46}Fe_{0.80}Cu_{0.20}O_{3-\delta}$ composition showed excellent performance and good conductivity.

Finally, materials used for cathode, anode, and electrolyte depend on SOFC application. Golkhatmi et al. [323] showed that in the case of oxide-ion SOFCs (O–SOFCs), electrolytes are composed of perovskite or fluorite structure with oxygen deficiency to provide oxygen pathways by oxygen vacancies. Zirconia-based (e.g., YSZ), ceria-based (e.g., GDC), and lanthanum gallate-based (e.g., $La_{0.8}Sr_{0.2}Ga_{0.8}Mg_{0.2}O_{3-\delta}$ (LSGM)) electrolytes are the most common examples for O–SOFCs. However, proton-conducting SOFCs (H–SOFCs) transport H^+ instead of O^{2-} and there is no generated water molecule at the anode side, which brings several advantages such as high performance at lower operating temperatures and better durability in using hydrocarbon fuels. For this type, $BaCeO_3$- and $BaZrO_3$-based perovskites, such as $BaCe_{0.7}Zr_{0.1}Y_{0.2}O_{3-\delta}$ (BCZY), are the most popular electrolytes [323]. Cathodes are also different for O–SOFCs and H–SOFCs. O–SOFC cathodes are mainly perovskite-based (ABX_3), such as $La_{1-x}Sr_xMnO_3$ (LSM) and $La_{0.6}Sr_{0.4}Co_{0.2}Fe_{0.8}O_3$ (LSCF), and layered perovskite-based ($AA'B_2O_{5+\delta}$), such as $NdBa_{1-x}Ca_xCo_2O_{5+\delta}$ (NBCaCO) cathodes, where A, A′, and B are cations but with different radius and X is an anion (mostly oxide) connected to cations [323]. Cathode function in H–SOFCs requires three charge carriers of O^{2-}, H^+, and e^- to show acceptable performance. Therefore, mixing the proton-conducting oxides with O^{2-} conductors can be the key to provide an effective electrode reaction [323]. $Ba_2YCu_3O_{6+\delta}$ (BYC), $Ba_xCo_{0.4}Fe_{0.4}Zr_{0.1}Y_{0.1}O_{3-\delta}$ (BCFZY), $BaFe_{0.8}Zn_{0.1}Bi_{0.1}O_{3-\delta}$ (BFZB), $Pr_2NiO_{3.9+\delta}F_{0.1}$ (PNOF), and $PrBaCo_{2-x}Ta_xO_{5+\delta}$ (PBCT) are some of these examples [323].

4.7.4 Applications of Nanomaterials for SOFCs

The nanoscience and nanotechnology play a key role to resolve many issues because of the modification of properties in SOFCs. Recently, researchers focus on methods and materials to prepare an excellent performance of nanostructures SOFCs. Nano-size of materials has unique properties for enhancing grain boundary ionic conductivity of electrolytes in SOFC devices. Arico et al. [342] reviewed nanostructured materials for energy storage devices, the development of low operating temperature FCs (<<200°C), fuel reforming and hydrogen (H_2) storage technology, the development and dispersion of non-precious and precious nonmetallic and metallic catalysts, and manufacturing of MEAs. Nanosized powders of ceria-based such as CGO, YDC, SDC, and YSZ (8% Y_2O_3-ZrO_2) induced the reduction in firing temperature in the fabrication of cells because of the different sintering properties of those polycrystalline powders. In addition, the mixed ionic and electronic conduction properties of nanosized Ce optimized the charge transfer reactions at the interface of electrodes and electrolytes. Nanostructure-based systems with larger surface area and grain boundaries also facilitate the enhancement of the number of mobile defects in the region of space charge.

Yuan et al. [343] focused on the preparation of Ce-based and controlled synthesis nanomaterials, crystal plan orientation, particle size, and tailor shape and assembled them in an effective way. Dong et al. [344] synthesized a nanosized electrolyte material of $Ce_{0.8}Gd_{0.2}O_{2-\delta}$ and $Ce_{0.79}Gd_{0.2}Cu_{0.01}O_{2-\delta}$, which is a cubic fluorite crystalline structure and porous foamy morphology. Bellino et al. [345] prepared a cobaltite nanotube cathode of $La_{0.6}Sr_{0.4}CoO_3$ showing low polarization resistance and high porosity. Martinelli et al. [346] synthesized NPs of $La_{0.8}Sr_{0.2}MnO_3$ cathode and investigated the effect of agglomeration and the variation of particle size in SOFCs.

Zhi et al. [347] synthesized LSCF nanofibers for the applications of the cathode in IT-SOFCs with YSZ electrolyte. The 3D nanofiber cathodic network has various advantages such as (1) high percolation, (2) high porosity, (3) continuous charge transportation, and (4) excellent thermal stability under a similar operating temperature. Ishihara et al. [348] introduced nanosized materials in their review article for the application of electrodes in IT-SOFCs. Yoon et al. [349] deposited a thin film of vertically aligned nanocomposite (VAN) structured between the GDC electrolyte, and the thin layer of $La_{0.5}Sr_{0.5}CoO_3$ (LSCO) cathode has achieved higher efficiency of thin-film SOFCs. Evans et al. [350] prepared a cathode by spin-coating suspension NPs of $La_{0.6}Sr_{0.4}CoO_{3-\delta}$ (LSC) followed by salt-assisted spray pyrolysis.

Nanocomposites are also used in SOFC. Recently, YSZ was replaced with much more effective Ni-GDC nanocomposite material [283]. Chavan et al. [351] studied the microstructure of NiO-GDC nanocomposite material, showing decreased activation energy with increased content of NiO. Fu et al. [352] also studied 65–35 wt% of NiO-GDC nanocomposite. The prepared NiO-GDC nanocomposite showed decreased PD with decreasing temperature from 650 to 500°C. Gil et al [353] synthesized NiO-GDC 50:50 wt% nanocomposite powder. Chen et al. [354] investigated NiO-GDC film deposited on GDC substrate by electrostatic-assisted ultrasonic spray pyrolysis technique. Ding et al. prepared a single cell of LSCF NPs printed over GDC-NiO-GDC by hydroxide coprecipitation method [355].

Guan et al. [356] successfully investigated a 3D porous type of microstructure of Ni-YSZ anode material by X-ray nano-tomography. Boldrin et al. [357] produced SOFC anodic material with $NiNO_3$ (nickel nitrate) solution by impregnating GDC scaffolds. In the heat treated, these scaffolds showed better performance at low operating temperature (<<700°C) in H_2 and excellent performance at all temperatures utilizing syngas with maximum PD of 0.15 W/cm^2 at $T=800$°C. Pelegrini et al. [358] reported that nanostructured Ni/Cu-YSZ material from nano-powders improves the TPB region for anodic applications in SOFCs. Kwon et al. [359] reported Co–Ni alloy NPs and synthesized $PrBaMn_{1.7}Co_{0.1}Ni_{0.2}O_{5+\delta}$, a double-layered perovskite by the Pechini method. Cavallaro et al. [360] investigated the deposition of $La_{0.8}Sr_{0.2}CoO_{3-\delta}$ dense films at various temperatures by pulsed laser deposition on silicon (Si) substrate for the applications of cathodic material. Depending on the different deposition temperatures, the amorphous or textured polycrystalline film was obtained. It was observed that O_2 diffusion coefficient had occurred in an amorphous film, which is four times more in comparison with crystalline materials and associated improvement of the surface exchange coefficient.

4.7.5 Mechanisms for Component Degradation in SOFC

The severe working conditions of SOFC have several diverse degradation processes, which arise from each component and their interactions, making it challenging to fulfill the long-term stability requirements. Degradation is commonly characterized as loss of performance, and the degradation rate is generally stated as the voltage loss per 1,000 hours, especially in stacks. Slodczyk et al. [361] showed that change in area-specific resistance (ASR) is another measure for reporting the degradation of single cells. Batfalsky et al. [362] noted that evaluating the degradation process in SOFC is quite complicated as long-term studies are needed, and the operation factors (temperature, fuel impurities, current density, etc.) affect the procedure. A summary of each cell component's main degradation mechanisms (cathode, electrolyte, anode) and a brief overview of stack elements (interconnects and sealants) are presented as follows [363–365].

4.7.5.1 Cathode

Cathode degradation mechanisms can be classified into three main groups [323], which include poisoning (by Cr, S, CO_2, and humidity), microstructural deformation, and chemical and thermal strains (delamination). The most rigorous degradation in LSCF cathodes is Cr poisoning, caused by the Cr evaporation from the unprotected metallic interconnects. Cr poisoning can happen in two potential ways for SOFC cathodes, chemical and electrochemical. In the chemical one, the volatile Cr species (CrO_3 or other gaseous kinds) directly counter the cathode surface and its segregated ions, which results in precipitated species. These species not only corrupt the electrical properties but also hinder the gas pathways of the cathode. This mechanism also increases the degradation effect of the cathode by material segregation. Operating conditions, such as temperature, humidity, and current density, can alter the Cr poisoning intensity. The lower temperature and higher humidity and current density enhance Cr poisoning and cathode degradation rate [323].

Sulfur (S) poisoning in cathodes was first determined by Yukokawa et al. [366]. They showed that the S deposition within the cathode material was strongly associated with the cathode high overpotentials. Even ppm level of SO_2 can affect the cell/stack performance and reduce the SOFC operation length [323]. In LSCF cathodes, S poisoning leads to fine-grain $SrSO_4$ precipitation in the grain boundaries of the cathode/electrolyte interface. This $SrSO_4$ deposition can also happen on the cathode surface with a homogenous distribution [323]. The decrease in Sr and Co components due to the formation of $SrSO_4$ and precipitation of $CoFe_2O_4$ also degrades ORR. The electrochemically active surface area is also reduced due to the formation of secondary phases such as $SrSO_4$, $La_2O_2SO_4$, and Co oxide due to the absorption of SO_2 on the cathode surface. Several parameters, including temperature, P_{SO2}, Sr content, and P_{O2}, affect the S poisoning process. Perovskite surface can also adsorb the CO_2, causing carbonate formation on the surface, increased polarization, reduction in O_2 adsorption, and ORR activity. The cathode's electrocatalytic properties are also impressed by the competition between O_2 and CO_2 adsorption on the cathode [323].

The change in cathodic microstructure caused by cathodic overpotential can also result in degradation. This phenomenon usually happens when the cathode has cations with much different mobility, leading to component separation, called "kinetic demixing" [323]. Sr segregation on the cathode surface affects the oxygen exchange kinetics and reduces ORR [323,367]. These Sr species can also react with surrounding gaseous phases such as Cr, CO_2, and humidity, forming insulating layers of $SrCrO_4$, $SrCO_3$, and $Sr(OH)_2$ [323]. The $SrCrO_4$ formation also causes Sr deficiency at the A-site, which reduces ORR activity. Sr reduction in perovskite lattice also degrades the electrical conductivity. Sr enrichment in the interface can react with YSZ and result in insulating phases such as $SrZrO_3$, which induces the increment in the cell's ohmic resistance [323].

Particle coarsening in high-temperature SOFC can also cause performance degradation because the reduced absorbent surface area increases polarization resistance. The chemical strain is also a degradation mechanism in cathode material due to the oxygen non-stoichiometry. With the formation of oxygen vacancies in the lattice, the B-site cations' overall valance number reduces. This reduction enlarges the B-site ionic radius and causes lattice expansion, resulting in a TEC mismatch between the cathode and the electrolyte. If the mismatch becomes too large, the electrolyte can be broken by bending. Besides, this difference in TEC can induce the applied thermal stress during the operation and result in component delamination. Delamination causes prolonged current pathway, hindering charge conduction, and destruction of reaction sites. When delamination happens, the current is localized in an intact area. This current localization contributes to higher cathode loss of activation and higher ohmic loss of electrolyte [323].

The poisoning effect can be reduced by different types of surface engineering measures. A protection layer against Cr diffusion and doping techniques are used to enhance the chemical and structural stability of the cathode material. A functional interlayer such as $Ce_{0.8}Gd_{0.2}O_{2-\delta}$ (CGO) can slow down the Sr diffusion from the cathode through the electrolyte if the barrier is dense enough [323]. The degradation process depends on the operating conditions. Moderate temperature and low polarization prevent any major overpotential and deterioration of the cathode. S content can be reduced by applying a chemical filter for the air inlet of the SOFC system. Moreover, a trapping layer on the cathode surface can catch the SO_2 and prevent degradation. Some additives such as $Sm_{0.5}Sr_{0.5}CoO_{3-\delta}$ (SSC) nanofibers, $BaCeO_3$, and BaO, can make the cathode more resistant to S poisoning [323].

4.7.5.2 Electrolyte

The main degradation mechanisms in the SOFC electrolyte are phase transition, impurities, and dopant diffusion and mechanical failures [368]. During the SOFC operation at high temperatures and in harsh atmosphere, a phase change in the electrolyte layer can affect the SOFC performance by reducing the ionic conductivity and phase stability, which eventually causes degradation. YSZ, as the most common electrolyte for SOFC, presents a competitive ion conductivity over a broad range of partial oxygen pressure, good stability under harsh operating conditions, and satisfying mechanical properties under elevated temperatures. However, several microstructural changes due to the long exposure at 1,000°C degrade the electrolyte conductivity and SOFC performance. Golkhatmi et al. [323] showed that the most notable phenomenon is the phase transformation from cubic to tetragonal zirconia, which strongly depends on the Y_2O_3 concentration in ZrO_2. Hattori et al. [369] showed that 9.5YSZ has high conductivity and excellent stability compared with 8, 8.5, 9, and 10YSZ electrolytes. Ionic conductivity faced a decreasing trend with higher dopant content, as the emergence of point defects lowers the defect mobility [323]. Phase transition is also a challenge in Sc_2O_3-stabilized ZrO_2 (ScSZ) electrolyte, a proper candidate for LT-IT operation, which experiences a cubic–rhombohedral–cubic phase transformation at lower temperatures. As the rhombohedral phase has a weaker ionic conductivity, the cell faces an increase in the ASR, causing lower performance [323]. Gao et al. [370] indicated that this transition causes unwanted residual stress in the SOFC stack as well.

Chemical interactions between the electrolyte and cathode are another source of degradation. The chemical interactions give rise to the formation interface and insulating secondary phases [371,372]. In the LSCF/YSZ systems, both the high-temperature sintering process and cathodic polarization cause Sr segregation and $SrZrO_3$

formation, which is undesirable for SOFC performance and durability. In the case of LSCF/GDC systems, the Sr segregation rate is much slower and less destructive than the one in YSZ. Limited and isolated Co accumulation happens in this case and is not as damaging as Sr segregation. Due to the better chemical stability of GDC, a barrier layer from this material may enhance the stability and electrochemical properties of the YSZ-based systems. Still, a highly resistive Ce–Zr solid solution phase forms at the GDC/YSZ interface in elevated sintering temperature above 1,300°C, inducing a severe degradation in SOFC [323]. The reaction between the LSGM electrolyte and cathode material is not similar to the one in YSZ or GCD electrolytes and generally takes place through the interdiffusion of cations rather than forming a secondary phase [323].

While a small amount of interdiffusion of Co, Fe, and Ni is not detrimental to electrolyte performance, extreme interdiffusion will cause degradation in both cathode and electrolyte performance. Therefore, applying a ceria protection layer between the LSGM electrolyte and LSC cathode may stop the Co interdiffusion, but then again the formation of an insulating phase can cause degradation [373]. Electrolyte/anode chemical reactions are less severe than electrolyte/cathode ones. Ni-YSZ, as the most common anode material, has no problems with the YSZ electrolyte. Nonetheless, the formation of a resistive layer may occur between LSGM and Ni-based anodes and cause SOFC degradation [373].

The last degradation mechanism in the SOFC electrolyte is mechanical failure resulting from thermal and chemical stresses. SOFCs are almost stress-free at high working temperatures, but cooling down to room temperature causes residual stresses due to the difference in TECs of cathode/electrolyte or anode/electrolyte. This residual stress introduces crack initiation or delamination, which eventually leads to mechanical failure [323]. Phase transformations are also responsible for residual stresses like the phase changes in ScSZ that experience a partial cubic to rhombohedral and back to cubic transformation through the heating range of 300–500°C. The difference in TEC between YSZ and GDC layers is also another example of delamination and mechanical degradation [323]. The oxidation of Ni-based anodes and the resulting volume change is probably the most crucial example of chemical stress. The penetration of unwanted oxygen to the anode, due to either system leakage or uncontrolled fuel utilization, causes irreversible expansion of Ni-based anodes. Since there is a significant difference between the Ni and NiO volumes, this oxidization creates internal stress, tension in the electrolyte, crack formation, and eventually system failure. The gas permeation through these cracks also speeds up the other degradation mechanism [323,374]. GDC reduction is another major cause of chemical stress as the GDC experiences a volume expansion and eventual cracking, along with the TEC mismatching between the other components [323,375].

4.7.5.3 Anode

Anode degradation mechanisms can be caused by microstructural changes, coking and poisoning, and delamination. The most common microstructural changes in Ni-based anodes are Ni coarsening, Ni migration, and Ni depletion, which are connected to each other [323]. Ni coarsening is known as the most detrimental degradation mechanism in SOFC anode electrodes. The primary reason is surface diffusion along with the interface and is generally related to a kind of "Ostwald ripening" mechanism [323]. Ni particle growth reduces both TBP sites and electrical conductivity, which weakens both performance and stability of the SOFC. Furthermore, the catalytic activity of the Ni decreases due to the loss of specific surface area in larger particles. This mechanism also deteriorates the Ni-YSZ contact and eventually causes the delamination of Ni from YSZ [323,376]. Ni coarsening can also result in Ni migration to the anode surface by evaporation/condensation process and diffusion. In the SOFC operating conditions, at high temperature and water pressure, Ni, O_2, and H_2O react together and form $Ni(OH)_2$, which is generally taking place near the TPB region. Since $Ni(OH)_2$ has a lower melting point than the operation temperature, it would be evaporated, transferred to the surface, and then condensed to Ni atomic form. This Ni migration to the surface brings inconsistency in the Ni content of the anode and causes Ni depletion around the electrolyte/anode interface. This redistribution may affect the TPB length, particle size distribution, porosity, and tortuosity and eventually cause the SOFC degradation. Further, Ni coarsening is responsible for Ni depletion, as the larger particles adsorb the smaller ones [323,377].

Coking due to a number of reactions occurring during reforming including the Boudourd reaction is a major problem for anode degradation. Several factors influence the coking rate, including steam/carbon ratio, anode composition, operation temperature, and current density. Anode coking covers the surface and blocks the TPBs and gas channels and causes mechanical and electrochemical degradation. As the carbon deposition increases, additional pressure can lead to the anode fraction [323]. Kan et al. [378] showed that carbon formation occurs at the beginning of the cell operation. After the carbon deposition, amorphous carbon changes to graphitic carbon, damaging the single cell's cohesive structure [378].

Ryan et al. [379] indicated that contaminants in hydrocarbon fuels including sulfur (S), phosphorous (P), arsenic (As), selenium (Se), chlorine (Cl), and antimony (Sb) may interfere with the anode and degrade the performance and stability of the SOFC. The type and amount of these elements in the hydrocarbon fuel depend on the coal's mine location and their process technique. Trembly et al. [380] indicated that the S poisoning from the hydrocarbon fuels in the anode is arisen from the interaction of H_2S with the anode, creating H_2 and elemental S. Ni particles have a strong tendency to adsorb this elemental S, which causes nickel sulfide (Ni_xS_y) deposition and blockage of active sites along with the redistribution of Ni at the interface. Temperature, polarization, cell configuration, and H_2S concentration impact the degree of S poisoning [323,380]. Zhi et al. [381] showed that P traces in coal lead to nickel phosphide's development at the anode/electrolyte interface and bring irreversible performance loss to SOFC as these species hinder the active sites. In addition to performance failure, the formation of NiP causes stress, resulting in the origination of microcracks in the Ni-free YSZ matrix and mechanical degradation in the anode [323,381].

Even less than 10 ppb amount of As can react with H2 to form AsH3, which is detrimental to SOFC. Furthermore, this element is a notable poison for Ni catalysts due to its strong tendency to react with Ni. The formation of Ni_5As_2 and $Ni_{11}As_8$, determined by temperature, As concentration, flow rate, and exposure time, causes Ni coarsening and Ni migration to the anode surface. These processes induce the loss of electrical connectivity in the anode support and, finally, result in sudden failure. Se (imbedded in some coals) poisoning, which is quite like S poisoning, originates from Se by Ni's surface adsorption and formation of Ni_3Se_2 at the anode/electrolyte interface. This solid phase inhibits the accessibility to active sites and causes SOFC degradation [323]. Cl poisoning is also a severe case since the SOFC practical fuels, such as biofuels, contain Cl compounds in high concentrations. Madi et al. [382] indicated that the presence of Cl, an electronegative species, can prevent the H_2 adsorption on the Ni surface, blocking the TBP region and limiting the electrochemical performance of SOFC. Haga et al. [383] noted a considerable microstructural change in Ni–Sc-stabilized Zr anode by Cl poisoning [323].

Sb is another coal impurity with a wide application as a passivating agent for Ni catalysts in refineries. Cell degradation from Sb poisoning results from two processes, depending on the exposure length, Sb concentration, and applied current density. Ni surface adsorbs Sb, and the electrocatalytic activity of the anode diminishes at the initial step, the same as S and Se poisoning. With a longer exposure duration, the late stage of the degradation, the severe one, begins with the broad formation of solid reaction products, especially NiSb. These products obstruct the electrical conduction pathways between particles (percolation loss), which irreversibly induces ohmic resistance. Furthermore, Sb poisoning leads to Ni coarsening, consumption, and migration to the surface, which are also unfavorable for anode performance [323,384].

Anode delamination is mainly due to oxidation cycling and thermal cycling. The first case causes volume changes in the anode, and the second one is because of the mismatch in TEC of the anode and the electrolyte [323]. Delamination is less common in anode than in cathode due to the similar TEC of anode and electrolyte, but it still exists. It should be noted that the Ni coarsening induces delamination because when the Ni particles grow bigger, the contact area with the electrolyte becomes smaller. Furthermore, delamination causes TPB reduction, which is one of the leading causes of degradation in SOFC [323].

4.7.5.4 Interconnects

The three degradation modes of interconnects are *corrosion, Cr vaporization, and mechanical failures.* Interconnects are a fundamental element in SOFC stacks as they provide electrically conductive pathways among the single cells and aid in separating one cell's anode side fuel from the cathode side air of the next cell in the stack [323]. High-temperature SOFCs use ceramic interconnects developed from semiconductor oxides. The most common of them are $LaCrO_3$-based interconnects, which are p-type semiconductor oxides. However, their application is restricted due to their challenging fabrication method, high price, and inadequate flexural strength. Current SOFCs with lower operating temperatures (500°C–800°C) use metallic interconnects instead of former ceramic ones owing to their lower cost, better electrical conductivity, and more straightforward fabrication processes [385]. Metal alloys such as Fe–Cr alloys, Cr alloys, Ni(Fe)–Cr-based heat-resistant alloys, and austenitic and ferritic SSs are widely utilized as metallic interconnects. However, most metals are affected by oxygen, which causes corrosion. The corrosion not only weakens the mechanical stability of the interconnect but also reduces its electrical conductivity due to the emergence of insulating oxide phases such as Cr_2O_3 and $(Mn, Cr)_3O_4$ [323,386]. Moreover, the simultaneous exposure of SOFC to fuel at one side and air at the other side in ferritic SSs is another cause of the corrosion, called the "dual atmosphere effect" [323]. Yang et a. [387] showed that this degradation is due to the different scale growth caused by the hydrogen transfer through the steel, accelerating the iron transfer and increasing its activity in growing iron oxide phases. "Metal dusting" phenomenon, a serious kind of corrosion in metals and alloys in the carbon-supersaturated gaseous atmosphere at high temperatures, is another hazard to the metallic interconnectors while using carbon-containing fuel gases. Metal dusting results in forming fine metal carbide or pure metal and carbon dust, causing a brittle structure and reducing the interconnector's mechanical strength [323].

Cr vaporization from interconnects causes Cr poisoning in the cathodes, one of the most severe degradation mechanisms in SOFCs responsible for a significant decrement in electrical conductivity by blocking the electrode's active TBP sites [323]. Furthermore, this Cr vaporization induces the Cr depletion in the interconnect, and this depletion below a specific threshold threatens its mechanical strength and structural integrity through the oxidation breakaway [388]. Compared to other metallic interconnects, superalloys such as Ni–Cr- or Ni–Fe–Cr-based alloys provide a lower-scale growth rate, leading to higher oxidation resistance behavior. Nevertheless, Ni-based alloys with enough Cr to obtain a high oxidation resistance show a high TEC, bringing the TEC mismatch between the other SOFC components and resulting in mechanical failure [323].

Zhang et al. [389] showed that the Cr vaporization issue can be addressed by modifying the alloy composition or surface conditions [389]. Stanislowski et al. [390] showed that surface modification is more favorable due to Cr release results from different common Ni-, Cr-, and Fe-based alloys. These results presented that the Cr-dominant oxide generators need to be protected by protective coatings to become free of Cr evaporation [388]. Generally, three types of coating materials are available for SOFC interconnects, including reactive element oxides (e.g., La_2O_3, Nd_2O_3, and Y_2O_3), rare earth perovskites (e.g., LSM and $La_{0.75}Sr_{0.25}Cr_{0.5}Mn_{0.5}O_{3-\delta}$ (LSCM)), and composite spinel oxides (e.g., Mn–Co-based, $NiFe_2O_4$, Cu-based). So far, composite spinel oxides found to be the most potential coating to save the interconnect from the Cr vaporization, enhance the electrical conductivity, and lower the TEC at the same time [323,391].

4.7.5.5 Sealant

In general, sealants face mechanical failure and leakage, corrosion, and poisoning. Bonding or rigid sealants, e.g., glass ceramics, are sensitive to thermomechanical stress during thermal cycles. This susceptibility causes nonlinear behavior in thermal properties, including TEC, viscosity, and porosity, and changes over time, leading to TEC mismatch between the sealant and other stack components, mechanical failure, and eventually leakage [323,392]. Typically, leakage indicates that the anode receives less fuel than planned, which increases the stack's fuel consumption, causing fuel shortage, operating limitations, performance loss, and anode oxidation [168]. Hence, quantifying any changes in the stack leakages throughout the operation is valuable to identify them from other causes of stack voltage degradation, such as increase in contact resistance, S poisoning, or measurement faults. Krainova et al. [393] showed that to lower the risk of mechanical failure and leakage rate, using non-crystallizing sealants, such as SiO_2–Al_2O_3–CaO–Na_2O–ZrO_2–Y_2O_3 systems, can be helpful since they are resistive to the considerable change in the TEC of material. Another solution can be hybrid sealants to develop a seal that takes mechanical characteristics from the compressible core but, unlike typical compressive seals, has meager interfacial leak rates due to the compliant surface coating [323,394]. In this regard, Rautanen et al. [395] developed a hybrid sealant, glass powder-coated Thermiculite 866, with leakage rates of 0.1–0.3 mL m/min, 60%–90% less than the uncoated sealant. This decreased leak rate was due to the conformability of the glass coating that covered the Thermiculite 866 surface defects and blockage of interfacial leak paths by the adjacent elements [395].

Although glass-based sealants show better leakage resistivity, their impurities, such as Si, can be poisonous to the cathode and anode and hinder their surface reactions, causing performance degradation in the SOFC stack. In Si poisoning, Si deposits on the TPBs and cathode surface, inhibiting the ORR [396]. Moreover, deposited Si can react with the cathode surface and change the surface composition by forming insulating phases. In fact, this form of poisoning is most commonly seen in Sr-containing perovskites at high temperatures, forming stable silicates (e.g., Sr_2SiO_4) and blocking the active surface sites [323,396]. In addition to impurities, their constructive cations, such as Ba^{2+} and Zr^{4+}, can become deleterious for the long-term performance of the SOFC. Ba^{2+} has a strong tendency to react with Cr from the ferritic interconnects and create unwanted and insulating $BaCrO_4$. However, Zr^{4+} leans to form bulk crystallization in the glass or glass–ceramic sealants, causing the growth of microcracks and weak mechanical stability in long-term performance [323]. Jiang et al. [397] showed that the presence of alkali cations (Li^+, Na^+, and K^+) in the glass- and glass–ceramic-based sealants makes them more likely to react with other cell components. They can induce cathode Cr poisoning by speeding the Cr vaporization of interconnects.

There are several techniques to control the sealant corrosion, such as lowering the Si content, making Ba^{2+}-, alkaline-, and alkaline earth metal-free sealants, and controlling the Zr^{4+} amount [323,398]. In this regard, Kiebach et al. [398] from DTU developed a CaO and ZnO-rich glass composed of 50 mol% CaO, 20 mol% ZnO, 20 mol% B_2O_3, and 10 mol% SiO_2, named CZBS. This CZBS glass showed no degradation or sealant-related leakage for over 400 hours under dual-phase atmospheres (air/H_2) at 750°C (for the first 100 hours) and 850°C for the rest of the operation.

4.7.5.6 Additional Perspectives on SOFC Durability Studies

SOFCs need to overcome durability and stability challenges to meet a commercial breakthrough. The characterization plays a critical role in better understanding the degradation mechanisms giving new insight to overcome the durability issues. The study tools can be categorized into electrochemical, chemical, and structural characterization. The first one contains I–V measurement, EIS, and calendar life tests. The chemical and structural characterization groups include Raman spectroscopy, Fourier-transform infrared (FT-IR) spectroscopy, X-ray diffraction (XRD), X-ray photoelectron spectroscopy (XPS), secondary ion mass spectrometry (SIMS), thermogravimetric analysis (TGA), differential scanning calorimetry (DSC), dilatometry, and microscopy techniques. These techniques are conducted in situ, ex situ, or both modes, depending on the characterization type, to provide the information as detailed as possible. Modeling is also another cost-effective method to understand SOFC performance better.

It is possible to achieve a better knowledge of the degradation of SOFC via modern characterization techniques as there is a lack of in-depth knowledge of degradation mechanisms. Degradation processes in SOFCs are complicated and depend on many parameters simultaneously. For instance, the role of temperature is still undefined. There are several reports about enhancing durability with increasing temperature, leading to uncertainty about the destructive act of high temperatures [323]. Besides, the short-term behavior cannot be interpolated into long-term behavior as the degradation process is nonlinear [323,399]. Furthermore, the absence of proper *in situ* characterization tools that can work during the electrochemical process of SOFC at high temperatures hinders a deeper understanding of the degradation mechanisms. Another challenge is the limited choice of materials for SOFC components due to the severe working conditions. The difficulty is finding a material with the desired properties for the cathode/electrolyte/anode (mentioned before) while having good compatibility with the SOFC harsh working conditions and other components at the same time [323,400,401]. There is also a lack of systematic standards to study the durability and stability of SOFCs. For instance, there is no set of standards for long-term/short-term operation durations or no particular range for the applied study techniques in current density or voltage amplitude on the laboratory scale. It is necessary to set guidelines and criteria to compare the degradation results accurately.

4.7.6 SOFC Configurations and Commercialization

Mahato et al. [287] indicated that among the different FCs, the SOFC is one of the most efficient technologies for power generation as it is flexible to fuel choice, noiseless, showing low CO_2 emissions, and has a potentially long lifetime of 40,000–80,000 hours. The high working temperature of SOFC, which is necessary to reach an adequate ionic conductivity, provides excellent heat by-products for combined cycle operations or cogeneration of energy. Another merit is their solid-state electrolyte, which is manageable and does not cause corrosion to the cell or handling issues. Furthermore, SOFCs are cost-effective for mass production since they do not use expensive noble metals [323,287]. Considering SOFC harsh operating conditions, such as high working temperatures, redox and thermal cycling, and poisonous atmosphere [323], they require several properties for their components (cathode, electrolyte, anode, interconnect, sealant), such as appropriate conductivity (electrolyte must be an electronic insulator and providing a good ionic conductivity, while electrodes should show a promising electronic and ionic conductivity); acceptable chemical, thermal, morphological, and mechanical stability and compatibility with other components, perfect gas tightness, and high resistivity against oxidation, sulfation, and carbon deposition for interconnects; porous structure for electrodes; dense electrolyte; and insulating nature for sealants [323,373]. Besides, these requirements of SOFC should be cost-effective and easy to fabricate.

Different companies have already begun the commercialization of SOFC for various purposes. For example, Bloom Energy Company (USA) has commercialized this technology for large stationary applications, while other companies such as JX Nippon Oil & Energy, Aisin (Japan), and Ceramic Fuel Cells Limited (CFCL) (Europe) have made the same effort on micro-combined heats and power (CHP) systems for small stationary applications [323]. In this regard, Ceres Company (Europe) reported a degradation rate of ~1%/1,000 hours for its first pre-commercial small-scale CHP [323]. However, high capital tariffs and expensive operating costs due to degradation issues are serious challenges for a commercial breakthrough of SOFC technology [323,402]. For instance, SOFC systems for stationary applications demand 40,000–80,000 hours of service for market launch [323]. US DOE has set targets toward system capital costs and degradation rates to overcome these challenges. The degradation rate, which is the electrical potential lowering rate, was targeted at 0.2%/1,000 hours for SOFC stacks in 2020. In this regard, scientists in the field expect average degradation rates of 0.5%/1,000 hours, 0.3%/1,000 hours, and 0.2%/1,000 hours for 2020, 2035, and 2050, respectively [323,402].

Golkhatmi et al. [323] point out that, so far, the longest SOFC operation belongs to F1002–97, a short stack from Forschungszentrum Jülich GmbH, which reached 93,000 hours of operation at 700°C under 0.5A/cm^2 constant current density with 40% fuel utilization of wet H_2 and compressed air as oxidant. This two-layer short stack consisted of a 500-µm-thick anode support (Ni-8YSZ; 8 mol% YSZ), a 7-µm-thick anode (Ni-8YSZ), a 10-µm-thick 8YSZ electrolyte, a 40-µm-thick LSCF cathode, and a 5-µm-thick GDC barrier layer. It also had a 5.5 mm IT metal interconnect with a MnO_x protective coating and glass sealants [323]. However, the average voltage degradation rate (0.5%/kh) crossed the given limit for SOFC commercialization [403], mainly due to the chromium (Cr)-poisoned cathodes and interconnector oxidation [323]. Besides, the stack was mostly run at a cell voltage of 0.7 V, which is much below the typical operating voltages for SOFCs [404]. There are also other stacks with long-term operations to study the durability performance of various SOFC components and design along with different working parameters [323], including ~ 40,000 hours and 0.5–1%/kh by Mai et al. [405], 6,000 hours and ~1.4%/kh by Chou et al. [406], 5,000 hours and 0.75%/kh by Ido et al. [407], and 1,000 hours without any noticeable degradation by Thaheem et al. [408].

Planar and tubular designs are the most common configurations for practical applications. The tubular cell consists of an array of sandwiched electrolyte and electrodes in a specific length and diameter. The planar design (radial or flat plate) includes a compact assembly of electrolyte and electrodes. The planar design has a simpler and cheaper fabrication procedure, higher PD, and low internal resistance due to its short current path. However, the tubular cell presents a more solid thermo-cycling performance, and it is easier to seal [323]. Both cell designs require sufficient mechanical strength to withstand the operation stresses provided by the support layer. The support layer has the largest thickness, and the thickness of other layers is minimized to avoid high internal resistance, enhance cell efficiency, and reduce costs. Generally, SOFCs have one support layer, and they can be designed as anode-, cathode-, or electrolyte-supported [409]. Planar designs are mostly anode-supported, while tubular ones are fabricated in electrolyte-supported configuration [410]. In the planar SOFCs, the reactant gases diffuse into the porous microstructure from the center to the circumference. Unlike the planar ones, the fuel flow runs outside and the oxidant inside in a cathode-supported tubular cell. For the anode-supported ones, the fuel flow goes inside, and the oxidant circulates outside the tube during the operation [323,410].

Electrolyte-supported SOFCs are the oldest design as YSZ provides a robust support layer and is easier to fabricate. However, a thick electrolyte layer causes higher ohmic losses, which degrades the SOFC PD output. Regarding the electrode-supported cells, an anode-supported design is more favorable than a cathode-supported one, owing to its higher power densities, particularly at lower temperatures. The second generation is the anode-supported cells, with a 200–1500 μm anode thickness and a thin electrolyte. This design decreases electrolytic resistance and leads to better conductivity at lower temperatures. Since the electrolyte no longer provides mechanical support, other materials with higher ionic conductivity and lower mechanical strength can be replaced with the YSZ to improve the cell's output. Furthermore, the anode-supported cells' fabrication process is simpler, and the anode microstructure is more controllable [323,409,411]. Although the industry is more interested in anode-supported design, both anode- and electrolyte-supported cells are used in laboratory experiments. For instance, electrolyte support allows for an easier independent analysis of each electrode process in a three-electrode operation, while anode support offers better output results [323].

Stacks also consist of several SOFC single cells joined to each other by interconnects. Interconnects act as a physical wall between the anode's reducing and cathode's oxidizing atmospheres. Moreover, planar design stacks require a sealant to avoid leakages or direct mixing of fuel and oxidant. Sealing, however, is typically not a major issue in tubular SOFCs [323]. It is worth noting that there is another geometry for SOFC stacks, the flat-tubular configuration, providing the features of both planar and tubular SOFCs into a single design, such as high PD, good thermal robustness, and ease of sealing [323]. The Siemens-Westinghouse SOFC Company invented this design to address the low PD of tubular cells [232,412]. Park et al. [413] reported a 5-cell stack with flat-tubular anode-supported cells without using metallic interconnect plates, showing a degradation rate of 0.69%/kh during 1,093 hours under a current load of 16 A at 750°C. However, there are no records of their durability performance on the cell level [413].

Finally, Kuterbekov et al. [414] presented a detailed classification of SOFC. In this review, an attempt is made to collect and structure all types of SOFC that exist today. Structural features of each type of SOFC have been described, and their advantages and disadvantages have been identified. To cover all SOFC concepts, the standard SOFC classification is supplemented by division according to such criteria as the presence/absence of electrolyte and gas space separation. Herewith, the types of SOFC that are usually not mentioned in the classifications (electrolyte-free FC and mixed-reactant SOFC) have been considered along with other types of SOFC from the standpoint of standard criteria: operating temperature, support types, and geometry. This has made it possible to compare the various designs. It is shown that the most developed group of SOFC is separate-reactant FCs with oxygen-ion conducting electrolytes. A comparison of the designs showed that among the well-studied dual-chamber SOFC with oxygen-ion conducting electrolyte, the anode-supported design is the most suitable for operation at temperatures below 800°C. Other SOFC types that are promising for LT operation are SOFC with proton-conducting electrolyte and electrolyte-free FCs. These have a greater potential for reducing operating temperatures than standard dual-chamber O-SOFC. All SOFC types have some drawbacks; therefore, further research and new ideas are necessary for the practical mass implementation of this technology. The recently developed technologies are still far from commercialization and require further research and development.

REFERENCES

1. Du, S., Pollet, B. G., Applications of nanomaterials in fuel cells. In David Rickerby (Ed), *Nanotechnology for Sustainable Manufacturing.* CRC Press, Boca Raton, 2014, pp. 113–152. DOI: 10.1201/b17046-8
2. Winter, M., Brodd, R. J., What are batteries, fuel cells, and supercapacitors? *Chem Rev* 2004;104:4245–4269. DOI: 10.1021/cr020730k
3. Garcia-Martinez, J. (Ed), *Nanotechnology for the Energy Challenge.* Wiley-VCH, Weinheim, Germany, 2010.
4. Minteer, S. D., Moehlenbrock, M. J., Extended lifetime biofuel cells. *Chem Soc Rev* 2008;37:1188–1196. DOI: 10.1039/b708013c
5. Zhao, F., Slade, R. C. T., Varcoe, J. R., Techniques for the study and development of microbial fuel cells: An electrochemical perspective. *Chem Soc Rev* 2009;38:1926–1939. DOI: 10.1039/b819866g
6. Barton, S. C., Gallaway, J., Atanassov, P., Enzymatic biofuel cells for implantable and microscale devices. *Chem Rev* 2004;104:4867–4886. DOI: 10.1021/cr020719k
7. Harnisch, F, Schröder, U., From MFC to MXC: Chemical and biological cathodes and their potential for microbial bioelectrochemical systems. *Chem Soc Rev* 2010;39:4433–4448. DOI: 10.1039/c003068f
8. Brandon, N. P., Skinner, S., Steele, B. C. H., Recent advances in materials for fuel cells. *Ann Rev Mat Res* 2003;33:183–213. DOI: 10.1146/annurev.matsci.33.022802.094122
9. Baxter, J., Bian, Z., Chen, G., Danielson, D., Dresselhaus, M. S., Fedorov, A. G., Fisher, T. S., Jones, C. W., Maginn, E., Kortshagen, U., Manthiram, A., Nozik, A., Rolison, D. R., Sands, T., Shi, L., Sholl, D., Wu, Y. Y., Nanoscale design to enable the revolution in renewable energy. *Energy Environ Sci* 2009;2:559–588. DOI: 10.1039/b821698c
10. Gong, K., Du, F., Xia, Z., Durstock, M., Dai, L., Nitrogen-doped carbon nanotube arrays with high electrocatalytic activity for oxygen reduction. *Science* 2009;323:760–764. DOI: 10.1126/science.1168049
11. Lefèvre, M., Proietti, E., Jaouen, F., Dodelet, J. P., Iron-based catalysts with improved oxygen reduction activity in polymer electrolyte fuel cells. *Science* 2009;324:71–74. DOI: 10.1126/science.1170051
12. Ma, Y, Wang, X, Li, S, Toprak, MS, Zhu, B, Muhammed, M., Samarium-doped ceria nanowires: Novel synthesis and application in low-temperature solid oxide fuel cells. *Adv Mater* 2010;22:1640–1644. DOI: 10.1002/adma.200903402.
13. Laberty-Robert, C., Long, J. W., Pettigrew, K. A., Stroud, R. M., Rolison, D. R., Ionic nanowires at 600°C: Using nanoarchitecture to optimize electrical transport in nano-crystal- line gadolinium-doped ceria. *Adv Mater* 2007;19:1734–1739 DOI: 10.1002/adma.200601840.
14. Jiang, C., Ma, J., Bonaccorso, A. D., Irvine, J. T. S., Demonstration of high power, direct conversion of waste-derived carbon in a hybrid direct carbon fuel cell. *Energy Environ Sci* 2012;5(5):6973–6980. DOI: 10.1039/c2ee03510c

15. Sazali, N., Wan Salleh, W. N., Jamaludin, A. S., Mhd Razali, M. N., New perspectives on fuel cell technology: A brief review, *Membranes* 2020;10:99; doi:10.3390/membranes10050099
16. Adamson, K. A., *Stationary Fuel Cells*. Elsevier, Amsterdam, the Netherlands, 2007.
17. Giorgi, L., Leccese, F., *Fuel Cells: Technologies and Applications. Open Fuel Cells J* 2013;6:1–20.
18. Bagotsky, V. S., *Fuel Cells: Problems and Solutions*. John Wiley & Sons, Hoboken, NJ, 2008.
19. FuelCellToday. Alkaline Fuel Cells (AFC). Available online: http://fuelcelltoday.com/technologies/afc (accessed on 10 June 2020).
20. Siwal, S., Thakur, S., Zhang, Q., Thakur, V., Electrocatalysts for electrooxidation of direct alcohol fuel cell: Chemistry and applications. *Mater Today Chem* 2019;14:100182.
21. Alhassan, M., Umar Garba, M., Design of an alkaline fuel cell. *Leonardo Electron J Pract Technol* 2006;5:99–106.
22. Office of Renewable Energy, *Types of Fuel Cells|Department of Energy*. U.S. Office of Energy Efficiency and Renewable Energy, Washington, DC, 2017.
23. Kiros, Y., Schwartz S., Pyrolyzed macrocycles on high surface area carbons for the reduction of oxygen in alkaline fuel cells. *J Power Sources* 1991;36(4):547–555.
24. Zhuang, Z., Giles, S. A., Zheng, J., Jenness, G. R., Caratzoulas, S., Vlachos, D. G., Yan, Y., Nickel supported on nitrogen-doped carbon nanotubes as hydrogen oxidation reaction catalyst in alkaline electrolyte. *Nat Commun* 2016;7(1):1–8.
25. Jukk, K., Alexeyeva, N., Ritslaid, P., Kozlova, J., Sammelselg, V., Tammeveski, K., Electrochemical reduction of oxygen on heat-treated pd nanoparticle/multi-walled carbon nanotube composites in alkaline solution. *Electrocatalysis* 2013;4(1):42–48.
26. Sun, Y, Wu, J, Tian J, Jin C, Yang R., Sulfur-doped carbon spheres as efficient metal-free electrocatalysts for oxygen reduction reaction. *Electrochim Acta* 2015;178:806–812.
27. Wu, J., Jin, C., Yang, Z., Tian, J., Yang, R., Synthesis of phosphorus- doped carbon hollow spheres as efficient metal-free electrocatalysts for oxygen reduction. *Carbon* 2015;82:562–571.
28. Qiu, Z., Huang, N., Ge, X., Xuan, J., Wang, P., Preparation of N- doped nano-hollow capsule carbon nanocage as ORR catalyst in alkaline solution by PVP modified F127. *Int J Hydrogen Energy* 2020;45(15):8667–8675.
29. Li, B, Sasikala, S. P., Kim, D. H., Bak, J., Kim, I. -D., Cho, E., Kim, S. O., Fe-N_4 complex embedded free-standing carbon fabric catalysts for higher performance ORR both in alkaline & acidic media. *Nano Energy* 2019;56:524–530.
30. Brandon, N. P., Brett, D. J., Engineering porous materials for fuel cell applications. *Phil Trans Math Phys Eng Sci* 2006;364(1838):147–59.
31. Tomantschger, K., Kordesch, K. V., Structural analysis of alkaline fuel cell electrodes and electrode materials. *J Power Sources* 1989;25(3):195–214.
32. Li, Y. S., Zhao, T. S., A high-performance integrated electrode for anion-exchange membrane direct ethanol fuel cells. *Int J Hydrogen Energy* 2011;36(13):7707–7713.
33. Bidault, F., Kucernak, A., Cathode development for alkaline fuel cells based on a porous silver membrane. *J Power Sources* 2011;196(11):4950–4956.
34. Tiwari, P., Tsekouras, G., Swiegers, G. F., Wallace, G. G., Gortex- based gas diffusion electrodes with unprecedented resistance to flooding and leaking. *ACS Appl Mater Interfaces* 2018;10(33):28176–28186.
35. Burchardt, T, Goue rec, P., Sanchez-Cortezon, E., Karichev, Z., Miners, J. H., Alkaline fuel cells: Contemporary advancement and limitations. *Fuel* 2002;81(17):2151–2155.
36. Ferriday, T. B., Middleton, P. H., Alkaline fuel cell technology - A review. *Int J Hydrogen Energy* 2021;46:18489–18510.

37. Akinyele, D., Olabode, E., Amole, A., Review of fuel cell technologies and applications for sustainable microgrid systems. *Inventions* 2020;5(3):42. DOI:10.3390/inventions5030042
38. Brown T., Gencell launches commercial alkaline fuel cell using cracked ammonia fuel. July 2018. Available online: https://www. ammoniaenergy.org/articles/gencell-launches-commercial- alkaline-fuel-cell-using-cracked-ammonia-fuel/ (access date Jan. 15, 2019).
39. Xu, F., Su, Y., Lin, B., Progress of alkaline anion exchange membranes for fuel cells: The effects of micro-phase separation. *Front Mater* 2020;7:4. DOI: 10.3389/fmats.2020.00004
40. Liu, L., Li, D. F., Xing, Y., Li, N. W., Mid-block quaternized polystyrene- b-polybutadiene-b-polystyrene triblock copolymers as anion exchange membranes. *J Membr Sci* 2018;564:428–435. DOI: 10.1016/j.memsci.2018.07.055
41. Zhang, Z., Wu, L., Varcoe, J., Li, C., Ong, A. L., Poynton, S., Xu, T., Aromatic polyelectrolytes via polyacylation of pre- quaternized monomers for alkaline fuel cells. *J Mater Chem A* 2013;1(7):2595–601.
42. Merle, G, Wessling, M, Nijmeijer, K., Anion exchange membranes for alkaline fuel cells: A review. *J Membr Sci* 2011;377(1–2):1–35.
43. Kim, D. J., Jeong, M. K., Nam, S. Y., Research trends in ion exchange membrane processes and practical applications. *Appl Chem Eng* 2015;26(1):1–16.
44. Varcoe, J. R., Atanassov, P., Dekel, D. R., Herring, A. M., Hickner, M. A., Kohl, PA, Kucernak, A. R., Mustain, W. E., Nijmeijer, K., Scott, K., Xu, T., Zhuang, L., Anion-exchange membranes in electrochemical energy systems. *Energy Environ Sci* 2014;7(10):3135–3191.
45. Huang, G., Mandal, M., Peng, X., Yang-Neyerlin, A. C., Pivovar, B. S., Mustain, W. E., Kohl, P. A., Composite poly (norbornene) anion conducting membranes for achieving durability, water management and high power (3.4 W/cm^2) in hydrogen/oxygen alkaline fuel cells. *J Electrochem Soc* 2019;166(10):F637.
46. Vijayalekshmi, V., Tae, Y. Son, Kwang, S.I., Ji, E.C., Hyoung, J. Kim, etc., Anion Exchange Composite membranes compound of quaternary ammonium-functionalized Poly (2,6-dimethyl-1,4-phenylene oxide) and silica for fuel cell application, ACS Omega 2021, 6, 15, 10168-10179, publication date April 6, 2021, https://doi.org/10.1021/acsomega.1c00247 (access date Nov. 3, 2022).
47. Alkaline anion exchange fuel cell, 2023. Wikipedia, The free encyclopedia, last edited 3 November 2022. https://en.wikipedia.org/wiki/Alkaline_anion_exchange_membrane_fuel_cell.
48. Vijayakumar, V., Nam, S. Y., Recent advancements in applications of alkaline anion exchange membranes for polymer electrolyte fuel cells. *J Ind Eng Chem* 2019;70:70–86.
49. Shin, D. W., Guiver, M. D., Lee, Y. M., Hydrocarbon-based polymer electrolyte membranes: Importance of morphology on ion transport and membrane stability. *Chem Rev* 2017;117(6):4759–805.
50. Mandal, M., Huang, G., Ul Hassan, N., Peng, X., Gu, T., Brooks- Starks, A. H., Bahar, B., Mustain, W. E., Kohl, P. A., The importance of water transport in high conductivity and high-power alkaline fuel cells. *J Electrochem Soc* 2019;167(5):054501.
51. Chen, N., Lee, Y. M., Anion exchange polyelectrolytes for membranes and ionomers. *Prog Polym Sci* 2021;113:101345.
52. Park, E. J., Maurya, S., Hibbs, M. R., Fujimoto, C. H., Kreuer, K. -D., Kim, Y. S., Alkaline stability of quaternized diels alder polyphenylenes. Macromolecules2019;52(14):5419–5428.
53. Omasta, T. J., Zhang, Y., Park, A. M., Peng, X., Pivovar, B., Varcoe, J. R., Mustain, W. E., Strategies for reducing the PGM loading in high power AEMFC anodes. J Electrochem Soc 2018;165(9):F710.

54. Gao, L., Wang, Y., Cui, C. Y., Zheng, W. J., Yan, X. M., Zhang, P., Anion exchange membranes with "rigid-side-chain" symmetric piperazinium structures for fuel cell exceeding 1.2W cm^{-2} at 60°C. *J Power Sources* 2019;438:227021. DOI: 10.1016/j.jpowsour.2019.227021
55. Jin, C. H., Zhang, S., Cong, Y. Y., Zhu, X. L., Highly durable and conductive poly(arylene piperidine) with a long heterocyclic ammonium side-chain for hydroxide exchange membranes. *Int J Hydrogen Energy* 2019;44:24954–24964. DOI: 10.1016/j.ijhydene.2019.07.184
56. Wang, L., Brink, J. J., Liu, Y., Herring, A. M., Ponce-Gonza lez, J., Whelligan, D. K., Varcoe, J. R., Non-fluorinated pre-irradiation- grafted (peroxidated) LDPE-based anion-exchange membranes with high performance and stability. *Energy Environ Sci* 2017;10(10):2154–2167.
57. Li, D., Chung, H. T., Maurya, S., Matanovic, I., Kim, Y. S., Impact of ionomer adsorption on alkaline hydrogen oxidation activity and fuel cell performance. *Curr Opin Electrochem* 2018;12:189–95.
58. Ul Hassan, N., Mandal, M., Huang, G., Firouzjaie, H. A., Kohl, P. A., Mustain, W. E., Achieving high-performance and 2000h stability in anion exchange membrane fuel cells by manipulating ionomer properties and electrode optimization. *Adv Energy Mat* 2020;10(40):2001986.
59. Leonard, D. P., Maurya, S., Park, E. J., Manriquez, L. D., Noh, S., Wang, X., Bae, C., Baca, E. D., Fujimoto, C., Kim, Y. S., Asymmetric electrode ionomer for low relative humidity operation of anion exchange membrane fuel cells. *J Mater Chem A* 2020;8(28):14135–14144.
60. Vincent, I., Lee, E. C., Kim, H. M., Comprehensive impedance investigation of low-cost anion exchange membrane electrolysis for large-scale hydrogen production. *Sci Rep* 2021;11:293. DOI: 10.1038/s41598-020-80683-6
61. Wang, Z. H., Wang, C. Y., Mathematical modeling of liquid-feed direct methanol fuel cells. *J Electrochem Soc* 2003;150:A508–A519.
62. Garc'ia, B. L., Weidner, J. W., Review of direct methanol fuel cells. In R. E. White, C. G. Vayenas, M.E. Gamboa-Aldeco (Eds), *Modern Aspects of Electrochemistry No. 40. Modern Aspects of Electrochemistry*. Springer, New York, 2010, vol 40. DOI: 10.1007/978-0-387-46106-9_5.
63. Feng, Y., Liu, H., Yang, J., A selective electrocatalyst–based direct methanol fuel cell operated at high concentrations of methanol. *Sci Adv* 2017;3(6):e1700580. DOI: 10.1126/sciadv.1700580
64. Kirubakaran, A., Jain, S., Nema, R. K., A review on fuel cell technologies and power electronic interface. *Renew Sustain Energy Rev* 2009;13:2430–2440.
65. Mekhilef, S., Saidur, R., Safari, A., Comparative study of different fuel cell technologies. *Renew Sustain Energy Rev* 2012;16:981–989.
66. Joghee, P., Malik, J. N., Pylypenko, S., O'Hayre, R., A review on direct methanol fuel cells—In the perspective of energy and sustainability. *MRS Energy Sustain* 2015;2:E3.
67. Mehmood, A., Scibioh, M. A., Prabhuram, J., An, M.-G., Ha, H. Y., A review on durability issues and restoration techniques in long-term operations of direct methanol fuel cells. *J Power Sources* 2015;297:224–241.
68. Zhao, T. S., Chen, R., Yang, W. W., Xu, C., Small direct methanol fuel cells with passive supply of reactants. *J Power Sources* 2009;191:185–202.
69. Li, X., Faghri, A., Review and advances of direct methanol fuel cells (DMFCs) part I: Design, fabrication, and testing with high concentration methanol solutions. *J Power Sources* 2013;226:223–240.
70. Zhao, T. S., Yang, W. W., Chen, R., Wu, Q. X., Towards operating direct methanol fuel cells with highly concentrated fuel. *J Power Sources* 2010;195:3451–3462.

71. de Sá, M. H., Pinto, A. M., Oliveira, V. B., Passive small direct alcohol fuel cells for low-power portable applications: Assessment based on innovative increments since 2018, *Energies* 2022;15:3787. DOI:10.3390/en15103787
72. Shaari, N., Kamarudin, S. K., Bahru, R., Osman, S. H., Md Ishak, N. A. I., Progress and challenges: Review for direct liquid fuel cell. *Int J Energy Res* 2021;45:6644–6688.
73. Sharaf, O. Z., Orhan, M. F., An overview of fuel cell technology: Fundamentals and applications. *Renew Sustain Energy Rev* 2014;32:810–853.
74. Alias, M., Kamarudin, S., Zainoodin, A., Masdar, M., Active direct methanol fuel cell: An overview. *Int J Hydrogen Energy* 2020;45:19620–19641.
75. de Sá, M., Pinto, A., Oliveira, V., Passive direct methanol fuel cells as a sustainable alternative to batteries in hearing aid devices–An overview. *Int J Hydrogen Energy* 2022;47:16552–16567.
76. Carneiro, L. P., Ferreira, N. S., Tavares, A. P., Pinto, A. M., Mendes, A., Sales, M. G. F., A passive direct methanol fuel cell as transducer of an electrochemical sensor, applied to the detection of carcinoembryonic antigen. *Biosens Bioelectron* 2020;175:112877.
77. Mohammed, H., Al-Othman, A., Nancarrow, P., Tawalbeh, M., Assad, M. E. H., Direct hydrocarbon fuel cells: A promising technology for improving energy efficiency. *Energy* 2019;172:207–219.
78. Pinto, A. M. F. R., Oliveira, V. B., Falcão, D. S., *Direct Alcohol Fuel Cells for Portable Applications Fundamentals, Engineering and Advances.* Elsevier Science, Amsterdam, the Netherlands, 2018.
79. Kjeang, E., Djilali, N., Sinton, D., Advances in microfluidic fuel cells. In *Micro Fuel Cells, Principles and Applications.* Academic Press, Cambridge, MA, 2009, pp. 99–139.
80. Falcão, D., Oliveira, V., Rangel, C., Pinto, A., Review on micro-direct methanol fuel cells. *Renew Sustain Energy Rev* 2014;34:58–70.
81. Zheng, Y., Wan, X., Cheng, X., Cheng, K., Dai, Z., Liu, Z. Advanced catalytic materials for ethanol oxidation in direct ethanol fuel cells. *Catalysts* 2020;10:166.
82. Mansor, M., Timmiati, S. N., Lim, K. L., Wong, W. Y., Kamarudin, S. K., Kamarudin, N. H. N., Recent progress of anode catalysts and their support materials for methanol electrooxidation reaction. *Int J Hydrogen Energy* 2019;44:14744–14769.
83. Ren, X., Lv, Q., Liu, L., Liu, B., Wang, Y., Liu, A., Wu, G., Current progress of Pt and Pt-based electrocatalysts used for fuel cells. *Sustain Energy Fuels* 2020;4:15–30.
84. Hu, C., Zhou, Y., Xiao, M., Yu, G., Precise size and dominant-facet control of ultra-small Pt nanoparticles for efficient ethylene glycol, methanol and ethanol oxidation electrocatalysts. *Int J Hydrogen Energy* 2020;45:4341–4354.
85. Du, L., Prabhakaran, V., Xie, X., Park, S., Wang, Y., Shao, Y., Low-PGM and PGM-free catalysts for proton exchange membrane fuel cells: Stability challenges and material solutions. *Adv Mater* 2020:33;1908232.
86. Kosmala, T., Bibent, N., Sougrati, M. T., Dražić, G., Agnoli, S., Jaouen, F., Granozzi, G., Stable, active, and methanol-tolerant PGM-free surfaces in an acidic medium: Electron tunneling at play in Pt/FeNC hybrid catalysts for direct methanol fuel cell cathodes. *ACS Catal* 2020;10:7475–7485.
87. Shi, Q., He, Y., Bai, X., Wang, M., Cullen, D. A., Lucero, M., Zhao, X., More, K. L., Zhou, H., Feng, Z., Liu, Y., Wu, G., Methanol tolerance of atomically dispersed single metal site catalysts: Mechanistic understanding and high-performance direct methanol fuel cells. *Energy Environ Sci* 2020;13:3544–3555.
88. Baricci, A., Bisello, A., Serov, A., Odgaard, M., Atanassov, P., Casalegno, A., Analysis of the effect of catalyst layer thickness on the performance and durability of platinum group metal-free catalysts for polymer electrolyte membrane fuel cells. *Sustain Energy Fuels* 2019;3:3375–3386.

89. Vecchio, C. L., Serov, A., Romero, H., Lubers, A., Zulevi, B., Aricò, A., Baglio, V., Commercial platinum group metal-free cathodic electrocatalysts for highly performed direct methanol fuel cell applications. *J Power Sources* 2019;437:226948.
90. Abdelkareem, M. A., Sayed, E., Mohamed, H.O., Obaid, M., Rezk, H., Chae, K. -J. Nonprecious anodic catalysts for low-molecular- hydrocarbon fuel cells: Theoretical consideration and current progress. *Prog Energy Combust Sci* 2020;77:100805.
91. Martinez, U., Babu, S. K., Holby, E. F., Chung, H. T., Yin, X., Zelenay, P., Progress in the development of Fe-based PGM-free electrocatalysts for the oxygen reduction reaction. *Adv Mater* 2019:31;e1806545.
92. Osmieri, L., Escudero-Cid, R., Armandi, M., Videla, A. M., Fierro, J. L. G., Ocón, P., Specchia, S., Fe-N/C catalysts for oxygen reduction reaction supported on different carbonaceous materials. Performance in acidic and alkaline direct alcohol fuel cells. *Appl Catal B Environ* 2017:205;637–653.
93. Osmieri, L., Zafferoni, C., Wang, L., Videla, A. H. A. M., Lavacchi, A., Specchia, S., Polypyrrole-derived Fe–Co–N–C catalyst for the oxygen reduction reaction: Performance in alkaline hydrogen and ethanol fuel cells. *ChemElectroChem* 2018;5:1954–1965.
94. Krewer, U., Weinzierl, C., Ziv, N., Dekel, D. R., Impact of carbonation processes in anion exchange membrane fuel cells. *Electrochim Acta* 2018;263:433–446.
95. Ziv, N., Mustain, W. E., Dekel, D. R., The effect of ambient carbon dioxide on anion-exchange membrane fuel cells. *ChemSusChem* 2018;11:1136–1150.
96. An, L., Zhao, T., Transport phenomena in alkaline direct ethanol fuel cells for sustainable energy production. *J Power Sources* 2017;341:199–211.
97. Borghei, M., Laocharoen, N., Kibena-Põldsepp, E., Johansson, L.-S., Campbell, J., Kauppinen, E., Tammeveski, K., Rojas, O. J., Porous N, P-doped carbon from coconut shells with high electrocatalytic activity for oxygen reduction: Alternative to Pt-C for alkaline fuel cells. *Appl Catal B Environ* 2017:204;394–402.
98. Yan, X., Zhao, T., An, L., Zhao, G., Shi, L., A direct methanol–hydrogen peroxide fuel cell with a Prussian Blue cathode. *Int J Hydrogen Energy* 2016;41:5135–5140.
99. An, L., Zhao, T., Zeng, L., Yan, X., Performance of an alkaline direct ethanol fuel cell with hydrogen peroxide as oxidant. *Int J Hydrogen Energy* 2014;39:2320–2324.
100. Pan, Z., An, L., Wen, C., Recent advances in fuel cells based propulsion systems for unmanned aerial vehicles. *Appl Energy* 2019;240:473–485.
101. Wang, J., Wang, H., Fan, Y., Techno-economic challenges of fuel cell commercialization. *Engineering* 2018;4:352–360.
102. Wang, J., System integration, durability and reliability of fuel cells: Challenges and solutions. *Appl Energy* 2017;189:460–479.
103. Sgroi, M. F., Zedde, F., Barbera, O., Stassi, A., Sebastián, D., Lufrano, F., Baglio, V., Aricò, A. S., Bonde, J. L., Schuster, M., Cost analysis of direct methanol fuel cell stacks for mass production. *Energies* 2016;9:1008.
104. Pedram, S., Batool, M., Yapp, K., Bonville, L., Jankovic, J., A review on bioinspired proton exchange membrane fuel cell: Design and materials. *Adv Energy Sustain Res* 2021;2(7):2000092 DOI: 10.1002/aesr.202000092
105. Lupatini, K. N., Schaffer, J. V., Machado, B., Silva, E. S., Ellendersen, L. S., Muniz, G. I., Ferracin, R. J., Alves, H. J.,, Development of chitosan membranes as a potential PEMFC electrolyte. *J Polym Environ* 2018;26:2964.
106. Fan, L., Tu, Z., Chan S. H., Recent development of hydrogen and fuel cell technologies: A review. *Energy Rep* 2021;7:8421–8446.
107. Adamson, K. A., Butler, J., Hugh, M., 2008. Fuel cell today industry review 2008: Fuel cells: Commercialization. *Platinum Met Rev* 2008;52(2):123.
108. Van Dao, D., Adilbish, G., Lee, I. -H., Yu, Y. -T., Enhanced electrocatalytic property of Pt/C electrode with double catalyst layers for PEMFC. *Int J Hydrogen Energy* 2019;44:24580–24590.

109. Hezarjaribi, M., Jahanshahi, M., Rahimpour, A., Yaldagard, M., Gas diffusion electrode based on electrospun Pani/CNF nanofibers hybrid for proton exchange membrane fuel cells (PEMFC) applications. *Appl Surf Sci* 2014;295:144–149.
110. Middelman, E., Improved PEM fuel cell electrodes by controlled self-assembly. *Fuel Cells Bull* 2002;2002:9–12.
111. Qu, L., Wang, Z., Guo, X., Song, W., Xie, F., He, L., Shao, Z., Yi, B., Effect of electrode Pt-loading and cathode flow-field plate type on the degradation of PEMFC. *J Energy Chem* 2019;35:95–103.
112. Su, H., Pasupathi, S., Bladergroen, B., Linkov, V., Pollet, B. G., Optimization of gas diffusion electrode for polybenzimidazole-based high temperature proton exchange membrane fuel cell: Evaluation of polymer binders in catalyst layer. *Int J Hydrogen Energy* 2013;38:11370–11378.
113. Sutradhar, S. C., Rahman, M. M., Ahmed, F., Ryu, T., Yoon, S., Lee, S., Kim, J., Lee, Y., Jin, Y., Kim, W., Thermally and chemically stable poly(phenylenebenzophenone) membranes for proton exchange membrane fuel cells by Ni (0) catalyst. *J Ind Eng Chem* 2019;76:233–239.
114. Neethu, B., Bhowmick, G. D., Ghangrekar, M. M., A novel proton exchange membrane developed from clay and activated carbon derived from coconut shell for application in microbial fuel cell. *Biochem Eng J* 2019;148:170–177.
115. Haragirimana, A., Ingabire, P. B., Zhu, Y., Lu, Y., Li, N., Hu, Z., Chen, S., Four-polymer blend proton exchange membranes derived from sulfonated poly(aryl ether sulfone)s with various sulfonation degrees for application in fuel cells. *J Membr Sci* 2019;583:209–219.
116. Teixeira, F. C., de Sá, A. I., Teixeira, A. P. S., Rangel, C. M., Nafion phosphonic acid composite membranes for proton exchange membranes fuel cells. *Appl Surf Sci* 2019;487:889–897.
117. Wang, J., Li, P., Zhang, Y., Liu, Y., Wu, W., Liu, J., Porous nafion nanofiber composite membrane with vertical pathways for efficient through-plane proton conduction. *J Membr Sci* 2019;585:157–165.
118. He, Y., Wang, J., Zhang, H., Zhang, T., Zhang, B., Cao, S., Liu, J., Polydopamine-modified graphene oxide nanocomposite membrane for proton exchange membrane fuel cell under anhydrous conditions. *J Mater Chem A* 2014;2(25):9548–9558. DOI: 10.1039/C3TA15301K
119. Lin, S. -Y., Chang, M. -H., Effect of microporous layer composed of carbon nanotube and acetylene black on polymer electrolyte membrane fuel cell performance. *Int J Hydrogen Energy* 2015;40:7879–7885.
120. Zhang, S., Yuan, X. -Z., Hin, J. N. C., Wang, H., Friedrich, K. A., Schulze, M., A review of platinum-based catalyst layer degradation in proton exchange membrane fuel cells. *J Power Sources* 2009;194:588–600.
121. Haan, J. L., Masel, R. I., The influence of solution pH on rates of an electrocatalytic reaction: Formic acid electrooxidation on platinum and palladium. *Electrochim Acta* 2009;54:4073–4078. DOI: 10.1016/j.electacta.2009.02.045
122. Wang, Q., Zhou, Z. -Y., Lai, Y. -J., You, Y., Liu, J. -G., Wu, X. -L., Terefe, E., Chen, C., Song, L., Rauf, M., Tian, N., Sun, S. -G., Phenylenediamine-based FeNx/C catalyst with high activity for oxygen reduction in acid medium and its active-site probing. *J Am Chem Soc* 2014;136:10882–10885.
123. Pillai, S. R., Sonawane, S. H., Gumfekar, S. P., Suryawanshi, P. L., Ashokkumar, M., Potoroko, I., Continuous flow synthesis of nanostructured bimetallic Pt- Mo/C catalysts in milli-channel reactor for PEM fuel cell application. *Mater Chem Phys* 2019;237:121854.
124. Wlodarczyk, R., Nanomaterials in low-temperatures fuel cells—the latest reports. *Mater Sci Appl* 2019;10(10);643–664. DOI: 10.4236/msa.2019.1010046

125. Zhang, J., Recent advances in cathode electrocatalysts for PEM fuel cells. *Front Energy* 2011;5:137–148. DOI: 10.1007/s11708-011-0153-y
126. Markovic, N. M., Gasteiger, H. A., Grgur, B. N., Ross F. N., Oxygen reduction reaction on Pt(111): Effects of bromide. *J Electroanal Chem* 1999;467:157–163. DOI: 10.1016/S0022-0728(99)00020-0
127. Antolini, E., Review formation, microstructural characteristics and stability of carbon supported platinum catalysts for low temperature fuel cells. *J Mater Sci* 2003;38, 2995–3005.
128. Rajalakshmi, N., Dhathathreyan, K. S., Nanostructured platinum catalyst layer prepared by pulsed electrodeposition for use in PEM fuel cells. *Int J Hydrogen Energy* 2008;33:5672–5677. DOI: 10.1016/j.ijhydene.2008.05.100
129. Santos, A. L., Profeti, D., Olivi, P., Electrooxidation of methanol on Pt microparticles dispersed on SnO_2 thin films. *Electrochim Acta* 2005;50:2615–2621. DOI: 10.1016/j.electacta.2004.11.006
130. Liu, K., Qiao, Z., Hwang, S., Liu, Z., Zhang, H., Su, D. Xu, H., Wu, G., Wang, G., Mn- and N- doped carbon as promising catalysts for oxygen reduction reaction: Theoretical prediction and experimental validation. *Appl Catal B* 2019;243:195–203.
131. de Bruijn, F. A., Papageorgopoulos, D. C., Sitters, E. F., Janssen, G. J. M., The influence of carbon dioxide on PEM fuel cell anodes. *J Power Sources* 2002;110:117–124. DOI: 10.1016/S0378-7753(02)00227-6
132. Bonilla, S. H., Zinola, C. F., Rodríguez, J., Díaz, V., Ohanian, M., Martínez, S., Giannetti, B. F., Catalytic effects of ruthenium and osmium spontaneous deposition on platinum surfaces toward methanol oxidation. *J Colloid Interface Sci* 2005;288:377–386. DOI: 10.1016/j.jcis.2005.03.039
133. Chen, A. C., Holt-Hindle, P., Platinum-based nanostructured materials: Synthesis, properties, and applications. *Chem Rev* 2010;110:3767–3804. DOI: 10.1021/cr9003902
134. Stamenkovic, V. R., Fowler, B., Mun, B. S., Wang, G. F., Ross, P. N., Lucas, C. A. and Markovic, N. M., Improved oxygen reduction activity on Pt3Ni(111) via increased surface site availability. *Science* 2007;315:493–497. DOI: 10.1126/science.1135941
135. Zhang, J., Sasaki, K., Sutter, E., Adzic, R. R., Stabilization of platinum oxygen-reduction electrocatalysts using gold clusters. *Science* 2007;315:220–222. DOI: 10.1126/science.1134569
136. Shao, Y. Y., Yin, G. P., Gao, Y. Z., Understanding and approaches for the durability issues of Pt-based catalysts for PEM fuel cell. *J Power Sources* 2007;171:558–566. DOI: 10.1016/j.jpowsour.2007.07.004
137. Morozan, A., Jousselme, B., Palacin, S., Low-platinum and platinum-free catalysts for the oxygen reduction reaction at fuel cell cathodes. *Energy Environ Sci* 2011;4:1238.
138. Fabian, T., O'Hayre, R., Litster, S., Prinz, F., Santiago, J., Water management at the cathode of a planar air-breathing fuel cell with an electroosmotic pump. *ECS Trans* 2006;3:949.
139. Zhao, Y. M., Yu, G. Q., Wang, F. F., Wei, P. J., Liu, J. G., Bioinspired transition-metal complexes as electrocatalysts for the oxygen reduction reaction. *Chem Eur J* 2019;25:3726.
140. Chen, Z., Higgins, D., Yu, A., Zhang, L., Zhang, J., A review on non-precious metal electrocatalysts for PEM fuel cells. *Energy Environ Sci* 2011;4:3167.
141. Le Goff, A., Artero, V., Jousselme, B., Tran, F. D., Guillet, N., Métayé, R., Fihri, A., Palacin, S., Fontecave, M., From hydrogenases to noble metal–free catalytic nanomaterials for H_2 production and uptake. *Science* 2009;326:1384.
142. Rodriguez-Maciá, P., Dutta, A., Lubitz, W., Shaw, W. J., Rüdiger, O., Direct comparison of the performance of a bio-inspired synthetic nickel catalyst and a [NiFe]-hydrogenase, both covalently attached to electrodes. *Angew Chem Int Ed* 2015;54:12303.

143. Wei, X., Li, D., Jiang, W., Gu, Z., Wang, X., Zhang, Z., Sun, Z., 3D printable graphene composite. *Sci Rep* 2015;5:1.
144. Wang, R., Higgins, D. C., Lee, D. U., Prabhudev, S., Hassan, F. M., Chabot, V., Lui, G., Jiang, G., Choi, J.Y., Rasenthiram, L., Fu, J., Botton, G., Chen, Z., Biomimetic design of monolithic fuel cell electrodes with hierarchical structures. *Nano Energy* 2016;20:57.
145. Kong, C. S., Zhang, H. L., Somodi, F., Morse, D. E., Bio-inspired synthesis of high-performance nanocomposite catalysts for hydrogen oxidation. *Adv Funct Mater* 2013;23:4585.
146. Yao, Y., You, Y., Zhang, G., Liu, J., Sun, H., Zou, Z., Sun, S., Highly functional bioinspired Fe/N/C oxygen reduction reaction catalysts: Structure-regulating oxygen sorption. *ACS Appl Mater Interfaces* 2016;8:6464.
147. Kahraman, H., Orhan, M. F., Flow field bipolar plates in a proton exchange membrane fuel cell: Analysis & modeling. *Energy Convers Manage* 2017;133:363–384.
148. Tüber, K., Oedegaard, A., Hermann, M., Hebling, C., Investigation of fractal flow-fields in portable proton exchange membrane and direct methanol fuel cells. *J Power Sources* 2004;131:175.
149. Lorenzini-Gutierrez, D., Hernandez-Guerrero, A., Ramos-Alvarado, B., Perez-Raya, I., Alatorre-Ordaz, A., Performance analysis of a proton exchange membrane fuel cell using tree-shaped designs for flow distribution. *Int J Hydrogen Energy* 2013;38:14750.
150. Wang, X. D., Yan, W. M., Duan, Y. Y., Weng, F. B., Jung, G. B., Lee, C. Y., Numerical study on channel size effect for proton exchange membrane fuel cell with serpentine flow field. *Energy Convers Manage* 2010;51:959.
151. Marappan, M., Palaniswamy, K., Velumani, T., Chul, K. B., Velayutham, R., Shivakumar, P., Sundaram, S., Performance studies of proton exchange membrane fuel cells with different flow field designs – review. *Chem Rec* 2021;21(4):663–714. DOI: 10.1002/tcr.202000138
152. Park, Y. H., Caton, J. A., An experimental investigation of electro-osmotic drag coefficients in a polymer electrolyte membrane fuel cell. *Int J Hydrogen Energy* 2008;33:7513–7520.
153. Wakita, H., Kawabata, N., Kani, Y., Measurement of water permeation through membranes from extremely high hydraulic pressure to atmospheric pressure. *Int J Hydrogen Energy* 2019;44:31257–31262.
154. Nikiforow, K., Ihonen, J., Keränen, T., Karimäki, H., Alopaeus, V., Modeling and experimental validation of H_2 gas bubble humidifier for a 50 kW stationary PEMFC system. *Int J Hydrogen Energy* 2014;39:9768–9781.
155. Wu, H., Li, X., Berg, P., On the modeling of water transport in polymer electrolyte membrane fuel cells. *Electrochim Acta* 2009;54:6913–6927.
156. Srinivasan, V., Higuchi, W. I., Su, M. -H., Baseline studies with the four- electrode system: The effect of skin permeability increase and water transport on the flux of a model uncharged solute during iontophoresis. *J Control Release* 1989;10:157–165.
157. Chen, B., Wang, M., Tu, Z., Gong, X., Zhang, H., Pan, M., Cai, Y., Wan, Z., Moisture dehumidification and its application to a 3 kW proton exchange membrane fuel cell stack. *Int J Hydrogen Energy* 2015;40:1137–1144.
158. Subianto, S., Recent advances in polybenzimidazole/phosphoric acid membranes for high-temperature fuel cells. *Polym Int* 2014;63(7):1134–1144. DOI: 10.1002/pi.4708
159. Proton exchange membrane fuel cell, a website report by Fuel cell works, 2023.
160. Authayanun, S., Mamlouk, M., Scott, K., Arpornwichanop, A., Comparison of high-temperature and low-temperature polymer electrolyte membrane fuel cell systems with glycerol reforming process for stationary applications. *Appl Energy* 2013;109:192–201. DOI: 10.1016/j.apenergy.2013.04.009
161. High temperature proton exchange membrane fuel cell, 2023. Wikipedia, the free encyclopedia, last edited 23 July 2023. https://en.wikipedia.org/wiki/High_Temperature_Proton_Exchange_Membrane_fuel_cell.

162. Guo, Z., Perez-Page, M., Chen, J., Ji, Z., Holmes, S. M., Recent advances in phosphoric acid–based membranes for high–temperature proton exchange membrane fuel cells, *J Energy Chem* 2021;63:393–429.
163. Williams, M.C., Fuel cells. In *Fuel Cells: Technologies for Fuel Processing*. National Energy Technology Laboratory, Morgantown, WV, 2011.
164. Bhosale, A. C., Suseendiran S. R., Ramkumar, R., Choudhury, S. R., Raghunathan, R. Phosphoric acid fuel cells. In *Reference Module in Earth Systems and Environmental Sciences*, 2021. DOI: 10.1016/B978-0-12-819727-1.00006-6
165. Phosphoric acid fuel cell, 2023. Wikipedia, The free encyclopedia, last edited 27 June 2023. https://en.wikipedia.org/wiki/Phosphoric_acid_fuel_cell.
166. Wang, H. S., Chang, C. P., Huang, Y. J., Su, Y. C., Tseng, F. G., A high-yield and ultra-low-temperature methanol reformer integratable with phosphoric acid fuel cell (PAFC), *Energy* 2017;133:1142–1152. DOI: 10.1016/j.energy.2017.05.140.
167. Behling, N. H., Fuel cells and the challenges ahead. In *Fuel Cells*, 2013, pp. 7–36. DOI: 10.1016/B978-0-444-56325-5.00002-8
168. O'Hayre, R. P., Cha, S. -W., Colella, W., Prinz, F. B., *Fuel Cell Fundamentals*. John Wiley & Sons, Inc, Hoboken, NJ, 2006.
169. Nayak, R., Sundarraman, M., Ghosh, P. C., Bhattacharyya, A. R., Doped poly (2, 5-benzimidazole) membranes for high temperature polymer electrolyte fuel cell: Influence of various solvents during membrane casting on the fuel cell performance. *Eur Polym J* 2018;100:111–120. DOI: 10.1016/j.eurpolymj.2017.08.026
170. Modestov, A. D., Tarasevich, M. R., Filimonov, V. Y., Davydova, E. S., CO tolerance and CO oxidation at Pt and Pt-Ru anode catalysts in fuel cell with polybenzimidazole-H_3PO_4 membrane. *Electrochim Acta* 2010;55(20):6073–6080. DOI: 10.1016/j.electacta.2010.05.068
171. Liang, H., Su, H., Pollet, B. G., Linkov, V., Pasupathi, S., Membrane electrode assembly with enhanced platinum utilization for high temperature proton exchange membrane fuel cell prepared by catalyst coating membrane method. *J Power Sources* 2014;266:107–113. DOI: 10.1016/j.jpowsour.2014.05.014
172. Sebastián, D., Ruíz, A. G., Suelves, I., Moliner, R., Lázaro, M. J., Baglio, V., Stassi, A., Aricò, A. S., Enhanced oxygen reduction activity and durability of Pt catalysts supported on carbon nanofibers. *Appl Catal B: Environ* 2012;115–116:269–275. DOI: 10.1016/j.apcatb.2011.12.041.
173. Galbiati, S., Morin, A., Pauc, N., Nanotubes array electrodes by Pt evaporation: Half- cell characterization and PEM fuel cell demonstration. *Appl Catal B: Environ* 2015;165:149–157. DOI: 10.1016/j.apcatb.2014.09.075.
174. Cunningham, B. D., Baird, D. G., Development of bipolar plates for fuel cells from graphite filled wet-lay material and a compatible thermoplastic laminate skin layer. *J Power Sources* 2007;168(2):418–425. DOI: 10.1016/j.jpowsour.2007.03.036.
175. Feng, K., Li, Z., Sun, H., Yu, L., Cai, X., Wu, Y., Chu, P. K., C/CrN multilayer coating for polymer electrolyte membrane fuel cell metallic bipolar plates. *J Power Sources* 2013;222:351–358.
176. Yan, W. M., Chen, C. Y., Liang, C. H., Comparison of performance degradation of high temperature PEM fuel cells with different bipolar plates. *Energy* 2019;186:115836. DOI: 10.1016/j.energy.2019.07.166
177. Wu, M., Zhang, H., Zhao, J., Wang, F., Yuan, J., Performance analyzes of an integrated phosphoric acid fuel cell and thermoelectric device system for power and cooling cogeneration. *Int J Refrig* 2018;89:61–69. DOI: 10.1016/j.ijrefrig.2018.02.018
178. Yang, P., Zhang, H., Hu, Z., Parametric study of a hybrid system integrating a phosphoric acid fuel cell with an absorption refrigerator for cooling purposes. *Int J Hydrogen Energy* 2016;41(5):3579–3590. DOI: 10.1016/j.ijhydene.2015.10.149

179. Açıkkalp, E., Ahmadi, M. H., Parametric investigation of phosphoric acid fuel cell - Thermally regenerative electro chemical hybrid system. *J Clean Prod* 2018;203:585–600. DOI: 10.1016/j.jclepro.2018.07.231
180. Ghaswalla, A. N., *Homegrown AIP tech is best fit for Scorpene submarines: DRDO.* The Hindu BusinessLine, 2017. Available online: https://www.thehindubusinessline.com/news/national/homegrown-aip-tech-is-best-fit-for- scorpene-submarines-drdo/article9973208.ece (accessed on 4 April, 2020).
181. Cui, C., Li, S., Gong, J., Wei, K., Hou, X., Jiang, C., Yao, Y., Ma, J., Review of molten carbonate-based direct carbon fuel cells. *Mater Renew Sustain Energy* 2021;10:12. DOI: 10.1007/s40243-021-00197-7
182. Edison, T. A., Process of and apparatus for generating electricity. US Patent No 460, 122, 1891.
183. Jacques, W.W., Method of converting potential energy of carbon into electrical energy. US Patent No 555511 A, 1896.
184. Hackett, G. A., Zondlo, J. W., Svensson, R., Evaluation of carbon materials for use in a direct carbon fuel cell. *J Power Sources* 2007;168(1):111–118. DOI: 10.1016/j.jpowsour.2007.02.021
185. Dicks, A. L., The role of carbon in fuel cells. *J Power Sources* 2006;156(2):128–141. DOI: 10.1016/j.jpowsour.2006.02.054
186. Li, X., Zhu, Z., De Marco, R., Bradley, J., Dicks, A., Evaluation of raw coals as fuels for direct carbon fuel cells. *J Power Sources* 2010;195(13):4051–4058. DOI: 10.1016/j.jpowsour.2010.01.048
187. Lee, C. -G., Kim, W. -K., Oxidation of ash-free coal in a direct carbon fuel cell. *Int J Hydrogen Energy* 2015;40(15):5475–5481. DOI: 10.1016/j.ijhydene.2015.01.068
188. Howard, H., Direct generation of electricity from coal and gas (fuel cells). *Chem Coal Util* 1945;2:1568–1585.
189. Liebhafsky, H. A., Cairns, E. J., *Fuel Cells and Fuel Batteries: Guide to Their Research and Development.* Wiley, New York, 1969.
190. Cao, D., Sun, Y., Wang, G., Direct carbon fuel cell: Fundamentals and recent developments. *J Power Sources* 2007;167(2):250–257. DOI: 10.1016/j.jpowsour.2007.02.034
191. Cao, T., Huang, K., Shi, Y., Cai, N., Recent advances in high-temperature carbon–air fuel cells. *Energy Environ Sci* 2017;10(2):460–490. DOI: 10.1039/c6ee03462d
192. Giddey, S., Badwal, S. P. S., Kulkarni, A., Munnings, C., A comprehensive review of direct carbon fuel cell technology. *Prog. Energy Combust* 2012;38(3):360–399. DOI: 10.1016/j.pecs.2012.01.003
193. Cooper, J. F., Selman, R., Electrochemical oxidation of carbon for electric power generation: A review. *J Electrochem Soc* 2009;19(14):15–25.
194. Gür, T. M., Critical review of carbon conversion in “carbon fuel cells.” *Chem Rev* 2013;113(8):6179–6206. DOI: 10.1021/cr400072b
195. Rady, A. C., Giddey, S., Badwal, S. P. S., Ladewig, B. P., Bhattacharya, S., Review of fuels for direct carbon fuel cells. *Energy Fuel* 2012;26(3):1471–1488. DOI: 10.1021/ef201694y
196. Zhou, W., Jiao, Y., Li, S. D., Shao, Z., Anodes for carbon-fueled solid oxide fuel cells. *ChemElectroChem* 2016;3(2):193–203. DOI: 10.1002/celc.201500420
197. Jiang, C., Ma, J., Corre, G., Jain, S. L., Irvine, J. T. S., Challenges in developing direct carbon fuel cells. *Chem Soc Rev* 2017;46(10):2889–2912. DOI: 10.1039/c6cs00784h
198. Glenn, M. J., Allen, J. A., Donne, S. W., Carbon electro-catalysis in the direct carbon fuel cell utilising alkali metal molten carbonates: A mechanistic review. *J Power Sources* 2020;453:227662. DOI: 10.1016/j.jpowsour.2019.227662
199. Cisneros, S., Sánchez, C., Kinetic characterization and molten KOH fuel cell simulation for dissolved coal oxidation on nickel anodes. *J Electrochem Soc* 2014;161(14):F1330–F1339. DOI: 10.1149/2.1181412jes

200. Ibrahim, U., Ayub, A., Direct carbon fuel cell-cleaner and efficient future power generation technology, *Adv J Grad Res* 2019;6(1):14–30. DOI: 10.21467/ajgr.6.1.14-30
201. Badwal, S. P. S., Giddey, S., The holy grail of carbon combustion – The direct carbon fuel cell technology. *Mater Forum* 2010;34:181–185.
202. Selman, J. R., Molten-salt fuel cells—Technical and economic challenges. *J Power Sources* 2006;160(2):852–857.
203. Wilemski, G., Simple porous electrode models for molten carbonate fuel cells. *J Electrochem Soc* 1983;130:117–121.
204. Goret, J., Tremillon, B., Propriétés chimiques et électrochimiques en solution dans les hydroxydes alcalins fondus—IV. Comportement électrochimique de quelques métaux utilisés comme électrodes indicatrices. *Electrochimica Acta* 1967;12(8):1065–1083.
205. Zecevic, S., Patton, E. M., Parhami, P., Direct carbon fuel cell with hydroxide electrolyte: Cell performance during initial stage of a long term operation. In *International Conference on Fuel Cell Science, Engineering and Technology*, 2005, pp. 507–514.
206. McPhail, S., Simonetti, E., Moreno, A., Bove, R., 7- Molten carbonate fuel cells. In M. Gasik (Ed), *Materials for Fuel Cells*. Woodhead Publishing, Salt Lake City, 2008, pp. 248–279.
207. Hu, L., Lindbergh, G., Lagergren, C., Performance and durability of the molten carbonate electrolysis cell and the reversible molten carbonate fuel cell. *J Phys Chem C* 2016;120(25):13427–13433.
208. Rao, Y. K., Adjorlolo, A., Haberman, J. H., On the mechanism of catalysis of the Boudouard reaction by alkali-metal compounds, *Carbon* 1982;20(3):207–212.
209. Kapteijn, F., Abbel, G., Moulijn, J. A., CO_2 gasification of carbon catalysed by alkali metals: Reactivity and mechanism. *Fuel* 1984;63(8):1036–1042.
210. Chen, M., Wang, C., Niu, X., Zhao, S., Tang, J., Zhu, B., Carbon anode in direct carbon fuel cell. *Int J Hydrog Energy* 2010;35(7):2732–2736.
211. Dicks, A., Siddle, A., Assessment of commercial prospects of molten carbonate fuel cells. *J Power Sources* 2000;86(1–2):316–323.
212. Ahn, S. Y., Eom, S. Y., Rhie, Y. H., Sung, Y. M., Moon, C. E., Choi, G.M., Kim, D. J., Utilization of wood biomass char in a direct carbon fuel cell (DCFC) system. *Appl Energ* 2013;105:207–216. DOI: 10.1016/j.apenergy.2013.01.023
213. Glugla, P., De Carlo, V., The specific conductance of molten carbonate fuel cell tiles. *J Electrochem Soc* 1982;129(8):1745. DOI: 10.1149/1.2124263
214. McKee, D. W., Spiro, C. L., Kosky, P. G., Lamby, E. J., Catalysis of coal char gasification by alkali metal salts. *Fuel* 1983;62(2), 217–220. DOI: 10.1016/0016-2361(83)90202-8
215. Janz, G.J., Lorena, M.R., Solid-liquid phase equilibria for mixtures of lithium, sodium, and potassium carbonates. *J Chem Eng Data* 1961;6(3):321–323. DOI: 10.1021/je00103a001
216. Kojima, T., Yanagida, M., Tanimoto, K., Tamiya, Y., Matsumoto, H., MiyazakiI, Y., The surface tension and the density of molten binary alkali carbonate systems. *Electrochemistry* 1999;67(6):593–602. DOI: 10.5796/electrochemistry.67.593
217. Posypaiko, V., Alekseeva, E., Vasina, N., *Diagrammy plavkosti solevykh sistem. Troynyye sistemy [Charts of salt systems' fusion. Triple systems]*. Chemistry, Moscow, 1977.
218. Licht, S., Stabilization of STEP electrolyses in lithium-free molten carbonates. arXiv preprint arXiv:1209.3512, 2012.
219. Weaver, R. D., Leach, S. C., Bayce, A. E., Nanis, L., *Direct Electrochemical Generation of Electricity from Coal*. SRI International Corp., Menlo Park, CA, 1979.
220. Mamantov, G., Mamantov, C., *Advances in Molten Salt Chemistry 5*. Elsevier Science Publishers Science and Technology Div, Amsterdam, 1983.
221. Vutetakis, D., Skidmore, D., Byker, H., Electrochemical oxidation of molten carbonate-coal slurries. *J Electrochem Soc* 1987;134(12):3027–3035. DOI: 10.1149/1.2100334

222. Jiang, C., Ma, J., Arenillas, A., Irvinea, J. T. S., Application of ternary carbonate in hybrid direct coal fuel cells. *ECS Trans* 2014;59(1):281–288. DOI: 10.1149/05901.0281ecst
223. Baur, E., Brunner, R., Über die eisenoxyd-kathode in der kohle-luft-kette. *Zeitschrift für Elektrochemie und angewandte physikalische Chemie* 1937;43(9):725–727. DOI: 10.1002/bbpc.19370430902
224. Anbar, M., Methods and apparatus for the pollution-free generation of electrochemical energy. US Patent No 3,741,809, 1973.
225. McKee, D. W., Gasification of graphite in carbon dioxide and water vapor—the catalytic effects of alkali metal salts. *Carbon* 1982;20(1):59–66. DOI: 10.1016/0008-6223(82)90075-6
226. Cooper, J. F., Direct conversion of coal and coal-derived carbon in fuel cells. In *The 2nd International Conference on Fuel Cell Science, Engineering and Technology*, 2004.
227. Cherepy, N. J., Krueger, R., Fiet, K. J., Jankowski, A. F., Cooper, J. F., Direct conversion of carbon fuels in a molten carbonate fuel cell. *J Electrochem Soc* 2005;152(1):A80–A87. DOI: 10.1149/1.1836129
228. Li, X., Zhu, Z. H., Roland, D. M., Andrew, D., John, B., Liu, S., Lu, G. Q., Factors that determine the performance of carbon fuels in the direct carbon fuel cell. *Ind Eng Chem Res* 2008;47(23), 9670–9677. DOI: 10.1021/ie800891m
229. Vutetakis, D., Skidmore, D., Byker, H., Electrochemical oxidation of molten carbonate-coal slurries. *J Electrochem Soc* 1987;134(12):3027–3035. DOI: 10.1149/1.2100334
230. Gür, T. M., Huggins, R. A., Direct electrochemical conversion of carbon to electrical energy in a high temperature fuel cell. *J Electrochem Soc* 1992;139(10), L95. DOI: 10.1149/1.2069025
231. Zhang, J., Zhong, Z., Shen, D., Xiao, J., Fu, Z., Zhang, H., Zhao, J., Li, W., Yang, M., Characteristics of a fluidized bed electrode for a direct carbon fuel cell anode. *J Power Sources* 2011;196(6), 3054–3059. DOI: 10.1016/j.jpowsour.2010.11.130
232. Zhang, J., Jiang, X., Piao, G., Yang, H., Zhong, Z., Simulation of a fluidized bed electrode direct carbon fuel cell. *Int J Hydrogen Energy* 2015;40(8):3321–3331. DOI: 10.1016/j.ijhydene.2014.12.090
233. Ido, A., Kawase, M., Development of a tubular molten carbonate direct carbon fuel cell and basic cell performance. *J Power Sources* 2020;449(15):227483. DOI: 10.1016/j.jpowsour.2019.227483
234. Watanabe, H., Umehara, D., Hanamura, K., Impact of gas products around the anode on the performance of a direct carbon fuel cell using a carbon/carbonate slurry. *J Power Sources* 2016;329:567–573. DOI: 10.1016/j.jpowsour.2016.08.122
235. Lee, E. -K., Chun, H. H., Kim, Y. -T., Enhancing Ni anode performance via Gd_2O_3 addition in molten carbonate-type direct carbon fuel cell. *Int J Hydrogen Energy* 2014;39(29):16541–16547. DOI: 10.1016/j.ijhydene.2014.03.180
236. Lee, E. -K., Park, S. -A., Jung, H. -W., Kim, Y. -T., Performance enhancement of molten carbonate-based direct carbon fuel cell (MC-DCFC) via adding mixed ionic-electronic conductors into Ni anode catalyst layer. *J Power Sources* 2018;386(15):28–33. DOI: 10.1016/j.jpowsour.2017.03.078
237. Bie, K., Fu, P., Liu, Y., Muhammad, A., Comparative study on the performance of different carbon fuels in a molten carbonate direct carbon fuel cell with a novel anode structure. *J Power Sources* 2020. DOI: 10.1016/j.jpowsour.2020.228101
238. Kudo, T., Hisamitsu, Y., Kihara, K., Mohamedi, M., Uchida, I., X-ray diffractometric study of in situ oxidation of Ni in Li/K and Li/Na carbonate eutectic. *J Power Sources* 2002;104(2):272–280. DOI: 10.1016/S0378-7753(01)00962-4
239. Soler, J., Gonzalez, T., Escudero, M., Rodrigo, T., Daza, L., Endurance test on a single cell of a novel cathode material for MCFC. *J Power Sources* 2002;106(1–2):189–195. DOI: 10.1016/S0378-7753(01)01041-2

240. Liu, Z., Guo, P., Zeng, C., Effect of Dy on the corrosion of NiO/Ni in molten (0.62Li, 0.38K)$_2CO_3$. *J Power Sources* 2007;166(2):348–353. DOI: 10.1016/j.jpowsour.2007.01.063
241. Suski, L., Kołacz, J., Mordarski, G., Ruggiero, M., Determination of open-circuit potentials at gas/electrode/YSZ boundary versus molten carbonate reference electrode at medium temperatures. *Electrochim Acta* 2005;50(14):2771–2780. DOI: 10.1016/j.electacta.2004.11.023
242. Wade, J. L., Lee, C., West, A. C., Lackner, K. S., Composite electrolyte membranes for high temperature CO_2 separation. *J Membr Sci* 2011;369(1–2):20–29. DOI: 10.1016/j.memsci.2010.10.05
243. Xu, X., Zhou, W., Zhu, Z., Stability of YSZ and SDC in molten carbonate eutectics for hybrid direct carbon fuel cells. *RSC Adv* 2014;4(5):2398–2403. DOI: 10.1039/c3ra46600k
244. Ju, H., Eom, J., Lee, J. K., Choi, H., Lim, T. -H., Song, R. -H., Lee, J., Durable power performance of a direct ash-free coal fuel cell. *Electrochim Acta* 2014;115:511–517. DOI: 10.1016/j.electacta.2013.10.124
245. Tulloch, J., Allen, J., Wibberley, L., Donne, S., Influence of selected coal contaminants on graphitic carbon electro-oxidation for application to the direct carbon fuel cell. *J Power Sources* 2014;260(15):140–149. DOI: 10.1016/j.jpowsour.2014.03.026
246. Eom, S., Ahn, S., Kang, K., Choi, G., Correlations between electrochemical resistances and surface properties of acid-treated fuel in coal fuel cells. *Energy* 2017;140(2):885–892. DOI: 10.1016/j energy.2017.09.034
247. Xie, H., Zhai, S., Chen, B., Liu, T., Zhang, Y., Ni, M., Shao, Z., Coal pretreatment and Ag-infiltrated anode for high-performance hybrid direct coal fuel cell. *Appl Energy* 2020. DOI: 10.1016/j.apenergy.2019.114197
248. Cai, W., Zhou, Q., Xie, Y., Liu, J., Long, G., Cheng, S., Liu, M., A direct carbon solid oxide fuel cell operated on a plant derived biofuel with natural catalyst. *Appl Energy* 2016;179(2):1232–1241. DOI: 10.1016/j.apenergy.2016.07.068
249. Hao, W., Mi, Y., Evaluation of waste paper as a source of carbon fuel for hybrid direct carbon fuel cells. *Energy* 2016;107(15):122–130. DOI: 10.1016/j.energy.2016.04.012
250. Hong, S. -G., Selman, J. R., Wetting characteristics of carbonate melts under MCFC operating conditions. *J Electrochem Soc* 2004;151(1):A77–A84. DOI: 10.1149/1.1629094
251. Cao, D., Wang, G., Wang, C., Wang, J., Lu, T., Enhancement of electrooxidation activity of activated carbon for direct carbon fuel cell. *Int J Hydrogen Energy* 2010;35(4):1778–1782. DOI: 10.1016/j.ijhydene.2009.12.133
252. Watanabe, H., Kimura, A., Okazaki, K., Impact of ternary carbonate composition on the morphology of the carbon/carbonate slurry and continuous power generation by direct carbon fuel cells. *Energy Fuel* 2016;30(3):1835–1840. DOI: 10.1021/acs.energyfuels.5b02224
253. Lan, R., Tao, S., A simple high-performance matrix-free biomass molten carbonate fuel cell without CO_2 recirculation. *Sci Adv* 2016;2(8):868. DOI: 10.1126/sciadv.1600772
254. Predtechensky, M. R., Varlamov, Y. D., Ul'yankin, S. N., Dubov, Y. D., Direct conversion of solid hydrocarbons in a molten carbonate fuel cell. *Thermophys Aeromech* 2009;16:601–610.
255. Li, C., Yi, H., Lee, D., On-demand supply of slurry fuels to a porous anode of a direct carbon fuel cell: Attempts to increase fuel-anode contact and realize long-term operation. *J Power Sources* 2016;309:99–107.
256. Frangini, S., Scaccia, S., Sensitive determination of oxygen solubility in alkali carbonate melts. *J Electrochem Soc* 2004;151:A1251–A1256.
257. Claes, P., Moyaux, D., Peeters, D., Solubility and solvation of carbon dioxide in the molten $Li_2CO_3/Na_2CO_3/K_2CO_3$ (43.5:31.5:25.0 mol-%) eutectic mixture at 973 K. *Eur J Inorg Chem* 1999;1999:583–588.

258. Pranda, P., Prandová, K., Hlavacek, V., Yang, F., Combustion of fly-ash carbon: Part II: Thermodynamic aspects and calorimetric experiment. *Fuel Process Technol* 2001;72:227–233.
259. Mohan, D., Pittman Jr, C. U., Steele, P. H., Pyrolysis of wood/biomass for bio-oil: A critical review. *Energy Fuels* 2006;20:848–889.
260. Lahijani, P., Zainal, Z. A., Mohammadi, M., Mohamed, A. R., Conversion of the greenhouse gas CO_2 to the fuel gas CO via the Boudouard reaction: A review. *Renew Sust Energy Rev* 2015;41:615–632. DOI: 10.1016/j.rser.2014.08.034
261. Elleuch, A., Boussetta, A., Halouani, K., Li, Y., Experimental investigation of direct carbon fuel cell fueled by almond shell biochar: Part II. Improvement of cell stability and performance by a three-layer planar configuration. *Int J Hydrogen Energy* 2013;38(36):16605–16614. DOI: 10.1016/j.ijhydene.2013.07.061
262. Bian, W., Wu, W., Orme, C. J., Ding, H., Zhou, M., Ding, D., Dual 3D ceramic textile electrodes: Fast kinetics for carbon oxidation reaction and oxygen reduction reaction in direct carbon fuel cells at reduced temperatures. *Adv Funct Mater* 2020;30(19):1910096. DOI: 10.1002/adfm.201910096
263. Nabae, Y., Pointon, K. D., Irvine, J. T. S., Electrochemical oxidation of solid carbon in hybrid DCFC with solid oxide and molten carbonate binary electrolyte. *Energy Environ Sci* 2008;1(1):148–155. DOI: 10.1039/b804785e
264. Lee, A. C., Mitchell, R. E., Gür, T. M., Thermodynamic analysis of gasification-driven direct carbon fuel cells. *J Power Sources* 2009;194(2):774–785. DOI: 10.1016/j.jpowsour.2009.05.039
265. Jiang, C., Irvine, J. T. S., Catalysis and oxidation of carbon in a hybrid direct carbon fuel cell. *J Power Sources* 2011;196(17):7318–7322. DOI: 10.1016/j.jpowsour.2010.11.066
266. Deleebeeck, L., Hansen, K. K., HDCFC performance as a function of anode atmosphere (N_2-CO_2). *J Electrochem Soc* 2013;161(1):F33–F46. DOI: 10.1149/2.027401jes
267. Li, S., Pan, W., Wang, S., Meng, X., Jiang, C., Irvine, J. T., Electrochemical performance of different carbon fuels on a hybrid direct carbon fuel cell. *Int J Hydrogen Energy* 2017;42(25):16279–16287. DOI: 10.1016/j.ijhydene.2017.05.150
268. Ma, J., Zhang, B., Hou, X., Gong, J., Yu, H., Xu, R., Jiang, C., The function of carbonate in a hybrid direct carbon fuel cell. *Solid State Ion* 2020;344:115094–115099. DOI: 10.1016/j.ssi.2019.115094
269. Cantero-Tubilla, B., Xu, C., Zondlo, J. W., Sabolsky, K., Sabolsky, E. M., Investigation of anode configurations and fuel mixtures on the performance of direct carbon fuel cells (DCFCs). *J Power Sources* 2013;238(2), 227–235. DOI: 10.1016/j.jpowsour.2013.03.072
270. Rodriguez-Reinoso, F., Thrower, P. A., Walker Jr, P. L., Kinetic studies of the oxidation of highly oriented pyrolytic graphites. *Carbon* 1974;12(1):63–70.
271. Stevens, F., Kolodny, L. A., Beebe, T. P., Kinetics of graphite oxidation: Monolayer and multilayer etch pits in HOPG studied by STM. *J Phys Chem B* 1998;102(52):10799–10804.
272. Kinoshita, K., *Carbon: Electrochemical and Physicochemical Properties*. John Wiley & Sons, New York, 1988.
273. Palniandy, L. K., Yoon, L. W., Wong, W. Y., Yong, S. T., Pang, M. M., Application of biochar derived from different types of biomass and treatment methods as a fuel source for direct carbon fuel cells. *Energies* 2019:12;2477. DOI: 10.3390/en12132477
274. Schlögl, R., *Chemical Energy Storage*. Walter de Gruyter, Berlin, 2012.
275. Palma, L., Enjeti, P. N., A modular fuel cell, modular DC–DC converter concept for high performance and enhanced reliability. *IEEE Trans Power Electron* 2009;24(6):1437–1443.
276. Matheny, M. S., Erickson, P. A., Niezrecki, C., Roan, V. P., Interior and exterior noise emitted by a fuel cell transit bus. *J Sound Vib* 2002;251(5):937–943.
277. Cowey, K., Green, K. J., Mepsted, G. O., Reeve, R., Portable and military fuel cells. *Curr Opin Solid State Mater Sci* 2004;8(5):367–371.

278. Patil, A. S., Dubois, T. G., Sifer, N., Bostic, E., Gardner, K., Quah, M., Bolton, C., Portable fuel cell systems for America's army: Technology transition to the field. *J Power Sources* 2004;136(2):220–225.
279. Varkaraki, E., Lymberopoulos, N., Zachariou, A., Hydrogen based emergency back-up system for telecommunication applications. *J Power Sources* 2003;118(1–2):14–22.
280. Abdullah, M. O., Yung, V. C., Anyi, M., Othman, A. K., Hamid, K. A., Tarawe, J. Review and comparison study of hybrid diesel/solar/hydro/fuel cell energy schemes for a rural ICT Telecenter. *Energy* 2010;35(2):639–646.
281. Bauen, A., Hart, D., Chase, A., Fuel cells for distributed generation in developing countries—an analysis. *Int J Hydrog Energy* 2003;28(7):695–701.
282. Briguglio, N., Ferraro, M., Brunaccini, G., Antonucci, V., Evaluation of a low temperature fuel cell system for residential CHP. *Int J Hydrog Energy* 2011;36(13):8023–8029.
283. Hussain, S., Yangping, L., Review of solid oxide fuel cell materials: Cathode, anode, and electrolyte. *Energy Transit* 2020;4:113–126. DOI: 10.1007/s41825-020-00029-8
284. Irshad, M., Siraj, K., Raza, R., Ali, A., Tiwari, P., Zhu, B., Rafique, A., Ali, A., Kaleem Ullah, M., Usman, A., A brief description of high temperature solid oxide fuel cell's operation, materials, design, fabrication technologies and performance. *Appl Sci* 2016;6(3):75. DOI: 10.3390/app6030075
285. Dziurdzia, B., Magonski, Z., Jankowski, H., Commercialization of solid oxide fuel cells—opportunities and forecasts. *IOP Conf Ser Mater Sci Eng* 2016. DOI: 10.1088/1757-899X/104/1/012020
286. Ruiz-Morales, J. C., Canales-Vázquez, J., Savaniu, C., Marrero-López, D., Zhou, W., Irvine, J. T. S., Disruption of extended defects in solid oxide fuel cell anodes for methane oxidation. *Nature* 2006;439(7076):568–571. DOI: 10.1038/nature04438
287. Mahato, N., Banerjee, A., Gupta, A., Omar, S., Balani, K., Progress in material selection for solid oxide fuel cell technology: A review. *Prog Mater Sci* 2015;72:141–337. DOI: 10.1016/j.pmatsci.2015.01.001
288. Chelmehsara, M. E., Mahmoudimehr, J., Techno-economic comparison of anode-supported, cathode-supported, and electrolyte-supported SOFCs. *Int J Hydrogen Energy* 2018;43:15521–15530 . DOI: 10.1016/j.ijhydene.2018.06.114
289. Wilson, J. R., Kobsiriphat, W., Mendoza, R., Yi Chen, H., Hiller, J. M., Miller, D. J., Thornton, K., Voorhees, P. W., Adler, S. B., Barnett, S. A., Three-dimensional reconstruction of a solid-oxide fuel-cell anode. *Nat Mater* 2006;5:541–544. DOI: 10.1038/nmat1668
290. Papandrew, A. B., Chisholm, C. R. I., Elgammal, R. A., Özer, M. M., Zecevic, S. K., Advanced electrodes for solid acid fuel cells by platinum deposition on CsH2PO4. *Chem Mater* 2011;23:1659–1667. DOI: 10.1021/cm101147y
291. Bessler, W. G., Vogler, M., Stormer, H., Gerthsen, D., Utz. A., Weber, A., Tiffee, E. I., Model anodes and anode models for understanding the mechanism of hydrogen oxidation in solid oxide fuel cells. *Phys Chem Chem Phys* 2010;12:13888–13903. DOI: 10.1039/c0cp00541j
292. Laosiripojana, N., Wiyaratn, W., Kiatkittipong, W., Arpornwichanop, A., Soottitantawat, A., Assabumrungrat, S., Reviews on solid oxide fuel cell technology. *Eng J* 2009;13:65–83. DOI: 10.4186/ej.2009.13.1.65
293. Guo, X., Waser, R., Electrical properties of the grain boundaries of oxygen ion conductors: Acceptor-doped zirconia and ceria. *Prog Mater Sci* 2006;51:151–210. DOI: 10.1016/j.pmatsci.2005.07.001
294. Fergus, J. W., Electrolytes for solid oxide fuel cells. J Power Sources 2006;162:30–40. DOI: 10.1016/j.jpowsour.2006.06.062
295. Ramesh, S., Raju, K. C. J., Preparation and characterization of $Ce_{1-x}(Gd_{0.5}Pr_{0.5})_xO_2$ electrolyte for IT-SOFCs. *Int J Hydrogen Energy* 2012;37:10311–10317. DOI: 10.1016/j.ijhydene.2012.04.008

296. Huang, J., Gao, Z., Mao, Z., Effects of salt composition on the electrical properties of samaria-doped ceria/carbonate composite electrolytes for low-temperature SOFCs. *Int J Hydrogen Energy* 2010;35:4270–4275. DOI: 10.1016/j.ijhydene.2010.01.063
297. Zhu, B., Liu, X., Zhu, Z., Ljungberg, R., Solid oxide fuel cell (SOFC) using industrial grade mixed rare-earth oxide electrolytes. *Int J Hydrogen Energy* 2008;33:3385–3392. DOI: 10.1016/j.ijhydene.2008.03.065
298. Leng, Y., Chan, S. H., Liu, Q., Development of LSCF-GDC composite cathodes for low-temperature solid oxide fuel cells with thin film GDC electrolyte. *Int J Hydrogen Energy* 2008;33:3808–3817. DOI: 10.1016/j.ijhydene.2008.04.034
299. Ferreira, A. S. V., Soares, C. M. C., Figueiredo, F. M. H. L. R., Marques, F. M. B., Intrinsic and extrinsic compositional effects in ceria/carbonate composite electrolytes for fuel cells. *Int J Hydrogen Energy* 2011;36:3704–3711. DOI: 10.1016/j.ijhydene.2010.12.025
300. Jacobson, A. J., Materials for solid oxide fuel cells. *Chem Mater* 2010;22:660–674. DOI: 10.1021/cm902640j
301. da Silva, F. S., de Souza, T. M., Novel materials for solid oxide fuel cell technologies: A literature review. *Int J Hydrogen Energy* 2017;42:26020–26036. DOI: 10.1016/j.ijhydene.2017.08.105
302. Song, X., Dong, X., Li, M., Wang, H., Effects of adding alumina to the nickel-zirconia anode materials for solid oxide fuel cells and a two-step sintering method for half-cells. *J Power Sources* 2016;308:58–64. DOI: 10.1016/j.jpowsour.2016.01.070
303. Wang, F., Wang, W., Ran, R., Tade, M. O., Shao, Z., Aluminum oxide as a dual-functional modifier of Ni-based anodes of solid oxide fuel cells for operation on simulated biogas. *J Power Sources* 2014;268:787–793. DOI: 10.1016/j.jpowsour.2014.06.087
304. Tian, X. Y., Zhang, J., Zuo, W., Kong, X., Wang, J., Sun, K., Zhou, X., Enhanced electrochemical performance and carbon anti-coking ability of solid oxide fuel cells with silver modified nickel-yttrium stabilized zirconia anode by electroless plating. *J Power Sources* 2016;301:143–150. DOI: 10.1016/j.jpows our.2015.10.006
305. Marina, O.A., Pederson, L. R., Williams, M. C., Coffey, G. W., Meinhard, K. D., Nguyen, C. D., Thomsen, E. C., Electrode performance in reversible solid oxide fuel cells. *J Electrochem Soc* 2007;154:B452. DOI: 10.1149/1.2710209
306. Bastidas, D. M., Tao, S., Irvine, J. T. S., A symmetrical solid oxide fuel cell demonstrating redox stable perovskite electrodes. *J Mater Chem* 2006;16:1603–1605. DOI: 10.1039/b600532b
307. Liu, Y., Jia, L., Li, J., Chi, B., Pu, J., Li, J., High-performance Ni in-situ exsolved $Ba(Ce_{0.9}Y_{0.1})_{0.8}Ni_{0.2}O_{3-\delta}/Gd_{0.1}Ce_{0.9}O_{1.95}$ composite anode for SOFC with long-term stability in methane fuel. *Compos. Part B Eng* 2020;193:108033. DOI: 10.1016/j.compositesb.2020.108033
308. Lu, Z. G., Zhu, J. H., Bi, Z. H., Lu, X. C., A Co-Fe alloy as alternative anode for solid oxide fuel cell. *J Power Sources* 2008;180:172–175. DOI: 10.1016/j.jpowsour.2008.02.051
309. Wei, T., Ji, Y., Meng, X., Zhang, Y., $Sr_2NiMoO_{6-\delta}$ as anode material for $LaGaO_3$-based solid oxide fuel cell. *Electrochem Commun* 2008;10:369–1372. DOI: 10.1016/j.elecom.2008.07.005
310. Cowin, P. I., Petit, C. T. G., Lan, R., Irvine, J. T. S., Tao, S., Recent progress in the development of anode materials for solid oxide fuel cells. *Adv Energy Mater* 2011;1:314–332. DOI: 10.1002/aenm.201100108
311. Mumtaz, S., Ahmad, M. A., Raza, R., Khan, M. A., Ashiq, M. N., Abbas, G., Nanostructured anode materials for low temperature solid oxide fuel cells: Synthesis and electrochemical characterizations. *Ceram Int* 2019;45:21688–21697. DOI: 10.1016/j.ceramint.2019.07.169
312. Benamira, M., Thommy, L., Moser, F., Joubert, O., Caldes, M. T., New anode materials for IT-SOFC derived from the electrolyte $BaIn_{0.3}Ti_{0.7}O_{2.85}$ by lanthanum and manganese doping. *Solid State Ionics* 2014;265:38–45. DOI: 10.1016/j.ssi.2014.07.006

313. Moura, C. G., Paulo, J., Grilo, D. F., Maribondo, R., A brief review on anode materials and reactions mechanism in solid oxide fuel cells. *Front Ceram Sci* 2017;9:26–41. DOI: 10.2174/9781681084312117010007
314. Sinha, A., Miller, D. N., Irvine, J. T. S., Development of novel anode material for intermediate temperature SOFC (IT-SOFC). *J Mater Chem A* 2016;4:11117–11123. DOI: 10.1039/c6ta03404g
315. Rossmeisl, J., Bessler, W. G., Trends in catalytic activity for SOFC anode materials. *Solid State Ionics* 2008;178:1694–1700. DOI: 10.1016/j.ssi.2007.10.016
316. Yang, X., Chen, J., Panthi, D., Niu, B., Lei, L., Yuan, Z., Du, Y., Li, Y., Chen, F., He, T., Electron doping of $Sr_2FeMoO_{6-\delta}$ as high performance anode materials for solid oxide fuel cells. *J Mater Chem A* 2019;7:733–743. DOI: 10.1039/c8ta10061f
317. Verbraeken, M. C., Iwanschitz, B., Mai, A., Irvine, J. T. S., Evaluation of Ca doped $La_{0.2}Sr_{0.7}TiO_3$ as an alternative material for use in SOFC anodes. *J Electrochem Soc* 2012;159:757–762. DOI: 10.1149/2.001212jes
318. Steiger, P., Burnat, D., Madi, H., Mai, A., Holzer, L., Herle, J.V., Krocher, O., Heel, A., Ferri, D., Sulfur poisoning recovery on a solid oxide fuel cell anode material through reversible segregation of Nickel. *Chem Mater* 2019;31:748–758. DOI: 10.1021/acs.chemmater.8b03669
319. Zha, S., Cheng, Z., Liu, M., A sulfur-tolerant anode material for SOFCs $Gd_2Ti_{1.4}Mo_{0.6}O_7$. *Electrochem Solid-State Lett* 2005;8:406–408. DOI: 10.1149/1.1945370
320. Li, F., Zhang, J., Luan, J., Liu, Y., Han, J., Preparation of Bi_2O_3-doped NiO/YSZ anode materials for SOFCs. *Surf Rev Lett* 2017;24:1750092. DOI: 10.1142/s0218625x17500925
321. Dong, G., Yang, C., He, F., Jiang, Y., Ren, C., Gan, Y., Lee, M., Xue, X., Tin doped $PrBaFe_2O_{5+\delta}$ anode material for solid oxide fuel cells. *RSC Adv* 2017;7:22649–22661. DOI: 10.1039/c7ra03143b
322. Shaheen, K., Shah, Z., Gulab, H., Hanif, M. B., Faisal, S., Suo, H., Metal oxide nanocomposites as anode and cathode for low temperature solid oxide fuel cell. *Solid State Sci* 2020;102:106162.
323. Golkhatmi, S. Z., Asghar, M. I., Lund, P. D., A review on solid oxide fuel cell durability: Latest progress, mechanisms, and study tools. *Renew Sustain Energy Rev* 2022;161:112339.
324. Sun, J., Liu, X., Han, F., Zhu, L., Bi, H., Wang, H., Yu, S., Pei, L., $NdBa_{1-x}Co_2O_{5+\delta}$ as cathode materials for IT-SOFC. *Solid State Ionics* 2016;288:54–60. DOI: 10.1016/j.ssi.2015.12.023
325. Choi, S., Yoo, S., Kim, J., Park, S., Jun, A., Sengodan, S., Kim, J., Shin, J., Jeong, H. Y., Choi, Y., Kim, G., Liu, M., Highly efficient and robust cathode materials for low-temperature solid oxide fuel cells: $PrB\alpha_{0.5}Sr_{0.5}Co_{2-x}Fe_xO_{5+\delta}$ *Sci Rep* 2013;3:3–8: DOI: 10.1038/srep02426
326. Ma, J., Tao, Z., Kou, H., Fronzi, M., Bi, L., Evaluating the effect of Pr-doping on the performance of strontium-doped lanthanum ferrite cathodes for protonic SOFCs. *Ceram Int* 2020;46:4000–4005. DOI: 10.1016/j.ceramint.2019.10.017
327. Dai, H., Kou, H., Tao, Z., Liu, K., Xue, M., Zhang, Q., Bi, L., Optimization of sintering temperature for SOFCs by a co-firing method. *Ceram Int* 2020;46:6987–6990. DOI: 10.1016/j.ceramint.2019.11.134
328. Baumann, F. S., Fleig, J., Cristiani, G., Stuhlhofer, B., Habermeier, H. -U., Maier, J., Quantitative comparison of mixed conducting SOFC cathode materials by means of thin film model electrodes. *J Electrochem Soc* 2007;154:B931–B941. DOI: 10.1149/1.2752974
329. Kim, J., Sengodan, S., Kwon, G., Ding, D., Shin, J., Liu, M., Kim, G., Triple-conducting layered perovskites as cathode materials for proton-conducting solid oxide fuel cells. *Chemsuschem* 2014;7:2811–2815. DOI: 10.1002/cssc.201402351

330. Wang, B., Bi, L., Zhao, X. S., Liquid-phase synthesis of $SrCo_{0.9}Nb_{0.1}O_{3-\delta}$ cathode material for proton-conducting solid oxide fuel cells. *Ceram Int* 2018;44:5139–5144. DOI: 10.1016/j.ceramint.2017.12.116
331. Ding, H., Xue, X., Cobalt-free layered perovskite $GdBaFe_2O_{5+x}$ as a novel cathode for intermediate temperature solid oxide fuel cells. *J Power Sources* 2010;195:4718–4721. DOI: 10.1016/j.jpowsour.2010.02.027
332. Lee, S. J., Yong, S. M., Kim, D. S., Kim, D. K., Cobalt-free composite cathode for SOFCs: Brownmillerite-type calcium ferrite and gadolinium-doped ceria. *Int J Hydrogen Energy* 2012;37:17217–17224. DOI: 10.1016/j.ijhydene.2012.08.100
333. Zhou, Q., Zhang, L., He, T., Cobalt-free cathode material $SrFe_{0.9}Nb_{0.1}O_{3-\delta}$ for intermediate-temperature solid oxide fuel cells. *Electrochem Commun* 2010;12:285–287. DOI: 10.1016/j.elecom.2009.12.016
334. Jiang, S., Zhou, W., Niu, Y., Zhu, Z., Shao, Z., Phase transition of a cobalt-free perovskite as a high-performance cathode for intermediate-temperature solid oxide fuel cells. *Chemsuschem* 2012;5:2023–2031. DOI: 10.1002/cssc.201200264
335. Yu, X., Long, W., Jin, F., He, T., Cobalt-free perovskite cathode materials $SrFe_{1-x}Ti_xO_{3-\delta}$ and performance optimization for intermediate-temperature solid oxide fuel cells. *Electrochim Acta* 2014;123:426–434.
336. Ling, Y., Zhao, L., Lin, B., Dong, Y., Zhang, X., Meng, G., Liu, X., Investigation of cobalt-free cathode material $Sm_{0.5}Sr_{0.5}Fe_{0.8}Cu_{0.2}O_{3-\delta}$ for intermediate temperature solid oxide fuel cell. *Int J Hydrogen Energy* 2010;35:6905–6910. DOI: 10.1016/j.ijhydene.2010.04.021
337. Zhu, Z., Yan, L., Sun, W., Liu, H., Liu, T., Liu, W., A cobalt-free composite cathode prepared by a superior method for intermediate temperature solid oxide fuel cells. *J Power Sources* 2012;217:431–436. DOI: 10.1016/j.jpowsour.2012.06.049
338. Duan, C., Hook, D., Chen, Y., Tong, J., O'Hayre, R., Zr and Y co-doped perovskite as a stable, high performance cathode for solid oxide fuel cells operating below 500°C. *Energy Environ Sci* 2017;10:176–182. DOI: 10.1039/c6ee01915c
339. Zhang, Z., Zhu, Y., Zhong, Y., Zhou, W., Shao, Z., Anion doping: A new strategy for developing high-performance perovskite-type cathode materials of solid oxide fuel cells. *Adv Energy Mater* 2017;7:1–9. DOI: 10.1002/aenm.201700242
340. Li, M., Zhao, M., Li, F., Zhao, W., Peterson, V. K., Xu, X., Shao, Z., Gentle, I., Zhu, Z., A niobium and tantalum co-doped perovskite cathode for solid oxide fuel cells operating below 500°C. *Nat Commun* 2017;8:1–9. DOI: 10.1038/ncomm s13990
341. Hussain, M., Muneer, M., Abbas, G., Shakir, I., Iqbal, A., Javed, M. A., Iqbal, M., Ur Rehman, Z., Raza, R., Cobalt free $La_xSr_{1-x}Fe_{1-y}Cu_yO_{3-\delta}$ ($x = 0.54, 0.8, y = 0.2, 0.4$) perovskite structured cathode for SOFC. *Ceram Int* 2020;46:18208–18215. DOI: 10.1016/j.ceramint.2020.04.143
342. Salvatore Arico, A., Bruce, P., Scrosati, B., Tarascon, J. M., Van Schalkwijk, W., A siloxane-incorporated copolymer as an in situ cross-linkable binder for high performance silicon anodes in Li-ion batteries. *Nat Mater* 2005;4:366–377. DOI: 10.1039/c6nr01559j
343. Yuan, Q., Duan, H. H., Le Li, L., Sun, L. D., Zhang, Y. W., Yan, C. H., Controlled synthesis and assembly of ceria-based nanomaterials. *J Colloid Interface Sci* 2009;335:151–167. DOI: 10.1016/j.jcis.2009.04.007
344. Dong, Y., Hampshire, S., Zhou, J. E., Meng, G., Synthesis and sintering of Gd-doped CeO_2 electrolytes with and without 1 at.% CuO doping for solid oxide fuel cell applications. *Int J Hydrogen Energy* 2011;36:5054–5066. DOI: 10.1016/j.ijhydene.2011.01.030
345. Bellino, M. G., Sacanell, J. G., Lamas, D. G., Leyva, A. G., Walsöe De Reca, N. E., High-performance solid-oxide fuel cell cathodes based on cobaltite nanotubes. *J Am Chem Soc* 2007;129:3066–3067. DOI: 10.1021/ja068115b
346. Martinelli, H., Lamas, D. G., Leyva, A. G., Sacanell, J., Influence of particle size and agglomeration in solid oxide fuel cell cathodes using manganite nanoparticles. *Mater Res Express* 2018;5:075013–075025.

347. Zhi, M., Lee, S., Miller, N., Menzler, N. H., Wu, N., An intermediate-temperature solid oxide fuel cell with electrospun nanofiber cathode. *Energy Environ Sci* 2012;5;7066–7071. DOI: 10.1039/c2ee02619h
348. Ishihara, T., Nanomaterials for advanced electrode of low temperature solid oxide fuel cells (SOFCs). *J Korean Ceram Soc* 2016:53:469–477. DOI: 10.4191/kcers.2016.53.5.469
349. Yoon, J., Cho, S., Kim, J. H., Lee, J. -H., Bi, Z., Serquis, A., Zhang, X., Manthiram, A., Wang, H., Vertically aligned nanocomposite thin films as a cathode/electrolyte interface layer for thin-film solid oxide fuel cells. *Adv Funct Mater* 2009;19:3868–3873. DOI: 10.1002/adfm.200901338
350. Evans, A., Benal, C., Darbandi, A. J., Hahn, H., Martynczuk, J., Gauckler, L. J., Prestat, M., Integration of spin-coated nanoparticulate-based $La_{0.6}Sr_{0.4}CoO_{3-\delta}$ cathodes into micro-solid oxide fuel cell membranes. *Fuel Cells* 2013;13:441–444. DOI: 10.1002/fuce.201300020
351. Chavan, A., Jamale, A., Patil, S., Jadhav, A., Effect of variation of NiO on properties of NiO/GDC (gadolinium doped ceria) nano-composites. *Ceram Int* 2012;38(4):3191–3196. DOI: 10.1016/j.ceramint.2011.12.023
352. Fu, C., Chan, S. H., Liu, Q., Ge, X., Pasciak, G., Fabrication and evaluation of Ni-GDC composite anode prepared by aqueous-based tape casting method for low-temperature solid oxide fuel cell. *Int J Hydrogen Energy* 2010;35:301–307. DOI: 10.1016/j.ijhydene.2009.09.101
353. Gil, V., Moure, C., Tartaj, J., Sinterability, microstructures and electrical properties of Ni/Gd-doped ceria cermets used as anode materials for SOFCs. *J Eur Ceram Soc* 2007;27:4205–4209. DOI: 10.1016/j.jeurceramsoc.2007.02.119
354. Chen, J. C., Hwang, B. H., Microstructure and properties of the Ni-CGO composite anodes prepared by the electrostatic-assisted ultrasonic spray pyrolysis method. *J Am Ceram Soc* 2008;91:97–102. DOI: 10.1111/j.1551–2916.2007.02109.x
355. Ding, C., Lin, H., Sato, K., Hashida, T., Synthesis of NiO-$Ce_{0.9}Gd_{0.1}O_{1.95}$ nanocomposite powders for low-temperature solid oxide fuel cell anodes by co-precipitation. *Scr Mater* 2009;60:254–256. DOI: 10.1016/j.scriptamat.2008.10.020
356. Guan, Y., Li, W., Gong, Y., Liu, G., Zhang, X., Chen, J., Gelb, J., Yun, W., Xiong, Y., Tian, Y., Wang, H., Analysis of the three-dimensional microstructure of a solid-oxide fuel cell anode using nano X-ray tomography. *J Power Sources* 2011;196:1915–1919. DOI: 10.1016/j.jpowsour.2010.09.059
357. Boldrin, P., Trejo, E. R., Yu, J., Gruar, R. I., Tighe, C. J., Chang, K.C., Ilavsky, J., Darr, J.A., Brandon, N., Nanoparticle scaffolds for syngas-fed solid oxide fuel cells. *J Mater Chem A* 2015;3:3011–3018. DOI: 10.1039/c4ta06029f
358. Pelegrini, L., Neto, J. B. R., Hotza, D., Process and materials improvements on Ni/Cu-YSZ composites towards nanostructured SOFC anodes: A review. *Rev Adv Mater Sci* 2016;46:6–21.
359. Kwon, O., Kim, K., Joo, S., Jeong, H.Y., Shin, J., Han, J. W., Sengodan, S., Kim, G., Self-assembled alloy nanoparticles in a layered double perovskite as a fuel oxidation catalyst for solid oxide fuel cells. *J Mater Chem A* 2018;6 15947–15953. DOI: 10.1039/c8ta05105d
360. Cavallaro, A., Pramana, S. S., Trejo, E. R., Sherrel, P. C., Ware, E., Kilner, J. A., Skinner, S. J., Amorphous-cathode-route towards low temperature SOFC. *Sustain Energy Fuels* 2018;2:862–875. DOI: 10.1039/c7se00606c
361. Slodczyk, A., Torrell, M., Hornés, A., Morata, A., Kendall, K., Tarancón, A., Understanding longitudinal degradation mechanisms of large-area micro-tubular solid oxide fuel cells. *Electrochim Acta* 2018;265:232–243.
362. Batfalsky, P., Malzbender, J., Menzler, N. H., Post-operational characterization of solid oxide fuel cell stacks. *Int J Hydrogen Energy* 2016;41:11399–11411.

363. Peng, J., Zhao, D., Xu, Y., Wu, X., Li, X., Comprehensive analysis of solid oxide fuel cell performance degradation mechanism, prediction, and optimization studies. *Energies* 2023;16:788. DOI: 10.3390/en16020788
364. Corigliano, O., Pagnotta, L., Fragiacomo, P., On the technology of solid oxide fuel cell (SOFC) energy systems for stationary power generation: A review, *Sustainability* 2022:14;15276. DOI: 10.3390/su142215276
365. Andersson, M., Sundén, B., *Technology Review – Solid Oxide Fuel Cell.* Energiforsk, Stockholm, Sweden, 2017.
366. Yokokawa, H., Suzuki, M., Yoda, M., Suto, T., Tomida, K., Hiwatashi, K., Shimazu, M., Kawakami, A., Sumi, H., Ohmori, M., Ryu, T., Achievements of NEDO durability projects on SOFC stacks in the light of physicochemical mechanisms. *Fuel Cell* 2019;19:311–339. DOI: 10.1002/fuce.201800187
367. Anjum, U., Agarwal, M., Khan, T. S., Gupta, R. K., Haider, M. A., Controlling surface cation segregation in a nanostructured double perovskite $GdBaCo_2O_{5+\delta}$ electrode for solid oxide fuel cells. *Nanoscale* 2019;11:21404–21418.
368. Skafte, T. L., Hjelm, J., Blennow, P., Graves, C., Quantitative review of degradation and lifetime of solid oxide cells and stacks. In *Proceedings of the 12th European SOFC & SOE Forum*, Lucerne, Switzerland, 2016, pp. 8–27.
369. Hattori, M., Takeda, Y., Sakaki, Y., Nakanishi, A., Ohara, S., Mukai, K., Lee, J. H., Fukui, T., Effect of aging on conductivity of yttria stabilized zirconia. *J Power Sources* 2004;126:23–27. DOI: 10.1016/j.jpowsour.2003.08.018
370. Gao, P., Bolon, A., Taneja, M., Xie, Z., Orlovskaya, N., Radovic, M., Thermal expansion and elastic moduli of electrolyte materials for high and intermediate temperature solid oxide fuel cell. *Solid State Ionics* 2017;300:1–9. DOI: 10.1016/j. ssi.2016.11.015
371. Sun, Y., He, S., Saunders, M., Chen, K., Shao, Z., A comparative study of surface segregation and interface of $La_{0.6}Sr_{0.4}Co_{0.2}Fe_{0.8}O_{3-\delta}$ electrode on GDC and YSZ electrolytes of solid oxide fuel cells. *Int J Hydrogen Energy* 2021;46:2606–2616. DOI: 10.1016/j. ijhydene.2020.10.113
372. Brandon, N., *Solid Oxide Fuel Cell Lifetime and Reliability: Critical Challenges in Fuel Cells.* Academic Press, Cambridge, 2017.
373. Sun, C., Hui, R., Roller, J., Cathode materials for solid oxide fuel cells: A review. *J Solid State Electrochem* 2010;14:1125–1144.
374. Parhizkar, T., Roshandel, R., Long term performance degradation analysis and optimization of anode supported solid oxide fuel cell stacks. *Energy Convers Manag* 2017;133:20–30. DOI: 10.1016/j.enconman.2016.11.045
375. Bishop, S. R., Chemical expansion of solid oxide fuel cell materials: A brief overview. *Acta Mech Sin* 2013;29:312–317.
376. Holzer, L., Münch, B., Iwanschitz, B., Cantoni, M., Hocker, T., Graule, T., Quantitative relationships between composition, particle size, triple phase boundary length and surface area in nickel-cermet anodes for solid oxide fuel cells. *J Power Sources* 2011;196:7076–7089. DOI: 10.1016/j.jpowsour.2010.08.006
377. Xu, C., Zondlo, J. W., Finklea, H. O., Demircan, O., Gong, M., Liu, X., The effect of phosphine in syngas on Ni–YSZ anode-supported solid oxide fuel cells. *J Power Sources* 2009;193:739–746.
378. Kan, H., Lee, H., Sn-doped Ni/YSZ anode catalysts with enhanced carbon deposition resistance for an intermediate temperature SOFC. *Appl Catal B Environ* 2010;97:108–114. DOI: 10.1016/j.apcatb.2010.03.029
379. Ryan, E. M., Xu, W., Sun, X., Khaleel, M. A., A damage model for degradation in the electrodes of solid oxide fuel cells: Modeling the effects of sulfur and antimony in the anode. *J Power Sources* 2012;210:233–242. DOI: 10.1016/j. jpowsour.2012.02.091
380. Trembly, J. P., Marquez, A. I., Ohrn, T. R., Bayless, D. J., Effects of coal syngas and H_2S on the performance of solid oxide fuel cells: Single-cell tests. *J Power Sources* 2006;158:263–273. DOI: 10.1016/j.jpowsour.2005.09.055

381. Zhi, M., Chen, X., Finklea, H., Celik, I., Wu, N. Q., Electrochemical and microstructural analysis of nickel–yttria-stabilized zirconia electrode operated in phosphorus- containing syngas. *J Power Sources* 2008;183:485–490. DOI: 10.1016/j. jpowsour.2008.05.055
382. Madi, H., Lanzini, A., Papurello, D., Diethelm, S., Ludwig, C., Santarelli, M., Solid oxide fuel cell anode degradation by the effect of hydrogen chloride in stack and single cell environments. *J Power Sources* 2016;326:349–356. DOI: 10.1016/j. jpowsour.2016.07.003
383. Haga, K., Shiratori, Y., Ito, K., Sasaki, K., Chlorine poisoning of SOFC Ni-cermet anodes. *J Electrochem Soc* 2008;155:B1233.
384. Marina, O. A., Pederson, L. R., Coyle, C. A., Thomsen, E. C., Nachimuthu, P., Edwards, D. J., Electrochemical, structural and surface characterization of nickel/zirconia solid oxide fuel cell anodes in coal gas containing antimony. *J Power Sources* 2011; 196:4911–4922. DOI: 10.1016/j.jpowsour.2011.02.027
385. Karczewski, J., Dunst, K. J., Jasinski, P., Molin, S., High temperature corrosion and corrosion protection of porous Ni22Cr alloys. *Surf Coating Technol* 2015;261:385–390. DOI: 10.1016/j.surfcoat.2014.10.051
386. Talic, B., Falk-Windisch, H., Venkatachalam, V., Hendriksen, P. V., Wiik, K., Lein, H. L., Effect of coating density on oxidation resistance and Cr vaporization from solid oxide fuel cell interconnects. *J Power Sources* 2017;354:57–67. DOI: 10.1016/j. jpowsour.2017.04.023
387. Yang, Z., Walker, M. S., Singh, P., Stevenson, J. W., Norby, T., Oxidation behavior of ferritic stainless steels under SOFC interconnect exposure conditions. *J Electrochem Soc* 2004;151:B669.
388. Kim, B. K., Kim, D. I., Yi, K. W., Suppression of Cr evaporation by Co electroplating and underlying Cr retention mechanisms for the 22wt.% Cr containing ferritic stainless steel. *Corrosion Sci* 2018;130:45–55. DOI: 10.1016/j. corsci.2017.10.019
389. Zhang, W., Hua, B., Yang, J., Chi, B., Pu, J., Jian, L., Performance evaluation of a new Fe- Cr-Mn alloy in the reducing atmosphere of solid oxide fuel cells. *J Alloys Compd* 2018;769:866–872. DOI: 10.1016/j.jallcom.2018.08.002
390. Stanislowski, M., Froitzheim, J., Niewolak, L., Quadakkers, W. J., Hilpert, K., Markus, T., Singheiser, L., Reduction of chromium vaporization from SOFC interconnectors by highly effective coatings. *J Power Sources* 2007;164:578–589.
391. Zanchi, E., Sabato, A. G., Molin, S., Cempura, G., Boccaccini, A. R., Smeacetto, F., Recent advances on spinel-based protective coatings for solid oxide cell metallic interconnects produced by electrophoretic deposition. *Mater Lett* 2021;286:129229. DOI: 10.1016/j.matlet.2020.129229
392. Rautanen, M., Himanen, O., Saarinen, V., Kiviaho, J., Compression properties and leakage tests of mica-based seals for SOFC stacks. *Fuel Cell* 2009;9:753–759.
393. Krainova, D. A., Saetova, N. S., Kuzmin, A. V., Raskovalov, A. A., Eremin, V. A., Ananyev, M. V., Steinberger-Wilckens, R., Non-crystallising glass sealants for SOFC: Effect of Y_2O_3 addition. *Ceram Int* 2020;46:5193–5200. DOI: 10.1016/j. ceramint.2019.10.266.
394. Chou, Y. -S., Stevenson, J. W., Novel infiltrated phlogopite mica compressive seals for solid oxide fuel cells. *J Power Sources* 2004;135:72–78. DOI: 10.1016/j. jpowsour.2004.02.037
395. Rautanen, M., Thomann, O., Himanen, O., Tallgren, J., Kiviaho, J., Glass coated compressible solid oxide fuel cell seals. *J Power Sources* 2014;247:243–248. DOI: 10.1016/j. jpowsour.2013.08.085
396. Sreedhar, I., Agarwal, B., Goyal, P., Singh, S. A., Recent advances in material and performance aspects of solid oxide fuel cells. *J Electroanal Chem* 2019;848:113315.
397. Jiang, S. P., Christiansen, L., Hughan, B., Foger, K., Effect of glass sealant materials on microstructure and performance of Sr-doped $LaMnO_3$ cathodes. *J Mater Sci Lett* 2001;20:695–697.

398. Kiebach, R., Agersted, K., Zielke, P., Ritucci, I., Brock, M. B., Hendriksen, P. V., A novel SOFC/SOEC sealing glass with a low SiO_2 content and a high thermal expansion coefficient. *ECS Trans* 2017;78:1739.
399. Fang, Q., Blum, L., Stolten, D., Electrochemical performance and degradation analysis of an SOFC short stack following operation of more than 100,000 hours. *J Electrochem Soc* 2019;166:F1320.
400. Fang, Q., Packbier, U., Blum, L., Long-term tests of a Jülich planar short stack with reversible solid oxide cells in both fuel cell and electrolysis modes. *Int J Hydrogen Energy* 2013;38:4281–4290.
401. Lashtabeg, A., Skinner, S. J., Solid oxide fuel cells—a challenge for materials chemists? *J Mater Chem* 2006;16:3161–3170.
402. Whiston, M. M., Azevedo, I. M., Litster, S., Samaras, C., Whitefoot, K. S., Whitacre, J. F., Meeting US solid oxide fuel cell targets. *Joule* 2019;3:2060–2065.
403. Zhang, Y. -C., Jiang, W., Tu, S. -T., Wang, C. -L., Chen, C., Effect of operating temperature on creep and damage in the bonded compliant seal of planar solid oxide fuel cell. *Int J Hydrogen Energy* 2018;43:4492–4504. DOI: 10.1016/j. ijhydene.2018.01.048
404. Fang, Q., de Haart, U., Schäfer, D., Thaler, F., Rangel-Hernandez, V., Peters, R., Blum, L., Degradation analysis of an SOFC short stack subject to 10,000 h of operation. *J Electrochem Soc* 2020;167:144508.
405. Mai, A., Iwanschitz, B., Schuler, J. A., Denzler, R., Nerlich, V., Schuler, A., Hexis' SOFC system Galileo 1000 N–Lab and field test experiences. *ECS Trans* 2013;57:73.
406. Chou, Y. S., Stevenson, J. W., Choi, J. P., Long-term evaluation of solid oxide fuel cell candidate materials in a 3-cell generic short stack fixture, part I: Test fixture, sealing, and electrochemical performance. *J Power Sources* 2014;255:1–8. DOI: 10.1016/j. jpowsour.2013.12.067
407. Ido, A., Asano, K., Morita, H., Yamamoto, T., Mugikura, Y., Degradation analysis of SOFC performance (1)—severe operation with high fuel utilization. *ECS Trans* 2019;91:801.
408. Thaheem, I., Joh, D. W., Noh, T., Lee, K. T., Highly conductive and stable $Mn_{1.35}Co_{1.35}Cu_{0.2}Y_{0.1}O_4$ spinel protective coating on commercial ferritic stainless steels for intermediate-temperature solid oxide fuel cell interconnect applications. *Int J Hydrogen Energy* 2019;44:4293–4303. DOI: 10.1016/j.ijhydene.2018.12.173
409. Kluczowski, R., Kawalec, M., Krauz, M., Świeca, A., Types, fabrication, and characterization of solid oxide fuel cells. In J. Kupecki (Eds), *Modeling, Design, Construction, and Operation of Power Generators with Solid Oxide Fuel Cells: From Single Cell to Complete Power System*. Springer, Cham, 2018, pp. 21–47.
410. Fang, X., Zhu, J., Lin, Z., Effects of electrode composition and thickness on the mechanical performance of a solid oxide fuel cell. *Energies* 2018;11:1735.
411. Huang, K., Singhal, S. C. Cathode-supported tubular solid oxide fuel cell technology: A critical review. *J Power Sources* 2013;237:84–97.
412. Behling, N. H., History of solid oxide fuel cells. In N. H. Behling (Ed), *Fuel Cells: Current Technology Challenges and Future Research Needs*. Elsevier, Amsterdam, the Netherlands, 2013, pp. 223–421. DOI: 10.1016/B978-0-444-56325-5.00006-5
413. Park, S., Sammes, N. M., Song, K. -H., Kim, T., Chung, J. -S., Monolithic at tubular types of solid oxide fuel cells with integrated electrode and gas channels. *Int J Hydrogen Energy* 2017;42:1154–1160. DOI: 10.1016/j.ijhydene.2016.08.212
414. Kuterbekov, K. A., Nikonov, A. V., Bekmyrza, K. Z., Pavzderin, N. B., Kabyshev, A. M., Kubenova, M. M., Kabdrakhimova, G. D., Aidarbekov, N., Classification of solid oxide fuel cells, *Nanomaterials* 2022;12:1059. DOI: 10.3390/nano12071059

5 Advances in Multi-Functional and Hybrid Fuel Cells

5.1 INTRODUCTION

In previous chapter, we examined several types of fuel cells (FCs) that are being developed to generate electricity. These operate at low, medium, and high temperatures using hydrogen as a fuel or other liquid and solid fuels to generate hydrogen and CO which can be used for power generation. Over the last several decades, significant progress has been made to commercialize FC technology, and they will be useful both for static and mobile applications and for portable electronics and large-scale power production. FC has a very bright future because it can be used as both power generation and power storage device, and it does not emit CO_2 or other harmful pollutants.

Besides the ones described in the previous chapter, there are several fuels cells, in particular molten carbonate FC (MCFC), microbial FC, and regenerative FCs that provide multiple functions including power generation. MCFC can capture CO_2, produce hydrogen, and heat along with power generation. Microbial FC can remediate wastewater and treat other agriculture and forestry wastes along with power generation. Reversible FC can be used as an electrolyzer-generating hydrogen as well as FC-generating power. This examines the workings of these multi-functional FCs.

The global electricity systems are currently witnessing a paradigm shift from the traditional centralized to balanced centralized and distributed generation technologies [1,2]. There are a number of ways FC can be used for distributed operations. One way to engage FC technologies is by integrating them with the renewable energy resources, in which they operate as a storage device for harnessing relatively high renewable energy. While FC offers very clean and renewable power, it can also be hybridized with other FCs or renewable technologies like solar, wind, natural gas combined cycle (NGCC) to generate more efficiency, longer life and durability, better dispatchability, lesser pollution, and sustainable power for the grid. These hybrid systems can also provide more balancing and stability to the grid. This chapter examines various alternatives for the hybrid FC systems.

The distributed power supply is in recent years facilitated by the use of microgrids [3–5]. In order to make the microgrids more sustainable, it is necessary to develop more diversified electrical energy production resources beyond the current solar, wind, hydro, biomass, diesel, and battery technologies for microgrid systems. Interestingly, FC systems are considered as promising energy resources for this purpose on the basis of being dispatchable, clean, pollution-free, and efficient. FCs also have potential to store high-caloric-value hydrogen compared to the chemical energy that may be stored by using most other materials and systems [6–9], and they are capable of supplying

DOI: 10.1201/9781003429906-5

energy for a relatively longer time [10]. FCs are also useful for off-grid system and for the emergency needs. This chapter also examines all of these roles of FC.

5.2 MULTI-FUNCTIONAL MCFC SYSTEM

In the previous chapter, we examined the role of MCFC as direct carbon FC. MCFCs are the most versatile multi-functional FCs which can handle not only solid fuel as in DCFC but also gaseous fuels like hydrogen, carbon monoxide, and carbon dioxide. They can also be used to capture carbon dioxide and produce hydrogen as well as heat. In this section, we examine multi-functional role of MCFC. As indicated in the previous chapter, MCFCs [11,12] are devices designed to generate clean energy using molten carbonate as electrolyte, especially Li-Na and Li-K carbonate [13,14], suspended in a chemically inert ceramic matrix, commonly an alumina-type $LiAlO_2$ [15], and Beta-alumina solid electrolyte, which allows the transport of carbonate ions. For this reason, MCFCs are high-temperature cells, operating at temperatures above 600°C. Figure 5.1 shows a schematic representation of an MCFC that works using hydrogen as fuel [16].

As pointed out by Contreras et al. [16], MCFCs could work with an efficiency of up to 60% reaching an operating power of 100 MW [17,18]. Using waste heat from the system, fuel efficiency can be as high as 85%, a value above performance of phosphoric

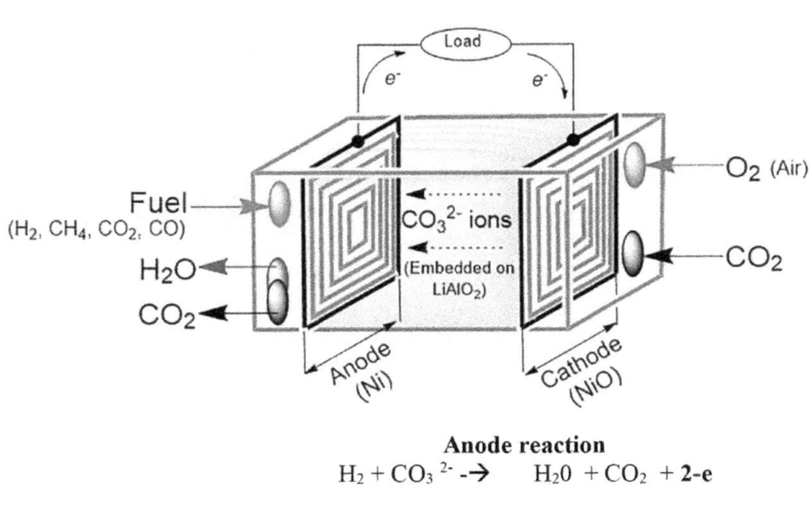

Anode reaction

$$H_2 + CO_3^{2-} \rightarrow H_20 + CO_2 + \mathbf{2\text{-}e}$$

Cathode reaction

$$^{1}/2 0_2 + CO_2 + 2e^- \rightarrow CO_3^{2-}$$

Anode Reformer Reactions

$$CO + H_2O \rightarrow H_2 + CO_2$$
$$CO_2 + 2H_2O \rightarrow CH_4 + 2\,O_2$$
$$CH4 + H_2O \rightarrow 3H_2 + CO$$

FIGURE 5.1 Schematic representation of a MCFC. Cathode chemical reactions are observed, anodic ones, including anodic reforming reactions [1].

acid fuel cells (PAFCs). Thus, MCFCs compete very well with PAFCs, proton-exchange membrane fuel cell (PEMFCs), and other types of FCs. They do not require an external reformer to convert other types of fuels into hydrogen, and due to the range of working temperatures, any hydrocarbon that is injected into the cell can be transformed into hydrogen through "anodic reforming reactions." The fuels can be traditional hydrocarbon fuels, flue gas, biofuels, natural gas (from coal), biomass, and in general syngas [16,19–22]. This flexibility results into a significant reduction in the operational costs.

MCFCs are not affected by carbon monoxide or carbon dioxide poisoning. On the contrary, as shown in the previous chapter, these cells produce CO_2 (+ CO) through an anodic oxidation reaction, which is used in anodic reformation transforming the injected hydrocarbons into the cell in combustible hydrogen [16]. In this sense, it is important to note that the MCFC could use carbon oxides themselves as fuel, which place them at the forefront of FCs in terms of the role they can play in capturing, separating, and concentrating CO_2 and producing hydrogen. MCFC can also be operated as a hybrid system by integrating with a thermoelectric power plant, forming a hybrid MCFC-GT system, with thermophotovoltaic (TPV) technology as hybrid MCFC–TPVC system, or with various solar thermal systems (PTC, LFR, and PDC) as well as with an organic Rankine cycle (ORC) forming an appropriate hybrid system [16,23–30]. All of these hybrid systems improve fuel efficiency of the cell. From a processing engineering point of view, the greatest disadvantage of MCFC technology is its low durability [16,13,14], which is a consequence of the range of high operating temperatures [31]. Furthermore, the corrosive nature of electrolytes accelerates the breakdown of cell components, reducing their useful life [32]. More research is needed to discover new materials that can be used in the construction of MCFCs [33,34].

MCFC can concentrate CO_2 up to around 75%–90% along with the production of water and small amount of hydrogen. Several studies [35–37] have investigated the use of MCFCs to produce power and simultaneously concentrating CO_2 coming out of coal plants and NGCC. A combination of NGCC and MCFC where FC is located between the gas turbine and the heat recovery steam generator has shown the largest reduction in energy required for CO_2 capture. In this system, the MCFC consumes ~20% of the total plant's fuel input and contributes to the electric power output by a similar fraction. Compared to a benchmark amine scrubbing process, the use of MCFC for concentrating CO_2 stream shows considerably better performance [23,38].

The typical configuration of a MCFC is based on a nickel alloy anode and a nickel oxide cathode, in contact through an electrolyte, a carbonate eutectic mixture of Li_2CO_3/Na_2CO_3 (52/48 mol %, melting point: 501°C), Li/Na–MCFC or Li_2CO_3/K_2CO_3 (62/38 mol %, 498°C), Li/Na–MCFC [39,40], confined in a reservoir or support made of γ-$LiAlO_2$. The electrochemical processes that occur are similar to those that take place in other FCs. However, the difference is the participation of CO_2 in the reactions of both the anode and the cathode. For example, CO_2 is produced and then consumed in the cathode chamber (Figure 5.1). Therefore, MCFCs need a CO_2 recycling system (CO_2 circuit) that allows CO_2 to be taken from the anode and reinjected into the cathode. In other words, the same amount of CO_2 consumed at the cathode will be regenerated by the anode. Consequently, the oxidizing agent at the cathode ends up being a mixture of O_2, N_2 and CO_2. For an MCFS- $H_2/O_{2(air)}$ system, H_2 is electrochemically oxidized at the anode and CO_2 is generated; at the cathode, O_2 is electrochemically reduced and CO_2 is consumed (Figure 5.2). In order for this electrochemical process to

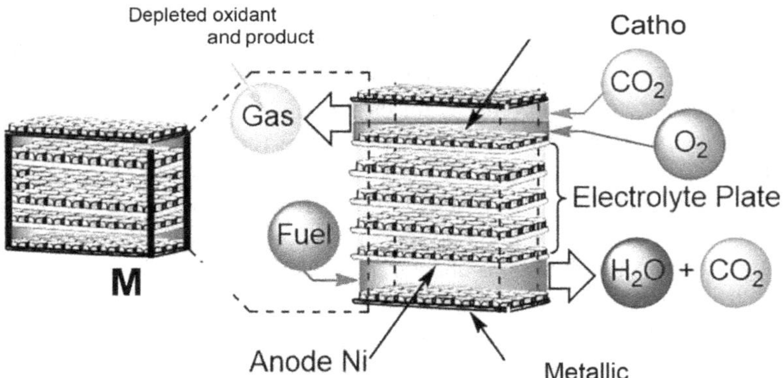

FIGURE 5.2 Diagram of the cross section of a MCFC. Fuel: H_2 (CH_4, CO_2, CO); electrolyte plate: CO_3^{2-}/$LiAlO_2$ [1].

be effective, transport of carbonate ions need to be efficient which is generally accomplished by γ-$LiAlO_2$ powder, a material commonly used to produce matrix electrolyte support plates in an MCFC. While γ-$LiAlO_2$ is thermally stable above 600°C, it is not very durable for the long term operation due to increase in its particle size and pore volume, which results in a loss of electrolyte that escapes from the support [41].

The γ-$LiAlO_2$ phase can be transformed into the α-$LiAlO_2$ phase, with a higher density (3.413 g/cm^3), which results in a change in the pore size distribution. Such a change can alter the distribution of the electrolyte concentration in the support across the length and width of the plate (panel) containing the $LiAlO_2 + CO_3^{2-}$ mixture, and change the transport phenomenon of ions of the molten carbonate. This can not only change electrochemical processes but cause leaks of the electrolytes from the support.

MCFC can run in reverse mode as a "molten carbonate electrolysis cell," with a Ni electrode working as "cathode," and producing H_2 or syngas (H_2 + CO), from CO_2 and water. Carbon dioxide as well as water must be present in the inlet gas, since CO_2 is a reactant when producing CO_3^{2-} ions. The electrochemical reaction on the "anode" (NiO electrode), is the electrolysis of carbonate ions producing O_2 and CO_2 and direct CO_2 electrolysis to generate CO which is carried out at the Ni electrode. However, Contreras et al. [16] reported that the kinetics of this reaction are much slower compared to water electrolysis on nickel-based electrodes. The use of MCFC as electrolysis cell to produce fuel gases and generate electricity (commonly known as "reversible MCFC," RMCFC) has been studied in details in the literature [42–44]. Reversible FCs are examined in more details later on in this chapter.

The materials from which an anode is made must have good electrical conductivity, structural stability under the working conditions of an MCFC and especially, be electrocatalytic toward the oxidation of the fuel. In MCFCs, the anode is manufactured as a plate composed of a matrix coated with a layer of metallic alloy. The most popular anodic material is a porous nickel alloy [45], which is an excellent electrochemical catalyst for the oxidation of hydrocarbons. Because of the fast oxidation reactions, the anode has a smaller surface (0.1–1.0 m^2/g) compared to the cathode, pore diameter in the range of 3–5 μm, the thickness between 0.5 and 0.8 mm, with a porosity of 55%–70% [43–47].

Under MCFC operating conditions, the porous plate is partially wetted by electrolyte which results in large three phase boundary (TPB) reaction surface area [48].

The use of Ni(0) as anode material can cause stress deformation or sintering resulting in its performance alterations. In order to address this issue various alloys such as Cu-Al, Ni–Cr, Ni–Cr protective by CeO_2 and ZrO_2 layer, Ni–Al with nano-ZrO_2 (3% wt.), Ni–Al with Ti (5% wt.), Ni–Ti, Ni–Fe–Cu, Ni–Nd or Ni–B, and construction of anodes based on an nickel alloy with dysprosium(III) oxides of the type Dy/Ni–Dy (Dy_2Ni_{17}) have been examined with a varying degree of success [49–58]. Even with these efforts, nickel(0) continues to be the preferred option in the manufacture of MCFC-anodes.

MCFC-cathodes are fabricated of porous nickel oxide (NiO) [59,60] and NiO-lithium, a material with good electrical conductivity, high porosity and low cost. The coating has an optimized pore diameter between 6 and 10 μm, and a thickness between 0.5 and 0.8 mm, with a porosity of 60%–80% [61]. The possibility of solubilizing Ni^{2+} ions of the MCFC-cathode in the electrolyte, facilitates the transport of these ions to the proximity of the anode, where it precipitates on the matrix and then meets the dissolved H_2 that reduces it to nickel(0) [62]. If this issue persists, it can ultimately result in the deposition of sufficient nickel metal in the middle of the cell, causing short-circuiting by contact cathode/Ni(0)/anode. It can also result in a loss of the integrity of the cathode and therefore, a decrease in the performance of the MCFC. One way to overcome this issue is to modify the cathode by coating the NiO with other materials less soluble in electrolytes, such as $LiCoO_2$, $LiFeO_2$, Li_2MnO_3 and $La_{1-x}Sr_xCoO_3$ [46,63–67]. Another approach is coating the surface of the anode–MCFC with rare earth oxides such as cerium(IV), CeO_2, La_2O_3, Pr_2O_3, Nd_2O_3 etc. Modifications of the micro-structure and the chemical composition of the material used for cathode can also influence the electrochemical behavior of MCFC- cathodes which can be done by, for example, by using a metallic foam support within electrodes [68–75]. Another strategy is to modify the Li_2CO_3-Na_2CO_3 electrolyte with the addition of complex mixtures of alkali metal carbonates [76].

Electrolytes used in MCFC are highly porous ceramic materials (50%–70% with pore size 0.1–0.3 μm), impregnated with an electrolyte based on alkali metal carbonates [77–79]. This matrix (1) provides the necessary electrical insulation so that the cell does not short- circuit, (2) allows ion transport phenomena to occur between the anode and the cathode and, (3) guarantees the separation of fuels or reagents. MCFC matrix is a solid at room temperature that transforms into a "paste" at the working temperature of 650°C and, additionally, must offer resistance to carbonates. This is achieved by isolating the electrolytic compartment. Based on these premises, several metal oxides including MgO and SrTiO3 [80] have been investigated, but γ-$LiAlO_2$ has been found to be the most desirable crystalline form as an MCFC matrix [81,82]. Unfortunately, in a temperature range of 650°C–700°C, a significant growth of the particle size, increase in volume and pore size with resulting cross-leakage and interaction between reactants or fuels causes drop in the cell voltage. When both alpha and gamma phases of $LiAlO_2$ are present, the proportion of the α-$LiAlO_2$ phase is expected to increase in a low temperature, high P_{CO2} environment. Although, the most stable phase under MCFC operating conditions is γ-$LiAlO_2$ [83], in terms of volume and pore size, the most recommended phase is α-$LiAlO_2$ [84,85]. At some point during the operation of the MCFC the process γ-$LiAlO_2$ → α-$LiAlO_2$ occurs.

Contreras et al. (1) point out that the performance and life span of an MCFC depend not only on the materials of construction of the anode, cathode, and electrolyte matrix or support, but on the composition of the electrolyte itself. Traditionally, the eutectic mixtures Li_2CO_3-Na_2CO_3 (52:48 mol%), $(Li_{0,52}Na_{0,48})_2CO_3$, y Li_2CO_3-K_2 CO_3 (62:38 mol%), $(Li_{0,62}K_{0,38})_2CO_3$, rich in lithium ions, have been used as standard electrolytes [86,87]. In general, the polarization of the anode changes only slightly when the composition varies between 40% and 70% in lithium ions, while the resistance to cathodic polarization decreases with increasing lithium content. Due to its higher ionic conductivity, Li-Na carbonates has other benefits. However, in the design of MCFCs with this electrolyte, it is necessary to foresee effects related to the loss of wettability and effects of temperature or losses due to vaporization [88]. The studies in relation to electrolytes for MCFC are fundamentally based on trying to change the carbonate ratio between Li:Na and Li:K, an aspect that is already quite optimized. While more research for better materials for anode, cathode and electrolyte is continuing, in future, new Ni-metal alloy anodes, alkaline earth metals or lanthanides added to NiO-based cathodes, optimizing the electrolytic compositions and improved porous electrolytic matrix of lithium aluminate through nanochemistry techniques are likely to be pursued [16].

5.2.1 Role of MCFC for CO_2 Capture

In recent years carbon capture and sequestration (CCS) from a fossil fuel based power plant is a much-discussed technology that energy industry require to achieve to reduce global GHG emissions [89–92]. While a wide variety of technologies are being pursued to address this challenge, including energy efficiency, biofuels, process intensification, hydrogen, CO_2 utilization, and many others, recent reviews [93,94] highlighted the potential for high temperature MCFCs for both CO_2 capture and H_2 generation along with power production, and the use of the H_2 as a low-GHG fuel in additional process heaters. The MCFCs are able to achieve this double or triple feat at high efficiency with a small, almost negligible, energy penalty.

CCS has been the subject of much research, development and a few demonstrations over the past few decades. The leading technology today is the amine scrubber, which has been deployed twice in post-combustion capture at commercially relevant sizes. Other CO_2 reducing technology has been deployed many times in natural gas processing, but those technologies are generally not relevant to combustion flue gas, which represents the overwhelming majority of global CO_2 emissions. Thus, there is a need for a new technology that is more cost effective and energy efficient than amines.

The use of hydrogen for combustion also offers the advantage of zero emissions at the point of consumption, but this requires a low-CO_2 method for its production, storage, and distribution. Most current H_2 production relies on the steam methane reforming process, which produces about 1 mol of CO_2 for each 2.3–2.5 mol of H_2. This CO_2 can be captured as demonstrated by Shell at the Quest project and by Air Products in Port Arthur, TX [95–99], although these two systems capture only about 60% of the total CO_2 content. For higher capture rates approaching 90%, amine systems on the furnace flue gas are required [98,99]. Alternatively, auto- thermal reformers can be integrated with CO_2 capture [98,99]. Recent work has reported advances in hydrogen production for methane conversion, water electrolysis, thermochemical water-splitting and various other methods [100]. However, these approaches are costly and still at relatively early technology stages.

MCFCs are capable to concentrate CO_2 from dilute sources [88–96]. MCFCs have been heavily investigated for distributed power generation as early as the 1980s, but their use in carbon capture applications has only more recently been investigated over the last roughly 15 years [89–100]. Most recently, efforts are also made to use MCFC for CO_2 capture, power production, and hydrogen generation [93,94]. Here, we briefly examine advances made in the use of MCFC for both carbon capture and hydrogen production along with power generation. Hydrogen can also be produced using unitized regenerative FC (URFC) system. We also evaluate an excellent study done in UK for techno-economic performances of combine cycle gas turbine (CCGT) with amine and MCFC for carbon capture.

In the use of MCFC for carbon capture, CO_2-containing flue gas is fed to the cathode, where CO_2 reacts with O_2 and electrons to form carbonate ions. These ions are transferred through the electrolyte to the anode, where they oxidize H_2 to produce electricity. Unlike traditional carbon capture technologies, such as amine scrubbing, in addition to capturing CO_2, MCFCs are able to produce additional power driven by the oxidation of fuel fed to the anode. The CO_2 concentration is upgraded from 4 to 10 vol % at the cathode inlet to around 70% (dry basis) at the anode outlet. Further processing, using conventional technology, converts the CO_2 to a high-purity, high-pressure stream, while making the H available for export or recycle to the front end of the system. In general, any combustion flue gas can be used, possibly supplemented with air to provide the required O_2 for the cathode reaction. A typical MCFC with possible reactions at anode and cathode is illustrated in Figure 5.3 [89].

FuelCell Energy (FCE)'s MCFC [101] concentrates flue-gas carbon by using carbon dioxide as almost a working fluid. At one end, a porous nickel cathode combines

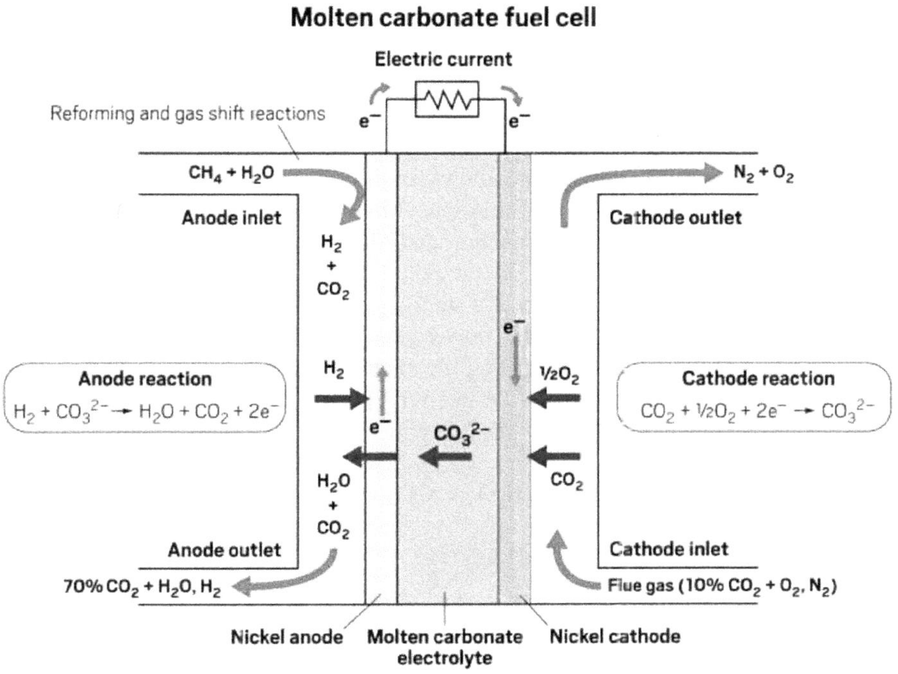

FIGURE 5.3 Typical MCFC [89]. (*Source*: Adapted from *Int. J. Hydrogen Energy*/IEA.)

CO_2, oxygen, and two electrons to make CO_3^{2-}, the carbonate ion. That CO_3^{2-} carries the charge through the electrolyte to a nickel anode on the other side, where it reacts with hydrogen to yield water, two electrons, and CO_2. The outlet gas is around 70% CO_2, and the balance is H_2 and H_2O—a mixture better for carbon capture, storage, and utilization than raw flue gas, which is 10% CO_2 diluted in nitrogen. FCE's MCFC differs from conventional cells in two crucial ways.

First, it can convert hydrocarbons to H_2 internally, using the water and waste heat generated by the cell. Its nickel electrode can catalyze reformation and gas shift reactions at the 600°C operating temperature inside. Because the H_2 is continuously consumed right there, Le Chatelier's principle pushes both reactions to completion. FCE claims that the cell can even consume alcohols as fuel, though the power plants currently in place run on methane or biogas. Second, the cell pumps CO_2 through the electrolyte as part of its electrochemical cycle, carrying it from the N_2-diluted air stream to the concentrated fuel stream. Stand-alone MCFCs have to recycle some CO_2 back to the air inlet to keep that cycle going. However, MCFCs attached to the end of a power plant can pull around 90% of the CO_2 out of the flue gas. Crucially for carbon capture, the gas flowing out of an MCFC is around 70% CO_2 along with H_2O and some extra H_2, and it's much easier to separate CO_2 from water than from the nitrogen that makes up around 70% of industrial flue gas.

The additional output of H_2 and water was a key selling point for Toyota when it was looking to install a fueling station for hydrogen-powered cars in Long Beach, California. An MCFC scheduled to come on line which will provide freshwater, 2.3 MW of electricity, and 1,200 kg of H_2 per day to fill Toyota's fuel-cell engines. It will be FCE's first commercial-scale plant with that configuration, which the firm calls trigeneration [101].

5.2.2 Chemistry within MCFC

As shown earlier, the MCFC consists of three active layers: the anode, the matrix, and the cathode, all of which contain molten salt electrolyte. The matrix is designed to be completely filled with electrolyte and allows for the conduction of the carbonate ions while being impervious to gas transmission. Additional repeat components include bipolar plate (BP), anode current collector, and cathode current collector which help distribute the electrical flow [16]. Understanding electrochemical phenomena which drive MCFC performance has been of a significant research focus. MCFCs are able to achieve simultaneous CO_2 capture, power generation, and H_2 production because of their unique chemistry. In an MCFC, the anode is fed natural gas, which is first reformed in the anode according to the steam reforming and water–gas shift (WGS) reactions, Eq. (5.1) and Eq. (5.2), to formulate the net Eq. (5.3):

$$CH_{4(a)} + H_2O_{(a)} = CO_{(a)} + 3H_{2(a)} \tag{5.1}$$

$$CO_{(a)} + H_2O_{(a)} = CO_{2(a)} + H_{2(a)} \tag{5.2}$$

$$CH_{4(a)} + 2H_2O_{(a)} = CO_{2(a)} + 4H_{2(a)} \tag{5.3}$$

At the anode, the hydrogen is oxidized by the carbonate ion from the electrolyte

$$H_{2(a)} + CO_3^{2-}{}_{(a)} = H_2O_{(a)} + CO_{2(a)} + 2e^- \tag{5.4}$$

the carbonate ion having been generated at the cathode by reduction of CO_2 and O_2 by electrons from the circuit:

$$CO_{2(c)} + 1/2\ O_{2(c)} + 2e^- = CO_3^{2-}{}_{(c)} \tag{5.5}$$

The net result of the electrochemical reactions in the FC is

$$H2_{(a)} + CO_{2(c)} + ½O_{2(c)} = H2O_{(a)} + CO_{2(a)} + \text{heat} + \text{electricity} \tag{5.6}$$

Combining all of these reactions, in appropriate relative amounts to reflect the actual FC operation, results in the same net result as for the normal combustion of methane:

$$CH_{4(a)} + 2H_2O_{(a)} + 4CO_{2(c)} + 2O_{2(c)} = 5CO_{2(a)} + 4H_2O_{(a)} + \text{heat} + \text{electricity} \tag{5.7}$$

Close inspection of Eq. (5.7) reveals how the MCFC can operate simultaneously as both a CO_2 capture device and a power generator: For each CH_4 oxidized to generate electricity, four CO_2 molecules are transferred from the cathode to the anode. The CO_2 that was generated by the oxidation of the CH_4 fed to the anode remains in the anode, leading to additional CO_2 capture. If all of the hydrogens generated in Eq. (5.3) are not consumed by the electrochemical reaction (Eq. (5.4)), then surplus H_2 will be generated at the same time as CO_2 is captured and power is generated. The observation of the simultaneous transfer of hydroxide ions and carbonate ions is scientifically interesting for FC operations. At standard power generation conditions, the hydroxide ion has almost no impact because of the high CO_2 concentrations. As the power output of the MCFCs is increased, the hydroxide ion conduction could become important, but it would take a fairly dramatic change in conditions for this to become true.

The use of MCFCs for CO_2 capture has multiple value propositions, as the MCFCs create additional power while simultaneously capturing CO_2 from dilute flue gas. The MCFCs and associated equipment are able to achieve both of these goals at very high efficiency; an operating cell potential of 0.75 V represents an electrical efficiency of 75% (which is to be compared with approximately 40% and 60% efficiency for conventional coal and natural gas power generation, respectively). A third value proposition exists in the hydrogen generation occurring on the anode side. Thus, MCFC has a triple value: (1) generation of CO_2 free power at high efficiency using hydroxide mechanism, (2) carbon capture along with generation of power, particularly as posttreatment of coal or natural gas combustion or combined cycle, and (3) carbon capture with both power and hydrogen generation.

In addition to chemistry described above, an unexpected non-carbonate ion transfer occurs in the molten carbonate electrolyte during FC operation at low CO_2 concentrations of <1% in the presence of water. Researchers have previously highlighted water's role on cathode performance, suggesting that hydroxide, the principal hydrated species in molten carbonates, acts as an intermediate reactant. The literature has shown that the operation above the limiting current for CO_2 diffusion was possible via direct reduction of O_2 to O^{2-} and that significant OH^- transport was possible in molten carbonate melts in the presence of large chemical driving forces [102–107]. The mass balance experiments reported in the literature [93,108] are consistent with hydroxide transport across the electrolyte, produced from the reduction of O_2 and H_2O at the cathode. In essence, when the cathode is starved of CO_2, reduction proceeds via hydroxide formation. This is largely caused by limited mass transport within the cathode, caused by the slow diffusion of the gas through the cathode pores to the gas–electrolyte interface. Reducing this bottleneck improves the CO_2 capture performance while minimizing the hydroxide current. At standard power generation conditions, the hydroxide ion has almost no impact because of the high CO_2 concentrations. As the power output of the MCFCs is increased, the hydroxide ion conduction could become important, but it would take a fairly dramatic change in conditions for this to become true.

5.2.3 CO_2 Capture from NGCC Flue Gas

CO_2 capture from flue gas of NGCC process is important because low CO_2 concentration from its flue gas represents the most challenging situation. In a typical NGCC–MCFC process, flue gas from the NGCC plant's heat recovery steam generation (HRSG) system is sent to the FC cathodes where CO_2 is removed. The resulting cathode exhaust is used to pre-heat the incoming gas. The FC anode is fed steam and a hydrocarbon fuel, such as methane, which is reformed to produce H_2 for the anode. The anode exhaust, consisting primarily of CO_2, H_2O, and unreacted H_2, is compressed, dehydrated, and subjected to a H_2/CO_2 separation. After separation and compression, the CO_2 is sent to sequestration (or other use), while the hydrogen by-product can either be recycled back to the FC as anode fuel or exported depending on the specific application and context. The addition of the MCFCs adds power output to the existing plant while capturing CO_2 from the existing NGCC plant, with an option for H_2 export as well [90,91].

Several technical challenges exist for the use of MCFCs as a CO_2 capture technology. From a process standpoint, the handling and treatment of significant flue gas streams at low pressure (i.e., high volumes) requires large equipment (ductwork, blowers, etc.). With respect to the FC electrochemistry, low CO_2 concentrations, typically less than 4%–10% by volume at the cathode entrance, affect cathode performance, the effects of which must be better understood and characterized. Practically speaking, the target in the cathode exit is less than 1% CO_2, with even lower concentrations preferable to maximize the CO_2 capture.

Past work has demonstrated MCFC operation for carbon capture where large fractions of the CO_2 are removed from cathode streams [90,91]. However, only a few studies have focused on very dilute CO_2 feeds, representative of natural-gas-fired flue gas, where inlet concentrations are less than 6 vol % CO_2. Even fewer investigations have demonstrated deep CO_2 capture (>90%) at these conditions. It has been recently noted that dilute CO_2 concentrations pose a significant challenge for the use of molten carbonate membranes in carbon capture and storage applications [97,98]. The challenge stems from significant excess air required for the combustion in the CO_2 source, which dilutes the CO_2 concentration to a very low level. The flue gas from a combined cycle gas turbine has a low CO_2:O_2 ratio of ≈1:2.5 by volume compared to the stoichiometric requirement of 2:1. This means that O_2 is in excess in the cathode, which serves to increase the cell potential slightly, although not enough to offset the substantial decline in potential by the low CO_2 concentration. This highlights the need for more experimental and process modeling work at these operating conditions to further develop, understand, and scale up MCFCs for CO_2 capture in NGCC applications.

In MCFC operation for power generation applications, the primary goal is to maximize power density. Therefore, designing MCFCs which operate with high voltage, high fuel utilization, and high current density is beneficial. In this mode, a high CO_2 utilization is actually disadvantageous to the FC performance, as it decreases CO_2 concentrations and therefore performance. The fuel utilization is a somewhat free parameter. The studies by Gatti et al. [90] and Campanari et al. [91] demonstrate experimentally that over 90% of the CO_2 can be removed from a simulated NGCC flue gas stream ($\leqslant$4 vol% CO_2) as well as from other flue gas sources. Electrochemical characterization shows that by modifying cathode conditions and FC operation, the CO_2 capture rates, cell potential, and power output can be controlled and tuned depending on the specific commercial embodiment.

Although feasible, higher capture rates may reduce overall process efficiency due to the decrease in theoretical Nernst voltage and increased contribution from the hydroxide pathway. Other flue gases, such as from nearly stoichiometric combustion devices, have a much higher CO_2 concentration, typically around 10%–15%. These flue gases are actually deficient in O_2 relative to what is required in the cathode reaction Eq. (5.7); therefore, supplemental air will need to be added to the flue gas prior to cathode, which will dilute the CO_2 concentration down to 8%–12%. In these cases, very high CO_2 capture rates without significant hydroxide current and Nernst voltage losses are possible unless the cell current density is increased to extremely high levels.

There are numerous studies on NGCC–MCFC combination reported in the literature. Here we briefly examine only two to illustrate the point. The study by Campanari et al. [91] examines the application of MCFCs in natural-gas-fired combined cycles to capture CO_2 from the exhaust of the gas turbine. The gas turbine flue gases are used as cathode feeding for a MCFC, based on Ansaldo FCs experience, fed with natural gas processed by an external reformer. The MCFC acts as an active CO_2 concentrator, where CO_2 is transferred and concentrated from the cathode to anode side, while producing additional power. The cell anode exhaust is sent to a CO_2 removal section consisting in an oxygen combustion of residual fuel compounds or a cryogenic CO_2 removal section, cooling the exhaust stream in the heat recovery steam generator. The plant shows the potential to achieve a CO_2 avoided ranging between 58% and 68% (depending on the configuration), while taking advantage from the introduction of the FC the final electric efficiency can be close or few points lower than the original combined cycle (0.2% lower heating value (LHV) in the most efficient configuration), and the power output increases by about 20%, giving a potentially relevant advantage with respect to competitive carbon capture technologies typically featuring a relevant net power output decay. Moreover, the role of the FC on the overall power balances is limited to 15%, leaving the majority of power to conventional components, improving the possibility of achieving a low plant-specific cost (€/kW). NGCC–MCFC combination plant examined in this study is illustrated in Figure 5.4.

The above discussions assume that all other aspects of the MCFC operation are not affected by the presence of the hydroxide ion, such as lifetime, corrosion, heat management, and so forth. Investigation of these topics is ongoing for carbon capture-specific applications. Further work is also focusing on optimizing the cathode design for CO_2 capture efficiency at very low CO_2 concentrations. The literature has thus shown that for NGCC flue gas upward of 90% CO_2 capture is possible with additional 1,000 W/m^2 power generation. At these conditions, significant amounts of non-carbonate, specifically hydroxide ions are able to conduct through the molten salt electrolyte. More research is needed to better understand the role of hydroxide ion conduction for the optimal FC design especially in carbon capture operation.

The study by Gatti et al. [90] examined performance and cost evaluation of four alternative technologies for post-combustion CO_2 capture in natural-gas-fired power plants. These include CO_2 permeable membranes, MCFCs, pressurized CO_2 absorption integrated with a multi-shaft gas turbine and heat recovery steam cycle, and supersonic flow-driven CO_2 anti-sublimation and inertial separation. The study indicates that MCFC seems to outperform the baseline from both performance (SPECCA of 0.31 MJ_{LHV} / $kg_{CO_2-avoided}$) and cost (CO_2 avoidance cost of 49 \$ / t_{CO_2}). CO_2 permeable membranes are affected by the large area and high capital costs which make this technology less attractive than amines for deep CO_2 removal from NGCC flue gas. To improve the performance and cost of this technology, a capture rate lower than the 90% threshold should be targeted. High-pressure solvent absorption from high-pressure exhaust gas, though a well-known and low-risk technology, does not outperform benchmark capture technology on performance. Although a decrease in specific investment cost appears to be achievable due to supplementary firing (which increases the power output), the cost of CO_2 avoided is affected by the lower net electric efficiency. Supersonic, flow-driven, CO_2 deposition is the technology with the lowest TRL, and the results of this study indicate that its energy performance can be similar to the MEA base case, only in case moderate flue gases compression ratios

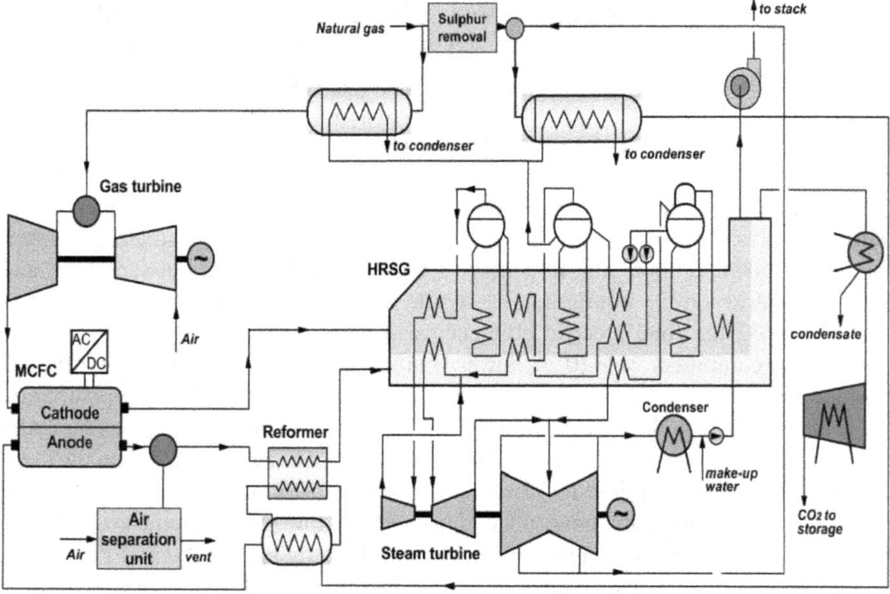

FIGURE 5.4 Plant layout with integration of the MCFC in a combined cycle, with CO_2 separation via oxygen combustion [91].

are achievable to ensure acceptable CO_2 removal levels via anti-sublimation. SSD application to post-combustion capture from NGCCs looks challenging due to the low CO_2 concentration in the flue gas. However, given the large uncertainty affecting the technology, it is difficult to draw conclusions on its economic performance. MCFC-NGCC integrated plant examined in this study is illustrated in Figure 5.5.

Since the energy and economic performance of CO_2 capture technologies are strongly dependent upon the application (e.g., flue gas composition, plant size, mass and energy integration with the power plant) and the modeling assumptions, the results and considerations found in this work cannot be generalized. As an example, for applications featuring higher CO_2 concentrations (coal-fired steam cycles, cement plants, refineries, smelters, etc.), technologies based on physical separation processes (e.g., HPS, SSD, and MEM) are expected to be much more competitive.

5.2.4 CO_2 Capture with Both Power and Hydrogen Productions

Most recent study by Barckholtz et al. [93] reviews advances made in the use of MCFC for CO_2 capture, power generation as well as hydrogen production. As indicated earlier, in MCFC, CO_2 is transferred from cathode to anode as well as generated from methane reforming. This CO_2 along with excess hydrogen is collected at the anode. The process simultaneously generates power by the electrochemical oxidation of a portion of the anode H_2. After compression and separation, the CO_2 can be sequestered while the H_2 can either be recycled as fuel or exported to a separate process for use as a chemical reagent, such as in ammonia synthesis or hydro-processing, or exported to be used as a low CO_2 fuel itself, such as in an industrial furnace. Due to the unique operation of the FC, all of this is accomplished at very high efficiency, often with little-to-no efficiency debit relative to the original CO_2-emitting process.

The technology has broad applicability to a wide variety of CO_2 sources, as highlighted in a recent review [15]. The study illustrates the utility of the technology in the power and heat generation sectors, focusing on gas turbines, steam boilers or process furnaces, and cogeneration units. MCFCs can also be useful for CO_2 removal in manned spacecraft, coal-fired power generation, power generation following coal or municipal solid waste gasification, natural-gas-fired power generation, steel, cement plants, oil sands production, and other systems. The conclusion of several of these studies [98,99,104,107–109] is that while MCFC concept can be considerably more cost-effective than present amine technology, more work for its commercialization is needed. Barckholtz et al. [93] point out that none of the previous studies have considered the creation of large volumes of H_2 in the MCFCs as a valuable co-product toward the decarbonization of additional CO_2 generators.

The flue gas can contain anywhere from 3 to 12 mol%, and sometimes higher, amounts of CO_2. There are several approaches to handle this CO_2. A common approach gas been to use liquid amine to capture CO_2, but this has been found to be very expensive. This post-combustion capture process causes a significant efficiency debit; for example, coal and natural-gas-fired power plants suffer from a 20% and 10% reduction, respectively, in power production (at constant fuel input), or require commensurate increases in fuel firing rate to achieve constant power production.

In another alternative of pre-combustion capture [97,98], relatively pure O_2 is used in the combustion process, producing a flue gas that is essentially only CO_2 and H_2O

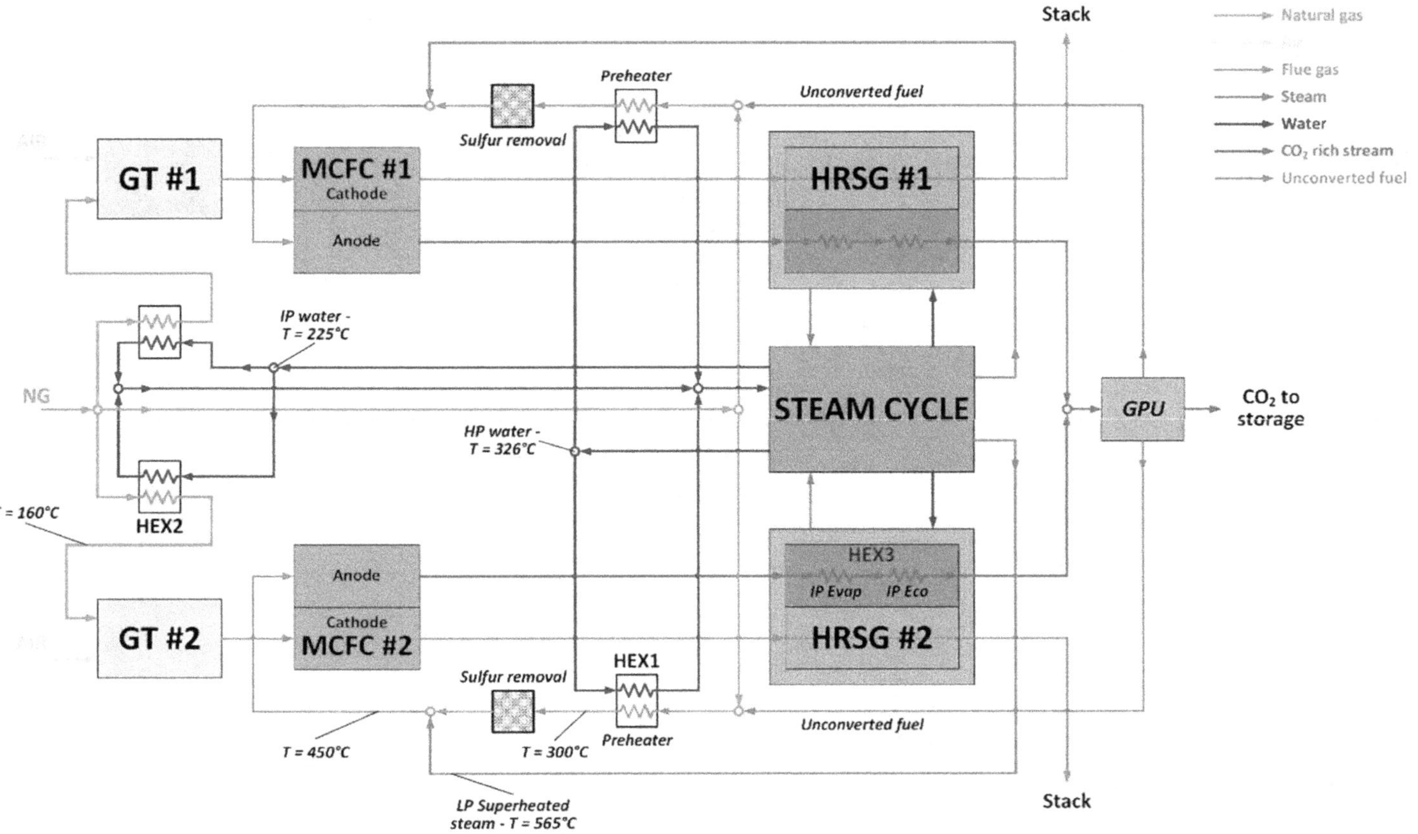

FIGURE 5.5 Process flow diagram of the NGCC–MCFC integrated plant [90].

which can be easily separated. Alternatively, the O_2 can be used to gasify coal, the resultant CO/H_2 shifted to CO_2 and H_2, which can then be separated, and H_2 can be used as a fuel in a power plant or other process. This technology is quite costly and inefficient, and multiple attempts at taking the technology to full scale have failed. Another approach is known as fuel switching, in which bioethanol for gasoline, natural gas for coal, and "green" hydrogen for carbon-based fuels are switched. The extent of decarbonization of each of these approaches depends on the GHG emissions intensity of the alternative fuel relative to the original fuel, and a complete life cycle analysis is generally required to make this assessment [95,102,108,110]. Finally, coal- or gas-fired power generation can also be replaced by clean solar and wind power generation systems.

The technology described in the study by Barckholtz et al. [93] is a novel approach that is a hybrid of each of these conventional approaches. The technology uses MCFCs to capture CO_2 in a post-combustion fashion from a conventional combustion process, while generating a co-product of H_2 that can be used to fuel switch other processes. The MCFCs also produce electricity while accomplishing the CO_2 capture and hydrogen generation and thus can be thought of to some degree as pre-combustion capture, too, in that the CO_2 generated in the anodes by the reforming and subsequent oxidation of the CH_4 is fully captured. The concept of using the MCFCs for H_2 production has been discussed in the literature [94,95,99] and has been deployed in a commercial test, but only the work of Rinaldi [110] develops the combination with CO_2 capture. None of them explore the potential of exceptionally large volumes of H_2 co-production. The use of MCFC thus has multiple value propositions. Furthermore, the presence of the hydroxide mechanism in MCFC generates power at high efficiency.

The study by Barckholtz et al. [93] examines cogeneration mode in which there are two cases: one in which the FCs synthesize only a small amount of co-product H_2, which is recycled to the fuel system, and another is a situation in which a large volume of co-product H_2 is generated, which is then exported to boilers that will use just H_2 as fuel. This series of cases illustrates the efficiency and effectiveness of the carbonate FC technology for CO_2 capture with co-production of power and hydrogen, creating a complete system with significantly reduced emissions. The study also shows that MCFCs can be used as effective CO_2 capture devices from a variety of flue gas sources, while producing additional power and possibly H_2. The system can be employed for NGCC, coal power generation, steel, hydrogen generation, cement, and many other applications. Hydrogen as fuel can be used as a chemical in hydro-processing or ammonia synthesis, or exported and used as a fuel for FC vehicles or many other applications. The IEA report modeled a 550-MW coal-fired power plant fitted with FCE's carbon-capturing MCFC running on a separate stream of natural gas. The FC in that model captured 90% of the coal plant's CO_2 emissions while putting out 351 MW of additional power.

5.2.5 Techno-Economic Comparison of CCGT with Post-Combustion CO_2 Capture Using Amine and MCFC

The study by Ferguson and Tarrant [92] presents the findings of the techno-economic assessment undertaken by Wood for the UK Government Department for Business, Energy and Industrial Strategy on the large-scale deployment of MCFCs for post-combustion CO_2 capture integrated with a new-build combined cycle gas turbine power plant for the generation of low-carbon electricity. The findings are compared with a state-of-the-art proprietary amine scrubbing technology. Based on a new-build power plant to be installed in the North East of England, with a power train comprising two trains of H-class gas turbines each with a dedicated steam turbine, the configuration presented utilizes MCFCs between the gas turbine exhausts and their heat recovery steam generators and cryogenic separation for unconverted fuel recycle and CO_2 purification. In this study, the process was summarized as follows:

1. Simulation of base CCGT power plant, cross-checked against public data such as Gas Turbine World.
2. CO_2 capture process cost and performance provided by Shell Cansolv for proprietary amine solvent case and various technical papers and public domain references for FCE's DCF3000 units for the MCFC case [109,111–117].
3. CO_2 compression and dehydration simulated in Hysys and cross-checked using recent vendor data for similar projects.
4. Heat and material balances and utilities summary provided the basis for thermal efficiency calculations, carbon balance, high-level equipment sizing, capital and operating cost estimates, and thus the calculation of LCOE.

While the CCGT with MCFC concept was expected to be capital intensive, its efficiency was expected to be high enough to make the scheme worthy of assessment to explore if use of this technology for post-combustion CO_2 capture had the potential to compete with, or exceed, the performance of current state-of-the-art proprietary amine solvents in this application once the technology has reached a suitable scale. The study included Case 1, a CCGT power plant with 90% proprietary amine solvent post-combustion CO_2 capture, and Case 2, which used MCFCs as the post-combustion CO_2 capture technology. In both cases, the CCGT power island was modeled using gate cycle, and non-proprietary elements of the CO_2 capture, compression, and dehydration systems were modeled using Aspentech Hysys. Overall process heat and material balances were then brought together along with FC performance calculations to perform the utility balance and hence describe the overall scheme.

Together, these key deliverables determined the efficiency and carbon balance of each case as well as providing the basis for a high-level equipment list from which the capital cost estimate was developed. At this level of study, this was done partly on the basis of costing individual equipment items in an in-house calibrated version of Aspentech Capital Cost Estimator, and partly using vendor quotations or public domain data for packaged units, such as the MCFC stacks on a per MW installed capacity basis. The material balance was also combined with the capital cost estimate and an estimate of manpower requirements to determine the variable portion of the plant operating costs. Once the capital and operating costs had been determined, it was possible to calculate illustrative overall project economics such as the levelized cost of electricity (LCOE).

- **Case 1—NGCC Gas Turbine with Post-Combustion Capture.** This case consists of a natural-gas-fired combined cycle power plant based upon 2 GE Frame 9HA.01 gas turbines each with a dedicated HRSG and steam turbine in a 2×2 configuration. The flue gas from both HRSGs is routed to a single-train Shell Cansolv proprietary post-combustion CO_2 capture unit, where it is cooled in a gas/gas heat exchanger, then boosted in pressure using a flue gas fan before entering a direct contact cooler. CO_2 is captured from the cooled flue gas using an amine-based solvent in an absorption column and is released from the solvent in the stripper. The captured CO_2 is then compressed in 4 stages, dehydrated and then compressed in a further stage to the required export pressure of 110 bar (abs).

- **Case 2- Combined Cycle Gas Turbines with MCFC Post-Combustion CO_2 Capture**. There are a number of different configurations in which MCFCs can be used for post-combustion CO_2 capture from CCGTs, with the following options defining many possibilities such as (1) internal (within the FC) or external (upstream) reforming; the study considered only internal reforming options, (2) locate FCs between GT and HRSG or downstream HRSG (greenfield or retrofit), (3) utilization of unconverted fuel species in anode exhaust, (4) oxy-combustion with heat integration, (5) recycle to GT, (6) recycle to FC and (7) CO_2 purification methodology.

MCFCs can be retrofitted as a bolt-on-the-back CO_2 capture technology analogous to an amine solvent post-combustion system. This would be the simplest configuration for retrofit to an existing CCGT or flue gas source. However, since the fuel cells operate at very high temperature, this configuration requires reheating of the gas turbine exhaust. A further potential configuration that was considered early in MCFC configuration development was location of the MCFCs downstream of the heat recovery steam generator. To make the most of the very high temperature between the gas turbine and its heat recovery steam generator, an alternative configuration, applicable particularly to new-build plants, would be to locate the MCFCs between these two. The unconverted fuel species in the MCFC fuel side exhaust being combusted with oxygen in a second heat recovery boiler scheme based on work undertaken by Politecnico di Milano [109,115,116]. A similar scheme presented by the same team showed the alternative of using cryogenic separation to separate the CO_2 product from the unconverted fuel species.

The scheme above has the anticipated advantage of being able to more directly control the purity of the CO_2 product. Subsequent work by the carbon capture project [112] found that slightly higher thermal efficiency of the overall scheme could be achieved by recycling the recovered unconverted fuel species to the FC rather than the gas turbine, and thus, this following configuration was selected as our basis for further techno-economic assessment as shown in Figure 5.6.

In summary, the above configuration was selected due to the study basis specifying a new-build plant, with an emphasis on a balance of controllability and maximum thermal efficiency.

The block flow diagram presents how the main process flows would be configured; however, physical integration of many FC units, two gas turbines, and their respective heat recovery steam generators connected by large cross sectional area

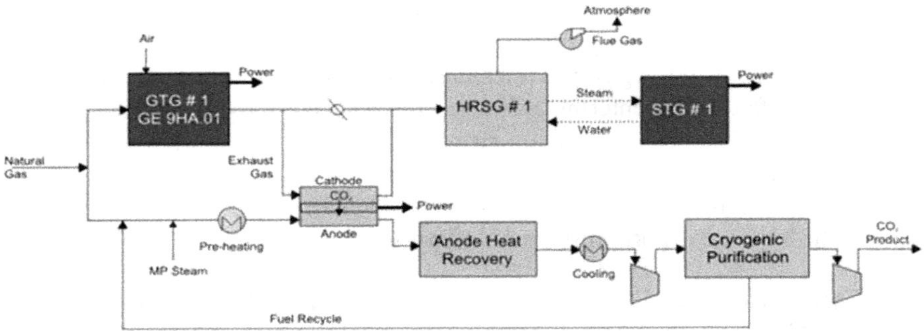

FIGURE 5.6 CCGT with MCFCs, cryogenic separation, and fuel recycle to MCFCs—selected configuration for study by Ferguson and Tarrant [92].

ductwork is logistically challenging. FCE has undertaken work separately investigating logistics of how best to arrange a large number of their stacks. This resulted in the following design which incorporates 208 individual FCs stacks into a single enclosure where each enclosure features a dedicated flue gas feed/exhaust gas heat recovery exchanger as shown in Figure 5.7.

To modify the above concept to fit the capacity required for larger gas turbines, Wood made the following adaptations:

1. Larger square ductwork with single inlet and single outlet of graduated cross-sectional area.
2. 5 × gas/gas heat exchangers per gas turbine train.
3. 5 × 208-stack enclosures per gas turbine train.

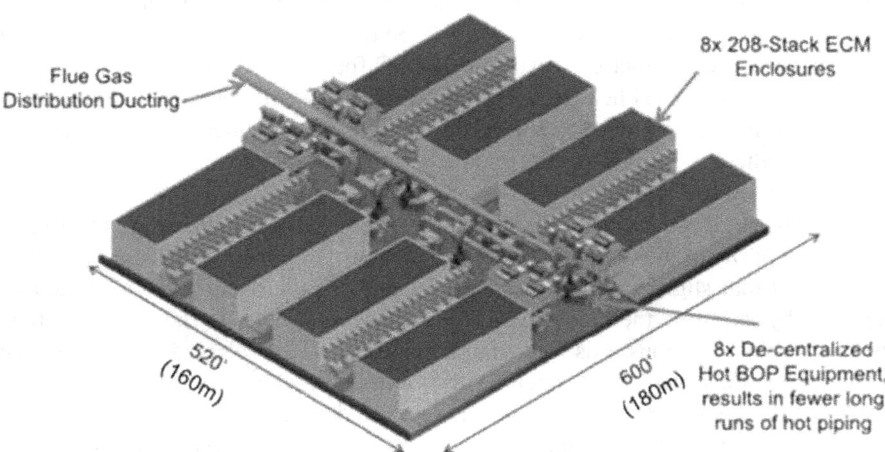

FIGURE 5.7 Sketch of a 350-MWe MCFC installation. (*Source*: Courtesy of FuelCell Energy Inc. based upon work supported by the US Department of Energy under Award Number DE-FE0026580) [92].)

The material of construction for equipment and ducting must be suitable for both high temperature and presence of carbon monoxide and hydrogen. This is a very challenging service for materials and required a high-grade nickel alloy (UNS N06696). Minimizing the quantity of such an expensive material of construction would be a key area in the design where costs could be reduced with further work, or substitution with alternative material as long as it is demonstrated to be able to withstand the duty without compromising safety. FC stack lifetime is another area which has significant impact on the overall cost of the scheme. FCE deems a 7-year lifetime to be an achievable target; therefore, the study assumed that the stacks would require replacement every 7 years. It is important to include this cost in the overall plant economics.

5.2.5.1 Comparative Techno-Economic Performance Results

The technical and economic performance results for Case 1, CCGT with amine solvent, and Case 2, MCFC post-combustion CO_2 capture were compared with a conventional CCGT without any CO_2 capture. The MCFC case captures 90% of the CO_2 from the GT exhausts and electrochemically consumes additional fuel, from which 100% of the CO_2 emitted is captured while producing additional power, the net effect of which is an increase in net power production and only a minimal 2.6% point net efficiency loss compared to an unabated CCGT plant. Ferguson and Tarrant [92] highlights the following points as basic differences in technical performance between the two cases:

1. Both cases use the same high-efficiency gas turbine power island configuration while the MCFC case adds the FCs, which have a gross LHV efficiency of ~75%. Thus, this case benefits from a very high efficiency underlying power production before any parasitic loads for carbon capture are applied.
2. The parasitic loads associated with the CO_2 capture and compression process result in a net exportable power from the CCGT with MCFC case of 444 MWe more than the amine solvent case, but with additional fuel electrochemically consumed. These balance each other to some extent, such that there is still a reduction in efficiency vs. the unabated case, but only of 2.6% lower net LHV efficiency.
3. Although the MCFCs require a significant amount of steam for the reforming and shift steps within the cell, this steam can be generated *via* heat recovery from the FC exhaust.
4. CO_2 compression power appears high compared to some schemes because the first compression stages are also compressing the unconverted hydrogen, CO, and water vapor prior to the cryogenic purification and fuel recycle step.
5. The carbon efficiency for the CCGT with MCFC case is 8% of the reference unabated case as this case captures 92% of the total CO_2 produced.

The study [92] also points out that the total project cost for the CCGT with MCFC case is 62% higher than the amine solvent case but produces 42% more net power output. The following additional points can be highlighted as basic differences in economic performance between the two cases with post-combustion CO_2 capture [92]:

1. The MCFC system includes several high capital cost elements, including the MCFC cells themselves and exotic materials required for heat exchangers downstream of the FC and high temperature ducting.
2. Operating costs are high for this case primarily because the MCFC stacks are assumed to be replaced every 7 years.
3. Despite the capital and operating costs being substantially higher for the MCFC case, the LCOE was almost identical to the amine case at £70.7/MWh, compared to the £69.9/MWh for the amine case. This was largely due to the very high thermal efficiency of this case combined with very low residual carbon emissions.

It should be noted that the UK has a penalty for CO_2 emissions which consequently rewards projects with lower residual CO_2 emissions, and has a very large cost impact on any project which does not feature carbon capture and storage. This is reflected in the estimations above, with both low-carbon electricity generation cases having a lower LCOE than the unabated fossil fuel power generation case. In this instance, where it can be considered that any new power plant with CCS is displacing an unabated plant from the grid, the cost of CO_2 avoided can be defined as the additional cost per unit of power produced divided by the delta carbon footprint between the plants with and without CO_2 capture. Since this study demonstrates lower LCOE for both abated plants, their cost of CO_2 avoided is negative.

While the results presented in the study by Ferguson and Tarrant [92] have drawn upon years of development undertaken by others investigating potential flow scheme configurations to integrate MCFCs with CCGTs, a further optimization of heat integration alternatives vs. capital and operating costs could result in a more cost-effective or more thermally efficient scheme by paying attention to heat recovery from the FC exhaust gas. All cases developed for the UK Government considered baseload power plants which would operate for the majority of the year, providing a base level of low-carbon power at all times. However, there is an increasing need for low-carbon power plants to be able to respond to changes in demand to balance production from variable renewables such as offshore wind. Therefore, an assessment of potential for flexible operation of the overall plant to meet grid demands for flexible/dispatchable low-carbon operation needs to be undertaken.

Most power plants can achieve increased overall thermal efficiency, as well as their value to society, by incorporating heat provision alongside power generation, particularly if value can be realized for low-grade heat. The scheme incorporating MCFCs may also be able to provide further potential combined heat and power benefits as the large-scale MCFC installation already in operation provides this. MCFCs generate hydrogen as part of the internal chemistry inherent to the FC. FCE has indicated that their stack can also be used to provide a pure hydrogen stream which is anticipated to be highly important as an energy vector for decarbonization. Anticipated uses for such a hydrogen stream include its use as a transport fuel, a low-carbon fuel for domestic and commercial space and water heating, and as an energy storage medium, whereby hydrogen is generated at times of low grid power demand and used for peak power generation, either *via* combustion in dedicated gas turbines or in hydrogen FCs. The ability to add alternative revenue streams *via* hydrogen sales can substantially improve the already competitive performance of this technology for post-combustion CO_2 capture compared to conventional state-of-the-art technologies.

In summary, the study by Ferguson and Tarrant [92] presents findings of a techno-economic assessment comparing the use of MCFCs and proprietary amine-based solvents for baseload low-carbon power generation using post-combustion CO_2 capture. It was found that incorporating MCFCs between each gas turbine and its respective heat recovery steam generator, using cryogenic separation to purify the CO_2 and recycling the unconverted fuel species back to the FC, could achieve 92% CO_2 capture by adding 440 MWe of FCs. Other findings included the following:

1. Net power production increased by 42% in MCFC case vs. amine case.
2. Thermal efficiency penalty improved from 7.4% points to 2.6% points in MCFC case.
3. Total CO_2 captured increased from 2.9 to 3.3 MTPA.
4. Total project cost increased by 65%, but specific project cost (per kW) increased by 14%.
5. Total operating cost (before fuel and carbon price) increased by 64%.
6. Income from electricity sales increased by 43%.

The increased capital and operating costs are balanced out by the increased power production, high thermal efficiency, and lower residual CO_2 emissions to result in an LCOE almost identical to that of the amine-based technology at £70.7/MWh and £69.9/MWh for the MCFC and amine technologies, respectively. Wood anticipates that this scheme may have significant additional advantages yet to be understood, such as flexibility to meet grid demand and ability to produce hydrogen as well as further potential for optimization of the design presented.

5.3 MULTI-FUNCTIONAL REGENERATIVE FC SYSTEMS

The regenerative FCs possess a dual operating mode. That is, they may be operated as an electrolyzer and alternately as FCs, thus providing the opportunity to operate either in the electrolysis or in the FC mode, respectively [118,119]. The electrolyzer consumes electrical energy to generate H_2.

In water electrolysis, produced hydrogen is fed to the hydrogen storage tank for operating a hydrogen-fueled generator [120]. The electrolyzer and the FC systems may be integrated to form a single-cell stack; the regenerative FC may also employ two separate cell stacks where one is being used as the FC while the other is used as an electrolyzer [118]. While the single-cell stack is referred to as a unitized system, the two separate cell stacks are regarded as a discrete system [119]. Figure 5.8 illustrates a microgrid based on solar PV modules and regenerative FCs with master and slave application of the FCs. Such a configuration is based on a dedicated controller for fuel supply and timing of operation for the two FCs.

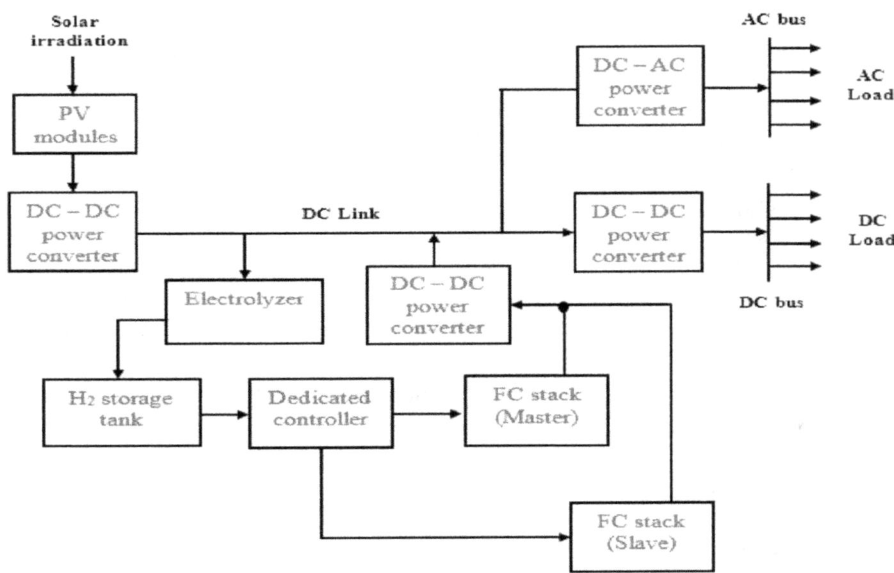

FIGURE 5.8 Microgrid system based on solar PV and regenerative FC [121].

The cell stack produces H_2 and oxygen from electrical energy in the electrolysis mode. However, it produces electrical energy from the stored H_2 (and O_2 that may be obtained from the air) in the FC mode [118]. By this characteristic (similar to a secondary battery), regenerative FC is an energy storage system that utilizes H_2 as the storage medium. It possesses the highest specific energy that may be reached of all rechargeable energy storage technologies. For instance, specific energy of 400–1,000 Wh kg/L (watt hour per kilogram) may be achieved for a practical system [118,122].

The technology may be used in applications where it is required to store a large quantity of energy such as [122] (1) used in combination with variable power sources such as solar and wind; (2) used as emergency power; (3) employed in marine systems, such as "unmanned" underwater vehicle; (4) used in spacecraft as hybrid energy storage or propulsion systems; and (5) used as solar rechargeable aircraft for the high-altitude, long-endurance purpose. The diagram illustrating the use of a regenerative FC system as storage in a solar photovoltaic (PV) microgrid was published in [122]. The overall specific energy of regenerative H_2/O_2 FCs that ranges from 400 to 1,000 Wh/kg is a multiple of the value achieved for any battery system [118]. For instance, specific energy values of 100–300 Wh/kg [123,124], 30–50 Wh/kg [125,126], 100–200 Wh/kg [125], and 10–50 Wh/kg [127] are for lithium-ion, lead-acid, sodium-nickel chloride also known as ZEBRA, and vanadium redox battery, respectively. These, including the fact that regenerative FCs are independent of hydrogen infrastructure like the other FC technologies, are important factors that attract their use in the mentioned applications.

Certain trade-offs are usually considered for planning and designing regenerative FCs. These include the choice between oxygen and air feed, single-stack and/or discrete systems, feed, and the choice of working pressure [119,128]. Each of the mentioned trade-offs may be weighed on different grounds, but the economic factor, efficiency, and the duty cycles are usually the driving factors for selecting an oxidant for a particular purpose.

Alia at NREL [129] points out that regenerative FCs offer a unique solution for grid energy storage. Unlike batteries, regenerative FCs can cost-effectively store a large amount of energy in the form of hydrogen. Energy in the form of hydrogen can be stored at a cost of about $35/kW-hr, versus $196/kW-hr for batteries. The regenerative FC systems could also provide an added benefit of hydrogen generation for FC vehicles. Regenerative FCs typically operate in either acidic (i.e., PEMFCs) or alkaline environments. The relevant reactions for regenerative FCs are shown as follows:

$$\text{Oxygen reduction reaction (ORR) in acid: } O_2 + 4\,H^- + 4\,e^- \rightarrow 2\,H_2O \tag{5.8}$$

$$\text{ORR in alkaline: } O_2 + 2\,H_2O + 4\,e^- \rightarrow 4\,OH^- \tag{5.9}$$

$$\text{Hydrogen Oxidation Reaction (HOR) in acid: } 2\,H_2 + 4\,e^- \rightarrow 4\,H^+ \tag{5.10}$$

$$\text{Hydrogen Oxidation Reaction in alkaline: } 2\,H_2 + 4\,OH^- \rightarrow 4\,H_2O + 4\,e^- \tag{5.11}$$

$$\text{Overall Energy Generation Reaction: } 2\,H_2 + O_2 \rightarrow 2\,H_2O \tag{5.12}$$

$$\text{Overall Energy Storage Reaction: } 2\,H_2O \rightarrow 2\,H_2 + O_2 \tag{5.13}$$

The overall reactions are the same in acid or alkaline conditions, the only difference is the ions being exchanged. The oxygen evolution reaction (OER) and hydrogen evolution reaction (HER), which occur during energy storage modes, are the reverse of the ORR and HOR, respectively.

5.3.1 Unitized Reversible FC

URFC is a highly developed and advanced FC technology [130–132]. In an URFC unit cell, the electricity is produced through the round-trip energy conversion of (1) the conventional FC mode and (2) the water electrolyzer (WE) mode [133–135]. The overall reaction mechanism of URFC is shown as follows:

$$\text{In FC mode: } 4H^+ + O_2 + 4e^- \rightarrow 2H_2O \tag{5.14}$$

$$\text{In WE mode: } 2H_2O \rightarrow 4H^+ + O_2 + 4e^- \tag{5.15}$$

In the FC mode, the hydrogen and oxygen are used to produce the electricity. The hydrogen and oxygen are passed to the anode and cathode, respectively. The ORR and electricity production process takes place through the FC mode. Initially, the hydrogen molecules dissociated into protons (H+) and electrons in the hydrogen electrode. The proton moves to another side of the electrode through the polymer-based proton-exchange membrane (PEM), and the electron moves through the external circuit. The end of the FC mode reaction is the production of water and electricity.

The WE mode is known as water-splitting process in the URFC unit cell. It involves the opposite reaction that undergoes in the FC mode. The water molecules are supplied to the oxygen electrode, and the reaction starts by splitting water into 4H+, and O_2 through the power supply. The produced oxygen and hydrogen gases can be stored and used in the FC mode as an energy carrier [75]. This is a most important achievement in FC applications.

5.3.1.1 Advantages and Limitations in URFC Unit Cell Systems

Along with the round-trip energy conversion, the URFC system has several potential advantages as follows: (1) It offers high specific energy density (packaged: 400–1,000 Wh/kg and theoretical: 3,660 Wh/kg), (2) it uses the abundant chemical compound H_2O as a fuel carrier, (3) it is a renewable and sustainable energy system, (4) no harmful emission occurs during the process, (5) it is lightweight, and (6) it is a high durable system [136,137]. Because of the above-mentioned potential properties, the URFC device is used in (1) spacecrafts, (2) zero-emission vehicles, (3) solar rechargeable aircrafts, (4) military applications, (5) on-site energy storage system, and (6) residential power sources, and is an especially prominent energy storage system for space applications [122,136,138–140]. Apart from the numerous advantages, there are a few drawbacks associated with the materials and components of the URFC unit cell. These are described later in this section. Figure 5.9 represents the schematic representation of the continual arrangement of a single URFC unit cell system.

The polymer electrolyte membrane (PEM) is the central part of the unit cell. The sequence of the components aligned from the center of the unit cell is the membrane, Nafion binder, electrode containing the electrocatalyst and supportive materials, gas diffusion backing (GDB) consisting of a micro- or meso-porous layer (MPL) and gas diffusion layer (GDL), and BP. A similar alignment is present on either side of the unit cell. The stability and durability of the unit cell are major challenges in the URFC system. For operating the WE mode in URFC system, it needs higher

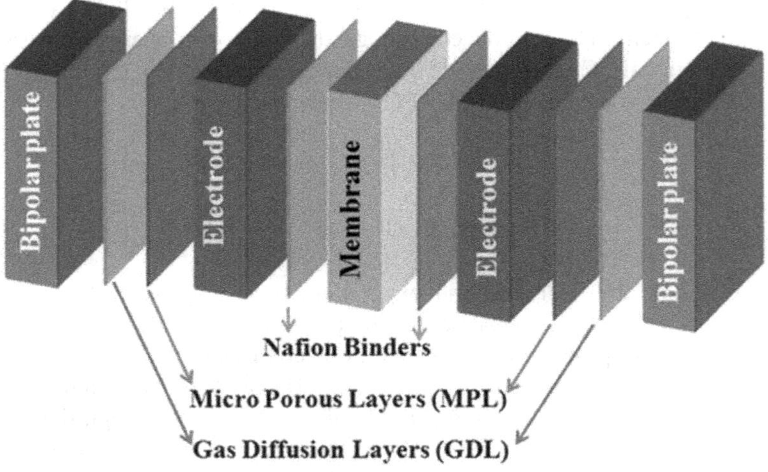

FIGURE 5.9 Schematic representation of the alignment of the URFC unit cell components [148].

resistant material, when compared to PEMFC system. The challenges and limitations associated with each component of the URFC device are as follows:

1. **Electrocatalysts**: the high cost of the Pt electrocatalyst, carbon corrosion of supportive materials, and aggregation and migration of electrocatalysts [134,140,141];
2. **Nafion binder**: dissolution or solubility of the Nafion binder solvents and cracking of the binder [128];
3. **PEM**: dimensional change and low proton conductivity [142];
4. **GDB layers and BP plates**: severe carbon corrosion [143–146]. In addition, the ultimate limitation is the cost effectiveness of each component in the URFC unit cell.

Table 5.1 shows the breakdown of the 2015 FC stack costs at 1000 and 500,000 systems per year, and the 2020 cost targets of URFC unit cell system are shown in Table 5.2, reprinted from the US-DOE hydrogen and FCs program record [147]. The US-DOE targets can be achieved through lower cost of the novel structured materials and composites using new technologies for advanced developments in the URFC unit cell components. To accomplish the higher performance, the limitations of each component should be overcome in URFC system.

TABLE 5.1
Breakdown of FC Stock Costs at 1,000 and 500,000 Systems per Year [148,147]

FC Component	Percentage for 1,000 sys/year	Percentage for 500,000 sys/year
BPs	20	5
Membranes	28	10
Catalyst + application	21	45
GDLs	20	5
MEA frame/gaskets	13	27
Balance of stack	5	8

TABLE 5.2
Projected cost status of each component at 500,000 systems per year compared with 2020-2024 cost targets [147,148]

Component	Cost Status	2020-2024 Cost Target
System	$\$53/kW_{net}$	$\$40/kW_{net}$
Stack	$\$26/kW_{net}$	$\$20/kW_{net}$
MEA	$\$17/kW_{net}$	$\$14/kW_{net}$
FC membrane	$\$17/m^2$	$\$20/m^2$
BPs	$\$7/kW_{net}$	$\$3/kW_{net}$
Air compressor (CEM)	$750/system	$500/system
Humidifier system	$81/system	$100/system
Humidifier membrane	$\$20/m^2$	$\$10/m^2$

5.3.1.2 Perspectives for Advanced Developments in URFC System

As mentioned earlier, the development of URFC faces several challenges and limitations. In order to address these challenges, the study by Sadhasivam et al. [148] has proposed new ideas for constructing advanced URFC systems. So far, high-cost Pt-based electrocatalysts have been considered as promising materials for enhanced electrochemical reactions. Various metals and their derivatives have been added with Pt electrocatalysts for reducing the Pt loading in the electrodes and to lower the manufacturing cost of the electrodes [149–155]. For making viable low-cost bifunctional electrocatalysts, novel structured or composite of non-Pt-based unsupported electrocatalysts such as metal and metal oxide (porous IrO_2, RuO_2) could provide efficient electrocatalyst materials for URFC applications.

The advanced structured materials should have the multiple functional properties of electrocatalysts and supportive materials. The novel electrocatalyst must contain the synergistic effect properties for enhancing the electrochemical reactions, protons, and electrical conductivity. The corrosion resistances of supportive materials should also contribute to enhancing the electrochemical performances [152,155,156]. Introducing sulfonation (-SO_3H functional group) on the supportive materials will be a priority choice for the advanced URFC systems, because sulfonation in the supportive materials can considerably enhance the electrochemical reactions due to the rapid proton transportation [157]. The severe corrosion in supportive materials can be controlled by the highly crystalline nature of novel structured materials. If the electrocatalyst support material contains the ORR and/or OER properties, it will be a more beneficial to the advanced URFC system.

In URFC technology, the next major challenge is to produce highly stable, low-cost membranes for replacing the conventional Nafion membrane. In the previous decade, various kinds of polymer-based membranes have been studied for different FC applications. However, no extensive analyses have been undertaken for URFC applications. Compared to the conventional Nafion membrane, the proposed organic–inorganic hybrid composite structure will be a promising membrane for advanced URFC applications. Sulfonated polyether ether ketone (SPEEK) and sulfonated poly phenylene oxide (SPPO) can be considered as efficient organic compounds instead of conventional Nafion due to their low cost, availability, and easy sulfonation processes [158–160]. The sulfonated polymer membrane can increase the proton conductivity, flexibility, and stability of the membrane. In order to augment the advanced properties in the membrane, functionalized inorganic materials must be dispersed in the polymer membrane matrix. The inorganic materials can enhance the mechanical and thermal stability of the polymer membrane due to strong interfacial interaction between the polymer main chain and inorganic functional properties. Furthermore, the incorporation of inorganic material can act as a barrier for the fuel crossover in the membrane. From the above-mentioned advanced properties, the hybrid organic–inorganic membrane will be a potential source for the further developments in the URFC systems. In URFC, carbon corrosion is also a major limitation in the BP plate and GDB. In order to prevent the carbon corrosion in GDB (GDL and MPL) and BP plates, the highly crystallite nature of graphitized carbon can be considered as a higher corrosion-resistant material [161].

The durability of graphitized carbon-based BP might be decreased during the long-time operation. At present, graphite-based metal composite plates [162,163] and metal-based PB plates, especially stainless steel and Ti [164–168], have the promising properties for enhancing the stability of the BP plate. For the advanced technologies, low cost of novel mesoporous or nanorod-structured metal particles such as Ti, IrO_2, and novel alloys incorporation on graphitized carbon will be promising BP plates for the URFC applications. The proposed materials will be more effective for the advanced developments in the near future for viable URFC applications.

The URFC is an optimum FC system owing to its specific high energy density. From the materials perspective, the degradation and poor stability of the materials in the unit cell is a significant challenge in the URFC device. Researchers have made noteworthy analyses and developments for commercializing the URFC system. In addition, novel scientific ideas have been suggested to develop and commercialize the URFC system with optimum operating conditions. For further sophisticated developments of URFCs, several other difficulties should be overcome in the near future. The major limitations associated with the MEA of URFCs are the high cost and poor stability of the materials. In MEA, the prominent Pt electrocatalyst and Nafion membranes have the highest production costs. Moreover, another critical challenge is the stability and durability of the MEA. In long-term operations, the electrocatalyst reaction has suffered owing to the agglomeration, migration, and sintering due to corrosion of supportive materials.

The membrane can be degraded through the dimensional change and inside penetration of electrocatalyst because of migration effect. The dissolution of Nafion binder has also hindered and caused the degraded performances in the unit cell because of electrocatalyst aggregation and diminished the interfacial interaction between electrodes and membrane. Apart from the major issues in MEA, carbon corrosion should be overcome in the GDB (MPL and GDL) and BPs. It is, however, possible to develop feasible URFC unit cell systems using the novel structured materials. The current trend shows several advantages for the novel structured materials, composites, and advanced techniques in the URFC unit cell. However, low-cost and highly durable materials and components need to be invented for creating advanced and optimum URFC systems.

5.3.1.3 Application of URFC to Aircraft Industry

One of the major applications of URFC system is for aircraft industry. Utz [169] explored URFC as potential power sources for multiple applications involving mass and volume minimization, including space and aerial vehicles. The study points out that much research and development has been performed on developing the technology from the cell level, which requires careful choice of materials that can survive the electrochemical environment in both FC and electrolysis modes, and flexible cell designs that handle reactants and products of each reaction without impeding the process. Developers have attempted to build their designs around either low-temperature PEM (LTPEM) or solid oxide FC (SOFC) technologies, each built upon several decades of maturation in FC applications. Demonstration hardware at the single-cell, short-stack, and full-stack levels has been manufactured and tested for performance. While both technologies have shown the ability to operate in a reversible mode, long-term, stable, and durable operation in practice requires addressing some challenges. With LTPEM technology, managing two-phase flow of water during transitions is problematic. SOFC technology requires large heat inputs and does not easily handle transitions between power levels or start–stop cycles. While both of these issues can be addressed by high-temperature PEM (HTPEM) technology, it is yet to be explored because of the low level of maturity of HTPEM.

The study points out that Teledyne has developed a conceptual design of a 125-kW HTPEM URFC system to be integrated into an aircraft. Potential applications include direct power for electric propulsion in small aircraft or a load-balancing device for larger aircraft, such as a 787 Dreamliner. The electric propulsion application was explored in detail in this study. The system consists of an open-cathode URFC stack with HTPEM technology stored within the oxygen reactant storage tank. This provides advantages by simplifying water management, eliminating oxygen-flow-control devices, and minimizing volume. This arrangement raises multiple safety issues that must be addressed in the detailed design, including the potential for hydrogen–oxygen gas mixtures due to leakage, product water sloshing, and electrical shorting of the stack. A hydrogen storage tank, thermal control system, pressure-/flow-control devices, and control logic were defined and described in detail to maintain safe and effective system operation over the length of the proposed flight. The concept of operations, system parameters, and alarm conditions were defined to provide a framework for system operation. Individual components were selected with performance data and physical specifications gathered.

The study [169] also developed a system model based upon the URFC system design. The model used power consumption data from all of the components required for operation, performance data from HTPEM technology during power production and electrolysis, thermal requirements defined by the stack mass and efficiency, and expected flight profile of 1-, 2-, and 5-hour flights. Performance was evaluated in both FC and electrolysis modes of operation. The system was shown to be effective in providing power for the electric propulsion application with enough reactant supply to handle the worst-case scenario. Refueling times were shown to be less than or equal to the mission duration. The major disadvantage of this system is the power and energy density of less than 100 and 200 Wh/kg, which is not competitive with lithium-ion battery technology of similar scale, largely due to the size and mass of the reactant storage tanks. Recommendations included scaling the system down to the kW or sub-kW scale to reduce the reactant tank mass, exploring failure modes and effects analysis studies to explore the safety implications and identify areas of improvement, and putting resources into improving the HTPEM technology itself to maximize efficiency and durability of the overall system.

5.4 HYBRID FC SYSTEMS

A hybrid power system consists of a combination of two or more power generation technologies to make best use of their operating characteristics and to obtain efficiencies higher than that could be obtained from a single power source. Hybrid fuel-cell systems are power generation systems in which a high-temperature FC is combined with another power generation technology [170,171]. The resulting system exhibits a synergism in which the combination has far greater efficiency than that could be provided by either system operating alone [172]. The efficiencies across a broad power range of various power generation technologies are shown in Figure 5.10. As an example, combining SOFC or MCFC with the gas turbine would increase the overall cycle efficiency while reducing per-kilowatt emissions. In some systems, combining FCs with wind or PV systems would extend the duration of the available power, which is of significance, rather than the overall efficiency. This type of system is used as a backup power or as an energy storage system. Getting higher efficiencies combined with low emissions, hybrid systems are likely to be the choice for the next generation of advanced power generation systems. These systems are not only used for stationary power generation, but also find application in transportation systems.

The literature [170,171,173–176] has classified the hybrid fuel-cell systems as Type-1 and Type-2 systems. In a Type-1 system, a high-temperature FC is combined with another power generation technology to increase the combined efficiency of the system. In Type-2 hybrid systems, a FC and another power generating system are combined to best make use of the operating characteristics of the individual units to either extend the duration of the availability of power or to supplement the fuel-cell power. Some examples of Type-1 systems are high-temperature SOFC–gas turbine system, SOFC–TPV system, FC with reciprocating (piston) engine, and designs that combine different fuel-cell technologies. Some examples of Type-2 systems are PEMFC–solar power hybrid system and PEMFC–wind power system. Here we examine advances made in these and several other novel hybrid FC systems.

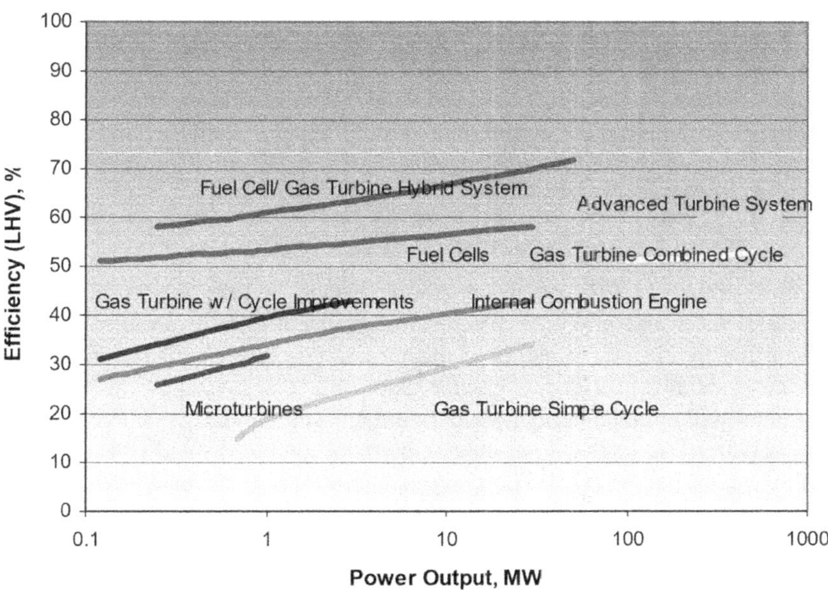

FIGURE 5.10 Estimated performance of power generation systems [173].

The hybrid system is a key to the Department of Energy's Vision 21 plants. The Vision 21 program has set power plant goals of achieving efficiencies greater than 75% (LHV) for natural gas.

The higher efficiencies play a key role in reducing emissions, another target in Vision 21 plants. As a comparison, conventional coal-burning power plants are typically 35% efficient, and natural-gas-fired plants are now 40%–50% efficient.

Hybrid systems can also involve cogeneration systems where excess heat is either used or converted to additional power in order to increase overall thermal efficiency. Here the choice of thermal energy converter can affect overall performance. Acikkalp et al. [177] compared the performance of an alkali metal thermal electric converter (AMTEC) and thermoelectric generator (TEG) as a subsystem for utilizing rejected heat from the MCFC. AMTEC and TEG have various advantages such as higher power density, no maintenance needed, silent, and generally lower cost. Performance parameters of systems are defined as power output density, exergy destruction rate density, and energy and exergy efficiencies. The study [177] examined thermodynamic performances of the MCFC-AMTEC and MCFC-TEG systems based on the first and second laws of thermodynamics. Performance analyses were carried out in terms of power density, energy efficiency, exergy efficiency, and exergy destruction rate density. Some important results obtained in this study were as follows:

1. The maximum power output densities for hybrids I and II are 2425.833 (W/m^2) and 1964.389 (W/m^2), respectively.
2. Energy efficiencies were 76.6% for system I and 76.4% for hybrid II.
3. Hybrid I is more advantageous than hybrid II.
4. Using AMTEC is more reasonable than using TEG to utilize waste heat from the MCFC. It is recommended that deeper analyses including economic and environmental approaches will be needed.

Hybrid energy management system design with renewable energy sources such as PV cells, wind energy, and fuel cells is reviewed in details by Agrawal et al. [178].

5.4.1 Hybrid FC–Gas Turbine Systems

Examples of hybrid FC power generation cycles include the combined high-temperature FCs and gas turbines, reciprocating engines, or another FC. These represent the hybrid power plants of the future. The conceptual systems have the potential to achieve efficiencies greater than 70%. The hybrid FC/turbine (FC/T) power plant will combine a high-temperature, conventional MCFC or a SOFC with a low-pressure-ratio gas turbine, air compressor, combustor, and in some cases, a metallic heat exchanger [173]. The synergistic effects of the hybrid FC/turbine technology will also provide the benefits of reduced greenhouse gas emissions. Nitrous (NOx) emissions will be an order of magnitude below those of non-FC power plants, and carbon monoxide emissions will be less than 2 parts per million (ppm) [174]. There will also be a substantial reduction in the amount of carbon dioxide produced as compared to conventional gas power plants.

FCs are inherently attractive for distributed power applications, but their use is limited by the need to package them in a system balance of plant (BOP) that allows them to function effectively. All FCs, especially high-temperature FCs, require spent fuel utilization/waste heat recovery subsystems. Low-temperature FCs also require fuel reforming subsystems. A common approach for providing these BOP functions has been to integrate the FC with another generating technology. There can be many different cycle configurations for the hybrid FC/turbine plant. In the direct mode, the FC serves as the combustor for the gas turbine while the gas turbine is the BOP for the FC, with some generation. In the indirect mode, the FC uses the gas turbine exhaust as air supply while the gas turbine is the BOP. In indirect systems, high-temperature heat exchangers are used [175].

The combination of the FC and turbine operates by using the rejected thermal energy and residual fuel from a FC to drive the gas turbine. The FC exhaust gases are mixed and burned, raising the turbine inlet temperature while replacing the conventional combustor of the gas turbine. Use of a recuperator, a metallic gas-to-gas heat exchanger, transfers heat from the gas turbine exhaust to the fuel and air used in the FC. Figure 5.11 illustrates an example of a proposed FC/turbine system. One item to note in Figure 5.11 is that the combustor shown prior to the turbine is for start-up purposes only.

Countries around the world are developing interest in the high-efficiency hybrid cycles. A 320-kW hybrid (SOFC and gas turbine) plant entered service in Germany in 2001, operated by a consortium under the leadership of RWE Energie AG. This was followed in 2002 by the first 1-MW plant, which was operated by Energie Baden-Wurttemberg AG (EnBW), Electricite de France (EDF), Gaz de France, and Austria's TIWAG [174]. The hybrid power systems based on high-temperature FCs and gas turbines have been extensively analyzed and studied by various universities, industries,

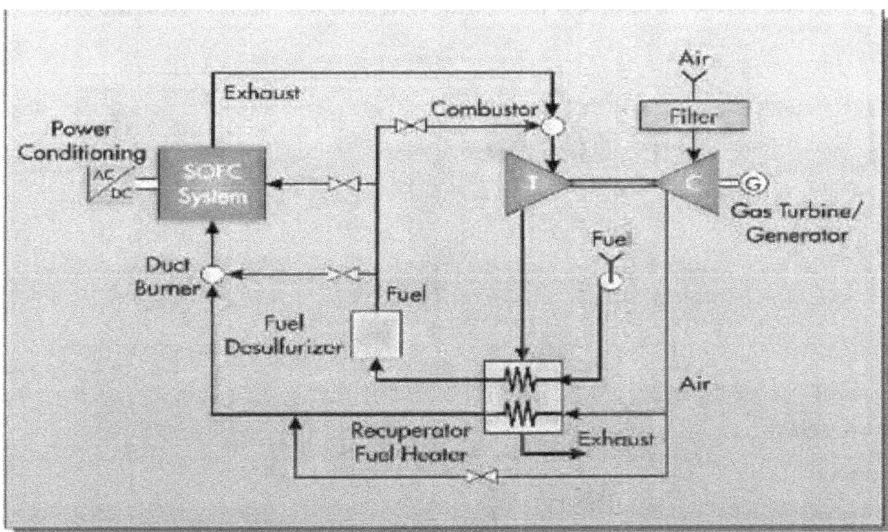

FIGURE 5.11 Diagram of a proposed Siemens–Westinghouse hybrid system. (Taken from DOE Project Fact Sheet—FC/ATS Hybrid Systems [173].)

and the US Department of Energy for stationary power generation applications [178–187]. Successful development and commercialization of FC/turbine hybrid power generation will allow the following: (1) extremely high efficiency compared to other fossil fuel systems; (2) ultralow emissions without additional cleanup; (3) siting flexibility with environmentally friendly energy systems; and (4) fuel flexibility.

The SOFC–micro-turbine hybrid systems are also being investigated for use as APUs in commercial airplanes to provide the power to all the electrical loads [188], and in railroad vehicles to provide the power to all the accessory loads. Combination of a high-temperature FC with a turbine/micro-turbine has several important ramifications to the energy and transportation industry. The SOFC systems are being developed in the range of 5 kW for automotive applications to several megawatts for power generation applications. The micro-turbines are being developed from 30 kW to 30 MW, and the gas turbines power is in the range of 100–1,000 MW. The study of Rajashekara [189,190] analyzes 500-kW SOFC–gas turbine, which can be used as an APU in cruise ships, airplanes, and trains. The same system could be used for distributed power generation applications. If more power is required, more of these systems could be paralleled or the individual systems can be distributed to meet the power demands of the local loads.

An SOFC/gas turbine hybrid system of 500-kW power is depicted in Figure 5.12. The fuel is first reformed to obtain the hydrogen-rich reformate and fed to the anode of the SOFC. The ambient air is drawn using a compressor and pressurized to about 300–400 kPa (3–4 atm). The compressed air is heated using the exhaust of the gas turbine with a heat exchanger and fed to the cathode. The cathode exhaust from the SOFC and the unused fuel from the anode are burned in a combustor to increase the temperature of the exhaust to about 1,000°C to meet the requirements of the turbine. The heat and the pressure difference drives the downstream turbine to generate more power without using additional fuel. The turbine exhaust after heating the compressor exit air is also used for heating the fuel that is going into the reformer. The turbine drives the generator

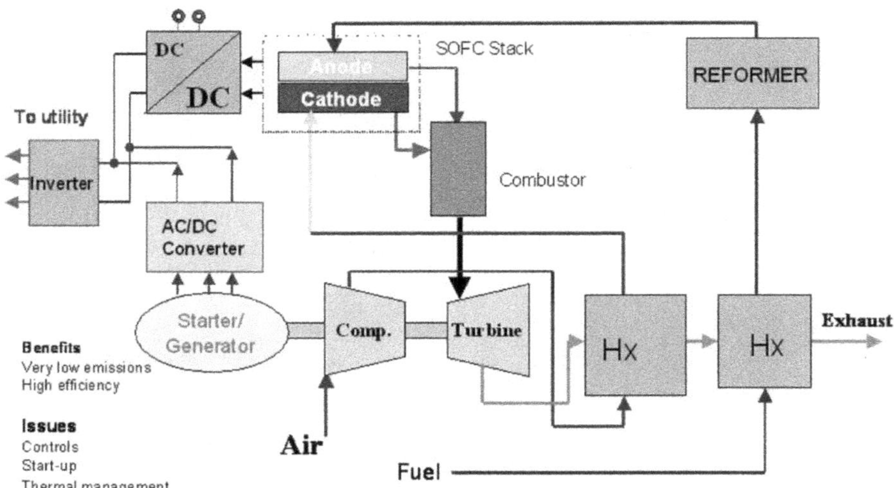

FIGURE 5.12 SOFC–gas turbine hybrid system [189,190].

and produces a three-phase AC output. This AC power is first converted to the DC power and then combined with the DC output from the FC using the power-conditioning system. This DC is converted to the AC output before feeding to the utility.

The gas turbine and the SOFC are tightly coupled in the system [191]. Also, operating with elevated pressure will yield increased power and efficiency for a given cycle. The use of a pressurized SOFC will also lead to optimum integration with the gas turbine. The gas turbine supplies heated compressed air to the SOFC. During normal operating conditions, no additional air or fuel is needed to the gas turbine unit. The reforming process could be endothermic, autothermal, or partial oxidation process. The fuel and air utilization could be varied to give the best system performance within the constraints of stack cooling and heat exchanger metrics. For stationary power generation applications, generally, natural gas is used as the fuel. For trains and ships, diesel fuel is used, and for airplanes, jet fuel is used. The system can use other types of fuels also. The reformer has to be capable of converting these fuels to hydrogen-rich fuel for the FC.

In the study by Rajashekara [189,190], the system was modeled using literature equations [192–194]. The selection was constrained by the operating pressures of the turbine and the compressor. For 500-kW power, the analysis showed that combining the turbine with the SOFC will result in higher efficiencies of the order of about 68%. If a larger stack size of 500,000 cm is used, the stack power density would be 0.89 W/cm and the combined efficiency would be 71.5% for a 500-kW hybrid power system. The DC output of the SOFC and the AC output of the turbine–generator system can be combined using several different power-conversion configurations. In Figure 5.13, the fuel-cell voltage is converted to a focusing of a boost converter and an inverter, and then combined with the three-phase AC output from the generator at the secondary of the three-phase transformers. Similar power-conversion techniques or variations of these are being used by different fuel-cell system manufacturers [195].

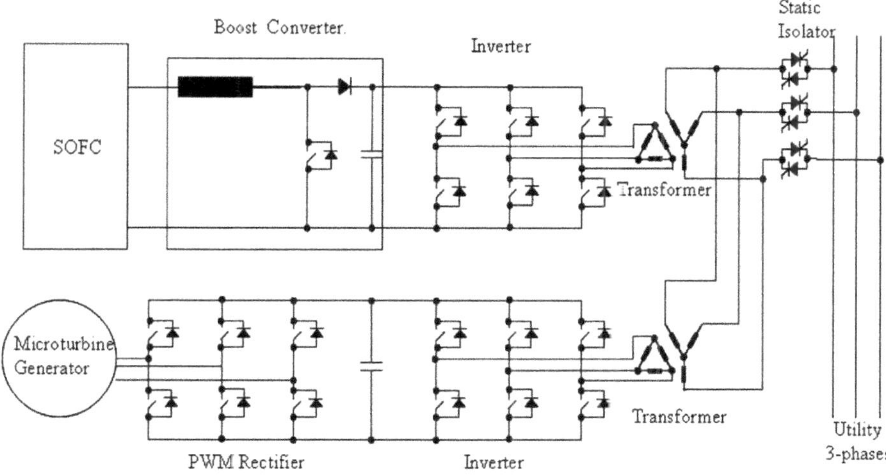

FIGURE 5.13 Electrical system diagram of the hybrid SOFC/turbine systems [189,190].

5.4.2 Hybrid SOFC–TPV System

With the improvements in the materials and fabrication technologies, thermoelectric and TPV devices are recently getting more attention to produce electric power [193–202]. These devices convert thermal or waste heat energy to electricity. Figure 5.14 shows the operating principle of a TPV unit. In a TPV system, the heated emitter produces electromagnetic radiation. A selective filter transmits that part of the radiation with photon energies above the bandgap of the photocells and reflects the lower energy radiation back to the emitter for recuperation. Based on the incident radiation, power is produced by the PV cells. The TPV device is similar to solar PV cells except that the source for TPV applications is much closer and has a temperature of around 1,500–1,800 K rather than 5,800 K for the sun. The heat source should have a temperature of about 1,200°C–1,500°C to achieve reasonable conversion efficiency.

The SOFC–TPV hybrid system proposed by Rajashekara [189,190] is shown in Figure 5.15. The hybrid SOFC–thermoelectric device uses the exhaust of the SOFC as a thermal source to produce the required electromagnetic radiation. The exhaust gases coming out of the SOFC are passed through a combustor to increase its temperature to the level required for operation of the TPV. The exhaust gases are heated to the required temperature in the combustor to meet the requirements of the TPV system. This will reduce the fuel requirements for the TPV thermal source. Once the SOFC unit starts operating, the unused fuel from the SOFC unit can be used for further heating the exhaust out of the SOFC unit, and depending on the power required, it is possible to

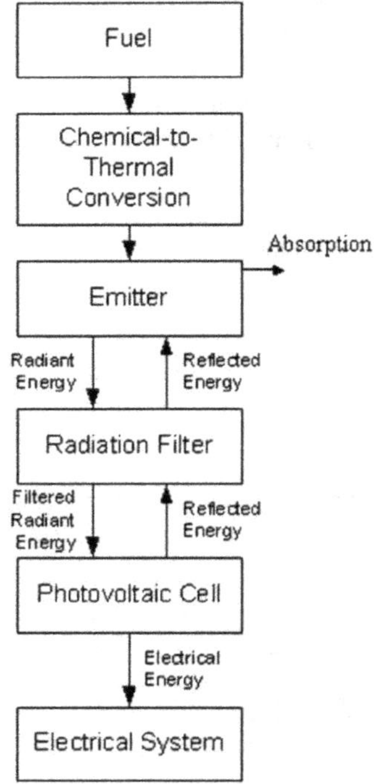

FIGURE 5.14 Operating principle of TPV power generation [189,190].

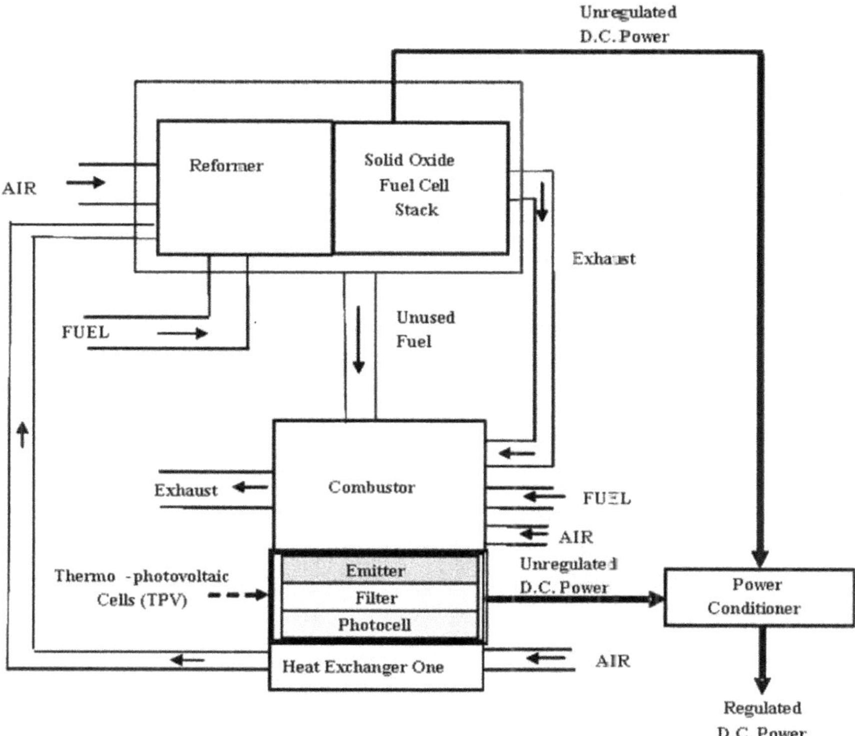

FIGURE 5.15 SOFC and TPV converter hybrid power generation system [189,190].

completely cut off the external fuel to the combustor of the TPV unit. The unused fuel from the SOFC itself may be sufficient. Depending on the power rating of the TPV unit, the DC power from the SOFC and the TPV could be combined using a power conditioner system to produce the required AC power. Instead of the TPV, it is possible to use other types of thermoelectric power-conversion devices in the fuel-cell hybrid system.

The TPV technology is still in the development stage and is not as efficient as a gas turbine. By using the exhaust of the SOFC as a thermal source, the TPV unit can be used to provide the electric power to the auxiliary loads of the SOFC unit. As the start-up time of the TPV is much faster than SOFC, the TPV unit output can provide the electric power to all the critical loads during the SOFC start-up process. It marginally contributes to the increase in the combined efficiency of the hybrid system. If the SOFC unit is 45% efficient and the thermoelectric system is 15%, and assuming 65% utilization of the waste heat from the SOFC, the combined efficiency of the hybrid system would be about 50%.

5.4.3 Hybrid SOFC–PEMFC Power Plant

PEMFCs are gaining importance as the FC for vehicular applications because of their low operating temperature, higher power density, specific power, longevity, efficiency, relatively high durability, and the ability to rapidly adjust to changes in power demand. The PEM is more suitable for automotive applications for a number of reasons [189]. PEMFCs can be started easily at ordinary temperatures and can operate at relatively low temperatures, below 100°C. Since they have relatively high power density, the size could be smaller. Hence, they could be easily packaged in the vehicles. Because of the simple structure compared to other types of FCs, their maintenance could be simpler. They can withstand the shock and vibrations of the automotive environment because of their composite structure. The PEM system also has some drawbacks. A PEMFC requires pure hydrogen as the fuel, thus complicating the design of the reformer system. Any small amount of carbon monoxide in the fuel will poison the electrodes, resulting in severe degradation of performance. As there is a continuous generation of water at the cathode and also the requirement of a certain level of humidification, a sophisticated water management system is required. Platinum metal is required to coat the electrodes to enhance the reactions. Because of the higher cost of platinum, the PEM system is relatively expensive.

High-temperature SOFCs are particularly suitable for automotive auxiliary power unit (APU) and stationary power generation applications. SOFCs possess numerous advantages [172,189,190]. The fuel processor requires a simple partial oxidation reforming process that eliminates the need for an external reformer. It has less stringent requirements for reformate quality and uses carbon monoxide directly as a fuel. Hence, a sophisticated reformer is not required. It operates at extremely high temperatures of the order of 700°C–1,000°C. As a result, it can tolerate relatively impure fuels, such as those obtained from the gasification of coal. Waste heat is high grade, allowing for smaller heat exchangers and the possibility of cogeneration to produce additional power. Water management is not a concern because the electrolyte is solid state and does not require hydration. The by-product is steam rather than liquid water, hence, no need for water management. The SOFC does not need precious metal catalysts. However, SOFC also has some drawbacks. Because of the high-temperature operation, the start-up is of the order of 20–30 minutes. Hence, the SOFC is not suitable for propulsion applications. Packaging of the low-temperature electronics and the high-temperature stack within the same enclosure is a major challenge. The electrical efficiencies of PEMFC and SOFC over various generation capacities are illustrated in Figure 5.16.

The power densities of both PEM and SOFC systems are of the order of about 500 mW/cm under typical operating conditions. The peak power densities under idealized conditions have been reported to be greater than 1,000 mW/cm. The relatively simple design (because of the solid electrolyte and fuel versatility), combined with the significant time required to reach operating temperature and to respond to changes in electricity demand, make the SOFC suitable for large-to-very large stationary power applications. The start-up time for the SOFC is of the order of 30–50 minutes, whereas the PEM system could be started in less than 1 minute. Hence, the SOFC is not suitable for propulsion applications. However, as an APU in transportation applications and in stationary power generation systems, the starting time of the SOFC is not a major issue.

For certain distributed power applications, fuel-cell/fuel-cell hybrids may be an effective approach to the BOP problem. Only certain combinations of FCs promise to synergize into a simplified BOP. Generally, the combination includes

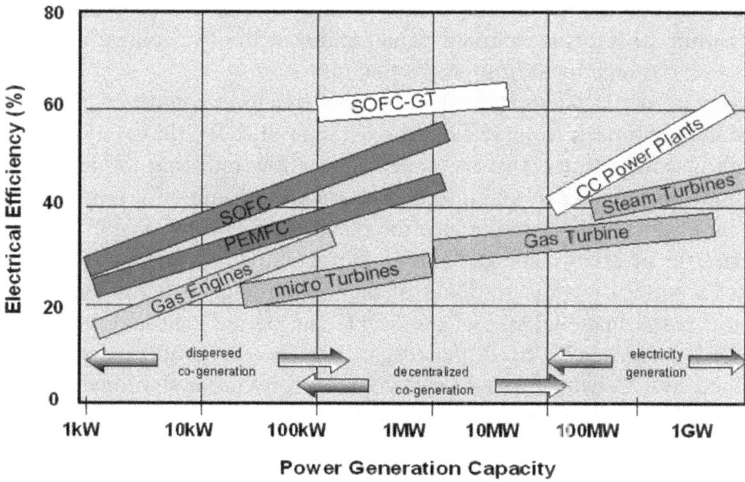

FIGURE 5.16 Efficiency over power generation capacity for large variables [189,190].

a high-temperature internally reforming FC with a low-temperature FC having a complementary electrolyte (anion-conducting versus cation-conducting electrolytes). The solid oxide FC/polymer electrolyte FC (SOFC/PEFC) hybrid is of particular interest due to several reasons. SOFC can be used as reformer for the PEFC. The fuel utilization in the SOFC is limited to a range necessary only to reform natural gas, but not completely oxidize it. PEFC can be used as chemical bottomer for the SOFC. The reformed fuel utilization is completed in the PEFC, where more favorable thermodynamics apply. The bulk air separation or acid gas removal is not required for isolation of CO_2, if sequestration is ultimately practiced in some way. Finally, the co-production of hydrogen, possibly in conjunction with a storage system, allows the SOFC to be base-loaded, while cycling is more effectively done with the PEFC.

The hybrid plants are projected to cost below comparably sized FCs, [170] and be capable of producing electricity at costs of 10%–20% below today's conventional turbine plants [172]. Operation of the plant is almost totally automatic. Therefore, it can be monitored and managed remotely with the possibility of controlling hundreds of the power plants from a single location [189]. Initial systems can be less than 20 MW, with typical system sizes of 1–10 MW. Future systems, in the megawatt size, can boost efficiency even further by combining two SOFC modules with more advanced gas turbines and introducing sophisticated cooling and heating procedures.

5.4.4 Hybrid Options for Dynamic Control of SOFC

Degradation and lifetime are among the main challenges for a wider adoption of SOFCs in the market [203]. Long-term operation of SOFC systems is particularly demanding because of the high level of integration within the system and of its vulnerability against electro-chemical, thermal, and mechanical stresses. Mueller et al. [204] suggested a list of precautions for limiting cell degradation in SOFC systems.

This included, among others, that the operating temperature of the cells should remain within 10 K of its nominal value, and that the FC voltage should be kept constant to avoid high local heat production rates.

While there is extensive interest in optimal dynamic control of SOFC systems, an alternative approach to increasing the lifetime of SOFC-based systems is based on including other elements in the system that allow reducing the load changes of the FC, both in rate and in magnitude. Batteries are commonly proposed as a way to mitigate load variations and reduce the installed size of the SOFC [205]. The use of batteries (or of other electrical energy storage devices, such as ultra-capacitors) allows for a better dynamic response of the system and for better system economics given that the installed size of the SOFC can be reduced and the average load of the SOFC increased [206]. In the case of vehicles, in addition, it was shown that a combined SOFC–battery power system could provide a significant reduction in the weight of the system (when compared to pure battery-powered vehicles) and in the investment cost (when compared to pure SOFC-powered systems) [207]. These systems have been also proposed for application in off-grid base transceiver stations [208]. However, systems that only rely on batteries for energy storage have limited capacity due to the high cost, weight, and volume of the batteries.

Different alternative systems were proposed in literature to deal with the required dynamic performance in SOFC-based systems. Tse et al. [209] suggested flywheel as an alternative to batteries to reduce load fluctuations for a marine application. Jia et al. [210] proposed a SOFC–GT hybrid where the design of the system and the sizing of its components were adapted to improve its overall transient performance, but at the cost of a decrease in system efficiency [210]. In residential applications, where the system is generally connected to the electric grid, the challenge is generally to follow the heating demand. In these regards, Wakui and Yokoyama [211] and Yang et al. [212] pointed out that heat storage is a potential solution for dealing with load fluctuations and avoiding over-dimensioning of different components of the system.

SOFCs can be used to produce a combination of electricity, hydrogen-rich synthesis gas, and high-temperature heat. This makes them particularly suitable both for stand-alone applications and integrated with other thermal cycles or energy conversion technologies [213]. An example of this approach was proposed by Obara [214], who suggested to exploit the waste heat from a SOFC in a reformer to generate hydrogen to be used as fuel for a PEMFC, with the introduction of a heat storage system to allow for a better handling of load fluctuations. This allows not only the integration of the reforming of the natural gas, but also the ability to shift it in time with respect to the operations of SOFC, which act as the source of waste heat for the reforming reaction.

More generally, several authors have suggested to exploit the flexibility of SOFC systems for the combined production of electric power and hydrogen. Perdikaris et al. [215] proposed a system for the production of electricity, heat, and hydrogen that could be operated in two modes: one for power generation, and one for hydrogen generation. The concept, involving the coupling of a SOFC and a SOEC as separate units, proved quite challenging to control. Becker et al. [216] introduced instead the concept of purifying the anode off-gas of a SOFC to produce hydrogen as a useful system output, achieving close to 70% efficiency in the combined generation of power and hydrogen, and over 85% efficiency when waste heat was also accounted. Similar results were obtained by Leal and Brouwer [217], who also showed that internal reforming is more appropriate for this type of systems, if estimated based on first-law efficiencies. The co-production of electricity and hydrogen based on a SOFC was also simulated by Shaffer and Brouwer [218] based on real data for the demand of a commercial building, using a 2D model that allowed investigating the effect of a highly dynamic load on the internal properties of the FC. Hemmes et al. [219] simulated these systems in detail, for different types of operational modes, showing that these systems can be operated flexibly, that is, varying the share of electric and hydrogen power. Pérez-Fortes et al. [220] proposed the design of an SOFC-based co-production system for electricity and hydrogen, based on the multi-stage optimization procedure proposed by Mian et al. [221], thus achieving a combined efficiency above 65% (excluding waste heat) that is maintained over a wide range of combinations of hydrogen and power generation.

Hydrogen can be used as fuel for other units. The combination of a SOFC with a PEM fuel-cell (PEMFC) system was first introduced by Vollmar et al. [222]. The integration of a hydrogen purification unit downstream of the SOFC allows operating the cells at a lower fuel utilization rate, thus resulting in a lower Nernst loss. This allows for a higher overall efficiency of the system, as the hydrogen not converted in the cell is recovered and used in the PEMFC downstream. In addition, using the hydrogen that leaves the SOFC stacks in a FC, instead of a burner, improves the overall efficiency of the system. This consideration is based on the assumption of using exergy to measure the combined quality and quantity of energy, according to which hydrogen is more valuable than waste heat regardless of the temperature [223].

Initial results based on simulations showed that SOFC–PEMFC hybrid systems can reach 61% net electric efficiency [224,225] and overall efficiency up to 90% [225,226], a significantly higher value compared to a reformer–PEMFC system and to the early stand-alone SOFC systems (efficiency of around 50%). In addition, the initial estimations presented by Dicks et al. [224] for the capital cost of the system suggested that the SOFC–PEMFC hybrid system can have a better economic performance compared to other systems of similar power output [224]. Similar results were obtained in the work of Yokoo and co-workers [227–229], who showed that the electrical efficiency of a hybrid SOFC–PEMFC can be up to 5% higher than that of a stand-alone SOFC system, a value that is comparable to SOFC–GT systems. Compared to the latter, the SOFC–PEMFC hybrid has the characteristic, typical of FCs, of an electric efficiency that does not depend on the size and on the load on the system. The proposed system is applied to two potential off-grid applications: an isolated dwelling and a cruise ship.

The proposed system by Baldi et al. [230] is based on the use of a SOFC as the main energy source of the poly-generation system. As SOFCs are not suitable to handle large and fast load changes, a proposed system added other components to make it more suitable for load-following system. These included a hydrogen purification system, as the main energy source of the system, a battery for fast transients and peak shaving, and a hydrogen storage system combined with a PEMFC for medium-slow load transients.

The hybrid SOFC (H-SOFC) is the main unit of the system and is able to generate both electrical power and waste heat for cogeneration purposes. The study assumed the use of the system structure proposed by Pérez-Fortes et al. [220], where the composition of the anode off-gas of the SOFC is adapted first by a two-stage WGS reactor to enhance the hydrogen content, and then by a pressure swing absorption (PSA) unit for achieving high H_2 purity (see Figure 5.17). The purified hydrogen flow from the PSA is sent to the hydrogen storage tanks. The unreacted gas after the PSA is combusted, and the generated heat is utilized within the system and for direct satisfaction of heat load. Part of the inlet natural gas flow can also be sent to the burner, depending on the operational conditions, to ensure the heat balance of the system. The system can operate in a wide range of combinations of electric power and hydrogen output. In addition to the electric power and hydrogen output, the H-SOFC also provides waste heat for cogeneration purposes, resulting from the intermediate cooling of the anode gas between the two WGS reactors.

In order to reduce load fluctuations and the installed size of the SOFC, the proposed system was also equipped with a PEMFC. PEMFCs operate at much lower temperatures compared to SOFCs and are, hence, more flexible in terms of load change [231,232]. The PEMFC is fueled using the hydrogen generated by the SOFC. The study proposed the use of high-temperature PEMFCs because of their better suitability to cogeneration purposes and of their higher tolerance of carbon monoxide impurities in the feed gas [231,232]. For the same purpose, the proposed system

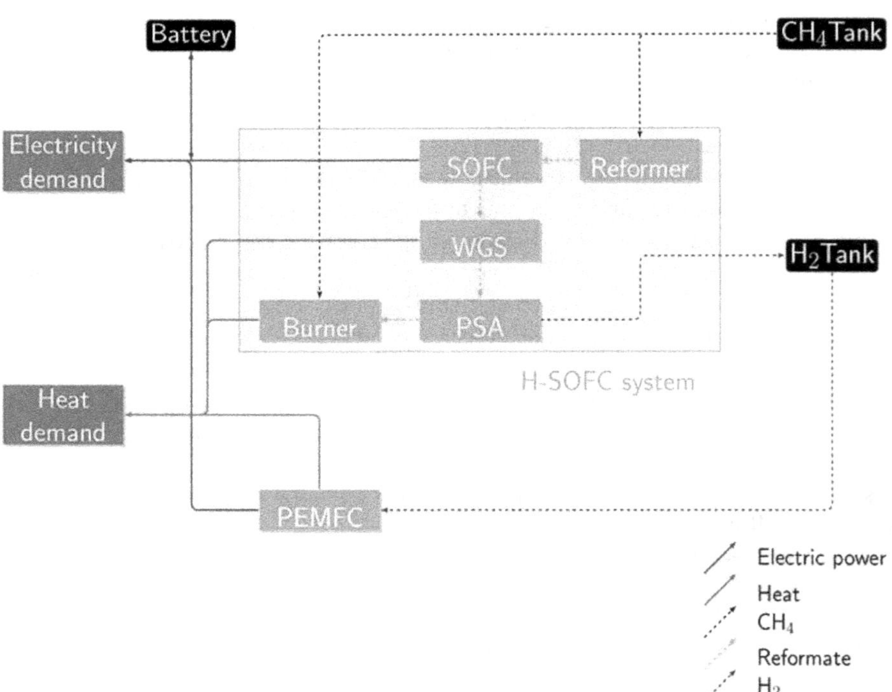

FIGURE 5.17 Graphical representation of the proposed energy system [230].

was also equipped with an electric energy storage device. While many types of EES could serve the purpose, the study focused on the use of batteries, and particularly of lithium-ion technology, that are becoming increasingly common for both mobile and stationary applications [233,234].

Thus, this poly-generation system was made of an H-SOFC, able to generate power, heat, and hydrogen, coupled with a PEMFC, a battery, and a hydrogen storage tank. The proposed system was shown to be economically feasible, and more convenient than alternative options under the constraint that the largest part of the fuel has to be used by the SOFC. In these conditions, the proposed system was expected to allow overcoming some of the major obstacles to a wider adoption of SOFC on the market. With reference to the cruise ship case study, for instance, van Biert et al. [232] report concerns in the maritime industry related to dynamic performance, system weight, and cost. The study achieved the promising results. The advantages of the proposed system, particularly in terms of investment cost and weight, would also prove beneficial in other applications related to the transport sector, such as cars, trucks, and airplanes, where the system's weight constitutes a significant constraint for the design of the power plant.

In the case of stationary applications, such as in the residential case, the main advantage came from the lower investment cost of the system. Furthermore, as shown by Pellegrino et al. [235] and Brunaccini et al. [208], the advantages of using energy storage to reduce the installed size of the SOFC, hence reducing the cost of the system, are known. In addition to the advantages, it provides to off-grid applications, and it should be noted that the proposed system could also provide beneficial in grid-connected systems, in presence of tariff schemes that encourage the usage of micro-CHP plants in self-consumption mode. In this case, the possibility of storing energy for later use at a low investment cost is crucial for reducing the payback time of the system, an aspect that makes the system proposed in this paper particularly suitable for these applications.

The results presented for the two case studies showed that the benefit of the proposed hybrid system is significantly based on the trade-off between investment and operational costs. Moret et al. [236] showed that using cost as the objective of the MILP-based optimization of energy systems forces a high degree of uncertainty on the problem: On the one hand, linear cost functions represent inherently a simplification of a model that is typically nonlinear, that is subject to high variability (the price of a component does not only depend on its size) and suffers from a general high uncertainty, particularly in the case of not-yet-mature technologies. In particular, the recent developments in battery and FC technology, both in relation to price and weight, may modify the optimal balance between hydrogen and electric storage. In addition, energy prices (in this case, the price of natural gas) are subjected to wide fluctuations. The linearization of the problem also prevents from taking into account the off-design performance of the installed components. While most of the main components considered in the proposed system maintain high conversion efficiency over a wide range of operations, this assumption might still lead to inaccurate results in the optimization.

Secondly, the benefits connected to the installation of the system proposed by Baldi et al. [230] are partly related to the choice of operating the SOFC close to constant load. While this is today widely accepted as a limitation of the operations of these systems, future technical developments, both in the system control and in material technology, are expected to mitigate this limitation, thus partly reducing the scope of the proposed system. Also the expected reduction in the specific cost of SOFCs may reduce its benefits. It should be noted, however, that the assumption of a specific investment cost of 1,600 EUR/kW for stand-alone SOFCs can be considered as optimistic. It should also be noted that the performance of the poly-generation SOFC unit was based on the optimization results reported in [220], in a case where a specific set of operational conditions were used for tailoring the performance on the system to the case under investigation, that is, a hydrogen refueling station. It can be expected that the system could perform more efficiently, from both energy and a cost perspective, if its energy integration was optimized for the specific case under study.

5.4.5 Hybrid PEMFC–Wind Power System

Recently, there has been a lot of emphasis on the electric power generation using wind energy. Wind turbines are being used not only for grid connection but also as stand-alone power generation systems. Wind power presents some challenges in producing continuous electric power. A significant problem is the intermittent nature of the wind, and the wind power generated depends on wind speed. Combining the wind power generation system with a fuel-cell system would solve some of the problems associated with wind power [237–239]. The grid-connected wind–hydrogen system provides off-peak hydrogen production and low-cost electricity. In Figure 5.18 is shown a Type-2 hybrid system based on wind power and PEMFC. The wind power is used for generating hydrogen using the electrolysis of water and is stored in cylinders at a certain pressure. This hydrogen is used as the fuel to the fuel-cell stack. The stored hydrogen can also be used to fuel the FC vehicles.

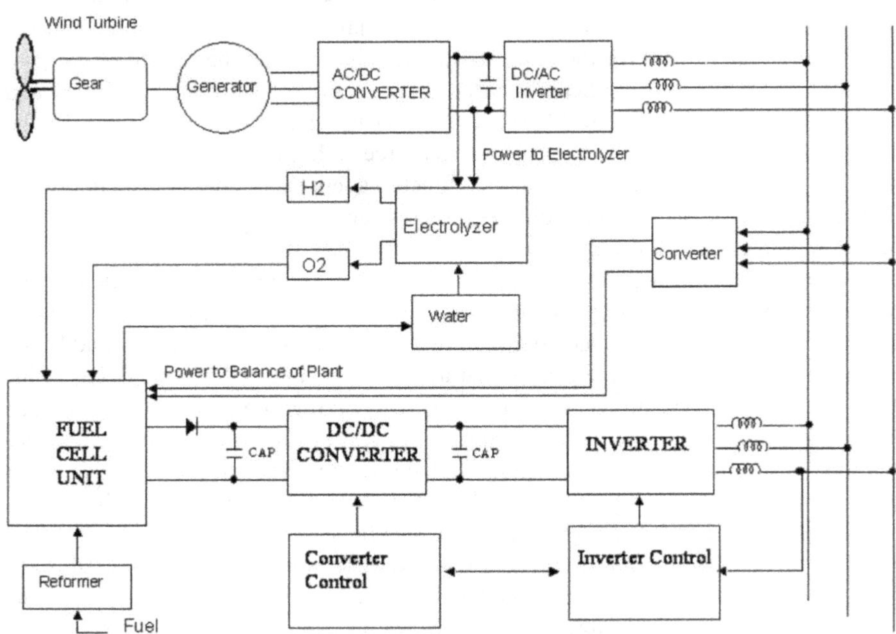

FIGURE 5.18 Wind–FC hybrid power system [189,190].

The hybrid system could be configured in several ways. The wind power could be used to supply the power to the balance of the plant of the fuel-cell system, particularly during start-up of the system, and the excess power could be used to supplement the power from the FC. This is a fuel-cell dominant system, and the wind generator supplements the fuel-cell power. In another configuration, hydrogen is generated using electrolysis and stored during the peak power availability from the wind power generation system. The stored hydrogen is used for generating power using the FC during the low-output-power operation of the wind unit. Electrolyzers can be used to reduce/eliminate surplus wind power generation. Fuel-cell power is generated only during daily peak load period to firm up the wind generation. This is a wind power dominant system, and the FC supplements the wind power.

As the hydrogen produced is from the electrolysis of water, it is free from any carbon monoxide. Hence, in this type of application, the PEMFC would be the most applicable choice. The hybrid wind power–fuel-cell system shown in Figure 5.18 can be modified to implement any of the above concepts. In addition, depending on the amount of hydrogen stored, the above scheme can be extended to transportation applications by transporting the hydrogen.

The systems shown in Figure 5.18 for the wind–fuel-cell hybrid system can be used for a FC–solar power hybrid system in stationary power generation applications. The difference is that solar power is generated only when there is a reasonable amount of availability of sunlight. It is applicable for systems that have enough hydrogen stored to provide the power for the loads during the time of darkness. For larger power generation systems, where the emphasis is on renewable energy, the wind energy could be combined with the solar and fuel-cell power as shown in Figure 5.19. This system, combining the fuel-cell power with wind and solar power, would be the ideal situation for reducing the emissions and dependency on fossil fuels [240].

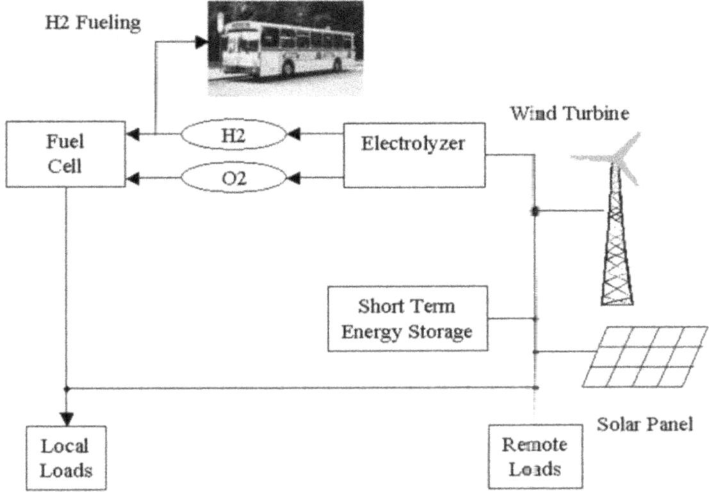

FIGURE 5.19 Hybrid fuel-cell system with wind and solar power [189,190].

5.4.6 Hybrid Fuel-Cell–PV Power System

Similar to the wind power, a FC–PV hybrid system produces hydrogen, stores it, and then converts its energy to electricity for further use. This type of hybrid system is particularly useful in spaceship applications, as shown in Figure 5.20 [241,242]. During the daytime, the PV cells convert the solar energy to electricity that is directly used for propelling the spaceship. The excess power is used for generating the hydrogen and oxygen through the electrolysis of water. During the night, the stored H and O are used as the fuel and oxygen supply for the FC to generate electric power to propel the spaceship. Hence, the duration of the power available to the critical loads is maximized. This system can also be used in land-based solar-powered vehicles. Again, for this type of application, considering the operating temperature and availability of pure hydrogen, the PEMFC is the right choice. In this type of system, the operating characteristics of the hybrid system components have to be optimized to provide the power to the load for a maximum amount of time. Hybrid power plant of the PV–FC was also evaluated by Herlambang et al. [243] to optimize the performance of PEMFC. Kim et al. [244] developed a PV–electrolysis–PEM hybrid model for a feasibility study, and simulations of several scenarios in Korea. Electricity load and solar irradiance information were used to test the performance model of the PV–electrolysis–PEM hybrid system for baseload and several peak load shave runs. When the baseload was set at 4,200 MW, the total capacity of the PV plants was 58.5 GW_p. In contrast, the hybrid system reduced the peak load more efficiently during daytime.

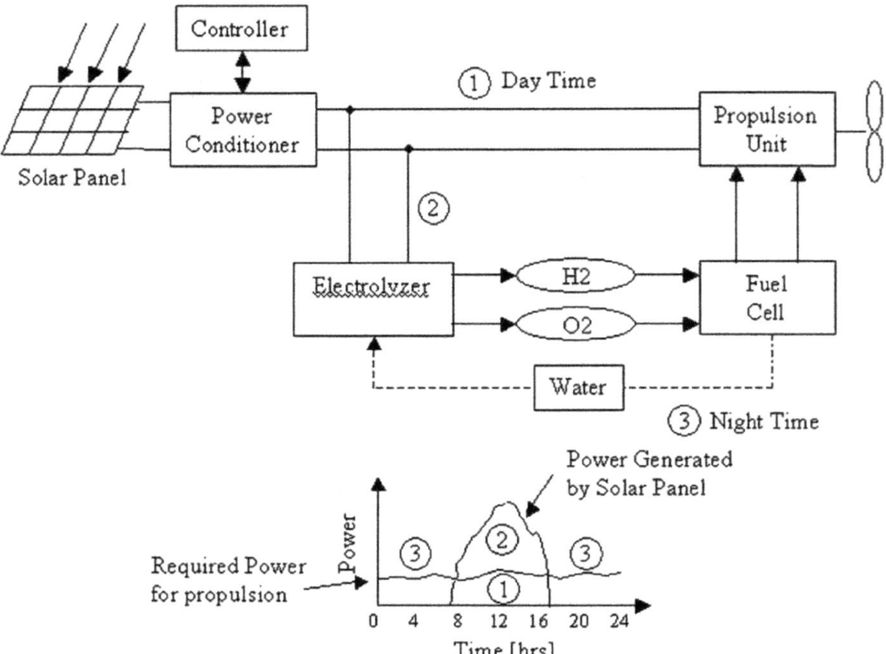

FIGURE 5.20 Hybrid PV–FC system for a spaceship propulsion application [189,190].

In particular, the capacity factor of the PEM system increased in winter because the solar irradiance is relatively weak in that season. These results provided useful insights for the development of control logic models for the PV–electrolysis–PEM system in microgrid setups.

A variation of the hybrid FC–PV system is hybrid FC–PV–SC system as proposed by Ferahtia et al. [245]. The typical system (see Figure 5.21) of this type consists of five parallel solar panels two series (Soltech model-245Wh), FC (PEMFC-6KW-45VDC), and a supercapacitor (80F-48VDC). They are connected to the DC bus via DC/DC converters to further control the power sharing and to stabilize the DC bus voltage (120 V). This microgrid has been designed to operate in islanded (off-grid). The performance is inspected by imposing deferent loads. On the whole, the FC stack and the PV source provide power to satisfy the demand power needs. The supercapacitor delivers the power to stabilize the DC bus voltage during the transient periods.

The structure of the studied system with the power electronic converters indicates that all the converters are connected in parallel. Two-phase parallel boost converters are used to control the FC stake, and this converter is controlled with interleaving technic; thus, the power quality is improved. A boost converter is utilized to control the produced power by the solar cell. The supercapacitor is always connected to the DC bus by a bidirectional boost converter, and this structure allows to charge or discharge the supercapacitor. Moreover, to ensure safe operating and soft dynamics, all the converters are controlled by the current regulation loop. This supercapacitor control loop is supposed to be much faster than the other control loops.

In this system, there are two energy variables to be controlled, the DC bus energy and the supercapacitor energy. In conclusion, this study's main objective is to ensure stable DC bus voltage by providing an effective system for DC microgrid powered by a FC and solar panel, and the backup storage unit (SC). The parallel employment of

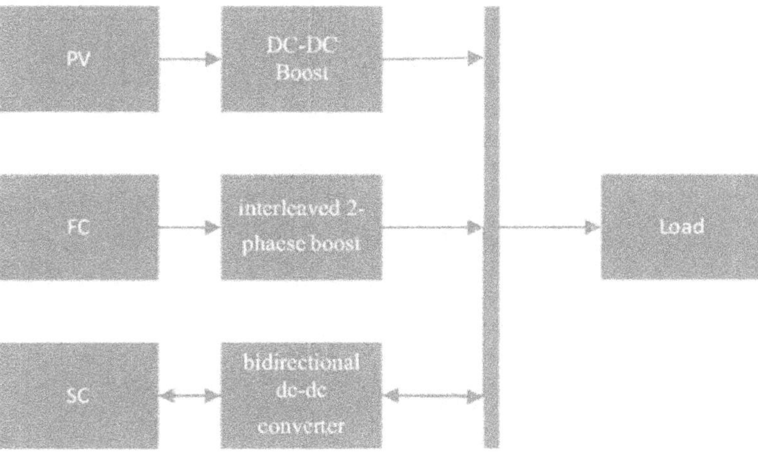

FIGURE 5.21 The proposed hybrid system with PV and FC as primary sources, and supercapacitor as an auxiliary source connected in parallel with the load side. Where the PV is connected via a boost converter, the FC is connected by an interleaved two-phase boost converter and the SC is connected via a bidirectional DC–DC converter [245].

PV and FC ensures to the load is sustainable supplying. The research focuses mostly on stabilizing the DC bus voltage, moreover employing the FC, solar panel and SC in the energy management approach, taking into account the energetic properties of these sources such as its power and energy density and its dynamics. Nonlinear differential flatness strategy for DC microgrid based on a renewable source with a FC provides excellent DC-link stabilization. The supercapacitor can move the load forward, regarding the characteristics of the main sources, by supplying a stronger power response to the load. During the important steps of the load, the supercapacitor offers the energy balance required during the transition of the load. Also, distributed power systems improve power quality and efficiency.

Thounthong et al. [246] presented an original control algorithm for a hybrid energy system (HES) with a renewable energy source, namely a PEMFC and a PV array and a single storage device, that is, a supercapacitor (ultracapacitor) module. The main weak point of FCs is slow dynamics because the power slope is limited to prevent fuel starvation problems, improve performance, and increase lifetime. The very fast power response and high specific power of a supercapacitor complement the slower power output of the main source to produce the compatibility and performance characteristics needed in a load. The energy in the system is balanced by DC bus energy regulation (or indirect voltage regulation). A supercapacitor module functions by supplying energy to regulate the DC bus energy. The FC, as a slow dynamic source in this system, supplies energy to the supercapacitor module in order to keep it charged. The PV array assists the FC during daytime. To verify the proposed principle, a hardware system is realized with analog circuits for the FC, solar cell and supercapacitor current control loops, and with numerical calculation (dSPACE) for the energy control loops. Experimental results with small-scale devices, namely a PEMFC (1,200 W, 46 A) manufactured by the Ballard Power System Company, a PV array (800 W, 31 A) manufactured by the Ekarat Solar Company, and a supercapacitor module (100 F, 32 V) manufactured by the Maxwell Technologies Company, illustrate the excellent energy management scheme during load cycles. Jayalakshmi et al. [247] presented modeling and control of PV/FC/supercapacitor hybrid power system for stand-alone applications. The hybrid power system used solar PV array and FC as the main sources. These sources share their power effectively to meet the load demand. The supercapacitor bank was used to supply or absorb the power during load transients. The main control system comprises controller for maximum power tracking from PV system, a DC–DC boost converter with controller for FC system for power management, and inverter controller to regulate voltage and frequency. The stand-alone hybrid system aims to provide quality power supply to the consumers with a constant voltage and frequency along with proper power management using simple control techniques. The modeling and control strategies of the hybrid system were realized in MATLAB®/SIMULINK.

5.4.7 Hybrid FC–Battery System

A hybrid power supply uses two or more sources that work together to deliver or store power to act as a single power delivery unit. One advantage that hybrid power supplies have over regular power supplies is that hybrid power supplies can store excess energy in one of its sources such as a battery or an ultracapacitor [248]. Another benefit is that hybrid power supplies can also be coupled with one source that has a slow dynamic response and one that has a fast dynamic response time in their power output. The system studied by Corcau et al. [249] contains a FC/battery hybrid power supply with two DC-to-DC boost converters (see Figure 5.22).

A hybrid fuel-cell system has higher power density and less weight and volume, providing a more sophisticated and reliable approach than batteries alone for military missions. Besides its application in transportation industry, FC–battery hybrid is very useful for portable electronics. Soldiers in the field require the most sophisticated portable electronic equipment to complete their missions. Meeting the power demands of such portable equipment requires technologically advanced energy systems. Conventional battery-based power systems require soldiers in the field to carry up to 30 lbs. of various batteries to autonomously operate their electronic equipment (e.g., night-vision goggles, laptops, communication devices, GPSs, and sensors). Batteries for different devices have to be repeatedly replaced or recharged, requiring frequent interruptions to the mission that can complicate logistics as well as add weight to the soldier's equipment supply. This has intensified the need for lightweight, reliable, mobile, and portable electrical power supply solutions.

As a result, FCs have become reliable power sources for mobile and portable defense applications. Direct methanol FCs (DMFCs) provide logistical, safety, and functionality advantages including virtually undetectable operation. They also are immune to extreme weather, generate power only when needed, and operate almost silently without producing exhaust. Unlike batteries, which store energy, DMFCs generate power by chemically converting methanol into electrical energy.

In DMFC, a mixture of methanol and water is introduced to the anode side, which is connected to the cathode by an electrical circuit. A patented water management system enables the use of 100% pure methanol with a very high energy density in the fuel cartridges. Ambient air is pumped into the stack on the cathode side. Upon contact with a platinum catalyst, methanol releases its electrons, which flow in the direction of the cathode, thus producing power. At the same time, protons are released and penetrate the membrane to the cathode. There, the oxygen reacts with the protons and electrons to form pure water. During this chemical process, the FC releases water in the form of water vapor and carbon dioxide. The process is environmentally friendly.

To address soldiers' growing power demands, SFC Smart FC (SFC) developed a fuel-cell/battery hybrid system solution that offers a lightweight alternative for non-stop equipment operation. The energy network—consisting of an SFC FC, fuel-cell cartridge, intelligent SFC Power Manager, and a rechargeable Li-ion battery—is also universally compatible with current and future power-consumption requirements.

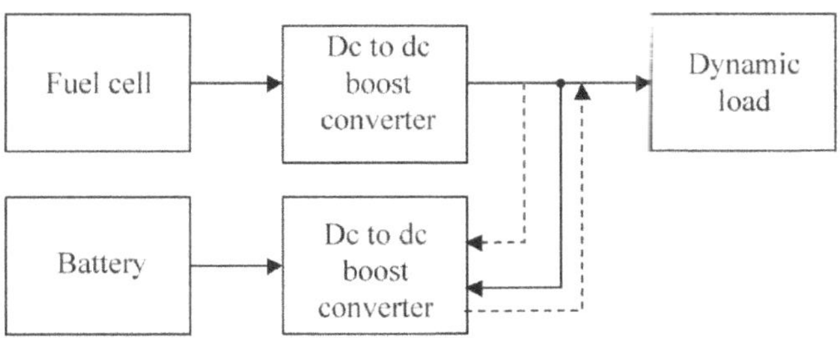

FIGURE 5.22 Block diagram of hybrid FC/battery [249].

The energy source of the network, the portable Jenny FC already in use by various military organizations, weighs only 3.7 lbs. and measures $10 \times 7 \times 3$ in. It provides 25 W continuous nominal power directly to electrical devices or for charging secondary batteries. Nominal voltage is 16.8 V and can be adapted to other voltages (output voltage is 10–30 Vdc). At 25 W, fuel consumption is less than one milliliter per Wh. The network produces power automatically as needed, continuously, as long as there is fuel. The FC itself remains maintenance-free throughout its entire life. The only maintenance required is occasional fuel-cartridge replacement.

Fuel for the Jenny FC comes in convenient 0.6-lb cartridges, each containing nearly 10 oz of fuel. Ten liters of methanol weighing approximately 18 lbs provides 10 kWh of power, a lot of energy at very low weight. Fuel cartridges can be easily exchanged during operation (hot swap). In operation, the FC is barely detectable due to its low surface temperature. It has been proven to operate reliably even in extreme temperatures in deserts or cold climates, as well as at high altitudes or fully submerged underwater.

The SFC Power Manager is the second central component of the network. It is an intelligent portable power-management device that assures continuous operation of any electrical equipment carried by special-operations soldiers, as well as for charging batteries. The Power Manager enables smart energy harvesting, including an option for assigning various priorities during charging/discharging of multiple batteries and powering several devices simultaneously. It automatically recognizes the voltage demands of the individual devices and adapts the output power accordingly. The SFC Power Manager also provides a constant indication of each battery's state of charge and other parameters (SMBus Level 3). The system readily hybridizes with conventional power sources, including vehicle power, solar, and FCs. Developed for defense organizations in North America and Europe, the fuel-cell/battery hybrid system is a flexible and intelligent solution that provides continuous operation of any electrical equipment. The combination of the FC and the power-management system enables versatility. For example, it can hybridize a solar energy system and provide lightweight and reliable energy for soldiers in the field.

5.4.8 Hybrid FC–Solar Panel–Battery System

In the fuel-cell/solar combination, the solar panel and the Jenny FC secure constant energy supply for power-consuming equipment. Based on 12 hours of sunshine and an average output rate of 50 W of the solar panel, the excess energy gets buffered in the battery. If the solar panel is unable to provide sufficient power, the Jenny FC with 25 W nominal power kicks in automatically. Connected to a rechargeable battery, the SFC Power Manager constantly monitors the battery's charge state. Once this drops below a predefined value, the FC automatically recharges the battery. When the battery is full, the FC returns to standby mode. In addition, the SFC Power Manager assures continuous operation of any electrical equipment carried by soldiers. It fully hybridizes available power sources, coordinates incoming energy, and can manage output up to 500 W. Moreover, it indicates energy levels of different power-consuming devices during a mission and evaluates power consumption, enabling soldiers to achieve corresponding optimization.

For example, a fuel-cell/solar combination with a Jenny FC weighing 3.7 lbs.—plus two 0.8-lb. fuel cartridges—would reduce the weight of the soldier's power supply in the 96-hour mission example with three power consumers (radio, notebook, and thermal-imaging device), by almost 70% compared to non-rechargeable batteries. The hybrid fuel-cell system is 31 lbs. lighter than batteries with equivalent power. Various system advantages include the following:

1. **Significant weight reduction**—With the use of the fuel-cell/battery hybrid system, the weight to be transported by soldiers in the field can be reduced up to 80% compared to conventional battery-based power systems;
2. **Improved logistics**—As a result, it simplifies logistics because conventional batteries must be discharged or replaced, which can be a challenge in critical situations, with the portable-power hybrid solution, soldiers' needs are optimized because the FC automatically recharges batteries fully and the rechargeable battery stores power, and fuel for several days in lightweight cartridges is easily packed and transported, making the soldier independent of logistics support;
3. **Ultra-high efficiency**—In combination with the SFC Power Manager, the energy network represents the modern way to efficiently supply power for all sorts of missions, and it is universally compatible with existing as well as anticipated future soldier equipment, and fully hybridizes available power sources;
4. **Increased soldier safety**—A fully integrated fuel-cell/battery hybrid system enables soldiers to take advantage of advanced electronic military equipment while reducing the weight of their load. Critical missions no longer have to be interrupted for battery replacement or recharging, and important systems do not fail due to drained batteries. Deploying a reliable power system can make a major difference for soldiers in the field by significantly increasing the soldier's safety, flexibility, mobility, and success.

5.4.9 Novel Hybrid Biomass-Driven FC–Solar Thermal System

Although low-temperature FCs powered by methanol or hydrogen have been well studied, existing low-temperature FC technologies cannot directly use biomass as a fuel because of the lack of an effective catalyst system for polymeric materials. Now, researchers (Yulin Deng) at the Georgia Institute of Technology have developed a new type of low-temperature FC that directly converts biomass to electricity with assistance from a catalyst activated by solar or thermal energy. The hybrid FC can use a wide variety of biomass sources, including starch, cellulose, lignin, and even switchgrass, powdered wood, algae, and waste from poultry processing. *A new solar-induced direct biomass-to-electricity hybrid FC can operate on a variety of fuels.* The device could be used in small-scale units to provide electricity for developing nations, as well as for larger facilities to provide power where significant quantities of biomass are available. The new solar-induced direct biomass-to-electricity hybrid FC was described in 7 February 2014, in the journal Nature Communications.

The challenge for biomass FCs is that the carbon–carbon bonds of the biomass—a natural polymer—cannot be easily broken down by conventional catalysts, including expensive precious metals. To overcome that challenge, scientists have developed microbial FCs in which microbes or enzymes break down the biomass. But that process has many drawbacks: Power output from such cells is limited, microbes or enzymes can only selectively break down certain types of biomass, and the microbial system can be deactivated by many factors. Deng and his research team got around those challenges by altering the chemistry to allow an outside energy source to activate the FC's oxidation–reduction reaction. In the new system, the biomass is ground up and mixed with a polyoxometalate (POM) catalyst in solution and then exposed to light from the sun—or heat. A photochemical and thermochemical catalyst, POM, functions as both an oxidation agent and a charge carrier. POM oxidizes the biomass under photo or thermal irradiation, and delivers the charges from the biomass to the FC's anode. The electrons are then transported to the cathode, where they are finally oxidized by oxygen through an external circuit to produce electricity.

The biomass and catalyst at room temperature will not react, but when you expose them to light or heat, the reaction begins. The POM introduces an intermediate step because biomass cannot be directly accessed by oxygen. The system provides major advantages, including combining the photochemical and solar–thermal biomass degradation in a single chemical process, leading to high solar conversion and effective biomass degradation. It also does not use expensive noble metals as anode catalysts because the fuel oxidation reactions are catalyzed by the POM in solution. Finally, because the POM is chemically stable, the hybrid FC can use unpurified polymeric biomass without concern for poisoning noble metal anodes.

The system can use soluble biomass, or organic materials suspended in a liquid. In experiments, the FC operated for as long as 20 hours indicates that the POM catalyst can be re-used without further treatment. The researchers reported a maximum power density of 0.72 mW/cm^2, which is nearly 100 times higher than cellulose-based microbial FCs, and near to that of the best microbial FCs. Deng believes the output can be increased five to ten times when the process is optimized. He believes that this type of FC could have an energy output similar to that of methanol FCs in the future. In order to improve the process, the researchers also need to compare operation of the system with solar energy and other forms of input energy, such as waste heat from other processes. Beyond the ability to directly use biomass as a fuel, the new cell also offers advantages in sustainability—and potentially lower cost compared to other FC types.

5.5 USE OF FC FOR SUSTAINABLE MICROGRID OPERATIONS

In recent years in order to balance centralized versus distributed power, hybrid utility versus microgrid transmissions have been extensively used. Sustainable microgrid operation requires balanced grid operation which can be obtained either by a dispatchable power generation scheme or by a hybrid scheme where intermittent power supply (like solar or wind) is balanced by dispatchable power supply or power storage. FC can play an important role in providing sustainable microgrid operation with or without utility grid operation. The application of microgrid technologies in electrical systems may be classified as grid-integrated and grid-independent systems. Whether the FC technologies are employed for grid-connected or off-grid purposes, it is necessary to establish the fact that an FC power plant consists of FC stack. This is due to the fact that a constant supply of reactants, that is, fuel and oxidant, is critical for achieving continuous production of electrical power [250]. Therefore, a connection between FC stack and microgrid is of interest. FC stack must meet certain requirements of fuel processing and admissible level of impurity, oxidant conditioning, electrolyte management, heat energy management unit, product removal, etc., to provide sustainable and dispatchable power supply to microgrid.

A typical FC microgrid connection can be illustrated as shown in Figure 5.23. This figure illustrates the additional subsystems required for processing the energy from the FC stack to the users. The FC microgrids are capable of producing electrical power from as low as <1 W to hundreds of kW [118].

The FC stack produces direct current but most of the appliances within the residential, commercial, and industrial premises, for example are alternating current (AC) powered. Therefore, a DC–AC converter is required. FCs also make use of DC–DC power converters in addition to DC–AC power converters for power-conditioning purposes [251]. The inverter, being a developed power electronic device, has efficiency as high as around 0.96 for MW-rated power generation systems [45]. FC power unit generates waste heat which can be used for CHP application or bottoming cycles for additional generation of electric power [252]. This further increases efficiency to as high as 0.85 or even more [250]. Through CHP, waste heat can also be used for water and space heating, food processing, drying, and preservation, including raising of steam for industrial applications.

In FC systems for microgrid operations, FC processors that convert fuels like coal, heavy oils, methanol, biomass, etc., to hydrogen and CO [250] are important, and they usually account for about 33% of the plant's weight and capital cost, especially for the one based on hydrocarbon [250]. The output power, efficiency, water balance, heat utilization, quick start-up, long dormancy, size, weight, and fuel supply are specific requirements for operating and managing FC technologies [118]. The expectation is that the FCs used in automobile systems are to have an operating life span between 3,000 and 5,000 hours, that is, less than 1 year, while those used for stationary applications have an operating life span between 40,000 and 80,000 hours, that is, about 5–10 years [118]. The applications of a stationary system with emphasis

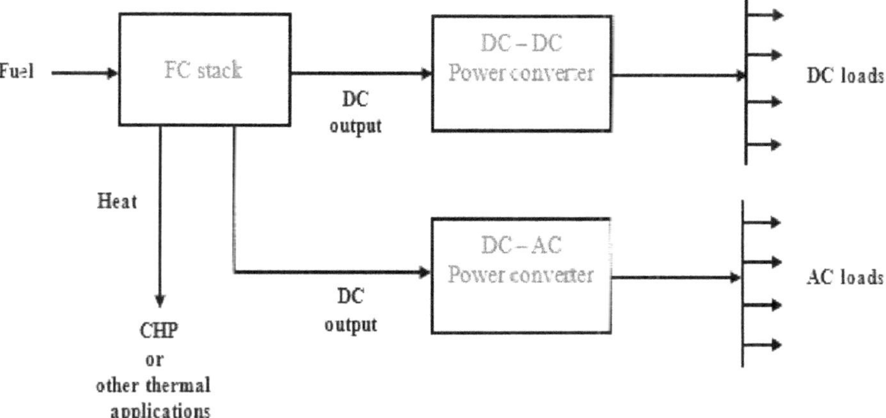

FIGURE 5.23 Schematic of the FC microgrid system [121].

on FC microgrid system include (1) grid-connected, (2) grid-parallel, (3) stand-alone power, (4) emergency or backup power, and (5) DC microgrid. HES containing FC stack can also be connected to microgrid.

The energy flow in grid-connected operation is allowed in three different paths such as from the electrical grid or network to the users' load, from the FC microgrid to the users' load, and from the FC microgrid to the electrical grid [126,253]. By using a "load-following" strategy, the microgrid system may be designed as a constant energy source to meet the users' maximum electricity consumption. In this case, the excess electricity production from the microgrid system may be exported to the electrical grid which can result in a reduction in the electricity bill from the utility [254]. A net metering system is also expected to be part of the electrical system for evaluation purposes [255]. Azmy and Erlich [256] examined a grid-tie hybrid power system based on FCs and micro-turbine or a parallel arrangement of three separate FC units for residential premise. A schematic of a grid-connected application of FCs is shown in Figure 5.24, where the heavy-current load is being served by the heat from the FCs and the electrical supply, which is different from the idea presented by Azmy and Erlich [256].

The microgrid system in Figure 5.24 was designed with the aim of satisfying the total load requirement in a typical household by the FC system and the grid which includes all accessories such as TV and DVD player.

Unlike in grid-connected mode, in grid-parallel application, energy may be purchased from the electrical grid to meet the users' load when required; however, the FC microgrid is not allowed to export any excess energy to the electrical grid [118]. This configuration also implies that two energy sources are available to the users, which are the existing electrical grid and the FC microgrid systems. Therefore, the energy flow is allowed in two different directions. As demand increases, the use can climb the energy ladder [257], but as shown in Figure 5.25, there is no energy storage and all the power is purchased from microgrid or grid [118] directly.

In stand-alone power application, there is no interaction between the users' load and the grid, nor is there any electrical connection between the FC microgrid system and the existing power grid [258]. Therefore, energy flow is only available from the on-site power system to the load. As shown in Figures 5.26 and 5.27, the on-site

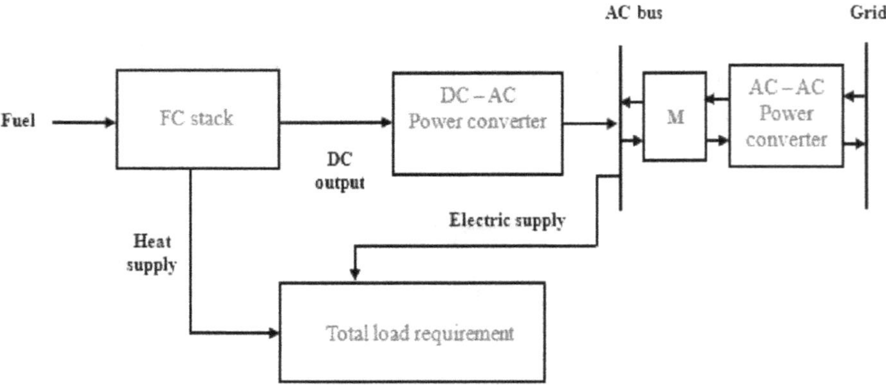

FIGURE 5.24 Grid-integrated application of FC power for residential loads [121].

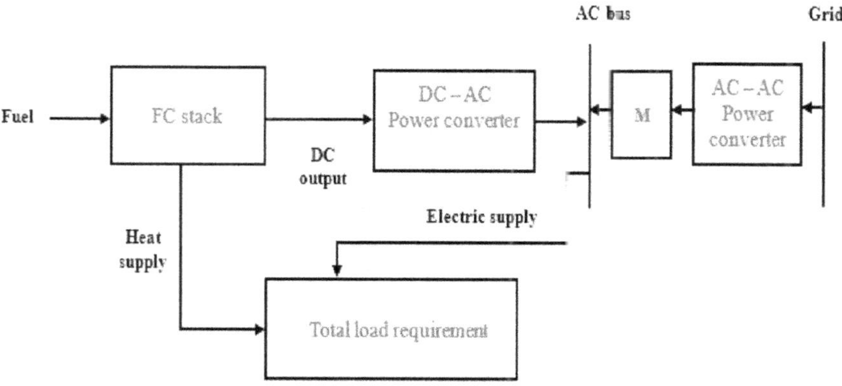

FIGURE 5.25 Grid-parallel application of FC power for residential loads [121].

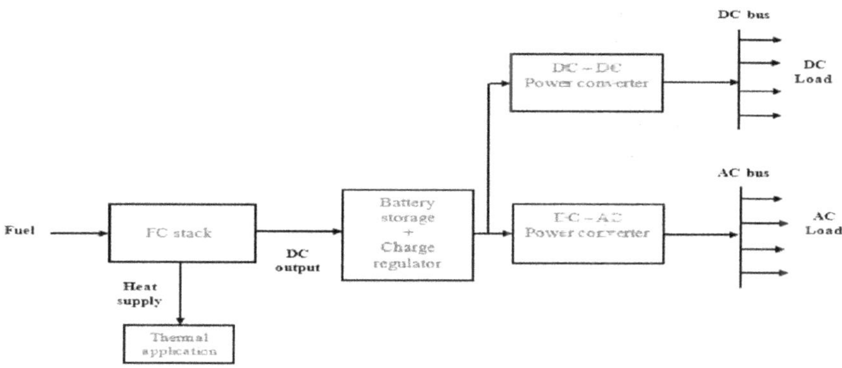

FIGURE 5.26 Stand-alone application based on FC power source [121].

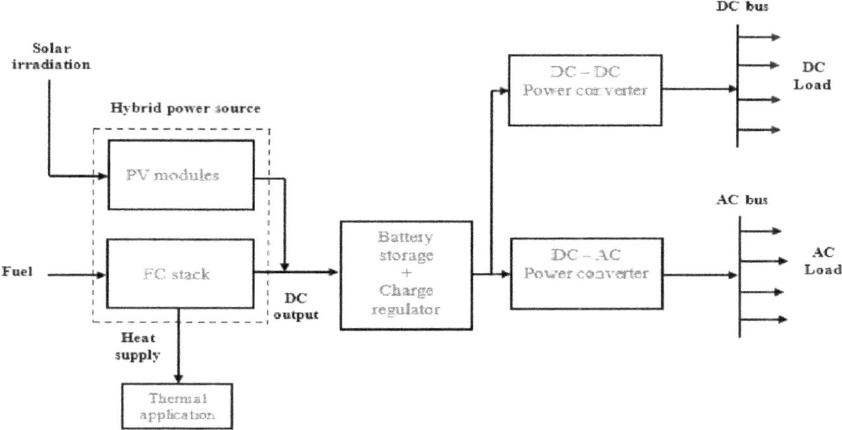

FIGURE 5.27 Stand-alone application based on the FC and PV power sources [121].

power system could be a single-source FC stack with energy storage or a hybrid system [259] of solar PV and FC stack with energy storage. The hybrid system can have other sources of power.

The stand-alone power has to have the technical capability to satisfy the start-up requirements of inductive loads such as water pumps, motors, and fans. This implies that the FC microgrid (FCs + battery storage) must be modeled to meet maximum continuous load demand for reliability purposes. In both single FC source and hybrid system, battery is charged when power load demand is low. Applications of energy storage system on renewable microgrid operation are discussed in detail by Gao [260].

In the emergency application, the microgrid system must not only possess the technical capability to meet a fast start-up but also be connected with a battery bank or another peaking plant [118]. The battery bank can offer a low backup power for a short duration (usually seconds to a few minutes); however, the FC stack can provide higher backup power in several kW with a longer duration of >30 minutes [118,126]. Both the electrolyzer and the hydrogen tank can be integrated with the FC microgrid used in backup power mode. The significance of this is that the system will be capable of producing its fuel (i.e., H_2) at those periods when electricity is imported from the grid [118,131]. This hydrogen can be used to operate hydrogen-fueled generator [120].

The DC generated by the FC stack may be used to power DC appliances, without the need for an inverter, or employed for developing a direct current (DC) microgrid. Figure 5.28 illustrates the DC microgrid with a connection with the existing power grid [261]. The DC microgrid is composed of a DC–DC voltage source, that is, an FC stack, which is connected to the source DC–DC converter for achieving a higher or lower voltage output depending on the design requirements. The common voltage levels that have been employed for telecom systems are 24 V and 48 V [261–263]. It is necessary to provide an interface between the existing power grid and the DC microgrid; this is achieved by using the bidirectional AC–DC converter [264]. Through this, it is possible to sell back the excess electricity from the DC microgrid to the existing power grid, as well as purchase energy from the power grid when the output of the microgrid system is lower than the load demand [265].

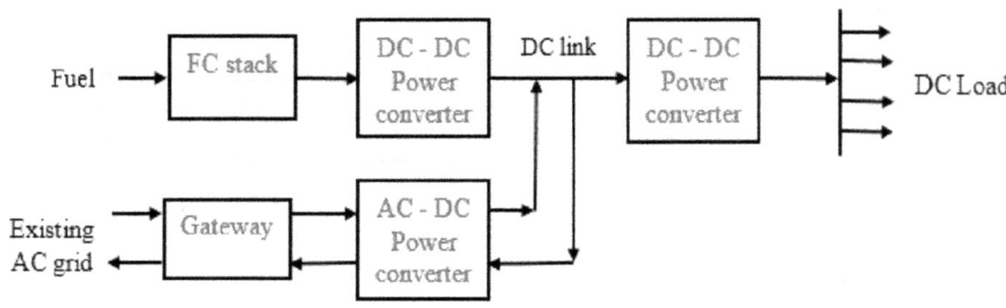

FIGURE 5.28 Schematic of a DC microgrid based on FC [121].

5.5.1 Microgrids for Hybrid FC Systems

As mentioned earlier, single FC system has limited applicability in microgrid operation [230]. In order to improve durability and stability to the grid, multiple energy systems with some control strategies work well [266]. In this section, we identify some case studies reported in the literature showing benefits of hybrid systems for microgrid operations.

Baldi et al. [230] considered a CHP system that is based on solid oxide and PEMFCs integrated with a hybrid storage configuration for off-grid purposes. In this case, SOFCs were used to satisfy the system's baseload, while the production and storage of pure H2 from the SOFC anode off-gas was stored and employed for driving PEMFCs when the electricity demand peaks. This combination thus improved durability of the overall system. Mudaliyar et al. [267] discussed the load-following capability of HES for microgrid operation, which is based on FCs and micro-turbine. Even PEMFC system, for instance, is one technology that demonstrates appreciable potential when employed in microgrids, but it requires the integration of energy storage such as the battery or the supercapacitor to realize a rapid electrical load following [268]. The control mechanism is such that the storage system responds to the transient load, while the FCs respond to the steady-state load [267]. However, the use of battery or ultracapacitor for load-following results in low life span and high capital cost. In light of this, a micro-turbine (microT) may also be integrated with the FCs to satisfy both the steady-state and the transient requirements of the load both in on-grid and off_-grid applications [267].

Hu et al. [269] discussed the application of FCs and heterogeneous energy storage systems for the energy management of data centers. The authors identified the "limited" load-following capability of FCs. This implies that the FC technology is slow in responding to a rapid increase in load demand due to a limitation placed on fuel delivery. The technology also has a delay in responding to and adjusting electrical power to a decrease in load, thus wasting energy. The combination of multiple batteries with FCs was introduced in the work to address the issue of limited load-following capability by providing electricity supply when the output of the FCs is less than the load requirement and employing the excess energy of the FCs to charge the batteries when the demand is low.

Asghari and Brouwer [270] showed that the dynamic power and cooling generation were achieved for household purposes by integrating a SOFC with an ORC and absorption chiller systems. The authors proposed a dynamic, physical model that integrates the SOFCs, ORC, and the absorption chiller to simulate the residential dynamic loads. It was reported that the hybrid design incorporating dynamic models of ORC and absorption chiller provided a capability to follow the residential dynamic load and the use of the SOFC dynamic waste heat to generate additional electrical power or cooling for the households.

Chang et al. [271] carried out the performance analysis of a cogeneration system based on PEMFCs and the lithium-ion battery system using a typical residential premise in North China as a case study. The authors recognized the good dynamic performance and the inherent charge/discharge properties of lithium-ion battery technology as a means to enhance the dynamic response when integrated with the

CHP system, especially in a situation where the electrical generation output does not match the load. The PEMFCs were employed as the prime mover in the study, and the entire system design had a coordinated thermal and electrical load-following strategy. It was reported that the hybrid design incorporating the lithium-ion battery produced an efficiency of 81.24% compared to the configuration without the battery that produced an efficiency of 70.22%.

Taleb et al. [272] used a direct hybrid design of polymer exchange membrane surface FC with small aqueous ultracapacitors. It was already established that the PEMFCs possess a high-power density, but this property may be degraded by the size and the weight of the associated supports or ancillaries required to manage and control the flow rates of air, H2, and the humidity. A hybrid design based on batteries or ultracapacitors was introduced by several researchers to overcome the challenges of limitation of FC dynamics by fuel supply mechanism consisting of the compressor, flow-control device, and/or the humidifiers. The authors, however, point out that the FC system's performance enhancement offered by a hybrid design with ultracapacitors raises a technical question about the best configurations and control approaches. This meant that it was desirable not to implement hybrid designs at the expense of the FC's durability and reliable operation. Although trade-offs may be entertained in certain situations, the system complexities and capital cost also need to be rationally moderated.

Proper control strategies are also important in hybrid systems. Yang et al. [273] achieved the load-following capability of SOFCs by using a time-delay control mechanism [273]. The authors also identified the limitation of load following, which affects the SOFC system's commercialization. In light of this, they introduced a time delay control with an observer in the gas supply system to improve the FC's dynamics—in terms of load following without curtailing or blocking the fuel supply. The initial approach of a time delay with an observer was enhanced by integrating a filter to forestall the unwanted effects, such as fuel disruptions. A 5-kW SOFC was developed in the work as a test case, and it was reported that the proposed control approach provided good dynamic performance when applied to the fuel supply part of the system design. Wu and Gao [274] introduced the optimal fault-tolerant control approach in the operation of SOFCs. The authors recognized that certain critical issues such as load tracking, thermal management, air excess ratio constraint, high efficiency, low capital cost, and fault detection are key factors required for developing SOFCs. It was also found that there are little or no research studies focused on the control strategies that are based on optimization and fault diagnosis in SOFCs. It was reported that the proposed control strategy can track the electrical, operating temperature, and the air excess ratio, realizing optimum efficiency, and cost under normal and abnormal (air compressor fault, for instance) conditions.

Energy management is also important for the optimization of hybrid systems. Barakat et al. [275] achieved the energy management of a HES based on the tidal turbine and hydrogen microgrid system. The idea of the work was to use an EM strategy to convert the energy produced by the tidal energy converter to H2 efficiently and ensuring the operation of the system components, and a reduction of losses in the system. Han et al. [276] proposed the hierarchical energy management system for the hybrid design of solar PV, H_2, battery island, and direct current microgrid systems. The authors first presented some of the advantages of DC microgrids such as the fact that they are not being affected by issues of power quality, synchronization, and the reactive power flow, which are common challenges of the AC microgrid design. These were part of the factors that motivated a strategy that was introduced to enhance the performance of the electrical system in terms of cost and robustness. The energy management system consists of the local control (LC) and the system control (SC) layers. Han et al. [277] also realized a two-level energy management (EM) strategy for a DC microgrid system based on solar PV, FCs, and battery systems. The authors mentioned the shortcoming of the conventional, distributed control technique, which is its limitation in satisfying the energy management requirements for operating the multiple-source DC microgrid system. A two-level EM approach was then proposed, which was similar to the idea presented in [276]. The strategy is segregated into two, viz. the device control (DC) and system control (SC), levels.

5.6 MULTI-FUNCTIONAL MFC

MFCs are relatively new and are one of the promising technologies that facilitate the simultaneous resolution of energy needs and environmental concerns. MFCs are equipped with the production of biohydrogen, biosensors, and in situ power sources that are utilized for bioremediation collectively with the treatment of wastewater facilities. The rationale that facilitates the use of MFCs for wastewater treatment includes the process of the direct conversion of energy obtained from substrate to electricity, production of controlled activated sludge, insensitivity to the operating environment at low temperatures, their ability to be used without treatment of gas and input of energy for aeration, and their utilization in areas with limited electrical infrastructures. The amount of energy generated through MFCs mainly depends on their design, the distance between electrodes, the electrode utilized, the PEM, the mediators, the substrate, and the microorganisms involved along with certain external influences. MFCs are composed of different designs, including single-chambered, double-chambered, stacked designs, etc. PEM, which is the main component in MFCs, plays a crucial role, as its area in comparison to the electrode surface area affects the power production [278].

PEM is composed of Nafion, cellophane, agar, etc. Mediators are the compounds that are involved in the transportation of electrons from microorganisms to the electrode surfaces and thus induce power density. A few examples of intracellular mediators include NADPH, NADH, cytochromes, and so forth. Certain synthetically obtained mediators include thionine, Meldola's blue, methylene blue (MB), neutral red (NR), hydroxy-1,4-naphthoquinone, and so forth [279]. Synthetic mediators might be integrated, but they have limited applications due to their toxicity. Recent studies suggest that the direct transfer of the electrons from the surface of the cell to the anode surface with increased stability and Coulombic efficiency (CE) plays an important role in the operation of MFCs. CE is governed by the transfer of electrons in a system to carry out an electrochemical reaction. In MFCs, CE measures the number of coulombs recovered as electrical current and is dependent on the types of microorganisms involved, the substrate used, the type of wastewater, the design of MFCs, and the experimental protocol [279]. The microbes that have been reported to transfer the electrons efficiently to the anode directly are Rhodoferax ferrireducens, Shewanella putrefaciens, Geobacter metallireducens, Geobacter sulfurreducens, and

Aeromonas hydrophila [279]. Microbes require nutrients to operate properly, and those nutrients can be provided using certain waste sources as substrates, such as swine waste, dairy waste, as well as combined industrial waste, and so forth [279].

The MFC system converts chemical energy to electrical energy supported by the metabolic activity of certain microbes. As a typical bio-electrochemical system, the MFC consists of an anode region and a cathode region separated by a PEM (see Figure 5.29). The electricity generation of MFCs relies on biological oxidation and oxygen reduction occurring in the anode and cathode regions, respectively. In the anode region, microbes act as biocatalysts to decompose substrates for the generation of electrons and protons through cellular respiration. These electrons transported through the external circuit and protons transported through the PEM result in a reduction reaction with oxygen to generate water in the cathode region [278]. Unlike other low-temperature FCs described earlier, microbial FCs serve multiple purposes of productions of electricity and generation of value-added products from biological wastes due to the diversity of strains and metabolic pathways. Microbes co-produce a variety of biofuels, volatile fatty acids, biopolymers, and other platform compounds through the fermentation process during the electricity generation of MFCs [278]. Furthermore, substrates for MFCs also extend from pure chemicals and organic wastewater to lignocellulosic biomass (LCB) due to the wide substrate availability of strains. Therefore, the MFC system is a promising sustainable technology for simultaneous energy production and waste valorization [278].

5.6.1 Multi-Functional Characteristics of MFC

While MFC plays a dual role as a power generation and a waste treatment system, in line with the theme of this book, here we mainly focus on its potential for power generation. According to Priya et al. [280], an ideal MFC should possess the highest

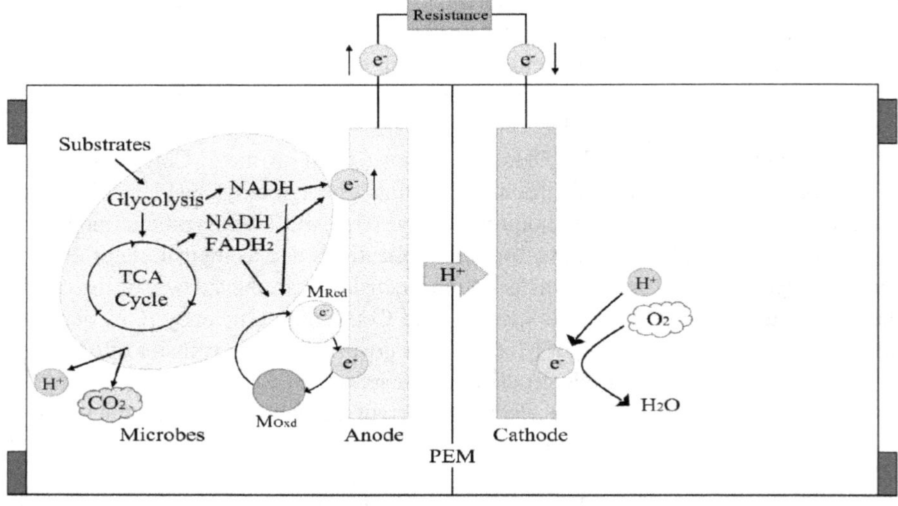

FIGURE 5.29 Typical MFC system [278].

power density of 8.98 W/m^2 under optimum conditions. While this theoretical power density has not been achieved by fabricated MFC, literature studies indicated that there are technical issues that restrict the commercialization of MFC. Major issues are relatively low power density due to poor electron transfer between microbes and anode, poor ORR that occurs at the cathode, and fouling effect in separator or PEM. Many challenges must be addressed which include the electron transfer process, scales of electrodes/PEM, and biofilm formation.

Based on the extensive research studies, MFC is believed to be a sustainable technology for the production of electricity. Many reviews emphasized designs and configurations of MFC, algae-based MFCs, and electrode materials [281–283]. The literature reviews also discuss recent improvements in anode, cathode, PEM of MFC. The review by Gajda et al. [284] was focused on changes in electrode materials and reactor designs with special emphasis on system performance and cost. The current challenges and opportunities for scaling up of MFC in various sectors were also discussed. Based on the literature studies, it appears that there are seven parameters that affect the role of MFC for electricity generation. These are anode, cathode, and membrane designs, constructions and operation, nature of microorganism, nature of substrate, cell operating conditions, and cell configuration. Here we examine the role of each of these parameters toward electricity generation and advances made by research in each case in some details.

5.6.2 Role of Anode Modification

Developing anode materials suitable for use in microbial FCs needs to meet the criteria of high electrical conductivity, high surface area, and biocompatibility that would not only allow for efficient electrochemical "wiring" (attachment) of living bacterial cells but also high conductivity, corrosion resistance, and chemical stability with cost as the primary driver. The modification of MFC anodes mainly focuses on improving the specific surface area and surface properties. Heat treatment and acid treatment are feasible surface treatment methods to increase the specific surface area of the anode. Electrochemical oxidation methods can increase the specific surface area of the anode and introduce new functional groups to the anode surface [278]. These methods are all conducive to facilitating electrical contacts of strain cells to form electron-donating biofilms. However, more studies have used different materials for electrode modification to enhance the adhesion of strain cells and promote electron transfer to the anode surface.

Anode component is critical in terms of the surface area available for the development of the microbial biofilm. In order to meet all criteria, granular carbon-based materials or fabric-based cloth, mesh, felt, etc., are generally chosen for anode construction. Research has shown that the cost-effective improvement in power density can be achieved by increasing the anodic surface area-to-volume ratio and anode packing. Metals and metal oxides have widely participated in anode modification. The development of new low-cost solutions such as stainless steel wool or carbon sponge is a promising anode alternative and deserves long-term testing [278].

Xu et al. [285] studied the electricity generation of dual-chamber MFCs (DCMFCs) with carbon cloth anodes modified with MnO_2, Pd, and Fe_3O_4, respectively. The maximum power densities achieved by anodes modified with such materials can reach 824,

782, and 728 mW/m^2, respectively. The authors also pointed out that anodes modified with different materials lead to the enrichment of different strains on the anode surface. Yu et al. [286] observed the maximum power density of 29.98 mW/m^2 using the anode modified with bentonite-Fe, and 18.28 mW/m^2 using the anode modified with Fe_3O_4. They also reported increases in the stable voltage and decreases in the internal resistance (IR) of the MFCs with modified anodes compared to the bare graphite felt anode. The carbon cloth modified with cobalt oxide and the nitrogen-doped carbon nanorods modified with Co-modified MoO_2 nanoparticles can also improve the electricity generation by MFCs [278].

Li et al. [287] reported that while zero-valent iron has a positive effect on power density, high concentration of zero-valent iron inhibited electricity production. Paul et al. [288] used the carbon felt (CF) anode modified with GO and zeolite to achieve a 3.6-times higher power density and 2.75-times higher CE than having used the bare CF anode. They indicated the higher biocompatibility that originated from the improved specific surface area by graphene oxide and enhanced microbe adhesion by zeolite. The power densities achieved by the CF anode modified with GO and Fe_2O_3 are 1.72 times and 2.59 times that of MFCs with the graphene anode and the unmodified anode, respectively [278]. Liang et al. [289] pointed out that anodes modified with graphene, GO, and CNT have higher electrochemically active surface areas and enriched microbial communities. Zhang et al. [290] indicated that the graphite felt modified with CNT can promote biofilm growth and enhance electron transfer. Often, composites of polymer and metal-type or carbon-type materials are used for anode modification. The power density achieved by the anode modified with composite polydopamine and reduced GO reaches 2.2 and 1.9 times that of MFCs with anodes modified with polydopamine and reduced GO individually [278]. The anode modified with composite polyaniline (PANI) and Au can also improve bioelectrochemical activity [278]. Mashkour et al. [291] pointed out the positive effect of PANI on biofilm growth. The CF anode modified with nitrogen-doped CNT, PANI, and MnO_2 can achieve a 2.76-times higher cell biomass content than that of the bare anode [278].

Exoelectrogens are vital and form a biofilm on the anode surface to decompose organic substrates like glucose, acetate, and waste materials to generate electrons and protons. Thus, this group is necessary for electricity generation. These microorganisms can transport electrons from electron donors (substrate) to electrode surfaces by various mechanisms, including direct contact, nanowires, and mediators produced by them [292]. Enhancing power generation in low-light settings and efficiently collecting electrons from photosynthetic bacteria on the electrode are also key difficulties in the development of a functional photosynthetic microbial FC (PMFC). In order to address these issues, an anode was developed in a dual-chambered PMFC with an abiotic cathode by casting a nanocomposite matrix that assisted biofilm development of the photo-catalyst *Synechococcus* sp., surging the bacterial photosystems (PS I and PS II) with suitable light ($\lambda_{650-750nm}$) at a broad excitation spectrum ($\lambda_{350-644\,nm}$) through fluorescence resonance energy transfer [292]. Researchers have established that an adequate microbial population structure and metabolic activity of the electrode biofilm communities are required for satisfying the performance of microbial electrochemical cells using this conversion technology [292].

The properties of the anode surface are of great importance for the colonization of exoelctrogenic bacteria by influencing the transfer of electrons and bacterial attachment [292]. An ideal MFC anode material has to expedite the flow of electrons and accelerate the attachment of microbes using conduction. Large conductivity, low active surface area, corrosion, and weak adhesion of microorganisms make the use of platinum, silver, gold, copper, and stainless steel [293] as electrode materials unsuitable for MFC [294]. In general, carbon as an anti-corrosion material with good conductivity is considered suitable for electrode fabrication in MFC. However, using conventional anodes, electron movement from microbial biofilm to the electrode surface is restricted due to bacteria's weak attachment resulting in low power output, making them impractical for MFC technology. The other significant shortcoming is high IR, resulting in a low rate of charge transfer and high-cost materials.

Inexpensive nanomaterials can provide an extended surface area for bacteria growth on the anode and enhance microbial attachment. This improvement (i.e., enhanced exoelectrogens colonization) through increased extracellular electron transfer (EET) and substrate metabolism lead to augmented function and effectiveness of the MFC [292]. The increase in the number of exoelectrogenic colonies boosts the redox activity of anodic biofilm, and hence, more electrons and protons are generated. Multi-walled carbon nanotubes (MWCNTs) are among the most widely used nanomaterials for anode modification. They can provide an excellent active surface for microbial attachment. Nambiar et al. [295] utilized MWCNTs for improving the growth of *Enterobacter cloacae* on a commercial carbon sheet anode in a DCMFC with a platinum-coated carbon sheet cathode. In this work, a maximum power density increase of 256% was achieved using the MWCNTs-modified anode rather than the bare carbon sheet. The authors attributed the improvement made to the extended highly conductive surface area on the carbon sheet obtained using a high concentration of MWCNTs ink (10 mg/mL) and chitosan as a green binder [295]. Despite the increased surface area for bacterial colonization, these nanomaterials have been reported as toxic at specific concentrations against bacteria. A recent work investigating the MWCNT's toxicity effect on microbial growth of *Geobacter sulfurreducens* biofilm revealed that the maximum MWCNTs concentration for improving anodic biofilm performance (i.e., highest power and current density compared to bare carbon sheet) was found to be 0.1 mg/mL [292]. However, further increments in MWCNTs concentration resulted in anodic biofilm's weak performance, suggesting that the MWCNTs concentration must have exceeded its toxic threshold [292]. The following reasons could explain the significantly different optimal concentrations reported by these two studies: (1) the differences in the biofilms developed and (2) the differences in the time of measurements. Mashkour et al. [291] tried to modify MWCNTs using biocompatible nanomaterials like PANI. The authors claimed that the modification could successfully mitigate the toxicity of the nanocomposite and consequently decrease the Rct (charge transfer resistance).

Other carbon-based nanomaterials have also been used for anode modification. For instance, reduced graphene oxide (rGO) nanosheets have been recently utilized for improving the exoelectrogenic biofilm activity of *Shewanella putrefaciens* in an MFC system [296]. Graphene oxide (GO) was dispersed in an anolyte solution of bacteria and was then reduced on carbon felt *in situ* biologically (br-GO). The power density generated by the modified anode stood at 240 mW/m^2, which was substantially higher than the value recorded for the unmodified carbon (0.015 mW/m^2) [296]. Moreover, br-GO was found highly biocompatible, which must have also contributed to the high affinity of the bacterial populations to grow at higher rates, more effectively colonize, and form a thicker layer of active biofilm on the anode [296].

Various other biocompatible nanomaterials have also been studied for anode modification to boost biofilm formation and activity. For instance, a mixture of nickel nanoparticles (NPs), poly(3,4-ethylenedioxythiophene) (PEDOT), and graphene was employed by Hernandez et al. [297] to modify a stainless steel (SS) anode with *E. Coli* as anodic bacteria in a DCMFC. The high surface area of graphene and its good conductivity on the one hand and the high redox activity of PEDOT, on the other hand, provided the SS anode with suitable sites for exoelectrogenic activity. Also, NiNPs accelerated the anodic biofilm formation because of their biocompatibility. The generated power density was improved by around twice compared with bare SS anode [297].

Gold NPs (AuNPs) were also used to modify anode electrode in an SCMFC by Duarte et al. [298]. Anodic biofilm activity of *Saccharomyces cerevisiae* on polyethyleneimine-functionalized carbon felt increased after the anode was coated by AuNPs. The modified anode led to an enhanced power density of 2,271 mW/m^2 vs. 381 mW/m^2 recorded for the control. Moreover, the AuNPs attachment onto *S. cerevisiae* membrane could result in direct electron transfer. Despite the promising results obtained using AuNPs, its high cost prohibits its commercial use.

Mashkour et al. [292] synthesized iron carbide NPs (FeCNPs) on carbon felt as an anode in an MFC with mixed-culture bacteria and showed that the power density of the modified anode increased by about 200%. FeCNPs could enrich the exoelectrogenic biofilm of mixed cultures and speed up the redox reaction. Moreover, the FeCNPs provided a much lower Rct than carbon felt. Nanocellulose is a low-cost group of useful materials for bacteria colonization. They can be coated and modified with a vast range of carbon, metal, and polymer-based materials to make this compound conductive for use in MFCs. Mashkour et al. [299,300] introduced a new generation of anode electrodes called hydrogel bio-anode. They fabricated porous hydrogel BC-based anodes coated by conductive polymers, that is, PANI [299] and polypyrrole [300]. The capillary effect between the BC nanofibers resulted in permanent access of bacteria to nutrients and prevented bacteria spoilage and clogging of pores [301]. The bio-anode improved the power density of the investigated MFCs significantly compared to the graphite plate [299,300]. The biocompatibility of both BC and the conductive polymers contributed to a more effective adhesion of the bacteria onto the anode, facilitating electron transfer. Moreover, the bio-anode led to higher redox activity. Low pH values of bio-anode used by Mashkour et al. [299,300] could have contributed to the excellent conductivity of the modified BC. However, the authors did not consider the effects of pH variations on the performance of the fabricated electrodes. Given the pH-dependent nature of conductive polymers [292], different results could have been obtained if different pH values had been taken into consideration. Redox properties, conductivity, surface area, biocompatibility, and cost of nanomaterials should be considered before their use for anode modification. From an economic perspective, carbon-based nanomaterials, conductive polymers, and inexpensive nanometals could be regarded as excellent choices. Chen et al. [302], however, point out that the effect of nanomaterials might differ for each species of exoelectrogens and should be taken into consideration during experimental designs.

The above discussions indicate that in general, improving the efficiency of MFCs using a nanocomposite nanomaterial electrode as a catalyst is critical for low-cost and efficient energy harvesting systems. Such improvement has a strong contribution to the assembly of strong nanocomposite materials hybridized from renewable forestry waste, two-dimensional layered nanomaterials, microscale and nanoscale electrode topography, their dimensional mesoporous electrodes, smart polymers, and conducting polymers with other nanomaterials as an electrode; in addition, membranes may offer performance improvement in MFC devices due to their synergistic effect [292]. Among nanomaterials, along with conducting polymers from PANI mentioned earlier, polypyrrole (PPy) is a good candidate [292]. According to the results of practical study, coating a single bacterial cell with a conjugated polymer/redox polymer results in a high-performance anode for MFC application [292]. These polymers due to their inexpensiveness, biocompatible nature, and flexibility in designing electrodes can increase efficiency in MFC devices. For example, a unique structural SS-based anode for high-performance MFCs was successfully modified by *in-situ* electrochemical deposition of PPy onto SS [292].

Many challenges still need to be addressed in order to modify effective bioelectricity generation in MFC. The type and nature of electrodes, bioreactor design, shape and size, and the types of active electro-microorganisms decide performance in MFC applications. While noble-based nanomaterials are effective in producing electricity even in small-device applications of MFCs, anodes modified by such nanomaterials are not cost-effective in scaling up MFC devices. Hence, to reduce use of such costly nanomaterials, novel, low-cost, biocompatible, and environmentally friendly nanocomposites are used. Although power generation by using nanocomposite-modified anodes yields low power production, factors governing MFC performance will be optimized in the future to advance effective anode materials. Among these factors, electroactive microorganisms/biocatalysts, the type and concentration of fuels (substrates), and pH are critical. In the future, all listed parameters should be optimized to reduce polarization loss, activation losses, concentration losses, and ohmic losses. All the losses will be minimized by fabricating a functional nanocomposite through low-cost metal oxide-conducting polymer integration.

The research on anode modification has resulted in few conclusions. High-surface-area/biocompatible nanomaterials are instrumentals in developing thick biofilms on electrodes' surface. Conductive NPs act as bridges between the biofilm and the electrode surface, and decrease electron transfer resistance. However, more investigation on the effect of nanoparticle morphology on electron transfer is still needed. Toxicity of CNTs to microbial biofilms would depend on CNTs' concentration. Also, the adverse effects associated with electrode modifications using CNTs could be moderated by adding biocompatible nanomaterials such as PANI.

5.6.3 Role of Cathode Catalyst Modification

Optimizing cathode components is based on the reduction reaction occurring in the cathodic half-cell. ORR is taking place at the MFC cathode which is often the limiting reaction and therefore a source of losses. The most feasible cathode configuration for large-scale application of MFCs is an air-breathing cathode; however, it shows limited performance under static operation mode. This is often connected to the cathode scaling with precipitated salts and its deactivation in time. It is then required to configure the cathodic chamber accordingly to avoid salt precipitation. One way is via the development of a catholyte-generating half cell which provides good long-term performance since the generated liquid washes away the deposits and provides additional electrochemical treatment through disinfection and bacterial killing [278].

The efficiency of cathode-based oxygen reduction directly affects the electricity generation of the MFC system. Appropriate cathode catalysts can improve the power output efficiency of MFC systems by promoting electron transfer and enhancing oxygen reduction. While platinum-based cathode catalysts can improve the activity of oxygen reduction [278], they are too costly for large-scale applications [303]. Currently, more studies are attempting to develop nanocomposite-based cathode catalysts to enhance the electrochemical activity of MFC systems. Liu et al. [304] focused on the cathode catalysts based on metals and metal oxides. They developed an activated carbon cathode modified with Cu_2O and Cu to achieve a peak power density of 16.12 W/m^2. Majidi et al. [305] observed a power density of 180 mW/m^2 using a carbon cloth cathode modified with α-MnO_2 nanowires and carbon Vulcan. Chiodoni et al. [306] reported the positive effect of manganese-oxide-based cathode catalysts on MFC performance. Rout et al. [307] focused on the cathode catalysts combined with metal oxides and non-metal materials. They developed a nanocomposite of MnO_2 and reduced GO to achieve a 2.7-times increase in volumetric power density. They also pointed out that this nanocomposite can provide a four-electron oxygen reduction pathway and enhance electron transfer.

Mecheri et al. [308] reported that the cathode catalyst based on FePc and GO can improve the electrochemical performance of the MFC. Lee et al. [309] considered the 3D composite of CNT and MoS_2 for cathode catalyst to achieve an efficient oxygen reduction. Li et al. [310] observed a maximum power density of 1,177.31 mW/m^2 and a current density of 6.73 A/m^2 using the cathode catalyst of bacterial cellulose doped with P and Cu. They indicated that more active sites in this cathode catalyst improve the catalytic activity of oxygen reduction. Kaur et al. [311] developed a composite catalyst of PANI and an iron-based metal–organic framework. It can achieve a power density of 680 mW/m^2 and a limiting current density of 3,500 mA/m^2. Lee et al. [312] considered the Ni-based metal–organic framework for an efficient cathode catalyst to promote oxygen reduction. In addition, the layered double hydroxide (LDH) has also participated in cathode catalyst development. Jiang et al. [313] developed a composite of Fe_3O_4 and NiFe-based LDH to achieve the maximum power density of 211.40 mW/m^2. They indicated the advantages of LDH in terms of electroactive site availability, rate capability, and cycling stability. In another study, the composite catalyst of NiFe-based LDH and Co_3O_4 achieved the maximum power density of 467.35 mW/m^2 [105]. Tajdid et al. [314] synthesized CoNiAl-based LDH. This material improved the performance of graphite cathode by working independently or combined with $NiCo_2O_4$. With the development of material science, co-composites based on various materials have become the choice for cathode catalyst development [278]. These co-composite cathode catalysts can exploit the specific advantages of each material. It is conducive to improving the comprehensive electrochemical performance of MFC systems, including electron transfer efficiency, oxygen reduction catalytic efficiency, and operation stability.

The effect of nanomaterials on electrotrophs' performance is generally similar to what is explained for exoelectrogens. In addition to those, nanomaterials can increase gas adsorption on the cathode side, leading to enhanced electrotrophic performance. The cathode has a major role in improving the efficiency and power density of MFC. Graphite or carbon cloth/paper/felt/brush and air cathode have been used in MFC to increase its efficiency. However, pristine cathodes showed insufficient activity results in poor performance. For example, although graphitic brush/rod exhibited high surface area, stability, and biocompatibility, they show poor performance due to insufficient catalytic active sites. Moreover, nanomaterials with photocatalytic properties might also benefit the electrotrophic activity by boosting the reaction taking place over the cathode.

For acetate production from CO_2 in an MES system, Bian et al. [315] modified the cathode surface with porous hollow fibers of nickel (Ni-PHF) and CNT in the presence of *Sporomusa ovata*. The porous structure of Ni-PHF could provide *S. ovata* with direct access to CO_2 on the cathode surface. At the same time, the high specific area of CNT could increase CO_2 adsorption by more than eleven times and improved charge transfer by boosting cathodic current density from 214 to 332 mA/m^2 [315]. In a different study but using the same electrotroph, a rGO sheet was used as a cathode for microbial colonization [292]. The authors claimed that acetate production rate and current density were increased by 7 and 8 times, respectively, compared to the carbon paper electrode. Compared to carbon cloth, the surface area of the cathodic biofilm of *Clostridium ljungdahlii* was expanded by 3.2-folds through incorporating graphene and CNT in a three-dimensional (3D) cathode structure [292]. The modified structure elevated the NADH/NAD+ ratio, and the excess NADH could be used in CO_2 fixation. Consequently, acetate production and cathodic current density increased by 5 and 4.4 times, respectively, compared to the values recorded for carbon cloth [292]. In a more recent study, carbon cloth was modified by GO and electro-polymerized conductive PEDOT to improve the interaction between *Methanobacterium* and CO_2 for a higher acetate production rate [292]. They also compared three types of modified carbon cloth, that is, GO, PEDOT, and GO-PEDOT modified carbon cloth and reported that the biofilm density increased due to the GO's high surface area, and a higher electron transfer rate from cathode to bacteria was obtained through the function of the conductive PEDOT [292].

Recently, a cathode modified by tungsten oxide (WO_3) and molybdenum oxide (MoO_3) NPs was reported used for a photo-assisted MES system. *Serratia marcescens* was utilized as an electrotroph to produce acetate from HCO_3. The photocurrent generated by the nanoparticle oxides-modified cathode was about five times higher than that of the bare carbon felt-based cathode. Photo-induced electrons placed on the conduction bands of the modified biocathode improved the HER. The produced hydrogen promoted acetate production by the electrotroph *S. marcescens* [292].

Overall, it could be concluded that the modification of cathode surface in MES systems using nanomaterials, by increasing surface area, could provide electrotrophs (responsible for producing chemicals) with more electrons and more gas molecules such as CO_2. The latter is ascribed to the higher gas adsorbtion rates of the nanomaterial-based modified cathodes. It should be noted that for these advantages to work, the nanomaterials used in cathode modification are bound to be biocompatible so that electrotrophs can grow densely, forming a thick biofilm. Also, the adsorption capacity of the utilized nanomaterials is vital in providing electrotrophs with additional gas molecules in reaction sites existing over the biocathode.

In MFC systems, ORR is desired to occur at the highest rates possible, signifying the critical role of catalysts (i.e., catalytic activity). On the other hand, the cost of the catalyst used for catalyzing ORR on the cathode is significantly important to achieve an economically viable system. As mentioned earlier, Pt-based catalysts are very expensive and have bacteria's poisonous effect to be useful on large scale [292]. Inexpensive nanomaterials are considered excellent choices to achieve the mentioned objective by extending the surface of the cathode. Metal oxides nanomaterials are among the promising alternatives to Pt for use as cathode catalysts. Tofighi et al. [316] coated carbon cloth with a mixture of α-MnO_2-GO and activated carbon and used it as the cathode in a DCMFC. The authors argued that the modified cathode possessed a substantially higher catalytic activity resulting in an increased power density of 140 μW/cm^2 *vs.* 0.6 μW/m^2 recorded for bare carbon cloth. A PANI/MnO_2 nanocomposite was introduced by Ansari et al. [317]. Both metal oxide nanomaterials and conductive polymers are categorized as pseudo-capacitors. Hence, the nanocomposite showed significant capacitance (525 F/g) and 76.9% cycling stability for 1,000 cycles. The maximum power density of 37.6 mW/m^2 was reported, which was considerably higher than the value recorded for carbon paper (approx. 0.003 mW/m^2). TiO_2 NPs were also utilized by Mashkour et al. [318] to modify a graphite-based cathode. The results indicated a higher power density (85 mW/m^2)^2 compared to 32 mW/m^2 recorded for the bare graphite [318].

In addition to metal oxides nanomaterials, nitrogen-doped carbonous nanostructures have also been used to synthesize high-performance catalysts for cathode modification. The higher ratio of nitrogen to oxygen, obtained by the nitrogen doping process, is an effective ORR rate parameter. He et al. [319] used nitrogen-doped CNT to modify carbon cloth cathode. The catalytic activity of the new composite electrode in ORR was higher than the Pt-coated carbon cloth. A maximum power density of 542 mW/m^3 was generated by the cathode, which was higher than that of the Pt-coated one (approx. 500 mW/m^3) [319]. In addition to nitrogen doping, phosphorus doping has also been reported by Liu et al. [320] who performed a dual-doping procedure by *in-situ* pyrolysis of cellulose in the presence of ammonium phosphate as nitrogen and phosphorus source. They claimed remarkable enhancement in power density compared to the Pt-coated cathode (2,293 mW/m^2 *vs.* 1,680 mW/m^2, respectively) [320]. Cobalt is another choice for doping. Liu et al. [321] and Zhong et al. [322] synthesized cobalt-doped carbon mixtures showing comparable performance in terms of power density with their nitrogen-doped counterparts. Moreover, to develop nitrogen- and cobalt-doped carbonous nanostructures, metal organic frameworks (MOFs) have been used recently. For instance, Zhao et al. [323] synthesized an ORR nanocatalyst by pyrolyzing a core–shell Co-MOF and adding rGO to the residue. The power density of the MFC equipped with the cathode modified by the MOF-based nanocatalyst increased to 2,350 mW/m^2 *vs.* 2,002 mW/m^2 measured for Pt [323]. In a different study, Zhong et al. [324] introduced a Zr-based MOF as an excellent template for nitrogen- and cobalt co-doped nanocomposite ORR catalyst. They showed that the MOF-based catalyst could enhance MFC output (power density) to near the value reached by Pt-coated cathode (300 mW/m^2 *vs.* 313 mW/m^2, respectively) [324].

Biofilm formation on the air-breathing cathode surface is a challenge in SCMFCs. The cathodic biofilm formed by anaerobic bacteria decreases the active sites of ORR on the cathode surface. So, hindering bacterial growth on air cathode might be considered a solution in the long-term operation of MFCs. There are numerous efforts to mitigate the biofouling of air cathode, including physical method, chemical cleaning, electrokinetic control, and surface modification [292]. Surface modification of air cathodes using nanomaterials can also assist with mitigating this unfavorable phenomenon. Ma et al. [325] used silver (Ag) and iron oxide (Fe_3O_4) NPs in the carbon composite of the cathode and managed to control biofouling. They argued that the recovered power density after 17 cycles was about 96% of the initial value compared to 60% for the bare cathode. Similarly, Noori et al. [326] also reported that the incorporation of silver in the catalyst layer of the composite helped recover 95% of the initial power density after 90 cycles. Both studies attributed the achievements made to the antibacterial properties of the silver NPs (AgNPs) [292].

In air-cathode single-chamber MFCs, electrode improvement and optimization are also ongoing [292]. EET of surface active microbes [292] as well as precise electron carrier events of electron-rich microbial species for electron transport functionality (e.g., biofilm formation, electron shuttles, swarming motility, dye decolorization, and bioelectricity generation) to MFCs aids in the generation of more stable and reliable power [292]. Literature studies have also resulted in the following conclusions: Nanoporous materials increase cathodic gas adsorption, leading to increased microbial electrosynthesis of chemicals; metal oxide- and MOF-based nanocomposites can efficiently catalyze the ORR and are considered inexpensive alternatives to Pt; they are also more stable in microbial environments; and nitrogen- and metal doping can extend ORR active sites of carbonous nanomaterials.

5.6.4 Role of Microorganism

Substrate oxidation by microbes in the anode is the only source of electron generation in MFC systems. *Geobacter* and *Shewanella* are electrogenic microbes commonly used in MFCs. Several yeast strains, such as *Saccharomyces cerevisiae*, *Candida melibiosica*, and *Kluyveromyces marxianus*, have also participated in the operation of MFC systems. In addition, archaebacteria, cyanobacteria, and proteobacteria are promising strains for electricity generation. However, the eukaryotic algae can act as both electron producers and acceptors in the anode and cathode, respectively [278]. The anodic inoculations of MFCs include pure cultures and mixed cultures. Pure cultures might achieve more efficient conversion of substrates to electricity because of simple and well-defined metabolic pathways. However, it also has higher requirements on the purity and concentration of the substrate. Therefore, selectivity for specific substrates might limit the ability of pure cultures to generate electricity using complex substrates such as wastewater and LCB hydrolysates [278].

Currently, pure cultures mainly participate in studies on electricity generation performance and electron transport mechanisms of specific strains. However, Pandit et al. [327] developed a pure culture-based bioaugmentation strategy to improve the volumetric current density and shorten the start-up time of MFCs. Mixed cultures are advantageous for the scale-up of MFC systems due to their higher adaptability to complex substrates. The synergistic effect of various strains in the mixed culture might also be conducive to the efficient operation of MFC systems. Activated sludge is the most representative mixed culture used for MFC systems. The sludge pretreated with acid and heat can further enhance electricity generation [278]. However, the microbe composition of activated sludge is very complex. It is difficult to determine the precise pathway of substrate conversion. Therefore, the co-culture of defined strains might enhance the performance of MFC systems through synergy based on their specific functions. Schmitz and Rosenbaum [328] developed a co-culture scheme of *Pseudomonas aeruginosa* and *Enterobacter aerogenes.* The electron mediator produced by *Pseudomonas aeruginosa* can improve the electron transfer efficiency to the anode. This co-culture system can achieve an over 400% increase in electrical current generation under an optimized oxygen supply.

Although various wild-type strains have successfully achieved electricity production, it is still necessary to further improve the electrochemical activity of strains through suitable methods. Several physical and chemical methods have been considered as potential ways to enhance the ability of strains to generate electricity [278]. Genetic engineering is also a promising strategy to improve the electrochemical activity of strains. It mainly involves gene modification related to metabolic activity, the electro-shuttle pathway, and substrate utilization. The enhancement of EET based on genetic engineering is an effective method to improve the performance of MFC systems. The modification of cytochrome c maturation can achieve a 77% increase in the current generation. A constructed hybrid system of cytochrome c maturation can also increase the overall current by 121% [278]. In addition, the cytochrome OmcZs expressed by Escherichia coli can enhance the current production by binding riboflavin. Synthetic biology has played a role in enhancing the EET efficiency of strains. Liu et al. [329] enhanced the electricity output of the *Pseudomonas aeruginosa* strain-based MFC by assembling type IV pili with high conductivity. Lin et al. [330] promoted the EET efficiency of *Shewanella oneidensis* by enhancing the biosynthesis and transportation of flavins. Kasai et al. [331] and Cheng et al. [332] also enhanced the EET of *Shewanella oneidensis* by improving the intracellular level of 3^0,5^0-cyclic adenosine monophosphate. Min et al. [333] developed an engineered *Shewanella oneidensis* strain with a gene cluster of flavin biosynthesis. This strain can achieve a 110% increase in the maximum current density of MFC systems. In addition, the increase in intracellular NAD(H/+) based on a modular synthetic biology strategy can improve both intracellular electron flux and EET efficiency [278]. McAnulty et al. [334] focused on the substrate expansion for power generation of MFC systems. They achieved the conversion of methane to electricity by developing a synthetic consortium with the main strains comprised of engineered *Methanosarcina acetivorans*, *Paracoccus denitrificans*, and *Geobacter sulfurreducens*. Li et al. [335] developed an engineered *Shewanella oneidensis* strain for electricity generation directly from xylose. It can achieve a maximum power density of 2.1 mW/m^2. Genetic-engineering-based strain modification for enhancing electricity generation has achieved desired results in laboratory-scale MFC systems. However, there is still a lack of specific progress in the scale-up of MFC systems.

5.6.5 Role of Substrate

One of the most crucial aspects of MFC is the substrates used due to their effect on the generation of electricity [278]. Various substrates can be utilized in the MFC technology to facilitate energy generation, either pure substances or complex mixture from organic materials present in wastewater. The most frequently used substrates are acetate, glucose, LCB, synthetic wastewater, brewery wastewater, dye wastewater, and other inorganic substrates. The aim of all treatment processes is mainly to reduce the number of pollutants in water in order for it to be safe for utilization and the environment. MFC performance is also determined by a unit, most often used, called current density that can signify the produced current per unit area of the surface area of anode (mA/cm^2) or the current produced per volume of the cell (mA/cm^3). The most frequently used substrates are acetate, glucose, LCB, synthetic wastewater, brewery wastewater, dye wastewater, and other inorganic substrates.

Currently, defined substrates for MFC systems mainly include sugars and organic acids. Glucose is a common substrate for MFC systems. Christwardana et al. [336] used glucose for a yeast-based MFC to achieve a maximum power density of 374.4 mW/m^2. However, Obileke et al. [337] pointed out that glucose might lead to low CE of MFC systems due to the electron loss caused by competition strains and the substrate consumption for fermentation. There are also studies using xylose as the substrate for MFC systems. Haavisto et al. [338] observed the highest power density of 333 mW/m^2 using xylose for an upflow MFC (UMFC) system. Li et al. [339] developed a microbial consortium consisting of engineered *Klebsiella pneumoniae* and *Shewanella oneidensis*. It can achieve a maximum power density of 104.7 mW/m^2 using co-substrates of xylose and glucose.

Several studies have compared the performance of MFC systems using different substrates. Ullah and Zeshan [340] studied the electricity generation of the DCMFC using glucose, acetate, and sucrose, respectively. They reported a maximum power density of 91 mW/m^2 using acetate as the most effective substrate. Jin et al. [341] also obtained similar results. They observed the best electricity production performance of the DCMFC using sodium acetate compared with glucose and lactose. In addition, acetates also have an advantage in the electrochemical performance of MFC compared to lactate and octanoate [278]. It might be related to the lower ohmic loss of the biofilm when using acetate as the substrate for MFC systems [278]. In recent years, acetate has also participated in the scheme of co-substrates with pollutants to improve the performance of MFCs in electricity generation and toxicant degradation. Shen et al. [342] obtained a voltage output of 389.0 mV with an initial phenol degradation of 78.8% using acetate as the co-substrate for the single-chamber MFC. Yu et al. [343] also used the DCMFC with acetate co-substrate to achieve increases of 4.3-fold in power generation and ~42% in removal efficiency of 4-chlorophenol. In addition, Ndayisenga et al. [344] obtained increases of 60.1% in CE and 64.7% in microcystin-LR using acetate co-substrate for the DCMFC. Mancilio et al. [345] used acetate co-substrate to achieve a power density of 398 mW/m^2 and a p-Coumaric acid degradation of 79%. However, they also reported higher potential and power density of the MFC using acetate as the single substrate.

As a renewable source rich in carbon, LCB exists mainly in the form of agricultural and forestry waste. These LCB hydrolysates containing a variety of hexoses and pentose have been considered promising substrates for cell growth and metabolism. Catal et al. [346] used the sulfuric acid hydrolysate of pinewood flour to achieve a voltage of 0.43 V at 1,000 Ω external resistance of single-chamber MFC. Jablonska et al. [347] obtained a power density of 54 mW/m^2 using rapeseed straw hydrolysates produced by hydrothermal pretreatment and enzymatic hydrolysis. Gurav et al. [348]

compared the electricity generation of a *Shewanella marisflavi* BBL25 strain-based MFC using hydrolysates of barley straw, Miscanthus, and pine, respectively. As the most effective substrate, barley straw hydrolysate can achieve the maximum current output density of 6.850 mA/cm^2 and the maximum power density of 52.80 mW/cm^2. Flimban et al. [349] studied the electricity generation of a DCMFC using the direct substrates of potato peels and rice straw, respectively. The power densities obtained from potato peels and rice straw can reach 152.55 and 119.35 mW/m^2, respectively. Mohd Zaini Makhtar and Tajarudin [350] compared the electricity generation of a membrane-less MFC system using banana peel, corn bran, and POME. They observed a voltage generation of 237.1 mV with a power density of 23.75 mW/m^2 achieved using the banana peel as the most effective substrate. Yoshimura et al. [351] developed a hydrodynamic cavitation system for the pretreatment of rice bran. They reported an increase of 26% in the total electricity generation using such pretreated rice bran because of the efficient substrate utilization. In addition, Jenol et al. [352] compared the electricity generation of a *Clostridium beijerinckii* SR1 strain-based MFC using the direct substrate and hydrolysate substrate of sago hampas. The power density achieved from these two substrate forms of sago hampas can reach 73.8 and 56.5 mW/cm^2, respectively. Despite the potential of LCB for MFC-based biomass valorization, difficulties in collection and transportation limit the large-scale application of LCB. Figure 5.30 illustrates the utilization of LCB substrates for MFC systems [278].

MFC is also facing constriction in producing energy because energy generation through the MFC system is dependent on the concentration of the substrate. If the concentration of the substrates present is larger than a certain value, production of power will be obstructed. The organic substrate is the most critical challenge.

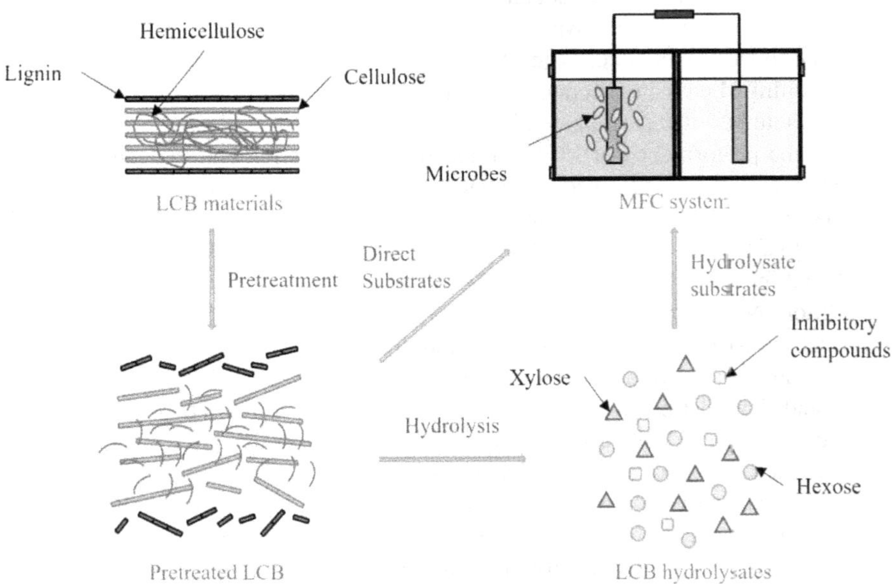

FIGURE 5.30 The utilization of LCB substrates for MFC systems [278].

The lower stability of the organic substrate may have had an impact on remediation efficiency and energy generation because the organic substrate did not provide enough power to the bacterial population, resulting in poor MFC output. The long-term stability of organic substrates in MFC for industrial use should be the subject of future research. According to the literature survey, there is no similar information available on the interaction of electrode and bacteria, and electrode effect in the presence of different organic substrates in MFC [278].

5.6.6 Role of Membrane

Although membrane-less systems can be characterized by less complex design, decreased IR, and lower cost due to the absence of membrane, they nevertheless lose efficiency due to the occurrence of ionic species crossover and side reactions. Due to simplicity and low cost, they can be implemented in water bodies for bioremediation and environmental sensing, or to provide a power source for charging mobile phones. For the purpose of the anolyte/catholyte separation, however, a membrane is required and this needs to be chosen appropriately to the application based on cost and simplicity of assembly/manufacturing. The choice of robust and low-cost membrane materials should be considered to meet the criteria of mechanical strength and longevity under various operating conditions in real-world applications [278,292].

Membranes are among the main factors affecting the performance of MFCs. In MFC, an ideal PEM should possess the following properties: (1) It should reduce the substrate flux effectively; (2) it should evade the back diffusion of H^+ ions from the cathode to the anode; (3) it should impede the movement of other ions across the membrane which may pull down the working efficiency of MFC; and (4) it should prevent oxygen transfer from cathode to anode [278,292]. With these requirements, Nafion 117 membrane is generally used as the separator. PEMs transfer protons from the anode to the cathode, prevent substrate leakage from the anode to the cathode, and stop oxygen crossover between the cathode and the anode [353]. In MFCs, typical commercial membranes, including Ultrex and Nafion, are generally used [292]. However, these membranes suffer from several disadvantages, including high prices (e.g., 1,500 USD/m^2 for Nafion 115), biofouling, and oxygen and substrate crossover [292]. The disadvantages of the commercial membranes have motivated the synthesis of diverse materials, particularly in their nanosized forms, and their incorporation into membrane structures to address the mentioned challenges.

Bazrgar Bajestani and Mousavi [354] modified a conventional Nafion using TiO_2 NPs. The formation of Ti-OH groups in the TiO_2-based nanocomposite membrane increased the exchange sites over the membrane, leading to higher ion conductivity than Nafion. The nanocomposite could also provide a higher water uptake comparatively [354]. Sulfonated TiO_2 ($sTiO_2$) NPs were also reported as a modifier for sulfonated polystyrene ethylene butylene polystyrene (SPSEBS). The nanocomposite membrane showed higher proton conductivity, much lower oxygen crossover, and much lower cost (200 USD/m^2) than commercial Nafion 117 [292]. Due to the mentioned improved features, power density and CE of the MFC with the $sTiO_2$-SPSEBS membrane increased.

SPEEK is also a typical polymer used to synthesize nanocomposite membranes for MFCs and is a low-cost polymer compared to Nafion. A nanocomposite made by SPEEK and sulfonated SiO_2 has been reported as a PEM for an SCMFC [355]. The SiO_2-modified membrane had an increased water uptake and proton conductivity compared to the bare SPEEK. Also, as a low-cost membrane, it showed much higher power density than Nafion 115. Moreover, the nanocomposite membrane showed a lower oxygen mass transfer coefficient and reduced substrate leakage. The high performance of the SPEEK nanocomposite was obtained using 7.5% SiO_2 in the polymer matrix [355]. For improving SPEEK, Shabani et al. [356] suggested adding GO nanosheets to the SPEEK mixture during synthesis. They claimed that oxygen crossover of the nanocomposite membrane was about 50% lower than that of Nafion 117 while its conductivity was 25% higher.

Contrary to the favorable results obtained using TiO_2 and SiO_2 NPs in nanocomposite membranes, using $Fe3O_4$ NPs in a polyethersulfone (PES) matrix led to lower performance than Nafion, that is, lower conductivity and higher oxygen crossover [292]. Moreover, increasing the NPs content in the nanocomposite resulted in reduced water uptake and increased oxygen transfer coefficient owing to the lack of sulfonated NPs and increased void space in the matrix, respectively. Therefore, the adverse effects associated with using nanocomposite membranes in MFCs should also be considered by future studies. Ceramic membranes are also used in MFCs owing to their low cost. Ahilan et al. [357] synthesized a polysiloxane-derived ceramic nanocomposite membrane with MWCNTs and GO and compared them with Nafion as control. The maximum power density obtained using the investigated membranes was in the following order: GO-modified membrane > Nafion > MWCNTs-modified membrane. However, the fresh nanomaterial-modified ceramic membranes showed less ion conductivity and higher oxygen crossover than the Nafion [357]. Nevertheless, it should be emphasized that the ion conductivity of the modified membrane might be sustained throughout time, whereas that of Nafion's is likely to diminish due to biofouling.

In addition to membrane composition, membrane electrode assembly (MEA) is also of great importance for the performance of SCMFCs. Nanocellulosic materials, by providing 3D nanoporous structures [292], are among the most exciting substrates for fabricating MEAs. In a recent investigation, a BC-based MEA was reported to outperform a carbon cloth-based gas diffusion electrode (GDE) [292]. In this study, one side of the BC was coated with MWCNTs for ORR. In contrast, the other side was coated by the hydrophobic nano-zycosil (NZ) to moderate the highly hydrophilic nature of the BC. This strategy would assist with achieving a favorable proton transfer driven by the hydrophilic nature of the BC and mitigating substrate leakage caused by the hydrophobic nature of NZ. Interestingly, despite the existing BC membrane resistance, the cell's IR with BC MEA was lower than that of the cell harboring the carbon cloth-based GDE [292]. This phenomenon could be attributed to the high conductivity of MWCNTs, homogeneous coating, and excellent behavior of NZ as a low-cost hydrophobic agent. Also, oxygen crossover was lower in the BC MEA than in the carbon cloth-based GDE, which could be explained by the barrier properties of nanocellulosic fibers [358]. In conclusion, inexpensive and high-performance alternatives to conventional PEMs could be obtained by modifying polymeric membranes with hydrophilic and antimicrobial NPs. The modified membranes are also less prone to biofouling.

5.6.7 Role of Operating Conditions and Mediators

Various operating conditions are conducive to improving the performance of MFCs. The increased temperature can increase power density and reduce IR due to an improved conductivity [278]. However, higher temperatures also negatively affect microbial activity, membrane stability, and partial pressure of oxygen [278]. The temperature range of 30°C–45°C is conducive to maintaining growth efficiency and electrochemical activity of microbes in the MFC system [278]. Environment or room temperature is generally the operating temperature of MFCs, although it might reduce the efficiency of electrical power generation. Heidrich et al. [125] reported a minor effect of low temperature on the power density of MFCs due to the potential self-heating performance of MFC biofilms. Gonzalez-Martínez et al. [126] studied the performance of the MFC system at 25°C and 8°C, respectively. They observed a difference in bacterial communities but similar voltage at these temperatures. It indicates the potential for stable MFC operation at lower temperatures.

The pH is the other primary condition affecting the performance of MFC systems. The promotion of proton transfer from anode to cathode in MFC systems usually depends on the different pH of the anode and the cathode. However, the limited proton transfer efficiency of the PEM might lead to a decrease in anodic pH due to proton accumulation and an increase in cathodic pH due to a lack of protons. The lower anolyte pH might result in lower electron generation efficiency due to the inhibition of the growth and metabolism of microbes. The higher catholyte pH might decrease oxygen reduction efficiency [278]. Therefore, the unstable anodic and cathodic pH might reduce the power production efficiency of MFC systems. Phosphate- and borate-based buffers can effectively maintain the electrolyte pH of MFCs. HCO_3^-/H_2CO_3 buffer systems based on anolyte or catholyte recirculation are promising alternatives to phosphate-based buffers [278]. There are also studies focusing on effects of initial substrate concentration, aeration rate, and hydraulic retention time [132,133]. Promoting cell growth and metabolism is still the primary solution to enhance MFC performance by controlling MFC operating conditions. In addition, the parameter optimization for the operating conditions of MFC systems can further improve power generation efficiency.

EET in MFC systems includes direct electron transfer and mediated electron transfer. Several strains, such as *Geobacter* and *Shewanella*, can transfer electrons directly to the anode surface via intricate networks of outer membrane cytochromes [278]. However, due to the lack of electrochemically active surface proteins, more strains need redox mediators for electron transfer [278]. Electron transfer mediators acquire electrons within strain cells and transfer the electrons to the cathode surface. They can achieve continuous electron transfer via the conversion of oxidized and reduced states. A number of organic compounds are common artificial exogenous mediators for electron transfer in MFC systems. Pal and Sharma [359] studied the electricity generation of a *Pichia fermentans* strain-based MFC with the mediator of MB. They reported higher maximum power densities of both single- and dual-chamber MFCs containing MB than those without mediators. MB can also achieve a 1.22-fold increase in the steady-state voltage of a DCMFC [278]. Christwardana et al. [360] compared the electricity generation of a *Saccharomyces cerevisiae* strain-based MFC with the addition of MB and methyl red. They pointed out a more efficient electron transfer of MB due to more effective capture by yeast and higher electron collection. MB also has advantages in enhancing the electricity generation of MFC systems compared to Congo red and crystal violet [278].

Chauhan et al. [361] reported positive effects of both MB and methyl orange on the electricity generation of a DCMFC. Chen et al. [117] reported a ~400% increase in CE of a DCMFC using NR as the mediator. They pointed out that the proper concentration of NR can improve electricity transfer efficiency and promote the growth of the exoelectrogens. Moreno et al. [362] also observed the positive effect of NR on the electricity generation in continuous flow MFCs. The addition of NR can improve the maximum power density from 777.8 to 1,428.6 mW/m^3 and the maximum current density from 3,444.4 to 5,714.3 mA/m^3. Marcílio et al. [363] used methylene green as the mediator to achieve a 20% increase in the voltage of an acetate-fed MFC with a stable operation for about 6 days. In addition, metabolites of specific microbes can also act as endogenous mediators to participate in the electron transfer [278]. Ajunwa et al. [364] determined the electricity generation of the glucose-fed MFC with flavins and pyocyanin as mediators. These endogenous mediators can improve the power production efficiency of MFC systems by simplifying the electron transfer process.

5.6.8 Role of Cell Configuration

Basic MFC cell can be housed in a single chamber (Figure 5.31) or a double chamber (Figure 5.32). Single-chamber MFCs combine both electrodes, the anode and the cathode, without a PEM. They normally possess only an anodic chamber without the requisite of aeration in a cathodic chamber. The single-chambered MFC is a simple and cost-effective design and may have various shapes of compartments, which have been reviewed in various studies. Two-chambered MFC is a typical and

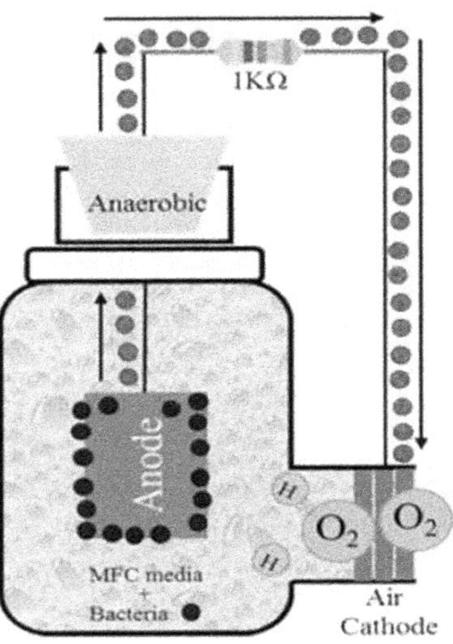

FIGURE 5.31 Single-chamber microbial FC [365].

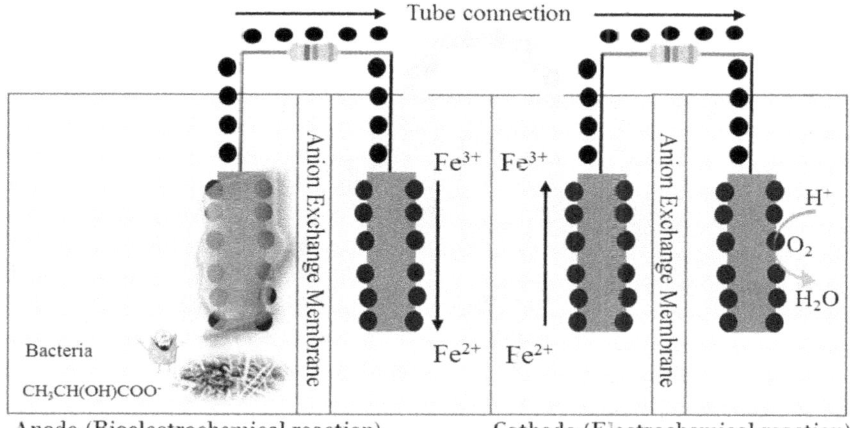

FIGURE 5.32 Schematic design of two-chamber configuration [365].

preferred MFC design, which mainly consists of anode and cathode compartment separated PEM or by a salt bridge, through which the protons move toward the cathode. The anode chamber remains free of oxygen for the occurrence of the anaerobic breakdown process, which is usually purged with nitrogen. Anode chamber contains microbes, substrate, and electrode whereas the cathode chamber contains electrode, freshwater, and oxygen supply [278,292,365] (Figure 5.32).

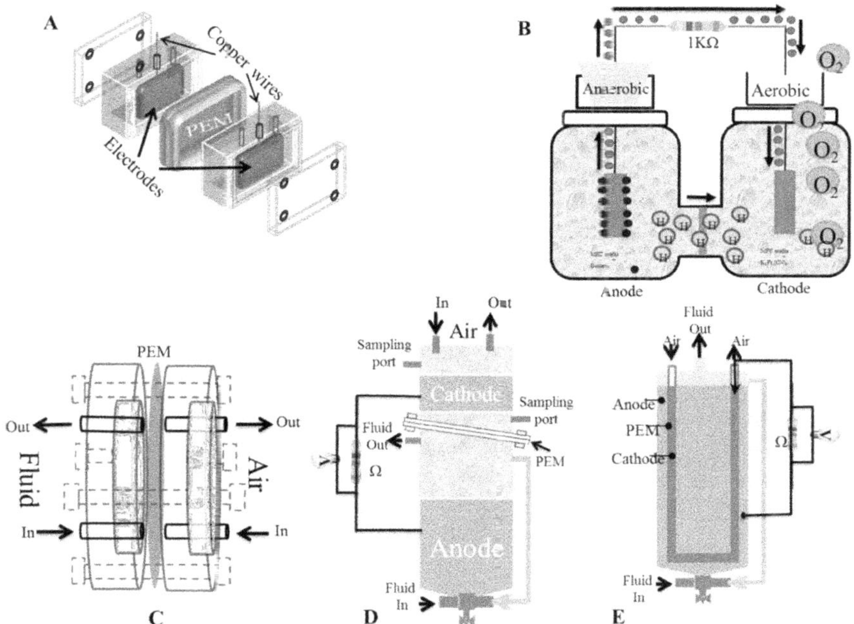

FIGURE 5.33 Schematic design of microbial FC configuration [365].

MFC configuration often depends on its application. Various types of MFC for different applications are illustrated in Figure 5.33 [365]. Figure 5.33a illustrates a cubed, mini square-shaped compact flat plate MFC (FPMFC), which resembles that of a conventional chemical FC having only a single electrode/PEM assembly, where the cathode is hot pressed to a Nafion PEM and is in contact with an anode to form an electrode/PEM assembly. The FPMFC with two non-conductive polycarbonate plates is bolted together. The PEM links the anodic and the cathodic chambers as shown in Figure 5.33a. The configuration in Figure 5.33b separates anaerobic and aerobic functions of anode and cathode, respectively. The miniature microbial FC (mini-MFC) of about 2 cm in diameter, shown in Figure 5.33c, provides a high-volume power density of 24 and 10 mW/m^2 using reticulated vitreous carbon and graphite felt electrodes, respectively, without the addition of exogenous mediators in the anolyte as reported by Ringeisen et al. [366]. They can be useful in powering self-directed sensors for long-term operations in less accessible regions. The compartments can take various practical shapes.

UMFC configuration (Figure 5.33d and e) can be suitably designed to scale up to treat large volumes of wastewater and various carbon sources. He et al. [366,367] employed two different UMFC configurations in their experiments. One generated a maximum power density of 170 mW/m^2 with 84 W internal resistance (Figure 5.33d) [368]; the other, a power density of 29.2 W/m^3 with 17.13 W internal resistance (Figure 5.33e) [367]. Since pumping fluids and recirculation are required in both configurations, it expends much greater energy than their power outputs. Therefore, the primary function of upflow configuration is wastewater treatment, not power generation. However, this configuration falls between single-chambered and double-chambered MFCs.

Power and current densities significantly decrease with the enlargement of the physical (geometrical) size of the reactor. For example, a module with a total volume of 250 L consisting of two MFC units achieved a relatively low power density of 0.47 W/m^3 [369]. This is because of the increase in the IR in the anodic, cathodic, membrane, and electrolyte components. Miniaturization of MFCs is one direction that

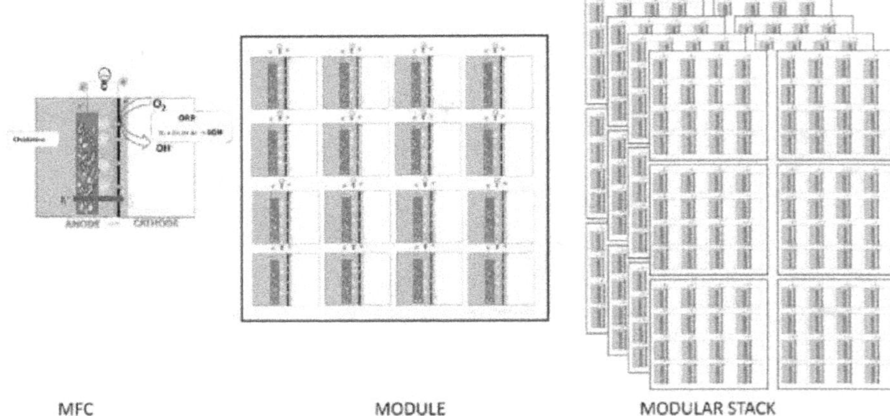

FIGURE 5.34 Stacked microbial FC [370].

allows for increased power densities and can be implemented in MFC stacks [365,370]. In order to overcome the practical challenges, the reactor should consist of modules involving multiple electrodes and/or multiple MFC units. Division into modules (parts of the whole system), hydraulic isolation of these modules, and multiplication of units are necessary for stepping-up the voltage values, when connected in series, in order to avoid short-circuiting. The modularity is here represented by the components of the module (anodes and cathodes) that can be connected electrically in parallel due to the fact that they share the same electrolyte; therefore, they form a group of multiple MFC units. We briefly examine here different types of MFC stacks that have been explored in the literature and their effects on the power production [365,370].

5.6.8.1 Stacked MFC

It is a type of configuration where FCs are stacked to form a battery of FC. This kind of construction does not affect each cell's distinct CE, but together it increases the output of the overall battery to be comparable to conventional power sources as shown in Figure 5.34. These can be either stacked in series or stacked in parallel. Both have their importance, are high in power efficiency, and can be practically utilized as a power source. Stack yields higher voltage and current by the series connection and parallel connection, respectively; accordingly, the required voltage, current, and power in electronic devices can be satisfied. MFC can be connected in series, and parallel circuits hence called stacked MFC. Designing stacked MFC efficiently is a critical issue. The type of electrode, stack direction (horizontal or vertical type), shape of the reactor, determination of connection methods, and modulation should be properly considered [370].

5.6.8.1.1 Bipolar Electrode Stack

Initial MFC reactors for stacking consist of anode chamber, a bipolar electrode, membrane, a cathode chamber, and end plates that were externally similar with PEMFC stack (Figure 5.35A). The bipolar electrode for FC stack has several advantages such as (1) minimizing resistive losses (good electrical contacts), (2) minimizing IR losses in current collectors, and (3) ease of fabrication. Shin et al. [371] used five BPs stacked MFCs in series, resulted in total voltage equal to the sum of individual MFCs and no electrical degradation in performance. The bipolar stacked MFC is composed of an H-beam shape of bipolar graphite that has two compartments at each side (anodic and cathodic compartments), a Pt/C catalyst-coated Nafion membrane at one side (positioned between each BP), and end plates. The titanium (Ti) plate functioned as both anode and cathode can also be used for MFC stack [365]. However, the bipolar stacked MFC in series has often been observed in voltage reversal [365], wherein the polarity of voltages of some MFCs among the stacked MFC is suddenly reversed (from a positive value to a negative value), while the voltages of the other MFCs still remain positive values. When the reversed MFCs operate for long term, the carbon corrosion makes the biofilm on the anode damage; therefore, the whole MFC system fails [365]. This compact design of bipolar electrode would be difficult to separate and maintain individually [365]. In addition, even if the bipolar electrode is separated from whole stacked MFC, it cannot be guaranteed that the voltage reversal phenomenon will not appear in the other bipolar stacked MFCs.

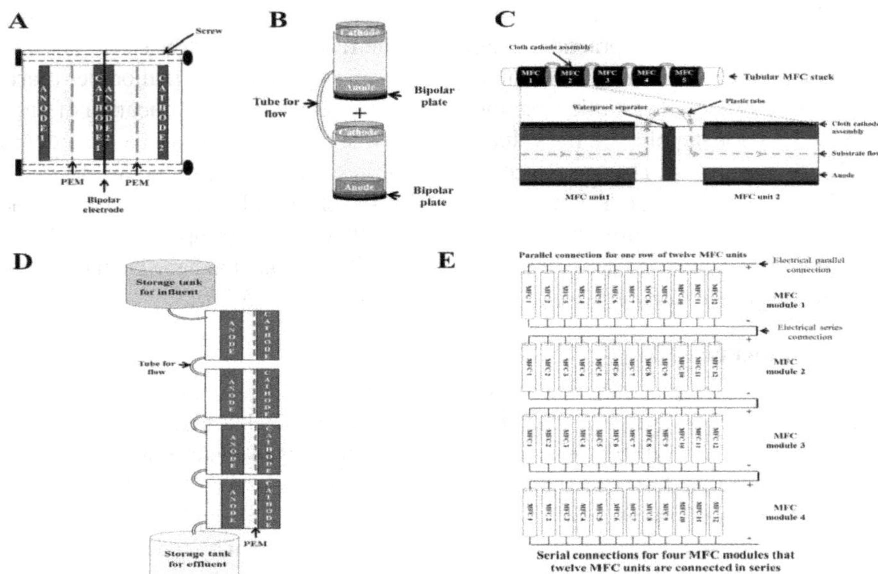

FIGURE 5.35 The schematic diagrams of stacked MFCs configuration. (a) A BP DCMFC horizontal stack; (b) a tubular BP membrane-less DCMFC vertical stack; (c) a tubular air-cathode MFC horizontal stack; (d) a cascade air-cathode MFC vertical stack; (e) the module MFC stack in combination of series and parallel connection [365].

An et al. [372,373] suggested membrane-less single-chambered MFC using a BP–electrode assembly (BEA) that can easily be separated and stackable in series (Figure 5.35B). An individual MFC unit was made of a cylindrical acrylic body, an anode (graphite felt) at the bottom, a cathode (graphite felt) at the top, and a rigid graphite plate under the anode to protect water leakage and air exposure. When two individual MFCs (MFC 1 unit and MFC 2 unit) are stacked in series, for example, the anode in MFC 1 unit with the graphite plate directly connects the cathode in MFC 2 unit without external wires. Whenever some MFCs show reversed polarity, the MFCs using BEA can easily be detached, and then other MFCs can be connected. Though the bipolar electrode distributes to reduce the IR during series stack, shortcut assembly and disassembly should be considered due to voltage reversal phenomenon. The stack using external metal wires such as copper (Cu) and Ti wire would be a better choice for easy connection of MFC units in series and in parallel rather than bipolar electrode stack.

5.6.8.1.2 Horizontal Stack & Vertical Stack

The MFC can be stacked horizontally and vertically. Many researchers have used a rectangular parallelepiped DCMFC for the horizontal stack. Initial MFC stacks containing bipolar electrode MFC stack were also arranged in the horizontal direction [365]. Individual MFC unit for horizontal stack without bipolar electrode consists of anode and cathode in the rectangular frames that have two ports (one for inlet and other for an outlet), and a CEM or PEM between two frames (anode chamber and cathode chamber) [365]. The MFC units are horizontally stacked with each other by connecting electrically metal wires to current collectors for making series or parallel circuits. The MFC units are also divided by rubber sheets or acryl sheets for not sharing anolyte and/or catholyte from other MFC units All the MFC units with the end plates and the sheets for separation of each MFC units are strongly tightened via several long screws, such called horizontal MFC stack. Aelterman et al. [374] reported a maximum power density of 308 W/m^3 for series connection and 263 W/m^3 for parallel connection based on the polarization curves using horizontal stacked MFC system consisting six individual MFC units. Recently, similar shape for pilot-scale horizontal MFC stack has been constructed and then operated. Wu et al. showed that the maximum power density of 51 W/m^3 and a COD removal efficiency of 97% were achieved using the scaled-up stacked MFC in parallel connection (total volume of 72 L) [365]. Vilajeliu-Pons et al. [375] presented the power density of between 2 and 4 W m^3 using a couple of six stacked MFCs in a combination of series and parallel connection from swine manure as a substrate for more than 6 months. The total volume of the six stacked MFC was 115 L.

The tubular air-cathode MFC design has also been used for the horizontal stack. Zhuang and Zhou reported the tubular air-cathode MFC stack horizontally [365]. The tubular MFC unit (void volume of 0.75 L) is constructed with polyvinyl chloride plastic tube serving as the frame of anode chamber and tubular membrane cathode assembly by hot-pressing carbon fiber cloth to CEM. The substrate hydraulically flows between anode chambers. They used Ti wires to serially connect each cell. This design could be adaptable for a drain pipe in the wastewater treatment processes and have continuous plug flow. Zhuang et al. [376] constructed 10-L serpentine-type MFC stack using forty tubular air-cathode MFC units and then operated it for 180 days (Figure 5.35C). The OCV of 23.0 V and maximum power density of 4.1 W/m^3 were obtained from brewery wastewater by the series stack. They concluded that each MFC unit would have different electrical performances due to different concentration of organic matter and ion conductivity faced with biofilm on the anode in continuous operation [365]. The low COD concentration near end MFC unit can reduce electrical performances or even produce negligible power compared to the initial MFC near inlet. In long-term operation, organic matter depletion near end MFC would result in extreme power loss and then may fail to operation due to inactivate microorganisms in the biofilm. Also, when multiple MFC units are connected in series, the voltage reversal could occur due to imbalance of substrate loading, causing a kinetic imbalance between MFC units or between anode and cathode [365].

The vertical cascade stacked MFC system (Figure 5.35D) can also be confronted with the voltage reversal phenomenon because of the substrate imbalances. However, the cascade stacked MFC can be used as inexpensive way because the fluid flow of electrolytes was mostly gravity driven without pumps. The horizontally stacked MFC system usually requires extra pumps for supplying substrate. It has been reported that the first self-sustainable MFC stack capable of self-maintenance using vertical cascade single-chamber MFC stacks consists of 40 identical 20-mL units (total volume of 0.8 L) [365]. Recently, the smartphone was successfully charged by scaled-up cascade MFC stack (10.5 L of whole stack volume) employing urine as substrate [365] though the MFC stack produced a low power of near 110 mW (maximum power density of ~10 W/m^3). The phone call was allowed for 1 hour 45 minutes after 3 hours of charge using the MFC stack. However, still, the methods for maintaining flow rate, reduced organic concentration near end MFC, the volume of the reactor, and height of the whole system would be considered for vertical cascade MFC stack to maintain electrical and biological performances (i.e., power generation and wastewater treatment) of MFC stack.

5.6.8.1.3 Modularized Multiple Electrodes MFC Stack

The electrolyte has not been shared from other electrolytes by rubber sheets, BP, thin inter-connecting tubes, and overflow [365] to prevent ionic cross-conduction, causing voltage reversal of stacked MFC in series connection. The ionic cross-conduction occurs when the same electrolyte (anolyte or catholyte) is shared by different MFCs in series connection [365]. These designs for separation of individual MFC reactors would require additional materials, leading to the complex structure of reactor and increase of capital costs for construction and maintenance. Modulation of MFC stack would be an efficient way to construct a stacked MFC system simply and reduce construction costs. He et al. [377] developed the stackable MFC with the anode and dual-cathode modules for simple construction, easy access, and maintenance of the electrodes. The highest power density (11.0 W/m^3) was achieved using raw domestic wastewater with the MFC module. After that, a larger MFC module (6.1 L) using four anode modules and three cathode modules were constructed and then operated, resulting in the maximum power density of 6.0 W/m^3 and COD removal of between 40% and 60% [365], but the researchers were not conducted to stack operation by electrical series and parallel connection.

A promising development direction for the modularized MFC stack is to combine electrical arrays with series connection and parallel connection of MFC units and/or MFC module. The MFC units in a module are connected in parallel first to be capable to prevent voltage reversal and increase current. Feng et al. [369] have shown that the increased high current (0.435 A) was obtained using modularized MFC stack in parallel connection of 32 carbon brush anodes. After series connection of modules in parallel-connected MFC units, the increased voltage and power can be satisfied with the required standard input voltage, and power of an electrical device depends on how many MFCs are parallelly and serially connected. Ge and He [378] developed a 200-L modularized MFC stack, which generated power of ~200 mW. The MFC system is comprised of eight MFC modules (Figure 5.35E), containing 12 tubular MFC units (2 L/each). The 12 MFC units are connected in parallel using external wires, and then eight MFC modules are connected in series (Figure 5.35E). They tried to avoid short-circuit connection between MFC modules because two modules (one set) are hydraulically connected for wastewater feeding.

5.6.9 MFC Commercialization and Future Perspectives

The commercialization of technology can be deemed successful if more people become aware of the product, and it is marketed in large amounts to a wide range of consumers. The commercialization of MFC will bring more benefits due to its function in energy production through the utilization of wastes. For instance, electricity can easily be generated at homes; energy generation with low expense can be accomplished all year long as waste materials. The xenobiotics are abundant in low-income countries like Africa; the MFC will be beneficial because of its low operating cost as compared to large foundations needed to set up energy generation plants that are unavailable. Lastly, MFC could be an alternative for the remediation process to remove hazardous substances from xenobiotics and wastes.

The performance of an MFC technology is dependent on the number of variables used to observe the system, for example, the utilized substrates, the setup, microbes present, catalyst, concentration, ideal membrane, and electrode materials. A large number of accessible reports have shown that MFC could be configured from sizes of a few milliliters to a few thousand liters. Based on these reports, the outcome is that the power generated is affected by the MFC scaling, which is a major weakness in marketing MFC. The space between anode and cathode is one of the components that can affect the power production for MFC. With bigger cell size, the size of electrodes used will also increase; however, the separation itself does not change to the same extent to avoid bulky cells, and this causes the low generation of power. Another reason that limits MFC scaling is the price of electrodes, which rationally should be inexpensive, but as the electrodes are bought instead of being produced in the industry, the price becomes higher, as well as because of the material itself. The membrane present in MFC is also created using high-cost material such as nylon. Substrates also play an important role for MFC, and depending on the type of substrate utilized, the power output will differ. As an example, the pure substrate will result in high power production, but by using wastes as substrates, the value of power generated decreased significantly. The reason for this is that microorganisms are incapable of metabolizing waste and pure sources of carbon. These are some of the limiting factors toward MFC commercialization. MFC is capable of utilizing various organic materials in producing energy.

Nevertheless, MFC still has few weaknesses that need to be addressed in order for the technology to be applied in a practical situation. The biggest challenge of an MFC is the poor density of power, which can be sorted out either by separating strong microbes that are capable of transferring electrons toward anode or by synthesizing engineered strains via DNA recombination that provides a higher transfer rate of electrons. Various consortia of bacteria are proven to be more efficient in transferring electrons than pure cultures, and more strains of bacteria can generate mediators for a more successful electron transfer. New types of mediators are recognized to enhance MFC performance. The small surface area of an MFC is also a big challenge as a limited number of microorganisms can adhere to it. Further studies reported new techniques that can improve the performance of MFC, which provide a more efficient configuration of small-scale MFC. Some of the techniques are stacked reactors, assemblies of cloth electrodes, and utilization of air cathodes. Utilization of air-cathode MFC is the most efficient method out of all techniques because it utilizes oxygen sources from air efficiently, in return removing the requirement of water aeration and employing chemical catholytes like ferricyanide that need to be regenerated. Distinct cell designs are used in evaluating the consequences of utilizing various shapes and positions for MFC improvement, and optimization of air cathode is also done for MFC utilization. Great results have been achieved through these attempts in which a power output of more than 1,000 W/m^3 was generated from an efficient small-scale MFC (~20 mL volume of anode). Nevertheless, producing a large-size

MFC that can provide increased energy generation and performance stability is still a difficult task. It appears from the literature that the highest value of power density 20 W/m^3 can be achieved by using MFC of 500 mL in volume. The final downside of the MFC system is in regard to wastewater treatment and MFC scaling up. These are crucial issues needing solutions as scaling up can result in MFC application for a large-scale setup that can help in thoroughly improving MFC performance, especially in the case of wastewater treatment that is abundant.

Various research papers and studies suggest that MFC technologies have been demonstrated as ecologically sustainable techniques to generate power while also removing pollutants from different forms of wastewater. However, there are a few significant issues, including economical aspects, development of designs that offer maximum output, and so forth, with MFC technologies. Thus, they have never been considered a significant competitor in the field of renewable energy or the wastewater treatment sector. Nevertheless, MFCs are probably sufficient in terms of net energy production, rather than utilizing energy obtained from the oxidation of organic materials using wastes and inorganic carbon under certain conditions. MFC systems have the benefit of being able to transform chemical energy directly into electrical energy through biological processes, allowing them to physiologically adapt to the treatment of a wide range of chemical substrates at different concentrations. MFC technology can be used by research groups to facilitate a better understanding of electrochemical, biochemical, microbial, and material surface responses under regulated conditions, which has had a favorable influence. The main focuses of their study are comprised of how materials, chemical compounds, and feedstock substrates, among other things, might affect them [379]. This strategy allows us to better understand the potential issues affecting MFCs' larger-scale applications.

To achieve commercial success of MFC technology, measures should be taken to minimize the high cost of operation and ameliorate poor power production [380]. Concerning configuration and treatment capabilities, MFCs have a capital cost that is 30 times greater than a standard activated sludge treatment system utilized for household wastewater [381]. The utilization of expensive electrode materials, such as catalyst, current collector, and separator materials, leads to the high-level capital cost of MFCs. In MFCs, bacteria can transfer electrons to the anode and protons into the solution, which causes a negative anode potential (approximately 0.2 V). The most promising cathode oxidants are oxygen and air, and both have a maximum theoretical potential of 0.805 V. Pt-catalyzed cathodes facilitate a maximum achievable potential of +0.3 V in MFC. The highest limits of power density in the case of MFCs were anticipated to be between 17 and 19 W/m^2, assuming negligible IR or first-order kinetic characteristics of microorganisms in biofilms.

The poor power density is due to a combination of factors, including solution conditions, high IR, substrate degradability, and the dynamics of biofilm. To improve currents and voltage output, the stacked MFCs could be used in parallel or series; however, parallel connections may enhance currents and power density, and CE is also substantially higher in parallel than in series. The stacked MFCs in a direct series can increase voltages, which is difficult to achieve with a chemical FC due to the impact of the external circuit on the microbial consortia. The voltage reversal in stacked MFCs and its impact on the enhancement of the performance in a direct series can be a topic of investigation for future study. The larger voltage could be accomplished by connecting arrays of MFCs to the charge capacitors in parallel and further discharging them in series. However, this technique can raise the cost and power consumption [381].

However, this technique can raise the cost and power consumption. Although MFCs are considered a new trend, further studies should concentrate on minimizing the limiting factors and comprehending the metabolic process involved to select high electrochemically active microorganisms. This entails creating a thick conductive biofilm and fine-tuning the operational parameters. The applicability of MFCs in wastewater treatment is directly determined by the design and architecture of the reactor [382,383]. Investigations on energy utilization and storage are required to develop power utilization and collecting system that would speed up the commercial deployment of MFCs as well as reduce the elimination of hazardous substances. Owing to remarkable advancements in electrode materials and inoculation patterns, MFC stacks can withstand the critical challenges of ionic short circuits and voltage reversal, which have been significant hurdles to practical use [384,385].

Further research is needed to address the following issues: enhanced power generation, regulated microbial performance in the unit, creation of new full-scale MFC models to reduce possible losses for optimal performance, and reduction of costs. Finally, the integration of MFCs with other wastewater treatment technologies can enhance treatment efficiency and hence reduce the overall power consumption to a great extent.

REFERENCES

1. Vezzoli, C., Ceschin, F., Osanjo, L., M'Rithaa, M.K., Moalosi, R., Nakazibwe, V., Diehl, J.C., *Designing Sustainable Energy for All: Sustainable Product-Service System Design Applied to Distributed Renewable Energy*. Springer, Berlin/Heidelberg, Germany, 2018.
2. Agüero, J. R., Takayesu, E., Novosel, D., Masiello, R., Grid modernization: Challenges and opportunities. *Electr J* 2017;30:1–6.
3. Salihu, T. Y., Akorede, M. F., Abdulkarim, A., Abdullateef, A. I., Off-grid photovoltaic microgrid development for rural electrification in Nigeria. *Electr J* 2020;33:106765.
4. Khodayar, M. E., Rural electrification and expansion planning of off-grid microgrids. *Electr J* 2017;30:68–74.
5. Moka, S., Pande, M., Rani, M., Gakhar, R., Sharma, M., Rani, J., Bhaskarwar, A. N., Alternative fuels: An overview of current trends and scope for future. *Renew Sustain Energy Rev* 2014, 32, 697–712.
6. Michaelides, E. E. S., *Alternative Energy Sources; Green Energy and Technology*. Springer, Berlin, Heidelberg, Germany, 2012.
7. Markowitz, M., *Hydrogen: An Energy Powerhouse with Unlimited Potential*. Open Access Government, 21 February 2019. Available online: https://www.openaccessgovernment.org/hydrogen-energypowerhouse/ 59506/#comment-167675 (accessed on 6 May 2020).
8. Squadrito, G., Andaloro, L., Ferraro, M., Antonucci, V., Hydrogen fuel cell technology. In A. Iulianelli and A. Basile (Eds.), *Advances in Hydrogen Production, Storage and Distribution*. Elsevier, Amsterdam, the Netherlands, 2014
9. Alaswad, A., Palumbo, A., Dassisti, M., Olabi, A. G., Fuel cell technologies, applications, and state of the art: A reference guide. In Saleem Hashmi (Ed.), *Reference Module in Materials Science and Materials Engineering*. Elsevier, Amsterdam, the Netherlands, 2016.

10. Theraja, B. L., Theraja, A. K., *A Textbook of Electrical Technology, Volume 2*. S. Chand, New Delhi, India, 2003.
11. Dicks, A. L., Molten carbonate fuel cells. *Curr Opin Solid State Mater Sci* 2004;8(5):379–383. DOI: 10.1016/j.cossms.2004.12.005
12. Jiang, S. P., Li, Q., Molten carbonate fuel cells. In *Introduction to Fuel Cells*. Singapore, Springer, 2022, pp. 673–693. DOI: 10.1007/978-981-10-7626-8_15
13. Lee, C.-W., Lee, M., Kang, M. G., Han, J., Yoon, S. P., Nam, S. W., Lee, K. B., Ham, H. C., Fabrication and operation characteristics of electrolyte impregnated matrix and cathode for molten carbonate fuel cells. *Int J Precis Eng Manuf-Green Technol* 2018;5(2):279–286. DOI: 10.1007/s40684-018-0029-2
14. Lee, K. -J., Kim, T. -K., Koomson, S., Lee, C. -G., Performance of molten carbonate fuel cell with Li-Na and Li-K carbonate electrolyte at extremely high-temperature condition. *Korean J Chem Eng* 2018;35(10):2010–2014. DOI: 10.1007/s11814-018-0098-5
15. Antolini, E., The stability of $LiAlO_2$ powders and electrolyte matrices in molten carbonate fuel cell environment. *Ceram Int* 2013;39(4):3463–3478. doi:10.1016/j.ceramint.2012.10.236
16. Contreras, R. R., Almarza, J., Rincón, L., Molten carbonate fuel cells: A technological perspective and review. *Energy Sources A: Recovery Util Environ Eff* 2021. DOI: 10.1080/15567036.2021.2013346
17. Sharaf, O. Z., Orhan, M. F., An overview of fuel cell technology: Fundamentals and applications. *Renew Sustain Energy Rev* 2014;32:810–853. DOI: 10.1016/j.rser.2014.01.012
18. Sundmacher, K., Kienle, A., Pesch, H. J., Berndt, J. F., Huppmann, G., *Molten Carbonate Fuel Cells*. Wiley-VCH, Weinheim, Germany, 2007.
19. Monforti Ferrario, A., Santoni, F., Della Pietra, M., Rossi, M., Piacente, N., Comodi, G., Simonetti, L., A system integration analysis of a molten carbonate electrolysis cell as an off-gas recovery system in a steam-reforming process of an oil refinery. *Front Energy Res* 2021;9:655915. DOI: 10.3389/fenrg.2021.655915
20. Chiodo, V., Zafarana, G., Maisano, S., Freni, S., Galvagno, A., Urbani, F., Molten carbonate fuel cell system fed with biofuels for electricity production. *Int J Hydrogen Energy* 2006;41(41):18815–18821. DOI: 10.1016/j.ijhydene.2016.05.119
21. Seo, H. -K., Park, W. S., Lim, H. C., The efficiencies of internal reforming molten carbonate fuel cell fueled by natural gas and synthetic natural gas from coal. *J Electrochem Energy Convers Storage* 2016;13(1):011005. DOI: 10.1115/1.4033255
22. McInerney, J., Ghezel-Ayagh, H. Sanderson, R., Hunt, J., Operation of carbonate fuel cell (MCFC) using syngas. In *Proceedings of the ASME 2011 9th International Conference on Fuel Cell Science, Engineering and Technology*, 2011, Vol. 54693, pp. 549–558. DOI: 10.1115/fuelcell2011-54959
23. Bettenhausen, C. A., A fuel cell that captures carbon. *C&EN Global Enterprise* 2021;99(11):18–19. DOI: 10.1021/cen-09911feature2
24. Audasso, E., Barelli, L., Bidini, G., Bosio, B., Discepoli, G., Molten carbonate fuel cell performance analysis varying cathode operating conditions for carbon capture applications. *J Power Sources* 2017;348:118–129. DOI: 10.1016/j.jpowsour.2017.02.081
25. Milewski, J., Lewandowski, J., Separating CO_2 from flue gases using a molten carbonate fuel cell. *IERI Procedia* 2012;1:23237. DOI: 10.1016/j.ieri.2012.06.036
26. Ansarinasab, H., Mehrpooya, M., Investigation of a combined molten carbonate fuel cell, gas turbine and Stirling engine combined cooling heating and power (CCHP) process by exergy cost sensitivity analysis. *Energy Convers Manag* 2018;165:291–303. DOI: 10.1016/j.enconman.2018.03.067
27. Jienkulsawad, P., Patcharavorachot, Y., Chen, Y. -S., Arpornwichanop, A., Energy and exergy analyses of a hybrid system containing solid oxide and molten carbonate fuel cells, a gas turbine, and a compressed air energy storage unit. *Int J Hydrogen Energy* 2021;46(70):34883–34895. DOI: 10.1016/j.ijhydene.2021.08.038

28. Mehrpooya, M., Khodayari, R., Ali-Moosavian, S. M., Dadak, A., Optimal design of molten carbonate fuel cell combined cycle power plant and thermo-photovoltaic system. *Energy Convers Manag* 2020;221:113177. DOI: 10.1016/j.enconman.2020.113177
29. Mehrpooya, M., Raeesi, M., Pourfayaz, F., Delpisheh, M., Investigation of a hybrid solar thermochemical water-splitting hydrogen production cycle and coal-fueled molten carbonate fuel cell power plant. *Sustain Energy Technol Assess* 2021;47:101458. DOI: 10.1016/j.seta.2021.101458
30. Mei, B., Qin, Y., Taghavi, M., Thermodynamic performance of a new hybrid system based on concentrating solar system, molten carbonate fuel cell and organic Rankine cycle with CO_2 capturing analysis. *Process Saf Environ Prot* 2021;146:531–551. DOI: 10.1016/j.psep.2020.12.001
31. Hu, L., Lindbergh, G., Lagergren, C., Performance and durability of the molten carbonate electrolysis cell and the reversible molten carbonate fuel cell. *J Phys Chem C* 2016;120(25):13427–13433. DOI: 10.1021/acs. jpcc.6b04417
32. Pachauri, R. K., Chauhan, Y. K., A study, analysis and power management schemes for fuel cells. *Renew Sustain Energy Rev* 2015;43:1301–1319. DOI: 10.1016/j.rser.2014.11.098
33. Kawase, M., Manufacturing method for tubular molten carbonate fuel cells and basic cell performance. *J Power Sources* 2015;285:260–265. DOI: 10.1016/j.jpowsour.2015.03.117
34. Kawase, M., Durability and robustness of tubular molten carbonate fuel cells. *Journal of Power Sources* 2017;371:106–111. DOI: 10.1016/j.jpowsour.2017.10.024
35. Spinelli, M., Campanari, S., Consonni, S., Romano, M. C., Kreutz, T., Ghezel-Ayagh, H., Jolly, S., Molten carbonate fuel cells for retrofitting postcombustion CO_2 capture in coal and natural gas power plants. *J Electrochem Energy Convers Storage* 2018;15(3):031001. DOI: 10.1115/1.4038601
36. Sugiura, K., Takei, K., Tanimoto, K., Miyazaki, Y., The carbon dioxide concentrator by using MCFC. *J Power Sources* 2003;113(1–2):218–227. DOI: 10.1016/s0378-7753(03)00084-3
37. Campanari, S., Chiesa, P., Manzolini, G., Bedogni, S., Economic analysis of CO_2 capture from natural gas combined cycles using molten carbonate fuel cells. *Appl Energy* 2014;130:562–573. DOI: 10.1016/j.apenergy.2014.04.011
38. Spinelli, M., Di Bona, D., Gatti, M., Martelli, E., Viganò, F., Consonni, S., Assessing the potential of molten carbonate fuel cell-based schemes for carbon capture in natural gas-red combined cycle power plants. *J Power Sources* 2020;448:227223. DOI: 10.1016/j.jpowsour.2019.227223
39. Frangini, S., Masi, A., Molten carbonates for advanced and sustainable energy applications: Part I. Revisiting molten carbonate properties from a sustainable viewpoint. *Int J Hydrogen Energy* 2016;41(41):18739–18746. DOI: 10.1016/j.ijhydene.2015.12.073
40. Frangini, S., Masi, A., Molten carbonates for advanced and sustainable energy applications: Part II. Review of recent literature. *Int J Hydrogen Energy* 2016;41(42):18971–18994. DOI: 10.1016/j.ijhydene.2016.08.076
41. Wong-Ng, W., McMurdie, H. F., Paretzkin, B., Hubbard, C. R., Dragoo, A. L., Stewart, J. M., Standard X-ray diffraction powder patterns of fifteen ceramic phases. *Powder Diffr* 1987;2(2):106–117. DOI: 10.1017/ S0885715600012495
42. Barelli, L., Bidini, G., Cinti, G., Milewski, J., High temperature electrolysis using molten carbonate electrolyzer. *Int J Hydrogen Energy* 2020;46(28):14922–14931. DOI: 10.1016/j.ijhydene.2020.07.220
43. Pérez-Trujillo, J. P., Elizalde-Blancas, F., McPhail, S. J., Della Pietra, M., Bosio, B., Preliminary theoretical and experimental analysis of a molten carbonate fuel cell operating in reversible mode. *Appl Energy* 2020;263:114630. DOI: 10.1016/j.apenergy.2020.114630

44. Hu, L., Lindbergh, G., Lagergren, C., Operating the nickel electrode with hydrogen-lean gases in the molten carbonate electrolysis cell (MCEC). *Int J Hydrogen Energy* 2016;41(41):18692–18698. DOI: 10.1016/j. ijhydene.2016.06.037
45. Cassir, M., Meléndez-Ceballos, A., Ringuedé, A., Lair, V., Molten carbonate fuel cells. In *Compendium of Hydrogen Energy.* Woodhead Publishing, 2016, pp. 71–87. DOI: 10.1016/b978-1-78242-363-8.00003-7
46. Kulkarni, A., Giddey, S., Materials issues and recent developments in molten carbonate fuel cells. *J Solid State Electrochem* 2012;16(10):3123–3146. DOI: 10.1007/s10008-012-1771-y
47. Zhu, X. -J., Huang, B., Molten carbonate fuel cells. *Electrochem Technol Energy Storage Conver* 2012;1:729–75. DOI: 10.1002/9783527639496.ch16
48. Janardhanan, V. M., Heuveline, V., Deutschmann, O., Three-phase boundary length in solid-oxide fuel cells: A mathematical model. *J Power Sources* 2008;178(1):368–372. DOI: 10.1016/j.jpowsour.2007.11.083
49. Hilmi, A., Yuh, C.-Y., Farooque, M., Carbonate fuel cell anode: A review. *ECS Transactions* 2014;61(1):245–253. DOI: 10.1149/06101.0245ecst
50. Özkan, G., Özkan, G., İyidir, U. C., Synthesis and characterization of molten carbonate fuel cell anode materials. *Energy Sources A: Recovery Util Environ Eff* 2015;37(22):2487–2495. DOI: 10.1080/15567036.2011.627415
51. Coppola, R., Rodriguez-Carvajal, J., Crystallographic characterization of a CeO_2 protective layer deposited on a Ni–Cr anode for molten carbonate fuel cell. *J Surf Investig: X-Ray Synchrotron Neutron Tech* 2020;14(1):S22–S24. DOI: 10.1134/S1027451020070095
52. Accardo, G., Frattini, D., Moreno, A., Yoon, S. P., Han, J. H., Nam, S. W., Influence of nano zirconia on NiAl anodes for molten carbonate fuel cell: Characterization, cell tests and post-analysis. *J Power Sources* 2017;338:74–81. DOI: 10.1016/j.jpowsour.2016.11.029
53. Frattini, D., Accardo, G., Moreno, A., Yoon, S. P., Han, J. H., Nam, S. W., A novel Nickel-Aluminum alloy with Titanium for improved anode performance and properties in molten carbonate fuel cells. *J Power Sources* 2017;352:90–98. DOI: 10.1016/j.jpowsour.2017.03.11
54. Belhomme, C., Gourba, E., Cassir, M., Tessier, C., Chemical and electrochemical behaviour of Ni–Ti in the cathodic conditions used in molten carbonate fuel cells. *J Electroanal Chem* 2001;503(1–2):69–77. DOI: 10.1016/s00220728(01)00375-8
55. Zheng, K., Du, K., Cheng, X., Jiang, R., Deng, B., Zhu, H., Wang, D., Nickel-iron-copper alloy as inert anode for ternary molten carbonate electrolysis at 650°C. *J Electrochem Soc* 2018;165(11):E572–E577. DOI: 10.1149/2.1211811jes
56. Fang, B., Chen, H., A new candidate material for molten carbonate fuel cell cathodes. *J Electroanal Chem* 2001;501(1–2):128–231. DOI: 10.1016/s0022-0728(01)00379-5
57. Başarir, E., Özkan, G., Özkan, G., Synthesis of nickel boride and investigation of availability as an additive in the molten carbonate fuel cell anode material. *Int J Energy Res* 2022;46(8):10088–10098. DOI: 10.1002/er.6626
58. Zeng, C. L., Zhang, T., Guo, P. Y., Wu, W. T., The corrosion behavior of two-phase Ni–Dy alloys in a eutectic $(Li, K)_2CO_3$ mixture at 650°C. *Corros Sci* 2004;46(9):2183–2189. DOI: 10.1016/j.corsci.2004.01.016
59. Czelej, K., Cwieka, K., Colmenares, J. C., Kurzydlowski, K. J., Catalytic activity of NiO cathode in molten carbonate fuel cells. *Appl Catal B: Environ* 2018;222:73–75. DOI: 10.1016/j.apcatb.2017.10.003
60. Kang, M. G., Song, S. A., Jang, S. C., Oh, I. H., Han, J., Yoon, S. P., Kim, S.H., Oh, S. G., Fabrication of electrolyte-impregnated cathode by dry casting method for molten carbonate fuel cells. *Korean J Chem Eng* 2012;29:876–885. DOI: 10.1007/s11814-011-0263-6

61. Lee, H., Hong, M., Bae, S., Lee, H., Park, E., Kim, K., A novel approach to preparing nano-size Co_3O_4-coated Ni powder by the Pechini method for MCFC cathodes. *J Mater Chem* 2003;13(10):2626–2632. DOI: 10.1039/ B303980C.
62. Ito, Y., Tsuru, K., Oishi, A., Miyazaki, Y., Kodama, T., Dissolution behavior of copper and nickel oxides in molten$Li_2CO_3/Na_2CO_3/K_2CO_3$. *J Power Sources* 1988;23(4):357–364. DOI: 10.1016/0378-7753(88)80080-6
63. Paoletti, C., Carewska, M., Presti, R. L., Mc Phail, S., Simonetti, E., Zaza, F., Performance analysis of new cathode materials for molten carbonate fuel cells. *J Power Sources* 2009;193(1):292–297. DOI: 10.1016/j. jpowsour.2008.12.094
64. Kim, M., Youn, J., Lim, J., Eom, K., Cho, E., Kwon, H., Corrosion-resistant coating for cathode current collector and wet-seal area of molten carbonate fuel cells. *Int J Hydrogen Energy* 2018;43(24):11363–11371. DOI: 10.1016/j.ijhydene.2018.02.196
65. Mustafa, K., Anwar, M., Akmal-Rana, M., Khan, Z. S., Development of cobalt doped lithiated NiO nano-composites and hot corrosion testing of $LiAlO_2$ matrices for molten carbonate fuel cell (MCFC) applications. *Mater Chem Phys* 2015;164:198–205. DOI: 10.1016/j.matchemphys.2015.08.44
66. Plomp, L., Veldhuis, J. B. J., Sitters, E. F., Van der Molen, S. B., Improvement of molten-carbonate fuel cell (MCFC) lifetime. *J Power Sources* 1992;39(3):369–373. DOI: 10.1016/0378-7753(92)80008-y
67. Ganesan, P., Colon, H., Haran, B., Popov, B. N., Performance of $La_{0.8}Sr_{0.2}CoO_3$ coated NiO as cathodes for molten carbonate fuel cells. *J Power Sources* 2003;115(1):12–18. DOI: 10.1016/s0378-7753(02)00621-3
68. Ibrahim, S. H., Wejrzanowski, T., Sobczak, P., Cwieka, K., Lysik, A., Skibinski, J., Oliver, G. J., Insight into cathode microstructure effect on the performance of molten carbonate fuel cell. *J Power Sources* 2021;491:229562. DOI: 10.1016/j. jpowsour.2021.229562
69. Wejrzanowski, T., Cwieka, K., Skibinski, J., Brynk, T., Ibrahim, S. H., Milewski, J., Xing, W., Metallic foam supported electrodes for molten carbonate fuel cells. *Mater Design* 2020;193:108864. DOI: 10.1016/j.matdes.2020.108864
70. Huang, B., Chen, G., Li, F., Yu, Q. C., Hu, K. A., Study of NiO cathode modified by rare earth oxide additive for MCFC by electrochemical impedance spectroscopy. *Electrochim Acta* 2004;49(28):5055–5068. DOI: 10.1016/j. electacta.2004.04.019
71. Okawa, H., Lee, J. H., Hotta, T., Ohara, S., Takahashi, S., Shibahashi, T., Yamamasu, Y., Performance of $NiO/MgFe_2O_4$ composite cathode for a molten carbonate fuel cell. *J Power Sources* 2004;131(1–2):251–255. DOI: 10.1016/j.jpowsour.2003.11.092
72. Escudero, M. J., Rodrigo, T., Mendoza, L., Cassir, M., Daza, L., Porous nickel MCFC cathode coated by potentiostatically deposited cobalt oxide: I. A structural and morphological study. *J Power Sources* 2005;140(1):81–87. DOI: 10.1016/j.jpowsour.2004.08.009
73. Kim, D., Li, H., Lee, M. H., Kim, K., Lim, S. N., Woo, J. Y., Han, H., Song, S. A., Improved performance and stability of low-temperature, molten-carbonate fuel cells using a cathode with atomic layer deposited ZrO_2. *J Power Sources* 2021;484:229254. DOI: 10.1016/j.jpowsour.2020.229254
74. Meléndez-Ceballos, A., Fernández-Valverde, S. M., Albin, V., Lair, V., Chávez-Carvayar, J. Á., Ringuedé, A., Cassir, M., Investigation on niobium oxide coatings for protecting and enhancing the performance of Ni cathode in the MCFC. *Int J Hydrogen Energy* 2016;41(41):18721–18731. DOI: 10.1016/j.ijhydene.2016.04.045
75. Cassir, M., Meléndez-Ceballos, A., Chavanne, M. H., Dallel, D., Ringuedé, A., ALD-processed oxides for high-temperaturefuel cells. In *Atomic Layer Deposition in Energy Conversion Applications* 2017, pp. 209–221. DOI: 10.1002/9783527694822.ch7
76. Tanimoto, K., Kojima, T., Yanagida, M., Nomura, K., Miyazaki, Y., Optimization of the electrolyte composition in a $(Li_{0.52}Na_{0.48})_{2-2x}AE_xCO_3$ (AE = Ca and Ba) molten carbonate fuel cell. *J Power Sources* 2004;131(1–2):256–260. DOI: 10.1016/j.jpowsour.2003.11.085

77. Kim, S. D., Hyun, S. H., Shin, M. Y., Lim, T. H., Hong, S. A., Lim, H. C., Phase and microstructure stabilities of $LiAlO_2$ in molten Li/Na carbonate for molten carbonate fuel cells. *J Power Sources* 2005;143(1–2):24–29. DOI: 10.1016/j.jpowsour.2004.12.008
78. Ottochian, A., Ricca, C., Labat, F., Adamo, C., Molecular dynamics simulations of a lithium/sodium carbonate mixture. *J Mol Model* 2016;22(3):61. DOI: 10.1007/s00894-016-2921-4
79. Terada, S., Nagashima, I., Higaki, K., Ito, Y., Stability of $LiAlO_2$ as electrolyte matrix for molten carbonate fuel cells. *J Power Sources* 1998;75(2):223–229. DOI: 10.1016/s0378-7753(98)00115-3
80. Arendt, R. H., Alternate matrix material for molten carbonate fuel cell electrolyte structures. *J Electrochem Soc* 1982;129(5):979–983. DOI: 10.1149/1.2124076
81. Isupov, V. P., Borodulina, I. A., Eremina, N. V., Bulina, N. V., Mechanically stimulated thermal synthesis of lithium aluminates. *Mater Today: Proc* 2019;12:44–47. DOI: 10.1016/j.matpr.2019.03.016
82. Tanimoto, K., Yanagida, M., Kojima, T., Tamiya, Y., Matsumoto, H., Miyazaki, Y., Long-term operation of the 100-cm^2 single cell of MCFC. In *Proceedings of the Fourth International Symposium on Carbonate Fuel Cell Technology*, 1997, pp. 79–84. DOI: 10.2172/460255
83. Baron, R., Wejrzanowski, T., Milewski, J., Szabłowski, Ł., Szczęśniak, A., Fung, K. Z., Manufacturing of γ-$LiAlO_2$ matrix for molten carbonate fuel cell by high-energy milling. *Int J Hydrogen Energy* 2018;43(13):6696–6700. DOI: 10.1016/j.ijhydene.2018.02.085
84. Gao, L., Selman, J. R., Nash, P., Wetting of porous α-$LiAlO_2$ by molten carbonate. *J Electrochem Soc* 2018;165(5):F324–F333. DOI: 10.1149/2.0771805jes
85. Hu, Q., Ma, S., Fang, L., Xia, Y., Lei, L., Kan, W. H., Sun, G., Peng, S., Coupling behavior between lattice dynamics and Li self-diffusion in layered α-$LiAlO_2$ ceramic. *Ceram Int* 2021;47(10):14587–14593. DOI: 10.1016/j. ceramint.2021.02.040
86. Breeze, P., The molten carbonate fuel cell. *Fuel Cells* 2017:53–62. DOI: 10.1016/B978-0-08-101039-6.00006-6
87. Kojima, T., Miyazaki, Y., Nomura, K., Tanimoto, K., Physical properties of molten Li_2CO_3-Na_2CO_3 (52:48 mol %) and Li_2CO_3-K_2CO_3 (62:38 mol%) containing additives. *J Electrochem Soci* 2013;160(10):H733–H741. DOI: 10.1149/2.073410jes
88. Sergeev, D., Yazhenskikh, E., Kobertz, D., Müller, M., Vaporization behavior of Na_2CO_3 and K_2CO_3. *Calphad* 2019;65:42–49. DOI: 10.1016/j.calphad.2019.02.004
89. Mohsin, I., Economic and environmental assessment of an integrated carbon capture and utilization pathway [M.S. thesis]. University of Calgary, Calgary, Canada, 2021.
90. Gatti, M., Martelli, E., Di Bona, D., Gabba, M., Scaccabarozzi, R., Spinelli, M., Viganò, F., Consonni, S., Preliminary performance and cost evaluation of four alternative technologies for post-combustion CO_2 capture in natural gas-fired power plants. *Energies* 2020;13:543. DOI: 10.3390/en13030543
91. Campanari, S., Chiesa, P., Manzolini, G., Giannotti, A., Federici, F., Bedont, P., Parodi, F., Application of MCFCs for active CO_2 capture within natural gas combined cycles. *Energy Procedia* 2011;4:1235–1242. DOI: 10.1016/j.egypro.2011.01.179
92. Ferguson, S., Tarrant, A., Molten carbonate fuel cells for 90% post combustion CO_2 capture from a new build CCGT. *Front Energy Res* 2021;9:668431. DOI: 10.3389/fenrg.2021.668431
93. Barckholtz, T. A., Taylor, K. M., Narayanan, S., Jolly, S., Ghezel-Ayagh, H., Molten carbonate fuel cells for simultaneous CO_2 capture, power generation, and H_2 generation. *Appl Energy* 2022;313:118553. DOI: 10.1016/j.apenergy.2022.118553
94. Shikhar, U., Hemmes, K., Woudstra, T., Exploring the possibility of using molten carbonate fuel cell for the flexible coproduction of hydrogen and power, *Front Energy Res* 2021;9:656490. DOI: 10.3389/fenrg.2021.656490

95. Mohsin, I., Al-Attas, T. A., Sumon, K. Z., Bergerson, J., McCoy, S., Kibria, M. G., Economic and environmental assessment of integrated carbon capture and utilization. *Cell Rep Phys Sci* 2020;1(7):100104. DOI: 10.1016/j.xcrp.2020.100104
96. Rossi, F., Nicolini, A., Palombo, M., Castellani, B., Morini, E., Filipponi, M., An innovative configuration for CO_2 capture by high temperature fuel cells. Sustainability 2014;6(10):6687–6695. DOI: 10.3390/su6106687
97. Kearns, D., Liu, H., Consoli, C., *Technology Readiness and Costs of CCS*, a report by Global CCS Institute, Washington, DC, March 2021.
98. Mirza, N., Kearns, D., *State of the Art: CCS Technologies*, Global CCS Institute report, Washington, DC, December 2022.
99. Martin-Roberts, E., Scott, V., Flude, S., Johnson, G., Haszeldine, R. S., Gilfillan, S., Carbon capture and storage at the end of a lost decade. *One Earth* 2021;4(11):1569–1584.
100. Shah, Y. T., *CO_2 Capture, Utilization, Sequestration Strategies*. CRC Press, Taylor and Francis Co., New York, 2022.
101. Wang, B., *Molten Carbonate High Temperature Fuel Cells Getting to Scale*, a website report on "Next Big Future", September, 2022.
102. Chen, S., Zhou, N., Wu, M., Chen, S., Xiang, W., Integration of molten carbonate fuel cell and chemical looping air separation for high-efficient power generation and CO_2 capture. *Energy* 2022;254:124184. DOI: 10.1016/j.energy.2022.124184
103. Evans, A, Xing, W., Norby, T., Electromotive force (EMF) determination of transport numbers for native and foreign ions in molten alkali metal carbonates. *J Electrochem Soc* 2015;162(10):1135–1143.
104. Weaver, D., Winnick, J., Sulfation of the molten carbonate fuel-cell anode. *J Electrochem Soc* 1989;136:1679–1686.
105. Weaver, R. D., Tietz, L., Cubicciotti, D., *Direct Use of Coal in a Fuel Cell: Feasibility Investigation*. SRI International Corp., Menlo Park, 1975.
106. Weaver, R., Leach, S., Nanis, L., Electrolyte management for the coal air fuel cell. In *Proceedings of the 16th Intersociety Energy Conversion Engineering Conference*, 1981, pp. 717–721.
107. Weaver, R. D., Nanis, L., Electrochemical oxidation of carbon in a molten carbonate coal-air fuel cell. *J Electrochem Soc* 1981;1981:316–333. DOI: 10.1149/198109.0316PV
108. Brouwer, J., Jabbari, F., Leal, E. M., Orr, T., Analysis of a molten carbonate fuel cell: Numerical modeling and experimental validation, *J Power Sources* 2006;158:213–224.
109. Consonni, S., Viganò, F., Martelli, E., Gatti, M., Bona, D. D., Capra, F., et al., *CCP Novel CO_2 Capture Technology Evaluation: WP1 MCFC Package*. LEAP, 2016, p. 151.
110. Rinaldi, G., McLarty, D., Brouwer, J., Lanzini, A., Santarelli, M., Study of CO_2 recovery in a carbonate fuel cell tri-generation plant. *J Power Sources* 2015;284:16–26. DOI: 10.1016/j.jpowsour.2015.02.147
111. Department for Business, Benchmarking state-of-the-art and next generation technologies, assessing the cost reduction potential and competitiveness of novel (next generation) UK carbon capture technology. In R. Ray, S. Ferguson, A. Tarrant (Eds), *Wood and Department for Business, Energy and Industrial Strategy*, 2018. Available online: https://www.gov.uk/government/publications/call-for-ccus-innovation-literature-review-benchmarking-report-and-calculator (accessed on March 1, 2019).
112. Forsyth, F., Lodge, S., Consonni, S., Bona, D. D., Gatti, M., Martelli, E., Scaccabarozzi, R., Viganò, F., Evaluation of five alternative CO_2 capture technologies with insights to inform further development. *Energy Procedia* 2017;114:2599–2610.
113. FuelcellEnergy. Publicity Material from FuelCellEnergy, 2018. Available online: http://www.fuelcellenergy.com/assets/DFC-CarbonCapture-White-Paper.pdf (accessed on June 21, 2018).
114. Fuel cell Energy. SureSource 3000 2.8 Megawatts Product Specification, 2017. Available online: https://www.fuelcellenergy.com/wp-content/uploads/2017/02/Product-Spec-SureSource-3000.pdf (accessed on April 15, 2017).

115. Fuel Cell, Molten carbonate fuel cells as means for postcombustion CO_2 capture: Retrofitting coal-fired steam plants and natural gas-fired combined cycles. In M. Spinelli, S. Campanari, M. C. Romano (Eds), *2015 Proceedings of the ASME 2015 13th International Conference on Fuel Cell Science, Engineering and Technology*. Thomas G. Kreutz of Princeton Environmental Institute, Hossein Ghezel-Ayagh, Stephen Jolly & Matthew Di Nitto of Fuel Cell Energy, Inc., 2015.
116. International Journal of Hydrogen Energy, *CO_2 Cryogenic Separation from Combined Cycles Integrated with Molten Carbonate Fuel Cells*. Paolo Chiesa, Stefano Campanari & Giampaolo Manzolini of Politecnico di Milano, Miami, 2010.
117. News and Large Stationary, Fuel cell energy and ExxonMobil to pilot MCFC carbon capture. *Fuel Cell Bull* 2016;2016(11):4–5. DOI:10.1016/s1464-2859(16)30306-6
118. Barbir, F., Fuel cell applications. In *PEM Fuel Cells*. AcademicPress, Cambridge, MA, 2013. Paperback ISBN: 9780128102398; Hardback ISBN: 9780123877109; eBook ISBN: 9780123983725.
119. Mittelsteadt, C., Norman, T., Rich, M., Willey, J., PEM electrolyzers and PEM regenerative fuel cells industrial view. In *Electrochemical Energy Storage for Renewable Sources and Grid Balancing*. Elsevier, Amsterdam, the Netherlands, 2015.
120. Lambert, T., Gilman, P., Lilienthal, P., Micropower system modeling with HOMER. *Integr Altern Sources Energy* 2006;1:379–418.
121. Akinyele, D., Olabode, E., Amole, A., Review of fuel cell technologies and applications for sustainable microgrid systems. *Inventions* 2020;5:42. DOI: 10.3390/inventions5030042
122. Barbir, F., Molter, T., Dalton, L., Efficiency and weight trade-off analysis of regenerative fuel cells as energy storage for aerospace applications. *Int J Hydrogen Energy* 2005;30:351–357.
123. Koohi-Fayegh, S., Rosen, M. A., A review of energy storage types, applications and recent developments. *J Energy Storage* 2020;27:101047.
124. Meishner, F., Sauer, D. U., Wayside energy recovery systems in DC urban railway grids. *eTransportation* 2019;1:100001
125. Chen, H., Cong, T. N., Yang, W., Tan, C., Li, Y., Ding, Y., Progress in electrical energy storage system: A critical review. *Prog Nat Sci* 2009;19:291–312.
126. Akinyele, D. O., Rayudu, R. K., Review of energy storage technologies for sustainable power networks. *Sustain Energy Technol Assess* 2014;8:74–91.
127. Sabihuddin, S., Kiprakis, A.E., Mueller, M., A numerical and graphical review of energy storage technologies. *Energies* 2015;8:172–216.
128. Jung, H. Y., Choi, J. H., The effect of a modified Nafion binder on the performance of a unitized regenerative fuel cell (URFC). *J Solid State Electrochem* 2012;16:1571–1576.
129. Alia, S., *Regenerative Fuel Cell System* (Technical report). NREL/TP-5900-76132 under CRADA number: CRD-16-635, contract no. DE-AC36-o8GO28308. National Renewable Energy Laboratory (NREL), Golden CO, 2020.
130. Wittstadt, U., Wagner, E., Jungmann, T., Membrane electrode assemblies for unitised regenerative polymer electrolyte fuel cells. *J Power Sources* 2005;145:555–562.
131. Pettersson, J., Ramsey, B., Harrison, D., A review of the latest developments in electrodes for unitised regenerative polymer electrolyte fuel cells. *J Power Sources* 2006;157:28–34.
132. Gabbasa, M., Sopian, K., Fudholi, A., Asim N., A review of unitized regenerative fuel cell stack: Material, design and research achievements. *Int J Hydrogen Energy* 2014;39:17765–17778.
133. Jung, H. Y., Park, S., Popov, B. N., Electrochemical studies of an unsupported PtIr electrocatalyst as a bifunctional oxygen electrode in a unitized regenerative fuel cell. *J Power Sources* 2009;191:357–361.

134. García, G., Roca-Ayats, M., Lillo, A., Galante, J. L., Peña, M. A., Martínez-Huerta, M. V., Catalyst support effects at the oxygen electrode of unitized regenerative fuel cells. *Catal Today* 2013;210:67–74.
135. Altmann, S., Kaz, T., Friedrich, K. A., Bifunctional electrodes for unitized regenerative fuel cells. *Electrochim Acta* 2011;56:4287–4293.
136. Mitlitsky, F., Myers, B., Weisberg, A. H., Regenerative fuel cell systems. *Energy Fuel* 1998;12:56–71.
137. Huang, S. Y., Ganesan, P., Jung, H. Y., Popov, B. N., Development of supported bifunctional oxygen electrocatalysts and corrosion-resistant gas diffusion layer for unitized regenerative fuel cell applications. *J Power Sources* 2012;198:23–29.
138. Smith W., The role of fuel cells in energy storage. *J Power Sources* 2000;86:74–83.
139. Maclay, J. D., Brouwer, J., Samuelsen, G. S., Dynamic analyses of regenerative fuel cell power for potential use in renewable residential applications. *Int J Hydrogen Energy* 2006;31:994–1009.
140. Zhang, Y., Wang, C., Wan, N., Mao, Z., Deposited RuO_2eIrO_2/Pt electrocatalyst for the regenerative fuel cell. *Int J Hydrogen Energy* 2007;32:400–404.
141. Vesborg, P. C. K., Jaramillo, T. F., Addressing the terawatt challenge: Scalability in the supply of chemical elements for renewable energy. *RSC Adv* 2012;2:7933–7947.
142. Lee, H., Kim, J., Park, J., Joe, Y., Lee, T., Performance of polypyrrole-impregnated composite electrode for unitized regenerative fuel cell. *J Power Sources* 2004;131:188–193.
143. Song, S., Zhang, H., Ma, X., Shao, Z. -G., Zhang, Y., Yi, B., Bifunctional oxygen electrode with corrosion-resistive gas diffusion layer for unitized regenerative fuel cell. *Electrochem Commun* 2006;8:399–405.
144. Gostick, J. T., Ioannidis, M. A., Fowler, M. W., Pritzker, M. D., On the role of the microporous layer in PEMFC operation. *Electrochem Commun* 2009;11:576–579.
145. Hermann, A., Chaudhuri, T., Spagnol, P., Bipolar plates for PEM fuel cells: A review. *Int J Hydrogen Energy* 2005;30:1297–1302.
146. Dihrab, S. S., Sopian, K., Alghoul, M. A., Sulaiman, M. Y., Review of the membrane and bipolar plates materials for conventional and unitized regenerative fuel cells. *Renew Sust Energy Rev* 2009;13:1663–1668.
147. U.S. Department of Energy Hydrogen and Fuel Cells Program, Record 15015: Fuel Cell System Cost e 2015. Available online: https://www.hydrogen.energy.gov/pdfs/15015_fuel_cell_ system_cost_2015.pdf (accessed on January 1, 2016).
148. Sadhasivam, T., Dhanabalan, K., Roh, S. H., Kim, T. H., Park, K. W., Jung, S., Kurkuri, M. D., Jung, H. Y., A comprehensive review on unitized regenerative fuel cells: Crucial challenges and developments, *Int J Hydrogen Energy* 2017;42(7):4415–4433. DOI:10.1016/j.ijhydene.2016.10.140
149. Di Blasi, A., D'urso, C., Baglio, V., Antonucci, V., Arico', A. S., Ornelas, R., Matteucci, F., Orozco, G., Beltran, D., Meas, Y., Arriaga, L. G., Preparation and evaluation of RuO_2eIrO_2, IrO_2ePt and $IrO_2eTa_2O_5$ catalysts for the oxygen evolution reaction in an SPE electrolyzer. *J Appl Electrochem* 2009;39:191–196.
150. Audichon, T., Mayousse, E., Morisset, S., Morais, C., Comminges, C., Napporn, T. W., Kokoh, K. B., Electroactivity of RuO_2eIrO_2 mixed nanocatalysts toward the oxygen evolution reaction in a water electrolyzer supplied by a solar profile. *Int J Hydrogen Energy* 2014;39:16785–16796.
151. Qu, L., Liu, Y., Baek, J. B., Dai, L., Nitrogen-doped graphene as efficient metal-free electrocatalyst for oxygen reduction in fuel cells. *ACS Nano* 2010;4:1321–1326.
152. Shahgaldi, S, Hamelin, J., The effect of low platinum loading on the efficiency of PEMFC's electrocatalysts supported on TiO_2eNb, and SnO_{2eNb}: An experimental comparison between active and stable conditions. *Energy Convers Manag* 2015;103:681–690.
153. Srirapu, V. K. V. P., Kumar, A., Srivastava, P., Singh, R. N., Sinha, A. S. K., Nanosized $CoWO_4$ and $NiWO_4$ as efficient oxygen-evolving electrocatalysts. *Electrochim Acta* 2016;209:75–84.

154. Guo, C., Wen, B., Liao, W., Li, Z., Sun, L., Wang, C., Wu, Y., Chen, J., Nie, Y., Liao, J., Chen, C., Template- assisted conversion of aniline nanopolymers into non- precious metal FeN/C electrocatalysts for highly efficient oxygen reduction reaction. *J Alloys Compd* 2016;686:874–882.
155. Lobato, J., Zamora, H., Plaza, J., Cañizares, P., Rodrigo, M. A., Enhancement of high temperature PEMFC stability using catalysts based on Pt supported on SiC based materials. *Appl Catal B* 2016;198:516–524.
156. Ejikeme, P. M., Makgopa, K., Ozoemena, K. I., Effects of catalyst- support materials on the performance of fuel cells. In K. Ozoemena, S. Chen (Eds), *Nanomaterials for Fuel Cell Catalysis*. Springer, Cham, 2016, pp. 517–550.
157. Roh, S. H., Sadhasivam, T., Kim, H., Park, J. H., Jung, H. Y., Carbon free SiO_2eSO_3H supported Pt bifunctional electrocatalyst for unitized regenerative fuel cells. *Int J Hydrogen Energy* 2016;41:20650–20659.
158. Du, L., Yan, X., He, G., Wu, X., Hu, Z., Wang, Y., SPEEK proton exchange membranes modified with silica sulfuric acid nanoparticles. *Int J Hydrogen Energy* 2012;37:11853–11861.
159. Bakangura, E., Ge, L., Muhammad, M., Pan, J., Wu, L., Xu T., Sandwich structure SPPO/BPPO proton exchange membranes for fuel cells: Morphology–electrochemical properties relationship. *J Membr Sci* 2015;475:30–38.
160. Molavian, M. R., Abdolmaleki, A., Eskandari, K., Theoretical investigation of proton-transfer in different membranes for PEMFC applications in low humidity conditions. *Comput Mater Sci* 2016;122:126–138.
161. Sadhasivam, T., Roh, S. H., Kim, T. H., Park, K. W., Jung, H. Y., Graphitized carbon as an efficient mesoporous layer for unitized regenerative fuel cells. *Int J Hydrogen Energy* 2016;41:18226–18230.
162. Cunningham, B. D., Huang, J., Baird, D. G., Development of bipolar plates for fuel cells from graphite filled wet-lay material and a thermoplastic laminate skin layer. *J Power Sources* 2007;165:764–773.
163. Huang, J., Baird, D. G., McGrath, J. E., Development of fuel cell bipolar plates from graphite filled wet-lay thermoplastic composite materials. *J Power Sources* 2005;150:110–119.
164. Lin, M. T., Wan, C. H., Wu, W., Comparison of corrosion behaviors between SS304 and Ti substrate coated with (Ti, Zr) N thin films as Metal bipolar plate for unitized regenerative fuel cell. *Thin Solid Films* 2013;544:162–169.
165. Zhang, H., Hou, M., Lin, G., Fu, Y., Sun S., Shao, Z., Yi, B., Performance of Ti-Ag-deposited titanium bipolar plates in simulated unitized regenerative fuel cell (URFC) environment. *Int J Hydrogen Energy* 2011;36:5695–5701.
166. Zhang, M., Hu, L., Lin, G., Shao, Z., Honeycomb-like nanocomposite Ti-Ag-N films prepared by pulsed bias arc ion plating on titanium as bipolar plates for unitized regenerative fuel cells. *J Power Sources* 2012;198:196–202.
167. Jung, H. Y., Huang, S. Y., Popov, B. N., High-durability titanium bipolar plate modified by electrochemical deposition of platinum for unitized regenerative fuel cell (URFC). *J Power Sources* 2010;195:1950–1956.
168. Jung, H. Y., Huang, S. Y., Ganesan, P., Popov, B. N. Performance of gold-coated titanium bipolar plates in unitized regenerative fuel cell operation. *J Power Sources* 2009;194:972–975.
169. Utz, R., *Study of Unitized Regenerative Fuel Cell Systems for Aircraft Applications*. DOT/FAA/TC-21/30 final report by US Dept. of Transportation, FAA, Washington DC, 2022.
170. *Hybrid Fuel Cell Technology Overview*, National Energy Technology Lab., U.S. DOE, Pittsburgh, PA, May 2001.

171. Samuelson, S., *Fuel Cell/Gas Turbine Hybrid Systems*. International Gas Turbine Institute, ASME, New York, 2004.
172. *Fuel Cell Handbook*, 6th ed., National Energy Technology Lab., U.S. DOE, Pittsburgh, PA, Nov. 2002., published by EG&G Technical Services, Inc. and Science Applications International Corporation, Under Contract No. DE-AM26-99FT40575.
173. Developing Power Systems for the 21st Century - Fuel Cell/ATS Hybrid Systems. U.S. Department of Energy, National Energy Technology Laboratory & Office of Industrial Technologies, Project facts for Advanced Clean/Efficient Power Systems, PS031.1099.
174. Eberl, U., *Fuel Cells and Gas Turbines: A Marriage of Efficiency*. Research and Innovation, January 2000.
175. Liese, E. A., Gemmen, R. S., Jabbari, F., Brouwer, J., Technical development issues and dynamic modeling of gas turbine and fuel cell hybrid systems. In *Proceedings of the 1999 Review Conference on Fuel Cell Technology*, August 1999.
176. Fuel Cells - Opening New Frontiers in Power Generation. U.S. Department of Energy, Office of Fossil Energy, National Energy Technology Laboratory, November 1999.
177. Açıkkalp, E., Chen, L., Ahmadi, M. H., Comparative performance analyses of molten carbonate fuel cell-alkali metal thermal to electric converter and molten carbonate fuel cell-thermo-electric generator hybrid systems. *Energy Rep* 2020;6:10–16. DOI: 10.1016/j.egyr.2019.11.108
178. Agrawal, S., Chourasiya, S., Palwalia, D. K., Hybrid energy management system design with renewable energy sources (fuel cells, PV cells and wind energy): A review *IJSET* 2018;6:174–177.
179. Uechi, H., Kimijima, S., Kasagi, N., Cycle analysis of gas turbine-fuel cell hybrid micro generation system. In *2001 International Joint Power Generation Conference*, New Orleans, LA, June 4–7, 2001.
180. Samuelsen, S., *Analyses and Technology Transfer for Fuel Cell Systems*. California Energy Commission, Sacramento, CA, 2000.
181. Veyo, S. E., Shockling, L. A., Dederer, J. T., Gillett, J. E., Lundberg, W. L., Tubular solid oxide fuel cell/gas turbine hybrid cycle power systems: Status. *J Eng Gas Turbines Power* 2002;124:845–849.
182. Rao, A. D., Samuelson, G. S., Analysis strategies for tubular solid oxide fuel cell based hybrid systems. *J Eng Gas Turbines Power* 2002;124:503–509.
183. Massardo, A. F., McDonald, C. F., Korakianitis, T., Microturbine/fuel-cell coupling for high-efficiency electrical- power generation. *J Eng Gas Turbines Power* 2002;124(1):110–116.
184. Pangalis, M. G., Martinez-Botas, R. F., Brandon, N. P., Integration of solid oxide fuel cells into gas turbine power generation cycles. Part 1: Fuel cell thermodynamic modeling. *Proc Inst Mech Eng A: J Power Energy* 2002;216:129–144.
185. Cunnel, C., Pangalis, M. G., Martinez-Botas, R. F., Integration of solid oxide fuel cells into gas turbine power generation cycles. Part 2: Hybrid model for various integration schemes. *Proc Inst Mech Eng A: J Power Energy* 2002;216:145–154.
186. Yi, Y., Rao, A. D., Brouwer, J., Samuelsen, G. S., Analysis and optimization of a solid oxide fuel cell and intercooled gas turbine (SOFC-ICGT) hybrid cycle. *J Power Sources* 2004;132(1–2):77–85.
187. Steinfeld, G., Maru, H. C., Sanderson, R. A., High efficiency carbonate fuel cell/turbine hybrid power cycle. In *Proc. 31st Intersoc. Energy Conversion Engineering Conf., IECEC 96*, Aug. 11–16, 1996, vol. 2, pp. 1123–1127.
188. Daggett, D., Eelman, S., Kristiansson, G., *Fuel Cell APU for Commercial Aircraft*. American Inst. Aeronautics Astronautics, Reston, VA, 2003.
189. Rajashekara, K., Propulsion system strategies for fuel cell vehicles, presented at the SAE 2000 World Congr., Paper 2000-01-0369, Detroit, MI, Mar. 6–9, 2000.

190. Rajashekara, K., Hybrid fuel cell strategies for clean power generation. In *Conference Record of the 2004 IEEE Industry Applications Conference, 2004. 39th IAS Annual Meeting*, Seattle, WA, 2004, pp. 2077–2083.
191. Li, W., Wang, Y., Liu, W., A review of solid oxide fuel cell application. *IOP Conf Ser: Earth Environ Sci* 2020;619:012012.
192. Keegan, K., Khaleel, M., Chick, L., Recknagle, K., Simner, S., Deibler, J., Analysis of planar solid oxide fuel cell based automotive auxiliary power unit, presented at the SAE 2002 World Congr., Paper 2002-01-0413, Detroit, MI, Mar. 4–7, 2002.
193. Jurado, F., Saenz, J. R., Adaptive control of a fuel cell-microturbine hybrid power plant, *IEEE Trans Energy Convers* 2003;18(2):342–347.
194. Ferrari-Trecate, G., Modeling and control of co-generation power plants: A hybrid system approach. *IEEE Trans Contr Syst Technol* 2004;12(5):694–705
195. Enjeti, P., Lai, J. S., Krein, P., NTU advanced power conditioning for fuel cell systems, presented at the Fuel Cell conference at National Energy Technology Laboratory at Morgantown, WV, Mar. 2002.
196. Rajashekara, K., Fattic, J., Husted, H., Comparative study of new on-board power generation technologies for automotive applications. In *Proc. IEEE Workshop Power Electronics in Transportation*, Auburn Hills, MI, Oct. 2002, pp. 3–10.
197. Coutts, T. J., Fitzegerald, M. C., Thermophotovoltaics. *Sci Am* 1998;279;90–95.
198. Coutts, T. J., A review of progress in thermo-photovoltaic generation of electricity. *Renew Sustain Energy Rev* 1999;3:77–184.
199. Black, R. E., Baldasaro, P. F., Charache, G. W., Thermophotovoltaics—Development status and parametric considerations for power applications. In *Eighteenth International Conference on Thermoelectrics. Proceedings, ICT'99 (Cat. No.99TH8407)*, 1999, pp. 639–644.
200. Mazzer, M., De Risi, A., Laforgia, D., Barnham, K., Rohr, C., High efficiency thermophotovoltaics for automotive applications, presented at the SAE 2000 World Congr., Paper 2000-01-0991, Detroit, MI, Mar. 6–9, 2000.
201. Morrison, O., Seal, M., West, E., Connelly, W., Use of a thermophotovoltaic generator in a hybrid electric vehicle. In *Proc. Fourth NREL Conf. Thermophotovoltaic Generation of Electricity*, Denver, CO, Oct. 1998, pp. 488–496.
202. Celanovic, I., Non-conventional electricity sources for motor vehicles, presented at the MIT/Industry Consortium Meeting Automotive Electrical/Electronic Components and Systems, Kyoto, Japan, Jun. 3, 2002.
203. Yokokawa, H., Tu, H., Iwanschitz, B., Mai, A., Fundamental mechanisms limiting solid oxide fuel cell durability. *J Power Sources* 2008;182, 400–412. DOI: 10.1016/j.jpowsour.2008.02.016
204. Mueller, F., Jabbari, F., Gaynor, R., Brouwer, J., Novel solid oxide fuel cell system controller for rapid load following. *J Power Sources* 2007;172, 308–323. DOI: 10.1016/j.jpowsour.2007.05.092
205. Wachsman, E., Marlowe, C., Lee, K., Role of solid oxide fuel cells in a balanced energy strategy. *Energy Environ Sci* 2012;5:5498–5509. DOI: 10.1039/C1EE02445K
206. Das, T., Snyder, S., Adaptive control of a solid oxide fuel cell ultra-capacitor hybrid system. *IEEE Trans Control Syst Technol* 2013;21:372–383. DOI: 10.1109/TCST.2011.2181514
207. Aguiar, P., Brett, D., Brandon, N., Feasibility study and techno-economic analysis of an SOFC/battery hybrid system for vehicle applications. *J Power Sources* 2007;171:186–197. DOI: 10.1016/j.jpowsour.2006.12.049
208. Brunaccini, G., Sergi, F., Aloisio, D., Ferraro, M., Blesznowski, M., Kupecki, J., Motylinski, K., Antonucci, V., Modeling of a SOFC-HT battery hybrid system for optimal design of off-grid base transceiver station. *Int J Hydrogen Energy* 2017;42:27962–27978. DOI: 10.1016/j.ijhydene.2017.09.062

209. Tse, L. K. C., Wilkins, S., McGlashan, N., Urban, B., Martinez-Botas, R., Solid oxide fuel cell/gas turbine trigeneration system for marine applications. *J Power Sources* 2011;196:3149–3162. DOI: 10.1016/j.jpowsour.2010.11.099
210. Jia, Z., Sun, J., Dobbs, H., King, J., Feasibility study of solid oxide fuel cell engines integrated with sprinter gas turbines: Modeling, design and control. *J Power Sources* 2015;275:111–125. DOI: 10.1016/j.jpowsour.2014.10.203
211. Wakui, T., Yokoyama, R., Optimal structural design of residential cogeneration systems in consideration of their operating restrictions. *Energy* 2014;64:719–733. DOI: 10.1016/j.energy.2013.10.002
212. Yang, W., Zhao, Y., Liso, V., Brandon, N., Optimal design and operation of a syngas-fuelled SOFC micro-CHP system for residential applications in different climate zones in China. *Ener Build* 2014;80:613–622. DOI: 10.1016/j.enbuild.2014.05.015
213. Zhang, X., Chan, S., Li, G., Ho, H., Li, J., Feng, Z., A review of integration strategies for solid oxide fuel cells. *J Power Sources* 2010;195:685–702. DOI: 10.1016/j.jpowsour.2009.07.045
214. Obara, S., Power generation efficiency of an SOFC-PEFC combined system with time shift utilization of SOFC exhaust heat. *Int J Hydrogen Energy* 2010;35:757–767. DOI: 10.1016/j.ijhydene.2009.11.032
215. Perdikaris, N., Panopoulos, K. D., Hofmann, P., Spyrakis, S., Kakaras, E., Design and exergetic analysis of a novel carbon free tri-generation system for hydrogen, power and heat production from natural gas, based on combined solid oxide fuel and electrolyser cells. *Int J Hydrogen Energy* 2010;35:2446–2456. DOI: 10.1016/j.ijhydene.2009.07.084
216. Becker, W. L., Braun, R. J., Penev, M., Melaina, M., Design and technoeconomic performance analysis of a 1MW solid oxide fuel cell polygeneration system for combined production of heat, hydrogen, and power. *J Power Sources* 2012;200:34–44. DOI: 10.1016/j.jpowsour.2011.10.040
217. Leal, E. M., Brouwer, J., A Thermodynamic analysis of electricity and hydrogen co-production using a solid oxide fuel cell. *J Fuel Cell Sci Technol* 2005;3:137–143. DOI: 10.1115/1.2173669
218. Shaffer, B., Brouwer, J., Feasibility of solid oxide fuel cell dynamic hydrogen coproduction to meet building demand. *J Power Sources* 2014;248, 58–69. DOI: 10.1016/j.jpowsour.2013.08.144
219. Hemmes, K., Patil, A., Woudstra, N., Flexible coproduction of hydrogen and power using internal reforming solid oxide fuel cells system. *J Fuel Cell Sci Technol* 2008;5:041010. DOI: 10.1115/1.2931459
220. Pérez-Fortes, M., Mian, A., Diethelm, S., Wang, L., Maréchal, F., Van Herle, J., Santhanam, S., Heddrich, M. P., Au, S. F., Varkaraki, E., Wuillemin, Z., Makkus, R., Mirabelli, I, Schoon, R., Grippa, M., Testi, M., Crema, L., Process optimization of a SOFC system for the combined production of hydrogen and electricity. In *Proceedings of the 13th European SOFC and SOE Forum 2018*, Lucerne, 2018.
221. Mian, A., Martelli, E., Maréchal, F., Framework for the multiperiod sequential synthesis of heat exchanger networks with selection, design, and scheduling of multiple utilities. *Indus Eng Chem Res* 2016;55:168–186. DOI: 10.1021/acs.iecr.5b02104
222. Vollmar, H. -E., Maier, C. -U., Nölscher, C., Merklein, T., Poppinger, M., Innovative concepts for the coproduction of electricity and syngas with solid oxide fuel cells. *J Power Sources* 2000;86:90–97. DOI: 10.1016/S0378-7753(99)00421-8
223. Kotas, T. J., *The Exergy Method of Thermal Plant Analysis*. Butterworths, London, 2013.
224. Dicks, A. L., Fellows, R. G., Martin Mescal, C., Seymour, C., A study of SOFC-PEM hybrid systems. *J Power Sources* 2000;86:501–506. DOI: 10.1016/S0378-7753(99)00492-9

225. Tan, L., Yang, C., Zhou, N., Performance of the solid oxide fuel cell (SOFC)/proton-exchange membrane fuel cell (PEMFC) hybrid system. *Chem Eng Technol* 2016;39:689–698. DOI: 10.1002/ceat.201500424
226. Subramanyan, K., Diwekar, U. M., Goyal, A, Multi-objective optimization for hybrid fuel cells power system under uncertainty. *J Power Sources* 2004;132:99–112. DOI: 10.1016/j.jpowsour.2003.12.053
227. Yokoo, M., Take, T., Simulation analysis of a system combining solid oxide and polymer electrolyte fuel cells. *J Power Sources* 2004;137:206–215. DOI: 10.1016/j.jpowsour.2004.06.007
228. Yokoo, M., Watanabe, K., Arakawa, M., Yamazaki, Y., The effect of fuel feeding method on performance of SOFC-PEFC system. *J Power Sources* 2006;159:836–845. DOI: 10.1016/j.jpowsour.2005.11.093
229. Yokoo, M., Watanabe, K., Arakawa, M., Yamazaki, Y., Influence of current densities in SOFC and PEFC stacks on a SOFC-PEFC combined system. *J Power Sources* 2007;163:892–899. DOI: 10.1016/j.jpowsour.2006.09.050
230. Baldi, F., Wang, L., Pérez-Fortes, M., Maréchal, F., A cogeneration system based on solid oxide and proton exchange membrane fuel cells with hybrid storage for off-grid applications. *Front Energy Res* 2019;6:139. DOI: 10.3389/fenrg.2018.00139
231. Nguyen, G., Sahlin, S., Andreasen, S. J., Shaffer, B., and Brouwer, J., Dynamic modeling and experimental investigation of a high temperature PEM fuel cell stack. *Int J Hydrogen Energy* 2016;41:4729–4739. DOI: 10.1016/j.ijhydene.2016.01.045
232. van Biert, L., Godjevac, M., Visser, K., Aravind, P., A review of fuel cell systems for maritime applications. *J Power Sources* 2016;327:345–364. DOI: 10.1016/j.jpowsour.2016.07.007
233. Dunn, B., Kamath, H., Tarascon, J.-M., Electrical energy storage for the grid: A battery of choices. *Science* 2011;334:928–935. DOI: 10.1126/science.1212741
234. Nykvist, B., Nilsson, M., Rapidly falling costs of battery packs for electric vehicles. *Nat Clim Change* 2015;5:329. DOI: 10.1038/nclimate2564
235. Pellegrino, S., Lanzini, A., Leone, P., Techno-economic and policy requirements for the market-entry of the fuel cell micro-CHP system in the residential sector. *Appl Energy* 2015;143:370–382. DOI: 10.1016/j.apenergy.2015.01.007
236. Moret, S., Gironès, V. C., Bierlaire, M., Maréchal, F., Characterization of input uncertainties in strategic energy planning models. *Appl Ener* 2017;202:597–617. DOI: 10.1016/j.apenergy.2017.05.106
237. Liu, E., Large scale wind hydrogen systems, presented at the U.S. DOE, Washington, DC, Sep. 2003.
238. Cengelci, E., Enjeti, P., Modular PM generator/converter topologies, suitable for utility interface of wind/micro turbine and flywheel type electromechanical energy conversion systems. In *Conference Record of the 2000 IEEE Industry Applications Conference. Thirty-Fifth IAS Annual Meeting and World Conference on Industrial Applications of Electrical Energy (Cat. No.00CH37129)*, Rome, Italy, 2000, Vol. 4, pp. 2269–2276.
239. Reicher, D., Renewable hydrogen, presented at the U.S. DOE, Washington, DC, Sep. 2003.
240. Eskander, M. N., El-Shatter, T. F., Energy flow and management of a hybrid wind/PV/fuel cell generation system. In *Proc. IEEE PESC'02*, Jun. 2002, pp. 347–353.
241. Eguchi, K., Research progress in solar RFC technology for SPF airship, presented at the Fuel Cell Seminar, Miami, FL, 2003.
242. Kato, N., Kurozumi, K., Hybrid power supply system composed of photovoltaic and fuel-cell systems. In *2001 Twenty-Third International Telecommunications Energy Conference INTELEC 2001*, Oct. 2001, pp. 631–635.
243. Herlambang, Y. D., Prasetyo, T., Roihatin, A., Safarudin, Y. M., Arifin, F., Hybrid power plant of the photovoltaic-fuel cell. *IOP Conf Ser: Mater Sci Eng* 2021;1108:012049.

244. Kim, C., Cho, H. S., Kim, C. H., Cho, W., Kim, H. G., A feasibility study of photovoltaic—electrolysis—PEM hybrid system integrated into the electric grid system over the Korean Peninsula. *Front Chem* 2021;9:732582. DOI: 10.3389/fchem.2021.732582
245. Ferahtia, S., Djerioui, A., Zeghlache, S., Houari, A., A hybrid power system based on fuel cell, photovoltaic source and supercapacitor. *SN Appl Sci* 2020;2:940. DOI:10.1007/s42452-020-2709-0
246. Thounthong, P., Chunkag, V., Sethakul, P., Sikkabut, S., Pierfederici, S., Davat, B., Energy management of fuel cell/solar cell/supercapacitor hybrid power source, *J Power Sources* 2011;196:313–324.
247. Jayalakshmi, N. S., Gaonkar, D. N., Nempu, P. B., Power control of PV/Fuel cell/supercapacitor hybrid system for stand-alone applications. *Int J Renew Energy Res* 2016;6(2):672–679.
248. Arregui, M. G., Theoretical study of a power generation unit based on the hybridization of a fuel cell stack and ultra capacitors, Doctorat de l'université de Toulouse, 2007.
249. Corcau, J. I., Dinca, L., Grigorie, T. L., Tudosie, A. N., Fuzzy energy management for hybrid fuel cell/battery systems for more electric aircraft. In *AIP Conference Proceedings*, 2017, vol. 1836, p. 020056. DOI: 10.1063/1.4981996
250. Li, X., FCs. In D.Y. Goswami and F. Keith (Eds.), *Energy Conversion*, 2nd ed. CRR Press, Boca Raton, FL, 2017.
251. Yu, X., Starke, M. R., Tolbert, L. M., Ozpineci, B., Fuel cell power conditioning for electric power applications: A summary. *IET Electr Power Appl* 2007;1:643–656.
252. Elmer, T., Worall, M., Wu, S., Riffat, S. B., Fuel cell technology for domestic built environment applications: State of-the-art review. *Renew Sustain Energy Rev* 2015;42:913–931.
253. Carrasco, J. M., Franquelo, L. G., Bialasiewicz, J. T., Galvan, E., Guisado, R. C., Prats, A. M., Leon, J. I., Moreno-Alfonso, N., Power-electronic systems for the grid integration of renewable energy sources: A survey. *IEEE Trans Ind Electron* 2006;53:1002–1016.
254. Emmanuel, M., Akinyele, D., Rayudu, R., Techno-economic analysis of a 10ákWp utility interactive photovoltaic system at Maungaraki school, Wellington, New Zealand. *Energy* 2017;120:573–583.
255. Basso, T. S., DeBlasio, R., IEEE 1547 series of standards: Interconnection issues. *IEEE Trans Power Electron* 2004;19:1159–1162.
256. Azmy, A. M., Erlich, I., Intelligent operation management of fuel cells and micro-turbines using genetic algorithms and neural networks. In *2004 New and Renewable Energy Technologies for Sustainable Development.* Wspc, Evora, Portugal, 2007.
257. Louie, H., Dauenhauer, P., Wilson, M., Zomers, A., Mutale, J., Eternal light: Ingredients for sustainable off-grid energy development. *IEEE Power Energy Mag* 2014;12:70–78.
258. IEEE. *IEEE Guide for Array and Battery Sizing in Stand-Alone Photovoltaic (PV) Systems; IEEEStd1562–2007.* IEEE, New York, 2008, p. 22.
259. Generation, D., Storage, E., *IEEE Guide for Optimizing the Performance and Life of Lead-Acid Batteries in Remote Hybrid Power Systems.* IEEE, New York, 2008.
260. Gao, D.W., Applications of ESS in renewable energy microgrids. In *Energy Storage for Sustainable Microgrid.* Academic Press, Cambridge, MA, 2015. Paperback ISBN: 9780128033746; eBook ISBN: 9780128033753.
261. Burmester, D., Rayudu, R., Seah, W., Akinyele, D., A review of nanogrid topologies and technologies. *Renew Sustain Energy Rev* 2017;67:760–775.
262. Cvetkovic, I., Dong, D., Zhang, W., Jiang, L., Boroyevich, D., Lee, F. C. Y., Mattavelli, P., A testbed for experimental validation of a low-voltage DC nanogrid for buildings. In *Proceedings of the 2012 15th International Power Electronics and Motion Control Conference and Exposition, EPE-PEMC 2012 ECCE Europe*, Novi Sad, Serbia, 4–6 September 2012.

263. Shwehdi, M. H., Mohamed, S. R., Proposed smart DC nano-grid for green buildings—A reflective view. In *Proceedings of the 3rd International Conference on Renewable Energy Research and Applications, ICRERA 2014*, Milwaukee, WI, USA, 19–22 October 2014.
264. Wu, H., Wong, S. C., Tse, C. K., Chen, Q., Control and modulation of a family of bidirectional AC-DC converters with active power compensation. In *Proceedings of the 2015 IEEE Energy Conversion Congress and Exposition, ECCE 2015*, Montreal, QC, Canada, 20–24 September 2015.
265. Ganesan, S. I., Pattabiraman, D., Govindarajan, R. K., Rajan, M., Nagamani, C., Control scheme for a bidirectional converter in a self-sustaining low-voltage dc nanogrid. *IEEE Trans. Ind. Electron* 2015;62:6317–6326.
266. Bizon, N., Oproescu, M., Raceanu, M., Efficient energy control strategies for a standalone renewable/fuel cell hybrid power source. *Energy Convers Manag* 2015;90:93–110.
267. Mudaliyar, S. R., Mishra, S., Sharma, R. K., Load following capability of fuel cell-microturbine based hybrid energy system for microgrid operation. In *Proceedings of the 6th International Conference on Computer Applications in Electrical Engineering—Recent Advances, CERA 2017*, Roorkee, India, 5–7 October 2017.
268. Wang, C., Nehrir, M. H., Load transient mitigation for stand-alone fuel cell power generation systems. *IEEE Trans Energy Convers* 2007;22:864–872.
269. Hu, X., Li, P., Wang, K., Sun, Y., Zeng, D., Guo, S., Energy management of data centers powered by fuel cells and heterogeneous energy storage. In *Proceedings of the 2018 IEEE International Conference on Communications*, Kansas City, MO, USA, 20–24 May 2018.
270. Asghari, M., Brouwer, J., Integration of a solid oxide fuel cell with an organic Rankine cycle and absorption chiller for dynamic generation of power and cooling for a residential application. *Fuel Cells* 2019;19:361–373.
271. Chang, H., Xu, X., Shen, J., Shu, S., Tu, Z., Performance analysis of a micro-combined heating and power system with PEM fuel cell as a prime mover for a typical household in North China. *Int J Hydrogen Energy* 2019;44:24965–24976.
272. Taleb, S. A., Brown, D., Dillet, J., Guillement, P., Mainka, J., Crosnier, O., Douard, C., Athouel, L., Brousse, L., Lottin, O., Direct hybridization of polymer exchange membrane surface fuel cell with small aqueous supercapacitors. *Fuel Cells* 2018;18:299–305.
273. Yang, J., Qin, S., Zhang, W., Ding, T., Zhou, B., Li, X., Jian, L., Improving the load-following capability of a solid oxide fuel cell system through the use of time delay control. *Int J Hydrogen Energy* 2017;42:1221–1236.
274. Wu, X., Gao, D., Optimal fault-tolerant control strategy of a solid oxide fuel cell system, *J. Power Sources* 2017;364:163–181.
275. Barakat, M., Tala-Ighil, B., Chaoui, H., Gualous, H., Hissel, D., Energy management of a hybrid tidal turbine-hydrogen micro-grid: Losses minimization strategy. *Fuel Cells* 2020;20:342–350.
276. Han, Y., Zhang, G., Li, Q., You, Z., Chen, W., Liu, H., Hierarchical energy management for PV/hydrogen/battery island DC microgrid. *Int J Hydrogen Energy* 2019;44:5507–5516
277. Han, Y., Chen, W., Li, Q., Yang, H., Zare, F., Zheng, Y., Two-level energy management strategy for PV-Fuel cell-battery-based DC microgrid. *Int J Hydrogen Energy* 2019;44:19395–19404.
278. Wang, J., Ren, K., Zhu, Y., Huang, J., Liu, S., A Review of recent advances in microbial fuel cells: Preparation, operation, and application. *BioTech* 2022;11:44. DOI: 10.3390/biotech11040044
279. Malik, S., Kishore, S., Dhasmana, A., Kumari, P., Mitra, T., Chaudhary, V., Kumari, R., Bora, J., Ranjan, A., Minkina, T., Rajput, V. D., A perspective review on microbial fuel cells in treatment and product recovery from wastewater. *Water* 2023;15:316. DOI: 10.3390/w15020316

280. Priya, A. K., Subha, C., Kumar, P. S., Suresh, R., Rajendran, S., Vasseghian, Y., Soto-Moscoso, M., Advancements on sustainable microbial fuel cells and their future prospects: A review. *Environ Res* 2022;210:112930.
281. Munoz-Cupa, C., Hu, Y., Xu, C., Bassi, A., An overview of microbial fuel cell usage in wastewater treatment, resource recovery and energy production. *Sci Total Environ* 2021;754:142429.
282. Kannan, N., Donnellan, P., Algae-assisted microbial fuel cells: A practical overview. *Bioresour Technol Rep* 2021;15:100747. DOI: 10.1016/j.biteb.2021.100747
283. Kaur, R., Singh, S., Chhabra, V. A., Marwaha, A., Kim, K. H., Tripathi, S. K., A sustainable approach towards utilization of plastic waste for an efficient electrode in microbial fuel cell applications. *J Hazard Mater* 2021;417:125992.
284. Gajda, I., Stinchcombe, A., Merino-Jimenez, I., Pasternak G., Sanchez-Herranz, D., Greenman, J., Ieropoulos, I. A., Miniaturized ceramic-based microbial fuel cell for efficient power generation from urine and stack development. *Front Energy Res* 2018;6:84. DOI: 10.3389/fenrg.2018.00084
285. Xu, H., Quan, X., Xiao, Z., Chen, L., Effect of anodes decoration with metal and metal oxides nanoparticles on pharmaceutically active compounds removal and power generation in microbial fuel cells. *Chem Eng J* 2018;335:539–547.
286. Yu, B., Li, Y., Feng, L., Enhancing the performance of soil microbial fuel cells by using a bentonite-Fe and Fe_3O_4 modified anode. *J Hazard Mater* 2019;377:70–77.
287. Li, C., Zhou, K., He, H., Cao, J., Zhou, S., Adding zero-valent iron to enhance electricity generation during MFC start-up. *Int J Environ Res Public Health* 2020;17:806.
288. Paul, D., Noori, M. T., Rajesh, P. P., Ghangrekar, M. M., Mitra, A., Modification of carbon felt anode with graphene oxide-zeolite composite for enhancing the performance of microbial fuel cell. *Sustain Energy Technol Assess* 2018;26:77–82.
289. Liang, Y., Zhai, H., Liu, B., Ji, M., Li, J., Carbon nanomaterial-modified graphite felt as an anode enhanced the power production and polycyclic aromatic hydrocarbon removal in sediment microbial fuel cells. *Sci Total Environ* 2020;713:136483.
290. Zhang, Y., Chen, X., Yuan, Y., Lu, X., Yang, Z., Wang, Y., Sun, J., Long-term effect of carbon nanotubes on electrochemical properties and microbial community of electrochemically active biofilms in microbial fuel cells. *Int J Hydrogen Energy* 2018;43:16240–16247.
291. Mashkour, M., Rahimnejad, M., Mashkour, M., Soavi, F., Electro-polymerized polyaniline modified conductive bacterial cellulose anode for supercapacitive microbial fuel cells and studying the role of anodic biofilm in the capacitive behavior. *J Power Sources* 2020;478:228822.
292. Mashkour, M., Rahimnejad, M., Raouf, F., Navidjouy, N., A review on the application of nanomaterials in improving microbial fuel cells. *Biofuel Res J* 2021;8:1400–1416. DOI: 10.18331/BRJ2021.8.2.5
293. Kumar, G. G., Sarathi, V. G., Nahm, K. S., Recent advances and challenges in the anode architecture and their modifications for the applications of microbial fuel cells. *Biosens Bioelectron* 2013;43:461–475. DOI: 10.1016/j.bios.2012.12.048
294. Huang, X., Duan, C., Duan, W., Sun, F., Cui, H., Zhang, S., Chen, X., Role of electrode materials on performance and microbial characteristics in the constructed wetland coupled microbial fuel cell (CW-MFC): A review. *J Clean Prod* 2021;301:126951.
295. Nambiar, S., Togo, C., Limson, J., Application of multi-walled carbon nanotubes to enhance anodic performance of an Enterobacter cloacae-based fuel cell. *Afr J Biotechnol* 2009;8(24): 6927–6932.
296. Zhu, W., Yao, M., Gao, H., Wen, H., Zhao, X., Zhang, J., Bai, H., Enhanced extracellular electron transfer between *Shewanella putrefaciens* and carbon felt electrode modified by bio-reduced graphene oxide. *Science Total Environ* 2019;691:1089–1097.

297. Hernández, L. A., Riveros, G., González, D. M., Gacitua, M., del Valle, M. A., PEDOT/graphene/nickel-nanoparticles composites as electrodes for microbial fuel cells. *J Mater Sci: Mater Electron* 2019;30(13):12001–12011.
298. Duarte, K. D., Frattini, D., Kwon, Y., High performance yeast- based microbial fuel cells by surfactant-mediated gold nanoparticles grown atop a carbon felt anode. *Appl. Energy* 2019;256:113912.
299. Mashkour, M., Rahimnejad, M., Mashkour, M., Bacterial cellulose-polyaniline nano-biocomposite: A porous media hydrogel bioanode enhancing the performance of microbial fuel cell. *J Power Sources* 2016;325:322–328.
300. Mashkour, M., Rahimnejad, M., Mashkour, M., Bakeri, G., Luque, R., Oh, S. E., Application of wet nanostructured bacterial cellulose as a novel hydrogel bioanode for microbial fuel cells. *ChemElectroChem* 2017;4(3):648–654.
301. Mashkour, M., Kimura, T., Kimura, F., Mashkour, M., Tajvidi, M., Tunable self-assembly of cellulose nanowhiskers and polyvinyl alcohol chains induced by surface tension torque. *Biomacromolecules* 2013;15(1):60–65.
302. Chen, Y., Yang, Z., Zhang, Y., Xiang, Y., Xu, R., Jia, M., Cao, J., Xiong, W., Effects of different conductive nanomaterials on anaerobic digestion process and microbial community of sludge. *Bioresour Technol* 2020;304:123016.
303. Priyadarshini, M., Ahmad, A., Das, S., Ghangrekar, M. M., Metal organic frameworks as emergent oxygen-reducing cathode catalysts for microbial fuel cells: A review. *Int J Environ Sci Technol* 2022;19:11539–11560.
304. Liu, P., Liu, X., Dong, F., Lin, Q., Tong, Y., Li, Y., Zhang, P., Electricity generation from banana peels in an alkaline fuel cell with a Cu_2O-Cu modified activated carbon cathode. *Sci Total Environ* 2018;631–632:849–856.
305. Majidi, M. R., Shahbazi Farahani, F., Hosseini, M., Ahadzadeh, I., Low-cost nanowired α-MnO_2/C as an ORR catalyst in air-cathode microbial fuel cell. *Bioelectrochemistry* 2019;125:38–45.
306. Chiodoni, A., Salvador, G.P., Massaglia, G., Delmondo, L., Muñoz-Tabares, J.A., Sacco, A., Garino, N., Castellino, M., Margaria, V., Ahmed, D., Pirri, C. F., Quaglio, M., MnxOy- based cathodes for oxygen reduction reaction catalysis in microbial fuel cells. *Int. J. Hydrogen Energy* 2018;44:4432–4441.
307. Rout, S., Nayak, A. K., Varanasi, J. L., Pradhan, D., Das, D., Enhanced energy recovery by manganese oxide/reduced graphene oxide nanocomposite as an air-cathode electrode in the single-chambered microbial fuel cell. *J Electroanal Chem* 2018;815:1–7.
308. Mecheri, B., Ficca, V. C. A., de Oliveira, M. A. C., D'Epifanio, A., Placidi, E., Arciprete, F., Licoccia, S., Facile synthesis of graphene-phthalocyanine composites as oxygen reduction electrocatalysts in microbial fuel cells. *Appl Catal B: Environ* 2018;237:699–707.
309. Lee, C., Ozden, S., Tewari, C. S., Park, O. K., Vajtai, R., Chatterjee, K., Ajayan, P. M., MoS_2 —Carbon nanotube porous 3 D network for enhanced oxygen reduction reaction. *ChemSusChem* 2018;11:2960–2966.
310. Li, H., Ma, H., Liu, T., Ni, J., Wang, Q., An excellent alternative composite modifier for cathode catalysts prepared from bacterial cellulose doped with Cu and P and its utilization in microbial fuel cell. *Bioresour Technol* 2019;289:121661.
311. Kaur, R., Singh, S., Chhabra, V. A., Marwaha, A., Kim, K. -H., Tripathi, S. K., A sustainable approach towards utilization of plastic waste for an efficient electrode in microbial fuel cell applications. *J Hazard Mater* 2021;417:125992.
312. Li, S., Zhu, X., Yu, H., Wang, X., Liu, X., Yang, H., Li, F., Zhou, Q., Simultaneous sulfamethoxazole degradation with electricity generation by microbial fuel cells using Ni-MOF-74 as cathode catalysts and quantification of antibiotic resistance genes. *Environ Res* 2021;197:111054.

313. Jiang, L., Chen, J., An, Y., Han, D., Chang, S., Liu, Y., Yang, R., Enhanced electrochemical performance by nickel-iron layered double hydroxides (LDH) coated on Fe3O4 as a cathode catalyst for single-chamber microbial fuel cells. *Sci Total Environ* 2020;745:141163.
314. Tajdid Khajeh, R., Aber, S., Zarei, M., Comparison of $NiCo_2O_4$, CoNiAl-LDH, and CoNiAl-LDH@$NiCo_2O_4$ performances as ORR catalysts in MFC cathode. *Renew Energy* 2020;154:1253–1271.
315. Bian, B., Alqahtani, M. F., Katuri, K. P., Liu, D., Bajracharya, S., Lai, Z., Rabaey, K., Saikaly, P. E., Porous nickel hollow fiber cathodes coated with CNTs for efficient microbial electrosynthesis of acetate from CO_2 using *Sporomusa ovata*. *J Mater Chem A* 2018;6(35):17201–17211.
316. Tofighi, A., Rahimnejad, M., Ghorbani, M., Ternary nanotube α- MnO_2/GO/AC as an excellent alternative composite modifier for cathode electrode of microbial fuel cell. *J Therm Anal Calorim* 2019;135(3):1667–1675.
317. Ansari, S. A., Parveen, N., Han, T. H., Ansari, M. O., Cho, M. H., Fibrous polyaniline@ manganese oxide nanocomposites as supercapacitor electrode materials and cathode catalysts for improved power production in microbial fuel cells. *Phys Chem Chem Phys* 2016;18(13):9053–9060.
318. Mashkour, M., Rahimnejad, M., Pourali, S., Ezoji, H., ElMekawy, A., Pant, D., Catalytic performance of nano-hybrid graphene and titanium dioxide modified cathodes fabricated with facile and green technique in microbial fuel cell. *Prog Nat Sci: Mater Int* 2017;27(6):647–651.
319. He, Y. R., Du, F., Huang, Y. X., Dai, L. M., Li, W. W., Yu, H. Q., Preparation of microvillus-like nitrogen-doped carbon nanotubes as the cathode of a microbial fuel cell. *J Mater Chem A* 2016;4(5):1632–1636.
320. Liu, Q., Zhou, Y., Chen, S., Wang, Z., Hou, H., Zhao, F., Cellulose- derived nitrogen and phosphorus dual-doped carbon as high performance oxygen reduction catalyst in microbial fuel cell. *J Power Sources* 2015;273:1189–1193.
321. Liu, Z., Ge, B., Li, K., Zhang, X., Huang, K., The excellent performance and mechanism of activated carbon air cathode doped with different type of cobalt for microbial fuel cells. *Fuel* 2016;176:173–180.
322. Zhong, M., Liang, B., Fang, D., Li, K., Lv, C., Leaf-like carbon frameworks dotted with carbon nanotubes and cobalt nanoparticles as robust catalyst for oxygen reduction in microbial fuel cell. *J Power Sources* 2021;482:229042.
323. Zhao, C. E., Qiu, Z., Yang, J., Huang, Z. D., Shen, X., Li, Y., Ma, Y., Metal–organic frameworks-derived core/shell porous carbon materials interconnected by reduced graphene oxide as effective cathode catalysts for microbial fuel cells. *ACS Sustain Chem Eng* 2020;8(37):13964–13972. DOI: 10.1021/acssuschemeng.0c03485
324. Zhong, K., Huang, L., Li, M., Dai, Y., Wang, Y., Zuo, J., Zhang, H., Zhang, B., Yang, S., Tang, J., Cobalt/nitrogen-Co-doped nanoscale hierarchically porous composites derived from octahedral metal-organic framework for efficient oxygen reduction in microbial fuel cells. *Int J Hydrogen Energy* 2019;44(57):30127–30140.
325. Ma, M., You, S., Gong, X., Dai, Y., Zou, J., Fu, H., Silver/iron oxide/graphitic carbon composites as bacteriostatic catalysts for enhancing oxygen reduction in microbial fuel cells. *J Power Sources* 2015;283:74–83.
326. Noori, M. T., Tiwari, B., Mukherjee, C., Ghangrekar, M., Enhancing the performance of microbial fuel cell using AgPt bimetallic alloy as cathode catalyst and anti-biofouling agent. *Int J Hydrogen Energy* 2018;43(42):19650–19660.
327. Pandit, S., Khilari, S., Roy, S., Ghangrekar, M. M., Pradhan, D., Das, D., Reduction of start-up time through bioaugmentation process in microbial fuel cells using an isolate from dark fermentative spent media fed anode. *Water Sci Technol* 2015;72:106–115.

328. Schmitz, S., Rosenbaum, M. A., Boosting mediated electron transfer in bioelectrochemical systems with tailored defined microbial cocultures. *Biotechnol Bioeng* 2018;115:2183–2193.
329. Liu, X., Wang, S., Xu, A., Zhang, L., Liu, H., Ma, L. Z., Biological synthesis of high-conductive pili in aerobic bacterium Pseudomonas aeruginosa. *Appl Microbiol Biotechnol* 2019;103:1535–1544.
330. Lin, T., Ding, W., Sun, L., Wang, L., Liu, C. -G., Song, H., Engineered *Shewanella oneidensis*-reduced graphene oxide biohybrid with enhanced biosynthesis and transport of flavins enabled a highest bioelectricity output in microbial fuel cells. *Nano Energy* 2018;50:639–648.
331. Kasai, T., Tomioka, Y., Kouzuma, A., Watanabe, K., Overexpression of the adenylate cyclase gene *cyaC* facilitates current generation by *Shewanella oneidensis* in bioelectrochemical systems. *Bioelectrochemistry* 2019;129:100–105.
332. Cheng, Z. H., Xiong, J. R., Min, D., Cheng, L., Liu, D. F., Li, W. W, Jin, F., Yang, M., Yu, H. Q., Promoting bidirectional extracellular electron transfer of *Shewanella oneidensis* MR-1 for hexavalent chromium reduction via elevating intracellular cAMP level. *Biotechnol Bioeng* 2020;117:1294–1303.
333. Min, D., Cheng, L., Zhang, F., Huang, X. -N., Li, D. -B., Liu, D.-F., Lau, T. -C., Mu, Y., Yu, H. -Q., Enhancing extracellular electron transfer of *Shewanella oneidensis* MR-1 through coupling improved flavin synthesis and metal-reducing conduit for pollutant degradation. *Environ Sci Technol* 2017;51:5082–5089.
334. McAnulty, M. J., Poosarla, V. G., Kim, K. -Y., Jasso-Chávez, R., Logan, B. E., Wood, T. K., Electricity from methane by reversing methanogenesis. *Nat Commun* 2017;8:15419.
335. Li, F., Li, Y., Sun, L., Li, X., Yin, C., An, X., Chen, X., Tian, Y., Song, H., Engineering *Shewanella oneidensis* enables xylose-fed microbial fuel cell. *Biotechnol Biofuels* 2017;10:196.
336. Christwardana, M., Frattini, D., Accardo, G., Yoon, S. P., Kwon, Y., Optimization of glucose concentration and glucose/yeast ratio in yeast microbial fuel cell using response surface methodology approach. *J Power Sources* 2018;402:402–412.
337. Obileke, K. C., Onyeaka, H., Meyer, E. L., Nwokolo, N., Microbial fuel cells, a renewable energy technology for bioelectricity generation: A mini-review. *Electrochem Commun* 2021;125:107003.
338. Haavisto, J., Dessì, P., Chatterjee, P., Honkanen, M., Noori, M. T., Kokko, M., Lakaniemi, A. -M., Lens, P. N. L., Puhakka, J. A., Effects of anode materials on electricity production from xylose and treatability of TMP wastewater in an up-flow microbial fuel cell. *Chem Eng J* 2019;372:141–150.
339. Li, F., An, X., Wu, D., Xu, J., Chen, Y., Li, W., Cao, Y., Guo, X., Lin, X., Li, C., Liu, S., Song, H., Engineering microbial consortia for high-performance cellulosic hydrolyzates-fed microbial fuel cells. *Front Microbiol* 2019;10:409.
340. Ullah, Z., Zeshan, S., Effect of substrate type and concentration on the performance of a double chamber microbial fuel cell. *Water Sci Technol* 2020;81:1336–1344.
341. Jin, Y. Z., Wu, Y. C., Li, B. Q., Zhu, H. D., Li, Y. P., Zhuang, M. Z., Fu, H. Y., Study on the electricity generation characteristics of microbial fuel cell with different substrates. *IOP Conf Series: Earth Environ Sci* 2020;435:012036.
342. Shen, J., Du, Z., Li, J., Cheng, F., Co-metabolism for enhanced phenol degradation and bioelectricity generation in microbial fuel cell. *Bioelectrochemistry* 2020;134:107527.
343. Yu, Y., Ndayisenga, F., Yu, Z., Zhao, M., Lay, C. -H., Zhou, D., Co-substrate strategy for improved power production and chlorophenol degradation in a microbial fuel cell. *Int J Hydrogen Energy* 2019;44:20312–20322.
344. Ndayisenga, F., Yu, Z., Yan, G., Phulpoto, I. A., Li, Q., Kumar, H., Fu, L., Zhou, D., Using easy-to-biodegrade co-substrate to eliminate microcystin toxic on electrochemically active bacteria and enhance bioelectricity generation from cyanobacteria biomass. *Sci Total Environ* 2021;751:142292.

345. Mancilio, L. B. K., Ribeiro, G. A., de Almeida, E. J. R., de Siqueira, G. M. V., Rocha, R. S., Guazzaroni, M. -E., De Andrade, A. R., Reginatto, V., Adding value to lignocellulosic byproducts by using acetate and p-coumaric acid as substrate in a microbial fuel cell. *Ind Crop Prod* 2021;171:113844.
346. Catal, T., Liu, H., Fan, Y., Bermek, H., A clean technology to convert sucrose and lignocellulose in microbial electrochemical cells into electricity and hydrogen. *Bioresour Technol Rep* 2019;5:331–334.
347. Jablonska, M. A., Rybarczyk, M. K., Lieder, M., Electricity generation from rapeseed straw hydrolysates using microbial fuel cells. *Bioresour Technol* 2016;208:117–122.
348 Gurav, R., Bhatia, S. K., Choi, T. R., Kim, H. J., Song, H. S., Park, S. L., Lee, S. M., Lee, H. S., Kim, S. H., Yoon, J. J., Yang, Y. H., Utilization of different lignocellulosic hydrolysates as carbon source for electricity generation using novel *Shewanella marisflavi* BBL25. *J Clean Prod* 2020;277:124084.
349. Flimban, S. G. A., Hassan, S. H. A., Rahman, M. M., Oh, S. E., The effect of Nafion membrane fouling on the power generation of a microbial fuel cell. *Int J Hydrogen Energy* 2020;45:13643–13651.
350. Mohd Zaini Makhtar, M., Tajarudin, H. A., Electricity generation using membrane-less microbial fuel cell powered by sludge supplemented with lignocellulosic waste. *Int J Energy Res* 2020;44:3260–3265.
351. Yoshimura, Y., Nakashima, K., Kato, M., Inoue, K., Okazaki, F., Soyama, H., Kawasaki, S., Electricity generation from rice bran by a microbial fuel cell and the influence of hydrodynamic cavitation pretreatment. *ACS Omega* 2018;3:15267–15271.
352. Jenol, M. A., Ibrahim, M. F., Bahrin, E. K., Kim, S. W., Abd-Aziz, S., Direct bioelectricity generation from sago hampas by *Clostridium beijerinckii* SR1 using microbial fuel cell. *Molecules* 2019;24:2397.
353. Shabani, M., Younesi, H., Pontié, M., Rahimpour, A., Rahimnejad, M., Zinatizadeh, A. A., A critical review on recent proton exchange membranes applied in microbial fuel cells for renewable energy recovery. *J Clean Prod* 2020;264:121446.
354. Bazrgar Bajestani, M., Mousavi, S. A., Effect of casting solvent on the characteristics of Nafion/TiO_2 nanocomposite membranes for microbial fuel cell application. *Int J Hydrogen Energy* 2016;41(1):476–482.
355. Sivasankaran, A., Sangeetha, D., Influence of sulfonated SiO_2 in sulfonated polyether ether ketone nanocomposite membrane in microbial fuel cell. *Fuel* 2015;159:689–696.
356. Shabani, M., Younesi, H., Rahimpour, A., Rahimnejad, M., Upgrading the electrochemical performance of graphene oxide-blended sulfonated polyetheretherketone composite polymer electrolyte membrane for microbial fuel cell application. *Biocatal Agric Biotechnol* 2019;22:101369.
357. Ahilan, V., de Barros, C. C., Bhowmick, G. D., Ghangrekar, M. M., Murshed, M. M., Wilhelm, M., Rezwan, K., Microbialfuel cell performance of graphitic carbon functionalized porous polysiloxane based ceramic membranes. *Bioelectrochemistry* 2019;129:259–269.
358. Aulin, C., Gällstedt, M., Lindström, T., Oxygen and oil barrier properties of microfibrillated cellulose films and coatings. *Cellulose* 2010;17(3):559–574.
359. Pal, M., Sharma, R. K., Exoelectrogenic response of *Pichia fermentans* influenced by mediator and reactor design. *J Biosci Bioeng* 2019;127:714–720.
360. Christwardana, M., Frattini, D., Accardo, G., Yoon, S. P., Kwon, Y., Effects of methylene blue and methyl red mediators on performance of yeast based microbial fuel cells adopting polyethylenimine coated carbon felt as anode. *J Power Sources* 2018;396:1–11.
361. Chauhan, S., Sharma, V., Varjani, S., Sindhu, R., Chaturvedi Bhargava, P., Mitigation of tannery effluent with simultaneous generation of bioenergy using dual chambered microbial fuel cell. *Bioresour Technol* 2022;351:127084.

362. Moreno, L., Nemati, M., Predicala, B., Biodegradation of phenol in batch and continuous flow microbial fuel cells with rod and granular graphite electrodes. *Environ Technol* 2018:39;144–156.
363. Marcílio, R., Neto, S. A., Ruvieri, B. M., Andreote, F. D., deAndrade, A.R., Reginatto, V., Enhancing the performance of an acetate-fed microbial fuel cell with methylene green. *Braz J Chem Eng* 2021;38:471–484.
364. Ajunwa, O. M., Odeniyi, O. A., Garuba, E. O., Marsili, E., Onilude, A. A., Influence of enhanced electrogenicity on anodic biofilm and bioelectricity production by a novel microbial consortium. *Process Biochem* 2021;104:27–38.
365. Flimban, S. G., Ismail, I. M., Kim, T., Oh, S. E., Overview of recent advancements in the microbial fuel cell from fundamentals to applications: Design, major elements, and scalability. *Energies* 2019;12:3390. DOI: 10.3390/en12173390
366. Ringeisen, B. R., Henderson, E., Wu, P. K., Pietron, J., Ray, R., Little, B., Biffinger, J. C., Jones-Meehan, J. M., High power density from a miniature microbial fuel cell using *Shewanella oneidensis* DSP10. *Environ Sci Technol* 2006;40:2629–2634.
367. He, Z., Wagner, N., Minteer, S. D., Angenent, L. T., An upflow microbial fuel cell with an interior cathode: Assessment of the internal resistance by impedance spectroscopy. *Environ Sci Technol* 2006;40:5212–5217.
368. He, Z., Minteer, S. D., Angenent, L. T., Electricity generation from artificial wastewater using an upflow microbial fuel cell. *Environ Sci Technol* 2005;39:5262–5267.
369. Feng, Y., He, W., Liu, J., Wang, X., Qu, Y., Ren, N., A horizontal plug flow and stackable pilot microbial fuel cell for municipal wastewater treatment. *Bioresour Technol* 2014;156:132–138.
370. Gajda, I., Greenman, J., Ieropoulos, I. A., Recent advancements in real-world microbial fuel cell applications. *Curr Opin Electrochem* 2018;11:78–83. DOI: 10.1016/j.coelec.2018.09.006
371. Shin, S. H., Choi, Y. J., Na, S. H., Jung, S. H., Kim, S. H., Development of bipolar plate stack type microbial fuel cells. *Bull Korean Chem Soc* 2006;27:281–285.
372. An, J., Kim, B., Jang, J. K., Lee, H. -S., Chang, I. S., New architecture for modulization of membraneless and single-chambered microbial fuel cell using a bipolar plate-electrode assembly (BEA). *Biosens Bioelectron.* 2014;59:28–34.
373. Aelterman, P., Rabaey, K., Pham, H. T., Boon, N., Verstraete, W., Continuous electricity generation at high voltages and currents using stacked microbial fuel cells. *Environ Sci Technol* 2006;40:3388–3394.
374. Vilajeliu-Pons, A., Puig, S., Salcedo-Dávila, I., Balaguer, M., Colprim, J., Long-term assessment of six-stacked scaled-up MFCs treating swine manure with different electrode materials. *Environ Sci* 2017;3:947–959.
375. Zhuang, L., Yuan, Y., Wang, Y., Zhou, S., Long-term evaluation of a 10-liter serpentine-type microbial fuel cell stack treating brewery wastewater. *Bioresour Technol* 2012;123:406–412.
376. An, J., Sim, J., Lee, H. -S., Control of voltage reversal in serially stacked microbial fuel cells through manipulating current: Significance of critical current density. *J Power Sources* 2015;283:19–23.
377. He, W., Wallack, M. J., Kim, K. -Y., Zhang, X., Yang, W., Zhu, X., Feng, Y., Logan, B. E., The effect of flow modes and electrode combinations on the performance of a multiple module microbial fuel cell installed at wastewater treatment plant. *Water Res* 2016;105:351–360.
378. Ge, Z., He, Z., Long-term performance of a 200 liter modularized microbial fuel cell system treating municipal wastewater: Treatment, energy, and cost. *Environ Sci Water Res Technol* 2016;2:274–281.
379. He, L., Du, P., Chen, Y., Lu, H., Cheng, X., Chang, B., Wang, Z., Advances in microbial fuel cells for wastewater treatment. *Renew Sustain Energy Rev* 2017;71:388–403.

380. Premier, G., Michie, I., Boghani, H., Fradler, K., Kim, J., Reactor design and scale-up. In K. Scott and E. Hao Yu (Eds.), *Microbial Electrochemical and Fuel Cells*. Elsevier, Amsterdam, the Netherlands, 2016, pp. 215–244.
381. Rahimnejad, M., Bakeri, G., Najafpour, G., Ghasemi, M., Oh, S. -E., A review on the effect of proton exchange membranes in microbial fuel cells. *Biofuel Res J* 2014;1:7–15.
382. Chin, M. Y., Phuang, Z. X., Woon, K. S., Hanafiah, M. M., Zhang Z., Liu,X., Life cycle-assessment of bioelectrochemical and integrated microbial fuel cell systems for sustainable wastewater treatment and resource recovery. *J Environ Manag* 2022;320:115778.
383. Ayol, A., Peixoto, L., Keskin, T., Abubackar, H. N., Reactor designs and configurations for biological and bioelectrochemical C1 gas conversion: A review. *Int J Environ Res Public Health* 2021:18:11683.
384. Jadhav, D. A., Park, S. -G., Eisa, T., Mungray, A. K., Madenli, E. C., Olabi, A. -G., Abdelkareem, M. A., Chae, K. -J., Current outlook towards feasibility and sustainability of ceramic membranes for practical scalable applications of microbial fuel cells. *Renew Sustain Energy Rev* 2022;167:112769.
385. Janicek, A., Fan, Y., Liu, H., Design of microbial fuel cells for practical application: A review and analysis of scale-up studies. *Biofuels* 2014;5:79–92.

6 Self-Powered Electrochemical Systems and Nanogenerators

6.1 INTRODUCTION

Electrochemistry, which refers to the interrelation of electrical and chemical effects, has played a crucial part in the sustainable advancement and innovation of industrial processes, including chemical industry, medicine, materials, energy, metal corrosion, and environmental science. The electrochemical process highly depends on external power supply, which aggravates the crisis of energy shortage and environmental pollution problems in modern society. To solve these issues, developing a self-powered (SP) electrochemical system (SPES), which can operate by integrating electrochemical system with energy harvesting technology for collecting energy from the environment, is one of the promising approaches. Due to the sufficient availability of mechanical energy, converting it from the environment to electricity has aroused broad attention. Recently, many technologies that extract mechanical energy have been reported, such as electromagnetic generator (EMG), triboelectric nanogenerator (TENG), and piezoelectric nanogenerator (PENG). However, the low energy conversion efficiency of EMG at low frequency and low output powers of PENG and TENG limits their practical applications [1–14].

The Internet of things (IoT) [15] and smart cities [16,17] are emerging themes in the literature that require sensors integration, based on recent advances in wearable electronics that are classified as epidermal devices and implantable devices [18], since these materials can be worn on or in the body, facilitating a plethora of applications such as the detection of toxic gases, volatile organic compounds, ultraviolet (UV) radiation on the skin, and temperature monitoring. Some critical properties for these devices (such as skin integration, stretchability, and long life cycle—which have been considered intrinsic drawbacks for conventional sensors and actuators) define the type of material to be explored. In addition to these required properties, there are critical factors relative to the access level of these devices: some implantable components (particularly depleted components in batteries for cardiac pacemakers, defibrillators, and deep brain stimulators) that require periodical substitution, representing a limiting aspect for implantables [19]. To circumvent the disadvantages of conventional energy storage devices, different mechanisms have been explored in the scavenge of energy from different movements (walking, jumping, running, or muscle contraction, relaxation, and cardiac motion) to reinforce the concept of SP sensors and power sources [20,21]. The overall mechanical energy involved in our daily activities is expected to reach overall power of up to 67 W [20] that can be conveniently harvested by nanogenerators, which are classified as triboelectric and piezoelectric devices [22].

DOI: 10.1201/9781003429906-6

The technology of nanogenerators favors plenty of new devices such as biosensors and health monitoring human–computer interfaces [23] in addition to the improvement in the performance of electrochemical devices (energy storage systems such as supercapacitors and batteries). In terms of sensors and biosensors, the incorporation of nanogenerators considers not only a primary role of generating energy to feed the electronic components (in prototypes of SP devices) but also for use as a sensing detection element that must be affected by the adsorption of the target molecules into nanogenerators. An example of this process is the creation of gate potential as reported by Selvarajan et al. [24] as a consequence of the interference of glucose on the flow of free electrons of nanogenerators, creating a direct relationship between the generated voltage and the amount of analyte (glucose). This process is facilitated by the modification of piezoelectric components with enzymes that interact with the analyte and modulate the generated power [18]. Consequently, the modulated potential can be explored as a potential parameter for the identification of specific components.

For monitoring physiological signals, nanogenerators represent relevant alternatives for diaphragmatic breathing monitoring in substitution into conventional sensors. The relation between output voltage from the nanogenerator and the respiration signal represents an important application for this SP prototype of the sensor [25]. In terms of application in energy power devices, the primary activity of energy generation has been added to an auxiliary role that refers to the association with supercapacitors in which the intermittent generation of electricity improves the energy density of wearable supercapacitors [26]. It is worth mentioning that these promising strategies to integrate nanogenerators and sensors, biosensors, and energy storage devices need to circumvent drawbacks from conventional materials such as comfort, flexibility, retention of properties under successive washing procedures, and wear-and-tear processes.

With the gradual rollout of the fifth-generation ((5G) mobile networks) technology across the world, the role of the IoT is becoming more and more essential in both industrial and commercial developments [27]. Entering the IoT era, wireless and portable electronics are undergoing explosive advancement, with the total number increasing tremendously and power consumption decreasing significantly. Affected by the nature of incredibly huge numbers, small power consumption, and ultra widely distributed location of IoT devices, energy supply in the new era should be changed from the centralized, immobile, ordered, and large-scale mode to the distributed, mobile, in situ, and small-scale mode. Yet, batteries as the primary choice of conventional energy supply exhibit apparent drawbacks such as the limited lifespan that needs frequent replacement or recharging, large volume and weight, rigidity, biological incompatibility, and environmental pollution. Thus, to meet the above requirements, the ideal energy supply units should possess the characteristics of portability, sustainability, miniaturization, wearability, and implantability depending on the applications [28,29]. In this regard, scavenging energy from the ambient surroundings by energy harvesters can provide a green, portable, and sustainable solution.

There exist various types of energy forms that are normally wasted in the environment, including mechanical energy associated with diverse nature vibrations and human activities, thermal energy, and solar energy. Accordingly, different types of energy transducing mechanisms and generators (i.e., energy harvesters) have been developed to scavenge these wasted energies, such as the EMG, PENG, and TENG for mechanical energy, the thermoelectric (TE) generator (TEG) and pyroelectric nanogenerator (PyENG) for thermal energy, and the solar cell (SC) for solar/light energy. In most scenarios, multitype energy forms coexist in the ambient environment, and the strength of a single-type energy form may vary significantly from time to time, such as solar energy varying with time and weather, random wind/wave energy, and different human activities. Therefore, generators employing a single-energy transducing mechanism may greatly suffer from unstable energy source, low energy utilization rate, low conversion efficiency, and low adaptability in different scenarios. However, hybrid generators utilizing integrated transducing mechanisms can be more effective in harvesting both single-type and multitype energies. Through synergetic designs, hybrid generators can compensate for the shortcomings of each mechanism, improve the space utilization efficiency, enhance the energy utilization rate, and thus can be applied in diverse scenarios as the energy supply units [30,31].

Among different energy forms, mechanical energy is one of the most ubiquitous energy sources, widely existing in water wave, wind, sound/ultrasound, machinery vibration, human motions, etc. Considering that most IoT devices are adopted in the environment- or human-relative applications where they exhibit abundant mechanical energy, generators with mechanical energy harvesting ability will be most desirable. Since the first invention in 2012 by Prof. Wang and his team [32], the TENG technology has been extensively explored for mechanical energy harvesting, due to its superior advantages of high output performance, versatile operation modes, broad material availability, wearable/implantable compatibility, simple fabrication, high scalability, and low cost [33,34]. Benefitting from these merits, integrating TENG with other transducing mechanisms yields a promising research direction for developing hybrid generators, which has received flourishing development in the past few years [35,36]. Furthermore, after integrating the TENG-based hybrid generators with power management circuitry, energy storage units, and functional components, a variety of hybridized systems with self-sustainability can be achieved for broad applications.

To scavenge different types of ambient available energies, generators based on different transducing mechanisms have been developed. As shown in Figure 6.1, TENG/EMG/PENG, PyENG/TEG, and SC are the most common generators employed for mechanical, thermal, and solar/light energy harvesting, respectively. Through synergetic integration, their hybrid generators can be further employed for complementary and effective energy harvesting of both single-type and multitype energies. In this chapter, we will mainly focus on TENG, EMG, PENG, and PyENG along with hybrid nanogenerators. TEG and SC were discussed extensively in my previous book on "Advanced power generation systems: Thermal sources." They will only be discussed here in the context of hybrid nanogenerators.

6.2 TENG NANOGENERATOR

A TENG is an energy harvesting device that converts external mechanical energy into electricity by a conjunction of triboelectric effect and electrostatic induction. The first TENG was invented in 2012 by Wang's group at the Georgia Institute of Technology [1]. The origin of TENG and other types of nanogenerators (e.g., PENG and PyENG) is Maxwell's displacement current theory, i.e., external current induced by a time-varying electric field in the nanogenerators [20]. In particular, when two dissimilar materials contact with each other, due to their different electron affinity, surface charges are generated at the contact interface. Then, upon separation, the built-up electric potential will drive electrons on the respective electrodes to flow in the external circuit until a new balance is achieved. In other words, in the inner circuit, a potential is created by the triboelectric effect due to the charge transfer between two thin organic/inorganic films that exhibit opposite tribo-polarity; in the outer circuit, electrons are driven to flow between two electrodes attached on the back sides of the films in order to balance the potential. If the two materials are brought into contact again, the electric potential will disappear and electrons will flow in a reverse direction. Thus, under periodic contact and separation, alternating current (AC) can be generated on the external load, and a rectification circuit is normally required for energy storage to convert the AC output into the direct current (DC) output [37,38]. There are also some advanced designs of DC-TENGs that do not need a rectification circuit before energy storage [39,40], but they exhibit more complicated structures and most developed TENGs are still AC-based. According to a material's ability to lose or accept electrons (electron affinity), different materials can be arranged into a sequence called the triboelectric series, from the most positive material to the most negative material [41,42].

Since the most useful materials for TENG are organic, it is also named organic nanogenerator, which is the first of using organic materials for harvesting mechanical energy. Ever since the first report of the TENG in January 2012, as shown in Figure 6.1, the output power density of TENG has been improved by five orders of magnitude within 12 months. The area power density reached 313 W/m^2, volume density reached 490 kW/m^3, and conversion efficiencies of ~60%–72% have been demonstrated. Besides the unprecedented output performance, this new energy technology also has a number of other advantages, such as low cost in manufacturing and fabrication, excellent robustness and reliability, and environmental friendliness. The TENG can be applied to harvest all kinds of mechanical energy that is available but wasted in our daily life, such as human motion, walking, vibration, mechanical triggering, rotating tire, wind, and flowing water.

During the last several years, different models including the electron-cloud-potential-well model have been developed to explain the origin of contact electrification among polymers, metals, semiconductors, and even liquids [43,44]. It is revealed that in most cases, electron transfer is the dominating effect of the contact electrification process. In particular, for the solid–liquid contact electrification, a two-step formation of the electric double layer (EDL) at the solid–liquid interface is proposed by Wang, which is also known as the Wang model [44,45]. The Wang model indicates that the electron transfer is required at the very first contact to create the first layer of electrostatic charges on the solid surface, and then, the ion transfer in solution dominates in the second step due to the electrostatic interactions with the charged solid surface. The surface charge density, as one of the most important parameters determining the output performance of TENGs, can be enhanced through several strategies, e.g., proper material selection (contact materials with a larger difference in electron affinity), surface modification (such as micro-/nanostructures, chemical treatment, and ion injection), structural optimization, middle layer insertion, and circuitry assistance [46–49].

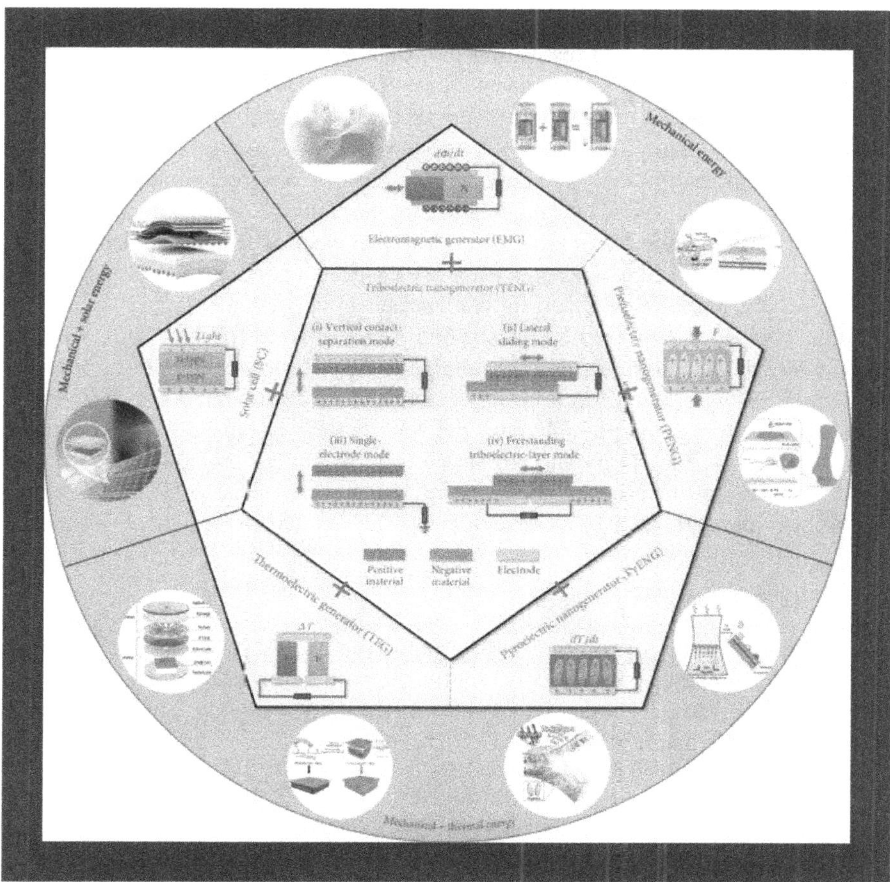

FIGURE 6.1 Overview of different transducing mechanisms for converting different forms of energy into electricity and their respective generators for hybrid generator integration [4].

In general, there are four basic operation modes of TENG, i.e., vertical contact–separation mode, lateral sliding mode, single-electrode mode, and freestanding triboelectric layer mode [50]. All these four modes share the same output equation. Based on these operation modes, a large variety of TENGs have been developed for both the mechanical energy harvesting and SP ambient parameter monitoring/intervention [51,52], such as micro-/nanopower sources [53], blue energy [54], physical/chemical sensors [55,56], human–machine interfaces [57,58], nerve/muscle/brain stimulators [59,60], air filters [61], droplet manipulation [62], and high-voltage applications [63,64], and wearable and implantable electronics. Hence, TENGs can contribute to not only power sources but also various functional components in self-sustainable hybridized systems.

TENG exhibits many unique advantages including light structural simplicity, diverse materials options, and high conversion efficiency. Moreover, TENG has demonstrated its potential applications in micro-/nanopower sources, SP sensors, large-scale blue energy, and direct high-voltage power sources. To promote the practical application of TENG, significant research efforts have been focused on improving its output performance via investigating the basic principles, enhancing the surface charge density, and power management; thus, 500 W/m^2 of area power density and >50% of conversion efficiency have been achieved, respectively. In view of these advantages, TENG can be served as a promising alternative energy harvesting power source to integrate with electrochemistry for various electrochemical operations. Currently, many SPESs based on TENG have been developed, which are elaborated in further detail as follows. Compared with traditional electrochemical systems, SPESs can drive the electrochemical process without the external power supply, which largely promotes the sustainable development of electrochemical systems.

Podila's group at Clemson University [65] demonstrated the first truly wireless TENGs, which were able to wirelessly charge energy storage devices (e.g., batteries and capacitors) without the need for any external amplification and boosters. These wireless generators could possibly pave the way for new systems that could be used to harvest mechanical energy and wirelessly transmit the generated energy for storage. A summary of the progress made in the output power density of TENGs within 12 months in the early days of its development is illustrated in Figure 6.2.

6.2.1 TENG Operation modes

The TENG has four basic operation modes: vertical contact–separation mode, in-plane sliding mode, single-electrode mode, and freestanding layer-based mode. They have different characteristics and are suitable for different applications. These modes are described in some detail as follows.

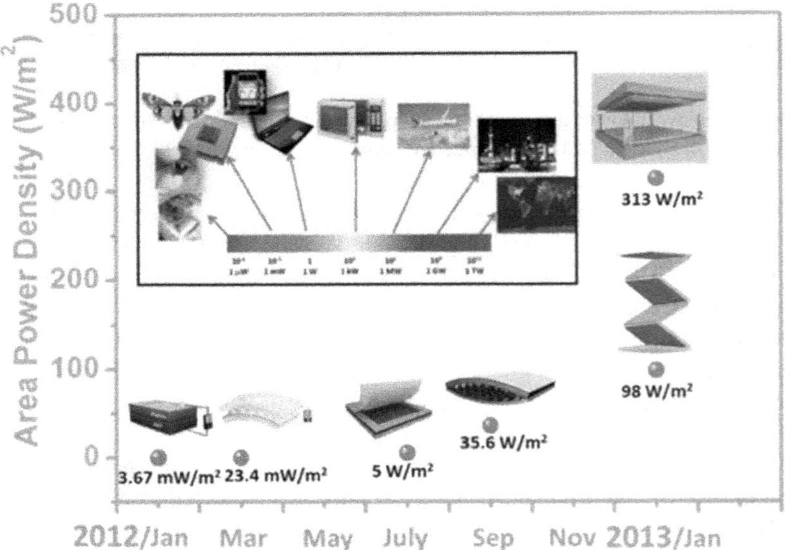

FIGURE 6.2 A summary of the progress made in the output power density of triboelectric nanogenerators within 12 months in the early days of its development [12].

6.2.1.1 Vertical Contact–Separation Mode

This mode is graphically illustrated in Figure 6.3. When mechanical agitation is applied onto the device to bend or press it, the inner surfaces of the two sheets will get into close contact and the charge transfer will begin, leaving one side of the surface with positive charges and the other with negative charges due to triboelectric effect. When the deformation is released, the two surfaces with opposite charges will separate automatically, so that these opposite triboelectric charges will generate an electric field in between and thus induce a potential difference across the top and bottom electrodes.

To screen this potential difference, the electrons will be driven to flow from one electrode to the other through the external load. The electricity generated in this process will continue until the potentials of the two electrodes get back to even again. Subsequently, when the two sheets are pressed toward each other again, the triboelectric charge-induced potential difference will begin to decrease to zero, so that the transferred charges will flow back through the external load, to generate another current pulse in the opposite direction. When this periodic mechanical deformation lasts, the AC signals will be continuously generated. As for the pair of materials getting in contact and generating triboelectric charges, at least one of them need to be an insulator, so that the triboelectric charges cannot be conducted away but will remain on the inner surface of the sheet. Then, these immobile triboelectric charges can induce AC electricity flow in the external load under the periodic distance change.

6.2.1.2 Lateral Sliding Mode

There are two basic friction processes: normal contact and lateral sliding. Here, a TENG is designed based on the in-plane sliding between the two surfaces in the lateral direction. With an intensive triboelectrification facilitated by sliding friction, a periodic change in the contact area between two surfaces leads to a lateral separation of the charge centers, which creates a voltage drop for driving the flow of electrons in the external load. The sliding-induced electricity generation mechanism is schematically depicted in Figure 6.4. In the original position, the two polymeric surfaces fully overlap and intimately contact with each other. Because of the large difference in the ability to attract electrons, the triboelectrification will leave one surface with net positive charges and the other with net negative charges with equal density. Since the

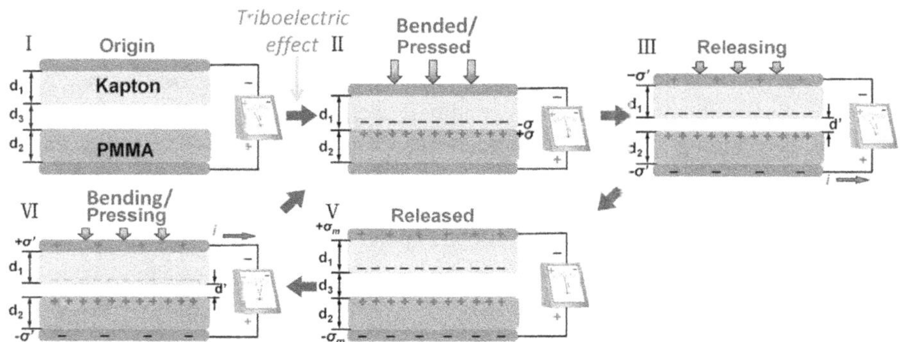

FIGURE 6.3 Vertical contact–separation mode of triboelectric nanogenerator [12].

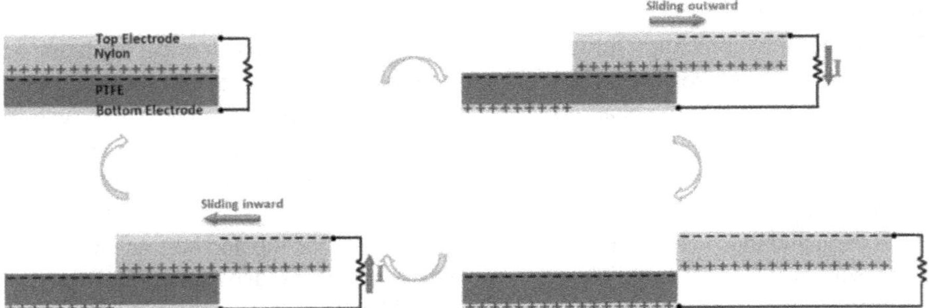

FIGURE 6.4 Lateral sliding mode of triboelectric nanogenerator [12].

tribo-charges on the insulators will only distribute in the surface layer and will not be leaked out for an extended period of time, the separation between the positively charged surface and the negatively charged surface is negligible at this overlapping position, and thus, there will be little electric potential drop across the two electrodes. Once the top plate with the positively charged surface starts to slide outward, the in-plane charge separation is initiated due to the decrease in contact surface area.

The separated charges will generate an electric field pointing from the right to the left almost parallel to the plates, inducing a higher potential at the top electrode. This potential difference will drive a current flow from the top electrode to the bottom electrode in order to generate an electric potential drop that cancels the tribo-charge-induced potential. Because the vertical distance between the electrode layer and the tribo-charged polymeric surface is negligible compared with the lateral charge separation distance, the amount of the transferred charges on the electrodes approximately equals the amount of the separated charges at any sliding displacement. Thus, the current flow will continue with the continuation of the ongoing sliding process that keeps increasing the separated charges, until the top plate fully slides out of the bottom plate and the tribo-charged surfaces are entirely separated. The measured current should be determined by the rate at which the two plates are being slid apart. Subsequently, when the top plate is reverted to slide backward, the separated charges begin to get in contact again but no annihilation due to the insulator nature of the polymer materials. The redundant transferred charges on the electrodes will flow back through the external load with the increase in the contact area, to keep the electrostatic equilibrium. This will contribute to a current flow from the bottom electrode to the top electrode, along with the second half cycle of sliding. Once the two plates reach the overlapping position, the charged surfaces get into full contact again. There will be no transferred charges left on the electrode, and the device returns to the first state. In this entire cycle, the processes of sliding outward and inward are symmetric, so a pair of symmetric AC peaks should be expected.

The mechanism of in-plane charge separation can work in either one-directional sliding between two plates or rotation mode. In the sliding mode, introducing linear grating or circular segmentation on the sliding surfaces is an extremely efficient means of energy harvesting. With such structures, two patterned triboelectric surfaces can get to a fully mismatching position through a displacement of only a grating unit length rather than the entire length of the TENG so that it dramatically increases the transport efficiency of the induced charges.

6.2.1.3 Single-Electrode Mode

This mode is graphically illustrated in Figure 6.5. A single-electrode-based TENG is introduced as a more practical and feasible design for some applications such as fingertip-driven TENG. The working principle of the single-electrode TENG is schematically shown in the figure by the coupling of contact electrification and electrostatic induction. In the original position, the surfaces of skin and polymethyl siloxane (PDMS) are fully contact with each other, resulting in charge transfer between them. According to the triboelectric series, electrons were injected from the skin to the PDMS since the PDMS is more triboelectrically negative than the skin, which is the contact electrification process. The produced triboelectric charges with opposite polarities are fully balanced/screened, leading to no electron flow in the external circuit. Once a relative separation between

PDMS and skin occurs, these triboelectric charges cannot be compensated. The negative charges on the surface of the PDMS can induce positive charges on the indium–tin–oxide (ITO) electrode, driving free electrons to flow from the ITO electrode to the ground. This electrostatic induction process can give an output voltage/current signal if the distance separating between the touching skin and the bottom PDMS is appreciably comparable to the size of the PDMS film. When negative triboelectric charges on the PDMS are fully screened from the induced positive charges on the ITO electrode by increasing the separation distance between the PDMS and skin, no output signals

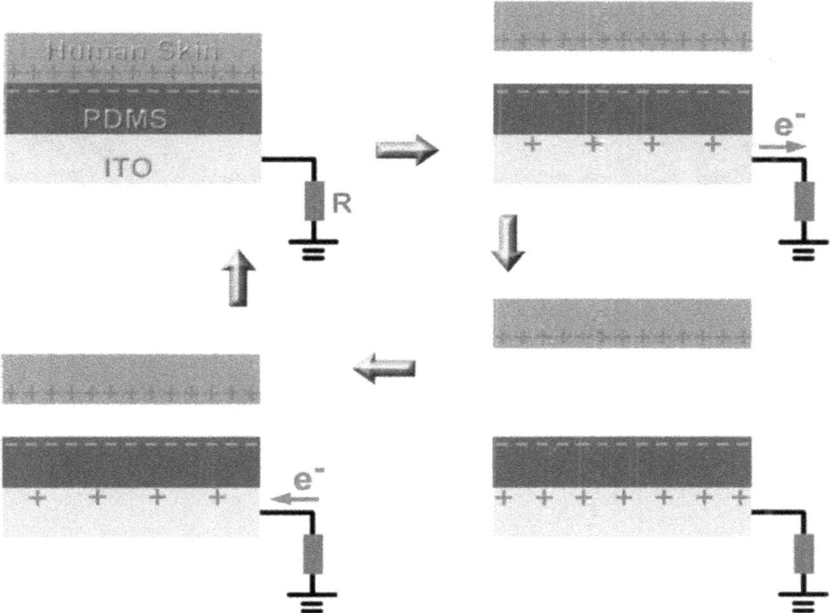

FIGURE 6.5 Single-electrode mode of triboelectric nanogenerator [12].

can be observed, as illustrated. Moreover, when the skin was reverted to approach the PDMS, the induced positive charges on the ITO electrode decreased and the electrons will flow from the ground to the ITO electrode until the skin and PDMS fully contact with each other again, resulting in a reversed output voltage/current signal. This is a full cycle of electricity generation process for the TENG in contact–separation mode.

6.2.1.4 Freestanding Triboelectric Layer-Based Nanogenerator

To prevent the electrostatic shield effect, the freestanding triboelectric layer-based nanogenerator (FTENG) is proposed with the paired-electrode structure [66]. In FTENG with contact mode (CFTENG), the freestanding layer is placed between the two opposite electrodes. In the CFTENG in Figure 6.6a, two Al electrodes are fixed in an acrylic frame with a fluorinated ethylene propylene (FEP) layer (the freestanding layer) supported by springs [14]. Compared with CFTENG, the FTENG with sliding mode (SFTENG) is more widely used with different structural designs. Firstly, SFTENG could be designed with a structure composed of a mover and a stator [67,68]. As the freestanding layer, the mover does not necessarily have to be attached with an electrode and a lead wire, and it could be fixed on moving objects such as humans and vehicles to monitor their motions [66]. As shown in Figure 6.6b, the mover is composed of two columns of copper strip coated on a polyimide (PI) film, which is attached to an acrylic sheet, while the stator is composed of a polytetrafluoroethylene (PTFE) film as the dielectric layer, and two interdigitated electrodes with a PI and an acrylic sheet as the base [68]. The interdigitated electrodes can lead to an alternating output when the mover slides linearly on the stator. Secondly, SFTENG could be designed

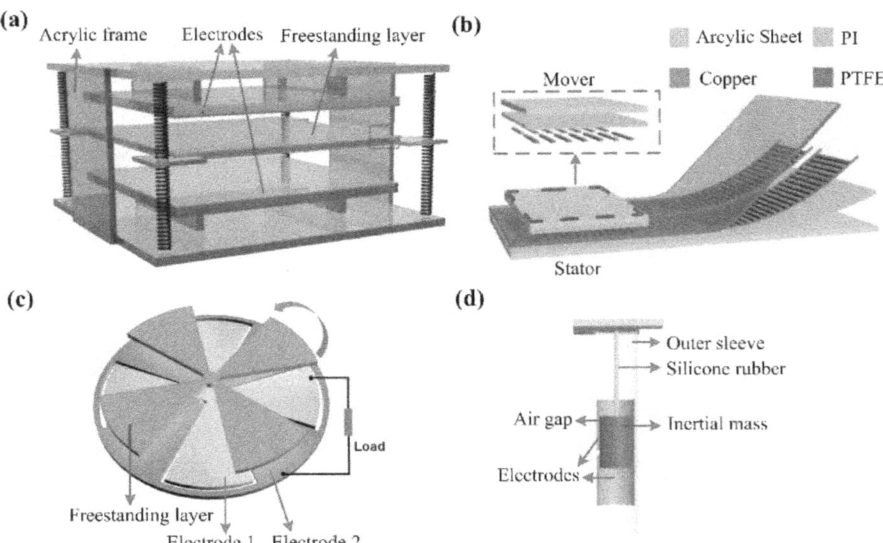

FIGURE 6.6 Structures of freestanding triboelectric layer-based TENGs: (a) an FTENG with contact mode for vibration sensing; (b) a FTENG with sliding mode for translational motion sensing; (c) a noncontact TENG with a disk-like structure; (d) a noncontact TENG with a tube-like structure [14].

with a noncontact structure, in which there is an air gap between the tribo-pair during the working process, to prevent energy loss from friction. Based on the noncontact structure, Lin et al. proposed a disk-like TENG as shown in Figure 6.6c [69]. The two electrodes of the device are fixed on the stationary base with the electricity generated in the rotation process of the freestanding layer. Yu et al. proposed a tube-like TENG, which is of a larger specific sensing area compared with planar devices [70]. As shown in Figure 6.6d, the device is composed of an outer sleeve tube coated with two Cu electrodes and an inner cylindrical inertial mass suspended by highly stretchable silicone rubber. They found that the output of the device decreases with the increase in the gap, which is controlled at 2 mm to obtain the best performance.

TENG is a physical process of converting mechanical agitation to an electric signal through triboelectrification (in the inner circuit) and electrostatic induction processes (in the outer circuit). This basic process has been demonstrated for two major applications. The first application is energy harvesting with a particular advantage of harvesting mechanical energy from various sources. The other application is to serve as a SP active sensor, because it does not need an external power source to drive. Here, we briefly examine both of these applications.

6.2.2 Use of TENG for Harvesting Mechanical Energy

Mechanical energy can be captured from a number of different sources such as energy from vibrations, energy from human body motions, energy from various sources in environment, and blue energy from distributed waves. Vibrations occur during walking or running, audible voices, shaking of engine, movements of transportation vehicles such as automobiles, bikes, train, aircraft, and gusty wind flow. This vibration energy can be harvested to power mobile electronics either individually or in a suitable combination. Various technologies based on the fundamental principles of TENGs have been demonstrated for harvesting vibration energy. These technologies can be classified into four categories. First, for microelectromechanical systems (MEMS), a cantilever-based technique can be used. By designing the contact surface of a cantilever with the top and bottom surfaces during vibration, TENG has been demonstrated for harvesting ambient vibration energy based on the contact–separation mode. Second, a rationally designed TENG with integrated rhombic gridding can effectively harvest energy from a backpack. This system greatly improves the total current output owing to structurally multiplied unit cells connected in parallel. Third, vibration energy from an automobile engine, a sofa, and a desk can be effectively captured by a harmonic resonator-based TENG with the use of four supporting springs that have been fabricated based on the resonance-induced contact–separation between the two triboelectric materials. Fourth, low-frequency, random vibration energy in multiple directions over a wide bandwidth can be harvested by a three-dimensional (3D)-TENG which is designed based on a hybridization mode of conjunction with the vertical contact–separation mode and the in-plane sliding mode. The last category allows the use of TENG in environmental/infrastructure monitoring, charging portable electronics, and IoT.

Mechanical energy generated from everyday motions of the human body can also be converted to electricity by TENG. This electricity can be used to charge portable electronics and biomedical devices and thereby expand their usage. A packaged power-generating insole with built-in flexible multilayered TENGs has been demonstrated. This insole acts as a direct power source by harvesting energy from mechanical pressure during normal walking. The TENG used here relies on the contact–separation mode and is effective in responding to the periodic compression of the insole. The insole can also be used for a fully packaged self-lighting shoe that has broad applications for display and entertainment purposes. A TENG can also be attached to the inner layer of a shirt for harvesting energy from body motion. The TENG with a single layer size of 2 cm × 7 cm × 0.08 cm sticking on the clothes was demonstrated as a sustainable power source that not only can directly light up 30 light-emitting diodes (LEDs), but also can charge a lithium-ion battery by persistently clapping clothes.

TENG is an ideal energy harvester, which can collect diverse mechanical energies from the environment, especially low-frequency mechanical energies, such as wind and water waves. Ever since 2013, Wang's group reported a rotary TENG for harvesting wind energy, various types of TENGs such as 3D spiral structure TENG to collect wave energy, fully enclosed TENGs for water, and harsh environment and multilayered disk nanogenerator for harvesting hydropower have been introduced. In these nanogenerators, however, the friction generated between layers of the TENG reduces the energy conversion efficiency and the durability of the device. Wang's group subsequently developed a frictionless electrostatic induction nanogenerator and showed that it exhibited high energy conversion efficiency and excellent durability. Circulation networks composed of such frictionless nanogenerators could be used to harvest water wave energy and continuously provide energy to some wireless devices. TENG has also been developed to harvest droplet energy from the environment. This nanogenerator relies on the contact electrification effect between liquid and solid to generate electricity. The advantage of this liquid–solid model is that it effectively avoids the wear of nanogenerators. Zi [71] reported a hybrid cell (HC) combining a silicon SC with a water droplet harvesting TENG. This HC had the capacity of harvesting both solar and raindrop energies.

For effectively harvesting distributed wave energy, the TENG is conceived to be organized in networks, which can have a hierarchical structure of modules [72]. The network structure also enables the device to be applied in different scales of harvesting, ranging from SP systems to large-scale clean energy (Figure 6.7a). In the development of TENGs for blue energy, efforts are mainly focusing on four aspects: TENG unit design, networking strategy, power management, and application system (Figure 6.7b). Of these four aspects, TENG unit design is the most important step and it is continuously improved by focusing on the structure, principle, and material to enhance its energy harvesting performance and to meet demands raised by various ocean environments, both on the water surface and beneath it. A networking strategy is then adopted to add outputs of single TENG units and expand them in a reliable way. It decides the connection pattern of massive TENG units and the coupling effect between TENG units, which could further enhance the performance. Before finally supplying electrical power to the application system, power management is required to manipulate the TENG output for a better match with appliances and improve the power efficiency with circuit approaches.

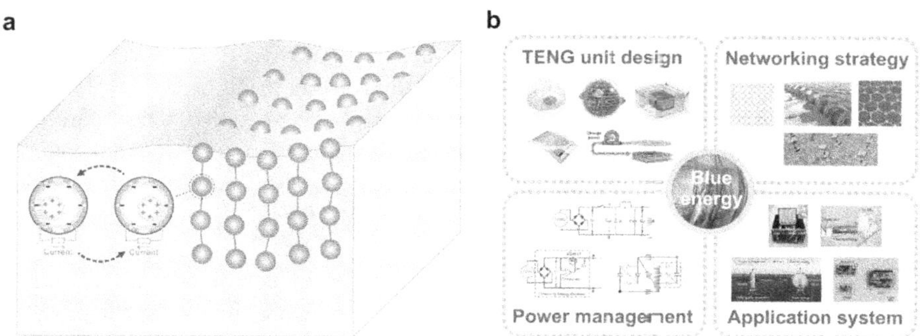

FIGURE 6.7 Schematics of blue energy harvesting based on TENGs. (a) Schematic diagram of the TENG network for harvesting wave energy. Reprinted with permission from ref. (b) Schematic diagram of major aspects of blue energy harvesting based on TENGs, including TENG unit design, networking strategy, power management, and application system [13].

6.2.3 Use of TENG for Various Types of Sensors

6.2.3.1 Self-Powered Active Strain/Force Sensors

A TENG automatically generates an output voltage and current once it is mechanically triggered. The magnitude of the output signal signifies the impact of the mechanical deformation and its time-dependent behavior. This is the basic principle of the TENG, which can be applied as a SP pressure sensor. The voltage output signal can reflect the applied pressure induced by a droplet of water. All types of TENGs have a high sensitivity and fast response to external force and show a sharp peak signal. Furthermore, the response to the impact of a piece of feather (20 mg, ~0.4 Pa in contact pressure) can be detected. The sensor signal can delicately show these details of the entire process.

The active pressure sensor has also been developed in the form of a composite. The term triboelectric composite refers to a sponge-shaped polymer with embedded wire. Applying pressure and impact on the composite in any direction causes charge separation between the soft polymer and the active wire because of the presence of composite air gap. Passive wire as the second electrode may be either embedded inside the sponge without any air gap or placed out of the composite allowing the sensor to work in single-electrode mode.

With the use of a matric array of the TENGs, a large area and SP pressure map applied on a surface can be obtained. The response of the TENG array with local pressure can be measured through a multichannel measurement system. There are two types of output signals from the TENG: open-circuit (OC) voltage and short-circuit (SC) current. The OC voltage is only dictated by the final configuration of the TENG after applying a mechanical triggering, so that it is a measure of the magnitude of the deformation, which is attributed to the static information to be provided by TENG. The output current depends on the rate at which the induced charge would flow, so that the current signal is more sensitive to the dynamic process of how the mechanical triggering is applied.

The active pressure sensor and the integrated sensor array based on the triboelectric effect have several advantages over conventional passive pressure sensors. First, the active sensor is capable of both static pressure sensing using the OC voltage and dynamic pressure sensing using the SC current, while conventional sensors are usually incapable of dynamic sensing to provide the loading rate information. Second, the prompt response of both static and dynamic sensing enables the revealing of details about the loading pressure. Third, the detection limit of the TENG for dynamic sensing is as low as 2.1 Pa, owing to the high output of the TENG. Fourth, the active sensor array may have no power consumption and could even be combined with its energy harvesting functionality for SP pressure mapping. Future works in this field involve the miniaturization of the pixel size to achieve higher spatial resolution, and the integration of the TEAS matrix onto fully flexible substrate for shape-adaptive pressure imaging.

6.2.3.2 Self-Powered Motion Sensors

The term SP sensors may reflect far beyond simple voltage output signal. It can refer to a system, which powers all the electronics responsible for measuring and demonstrating the detectable movement (see Figure 6.8). For example, the SP triboelectric encoder, integrated in smart belt–pulley system, converts friction into useful electrical energy by storing the harvested energy in a capacitor and fully powering the circuit, including a microcontroller and an LCD.

6.2.3.3 Self-Powered Active Chemical Sensors

As for TENGs, maximizing the charge generation on opposite sides can be achieved by selecting the materials with the largest difference in the ability to attract electrons and changing the surface morphology. In such a case, the output of the TENG depends on the type and concentration of molecules adsorbed on the surface of the triboelectric materials, which can be used for fabricating chemical and biochemical sensors. As an example, the performance of the TENG depends on the assembly of Au nanoparticles (NPs) onto the metal plate. These assembled Au NPs not only act as steady gaps between the two plates at strain-free condition, but also enable the function of enlarging the contact area of the two plates, which will increase the electrical output of the TENG. Through further modification of 3-mercaptopropionic acid (3-MPA) molecules on the assembled Au NPs, the high-output nanogenerator can become a highly sensitive and selective nanosensor toward Hg^{2+} ions detection because of the different triboelectric polarity of Au NPs and Hg^{2+} ions. With its high sensitivity, selectivity, and simplicity, the TENG holds great potential for the determination of Hg^{2+} ions in environmental samples. The TENG is a future sensing system for unreachable and access-denied extreme environments. As different ions, molecules, and materials have their unique triboelectric polarities, the TENG can become either an electrical turn-on or turn-off sensor when the analytes are selectively binding to the modified electrode surface. In the future, TENG may be developed to identify metal ions and biomolecules such as DNA and proteins.

6.2.3.4 Self-Powered Electrochemical Sensors

Based on TENG, lots of SP electrochemical sensors have been proposed to detect chemical substances. According to the mechanism of sensing, SP electrochemical sensor can be classified into two types including SP electrochemical passive sensor and

FIGURE 6.8 Smart belt–pulley system powers the encoder circuit by converting friction into electrical energy [12].

SP electrochemical active sensor [73]. SP electrochemical passive sensor is that the conventional sensors are driven by TENG for collecting mechanical energy from the environment. Zhang et al. proposed an SP glucose biosensor based on contact–separation type TENG combined with lithium-ion battery (LIB) [74]. Aiming to voluntarily supervise the water quality, Bai et al. reported an in situ SP sensing system that can convert the water wave energy to electricity based on tandem disk TENG [75]. Owing to the radial grating disk structure with swinging mass blocks, the tandem disk TENG driven by water waves can realize a conversion from low-frequency water wave motions into high-frequency output. Through rectification and energy storage, the electricity produced by TENG can be utilized to drive electronics for monitoring water quality.

As for the SP electrochemical active sensor, a TENG is designed to actively generate electrical signal for responding to the stimulation of chemical molecules or environmental factors such as ethanol, phenol, catechin, and pH, where the output performance of TENG and the target shows a linear relationship [76–80]. Zhang et al. demonstrated SP sensors based on the TENGs fabricated by polyamide (PA) film and PTFE film for detecting liquid/gaseous ethanol [76]. Lin et al. proposed a contact–separation-type TENG as an SP nanosensor toward catechin detection [77]. Wu et al. reported an SP triboelectric sensor based on the sliding-type TENG for detecting pH value from a periodic contact/separation motion [78]. The output voltage of the SP triboelectric sensor enhanced with increased pH value due to the increased ion concentration. Therefore, the pH value of the buffer solution can be actively monitored in real time by reading the output voltage. Li et al. proposed an SP active sensor for Hg ions monitoring based on TENG where 3-MPA-modified gold NPs were used as recognition element [79]. Jie et al. demonstrated an SP triboelectric sensor for monitoring dopamine in the alkaline condition on the bases of TENG, which was composed of PTFE with NP arrays and an Al film [80].

6.2.4 Practical Applications of TENG

6.2.4.1 Pollutant Treatment

As shown earlier, since TENG captures mechanical energy from the environment, SPES has been proposed as a candidate for environmental treatment [81]. Efforts are made to use TENG to remove environmental pollutants from water and air. While water may contain various inorganic, organic, and biological contaminants, the most formidable contaminant is heavy metal ions, which can affect the entire food chain [82]. Efforts have been made to remove this contaminant using TENG. Li et al. reported a water-driven TENG to extract the kinetic energy from wastewater flow, which is used to drive electrochemical reaction for Pb and Cu removal [83]. Through comparing the electrochemical property of Cr(VI) driven by continuous DC (CDC) and pulsed DC (PDC), Zhou et al. confirmed an enhanced efficiency of removing Cr(VI) under PDC than that of CDC due to the better utilization of ferrous ion, the lower electrode passivation, and the higher ion diffusion rate during the reaction process [84].

Organic pollutants in wastewater are toxic and carcinogenic and also need to be removed. SP electrooxidation is the common process for organic pollutants treatment using the produced chlorine and hypochlorite. Li et al. proposed a unique SP electrooxidation system for phenol removal by creatively employing β-cyclodextrin to increase triboelectrification [85]. Gao et al. reported a freestanding mode TENG integrated with electrocatalytic technology for degrading 4-aminoazobenzene [86]. Chen et al. fabricated an SP multifunctional system, which can simultaneously realize heavy metal ions and organic pollutant removal driven by a rotary TENG [87]. Yang et al. reported an SP electrocatalytic system containing a hybrid energy cell including TENG to degrade methyl orange, which can further achieve a higher performance [88]. SP electrochemical advanced oxidation system such as SP electro-Fenton process is proposed to remove organic pollutants from the water [89,90]. Feng et al. integrated a rotary TENG with an electrochemical cell to build an SP electro-Fenton system to remove dyes [91]. Combining with 3D printing techniques, Tian et al. prepared a 3D-printed elastic TENG to form an SP electro-Fenton system for removing methylene blue, where 97.0% of methylene blue was removed within 140 minutes [92]. Biological contaminants such as harmful bacteria and algal blooms caused by the uncontrolled discharge of wastewater are the other kinds of water pollutants. Jiang et al. reported an SP electrochemical water treatment system based on an arch-shaped TENG for cleaning sterilization and algae in wastewater [93].

Particulate matters and gaseous pollutants can cause pulmonary and cardiovascular disease [94]. To remove PM 2.5, Guo et al. reported an SP triboelectric negative air ion generator (MSNG) powered by TENG [95]. The harmful pollutants are also oxysulfide, oxynitride, and organic compounds [96]. For removing sulfur dioxide (SO_2) and dust, Chen et al. reported an SP air cleaning system based on a rotary TENG [97]. For cleaning oxynitride, Han et al. proposed an SP nitrous oxide (NOx) absorption and degradation system based on a radial engine-shaped TENG system [98]. For cleaning formaldehyde in indoor atmosphere, Feng et al. demonstrated an SP electrostatic filter by integrating TENG with photocatalysis technology [99].

6.2.4.2 Self-Powered Electrochemical Synthesis System

It has been known that the energy from the environment can be converted to storable chemical energy such as hydrogen and formic acid [100]. Generally, the process of clean fuel production needs an external input power, where renewable energies such as solar, wind, geothermal, and hydro have been applied along with carbon dioxide and water to generate clean fuels [101]. Owing to the ability of harvesting ambient mechanical energy, the TENG-based SP electrochemical system has been demonstrated as a promising technology for clean fuel generation [102,103]. Tang et al. proposed a fully SP water splitting system for producing hydrogen [104], which integrated a rotary TENG with a water splitting unit. Li et al. (105) boosted photoelectrochemical water splitting by TENG-charged Li-ion battery. Yang et al. designed an SP water splitting system based on a hybrid energy cell, where the hydrogen production speed had been improved to mL/s [106].

Formic acid with characteristics of high volumetric capacity, low toxicity, and flammability under ambient condition has also drawn wide attention [107]. Leung et al. presented an SP carbon dioxide reduction system that harvests energy from ocean wave for converting carbon dioxide into formic acid [108]. Ammonia (NH_3) acts a crucial role in food production, industrial manufacturing, and a predictable ideal energy carrier in the future [109]. In industry, Haber–Bosch process is commonly employed to produce NH_3 with the existence of hydrogen and external energy supply, where the challenge of grueling conditions increases the cost of production [110]. Due to a controllable operation under mild conditions utilizing mechanical energy by TENG, the SP electrocatalytic NH_3 synthesis system based on TENG provides a promising candidate for the conversion of N to NH_3. Gao et al. reported an SP sustainable metal-free NH_3 production system based on a multilayer as a way TENG by 3D printing technology, which can efficiently convert N into NH3 [111]. SPESs such as SP electrochemical oxidation system and SP electrodeposition system were also developed for electrochemical synthesis [112–114]. Zheng et al. reported an SP electrochemical oxidation system based on a designed cross-linked TENG for synthesizing polyaniline [113]. Wang et al. established a SP electrodeposition system for polypyrrole synthesis, where the polypyrrole as the electrode material of TENG was produced by TENG [114].

6.2.4.3 Self-Powered Electrochromic System

Cai et al. [115] showed that electrochromic devices can reversibly change their optical properties by the electrochemical redox reaction under an external electric field. Zhang et al. [116] and Sun et al. [117] showed that a more sustainable power solution can be adopted by replacing batteries with TENG as the electricity source to provide a constant voltage. Driven by a dual-mode TENG for harvesting wind and raindrop energy, Yeh et al. realized an SP smart window system [118]. The dual-mode TENG involved a single-electrode TENG on the top of the SP smart window for harvesting the energy from raindrop motions and a contact mode TENG assembled by elastic springs below the abovementioned single-electrode mode TENG for collecting energy from wind energy. Both of the two TENGs consisted of a PDMS thin film adhered to a conducting substrate that was adhered to the electrochromic device with a substrate to fabricate an SP system. Yang et al. reported a tungsten oxide (WO)-based electrochromic device integrating with TENG to fabricate an SP electrochromic device [119]. The SP electrochromic device had a multilayered structure. Commercial glass was used as the substrate, on which there was a layer of freedom to operate (FTO) thin films that acted as electrodes. Compared with the SP electrochromic device, the transmittance of the device greatly dropped. To effectively harvest acoustic energy, Qiu et al. integrated a sandwich-like structured TENG with an electrochromic device for reversible color changing [120]. In this system, reversible switches controlled the oxidation and reduction process and thus controlled the color change. Driven by the TENG, the color of electrochromic film changed between transparent white and dark black, which were controlled by the switching dote.

6.2.4.4 Self-Powered Anticorrosion System

Material corrosion is a long-standing challenge in many engineering applications [121]. To solve this issue, many works have been focused on the SP anticorrosion system by integrating TENG with chemical anticorrosion protection [122,123]. Wang et al. realized an SP anticorrosion system for iron sheet driven by a high-performance TENG [124]. The study indicated a good anticorrosion property of the SP system, which exhibited a good prospect to protect materials from rusting with low energy cost. Considering that metal corrosion is more likely to happen under ocean environmental conditions, Li et al. developed an SP system based on networking TENG and supercapacitor to convert water wave to electricity for metal anticorrosion [125]. The results of this study demonstrated that the SP system significantly decreased the corrosion rate, which could be adopted in marine corrosion. Zhu et al. designed an SP cathodic protection system based on a flexible TENG, which can harvest energy from natural raindrops and wind to drive the cathodic protection process [122]. The contact–separation-type TENG was mainly composed of PDMS film and ITO, which acted as a pair of triboelectric layers. The results of this study showed the successful elimination of corrosion.

6.2.5 Choice of Materials and Surface Structures for TENG

Almost all materials known exhibit the triboelectrification effect, from metal, to polymer, to silk, and to wood, almost everything. All of these materials can be candidates for fabricating TENGs, so the material choices for TENG are huge. However, the ability of a material for gaining/losing electron depends on its polarity. John Carl Wilcke published the first triboelectric series in 1757 on static charges. A material toward the bottom of the series, when touched with a material near the top of the series, will attain a more negative charge. The further away two materials are from each other on the series, the greater the charge transferred. Besides the choice of the materials in the triboelectric series, the morphologies of the surfaces can be modified by physical techniques with the creation of pyramid-, square-, or hemisphere-based micro- or nano-patterns, which are effective for enhancing the contact area and possibly the triboelectrification. However, the created bumpy structure on the surface may increase the friction force, which may possibly reduce the energy conversion efficiency of the TENG. Therefore, an optimization has to be designed for maximizing the conversion efficiency.

The surfaces of the materials can be functionalized chemically using various molecules, nanotubes, nanowires (NWs), or NPs, to enhance the triboelectrification effect. Surface functionalization can largely change the surface potential. The introduction of nanostructures on the surfaces can change the local contact characteristics, which may improve triboelectrification. This will involve a large number of studies for testing a range of materials and a range of available nanostructures.

Besides these pure materials, the contact materials can be made of composites, such as embedding NPs in polymer matrix. This changes not only the surface electrification, but also the permittivity of the materials so that they can be effective for electrostatic induction. Therefore, there are numerous ways for enhancing the performance of the TENG from the materials' point of view. This gives an excellent opportunity for chemists and material scientists to do extensive study both in basic science and in practical application. In contrast, material systems for SC and thermal electric, for example, are rather limited, and there are not very many choices for high-performance devices.

6.2.6 Direct Current Triboelectric Nanogenerator Based on Electrostatic Breakdown

The DC-TENG is a new-type TENG that can harvest mechanical energy and convert it into electrical energy through the conjunction of triboelectric effect and electrostatic breakdown effect. Different from the abovementioned traditional TENGs possessing standard AC output signal, in which the signal is generated by a conjunction of triboelectric effect and electrostatic induction effect, the DC-TENG produces a unidirectional current flow in the external circuit and can drive electronic devices and charge energy storage devices directly. When the DC-TENG is fabricated as rotary structure, it will produce continuous constant current, which can be utilized as a constant current power supply at a wide range of external load resistance. This new-type TENG was firstly invented at the Beijing Institute of Nanoenergy and Nanosystems in 2019. As a device for converting mechanical energy into electrical energy, this new-type TENG can be utilized in micro-/nanopower source or self-power sensing system.

The working mechanism (see Figure 6.9) of DC-TENG is the coupling between contact electrification and electrostatic breakdown effects. Its conventional structure is quite simple, which constituted friction electrode that is used for contact electrification with a friction layer and charge collecting electrode that collects the charges formed by electrostatic breakdown. During the sliding process, the friction electrode contacts with the friction layer, resulting in positive charges on the friction electrode and negative charges on the friction layer due to contact electrification. Due to the existence of charges on the friction layer, with the device sliding further, the following gap between charge collecting electrode and friction layer will build a strong electric field, which will lead to the air breakdown in the gap and generate DC in the external circuit if the electric field is strong enough. When the DC-TENG device slides to the end of the friction layer film, the air breakdown process will stop.

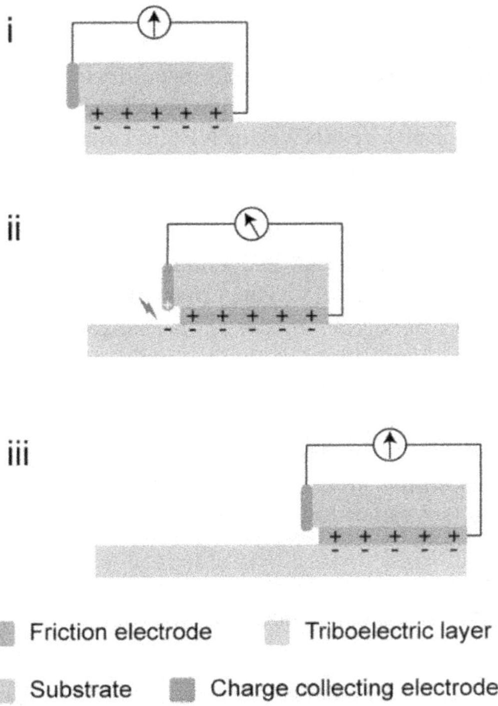

FIGURE 6.9 Mechanism of direct current triboelectric nanogenerator based on electrostatic breakdown [12]. The figure shows positions i, ii and iii, which indicate three levels of air breakdown in the gap and the gradual generation of electric current by the nanogenerator.

6.2.7 Output Enhancement Strategies for TENGs

For the TENG-based hybrid generators, the output performance of TENG components plays an important role in determining the overall output of the hybrid generators. Therefore, it is of great significance to enhance the performance of the integrated TENG components. Sadasivuni et al. [126] and Zi et al. [127] noted that the surface charge density of TENGs is limited by the air breakdown effect between two triboelectric surfaces, and thereby, it is severely detrimental to their practical applications. As the earliest strategy to enhance performance, TENG devices are improved by material selection [128], structure optimization [129], surface modification [130], ion injection [131,132], and environment control [133]. Since charge improvement from material modifications is finite and a vacuum strategy limits applications of TENGs, more effective methods are desired to improve the charge density in the air for broad applications of TENGs. Advanced mechanisms are also needed to improve output performance, which can be integrated into practical devices for different working environments in an effortless manner.

Recently, there are a few effective methods proposed to obtain an optimum contact structure and improved output performance. Liu et al. proposed a standard method to precisely evaluate the contact status of TENG to optimize the contact of two tribo-surfaces, as shown in Figure 6.10a [134]. They illuminate the strategies of enhancing the charge output for the charge excitation TENG, including the reduction

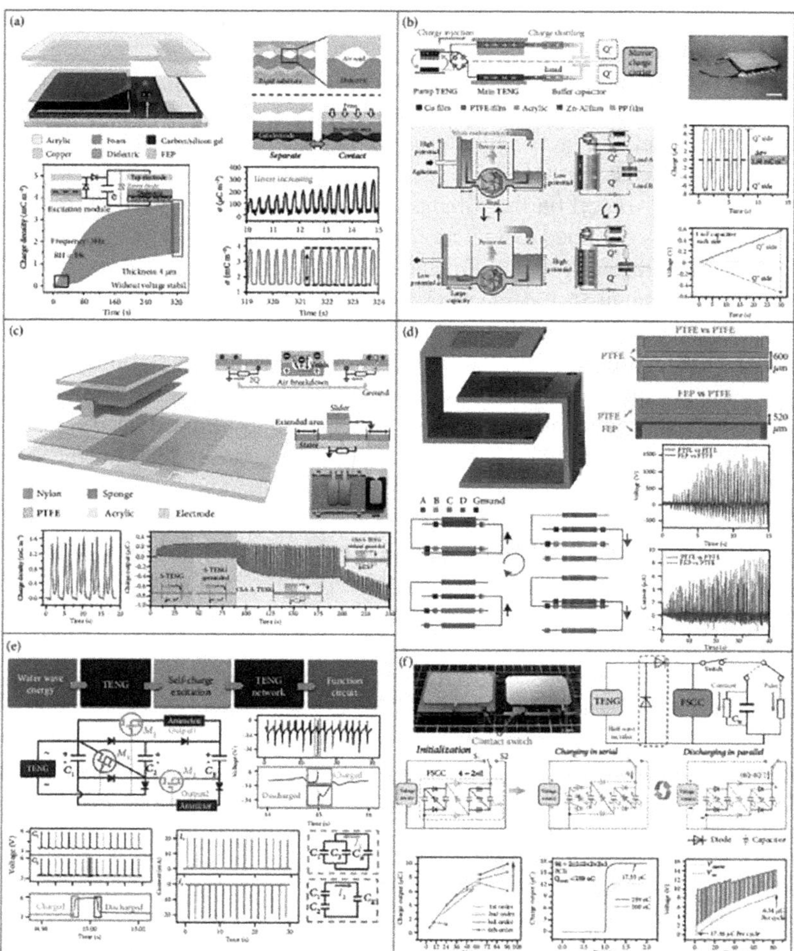

FIGURE 6.10 Output enhancement strategies for TENGs. (a) The air breakdown model of a charge excitation TENG to enhance charge density. (b) A high-performance TENG based on charge shuttling. (c) Boosting the output performance of a sliding mode TENG by charge space accumulation effect. (d) The out-of-plane design of a Bennet doubler-based TENG. (e) A high-output charge excitation TENG based on the voltage multiplying circuit. (f) The switched capacitor converters based on the fractal design for TENG output enhancement [4].

in the thickness of dielectrics, the increment of external capacitor, and the control of the atmospheric environment. As illustrated in Figure 6.10b, Wang et al. proposed a high-performance TENG based on the shuttling of charges [135]. The charge shuttling TENG consists of a pump TENG, a main TENG, and a buffer capacitor. The electrodes of the main TENG and the buffer capacitor form two conduction domains, presenting a quasi-symmetrical structure with a side and a side. Based on this mechanism, an integrated device for water wave energy harvesting shows the feasibility of the charge shuttling TENG as a fundamental device to be applied in complex structures for various practical applications.

TENG in the lateral sliding mode provides an effective approach for in-plane low-frequency mechanical energy harvesting. Bai et al. [136] and Kim et al. [137] showed that surface modification and charge excitation do not work well for this mode as an output enhancement strategy. As shown in Figure 6.10c, He et al. developed a new strategy with a shielding layer and alternative blank-tribo-area-enabled charge space accumulation (CSA) design for enormously improving the charge density of a sliding mode TENG [138]. Based on the principle of the linearly sliding TENG unit, CSA-TENG can be easily designed to a rotational working mode. Based on the CSA mechanism, Wang et al. presented an out-of-plane design to achieve high-performance TENG [139]. As shown in Figure 6.10d, electrodes B and D are arranged on the left part, while electrodes A and C are arranged on the right part of the designed structure.

Vaisband et al. [140] and Xie and Li [141] noted that another common method for increasing the output of TENGs is designing the external circuit. In this respect, the switched capacitor converter (SCC) plays an increasingly important role compared with the traditional transformer because it is easy to integrate and it is magnet-free and lightweight. This means that a power management system with higher transfer efficiency and multifunctional output mode is urgently needed and has great significance for practical applications of TENGs. As shown in Figure 6.10e, Liang et al. developed a new charge excitation system based on the voltage multiplying circuit (VMC) to achieve high-output TENGs for effectively harvesting the water wave energy [142]. Not only the output performance of a single TENG is increased by multiple times but also a scheme is proposed to realize a high-output TENG network through integrating with the charge excitation circuits (CECs). The CEC can improve both the output current and power of a TENG, which is the most important difference from an ordinary voltage doubler circuit that only increases the output voltage. Finally, the high-output TENG network is utilized to drive a thermometer to work continuously and realize wireless communication with a mobile phone for remote environmental monitoring. However, the high output impedance and switching loss largely reduce the SCC's power efficiency, due to the imperfect topology and transistors. To address this issue, as shown in Figure 6.10f, Liu et al. proposed fractal design-based switched capacitor converters (FSCCs) with characteristics including high conversion efficiency, minimum output impedance, and electrostatic voltage applicability [143].

6.2.8 Perspectives on Future Directions for TENG

The above discussions indicate that an integration of electrochemical system with TENG is the most promising strategy to break the limitation of external powers source for electrochemical operation. As shown earlier, SPESs can be very useful for five major fields such as pollutants treatment, electrochemical synthesis, electrochemical sensor, electrochromic reaction, and metal anticorrosion. While SPES has shown remarkable progresses, several issues should be addressed for promoting further development in this field.

The high performance of SPES requires high output power and durability of TENG. To improve the output power of TENG, further improvement in the surface charge density and integration of existing methods on a large scale are recommended. An introduction of interface liquid lubrication and the development of materials with the most robust mechanical durability and stability might be promising strategies to enhance the durability of TENG. TENG has the characteristics of high voltage output and low current output, while the performance of electrochemical process exhibits a positive correlation with current density and negative influence such as passive electrode effect and secondary reaction induced by high potential. To achieve a high performance of SPES, the power circuit management of TENG is highly desired to match the corresponding electrochemical process. TENG with a pulse output signal has been reported to reduce the electrode passive effect; however, the phase superposition of TENG caused by multiple parallel electrodes makes it hard to realize a full-waveform pulse current. Therefore, rational design of the structure of TENG, such as adjusting the rotation center angle ratio between each rotator and stator to obtain a full-waveform pulse current, will be a promising strategy to optimize the processes of SPES. Since AC exhibits lower energy consumption, improved mass transfer characteristics, and delayed electrode passivation compared with DC in the electrochemical system, utilizing the AC of TENG to build SP electrode system is the other method to reduce the electrode passive and improve the electrochemical performance.

Since electrodes play a key role in the electrochemical reaction, the materials used for electrodes play an important role in the properties of electrochemical reaction. Novel electrode materials such as nanoscale electrode materials and metal–organic framework materials with merits of high conductivity, high specific surface areas, high activity, and cycle stability should be prepared to further enhance the performance of SPES. With the improved performance of TENG, it can be utilized to substitute an external power supply for an electricity source to power some new electrochemical reactions such as electrocoagulation for the removal of oil in water, electrodialysis for desalination and water reuse, and electrophoresis for the separation of protein. The interrelation of electrical and chemical effects can also be realized by high-voltage electrostatic discharge, which can be easily achieved by utilizing TENG. Therefore, integrating TENG with the electrochemical system to build SP discharge electrochemical system for removing pollutants such as PM provides a new area for its application.

6.3 EMG

The mechanism of EMG is based on Faraday's law of electromagnetic induction, in which the voltage on a closed loop is proportionally induced by the loop's magnetic flux variation over time. Due to the high energy conversion efficiency, EMGs have been widely adopted in modern energy farms for centralized and large-scale electricity generation. Pan et al. [144] and Liu et al. [145,146] showed that EMG can also be used as distributed and small-scale energy sources for harvesting the in situ energy from various machinery vibrations and human motions. When there are relative movements between the magnets and coils in the EMGs, an induction current will be generated in the coil. According to the difference in relative movements, EMGs can be classified into two basic operation modes, i.e., movable magnet-fixed coil mode and movable coil-fixed magnet mode. Compared to TENG, EMG normally exhibits small impedance with large current but small voltage outputs. Due to these distinct output characteristics and the similar triggering forms of EMG and TENG (both by the relative movements of different components), EMG can be a good complement to TENG in a hybrid generator and hybridized system.

TENGs use a wide range of conventional materials and are currently known to be mechanically flexible, lightweight, cost-effective, and easily scalable with low operation frequencies and large bandwidths [147]. EMGs are well-established, efficient, versatile, reliable, effective at large scales, and have an easily controllable internal impedance and high frequency of operation. Electrically, TENGs behave as low current sources with high parallel internal impedance because of the electrostatic induction mechanism and the nature of the insulator-to-insulator or insulator-to-metal interface. Mesoscale devices typically have high output OC voltages (~ 1–1,000 V) and low output SC currents (~ 1–1,000 μA) and capacitive internal impedance characteristics. Their characteristics of high output voltage and low current and susceptibility to wear, ambient humidity and temperature, and low and unstable charge density on tribo-layers still limit the practical applications of TENGs [147]. EMGs, however, behave as low voltage sources with low series internal impedance due to the electromagnetic induction mechanism and the high conductivity of the coils. They typically have low output OC voltages (~ 1–1,000 mV) and high SC currents (~ 1–1,000 mA) and resistive and inductive internal impedance characteristics. As such, both technologies are complementary.

6.3.1 E-TENG

Depending on the environment of operation of SP electronic devices, they may require multiple types of energy sources. TENG is naturally suited to form HCs to simultaneously harvest energy from various sources, such as mechanical [148,149], solar [150–152], thermal [152], or chemical [150,153]. Bai et al. [154] have shown that each type of mechanical energy harvesters provides its own benefits and unique advantages/drawbacks. TENG/EMG HCs (E-TENGs) can be used to simultaneously scavenge vibrational energy by taking advantage of their complementarity: high voltage (TENG) and high current (EMG) or, alternatively, to use either of these to meet the requirements of particular applications [155,156]. Furthermore, they can be used to broaden the operating bandwidth of the nanogenerator due to TENG's high efficiency at low frequencies and amplitudes of excitation and increased performance of EMGs at high frequencies and amplitudes. In fact, early comparative studies between transduction mechanisms have suggested that piezoelectric and triboelectric energy harvesters provide superior performance in relation to EMGs at low frequencies and low dimensions [157,158], as well as for small displacement amplitudes of excitation [159]. By taking into account constitutive equations for their respective conversion mechanisms, scaling analysis of the output power of different transducer types as a function of effective material volume (V) has shown that it should be roughly proportional to V^2 and $V^{2/3}$ for the electromagnetic and electrostatic generators, respectively [156,160]. Thus, below a critical volume of ~ 0.5 cm^3 the triboelectric mechanism can become more attractive [160]. Nevertheless, technological difficulties may still be encountered at smaller size scales.

Recently, Zhao et al. [159] examined the remarkable merits of TENG and EMG for harvesting small-amplitude mechanical energy. Linear TENGs and EMGs, with similar geometry and size, were fabricated and their electrical output characteristics were systematically studied as a function of the amplitude and frequency of excitation, yielding significantly larger output powers for the TENG in the low-frequency ($\lesssim$1 Hz) and small-amplitude regime ($\lesssim$ 1 mm), as illustrated in Figure 6.11a and b [159]. The most important electromechanical characteristics of TENGs and EMGs are summarized in Figure 6.11c. Xu et al. [161] compared the applied torque and energy conversion efficiencies between rotational TENGs and EMGs. The input mechanical torque of the EMG was shown to be balanced by the friction and electromagnetic resisting torques, which increased with increasing rotation rate due to Ampère's force. The input torque of the TENG was balanced by the friction and electrostatic resisting torques, which were nearly constant with the rotation rate. The energy conversion efficiency of the EMG was observed to increase with increasing mechanical power inputs, while the one of the TENG remains nearly constant. These results suggest that the TENG can be superior to the EMG for harvesting mechanical energy with low input powers ($\lesssim$11.4 mW). Many designs of hybridized E-TENGs have already been proposed and tested in applications such as general vibration energy harvesting, wheel rotation energy, biomechanical energy, blue energy (wave energy and fluid flow), wind energy, and thermal energy. These generators achieved

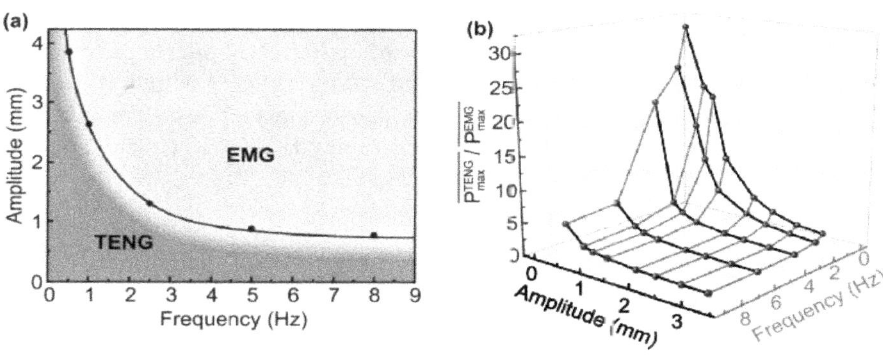

(c)

	EMG	TENG
Approximate equivalence	Voltage Source	Current Source
Internal impedance	Small	Large
Impedance characteristic	Resistive, inductive	Capacitive
Open-circuit voltage	Low	High
Short-circuit current	High	Low
Frequency	High	Low
Amplitude	Large	Small

FIGURE 6.11 (a) Domain of excitation amplitude and frequency values where the energy harvesting performance of a test EMG or TENG is superior. The light grey area denotes the dominant scope of the TENG in low frequency and small amplitude, while the dark grey area denotes that of the EMG. (b) Maximum average output power ratio of the EMG and TENG versus amplitude and frequency. (c) Summarized overall comparison table between the complementary characteristics of EMGs and TENGs [9].

maximum output peak powers up to ca. 100 mW [162], average powers around 1 mW [163–165], and peak power densities up to 1 mW/cm^3 [166,162].

Motion-driven electromagnetic–triboelectric energy generators (E-TENGs) hold a great potential to provide higher voltages, higher currents, and wider operating bandwidths than both electromagnetic and triboelectric generators standing alone. Therefore, they are promising solutions to autonomously supply a broad range of highly sophisticated devices. The study by Vidal et al. [9] provides a thorough review focused on major recent breakthroughs in the area of electromagnetic–triboelectric vibrational energy harvesting. A detailed analysis was conducted on various architectures including rotational, pendulum, linear, sliding, cantilever, flexible blade, multidimensional and magnetoelectric, and the following hybrid technologies. They enable highly efficient ways to harvest electric energy from many forms of vibrational, rotational, biomechanical, wave, wind, and thermal sources. OC voltages up to 75 V, SC currents up to 60 mA, and instantaneous power of up to 144 mW were already achieved by these nanogenerators. Their transduction mechanisms, including proposed models to make intelligible the involved physical phenomena, are also overviewed in this study. The review also presents a comprehensive analysis to compare their respective construction designs, external excitations, and electric outputs. The results highlight the potential of hybrid E-TENGs to convert unused mechanical motion into electric energy for both large- and small-scale applications. The review also proposes future research directions toward optimization of energy conversion efficiency, power management, durability and stability, packaging, energy storage, operation input, research of transduction mechanisms, quantitative standardization, system integration, miniaturization, and multi-energy HCs. A schematic of hybridized E-TENG nanogenerator is illustrated in Figure 6.12.

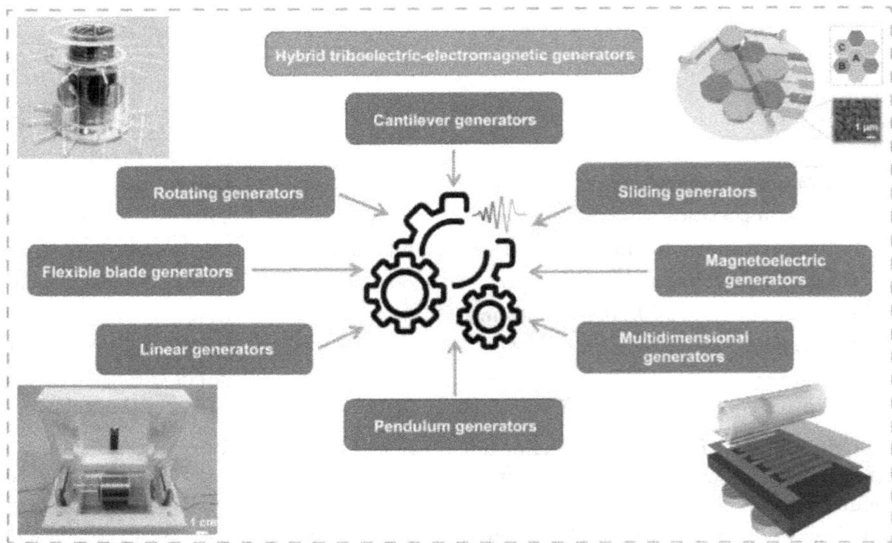

FIGURE 6.12 A schematic of hybridized E-TENG nanogenerator [9].

Hybridized E-TENGs can offer a more efficient vibration energy conversion by taking advantage of both of their desirable complementary high-voltage and high-current characteristics and wider operating bandwidths. As mentioned earlier, the TENGs are effectively able to harvest electric energy from low-frequency (<1 Hz) and low amplitude (<1 mm) kinetic energy, providing large output voltages. The EMGs are effective technologies to harvest at high frequencies and amplitudes of excitation, providing high output currents. E-TENGs can also be connected in series or in parallel, or used independently, to output high voltages or currents to fulfill customized requirements and suitability for particular applications. Thus, by yielding peak output powers higher than 100 mW, they might find use as micro-/nanopower sources or in SP sensors by scavenging general forms of vibration energy, wheel rotation energy, biomechanical energy, blue energy, wind energy, thermal energy, and more. Therefore, E-TENG generators hold the potential to power captivating technologies, such as those related to the IoT, wireless sensor networks, portable electronics, and implantable biomedical devices. Despite the observed impressive progress, according to Vidal et al. [9] energy conversion efficiency, power management, durability and stability of device, packaging, energy storage, operation input, understanding of the conversion mechanism, quantitative standardization, system integration, miniaturization, large-scale manufacturing, and use of multi-energy HCs are some of the issues that still needs to be addressed in the future. These issues as outlined by Vidal et al. [9] are briefly summarized as follows.

The energy scavenging capabilities of TENGs are still not sufficient to continuously power most conventional electronic devices. Since its output power increases with the square of the triboelectric surface charge density, enhancing this value is mandatory. At least five different strategies, modification in material composition, improvement in the effective contact area, surface charge pumping, ionized-air injection, and control of environmental conditions, are being explored. Coupling the output powers from the TENG and EMG units is a tremendous issue due to the huge impedance mismatch between them and between TENGs and storage devices. A better understanding of the combined energy harvesting efficiency in hybrid systems is required. Energy transfer maximization has been explored through a rationally modulated charging cycle controlled by a motion-triggered switch placed in parallel to the energy storage unit [167]. The undesirable small output current from TENGs can also be enhanced, e.g., using sliding micro-grating [168], radial-arrayed [169], or multilayered stacked structures [170]. Stable and reliable materials and devices are required for real-life applications, which could constitute a problem, mainly in the case of high-friction lateral sliding TENGs. More durable composite structures must be investigated for TENGs or using a conjunction of working modes. Strategies already tested have included mechanically switchable structures with the automatic transition between contact (triboelectric charge buildup) and noncontact working states [171] and low-friction freestanding structures with intermediate rolling rods [172]. The performance of the nanogenerators, especially TENGs, sensibly depends on the moisture and temperature. Thus, good packaging methods for devices with moving parts are required to guarantee thorough water sealing, while raising the working temperature of the devices could also prove useful [173]. Humidity-resistive TENGs were also fabricated making use of metal organic framework composites. Energy storage of irregular electrical outputs is required. Improvements must be carried out regarding the leakage problem in supercapacitors, and new designs of Li-ion batteries should be tested [174]. Wider bandwidths of input frequencies and amplitudes are valuable and they could be achieved, e.g., by periodic tuning of the resonant frequency, mechanical stoppers, nonlinear springs, or bistable structures [175]. Obtaining a better understanding of the triboelectric effect and its relation to environmental conditions could be useful to optimize the performance of the harvesters. New computational models to predict the electromechanical output for realistic irregular tridimensional motions of combined hybrid harvesters should also be developed.

Quantitative experimental standards should be employed for comparing and calibrating the performance of developed E-TENG prototypes, since testing conditions vary widely in the literature. Rational designs should be developed for system integration with different kinds of vibration environments. Biomechanical and hydrodynamic fluid flow studies could be important. Economic studies should be carried out to estimate the maintenance and production costs of hybrid nanogenerators to evaluate their competitiveness in the energy market. The possible impact of relatively large stray magnetic and electric fields associated with EMGs and TENGs, respectively, should also be addressed. Miniaturization of E-TENGs for the integration with wearable equipment or incorporation within innovative implantable bioelectronic medical devices should be explored [176]. Significant output powers from biomechanical sources and the ability to work in biological environment are relevant requirements for future advanced applications of miniaturized E-TENGs. While large-scale EMGs are easier to manufacture, as only conventional machining technology is required, large-scale triboelectric generators require special attention, because upscaling of the laboratory production techniques such as pulsed laser deposition (PLD) or atomic layer deposition (ALD) to a mass production level is difficult. Moreover, standard techniques reported in the literature to boost the performance of TENGs involve nano-/micro-patterning of the electrodes, which is complex, tedious, and expensive. The solutions for large-scale patterning could be soft lithography and surface treatment processes, including different kinds of plasma treatment processes and chemical synthesis methods suitable for large-scale manufacturing [177]. HCs to simultaneously harvest energy from multiple available types of environmental energy, including solar, thermal, and mechanical, will most likely be very useful in the future.

6.4 PENG NANOGENERATOR

A PENG is an energy harvesting device capable of converting external kinetic energy into electrical energy via action by a nanostructured piezoelectric material. Although its definition may include any type of energy harvesting devices using nanostructures to convert various types of ambient energy (e.g., solar power and thermal energy), it is generally used to indicate kinetic energy harvesting devices utilizing nanoscaled piezoelectric material like in thin-film bulk acoustic resonators. Although still in the early stages of development, the technology has been regarded as a potential breakthrough toward further miniaturization of conventional energy harvesters, possibly leading to facile integration with other types of energy harvesters and the independent operation of mobile electronic devices with reduced concern for sources of energy.

As illustrated by Wang [178], Wang et al. [179], and Xu et al. [180], the fundamental mechanism of PENG is the appearance of an electric potential (electric dipole moment) on a piezoelectric material when it undergoes external pressure, which is also known as the direct piezoelectric effect. Liu et al [181] showed that the common piezoelectric materials can be categorized into two classifications—inorganic piezoelectric materials and organic piezoelectric materials. Popular inorganic materials include piezoelectric ceramics and crystals, such as lead zirconate titanate (PZT), aluminum nitride (AlN), barium titanate ($BaTiO_3$), lithium niobate ($LiNbO_3$), zinc oxide (ZnO), and quartz. Meanwhile, the representative organic piezoelectric materials are polyvinylidene fluoride (PVDF) fiber/thin film and its copolymers, with good flexibility and suitability for wearable electronics. One of the key parameters determining the output performance of piezoelectric materials is the piezoelectric coefficient, which is the generated charge density normalized by the applied stress. Denoting "3" as the polar axis direction of the material, and denoting "1" as one of the directions that are perpendicular to the polar axis due to symmetry, depending on the direction of the applied stress, PENGs can be classified into two modes, 33-mode (stress along the polar axis) and 31-mode (stress perpendicular to the polar axis). Dagdeviren et al. [182] and Park et al. [183] observed that benefitting from the advanced transferring technology in recent years, rigid piezoelectric materials with normally higher piezoelectric coefficient can also be transferred to flexible substrates, greatly improving the output performance of PENGs in wearable and implantable applications.

The working principle of PENG is illustrated in Figures 6.13 and 6.14 for two different cases: the force exerted perpendicular and parallel to the axis of the NW. The working principle for the first case is explained by a vertically grown NW subjected to the laterally moving tip. When a piezoelectric structure is subjected to external force by the moving tip, the deformation occurs throughout the structure. The piezoelectric effect will create the electrical field inside the nanostructure; the stretched

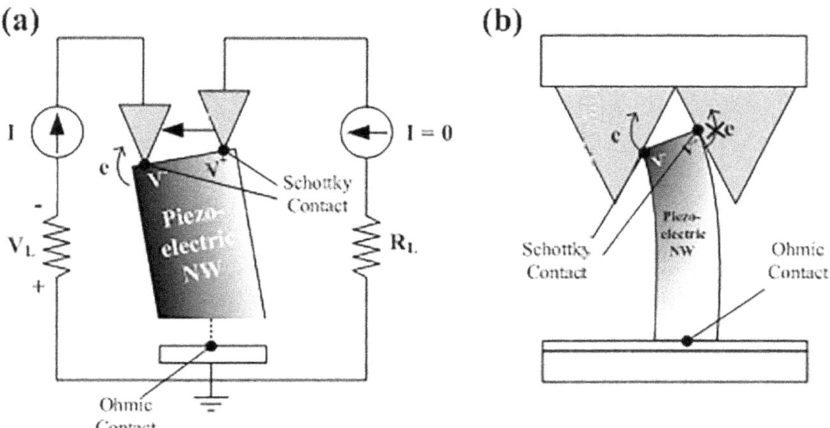

FIGURE 6.13 Working principle of nanogenerator where an individual nanowire is subjected to the force exerted perpendicular to the growing direction of nanowire. (a) An AFT tip is swept through the tip of the nanowire. Only the negatively charged portion will allow the current to flow through the interface. (b) The nanowire is integrated with the counter electrode with AFT tip-like grating. As of (a), the electrons are transported from the compressed portion of nanowire to the counter electrode because of Schottky contact [12].

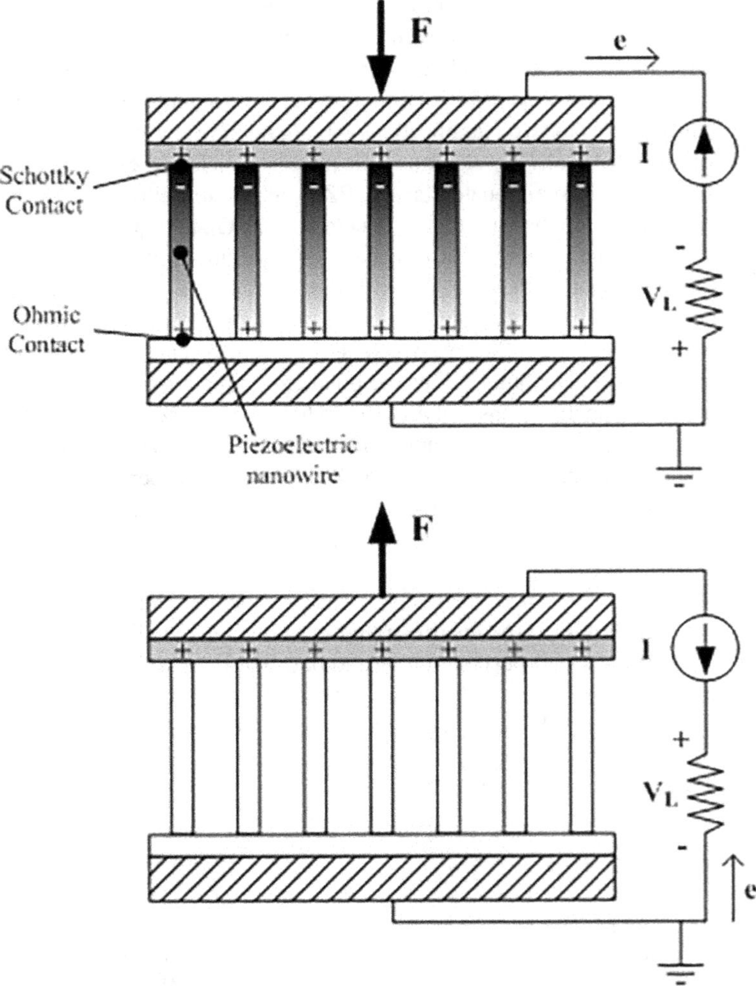

FIGURE 6.14 Working principle of nanogenerator where an individual nanowire is subjected to the force exerted parallel to the growing direction of nanowire [12].

part with the positive strain will exhibit positive electrical potential, whereas the compressed part with the negative strain will show negative electrical potential. This is due to the relative displacement of cations with respect to anions in their crystalline structure. As a result, the tip of the NW will have an electrical potential distribution on its surface, while the bottom of the NW is neutralized since it is grounded.

6.4.1 Mechanism

The electrical contact plays an important role to pump out charges in the surface of the tip. The Schottky contact must be formed between the counter electrode and the tip of the NW since the ohmic contact will neutralize the electrical field generated at the tip. To form an effective Schottky contact, the electron affinity (E_a) must be smaller than the work function (φ) of the metal composing the counter electrode. For the case of ZnO-NW with the electron affinity of 4.5 eV, Pt (φ = 6.1 eV) is a suitable metal to construct the Schottky contact. By constructing the Schottky contact, the electrons will pass to the counter electrode from the surface of the tip when the counter electrode is in contact with the regions of the negative potential, whereas no current will be generated when it is in contact with the regions of the positive potential, in the case of n-type semiconductive nanostructure (p-type semiconductive structure will exhibit the reversed phenomenon since the hole is mobile in this case). The formation of the Schottky contact also contributes to the generation of DC output signal consequently.

For the second case, a model with a vertically grown NW stacked between the ohmic contact at its bottom and the Schottky contact at its top is considered. When the force is applied toward the tip of the NW, the uniaxial compressive is generated in the NW. Due to the piezoelectric effect, the tip of the NW will have a negative piezoelectric potential, increasing the Fermi level at the tip. Since the electrons will then flow from the tip to the bottom through the external circuit as a result, the positive electrical potential will be generated at the tip. The Schottky contact will barricade the electrons being transported through the interface, therefore maintaining the potential at the tip. As the force is removed, the piezoelectric effect diminishes, and the electrons will be flowing back to the top to neutralize the positive potential at the tip. The second case will generate an AC output signal.

6.4.2 Geometrical Configuration

Depending on the configuration of the piezoelectric nanostructure, most of the nanogenerators can be categorized into three types: vertical nanowire integrated nanogenerator (*VING*), lateral nanowire integrated nanogenerator (*LING*), and "*NEG*." Still, there is a configuration that does not fall into the aforementioned categories, as stated in other types.

6.4.2.1 Vertical Nanowire Integrated Nanogenerator (VING)

VING is a 3D configuration consisting of a stack of three layers in general, which are the base electrode, the vertically grown piezoelectric nanostructure, and the counter electrode (see Figure 6.15). The piezoelectric nanostructure is usually grown from the base electrode by various synthesizing techniques, which are then integrated with the counter electrode in full or partial mechanical contact with its tip. After Professor Wang of the Georgia Institute of Technology introduced a basic configuration of VING in 2006 where he used a tip of atomic force microscope (AFM) to induce the deformation of a single vertical ZnO-NW, the first development of VING is followed in 2007. The first VING utilizes the counter electrode with the periodic surface grating resembling the arrays of AFM tip as a moving electrode. Since the counter electrode is not in full contact with the tips of the piezoelectric NW, its motion in-plane or out-of-plane occurred by the external vibration induces the deformation of the piezoelectric nanostructure, leading to the generation of the electrical potential distribution inside each individual NW. The counter electrode is coated with the metal forming the Schottky contact with the tip of the NW, where only the compressed

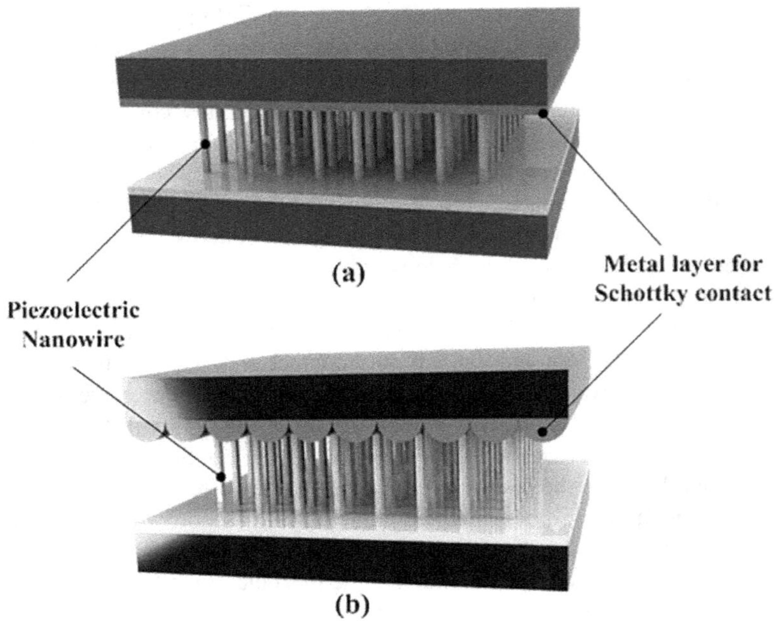

FIGURE 6.15 Schematic view of typical vertical nanowire integrated nanogenerator, (a) with full contact, and (b) with partial contact. Note that the grating on the counter electrode is important in the latter case [12].

portion of piezoelectric NW would allow the accumulated electrons to pass through the barrier between its tip and the counter electrode, in the case of n-type NW. The switch-on and switch-off characteristic of this configuration shows its capability of generating DC generation without any requirement for the external rectifier.

In VING with partial contact, the geometry of the counter electrode plays an important role. The flat counter electrode would not induce sufficient deformation of the piezoelectric nanostructures, especially when the counter electrode moves by in-plane mode. After the basic geometry resembling the array of AFM tips, a few other approaches have been followed for the facile development of the counter electrode. Professor Wang's group has generated a counter electrode composed of ZnO nanorods utilizing a similar technique used for synthesizing ZnO-NW array. Professor Sang-Woo Kim's group at Sungkyunkwan University (SKKU) and Dr. Jae-Young Choi's group at Samsung Advanced Institute of Technology (SAIT) in South Korea introduced bowl-shaped transparent counter electrode by combining anodized aluminum and electroplating technology. They also have developed other types of counter electrodes using a networked single-walled carbon nanotube (SWNT) on the flexible substrate, which is not only effective for energy conversion but also transparent.

The other type of VING has been also suggested. While it shares the identical geometric configuration with the aforementioned, such a VING has full mechanical contact between the tips of the NWs and the counter electrode. This configuration is effective for application where the force is exerted in the vertical direction (toward the c-axis of the piezoelectric NW), and it generates AC unlike VINGs with partial contact.

6.4.2.2 Lateral Nanowire Integrated Nanogenerator (LING)

LING is a two-dimensional configuration consisting of three parts: the base electrode, the laterally grown piezoelectric nanostructure, and the metal electrode for Schottky contact (see Figure 6.16). In most cases, the thickness of the substrate film is much thicker than the diameter of the piezoelectric nanostructure, so the individual nanostructure is subjected to the pure tensile strain. LING is an expansion of a single wire generator (SWG), where a laterally aligned NW is integrated on the flexible substrate. SWG is rather a scientific configuration used for verifying the capability of electrical energy generation of a piezoelectric material and is widely adopted in the early stage of development.

As for VINGs with full mechanical contact, LING generates AC electrical signal. The output voltage can be amplified by constructing an array of LING connected in series on the single substrate, leading to the constructive addition of the output voltage. Such a configuration may lead to the practical application of LING for scavenging large-scale power, for example, wind or ocean waves.

6.4.2.3 Nanocomposite Electrical Generators (NEG)

"NEG" is a 3D configuration consisting of three main parts: the metal plate electrodes, the vertically grown piezoelectric nanostructure, and the polymer matrix, which fills in between the piezoelectric nanostructure (see Figure 6.17). NEG was introduced by Momeni et al. [184]. It was shown that NEG has a higher efficiency compared with the

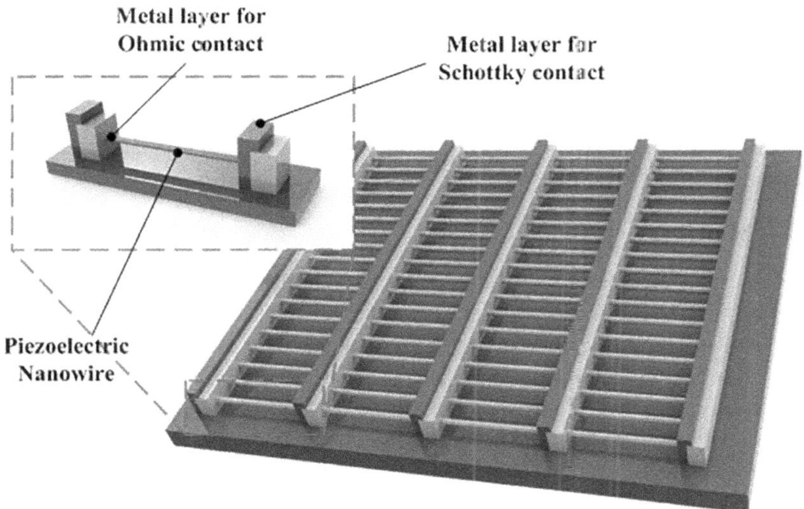

FIGURE 6.16 Schematic view of typical lateral nanowire integrated nanogenerator [12].

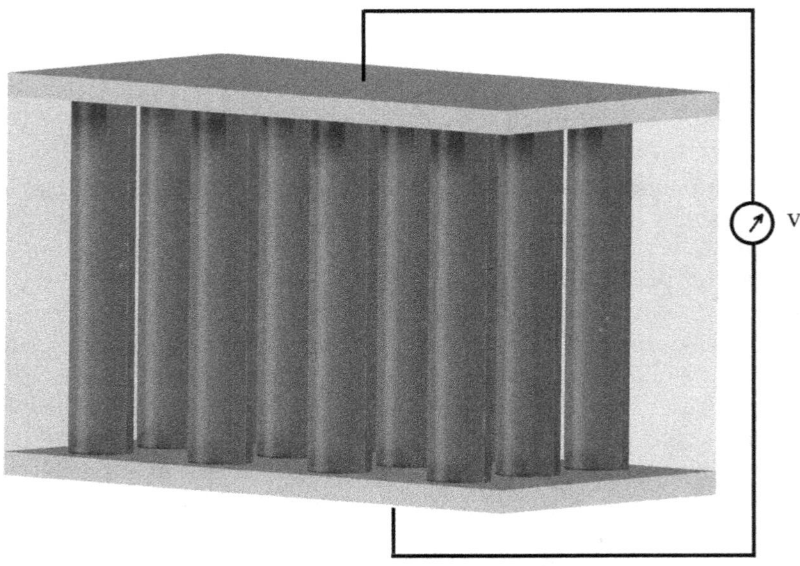

FIGURE 6.17 Schematic view of typical nanocomposite electrical generator [12].

original nanogenerator configuration, in which a ZnO-NW will be bent by an AFM tip. It is also shown that it provides an energy source with higher sustainability.

6.4.2.4 Other Types

The fabric-like geometrical configuration has been suggested by Professor Zhong Lin Wang in 2008. The piezoelectric NW is grown vertically on the two microfibers in its radial direction, and they are twined to form a nanogenerator. One of the microfibers is coated with the metal to form a Schottky contact, serving as the counter electrode of VINGs. As the movable microfiber is stretched, the deformation of the nanostructure occurs on the stationary microfiber, resulting in voltage generation. Its working principle is identical to VINGs with partial mechanical contact, thus generating DC electrical signal.

6.4.3 Materials

Among various piezoelectric materials studied for the nanogenerator, many of the researches have been focused on the materials with wurtzite structures such as ZnO, CdS, and GaN. The greatest advantage of these materials arises from the facile and cost-effective fabrication technique, hydrothermal synthesis. Since the hydrothermal synthesis can be conducted in a low-temperature environment under 100°C in addition to vertical and crystalline growth, these materials can be integrated into various substrates with reduced concern for their physical characteristics such as melting temperature.

Endeavors for enhancing the piezoelectricity of the individual NW also led to the development of other piezoelectric materials based on the wurtzite structure. Professor Wang of Georgia Institute of Technology introduced p-type ZnO-NW.

Unlike the n-type semiconductive nanostructure, the mobile particle in p-type is a hole; thus, the Schottky behavior is reversed from that of n-type case; the electrical signal is generated from the portion of the nanostructure where the holes are accumulated. It is experimentally proved that p-type ZnO-NW can generate an output signal nearly ten times that of n-type ZnO-NW.

From the idea that the material with perovskite structure is known to have more effective piezoelectric characteristics compared with that with wurtzite structure, $BaTiO_3$ NW has been also studied by Professor Yu of the University of Illinois at Urbana Champaign. The output signal was found to be more than 16 times that from a similar ZnO-NW. Professor Lin of the University of California, Berkeley, has suggested that PVDF can be also applied to form a nanogenerator. Being a polymer, PVDF utilizes near-field electrospinning for its fabrication, which is rather a different technique compared with other materials. The nanofiber can be directly written on the substrate controlling the process, and this technique is expected to be applied for forming SP textile based on nanofiber. Researchers from SUTD presented the successful synthesis of ultra-long potassium niobate ($KNbO_3$) nanofibers using a sol–gel-assisted far-field electrospinning process and utilized them to develop a high-output voltage flexible nanogenerator.

Considering that the piezoelectric constant plays a critical role in the overall performance of a PENG, another research direction to improve device efficiency is to find new material of large piezoelectric response. Lead magnesium niobate–lead titanate (PMN-PT) is a next-generation piezoelectric material with super high piezoelectric constant when ideal composition and orientation are obtained. In 2012, PMN-PT NWs with a very high piezoelectric constant were fabricated by a hydrothermal approach and then assembled into an energy harvesting device. The record-high piezoelectric constant was further improved by the fabrication of a single-crystal PMN-PT nanobelt, which was then used as the essential building block for a PENG. A comparison of the reported materials by 2010 is given in Table 6.1

6.4.4 Applications of PENG

Nanogenerator is expected to be applied for various applications where periodic kinetic energy exists, such as wind and ocean waves on a large scale to the muscle movement by the beat of a heart or inhalation of lung on a small scale. The further feasible applications are as follows.

6.4.4.1 Self-Powered Nano-/Micro-Devices

One of the feasible applications of nanogenerator is an independent or a supplementary energy source to nano-/micro-devices consuming a relatively low amount of energy in a condition where the kinetic energy is supplied continuously. One example was introduced by Professor Wang's group in 2010 by the SP pH or UV sensor-integrated VING with an output voltage of 20~40 mV onto the sensor. Unfortunately, since the converted electrical energy is relatively small for operating nano/micro-devices, a supplementary energy source of battery is often required. The breakthrough is being sought by combining the nanogenerator with other types of energy harvesting devices, such as SC or biochemical energy harvester. This approach is

TABLE 6.1
Comparison of Reported Materials for PENG [12]

Material	Type	Geometry	Output Voltage	Output Power	Synthesis	Researched at
ZnO (n-type)	Wurtzite	D: ~100 nm, L: 200~500 nm	V_P = ~9 mV @ R = 500 MΩ	~0.5 pW per cycle (estimated)	CVD, hydrothermal process	Georgia Tech.
ZnO (p-type)	Wurtzite	D: ~50 nm, L: ~600 nm	V_P = 50~90 mV @ R = 500 MΩ	5~16.2 pW per cycle (calculated)	CVD	Georgia Tech.
ZnO-ZnS	Wurtzite (heterostructure)	Not stated	V_P = ~6 mV @ R = 500 MΩ	~0.1 pW per cycle (calculated)	Thermal evaporation and etching	Georgia Tech.
GaN	Wurtzite	D: 25~70 nm, L: 10~20 μm	V_{avg} = ~20 mV, V_{max} = ~0.35 V@ R = 500 MΩ	~0.8 pW per cycle (average, calculated)	CVD	Georgia Tech.
CdS	Wurtzite	D: ~100 nm, L: 1 μm	V_P = ~3 mV	Not stated	PVD, hydrothermal process	Georgia Tech.
$BaTiO_3$	Perovskite	D: ~280 nm, L: ~15 μm	V_P = ~25 mV @ R = 100 MΩ	~0.3 aJ per cycle (stated)	High-temperature chemical reaction	UIUC
PVDF	Polymer	D: 0.5~6.5 μm, L: 0.1~0.6 mm	V_P = 5~30 mV	2.5 pW~90 pW per cycle (calculated)	Electrospinning	UC Berkeley
$KNbO_3$	Perovskite	D: ~100 nm; L: few cm	V_p = ~16V @ R = 100MΩ		Electrospinning	SUTD/MIT

expected to contribute to the development of the energy source suitable for the application where independent operation is crucial, such as smartdust.

6.4.4.2 Smart Wearable Systems

The outfit integrated or made of the textiles with the piezoelectric fiber is one of the feasible applications of the nanogenerator. The kinetic energy from the human body is converted to electrical energy through the piezoelectric fibers, and it can be possibly applied to supply portable electronic devices such as health monitoring system attached to the smart wearable systems. The nanogenerator such as VING can be also easily integrated into the shoe employing the walking motion of the human body. Another similar application is a power-generating artificial skin. Professor Wang's group has shown the possibility of generating an AC voltage of up to 100 mV from the flexible SWG attached to the running hamster.

6.4.4.3 Transparent and Flexible Devices

Some of the piezoelectric nanostructures can be formed in various kinds of substrates, such as flexible and transparent organic substrates. The research groups in

SKKU and SAIT have developed the transparent and flexible nanogenerator, which can be possibly used for SP tactile sensor, and anticipated that the development may be extended to energy-efficient touchscreen devices. Their research focus is being extended to enhance the transparency of the device and the cost-effectiveness by substituting ITO electrode with a graphene layer.

6.4.4.4 Implantable Telemetric Energy Receiver

The nanogenerator based on ZnO-NW can be applied for implantable devices since ZnO not only is biocompatible but also can be synthesized upon the organic substrate, rendering the nanogenerator biocompatible overall. The implantable device integrated with the nanogenerator can be operated by receiving the external ultrasonic vibration outside the human body, which is converted to electrical energy by the piezoelectric nanostructure.

6.5 PyENG

Yang et al. [185] and Shin et al. [186] showed that the operation mechanism of PyENG is based on the pyroelectric effect of a material, referring to the spontaneous polarization change under the temperature variation over time. Generally, the spontaneous polarization intensity of a pyroelectric material will remain unchanged (no pyroelectric current) when there are no temperature variations over time, no matter how high/low the temperature is or with/without a temperature gradient over space. Once there is an ascent or descent in temperature over time, the spontaneous polarization intensity of a pyroelectric material will then change accordingly, generating a pyroelectric current in the external circuit until a new equilibrium is achieved. Since most piezoelectric materials also exhibit pyroelectric property, thus the same material can be used to harvest both mechanical and thermal energies under different usage scenarios, based on the coexisted piezoelectric and pyroelectric properties.

A PyENG is an energy harvesting device converting the external thermal energy into electrical energy using nanostructured pyroelectric materials. Usually, harvesting TE energy mainly relies on the Seebeck effect that utilizes a temperature difference between two ends of the device for driving the diffusion of charge carriers. However, in an environment where the temperature is spatially uniform without a gradient, such as in the outdoors, the Seebeck effect cannot be used to harvest thermal energy from a time-dependent temperature fluctuation. In this case, the pyroelectric effect has to be the choice, which is about the spontaneous polarization in certain anisotropic solids as a result of temperature fluctuation. The first PyENG was introduced by Prof. Wang at the Georgia Institute of Technology in 2012. By harvesting the waste heat energy, this new type of nanogenerator has potential applications such as wireless sensors, temperature imaging, medical diagnostics, and personal electronics.

6.5.1 Mechanism for PyENG

The working principle of PyENG will be explained in two different cases: the primary pyroelectric effect and the secondary pyroelectric effect (see Figure 6.18). The working principle for the first case is explained by the primary pyroelectric effect, which describes the charge produced in a strain-free case. The primary pyroelectric effect dominates the pyroelectric response in PZT, BTO, and some other ferroelectric materials. The mechanism is based on the thermally induced random wobbling of the electric dipole around its equilibrium axis, the magnitude of which increases with increasing temperature. Due to thermal fluctuations under room temperature, the electric dipoles will randomly oscillate within a degree from their respective aligning axes. Under a fixed temperature, the total average strength of the spontaneous polarization from the electric dipoles is constant, resulting in no output of the PyENG. If we apply a change in temperature in the nanogenerator from room temperature to a higher temperature, the increase in temperature will cause the electric dipoles to oscillate within a larger degree of spread around their respective aligning axes. The total average spontaneous polarization is decreased due to the spread of the oscillation angles. The quantity of induced charges in the electrodes is thus reduced, resulting in a flow of electrons. If the nanogenerator is cooled instead of heated, the spontaneous polarization will

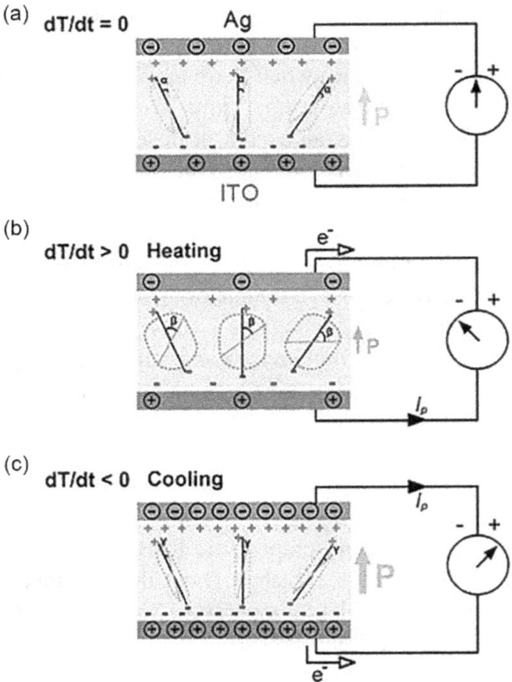

FIGURE 6.18 Mechanism of the pyroelectric nanogenerator based on a composite structure of pyroelectric nanowires. (a–c) Schematic diagrams of the pyroelectric nanogenerator with negative electric dipoles under room temperature (a), heated (b), and cooled (c) conditions. The angles marked in the diagrams represent the degrees to which the dipole would oscillate as driven by statistical thermal fluctuations [12].

be enhanced since the electric dipoles oscillate within a smaller degree of spread angles due to the lower thermal activity. The total magnitude of the polarization is increased and the amount of induced charges in the electrodes is increased. The electrons will then flow in an opposite direction.

For the second case, the obtained pyroelectric response is explained by the secondary pyroelectric effect, which describes the charge produced by the strain induced by thermal expansion. The secondary pyroelectric effect dominates the pyroelectric response in ZnO, CdS, and some other wurtzite-type materials. The thermal deformation can induce a piezoelectric potential difference across the material, which can drive the electrons to flow in the external circuit. The output of the nanogenerator is associated with the piezoelectric coefficient and the thermal deformation of the materials.

6.5.2 Applications of PyENG

PyENG is expected to be applied for various applications where time-dependent temperature fluctuation exists. One of the feasible applications of the PyENG is used as an active sensor, which can work without a battery. One example introduced by Wang's group [1] is the use of a PyENG as the SP temperature sensor for detecting a change in temperature, where the response time and reset time of the sensor are about 0.9 and 3 seconds, respectively. In general, the PyENG gives a high output voltage, but the output current is small. It not only can be used as a potential power source, but also as an active sensor for measuring temperature variation.

6.6 HYBRID CELLS FOR THE CONCURRENT HARVESTING OF MULTIPLE ENERGY TYPES

Rationally designed materials and technologies have been developed in the past decades for the conversion of various types of energy, such as solar, thermal, mechanical, and chemical energy, into electricity. These existing approaches, however, were investigated and developed based on drastically different physical principles and diverse engineering approaches to specifically harvest a certain type of energy, while the other types of energy were wasted. Innovative approaches have to be developed for the conjunctional harvesting of multiple types of energy through the use of integrated structures/materials, so that all available energy resources can be effectively and complementarily utilized [1]. On a smaller scale, the temporal/spatial distribution and availability of energy sources for driving MNSs vary drastically. The concurrent harvesting of multiple energy types from the ambient environment by a single integral device has therefore emerged as a promising approach toward the sustainable and maintenance-free operation of MNSs. Ever since the first demonstration of a nanotechnology-enabled HC by the Wang research group for the simultaneous harvesting of multiple types of energy with a single device, [128] this technology has been advancing at an increasing pace.

6.6.1 Hybrid Cells for the Harvesting of Solar and Mechanical Energy

The first nanotechnology-enabled HC was developed by the Wang research group in 2009 for harvesting solar and mechanical energy with a single-energy harvester. The device essentially integrates a dye-sensitized SC (DSSC) and a piezoelectric NG, both of which are based on an array of ZnO-NWs, on a common substrate [187]. The cathode of the NG and the anode of the DSSC were integrated on the same silicon substrate to form a serial connection between the DSSC and the NG. The DSSC and NG units in the HC can work independently when a source of either solar or mechanical energy is available. It has also been demonstrated that the HC can harvest both solar energy and mechanical energy simultaneously and synergistically [1].

The practical application of the original HC was limited due to low power output and unreliable performance due to solvent leakage and evaporation. Xu and Wang [188] developed a prototype of a compact HC in which an NG based on a ZnO-NW array was integrated with a solid-state DSSC, which showed enhanced performance and durability owing to the introduction of a solid-state electrolyte and convolute structures formed between the NG and the DSSC. The ZnO-NW array in this flexible HC not only serves as the NG, but also acts simultaneously as the SC part of the device by integrating with an infiltrated organic polymer. Owing to the controllability of the output behavior, the performance of the HC can be synergistically enhanced by the contribution of the NG part. Lee et al. demonstrated a conceptually similar HC prototype based on the integration of a ZnO-NW NG with infiltrated quantum dots (QDs), which surrounded the NWs. This HC was specifically developed for harvesting sound and solar energy simultaneously [189]. Bae et al. [190] integrated multiple energy harvesters together with a storage device along a single fiber, which involved the use of ZnO-NWs and graphene. This approach allows simultaneous harvesting of solar and mechanical energy and in situ storage of this harvested energy for potential applications in flexible and wearable electronics.

6.6.2 Hybrid Cells for the Harvesting of Biomechanical and Biochemical Energy

Sustainably SP and implantable wireless micro-/nanodevices for in vivo biomedical applications are becoming increasingly important. One viable approach is to concurrently harvest energy from multiple energy sources such as mechanical and biochemical energy due to body motion, muscle stretching, and metabolic processes abound in the biological entity. Hansen et al. [191] developed a prototype hybrid energy scavenging device to address the above application needs through the direct harvesting of mechanical and biochemical energies in a biofluid environment. This hybrid energy scavenger consists of a piezoelectric PVDF nanofiber NG for harvesting mechanical energy, such as from respiration and blood flow in the vessels, integrated with a flexible enzymatic biofuel cell (BFC) for harvesting the biochemical energy from the chemical processes between glucose and O_2 in biofluid. These two energy harvesting approaches, integrated within one single device, can work either individually or synergistically. This HC for harvesting biomechanical and biochemical energy has similar disadvantages to previous HCs: The separate arrangement of the two components on the substrate without sophisticated integration leads to engineering problems, such as cross talk, and hence deteriorated overall performance. To solve this problem, Pan et al. [192] developed a compact structure by the integration of a ZnO-NW NG and a BFC on single carbon fiber. The NG for harvesting mechanical energy is based on a textured ZnO-NW film and is grown radially on the carbon fiber, which serves as both the core electrode and the substrate for ZnO growth. The BFC for converting chemical energy from the ambient biofluid is fabricated at the other end of the same carbon fiber. Elimination of the separating membrane and mediator significantly reduced the size of the BFC relative to that of conventional BFCs. The integrated structure improved the performance and the adaptability of the HC for harvesting biomechanical and biochemical energy.

6.6.3 Hybrid Cells for the Harvesting of Solar and Thermal Energy

During the solar photovoltaic (PV) process, a big proportion of the wasted energy is converted into heat, which leads to a temperature rise in the SCs. Furthermore, incident photons with longer wavelengths, which cannot participate in PV conversion, may also be converted into heat. To improve the conversion efficiency and fully utilize the solar spectrum, Guo et al. [193] designed an HC to harvest solar energy and the concurrently generated heat. The two-compartment hybrid tandem cell consists of a DSSC and a TE cell (TC). Solar energy is first converted into electricity in the DSSC, and the heat induced during this process is then transmitted to the TC for subsequent TE conversion. This HC is more efficient than a single harvester and fully utilizes the energy from the solar spectrum. Recently, Wang et al. [194] developed a novel PV–TE hybrid device composed of a series-connected DSSC, a solar-selective absorber (SSA), and a TE generator. The study reported a significantly enhanced efficiency of 13% by this device.

Although the concept of hybrid energy harvesting and the proposed approaches described above are promising, several practical issues need to be addressed before real applications of these prototypes are possible. One of the biggest issues is network matching between different energy harvesters. The power output from different harvesters differs significantly. Strategic approaches for matching and reconciling the different outputs should therefore be implemented. However, solutions to the current problems might also impart increased cost and difficulties in manufacturing. Overall, it can be anticipated that the concept of hybrid energy harvesting will play a critical role in the implementation of novel sustainable micro-/nanotechnology with more flexibility and adaptability. It can also be expected that more sophisticated hybrid energy scavenging devices that are capable of concurrently harvesting even more types of energy may be developed.

6.7 HYBRID NANOGENERATORS

6.7.1 TENG-Based Hybrid Generators for Outdoor Applications

Previously, the most common method for water energy harvesting is using EMG to convert the water energy into electric power by a turbine, but this kind of method means low energy conversion efficiency, large device volume, and high cost for equipment maintenance. Considering that the optimal operation frequency of EMG is higher than 50 Hz, not all water motions are suitable to be harvested by EMG, especially for those rectilinear motions operating below 10 Hz, i.e., wave energy. In this regard, TENGs that are capable of scavenging low-frequency (<5 Hz) mechanical energy with the advantages of low cost, high voltage, and simple fabrication have been explored globally in recent years for low-frequency wave energy harvesting [195–199]. Furthermore, as shown earlier, the energy conversion efficiency of TENG can be further improved by integrating with EMG to form a hybrid generator, where the low output current of TENG can be compensated by the high output current of EMG, and the unsatisfactory performance of EMG in the low-frequency energy scavenging can also be made up by TENG [200,201]. Shi et al. [4] have outlined numerous TENG-EMG hybrid nanogenerators that are useful for wave energy applications.

A teeterboard-like hybrid nanogenerator was developed by Wu et al. [202] with a multilayered TENG floating on water at one end and an EMG driven by a lever, which moved vertically at the other end. With the novel separate design, the lightweight TENG end can be easily triggered by low-frequency ocean waves, resulting in the freestanding Al tubes rolling on the PTFE surface and generating desirable triboelectric outputs. A rotational pendulum-based TENG-EMG hybrid system was proposed by Hou et al. [203]. The unrestricted rotational movement of the pendulum enables the hybrid generator with wide applicability to low frequency (<5 Hz) and irregular vibration. When the pendulum rotor rotates, the magnetic flux across each coil will change and generate the electromagnetic output. This hybrid generator is successfully demonstrated to be integrated into a buoy, which is able to utilize the energy from waves to directly drive IoT sensors. A TENG-EMG hybrid generator with a cubic structure, which works in contact freestanding mode, was reported by Wang et al. [204] to further improve energy conversion efficiency. The external oscillation enables the reciprocating contact and separation of the PTFE board between two electrodes, altering the electrodes' potential difference and thus generating the triboelectric output. This design has the prospect for future large-scale blue energy harvesting.

To efficiently harvest wind energy under different frequencies, the combination of TENG and EMG has been explored and proven as an effective approach with high energy conversion efficiency benefitting from the complementary properties of the two energy harvesting mechanisms. However, most triboelectric-integrated wind energy harvesters are based on rotational structures, resulting in inevitable abrasion between two friction layers and reduction in device life [205,206]. One way to solve this problem is to change the working mode of TENG from sliding-based to contact–separation-based. A TENG-EMG hybrid wind generator was proposed by Fan et al. [207] in which for the EMG part, the output is generated by the relative rotating motion between the upper magnet rotator and the bottom coil stator. For the TENG part, the novel structural design of the slider can help to convert the rotating motion of the rotator into the driving force of its own reciprocating motion, thus resulting in the contact and separation between the aluminum and the silicone rubber for the triboelectric output generation. Although the friction of the TENG part has been reduced by using the contact–separation mode, friction still exists at the rotational EMG, which requires the slowest driving wind speed to exceed 4 m/s, greatly limiting its application for low-speed breeze energy harvesting. Zhang et al. reported a windmill-like hybrid generator applicable to low-speed wind [208]. The spring steel sheet and the magnets mounted on the fan can effectively reduce the rotation resistance.

Other than the TENG-EMG hybrid generator, PENG is another suitable mechanism to be easily hybridized with TENG due to the similar thin-film nature of their active layers [209,210]. The advantages of TENG-PENG hybridization are briefly illustrated in Figure 6.19. Zhao et al. [211] reported a TENG-PENG hybrid nanogenerator for wind energy harvesting. The rotator is driven to rotate by wind and force the polyethylene terephthalate (PET)/Al sheet to contact the PTFE surface, thus generating triboelectric output in the Al/Au electrodes. When the PTFE film fully contacts with the Al layer, the further rotation will induce tensile stress on the integrated PENG and a 31-mode piezoelectric potential is then produced in the PVDF-TrFE layer. Rahman et al. [212] and Gong et al. [213] showed that the vibration generated by machinery and human activities, e.g., suspension systems in vehicles and human walking, can also be harvested for specific applications.

A broadband TENG-PENG hybrid nanogenerator with a bistable structure was developed by Deng et al. [215] for ultra-low-frequency rectilinear vibration. This device is successfully demonstrated to charge a 10 μF capacitor to 0.12 V within 60 seconds under the excitation frequency of 0.1 Hz, proving its applicability to ultra-low-frequency motions. By integrating into the suspension system of vehicles, the hybrid nanogenerator can be utilized to harvest the vibration energy of cars, which can be further stored to power electronics to reduce fuel consumption. Another hybrid nanogenerator that could be used to scavenge the energy of vehicles was reported by Yang et al. [216]. The crankshaft piston-based TENG-PENG hybrid nanogenerator could be mounted on the shafts of cars and driven by shafts' rotation. The rotational motion of the crankshaft drives the contact–separation motion between the Al and the triboelectric layers, as well as the relative motion between the magnet and copper,

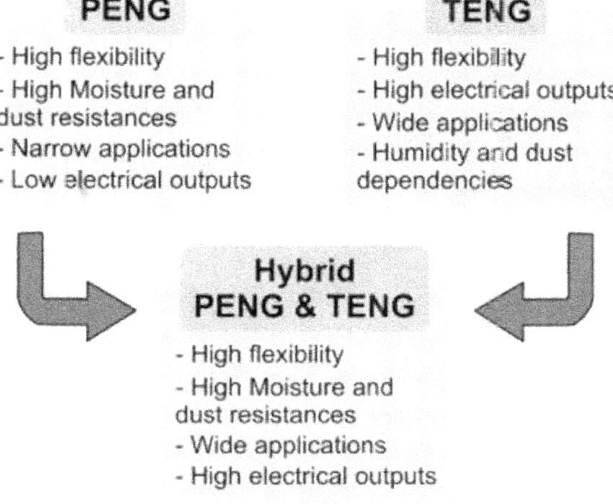

FIGURE 6.19 Advantages of using the hybrid PENG and TENG concept [214].

for generating the triboelectric and electromagnetic outputs. Many other applications of TENG-PENG hybridization is illustrated in Figure 6.20.

Hybridization of SCs with other energy harvesting devices can achieve continuous energy harvesting in varying weather conditions. TENG has been proven as a reliable energy harvester for raindrop energy scavenging based on the mechanism of triboelectrification at the liquid/solid interfaces [217–220]. Hence, TENG can be further integrated with SCs as a complement to insufficient solar energy on rainy days. Though several works have successfully demonstrated the feasibility of such kind of hybridization [221–223], the insufficient transparency of TENG top layers and the high cost of the antireflective process limit their applications for efficient and large-scale energy harvesting. To address this issue, Wang et al. [224] proposed an HC consisting of a silicone SC and a TENG with a highly transparent ionic conductor using a carbon dot composite film. This hybrid generator provides a promising method to simultaneously scavenge solar energy and raindrop energy with high power conversion efficiency. An ultrathin hybrid generator capable of harvesting raindrop, wind, and solar energies in varying weather conditions was also developed by Roh et al. [225]. The hybrid device is composed of two TENGs and one SC. The top TENG

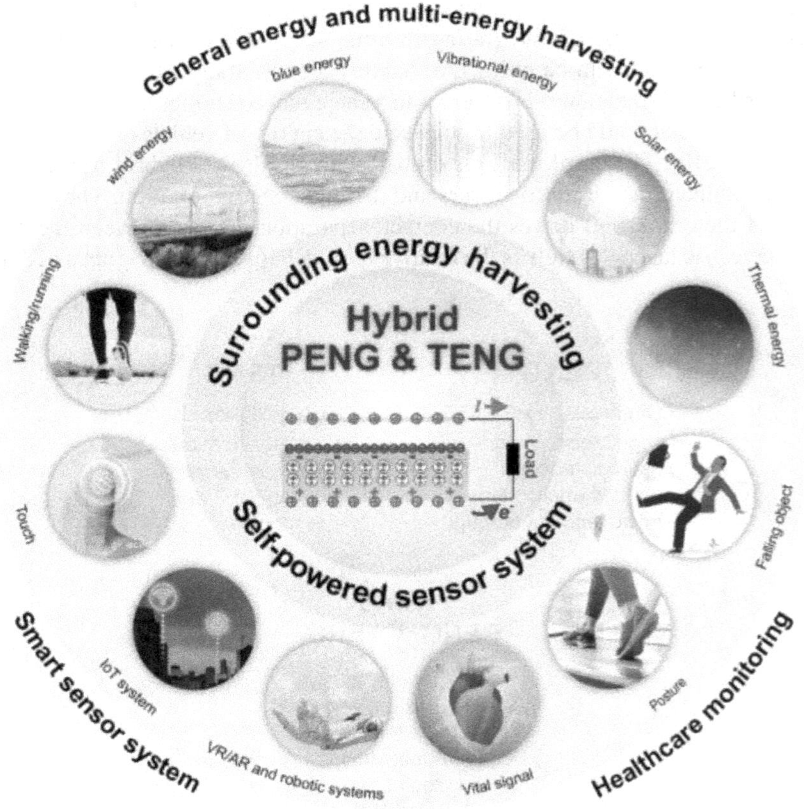

FIGURE 6.20 Outline illustration for hybrid PENG-TENG applications [214].

consisting of two ITO electrodes and a transparent FEP film is utilized for raindrop energy harvesting based on the triboelectrification on the FEP surface and electric potential changes in the two ITO electrodes when the droplets slide over the FEP film. Another novel all-in-one hybrid generator for wind, raindrop, and sunlight energy harvesting was proposed by Xu et al. [226]. The hybrid generator consists of four spherical TENG units covered by truncated cone-shaped capture rims for wind and rain fluid energy harvesting and four SCs mounted on two sides of the acrylic frame for solar energy collecting. This work utilizes TENGs to scavenge the weather-related mechanical energies to complement the output of SCs from solar energy, forming the hybridization of two energy harvesting mechanisms.

To further improve energy harvesting efficiency, TENG, EMG, and SC can be integrated together to form a tri-mechanism-based hybrid generator with the merits of enhanced energy scavenging ability for mechanical energies brought by the complementary properties of TENG and EMG as mentioned previously. Chandrasekhar et al. [227] reported a fully packed spheroidal hybrid generator consisting of a TENG and an EMG for water wave energy harvesting and a solar panel as a backup power source for low wave condition. When the hybrid generator is driven to sway by the water wave, the cylinder tube will slide back and forth in the device shell, resulting in the alternative contact and separation between the PDMS and the Al layers of two TENGs and producing the triboelectric output. Based on the same tri-mechanism hybrid strategy, a nonencapsulated pendulum-like paper-based hybrid generator was developed by Yang et al. [228]. There are three components in this device: one solar panel on the mover's top, two paper-based multilayered TENG units, and three EMG units. One of the two magnets mounted on the top portion of the mover is used to drive the device to wobble based on the repulsive force between magnets when the varying magnetic field is induced by water waves, and the other one is used to maintain the balance of the device. By integrating an SC for additional energy harvesting, the volume of the device could be effectively utilized, presenting a new strategy of hybrid generators to realize multisource energy harvesting for a SP navigation system.

Besides solar energy, thermal energy is another valuable static energy in the outdoor environment that is also desirable to be harvested considering the wide distribution of heat resources existing in geothermal and industrial factories, e.g., power plants, chemical plants, and metal productions [229]. Moreover, most thermal energy appears in the form of hot water, which contains not only heat but also kinetic energy. TENGs have been proven as reliable energy harvesters to scavenge the kinetic energy from water drops based on the solid–liquid triboelectrification, so it is a good strategy to hybridize PyENGs or TEGs with TENGs to fully utilize both the thermal and kinetic energies for maximum energy conversion efficiency. Jiang et al. proposed a TENG-PyENG hybrid generator aiming at collecting energy from low-grade waste fluids [230]. When a hot droplet slides across the surface of the hydrophobic layer, an electric potential difference in the attached two Ag electrodes arises, thus producing the triboelectric output. Wu et al. [231] integrated a TEG into a 2D rotary TENG to harvest the extra thermal energy induced by the TENG's friction. The triboelectric output is generated by the relative friction between the stator and rotator, while the TE output is produced because of the temperature difference between the friction-induced hot side and the bottom cold side.

6.7.2 TENG-Based Hybrid Generators for Indoor Applications

In recent years, advanced energy harvesting technologies have been developed to effectively scavenge indoor energies, e.g., biomechanical energy from human activities [232,233], waste heat energy from appliances [234], and indoor light [235], to directly power smart electronic devices and realize sustainable systems without too much maintenance operations. TENG, due to its wide selection of materials and high energy conversion efficiency for low-frequency excitation, is quite suitable to be designed as different household products, e.g., floor [236–240] and bed sheet [241], and wearable devices [242–244] for indoor biomechanical energy harvesting and sensing. Moreover, by integrating TENGs with other energy harvesting mechanisms such as EMG, PENG, PyENG, TEG, and SC, hybrid generators could be implemented toward effective indoor energy harvesting applications.

Islam et al. [245] reported a triboelectric–electromagnetic hybrid energy tile for stepping energy harvesting. The TENG part is composed of a layer of Al and a layer of high-polarized Kapton serving as the two opposite triboelectric materials, and the EMG part consists of multiple pairs of coils and magnets. When the smart tile is compressed by human footsteps, there will be contact–separation motions induced between the Kapton and the Al layer, as well as relative vertical motions in the coil and magnet pairs, thus producing the triboelectric and electromagnetic outputs. This hybrid generator is successfully demonstrated to charge a LIB with high speed and efficiency, illustrating its potential for powering indoor electronics.

In the information age, the usage of computers has become an indispensable part of our daily activities, where the biomechanical energies generated from hand motions could also be utilized for SP smart electronic purpose. A TENG-EMG hybrid smart mouse was developed by Rana et al. [246]. The TENG installed in the smart mouse consists of two parts: a sliding part composed of a PTFE film with two pairs of interdigitated electrodes and an immobile part made up of the nylon-11 film as the mouse pad. When the mouse slides on the mouse pad, the electric potential in the electrode pairs will change, thus driving the electrons to flow and generating the triboelectric output. The energy is successfully utilized to power portable electronics including a Bluetooth mouse, smartwatches, and smartphones. In addition to the mouse, the keyboard is also a commonly used human–computer interaction device, which can be combined with a hybrid generator to collect the mechanical energy of finger tapping. A triboelectric–pyroelectric–piezoelectric hybrid nanogenerator designed as the computer keyboard cover was proposed by Zhang et al. [247]. The top and bottom rubber layers serve as the negative triboelectric material in the device, and the triboelectric output is produced in the embedded Ag-coated fabric fiber when the finger taps on the rubber surface. The pressure of finger tapping induces piezoelectric polarization charges of the pyroelectric– PENG (PPENG), which is more sensitive to strain compared with the TENG unit. Besides, the PPENG can also harvest the thermal energy of the human body based on the pyroelectric effect to enhance the output performance of the hybrid nanogenerator. The hybrid nanogenerator is able to charge a capacitor to 3 V within 200 seconds by tapping, which can be directly used to power portable devices.

A TENG-EMG hybrid generator was proposed by Zhong et al. [248] for collecting energies from multiple sources. The proposed hybrid generator consists of a cylindrical stator with multiple attached electrodes and a cylindrical rotator with several FEP films mounted on the outer surface. When the top driver of the device is driven to rotate by wind or water flow, the rotator will spin due to the magnetic coupling force, resulting in the contact–separation motion between the FEP film blades and the fixed Cu electrodes for triboelectric output generation. The combined output power is sufficient to directly drive small electronics, e.g., humidity sensor and thermometer, indicating that the practical issues of sustainable power supply in certain applications could be solved by such hybridizing strategy for smart home, smart building, and smart city.

6.7.3 TENG-Based Hybrid Generators on Human Bodies

Along with the rapid development of TENG-based hybrid generators for outdoor and indoor applications, they are also extensively explored in different aspects of wearable and implantable applications, such as sustainable power sources, SP sensors, human–machine interfaces, and healthcare monitoring.

6.7.3.1 Wearable Hybrid Generators

Due to the capability of converting mechanical energy into electricity, wearable TENGs are suitable for acting as energy harvesters and SP sensors, especially to detect human daily activities and physiological status [249–253]. Even though TENGs hold great promise as wearable electronics, they still inevitably have limitations in power generation, sensing range, sensitivity, and also the sensing domain for the intrinsic limitations of electrification [254–256]. Therefore, hybrid generator systems that combine multiple energy harvesting units with energy storage units are widely studied for wearable electronics.

In most situations, PENGs are extensively investigated as a common compensation of energy harvesters in hybrid integrated systems, and the use of a piezoelectric element as a sensor has also been studied for the detection of more diversified signals [257–259]. Gong et al. [260] proposed a hybrid generator by combining the high voltage from a nonpiezoelectric meso-poly(lactic acid) (meso-PLA) electret-based TENG (E-TENG) and the relatively high current from a double-layered poly(l-lactic acid) (PLLA)-based PENG. This hybrid generator can also be woven into the shape of a fabric format as E-skin for wearable applications. Zou et al. [261] presented a spacer-free hybrid generator with a self-arched structure based on the effect of stress mismatch at the interface of two polymers. The self-arched nanogenerator (SANG) consists of a self-arched layer and a flat layer. The Ecoflex film in the self-arched layer and the Al film in the flat layer form a TENG, and the PVDF film with Ag electrodes forms the core part of the PENG. With different mass ratios of PDMS and Ecoflex, the bending degrees of the self-arched structure of the SANG will be different. More PDMS in the self-arched structure leads to a larger bending degree.

Syu et al. [262] developed a biomimetic hybrid SP sensor (BHSS), in which piezoelectric PVDF nano-microfiber (NMF) is deposited on the prepatterned printed circuit board (PCB) substrate in the in situ poled fashion of aligned dipoles, forming the PENG part of the BHSS. The SP wearable sensor can be achieved through the hybridization of piezoelectric and triboelectric mechanisms. Zhu et al. [263] reported a SP and self-functional sock (S2-sock) based on a poly(3,4-ethylenedioxythiophene) polystyrene-sulfonate (PEDOT:PSS)-coated textile-TENG integrated with a PZT-based piezoelectric sensor. PEDOT:PSS is adopted as the electrode material to incorporate good conductivity and high stability into a normal fabric. Lee et al. [264] reported another hybrid generator by electric polarization-controlled PVDF-based TENG and PENG for effectively harvesting the vibrational energy from human footsteps. The proposed TENG-PENG hybrid generator consists of two PVDF films, three Al electrodes, and two acrylic supports. The upper TENG component is vertically stacked on and electrically connected to the lower PENG component. When three hybrid generators were embedded at the forefoot, arch, and heel positions in a shoe insole, the whole insole device can generate enough energy to light up LEDs and drive a wireless pressure sensor during normal walking, which can be further applied for diagnostic health care.

In addition to the integration with PENGs, TENGs can also be combined with other mechanisms for wearable application [265,266]. Among these mechanisms, EMGs normally exhibit high current outputs that would be a good complement to the TENG's high voltage outputs [267,268]. Hybridizing multiple mechanisms such as TENG and EMG into an energy harvester is a suitable approach for improving the power conversion efficiency of human-induced mechanical excitations. Rahman et al. [269] reported a highly miniaturized freestanding kinetic impact-based hybrid generator as an effective energy harvester for various human-induced vibrations. The rational integration of EMG and TENG into a common mechanical system can improve the power generation capability of the hybrid generator under the same mechanical input. With the aid of a customized power management circuit (PMC), the output can be used to power modern electronics such as smartphones, smartwatches, and wireless temperature sensors.

In terms of miniaturized energy harvesting systems, integrating TENG with organic SC becomes a significant approach to collect solar energy owing to its flexibility that can be seamlessly integrated with human and the compatibility with large-scale and low-cost manufacturing techniques [270]. The integration of TENGs and flexible SCs enables the simultaneous collection of human mechanical and solar energy [271]. Ren et al. [272] developed a novel self-cleaning flexible hybrid energy harvesting system (HEHS), which includes a groove-shaped micro-/nanostructured haze thin film (GHF), a flexible PMC, and a hybrid generator composed of a flexible organic SC (F-OSC) and an autonomous single-electrode TENG (AS-TENG) via one common electrode. This hybrid system would collect both solar and mechanical energies through the top F-OSC and the bottom AS-TENG that can simultaneously utilize the large current of the SC and the high voltage of the TENG by the flexible PMC. The hybrid generator can be integrated into the garment in an embedded manner as a high-performance wearable HEHS.

Integrating TENG with other mechanisms could provide a feasible solution to realize skin-like sensing systems. Wang et al. [273] presented a multifunctional, tactile, and SP sensor that enables pressure, temperature, and material sensing. The sensor exhibits the form of a multilayer stack: a hydrophobic PTFE film as the triboelectrification layer, two Cu sheets coated with the Ag NW film as electrodes, and a sponge-like graphene/PDMS composite as the responsive component for pressure and temperature sensing based on the piezoresistive and TE effects. The proposed device can infer ten different flat materials, which opens a new path for using SP sensors in tactile sensing and material identification. As a result, this hierarchically patterned SP sensor offers a promising approach for multifunctional sensing with potential applications in wearable electronics and robotics.

6.7.3.2 Implantable Hybrid Generators

Since the motion energy of organs could be detected and harvested by nanogenerators, their outputs are strongly related to many biomedical signals, such as electrocardiogram (ECG) [274], heart rate [275], blood pressure [276], and respiratory motion [277]. Therefore, nanogenerators show great potential in implantable electrical stimulation systems, such as the brain, muscle, and peripheral nervous system [278–281]. TENGs with appropriate material and structural engineering have been extensively investigated for diversified implantable applications [282,283]. By harvesting the mechanical deformation energy from the contraction of the heart, brain, and vagus nerve, stimulation has been successfully realized by PENGs and TENGs, respectively [284,285]. The integration of neural interfaces, sensors, and other functional devices is emerging to form a SP implantable system with a variety of applications. Based on TENG, Liu et al. [286] reported a miniaturized, flexible, ultrasensitive, and SP endocardial pressure (EP) sensor (SEPS) based on TENG for real-time EP monitoring. SEPS can convert the energy of blood flow within the heart chambers into electricity. The electric outputs of the device can indicate the physiological and pathological cardiovascular status, including EP, ventricular fibrillation, and ventricular premature contraction. The device promotes the development of miniature implantable medical sensors for monitoring and diagnosis of cardiovascular diseases. With the robust SP capability, the SEPS exempts the necessity of onboard batteries, showing great potential in monitoring and diagnosing cardiovascular diseases.

Recently, the integration of TENGs and PENGs offers an efficient approach to enhance energy conversion efficiency. Shi et al. presented a piezoelectric and triboelectric hybrid nanogenerator (PTNG) to realize a packaged SP system (PSNGS) [287]. The triboelectric part is working based on the vertical contact–separation mode, whereas the contact layers are the $BaTiO_3$-doped PDMS film and the Al film. Besides, the polarized $BaTiO_3$-doped PDMS film can generate piezoelectric potential under stretched status, which can be used to enhance the triboelectric output through a synergistic effect. The PSNGS mainly consists of four parts: the PTNG as a power source, a rectifier, a micro-battery, and a flexible substrate. Huang et al. [288] reported a "self-matched" TENG-PENG hybrid nanogenerator using vapor-induced phase-separated poly(vinylidene fluoride) and recombinant spider silk to harvest the mechanical energy more efficiently. The electrons occupying the specific protein molecular orbits tend to transfer to the empty orbits of PET when they contact.

Li et al. [289] reported a HEHS consisting of a TENG and a glucose fuel cell (GFC) on a flexible PET substrate for simultaneously harvesting the biomechanical energy and the biochemical energy in simulated body fluid. The body fluid containing glucose molecules penetrates the active materials and then participates in the redox reaction around the anode electrode of the GFC. The lost electrons of glucose migrate from the anode to the cathode and are captured by dissolved oxygen in the body fluid. This process converts the biochemical energy in glucose into electric energy. Thus, implantable hybrid generators exhibit advantages for powering implantable medical electronics with excellent output performance, high power density, and good durability.

6.8 POWER MANAGEMENT AND ENERGY STORAGE

Accounted for the high-voltage–low-current and large internal impedance characteristics of conventional TENGs, different power management strategies have been developed for TENGs to productively improve energy conversion and energy storage efficiency [290]. For the hybrid generators, specific consideration in a PMC should be taken into account due to the more complicated configuration and output characteristics, such as the impedance match of different generator components, the mixing AC/DC outputs from different components, and the impedance match of the hybrid generators and storage units. Therefore, achieving high efficiency power management and energy storage is essential to realize high-performance and self-sustainable systems.

6.8.1 Power Management

One of the most common TENG-based hybrid generators is the TENG and EMG hybrid, in which the power management strategy is highly important due to the significant impedance mismatch of TENG and EMG [291–293]. Cao et al. [294] developed a PMC with an impedance-matching strategy for a TENG-EMG hybrid generator. The PMC consists of two commercial transformers and a bridge rectifier. One transformer with a ratio of 1 : 10 is connected to the EMG, and the other transformer with a ratio of 12 : 1 is connected to the TENG, to achieve an impedance match between the two components. After that, the two outputs are connected in series into the bridge rectifier for capacitor charging. Another common strategy for power management of TENG-EMG hybrid generators is using parallel–series switchable capacitors. Chen et al. [295] proposed a transistor-controlled PMC for automatic parallel–series capacitor transformation to overcome the huge impedance mismatch between TENG, EMG, and the energy storage unit. In the PMC, automatic switchability is designed by combining a transistor with transition capacitors (small capacitance) and diodes for the rectified triboelectric output. In this way, the impedance match between the TENG and the energy storage capacitor can be greatly improved. While TENG, EMG, PENG, and PyENG normally show AC output characteristics, TEG and SC exhibit DC characteristics that do not require rectifiers for output regulation. Ren et al. [296] presented a PMC on a flexible substrate for a hybrid generator composed of an AS-TENG and a F-OSC. A bridge rectifier is adopted to convert the AC outputs from the TENG into DC outputs before charging a capacitor, while the DC outputs from the SC can be directly connected. In the PMC, a diode is used to prevent the TENG currents to go through the SC.

In the practical application of powering small electronics, continuous and stable DC output is always required, which exerts higher demand for traditional power management. Rasel et al. [297] developed a customized PMC with a universal serial bus (USB) port as the stable DC output for a TENG-PENG hybrid generator. The PMC consists of two respective bridge rectifiers for the TENG and the PENG, an energy storage capacitor, a DC/DC converter, and a USB port for stable output. To achieve quick response time and minimize the power dissipation in the PMC, a high conversion efficiency DC/DC voltage converter is employed.

6.8.2 Energy Storage

To efficiently store the converted electrical energy from hybrid generators, flexible energy storage units such as supercapacitors and batteries have been given serious attention for wearable applications [298–301]. To construct functional and self-sustainable systems, energy storage units that are developed under the same platform as the hybrid generators are more favorable and are attracting increasing research attention. Qin et al. [302] demonstrated a self-charging power package through the integration of a TENG-PENG hybrid generator and an electrochromic microsupercapacitor array. Benefitting from the electrochromic property, the power package is able to indicate its charging status with color changes, which offers more convenience in monitoring the charging process. The supercapacitors as the energy storage units are fabricated under the same platform to store the energy from the TENG-SC hybrid generator.

With the extensive innovation in wearable electronics in the past few years, textile-based generators and energy storage units have also received rapid development. Wen et al. [303] demonstrated a self-charging power textile through the integration of a fiber-shaped TENG-SC hybrid generator and multiple fiber-shaped supercapacitors, for simultaneous biomechanical/solar energy collection and storage. Since both the hybrid generator and the supercapacitor are developed under the same fiber platform, the self-charging power textile offers great convenience in systemic integration and can be easily woven into normal clothes to enable the application in smart/power wearables.

Pu et al. [304] developed another power textile by integrating a fabric TENG (grating structure) and fiber-shaped DSSCs (FDSSCs) with a LIB as the energy storage unit. The TENG is designed with the grating structure and fabricated by Ni plating, to convert the common low-frequency human motions into high-frequency electrical outputs. Then, the outputs from the TENG are first connected to a bridge rectifier and then in parallel with seven FDSSCs (with an output voltage of ~5 V in series) to charge the LIB. To compare the charging performance, the LIB is charged by the fabric TENG, the FDSSCs, and the hybrid generator for 10 minutes, respectively, followed by a constant discharge at 1 μA. The results demonstrate that the discharge curves can last for 28, 59, and 98 minutes, respectively, showing the synergetic and improved charging effect of the hybrid generator.

6.9 FUNCTIONAL AND SELF-SUSTAINABLE HYBRIDIZED SYSTEMS

With the proper power management and energy storage discussed in the above section, functional and self-sustainable hybridized systems without external power supplies can be eventually realized by further combining the hybrid generator-based power packages with functional components [305–310]. Other than just serving as the power sources, the energy harvesting components in hybrid generators can also function as SP active sensors to reduce the overall power consumption of the hybridized system.

Rahman et al. [311] demonstrated a wireless SP environmental monitoring system by harvesting wind energy from the surroundings. The SP system is developed from a windmill-shaped hybrid generator that comprises a TENG, a PENG, and an EMG. All three generators are integrated into a common 3D-printed cylinder structure to convert the wind-induced rotation energy into electricity. The hybrid generator is further combined with a customized PMC (composed of transformers, rectifiers, a supercapacitor, a voltage regulator, and a switch), a microcontroller unit, an environmental sensor, and a Bluetooth module for constructing the wireless SP system.

To enable SP disaster monitoring, Qian and Jing [312] developed a SP system that can harvest the ambient mechanical wind and solar energy simultaneously. The wind energy harvester has a rotating structure that integrates sliding TENGs and EMGs. Next, a waterproof and flexible SC is integrated into the outer frame of the structure for solar energy harvesting. The developed SP disaster monitoring network includes three stages: energy harvesting-enabled SP monitoring, information collection and data transmission, and data receiving and response. Due to the complicated and offshore environment, SP systems are of great importance in ocean-relative applications. Gao et al. [313] presented a SP tracking system for monitoring marine equipment's position and attitude, by combining an EMG-powered global position system (GPS) module with a TENG-based inertial sensor. The TENG and EMG are integrated into a rotating gyro-structure, which shows good performance in energy harvesting and multiple-direction sensing. Furthermore, the SP system is packaged inside an autonomous underwater vehicle (AUV) and validated in the Huanghai Sea under actual working scenarios, showing the great potential of hybrid generators in blue energy applications.

To solve the limited working distance of the traditional radio frequency identification (RFID) tags while still maintaining battery-less characteristics, Chen et al. [314] proposed a SP RFID tag by integration with a TENG-EMG hybrid generator. The hybrid generator can effectively harvest human biomechanical energy with a maximum power density of $6.79\,W/m^2$. In addition, a PMC with an impedance-matching strategy by the parallel–series transformation of capacitors is developed, which can improve the capacitor charging efficiency by approximately 50%. With the harvested energy from human working, the hybrid generator-driven SP RFID tag shows a significantly enhanced working distance. Furthermore, based on the received signal strength indicator (RSSI) on the receiver side, the distance from the tag to the receiver can also be calculated and used for automatic door control in smart home relative applications.

In the rapidly expanded IoT era, developing self-sustainable human–machine interaction systems is of great significance for information exchange and communication. Qiu et al. [315] demonstrated a self-sustainable control system for smart home interactions. The control system consists of a photovoltaic cell as the power source, a sliding TENG as the control interface, and a PMC for power management, signal processing, and wireless communication. In the TENG control interface, a 3-bit binary-reflected gray code (BRGC) pattern is designed with two sensing electrodes, where one of them represents the bit "0" and the other one represents the bit "1." A third electrode located in the middle of the control interface is adopted to differentiate the inward and outward sliding directions. When the finger slides across different patterns, the unique output signals can be detected for different appliances' control.

6.10 FUTURE PROSPECTS

In the rapidly developing IoT era, TENG-based hybrid generators (e.g., TENG integration with one or more of the generators such as EMG, PENG, PyENG, TEG, and SC) provide a promising solution to enable the realization of functional and self-sustainable systems, by synergistically collecting the single-source and multisource energies. The review by Shi et al. [4] systematically summarize the recent development of TENG-based hybrid generators and hybridized systems in terms of outdoor, indoor, wearable, and implantable applications. As illustrated in Figure 6.21, various forms of energy in the environment (such as mechanical, thermal, and solar) can be effectively converted into electrical energy through employing the appropriate transducing mechanisms and their synergistically integrated hybrids [316–320]. For the single-source mechanical energy harvesting, TENG can be integrated with EMG and PENG under the same actuation scheme for compensating for the shortcomings of each generator and improving the energy conversion efficiency. As for the multisource energy harvesting, TENG-integrated PyENG, TEG, and SC under the same platform will be more favorable in terms of facile systemic integration and structural complexity reduction. By further integrating the hybrid generators with efficient PMCs, energy storage units, and other functional components, self-sustainable hybridized systems can be realized for a wide range of applications,

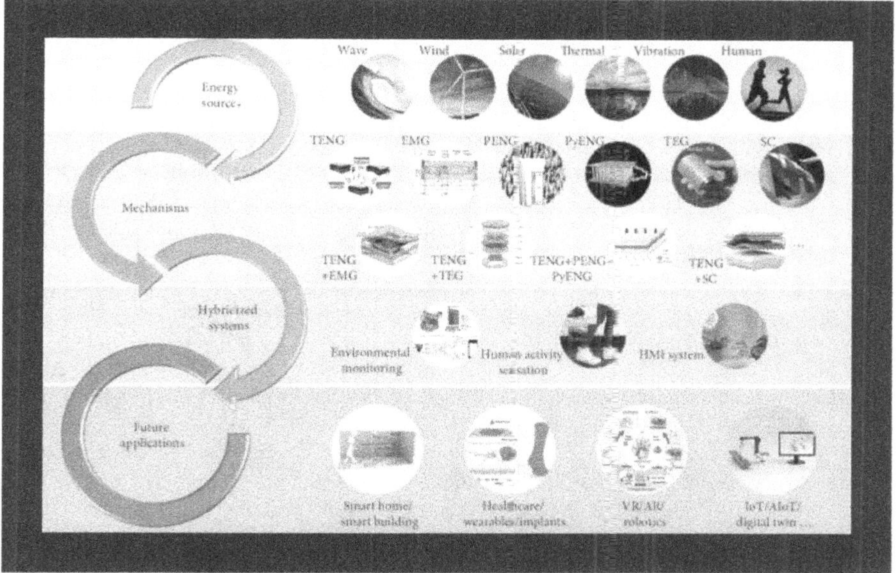

FIGURE 6.21 Prospects on the development and applications of hybridized systems [4].

e.g., environmental monitoring, human activity sensation, human–machine interaction, smart home, healthcare, wearables, implants, virtual/augmented reality (VR/AR), robotics, IoT, and artificial intelligence of things (AIoT) [320–324].

Although the future is promising, there are still some challenges remaining to be further addressed in the currently developed TENG-based hybrid generators and hybridized systems. First, further power enhancement of the hybrid generators is still inevitably desired, through maximizing the output performance of each generator and synergic integration. In most of the demonstrated SP systems, the whole system can only function for a short period of time but requires a much longer charging period. This intermittent operation mode highly limits their applicability in some crucial applications such as toxic matter monitoring, where intensive and continuous operations are required. Second, the long-term durability and performance robustness under different ambient conditions are major concerns for TENG-based systems. The involved periodical contact/friction of triboelectric materials and the high susceptibility to the ambient humidity of TENGs raise grand challenges for the material and structural design. Third, due to the large impedance mismatch between different energy harvesting components and energy storage units, effective power management is of great importance in achieving higher conversion efficiency. Moreover, TENG enhancement through optimized material/structural/circuitry design can be implemented before the power management for the hybrid generator to obtain a higher output performance. Last but not least, for the long-term functionality and stability of the hybridized systems, energy storage units such as supercapacitors and batteries should possess high energy storage density and reliability. Besides, energy storage units that are developed under the same platform as the hybrid generators (e.g., flexible, stretchable, and textile) will be more favorable for the convenient integration of the whole hybridized system. In this regard, the bright future of the IoT era will be greatly beneficial by the realization of functional and self-sustainable hybridized systems in diverse application areas.

REFERENCES

1. Wang, L. W., Wu, W., Nanotechnology-enabled energy harvesting for self-powered micro-/nanosystems. *Angew Chem Int Ed* 2012;51:2–24. DOI: 10.1002/anie.201201656
2. Zhou, L., Liu, D., Liu, L., He, L., Cao, X., Wang, J., Wang, Z., Recent advances in self-powered electrochemical systems. *Research* 2021;2021:4673028. DOI: 10.34133/2021/4673028
3. Yan, J., Mei, N., Zhang, D., Zhong, Y., Wang, C., Review of wave power system development and research on triboelectric nano power systems. *Front Energy Res* 2022. DOI: 10.3389/fenrg.2022.966567
4. Shi, Q., Sun, Z., Zhang, Z., Lee, C., Triboelectric nanogenerators and hybridized systems for enabling next-generation IoT applications. *Research* 2021;2021:6849171. DOI: 10.34133/2021/6849171
5. Wu, Z., Cheng, T., Wang, Z. L., Self-powered sensors and systems based on nanogenerators. *Sensors* 2020;20;2925. DOI: 10.3390/s20102925
6. Oliveira, H. P. D., Wearable nanogenerators: Working principle and self-powered biosensors applications. *Electrochem* 2021;2:118–134. DOI: 10.3390/electrochem2010010
7. Zhang, J., Yang, Z., Liang, X., Development and prospects of triboelectric nanogenerators in sports and physical state monitoring. *Front Mater* 2022;9:902499. DOI: 10.3389/fmats.2022.902499

8. Walden, R., Kumar, C., Mulvihill, D. M., Pillai, S. C., Opportunities and challenges in triboelectric nanogenerator (TENG) based sustainable energy generation technologies: A mini-review. *Chem Eng J Adv* 2022;9:100237. DOI: 10.1016/j.ceja.2021.100237
9. Vidal, J. V., Slabov, V., Kholkin, A. L., Dos Santos, M. P. S., Hybrid triboelectric-electromagnetic nanogenerators for mechanical energy harvesting: A review. *Nano-Micro Lett* 2021;13:1–58. DOI: 10.1007/s40820-021-00713-4
10. Xu, L., Hasan, M. A. M., Wu, H., Yang, Y., Electromagnetic–triboelectric hybridized nanogenerators. *Energies* 2021;14(19):6219. DOI: 10.3390/en14196219
11. Han, C., Cao, Z., Yuan, Z., Zhang, Z., Huo, X., Zhang, L. A., Wu, Z., Wang, Z. L., Hybrid triboelectric-electromagnetic nanogenerator with a double-sided fluff and double Halbach array for wave energy harvesting. *Adv Funct Mater* 2022;32(36):2205011. DOI: 10.1002/adfm.20220501
12. Nanogenerators, 2023. Wikipedia, The free encyclopedia, last edited 8 September 2023. https://en.wikipedia.org/wiki/Nanogenerator.
13. Wang, H., Xu, L., Wang, Z., Advances of high-performance triboelectric nanogenerators for blue energy harvesting, *Nanoenergy Adv* 2021;1:32–57. DOI: 10.3390/nanoenergyadv1010003
14. Yao, L., Zhang, H., Jiang, J., Zhang, Z., Zheng, X., Recent progress in sensing technology based on triboelectric nanogenerators in dynamic behaviors. *Sensors* 2022;22(13):4837. DOI: 10.3390/s22134837
15. Miorandi, D., Sicari, S., De Pellegrini, F., Chlamtac, I., Internet of things: Vision, applications and research challenges. *Ad HocNetw* 2012;10:1497–1516.
16. Batty, M., Axhausen, K. W., Giannotti, F., Pozdnoukhov, A., Bazzani, A., Wachowicz, M., Ouzounis, G., Portugali, Y., Smart cities of the future. *Eur Phys J Spec Top* 2012;214:481–518.
17. Zanella, A., Bui, N., Castellani, A., Vangelista, L., Zorzi, M., Internet of things for smart cities. *IEEE Internet Things J* 2014;1:22–32.
18. Al Mamun, M. A., Yuce, M. R., Recent progress in nanomaterial enabled chemical sensors for wearable environmental monitoring applications. *Adv Funct Mater* 2020;30:2005703.
19. Ho, D. H., Han, J., Huang, J., Choi, Y. Y., Cheon, S., Sun, J., Lei, Y., Park, G. S., Wang, Z. L., Sun, Q., Cho, J. H., β-phase-preferential blow-spun fabrics for wearable triboelectric nanogenerators and textile interactive interface. *Nano Energy* 2020;77:105262.
20. Zou, Y., Raveendran, V., Chen, J., Wearable triboelectric nanogenerators for biomechanical energy harvesting. *Nano Energy* 2020;77:105303
21. Lin, Z., Chen, J., Yang, J., Recent progress in triboelectric nanogenerators as a renewable and sustainable power source. *J Nanomater* 2016 2016:5651613.
22. Zou, Y., Liao, J., Ouyang, H., Jiang, D., Zhao, C., Li, Z., Qu, X., Liu, Z., Fan, Y., Shi, B., Zheng, L., Li, Z., A flexible self-arched biosensor based on combination of piezoelectric and triboelectric effects. *Appl Mater Today* 2020;20:100699.
23. Li, Z., Zheng, Q., Wang, Z. L., Li, Z., Nanogenerator-based self-powered sensors for wearable and implantable electronics. *Research* 2020;2020:8710686.
24. Selvarajan, S., Alluri, N. R., Chandrasekhar, A., Kim, S. J. Unconventional active biosensor made of piezoelectric $BaTiO_3$ nanoparticles for biomolecule detection. *Sens Actuators B Chem* 2017;253:1180–1187.
25. Liu, Z., Zhang, S., Jin, Y. M., Ouyang, H., Zou, Y., Wang, X. X., Xie, L. X., Li, Z., Flexible piezoelectric nanogenerator in wearable self-powered active sensor for respiration and healthcare monitoring. *Semicond Sci Technol* 2017;32:064004.
26. Xiong, W., Hu, K., Li, Z., Jiang, Y., Li, Z., Li, Z., Wang, X., A wearable system based on core-shell structured peptide-Co9S8 supercapacitor and triboelectric nanogenerator. *Nano Energy* 2019;66:104149.

27. Li, S., Ni, Q., Sun, Y., Min, G., Al-Rubaye, S., Energy-efficient resource allocation for industrial cyber-physical IoT systems in 5G era. *IEEE Transactions on Industrial Informatics.* 2018;14(6):2618–2628. DOI: 10.1109/TII.2018.2799177
28. Liu, J., Gu, L., Cui, N., Xu, Q., Qin, Y., Yang, R., Fabric-based triboelectric nanogenerators. *Research* 2019;2019:1091632. DOI: 10.34133/2019/1091632
29. Li, Z., Zheng, Q., Wang Z. L., Li, Z., Nanogenerator-based self-powered sensors for wearable and implantable electronics. *Research* 2020;2020:8710686. DOI: 10.34133/2020/8710686
30. Ryu, H., Yoon, H., Kim, S., Hybrid energy harvesters: Toward sustainable energy harvesting. *Adv Mater* 2019;31(34):1802898. DOI: 10.1002/adma.201802898
31. Khan, A. A., Mahmud, A., Ban, D., Evolution from single to hybrid nanogenerator: A contemporary review on multimode energy harvesting for self-powered electronics. *IEEE Trans Nanotechnol* 2019;18:21–36. DOI: 10.1109/TNANO.2018.2876824
32. Fan, F. R., Tian, Z. Q., Lin Wang, Z., Flexible triboelectric generator. *Nano Energy* 2012;1(2):328–334. DOI: 10.1016/j.nanoen.2012.01.004
33. Wu, C., Wang A. C., Ding, W., Guo, H., Wang, Z. L., Triboelectric nanogenerator: A foundation of the energy for the new era. *Adv Energy Mater* 2019;9(1):1802906. DOI: 10.1002/aenm.201802906
34. Shi, Q., Dong B., He T., Sun, Z., Zhu, J., Zhang, Z., Lee, C., Progress in wearable electronics/photonics—moving toward the era of artificial intelligence and Internet of things. *InfoMat* 2020;2(6):1131–1162. DOI: 10.1002/inf2.12122
35. Karan, S. K., Maiti, S., Lee, J. H., Mishra, Y. K., Khatua, B. B., Kim, J. K., Recent advances in self-powered tribo-/piezoelectric energy harvesters: All-in-one package for future smart technologies. *Adv Funct Mater* 2020;30(48):2004446. DOI: 10.1002/adfm.202004446
36. Zhang, Q., Zhang, Z., Liang, Q., Liang, Q., Gao, F., Yi, F., Ma, M., Liao, Q., Kang, Z., Zhang, Y., Green hybrid power system based on triboelectric nanogenerator for wearable/portable electronics. *Nano Energy* 2019;55:151–163. DOI: 10.1016/j.nanoen.2018.10.078
37. Wang, Z., Liu, W., Hu, J., He, W., Yang, H., Ling, C., Xi, Y., Wang, X., Liu, A., Hu, C., Two voltages in contact-separation triboelectric nanogenerator: From asymmetry to symmetry for maximum output. *Nano Energy* 2020;69:104452. DOI: 10.1016/j.nanoen.2020.104452
38. He, T., Wang, H., Wang, J., Yazici, M. S., Zhu, M., Ma, Y., Dong, B., Liu, Y., Lee, C., Self-powered multifunctional monitoring system using hybrid integrated triboelectric nanogenerators and piezoelectric microsensors. *Nano Energy* 2019;58:612–623. DOI: 10.1016/j.nanoen.2019.01.096
39. Qin, H., Cheng, G., Zi, Y., Gu, G., Zhang, B., Shang, W., Yang, F., Yang, J., Du, Z., Wang, Z. L., High energy storage efficiency triboelectric nanogenerators with unidirectional switches and passive power management circuits. *Adv Funct Mater* 2018;28(51):1805216. DOI: 10.1002/adfm.201805216
40. Liu, D., Yin, X., Guo, H., Zhou, L., Li, X., Zhang, C., Wang, J., Wang, Z. L., A constant current triboelectric nanogenerator arising from electrostatic breakdown. *Sci Adv* 2019;5(4):eaav6437. DOI: 10.1126/sciadv.aav6437
41. Zou, H., Zhang, Y., Guo, L., Wang, P., He, X., Dai, G., Zheng, H., Chen, C., Wang, A. C., Xu, C., Wang, Z. L., Quantifying the triboelectric series. *Nat Commun* 2019;10(1):1427. DOI: 10.1038/s41467-019-09461-x
42. Zou, H., Guo, L., Xue, H., Zhang, Y., Shen, X., Liu, X., Wang, P., He, X., Dai, G., Jiang, P., Zheng, H., Quantifying and understanding the triboelectric series of inorganic nonmetallic materials. *Nat Commun* 2020;11(1):2093. DOI: 10.1038/s41467-020-15926-1
43. Wang, Z. L., Wang, A. C., On the origin of contact-electrification. *Mater Today* 2019;30:34–51. DOI: 10.1016/j.mattod.2019.05.016

44. Lin, S., Xu, L., Chi Wang, A., Wang, Z. L., Quantifying electron-transfer in liquid-solid contact electrification and the formation of electric double-layer. *Nat Commun* 2020;11(1):399. DOI: 10.1038/s41467-019-14278-9
45. Lin, S., Zheng, M., Luo, J., Wang, Z. L., Effects of surface functional groups on electron transfer at liquid–solid interfacial contact electrification. *ACS Nano* 2020;14(8):10733–10741. DOI: 10.1021/acsnano.0c06075
46. Fan, F. -R., Lin, L., Zhu, G., Wu, W., Zhang, R., Wang, Z. L., Transparent triboelectric nanogenerators and self-powered pressure sensors based on micropatterned plastic films. *Nano Lett* 2012;12(6):3109–3114. DOI: 10.1021/nl300988z
47. Chen, L., Shi, Q., Sun, Y., Nguyen, T., Lee, C., Soh, S., Controlling surface charge generated by contact electrification: Strategies and applications. *Adv Mater* 2018;30(47):1802405. DOI: 10.1002/adma.201802405
48. Xie, X., Chen, X., Zhao, C., Liu, Y., Sun, X., Zhao, C., Wen, Z., Intermediate layer for enhanced triboelectric nanogenerator. *Nano Energy* 2021;79:105439. DOI: 10.1016/j.nanoen.2020.105439
49. Niu, S., Wang, Z. L., Theoretical systems of triboelectric nanogenerators. *Nano Energy* 2015;14:161–192. DOI: 10.1016/j.nanoen.2014.11.034
50. Wang, Z. L., Triboelectric nanogenerators as new energy technology and self-powered sensors – principles, problems and perspectives. *Faraday Discussions*. 2014;176:447–458. DOI: 10.1039/C4FD00159A
51. Zhu, J., Zhu, M., Shi, Q., Wen, F., Liu, L., Dong, B., Haroun, A., Yang, Y., Vachon, P., Guo, X., He, T., Progress in TENG technology—a journey from energy harvesting to nanoenergy and nanosystem. *EcoMat* 2020;2(4):eom2.12058. DOI: 10.1002/eom2.12058
52. Shi, Q., He, T., Lee, C., More than energy harvesting – combining triboelectric nanogenerator and flexible electronics technology for enabling novel micro-/nano-systems. *Nano Energy* 2019;57:851–871. DOI: 10.1016/j.nanoen.2019.01.002
53. Ouyang, H., Liu, Z., Li, N., Shi, B., Zou, Y., Xie, F., Ma, Y., Li, Z., Li, H., Zheng, Q., Qu, X., Symbiotic cardiac pacemaker. *Nat Commun* 2019;10(1):1821. DOI: 10.1038/s41467-019-09851-1
54. Liu, L., Shi, Q., Ho, J. S., Lee, C., Study of thin film blue energy harvester based on triboelectric nanogenerator and seashore IoT applications. *Nano Energy* 2019;66:104167. DOI: 10.1016/j.nanoen.2019.104167
55. Shi, Q., Wang, H., Wang, T., Lee, C., Self-powered liquid triboelectric microfluidic sensor for pressure sensing and finger motion monitoring applications. *Nano Energy* 2016;30:450–459. DOI: 10.1016/j.nanoen.2016.10.046
56. Wang, H., Wu, H., Hasan, D., He, T., Shi, Q., Lee, C., Self-powered dual-mode amenity sensor based on the water-air triboelectric nanogenerator. *ACS Nano* 2017;11(10):10337–10346. DOI: 10.1021/acsnano.7b05213
57. Ding, W., Wang, A. C., Wu, C., Guo, H., Wang, Z. L., Human-machine interfacing enabled by triboelectric nanogenerators and tribotronics. *Adv Mater Technol* 2019;4(1):1800487. DOI: 10.1002/admt.201800487
58. Zhu, M., He, T., Lee, C., Technologies toward next generation human machine interfaces: from machine learning enhanced tactile sensing to neuromorphic sensory systems. *Appl Phys Rev* 2020;7(3):031305. DOI: 10.1063/5.0016485
59. Lee, S., Wang, H., Shi, Q., Dhakar, L., Wang, J., Thakor, N. V., Yen, S. C., Lee, C., Development of battery-free neural interface and modulated control of tibialis anterior muscle via common peroneal nerve based on triboelectric nanogenerators (TENGs) *Nano Energy* 2017;33:1–11. DOI: 10.1016/j.nanoen.2016.12.038
60. Wang, J., He, T., Lee, C., Development of neural interfaces and energy harvesters towards self-powered implantable systems for healthcare monitoring and rehabilitation purposes. *Nano Energy* 2019;65:104039. DOI: 10.1016/j.nanoen.2019.104039

61. Bai, Y., Han, C. B., He, C., Gu, G. Q., Nie, J. H., Shao, J. J., Xiao, T. X., Deng, C. R., Wang, Z. L., Washable multilayer triboelectric air filter for efficient particulate matter PM 2.5 removal. *Adv Funct Mater* 2018;28(15):1706680. DOI: 10.1002/adfm.201706680
62. Nie, J., Ren, Z., Shao, J., Deng, C., Xu, L., Chen, X., Li, M., Wang, Z. L., Self-powered microfluidic transport system based on triboelectric nanogenerator and electrowetting technique. *ACS Nano* 2018;12(2):1491–1499. DOI: 10.1021/acsnano.7b08014
63. Xu, L., Wu, H., Yao, G., Chen, L., Yang, X., Chen, B., Huang, X., Zhong, W., Chen, X., Yin, Z., Wang, Z. L., Giant voltage enhancement via triboelectric charge supplement channel for self-powered electroadhesion. *ACS Nano* 2018;12(10):10262–10271. DOI: 10.1021/acsnano.8b05359
64. Cheng, J., Ding, W., Zi, Y., Lu, Y., Ji, L., Liu, F., Wu, C., Wang, Z. L., Triboelectric microplasma powered by mechanical stimuli. *Nat Commun* 2018;9(1):3733. DOI: 10.1038/s41467-018-06198-x
65. Pacha, A., Nanogenerators go wireless. *The Hindu*, 2017.
66. Niu, S., Liu, Y., Chen, X., Wang, S., Zhou, Y. S., Lin, L., Xie, Y., Wang, Z. L., Theory of freestanding triboelectric-layer-based nanogenerators. *Nano Energy* 2015;12:760–774. DOI: 10.1016/j.nanoen.2015.01.013
67. Zhu, J., Guo, X., Meng, D., Cho, M., Park, I., Huang, R., Song, W., A flexible comb electrode triboelectric–electret nanogenerator with separated microfibers for a self-powered position, motion direction and acceleration tracking sensor. *J Mater Chem A* 2018;6:16548–16555. DOI: 10.1039/C8TA04443K
68. Li, W., Liu, G., Jiang, D., Wang, C., Li, W., Guo, T., Zhao, J., Xi, F., Liu, W., Zhang, C., Interdigitated electrode-based triboelectric sliding sensor for security monitoring. *Adv Mater Technol* 2018;3:1800189. DOI: 10.1002/admt.201800189
69. Lin, L., Wang, S., Niu, S., Liu, C., Xie, Y., Wang, Z. L., Noncontact free-rotating disk triboelectric nanogenerator as a sustainable energy harvester and self-powered mechanical sensor. *ACS Appl Mater Interfaces* 2014;6:3031–3038. DOI: 10.1021/am405637s
70. Yu, H., He, X., Ding, W., Hu, Y., Yang, D., Lu, S., Wu, C., Zou, H., Liu, R., Lu, C., Wang, Z. L., A self-powered dynamic displacement monitoring system based on triboelectric accelerometer. *Adv Energy Mater* 2017;7:1700565. DOI: 10.1002/aenm.201700565
71. Wang, L., Wang, Y., Wang, H., Xu, G., Döring, A., Daoud, W. A., Xu, J., Rogach, A. L., Xi, Y., Zi, Y., Carbon dot-based composite films for simultaneously harvesting raindrop energy and boosting solar energy conversion efficiency in hybrid cells. *ACS Nano* 2020;14(8):10359–10369. DOI: doi:10.1021/acsnano.0c03986.s001
72. Wang, Z. L., Triboelectric nanogenerators as new energy technology and self-powered sensors—Principles, problems and perspectives. *Faraday Discuss* 2014;176:447–458.
73. Wang, S., Lin, L., Wang, Z. L., Triboelectric nanogenerators as self- powered active sensors. *Nano Energy* 2015;11:436–462.
74. Zhang, H., Yang, Y., Hou, T. C., Su, Y., Hu, C., Wang, Z. L., Triboelectric nanogenerator built inside clothes for self-powered glucose biosensors. *Nano Energy* 2013;2(5):1019–1024.
75. Bai, Y., Xu, L., He, C., Zhu, L., Yang, X., Jiang, T., Nie, J., Zhong, W., Wang, Z. L., High-performance triboelectric nanogenerators for self-powered, in-situ and real-time water quality mapping. *Nano Energy* 2019;66:104117.
76. Zhang, H., Yang, Y., Su, Y., Chen, J., Hu, C., Wu, Z., Liu, Y., Wong, C. P., Bando, Y., Wang, Z. L., Triboelectric nanogenerator as self-powered active sensors for detecting liquid/gaseous water/ethanol. *Nano Energy* 2013;2(5):693–701.
77. Lin, Z. H., Xie, Y., Yang, Y., Wang, S., Zhu, G., Wang, Z. L., Enhanced triboelectric nanogenerators and triboelectric nanosensor using chemically modified TiO_2 nanomaterials. *ACS Nano* 2013;7(5):4554–4560.
78. Wu, Y., Su, Y., Bai, J., Zhu, G., Zhang, X., Li, Z., Xiang, Y., Shi, J., A self-powered triboelectric nanosensor for PH detection," *J Nanomater* 2016;2016:5121572.

79. Lin, Z. H., Zhu, G., Zhou, Y. S., Yang, Y., Bai, P., Chen, J., Wang, Z. L., A self-powered triboelectric nanosensor for mercury ion detection. *Angew Chem Int Ed* 2013;52(19):5065–5069.
80. Jie, Y., Wang, N., Cao, X., Xu, Y., Li, T., Zhang, X., Wang, Z. L., Self-powered triboelectric nanosensor with poly(tetrafluoroethylene) nanoparticle arrays for dopamine detection. *ACS Nano* 2015;9(8):8376–8383.
81. Cao, X., Jie, Y., Wang, N., Wang, Z. L., Triboelectric nanogenerators driven self-powered electrochemical processes for energy and environmental science. *Adv Energy Mater* vol. 6(23):1600665, 2016.
82. Bolisetty, S., Peydayesh, M., Mezzenga, R., Sustainable technologies for water purification from heavy metals: Review and analysis. *Chem Soc Rev* 2019;48(2):463–487.
83. Li, Z., Chen, J., Guo, H., Fan, X., Wen, Z., Yeh, M. H., Yu, C., Cao, X., Wang, Z. L., Triboelectrification-enabled self-powered detection and removal of heavy metal ions in wastewater. *Adv Mater* 2016;28(15):2983–2991.
84. Zhou, L., Liu, D., Li, S., Yin, X., Zhang, C., Li, X., Zhang, C., Zhang, W., Cao, X., Wang, J., Wang, Z. L., Effective removing of hexavalent chromium from wasted water by triboelectric nanogenerator driven self-powered electrochemical system - Why pulsed DC is better than continuous DC? *Nano Energy* 2019;64:103915.
85. Li, Z., Chen, J., Yang, J., Su, Y., Fan, X., Wu, Y., Yu, C., Wang, Z. L., β-cyclodextrin enhanced triboelectrification for self-powered phenol detection and electrochemical degradation. *Energy Environ Sci* 2015;8(3):887–896.
86. Gao, S., Su, J., Wei, X., Wang, M., Tian, M., Jiang, T., Wang, Z. L., Self-powered electrochemical oxidation of 4- aminoazobenzene driven by a triboelectric nanogenerator. *ACS Nano* 2017;11(1):770–778.
87. Chen, S., Wang, N., Ma, L., Li, T., Willander, M., Jie, Y., Cao, X., Wang, Z. L., Triboelectric nanogenerator for sustainable wastewater treatment via a self-powered electrochemical process. *Adv Energy Mater* 2016;6(8):1501778.
88. Yang, Y., Zhang, H., Lee, S., Kim, D., Hwang, W., Wang, Z. L., Hybrid energy cell for degradation of methyl orange by self-powered electrocatalytic oxidation. *Nano Lett* 2013;13(2):803–808.
89. Gao, S., Wang, M., Chen, Y., Tian, M., Zhu, Y., Wei, X., Jiang, T., An advanced electro-Fenton degradation system with triboelectric nanogenerator as electric supply and biomass- derived carbon materials as cathode catalyst. *Nano Energy* 2018;45:21–27.
90. Tian, M., Zhu, Y., Zhang, D., Wang, M., Chen, Y., Yang, Y., Gao, S., Pyrrolic-nitrogen-rich biomass-derived catalyst for sustainable degradation of organic pollutant via a self-powered electro-Fenton process. *Nano Energy* 2019;64:103940.
91. Feng, Y., Han, K., Jiang, T., Bian, Z., Liang, X., Cao, X., Li, H., Wang, Z. L., Self-powered electrochemical system by combining Fenton reaction and active chlorine generation for organic contaminant treatment. *Nano Research* 2019;12(11):2729–2735.
92. Tian, M., Zhang, D., Wang, M., Zhu, Y., Chen, C., Chen, Y., Jiang, T., Gao, S., Engineering flexible 3D printed triboelectric nanogenerator to self-power electro-Fenton degradation of pollutants. *Nano Energy* 2020;74:104908.
93. Jiang, Q., Jie, Y., Han, Y., Gao, C., Zhu, H., Willander, M., Zhang, X., Cao, X., Self-powered electrochemical water treatment system for sterilization and algae removal using water wave energy. *Nano Energy* 2015;18:81–88.
94. Frye, R. E., Cakir, J., Rose, S., Delhey, L., Bennuri, S. C., Tippett, M., Melnyk, S., James, S. J., Palmer, R. F., Austin, C., Curtin, P., Arora, M., Prenatal air pollution influences neurodevelopment and behavior in autism spectrum disorder by modulating mitochondrial physiology. *Molecular Psychiatry* 2021;26(5):1561–1577.
95. Guo, H., Chen, J., Wang, L., Wang, A. C., Li, Y., An, C., He, J. H., Hu, C., Hsiao, V. K., Wang, Z. L., A highly efficient triboelectric negative air ion generator. *Nat Sustain* 2021;4(2):147–153.

96. Liu, Y. Z., Guo, R. T., Duan, C. P., Wu, G. L., Miao, Y. F., Gu, J. W., Pan, W. G., Removal of gaseous pollutants by using 3DOM-based catalysts: A review. *Chemosphere* 2021;262:127886.
97. Chen, S., Gao, C., Tang, W., Zhu, H., Han, Y., Jiang, Q., Li, T., Cao, X., Wang, Z., Self-powered cleaning of air pollution by wind driven triboelectric nanogenerator. *Nano Energy* 2015;14:217–225.
98. Han, K., Luo, J., Feng, Y., Lai, Q., Bai, Y., Tang, W., Wang, Z. L., Wind-driven radial-engine-shaped triboelectric nanogenerators for self-powered absorption and degradation of NO_X. *ACS Nano* 2020;14(3):2751–2759.
99. Feng, Y., Ling, L., Nie, J., Han, K., Chen, X., Bian, Z., Li, H., Wang, Z. L., Self-powered electrostatic filter with enhanced photocatalytic degradation of formaldehyde based on built-in triboelectric nanogenerators. *ACS Nano* 2017;11(12):12411–12418.
100. Fukuzumi, S., Production of liquid solar fuels and their use in fuel cells. *Joule* 2017;1(4):689–738.
101. Gao, P., Zhang, L., Li, S., Zhou, Z., Sun, Y., Novel heterogeneous catalysts for CO_2 hydrogenation to liquid fuels. *ACS Central Science* 2020;6(10):1657–1670.
102. Leung, S. F., Fu, H. C., Zhang, M., Hassan, A. H., Jiang, T., Salama, K. N., Wang, Z. L., He, J. H., Blue energy fuels: Converting ocean wave energy to carbon-based liquid fuels via CO_2 reduction. *Energy Environ Sci* 2020;13(5):1300–1308.
103. Jiang, Q., Han, Y., Tang, W., Zhu, H., Gao, C., Chen, S., Willander, M., Cao, X., Wang, Z. L., Self-powered seawater desalination and electrolysis using flowing kinetic energy. *Nano Energy* 2015;15:266–274.
104. Tang, W., Han, Y., Han, C. B., Gao, C. Z., Cao, X., Wang, Z. L., Self-powered water splitting using flowing kinetic energy. *Adv Mater* 2015;27(2):272–276.
105. Li, T., Xu, Y., Xing, F., Cao, X., Bian, J., Wang, N., Wang, Z. L., Boosting photo-electrochemical water splitting by TENG-charged Li-ion battery. *Adv Energy Mater* 2017;7(15):1700124.
106. Yang, Y., Zhang, H., Lin, Z. H., Liu, Y., Chen, J., Lin, Z., Zhou, Y. S., Wong, C. P., Wang, Z. L., A hybrid energy cell for self-powered water splitting. *Energy Environ Sci* 2013;6(8):2429.
107. Wang, W. H., Himeda, Y., Muckerman, J. T., Manbeck, G. F., Fujita, E., CO_2 hydrogenation to formate and methanol as an alternative to photo- and electrochemical CO_2 reduction. *Chem Rev* 2015;115(23):12936–12973.
108. Zhai, N., Wen, Z., Chen, X., Wei, A., Sha, M., Fu, J., Liu, Y., Zhong, J., Sun, X., Blue energy collection toward all-hours self- powered chemical energy conversion. *Adv Energy Mater* 2020;10(33):2001041.
109. Duan, G., Chen, Y., Tang, Y., Gasem, K.A., Wan, P., Ding, D., Fan, M., Advances in electrocatalytic ammonia synthesis under mild conditions. *Prog Energy Combust Sci* 2020;81:100860.
110. Chen, A., Xia, B. Y., Ambient dinitrogen electrocatalytic reduction for ammonia synthesis. *J Mater Chem A* 2019;7(41):23416–23431.
111. Gao, S., Zhu, Y., Chen, Y., Tian, M., Yang, Y., Jiang, T., Wang, Z. L., Self-power electroreduction of N_2 into NH_3 by 3D printed triboelectric nanogenerators. *Mater Today* 2019;28:17–24.
112. Zhu, H., Xu, Y., Han, Y., Chen, S., Zhou, T., Willander, M., Cao, X., Wang, Z., Self-powered electrochemical anodic oxidation: A new method for preparation of mesoporous Al_2O_3 without applying electricity. *Nano Res* 2015;8(11):3604–3611.
113. Zheng, X., Su, J., Wei, X., Jiang, T., Gao, S., Wang, Z. L., Self-powered electrochemistry for the oxidation of organic molecules by a cross-linked triboelectric nanogenerator. *Adv Mater* 2016;28(26):5188–5194.
114. Wang, J., Wen, Z., Zi, Y., Lin, L., Wu, C., Guo, H., Xi, Y., Xu, Y., Wang, Z. L., Self-powered electrochemical synthesis of polypyrrole from the pulsed output of a triboelectric nanogenerator as a sustainable energy system. *Adv Funct Mater* 2016;26(20):3542–3548.

115. Cai, G., Wang, J., Lee, P. S., Next-generation multifunctional electrochromic devices. *Acc Chem Res* 2016;49(8):1469–1476.
116. Zhang, M., Jie, Y., Cao, X., Bian, J., Li, T., Wang, N., Wang, Z. L., Robust design of unearthed single-electrode TENG from three-dimensionally hybridized copper/polydimethylsiloxane film. *Nano Energy* 2016;30:155–161.
117. Sun, J. G., Yang, T. N., Kuo, I. S., Wu, J. M., Wang, C. Y., Chen, L. J., A leaf- molded transparent triboelectric nanogenerator for smart multifunctional applications. *Nano Energy* 2017;32:180–186.
118. Yeh, M. H., Lin, L., Yang, P. K., Wang, Z. L., Motion-driven electrochromic reactions for self-powered smart window system," *ACS Nano* 2015;9(5):4757–4765.
119. Yang, X., Zhu, G., Wang, S., Zhang, R., Lin, L., Wu, W., Wang, Z. L., A self-powered electrochromic device driven by a nanogenerator. *Energy Environ Sci* 2012;5(11):9462–9466.
120. Qiu, W., Feng, Y., Luo, N., Chen, S., Wang, D., Sandwich-like sound-driven triboelectric nanogenerator for energy harvesting and electrochromic based on Cu foam. *Nano Energy* 2020;70:104543.
121. Sabel, C. F., Victor, D. G., Governing global problems under uncertainty: Making bottom-up climate policy work. *Clim Change* 2015;144(1):15–27.
122. Zhu, H. R., Tang, W., Gao, C. Z., Han, Y., Li, T., Cao, X., Wang, Z. L., Self-powered metal surface anti- corrosion protection using energy harvested from rain drops and wind. *Nano Energy* 2015;14:193–200.
123. Guo, W., Li, X., Chen, M., Xu, L., Dong, L., Cao, X., Tang, W., Zhu, J., Lin, C., Pan, C., Wang, Z. L.,Electrochemical cathodic protection powered by triboelectric nanogenerator. *Adv Funct Mater* 2014;24(42):6691–6699.
124. Z. Wang, L. Cheng, Y. Zheng, Y. Qin, and Z. L. Wang, "Enhancing the performance of triboelectric nanogenerator through prior-charge injection and its application on self-powered anticorrosion," *Nano Energy* 2014;10:37–43.
125. Li, X., Tao, J., Guo, W., Zhang, X., Luo, J., Chen, M., Zhu, J., Pan, C., A self-powered system based on triboelectric nanogenerators and supercapacitors for metal corrosion prevention. *J Mater Chem A* 2015;3(45):22663–22668.
126. Sadasivuni, K. K., Deshmukh, K., Ahipa, T. N., Muzaffar, A., Ahamed, M. B., Pasha, S. K., Al-Maadeed, M. A. A., Flexible, biodegradable and recyclable solar cells: A review. *J Mater Sci: Mater Electron* 2019;30(2):951–974. DOI: 10.1007/s10854-018-0397-y
127. Zi, Y., Niu, S., Wang, J., Wen, Z., Tang, W., Wang, Z. L., Standards and figure-of-merits for quantifying the performance of triboelectric nanogenerators. *Nat Commun* 2015;6(1):8376. DOI: 10.1038/ncomms9376
128. Kim, M., Park, D., Alam, M. M., Lee, S., Park, P., Nah, J., Remarkable output power density enhancement of triboelectric nanogenerators via polarized ferroelectric polymers and bulk MoS_2 composites. *ACS Nano* 2019;13(4):4640–4646. DOI: 10.1021/acsnano.9b00750
129. Wang, J., Pan, L., Guo, H., Zhang, B., Zhang, R., Wu, Z., Wu, C., Yang, L., Liao, R., Wang, Z. L., Rational structure optimized hybrid nanogenerator for highly efficient water wave energy harvesting. *Adv Energy Mater* 2019;9(8):1802892. DOI: 10.1002/aenm.201802892
130. Chen, B. D., Tang, W., Zhang, C., Xu, L., Zhu, L. P., Yang, L. J., He, C., Chen, J., Liu, L., Zhou, T., Wang, Z. L., Au nanocomposite enhanced electret film for triboelectric nanogenerator. *Nano Research*. 2018;11(6):3096–3105. DOI: 10.1007/s12274-017-1716-y
131. Wang, S., Xie, Y., Niu, S., Lin, L., Liu, C., Zhou, Y. S., Wang, Z. L., Maximum surface charge density for triboelectric nanogenerators achieved by ionized-air injection: Methodology and theoretical understanding. *Adv Mater* 2014;26(39):6720–6728. DOI: 10.1002/adma.201402491.
132. Chung, J., Heo, D., Shin, G., Choi, D., Choi, K., Kim, D., Lee, S, Ion-enhanced field emission triboelectric nanogenerator. *Adv Energy Mater* 2019;9(37):1901731. DOI: 10.1002/aenm.201901731

133. Wang, J., Wu, C., Dai, Y., Zhao, Z., Wang, A., Zhang, T., Wang, Z. L., Achieving ultrahigh triboelectric charge density for efficient energy harvesting. *Nat Commun* 2017;8(1):p. 88. DOI: 10.1038/s41467-017-00131-4
134. Liu, Y., Liu, W., Wang, Z., He, W., Tang, Q., Xi, Y., Wang, X., Guo, H., Hu, C., Quantifying contact status and the air-breakdown model of charge-excitation triboelectric nanogenerators to maximize charge density. *Nat Commun* 2020;11(1):1599. DOI: 10.1038/s41467-020-15368-9
135. Wang, H., Xu, L., Bai, Y., Wang, Z. L., Pumping up the charge density of a triboelectric nanogenerator by charge-shuttling. *Nat Commun* 2020;11(1):4203. DOI: 10.1038/s41467-020-17891-1
136. Bai, Y., Xu, L., Lin, S., Luo, J., Qin, H., Han, K., Wang, Z. L., Charge pumping strategy for rotation and sliding type triboelectric nanogenerators. *Adv Energy Mater* 2020;10(21):2000605. DOI: 10.1002/aenm.202000605
137. Kim, J., Ryu, H., Lee, J. H., Khan, U., Kwak, S. S., Yoon, H. J., Kim, S. W., High permittivity $CaCu_3Ti_4O_{12}$ particle-induced internal polarization amplification for high performance triboelectric nanogenerators. *Adv Energy Mater* 2020;10(9):1903524. DOI: 10.1002/aenm.201903524
138. He, W., Liu, W., Chen, J., Wang, Z., Liu, Y., Pu, X., Yang, H., Tang, Q., Yang, H., Guo, H., Hu, C., Boosting output performance of sliding mode triboelectric nanogenerator by charge space-accumulation effect. *Nat Commun* 2020;11(1):4277. DOI: 10.1038/s41467-020-18086-4
139. Wang, H., Zhu, J., He, T., Zhang, Z., Lee, C., Programmed-triboelectric nanogenerators—a multi-switch regulation methodology for energy manipulation. *Nano Energy* 2020;78:105241. DOI: 10.1016/j.nanoen.2020.105241
140. Vaisband, I., Saadat, M., Murmann, B., A closed-loop reconfigurable switched-capacitor DC-DC converter for sub-mW energy harvesting applications. *IEEE Trans Circuits Syst I: Regul Pap* 2015;62(2):385–394. DOI: 10.1109/TCSI.2014.2362971
141. Xie, H., Li, R., A novel switched-capacitor converter with high voltage gain. *IEEE Access* 2019;7:107831–107844. DOI: 10.1109/ACCESS.2019.2931562
142. Liang, X., Jiang, T., Feng, Y., Lu, P., An, J., Wang, Z. L., Triboelectric nanogenerator network integrated with charge excitation circuit for effective water wave energy harvesting. *Adv Energy Mater* 2020;10(40):2002123. DOI: 10.1002/aenm.202002123
143. Liu, W., Wang, Z., Wang, G., Zeng, Q., He, W., Liu, L., Wang, X., Xi, Y., Guo, H., Hu, C., Wang, Z. L., Switched-capacitor-convertors based on fractal design for output power management of triboelectric nanogenerator. *Nat Commun* 2020;11(1):1883. DOI: 10.1038/s41467-020-15373-y
144. Pan, Y., Liu, F., Jiang, R., Tu, Z., Zuo, L., Modeling and onboard test of an electromagnetic energy harvester for railway cars. *Appl Energy* 2019;250:568–581. DOI: 10.1016/j.apenergy.2019.04.182
145. Liu, H., Hou, C., Lin, J., Li, Y., Shi, Q., Chen, T., Sun, L., Lee, C., A non-resonant rotational electromagnetic energy harvester for low-frequency and irregular human motion. *Appl Phys Lett* 2018;113(20):203901. DOI: 10.1063/1.5053945
146. Liu, H., How Koh, K., Lee, C., Ultra-wide frequency broadening mechanism for micro-scale electromagnetic energy harvester. *Appl Phys Lett* 2014;104(5):053901. DOI: 10.1063/1.4863565
147. Liu, W., Wang, Z., Wang, G., Liu, G., Chen, J., Pu, X., Xi, Y., Wang, X., Guo, H., Hu, C., Wang, Z. L., Integrated charge excitation triboelectric nanogenerator. *Nat Commun* 2019;10(1):1426. DOI: 10.1038/ s41467-019-09464-8
148. Hu, Y., Yang, J., Niu, S., Wu, W., Wang, Z. L., Hybridizing triboelectrification and electromagnetic induction effects for high-efficient mechanical energy harvesting. *ACS Nano* 2014;8(7):7442. DOI: 10.1021/nn502684f

149. Wang, X., Wang, S., Yang, Y., Wang, Z. L., Hybridized electro-magnetic–triboelectric nanogenerator for scavenging air-flow energy to sustainably power temperature sensors. *ACS Nano* 2015;9(4):4553. DOI: 10.1021/acsnano.5b01187
150. Wu, Y., Zhong, X., Wang, X., Yang, Y., Wang, Z. L., Hybrid energy cell for simultaneously harvesting wind, solar, and chemical energies. *Nano Res* 2014;7(11):1631. DOI: 10.1007/s12274-014-0523-y
151. Zheng, L., Lin, Z. H., Cheng, G., Wu, W., Wen, X., Lee, S., Wang, Z. L., Silicon- based hybrid cell for harvesting solar energy and raindrop electrostatic energy. *Nano Energy* 2014;9:291. DOI: 10.1016/j.nanoen.2014.07.024
152. Yang, Y., Zhang, H., Lin, Z.H., Liu, Y., Chen, J., Lin, Z., Zhou, Y. S., Wong, C. P., Wang, Z. L., A hybrid energy cell for self-powered water splitting. *Energy Environ Sci* 2013;6(8):2429. DOI: 10.1039/C3EE41485J
153. Yang, Y., Zhang, H., Chen, J., Lee, S., Hou, T. C., Wang, Z. L., Simultaneously harvesting mechanical and chemical energies by a hybrid cell for self-powered biosensors and personal electronics. *Energy Environ Sci* 2013;6(6):1744. DOI: 10.1039/C3EE40764K
154. Bai, Y., Jantunen, H., Juuti, J., Hybrid, multi-source, and integrated energy harvesters. *Front Mater* 2018;5:65. DOI: 10.3389/fmats.2018 00065
155. Zhang, T., Yang, T., Zhang, M., Bowen, C. R., Yang, Y., Recent progress in hybridized nanogenerators for energy scavenging. *iScience* 2020;23(11):101689. DOI: 10.1016/j.isci.2020.101689
156. Zhang, T., Wen, Z., Liu, Y., Zhang, Z., Xie, Y., Sun, X., Hybridized nanogenerators for multifunctional self-powered sensing: Principles, prototypes, and perspectives. *iScience* 2020;23(12):101813. DOI: 10.1016/j.isci.2020.101813
157. Mitcheson, P. D., Reilly, E. K., Toh, T., Wright, P. K., Yeatman, E. M., Performance limits of the three MEMS inertial energy generator transduction types. *J Micromech Microeng* 2007;17(9):S211. DOI: 10.1088/0960-1317/17/9/s01
158. Chen, H., Xing, C., Li, Y., Wang, J., Xu, Y., Triboelectric nanogenerators for a macro-scale blue energy harvesting and self-powered marine environmental monitoring system. *Sustain Energy Fuels* 2020;4(3):1063. DOI: 10.1039/C9SE01184F
159. Zhao, J., Zhen, G., Liu, G., Bu, T., Liu, W., Fu, X., Zhang, P., Zhang, C., Wang, Z.L., Remarkable merits of triboelectric nanogenerator and electromagnetic generator for harvesting small-amplitude mechanical energy. *Nano Energy* 2019;61:111. DOI: 10.1016/j.nanoen.2019.04.047
160. Marin, A., Bressers, S., Priya, S., Multiple cell configuration electromagnetic vibration energy harvester. *J Phys D-Appl Phys* 2011;44(29):295501. DOI: 10.1088/0022-3727/44/29/295501
161. Xu, S., Fu, X., Liu, G., Tong, T., Bu, T., Wang, Z. L., Zhang, C., Comparison of applied torque and energy conversion efficiency between rotational triboelectric nanogenerator and electromagnetic generator. *iScience* 2021;24(4):102318. DOI: 10.1016/j.isci.2021.102318
162. Hou, C., Chen, T., Li, Y., Huang, M., Shi, Q., Liu, H., Sun, L., Lee, C., A rotational pendulum based electromagnetic/triboelectric hybrid-generator for ultra-low-frequency vibrations aiming at human motion and blue energy applications. *Nano Energy* 2019;63:103871. DOI: 10.1016/j.nanoen.2019.103871
163. Zhang, C., Tang, W., Han, C., Fan, F., Wang, Z. L., Theoretical comparison, equivalent transformation, and conjunction operations of electromagnetic induction generator and triboelectric nanogenerator for harvesting mechanical energy. *Adv Mater* 2014;26(22):3580. DOI: 10.1002/adma.201400207
164. Shao, H., Cheng, P., Chen, R., Xie, L., Sun, N., Shen, Q., Chen, X., Zhu, Q., Zhang, Y., Liu, Y., Wen, Z., Triboelectric–electromagnetic hybrid generator for harvesting blue energy. *Nano-Micro Lett* 2018;10(3):54. DOI: 10.1007/s40820-018-0207-3

165. Wen, Z., Guo, H., Zi, Y., Yeh, M. H., Wang, X., Deng, J., Wang, J., Li, S., Hu, C., Zhu, L., Wang, Z. L., Harvesting broad frequency band blue energy by a triboelectric– electromagnetic hybrid nanogenerator. *ACS Nano* 2016;10(7):6526. DOI: 10.1021/acsnano.6b03293
166. Zhong, Y., Zhao, H., Guo, Y., Rui, P., Shi, S., Zhang, W., Liao, Y., Wang, P., Wang, Z. L., An easily assembled electromagnetic-triboelectric hybrid nanogenerator driven by magnetic coupling for fluid energy harvesting and self-powered flow monitoring in a smart home/city. *Adv Mater Technol* 2019;4(12):1900741. DOI: 10.1002/admt.201900741
167. Zi, Y., Wang, J., Wang, S., Li, S., Wen, Z., Guo, H., Wang, Z. L., Effective energy storage from a triboelectric nanogenerator. *Nat Commun* 2016;7(1):10987. DOI: 10.1038/ncomms10987
168. Zhu, G., Zhou, Y. S., Bai, P., Meng, X. S., Jing, Q., Chen, J., Wang, Z. L., A shape-adaptive thin-film-based approach for 50% high-efficiency energy generation through micro-grating sliding electrification. *Adv Mater* 2014;26(23):3788. DOI: 10.1002/adma.201400021
169. Zhu, G., Chen, J., Zhang, T., Jing, Q., Wang, Z. L. Radial-arrayed rotary electrification for high performance triboelectric generator. *Nat Commun* 2014;5(1):3426. DOI: 10.1038/ncomms4426
170. Yang, W., Chen, J., Jing, Q., Yang, J., Wen, X., Su, Y., Zhu, G., Bai, P., Wang, Z. L. 3D stack integrated triboelectric nanogenerator for harvesting vibration energy. *Adv Funct Mater* 2014;24(26):4090. DOI: 10.1002/adfm.201304211
171. Li, S., Wang, S., Zi, Y., Wen, Z., Lin, L., Zhang, G., Wang, Z. L., Largely improving the robustness and lifetime of triboelectric nanogenerators through automatic transition between contact and noncontact working states. *ACS Nano* 2015;9(7):7479. DOI: 10.1021/acsnano.5b02575
172. Lin, L., Xie, Y., Niu, S., Wang, S., Yang, P. K., Wang, Z. L., Robust triboelectric nanogenerator based on rolling electrification and electrostatic induction at an instantaneous energy conversion efficiency of ~ 55%. *ACS Nano* 2015;9(1):922. DOI: 10.1021/nn506673x
173. Xu, C., Wang, A. C., Zou, H., Zhang, B., Zhang, C., Zi, Y., Pan, L., Wang, P., Feng, P., Lin, Z., Wang, Z. L., Raising the working temperature of a triboelectric nanogenerator by quenching down electron thermionic emission in contact-electrification. *Adv Mater* 2018;30(38):1803968. DOI: 10.1002/adma.201803968
174. Li, S., Zhang, D., Meng, X., Huang, Q. A., Sun, C., Wang, Z. L., A flexible lithium-ion battery with quasi-solid gel electrolyte for storing pulsed energy generated by triboelectric nanogenerator. *Energy Storage Mater* 2018;12:17. DOI: 10.1016/j.ensm.2017.11.013
175. Zhu, D., Tudor, M. J., Beeby, S. P., Strategies for increasing the operating frequency range of vibration energy harvesters: A review. *Meas Sci Technol* 2009;21(2):022001. DOI: 10.1088/0957-0233/21/2/022001
176. Zheng, Q., Shi, B., Fan, F., Wang, X., Yan, L., Yuan, W., Wang, S., Liu, H., Li, Z., Wang, Z. L., In vivo powering of pacemaker by breathing-driven implanted triboelectric nanogenerator. *Adv Mater* 2014;26(33):5851. DOI: 10.1002/adma.201402064
177. Meng, B., Fabrication of triboelectric nanogenerators. In M. Han, X. Zhang, H. Zhang (Eds), *Flexible and Stretchable Triboelectric Nanogenerator Devices: Toward Self-Powered Systems.* Wiley-VCH Verlag, 2019, pp. 41–57. DOI: 10.1002/9783527820153.ch3
178. Wang, Z. L., Piezoelectric nanogenerators based on zinc oxide nanowire arrays. *Science* 2006;312(5771):242–246. DOI: 10.1126/science.1124005
179. Wang, X., Song, J., Liu, J., Wang, Z. L., Direct-current nanogenerator driven by ultrasonic waves. *Science* 2007;316(5821):102–105. DOI: 10.1126/science.1139366
180. Xu, S., Qin, Y., Xu, C., Wei, Y., Yang, R., Wang, Z. L., Self-powered nanowire devices. *Nat Nanotechnol* 2010;5(5):366–373. DOI: 10.1038/nnano.2010.46

181. Liu, H., Zhong, J., Lee, C., Lee, S.-W., Lin, L., A comprehensive review on piezoelectric energy harvesting technology: materials, mechanisms, and applications. *Appl Phys Rev* 2018;5(4):041306. DOI: 10.1063/1.5074184
182. Dagdeviren, C., Yang, B. D., Su, Y., Tran, P. L., Joe, P., Anderson, E., Xia, J., Doraiswamy, V., Dehdashti, B., Feng, X., Lu, B., Conformal piezoelectric energy harvesting and storage from motions of the heart, lung, and diaphragm. *Proc Natl Acad Sci U S A* 2014;111(5):1927–1932. DOI: 10.1073/pnas.1317233111
183. Park, K. -I., Son, J. H., Hwang, G. -T., Jeong, C. K., Ryu, J., Koo, M., Choi, I., Lee, S. H., Byun, M., Wang, Z. L., Lee, K. J., Highly-efficient, flexible piezoelectric PZT thin film nanogenerator on plastic substrates. *Adv Mater* 2014;26(15):2514–2520. DOI: 10.1002/adma.201305659
184. Momeni, K., Odegard, G. M., Yassar, R. S., Nanocomposite electrical generator based on piezoelectric zinc oxide nanowires. *J Appl Phys* 2010;108(11):114303. DOI: 10.1063/1.3517095
185. Yang, Y., Guo, W., Pradel, K. C., Zhu, G., Zhou, Y., Zhang, Y., Hu, Y., Lin, L., Wang, Z. L., Pyroelectric nanogenerators for harvesting thermoelectric energy. *Nano Lett* 2012;12(6):2833–2838. DOI: 10.1021/nl3003039.
186. Shin, Y. E., Sohn, S. D., Han, H., Park, Y., Shin, H. J., Ko, H., Self-powered triboelectric/pyroelectric multimodal sensors with enhanced performances and decoupled multiple stimuli. *Nano Energy* 2020;72:104671. DOI: 10.1016/j.nanoen.2020.104671
187. Xu, C., Wang, X., Wang, Z. L., Nanowire structured hybrid cell for concurrently scavenging solar and mechanical energies. *J Am Chem Soc* 2009;131:5866–5872.
188. Xu, C., Wang, Z. L., Compact hybrid cell based on a convoluted nanowire structure for harvesting solar and mechanical energy. *Adv Mater* 2011;23:873–877.
189. Lee, M., Yang, R., Li, C., Wang, Z. L., Nanowire– quantum dot hybridized cell for harvesting sound and solar energies. *J Phys Chem Lett* 2010;1:2929–2935.
190. Bae, J., Park, Y. J., Lee, M., Cha, S. N., Choi, Y. J., Lee, C. S., Kim, J. M., Wang, Z. L., 2011. Single-fiber-based hybridization of energy converters and storage units using graphene as electrodes. *Adv Mater* 2011;23:3446–3449.
191. Hansen, B. J., Liu, Y., Yang, R., Wang, Z. L., Hybrid nanogenerator for concurrently harvesting biomechanical and biochemical energy. *ACS Nano* 2010;4:3647–3652.
192. Pan, C., Li, Z., Guo, W., Zhu, J., Wang, Z. L., Fiber-based hybrid nanogenerators for/as self-powered systems in biological liquid. *Angew Chem* 2011;123:11388–11392.
193. Guo, X. Z., Zhang, Y. D., Qin, D., Luo, Y. H., Li, D. M., Pang, Y. T., Meng, Q. B., Hybrid tandem solar cell for concurrently converting light and heat energy with utilization of full solar spectrum. *J Power Sources* 2010;195:7684–7690.
194. Wang, N., Han, L., He, H., Park, N. H., Koumoto, K., A novel high-performance photovoltaic–thermoelectric hybrid device. *Energy Environ Sci* 2011;4:3676–3679.
195. Wang, X., Niu, S., Yin, Y., Yi, F., You, Z., Wang, Z. L., Triboelectric nanogenerator based on fully enclosed rolling spherical structure for harvesting low-frequency water wave energy. *Adv Energy Mater* 2015;5(24):1501467. DOI: 10.1002/aenm.201501467
196. Xiong J., Lin M.-F., Wang J., Gaw S. L., Parida K., Lee P. S. Wearable all-fabric-based triboelectric generator for water energy harvesting. *Adv Energy Mater* 2017;7(21):1701243. DOI: 10.1002/aenm.201701243
197. Shi, Q., Wang, H., Wu, H., Lee, C., Self-powered triboelectric nanogenerator buoy ball for applications ranging from environment monitoring to water wave energy farm. *Nano Energy* 2017;40:203–213. DOI: 10.1016/j.nanoen.2017.08.018
198. Chen, G., Xu, L., Zhang, P., Chen, B., Wang, G., Ji, J., Pu, X., Wang, Z. L., Seawater degradable triboelectric nanogenerators for blue energy. *Adv Mater Technol* 2020;5(9):2000455. DOI: 10.1002/admt.202000455
199. Xia, K., Fu, J., Xu, Z., Multiple-frequency high-output triboelectric nanogenerator based on a water balloon for all-weather water wave energy harvesting. *Adv Energy Mater* 2020;10(28):2000426. DOI: 10.1002/aenm.202000426

200. Wang, P., Liu, R., Ding, W., Zhang, P., Pan, L., Dai, G., Zou, H., Dong, K., Xu, C., Wang, Z. L., Complementary electromagnetic-triboelectric active sensor for detecting multiple mechanical triggering. *Adv Funct Mater* 2018;28(11):1705808. DOI: 10.1002/adfm.201705808
201. Zi, Y., Guo, H., Wen, Z., Yeh, M. H., Hu, C., Wang, Z. L., Harvesting low-frequency (<5 Hz) irregular mechanical energy: A possible killer application of triboelectric nanogenerator. *ACS Nano* 2016;10(4):4797–4805. DOI: 10.1021/acsnano.6b01569
202. Wu, Y., Zeng, Q., Tang, Q., Liu, W., Liu, G., Zhang, Y., Wu, J., Hu, C., Wang, X., A teeterboard-like hybrid nanogenerator for efficient harvesting of low-frequency ocean wave energy. *Nano Energy* 2020;67:104205. DOI: 10.1016/j.nanoen.2019.104205
203. Hou, C., Chen, T., Li, Y., Huang, M., Shi, Q., Liu, H., Sun, L., Lee, C., A rotational pendulum based electromagnetic/triboelectric hybrid-generator for ultra-low-frequency vibrations aiming at human motion and blue energy applications. *Nano Energy* 2019;63:103871. DOI: 10.1016/j.nanoen.2019.103871
204. Wang, J., Pan, L., Guo, H., Zhang, B., Zhang, R., Wu, Z., Wu, C., Yang, L., Liao, R., Wang, Z. L., Rational structure optimized hybrid nanogenerator for highly efficient water wave energy harvesting. *Adv Energy Mater* 2019;9(8):1802892. DOI: 10.1002/aenm.201802892
205. Chen, S., Gao, C., Tang, W., Zhu, H., Han, Y., Jiang, Q., Li, T., Cao, X., Wang, Z., Self-powered cleaning of air pollution by wind driven triboelectric nanogenerator. *Nano Energy* 2015;14:217–225. DOI: 10.1016/j.nanoen.2014.12.013
206. Xie, Y., Wang, S., Lin, L., Jing, Q., Lin, Z. H., Niu, S., Wu, Z., Wang, Z. L., Rotary triboelectric nanogenerator based on a hybridized mechanism for harvesting wind energy. *ACS Nano* 2013;7(8):7119–7125. DOI: 10.1021/nn402477h
207. Fan, X., He, J., Mu, J., Qian, J., Zhang, N., Yang, C., Hou, X., Geng, W., Wang, X., Chou, X., Triboelectric-electromagnetic hybrid nanogenerator driven by wind for self-powered wireless transmission in Internet of things and self-powered wind speed sensor. *Nano Energy* 2020;68:104319. DOI: 10.1016/j.nanoen.2019.104319
208. Zhang, Y., Zeng, Q., Wu, Y., Wu, J., Yuan, S., Tan, D., Hu, C., Wang, X., An ultra-durable windmill-like hybrid nanogenerator for steady and efficient harvesting of low-speed wind energy. *Nano-Micro Lett* 2020;12(1):175. DOI: 10.1007/s40820-020-00513-2
209. Chen, S., Tao, X., Zeng, W., Yang, B., Shang, S., Quantifying energy harvested from contact-mode hybrid nanogenerators with cascaded piezoelectric and triboelectric units. *Adv Energy Mater* 2017;7(5):1601569. DOI: 10.1002/aenm.201601569.
210. Han, M., Zhang, X. S., Meng, B., Liu, W., Tang, W., Sun, X., Wang, W., Zhang, H., R-shaped hybrid nanogenerator with enhanced piezoelectricity. *ACS Nano* 2013;7(10):8554–8560. DOI: 10.1021/nn404023v
211. Zhao, C., Zhang, Q., Zhang, W., Du, X., Zhang, Y., Gong, S., Ren, K., Sun, Q., Wang, Z. L., Hybrid piezo/triboelectric nanogenerator for highly efficient and stable rotation energy harvesting. *Nano Energy* 2019;57:440–449. DOI: 10.1016/j.nanoen.2018.12.062
212. Rahman, M. T., Rana, S. S., Salauddin, M., Maharjan, P., Bhatta, T., Park, J. Y., Biomechanical energy-driven hybridized generator as a universal portable power source for smart/wearable electronics. *Adv Energy Mater* 2020;10(12):1903663. DOI: 10.1002/aenm.201903663
213. Gong, S., Zhang, B., Zhang, J., Wang, Z. L., Ren, K., Biocompatible poly(lactic acid)-based hybrid piezoelectric and electret nanogenerator for electronic skin applications. *Adv Funct Mater* 2020;30(14):1908724. DOI: 10.1002/adfm.201908724
214. Sriphan, S., Vittayakorn, N., Hybrid piezoelectric-triboelectric nanogenerators for flexible electronics: Recent advances and perspectives, *J Sci: Adv Mater Devices* 2022;7(3):100461. DOI: 10.1016/j.jsamd.2022.100461
215. Deng, H., Ye, J., Du, Y., Zhang, J., Ma, M., Zhong, X., Bistable broadband hybrid generator for ultralow-frequency rectilinear motion. *Nano Energy* 2019;65:103973. DOI: 10.1016/j.nanoen.2019.103973

216. Yang, H., Yang, H., Lai, M., Xi, Y., Guan, Y., Liu, W., Zeng, Q., Lu, J., Hu, C., Wang, Z. L., Triboelectric and electromagnetic hybrid nanogenerator based on a crankshaft piston system as a multifunctional energy harvesting device. *Adv Mater Technol* 2019;4(2):1800278. DOI: 10.1002/admt.201800278
217. Zhao, X. J., Kuang, S. Y., Wang, Z. L., Zhu, G., Highly adaptive solid-liquid interfacing triboelectric nanogenerator for harvesting diverse water wave energy. *ACS Nano* 2018;12(5):4280–4285. DOI: 10.1021/acsnano.7b08716
218. Xu, M., Zhao, T., Wang, C., Zhang, S. L., Li, Z., Pan, X., Wang, Z. L., High power density tower-like triboelectric nanogenerator for harvesting arbitrary directional water wave energy. *ACS Nano* 2019;13(2):1932–1939. DOI: 10.1021/acsnano.8b08274
219. Xu, W., Zheng, H., Liu, Y., Zhou, X., Zhang, C., Song, Y., Deng, X., Leung, M., Yang, Z., Xu, R. X., Wang, Z. L., A droplet-based electricity generator with high instantaneous power density. *Nature* 2020;578(7795):392–396. DOI: 10.1038/s41586-020-1985-6
220. Tang, W., Chen, B. D., Wang, Z. L., Recent progress in power generation from water/liquid droplet interaction with solid surfaces. *Adv Funct Mater* 2019;29(41):1901069. DOI: 10.1002/adfm.201901069
221. Zheng, L., Cheng, G., Chen, J., Lin, L., Wang, J., Liu, Y., Li, H., Wang, Z. L., A hybridized power panel to simultaneously generate electricity from sunlight, raindrops, and wind around the clock. *Adv Energy Mater* 2015;5(21):1501152. DOI: 10.1002/aenm.201501152
222. Liu, Y., Sun, N., Liu, J., Wen, Z., Sun, X., Lee, S. T., Sun, B., Integrating a silicon solar cell with a triboelectric nanogenerator via a mutual electrode for harvesting energy from sunlight and raindrops. *ACS Nano* 2018;12(3):2893–2899. DOI: 10.1021/acsnano.8b00416
223. Zheng, L., Lin, Z. H., Cheng, G., Wu, W., Wen, X., Lee, S., Wang, Z. L., Silicon-based hybrid cell for harvesting solar energy and raindrop electrostatic energy. *Nano Energy* 2014;9:291–300. DOI: 10.1016/j.nanoen.2014.07.024
224. Wang, L., Wang, Y., Wang, H., Xu, G., Döring, A., Daoud, W. A., Xu, J., Rogach, A. L., Xi, Y., Zi, Y., Carbon dot-based composite films for simultaneously harvesting raindrop energy and boosting solar energy conversion efficiency in hybrid cells. *ACS Nano* 2020;14(8):10359–10369. DOI: 10.1021/acsnano.0c03986
225. Roh, H., Kim, I., Kim, D., Ultrathin unified harvesting module capable of generating electrical energy during rainy, windy, and sunny conditions. *Nano Energy* 2020;70:104515. DOI: 10.1016/j.nanoen.2020.104515
226. Xu, L., Xu, L., Luo, J., Yan, Y., Jia, B. E., Yang, X., Gao, Y., Wang, Z. L., Hybrid all-in-one power source based on high-performance spherical triboelectric nanogenerators for harvesting environmental energy. *Adv Energy Mater* 2020;10(36):2001669. DOI: 10.1002/aenm.202001669
227. Chandrasekha,r A., Vivekananthan, V., Kim, S. J., A fully packed spheroidal hybrid generator for water wave energy harvesting and self-powered position tracking. *Nano Energy*. 2020;69:104439. DOI: 10.1016/j.nanoen.2019.104439
228. Yang, H., Deng, M., Tang, Q., He, W., Hu, C., Xi, Y., Liu, R., Wang, Z. L., A nonencapsulative pendulum-like paper–based hybrid nanogenerator for energy harvesting. *Adv Energy Mater* 2019;9(33):1901149. DOI: 10.1002/aenm.201901149
229. Chu, S., Majumdar, A., Opportunities and challenges for a sustainable energy future. *Nature* 2012;488(7411):294–303. DOI: 10.1038/nature11475
230. Jiang, D., Su, Y., Wang, K., Wang, Y., Xu, M., Dong, M., Chen, G., A triboelectric and pyroelectric hybrid energy harvester for recovering energy from low-grade waste fluids. *Nano Energy* 2020;70:104459. DOI: 10.1016/j.nanoen.2020.104459
231. Wu, Y., Kuang, S., Li, H., Wang, H., Yang, R., Zhai, Y., Zhu, G., Wang, Z. L., Triboelectric-thermoelectric hybrid nanogenerator for harvesting energy from ambient environments. *Adv Mater Technol* 2018;3(11):1800166. DOI: 10.1002/admt.201800166

232. Rahman, M. T., Rana, S. S., Salauddin, M., Maharjan, P., Bhatta, T., Kim, H., Cho, H., Park, J. Y., A highly miniaturized freestanding kinetic-impact-based non-resonant hybridized electromagnetic-triboelectric nanogenerator for human induced vibrations harvesting. *Appl Energy* 2020;279:115799. DOI: 10.1016/j.apenergy.2020.115799
233. Zhu, M., Shi, Q., He, T., Yi, Z., Ma, Y., Yang, B., Chen, T., Lee, C., Self-powered and self-functional cotton sock using piezoelectric and triboelectric hybrid mechanism for healthcare and sports monitoring. *ACS Nano* 2019;13:1940–1952. DOI: 10.1021/acsnano.8b08329
234. Ding, L. C., Meyerheinrich, N., Tan, L., Rahaoui, K., Jain, R., Akbarzadeh, A., Thermoelectric power generation from waste heat of natural gas water heater. *Energy Procedia* 2017;110:32–37. DOI: 10.1016/j.egypro.2017.03.101
235. Arai, R., Furukawa, S., Hidaka, Y., Komiyama, H., Yasuda, T., High-performance organic energy-harvesting devices and modules for self-sustainable power generation under ambient indoor lighting environments. *ACS Appl Mater Interfaces* 2019;11(9):9259–9264. DOI: 10.1021/acsami.9b00018
236. Jeon, S. B., Nho, Y. H., Park, S. J., Kim, W. G., Tcho, I. W., Kim, D., Kwon, D. S., Choi, Y. K., Self-powered fall detection system using pressure sensing triboelectric nanogenerators. *Nano Energy.* 2017;41:139–147. DOI: 10.1016/j.nanoen.2017.09.028
237. Ma, J., Jie, Y., Bian, J., Li, T., Cao, X., Wang, N., From triboelectric nanogenerator to self-powered smart floor: A minimalist design. *Nano Energy* 2017;39:192–199. DOI: 10.1016/j.nanoen.2017.06.025
238. Cheng, X., Song, Y., Han, M., Meng, B., Su, Z., Miao, L., Zhang, H., A flexible large-area triboelectric generator by low-cost roll-to-roll process for location-based monitoring. *Sens Actuators A Phys* 2016;247:206–214. DOI: 10.1016/j.sna.2016.05.051
239. He, C., Zhu, W., Chen, B., Xu, L., Jiang, T., Han, C. B., Gu, G. Q., Li, D., Wang, Z. L., Smart floor with integrated triboelectric nanogenerator as energy harvester and motion sensor. *ACS Appl Mater Interfaces* 2017;9(31):26126–26133. DOI: 10.1021/acsami.7b08526
240. Shi, Q., Zhang, Z., He, T., Sun, Z., Wang, B., Feng, Y., Shan, X., Salam, B., Lee, C., Deep learning enabled smart mats as a scalable floor monitoring system. *Nat Commun* 2020;11(1):4609. DOI: 10.1038/s41467-020-18471-z.
241. Lin, Z., Yang, J., Li, X., Wu, Y., Wei, W., Liu, J., Chen, J., Yang, J., Large-scale and washable smart textiles based on triboelectric nanogenerator arrays for self-powered sleeping monitoring. *Adv Funct Mater* 2018;28(1):1704112. DOI: 10.1002/adfm.201704112
242. Zhang, Z., He, T., Zhu, M., Sun, Z., Shi, Q., Zhu, J., Dong, B., Yuce, M. R., Lee, C., Deep learning-enabled triboelectric smart socks for IoT-based gait analysis and VR applications. *npj Flex Electron* 2020;4(1):29. DOI: 10.1038/s41528-020-00092-7
243. Wen, F., Sun, Z., He, T., Shi, Q., Zhu, M., Zhang, Z., Li, L., Zhang, T., Lee, C., Machine learning glove using self-powered conductive superhydrophobic triboelectric textile for gesture recognition in VR/AR applications. *Adv Sci* 2020;7(14):2000261. DOI: 10.1002/advs.202000261
244. He, T., Shi, Q., Wang, H., Wen, F., Chen, T., Ouyang, J., Lee, C., Beyond energy harvesting - multi-functional triboelectric nanosensors on a textile. *Nano Energy.* 2019;57:338–352. DOI: 10.1016/j.nanoen.2018.12.032.
245. Islam, E., Abdullah, A. M., Chowdhury, A. R., Tasnim, F., Martinez, M., Olivares, C., Lozano, K., Uddin, M. J., Electromagnetic-triboelectric-hybrid energy tile for biomechanical green energy harvesting. *Nano Energy* 2020;77:105250. DOI: 10.1016/j.nanoen.2020.105250
246. Rana, S. S., Rahman, M. T., Salauddin, M., Maharjan, P., Bhatta, T., Cho, H., Park, J. Y., A human-machine interactive hybridized biomechanical nanogenerator as a self-sustainable power source for multifunctional smart electronics applications. *Nano Energy* 2020;76:105025. DOI: 10.1016/j.nanoen.2020.105025

247. Zhang, Q., Liang, Q., Zhang, Z., Kang, Z., Liao, Q., Ding, Y., Ma, M., Gao, F., Zhao, X., Zhang, Y., Electromagnetic shielding hybrid nanogenerator for health monitoring and protection. *Adv Funct Mater* 2018;28(1):1703801. DOI: 10.1002/adfm.201703801
248. Zhong, Y., Zhao, H., Guo, Y., Rui, P., Shi, S., Zhang, W., Liao, Y., Wang, P., Wang, Z. L., An easily assembled electromagnetic-triboelectric hybrid nanogenerator driven by magnetic coupling for fluid energy harvesting and self-powered flow monitoring in a smart home/city. *Adv Mater Technol* 2019;4(12):1900741. DOI: 10.1002/admt.201900741
249. Shi, Q., Qiu, C., He, T., Wu, F., Zhu, M., Dziuban, J. A., Walczak, R., Yuce, M. R., Lee, C., Triboelectric single-electrode-output control interface using patterned grid electrode. *Nano Energy* 2019;60:545–556. DOI: 10.1016/j.nanoen.2019.03.090
250. Dong, B., Shi, Q., He, T., Zhu, S., Zhang, Z., Sun, Z., Ma, Y., Kwong, D. L., Lee, C., Wearable triboelectric/aluminum nitride nano-energy-nano-system with self-sustainable photonic modulation and continuous force sensing. *Adv Sci* 2020;7(15):1903636. DOI: 10.1002/advs.201903636
251. Chen, T., Shi, Q., Zhu, M., He, T., Yang, Z., Liu, H., Sun, L., Yang, L., Lee, C., Intuitive-augmented human-machine multidimensional nano-manipulation terminal using triboelectric stretchable strip sensors based on minimalist design. *Nano Energy* 2019;60:440–448. DOI: 10.1016/j.nanoen.2019.03.071
252. Chen, T., Shi, Q., Zhu, M., He, T., Sun, L., Yang, L., Lee, C., Triboelectric self-powered wearable flexible patch as 3D motion control interface for robotic manipulator. *ACS Nano* 2018;12(11):11561–11571. DOI: 10.1021/acsnano.8b06747
253. Shi, Q., Lee, C., Self-powered bio-inspired spider-net-coding interface using single-electrode triboelectric nanogenerator. *Adv Sci* 2019;6(15):1900617. DOI: 10.1002/advs.201900617
254. Chen, C., Guo, H., Chen, L., Wang, Y. C., Pu, X., Yu, W., Wang, F., Du, Z., Wang, Z. L., Direct current fabric triboelectric nanogenerator for biomotion energy harvesting. *ACS Nano* 2020;14(4):4585–4594. DOI: 10.1021/acsnano.0c00138
255. Lin, Z., Chen, J., Li, X., Zhou, Z., Meng, K., Wei, W., Yang, J., Wang, Z. L., Triboelectric nanogenerator enabled body sensor network for self-powered human heart-rate monitoring. *ACS Nano* 2017;11(9):8830–8837. DOI: 10.1021/acsnano.7b02975
256. He, T., Wang, H., Wang, J., Tian, X., Wen, F., Shi, Q., Ho, J. S., Lee, C., Self-sustainable wearable textile nano-energy nano-system (NENS) for next-generation healthcare applications. *Adv Sci* 2019;6(24):1901437. DOI: 10.1002/advs.201901437
257. Zhang, G., Li, M., Li, H., Wang, Q., Jiang, S., Harvesting energy from human activity: Ferroelectric energy harvesters for portable, implantable, and biomedical electronics. *Energy Technol* 2018;6(5):791–812. DOI: 10.1002/ente.201700622
258. Wang, Y., Wang, H., Xuan, J., Leung, D. Y. C., Powering future body sensor network systems: A review of power sources. *Biosens Bioelectron* 2020;166:112410. DOI: 10.1016/j.bios.2020.112410
259. Tang, G., Shi, Q., Zhang, Z., He, T., Sun, Z., Lee, C., Hybridized wearable patch as a multi-parameter and multi-functional human-machine interface. *Nano Energy* 2021;81:105582. DOI: 10.1016/j.nanoen.2020.105582
260. Gong, S., Zhang, B., Zhang, J., Wang, Z. L., Ren, K., Biocompatible poly(lactic acid)-based hybrid piezoelectric and electret nanogenerator for electronic skin applications. *Adv Funct Mater* 2020;30(14):1908724. DOI: 10.1002/adfm.201908724
261. Zou, Y., Liao, J., Ouyang, H., Jiang, D., Zhao, C., Li, Z., Qu, X., Liu, Z., Fan, Y., Shi, B., Zheng, L., A flexible self-arched biosensor based on combination of piezoelectric and triboelectric effects. *Appl Mater Today* 2020;20:100699. DOI: 10.1016/j.apmt.2020.100699
262. Syu, M. H., Guan, Y. J., Lo, W. C., Fuh, Y. K., Biomimetic and porous nanofiber-based hybrid sensor for multifunctional pressure sensing and human gesture identification via deep learning method. *Nano Energy* 2020;76(300):105029. DOI: 10.1016/j.nanoen.2020.105029

263. Zhu, M., Shi, Q., He, T., Yi, Z., Ma, Y., Yang, B., Chen, T., Lee, C., Self-powered and self-functional cotton sock using piezoelectric and triboelectric hybrid mechanism for healthcare and sports monitoring. *ACS Nano* 2019;13:1940–1952. DOI: 10.1021/acsnano.8b08329
264. Lee, D. W., Jeong, D. G., Kim, J. H., Kim, H. S., Murillo, G., Lee, G. H., Song, H. C., Jung, J. H., Polarization-controlled PVDF-based hybrid nanogenerator for an effective vibrational energy harvesting from human foot. *Nano Energy* 2020;76:105066. DOI: 10.1016/j.nanoen.2020.105066
265. Koh, K. H., Shi, Q., Cao, S., Ma, D., Tan, H. Y., Guo, Z., Lee, C., A self-powered 3D activity inertial sensor using hybrid sensing mechanisms. *Nano Energy* 2019;56:651–661. DOI: 10.1016/j.nanoen.2018.11.075
266. Liu, H., Fu, H., Sun, L., Lee, C., Yeatman, E. M., Hybrid energy harvesting technology: From materials, structural design, system integration to applications. *Renew Sustain Energy Rev* 2020;137:110473. DOI: 10.1016/j.rser.2020.110473
267. Tan, P., Zheng, Q., Zou, Y., Shi, B., Jiang, D., Qu, X., Ouyang, H., Zhao, C., Cao, Y., Fan, Y., Wang, Z. L., A battery-like self-charge universal module for motional energy harvest. *Adv Energy Mater* 2019;9(36):1901875. DOI: 10.1002/aenm.201901875
268. Zhang, S. L., Jiang, Q., Wu, Z., Ding, W., Zhang, L., Alshareef, H. N., Wang, Z. L., Energy harvesting-storage bracelet incorporating electrochemical microsupercapacitors self-charged from a single hand gesture. *Adv Energy Mater* 2019;9(18):1900152. DOI: 10.1002/aenm.201900152
269. Rahman, M. T., Rana, S. S., Salauddin, M., Maharjan, P., Bhatta, T., Kim, H., Cho, H., Park, J. Y., A highly miniaturized freestanding kinetic-impact-based non-resonant hybridized electromagnetic-triboelectric nanogenerator for human induced vibrations harvesting. *Appl Energy* 2020;279:115799. DOI: 10.1016/j.apenergy.2020.115799
270. Kwak, S. S., Yoon, H. -J., Kim, S. -W. Textile-based triboelectric nanogenerators for self-powered wearable electronics. *Adv Funct Mater* 2019;29(2):1804533. DOI: 10.1002/adfm.201804533
271. Varma, S. J., Sambath Kumar, K., Seal, S., Rajaraman, S., Thomas, J., Fiber-type solar cells, nanogenerators, batteries, and supercapacitors for wearable applications. *Adv Sci* 2018;5(9):1800340. DOI: 10.1002/advs.201800340
272. Ren, Z., Zheng, Q., Wang, H., Guo, H., Miao, L., Wan, J., Xu, C., Cheng, S., Zhang, H., Wearable and self-cleaning hybrid energy harvesting system based on micro/nanostructured haze film. *Nano Energy* 2020;67:104243. DOI: 10.1016/j.nanoen.2019.104243
273. Wang, Y., Wu, H., Xu, L., Zhang, H., Yang, Y., Wang, Z. L., Hierarchically patterned self-powered sensors for multifunctional tactile sensing. *Sci Adv* 2020;6(34):eabb9083. DOI: 10.1126/sciadv.abb9083
274. Ma, Y., Zheng, Q., Liu, Y., Shi, B., Xue, X., Ji, W., Liu, Z., Jin, Y., Zou, Y., An, Z., Zhang, W., Self-powered, one-stop, and multifunctional implantable triboelectric active sensor for real-time biomedical monitoring. *Nano Lett* 2016;16(10):6042–6051. DOI: 10.1021/acs.nanolett.6b01968
275. Zheng, Q., Zhang, H., Shi, B., Xue, X., Liu, Z., Jin, Y., Ma, Y., Zou, Y., Wang, X., An, Z., Tang, W., *In vivo* self-powered wireless cardiac monitoring via implantable triboelectric nanogenerator. *ACS Nano* 2016;10(7):6510–6518. DOI: 10.1021/acsnano.6b02693
276. Tang, W., Tian, J., Zheng, Q., Yan, L., Wang, J., Li, Z., Wang, Z. L., Implantable self-powered low-level laser cure system for mouse embryonic osteoblasts' proliferation and differentiation. *ACS Nano* 2015;9(8):7867–7873. DOI: 10.1021/acsnano.5b03567
277. Zheng, Q., Zou, Y., Zhang, Y., Liu, Z., Shi, B., Wang, X., Jin, Y., Ouyang, H., Li, Z., Wang, Z. L., Biodegradable triboelectric nanogenerator as a life-time designed implantable power source. *Sci Adv* 2016;2(3):e1501478. DOI: 10.1126/sciadv.1501478
278. Wang, J., Wang, H., Thakor, N. V., Lee, C., Self-powered direct muscle stimulation using a triboelectric nanogenerator (TENG) integrated with a flexible multiple-channel intramuscular electrode. *ACS Nano* 2019;13(3):3589–3599. DOI: 10.1021/acsnano.9b00140

279. Long, Y., Wei, H., Li, J., Yao, G., Yu, B., Ni, D., Gibson, A. L., Lan, X., Jiang, Y., Cai, W., Wang, X., Effective wound healing enabled by discrete alternative electric fields from wearable nanogenerators. *ACS Nano* 2018;12(12):12533–12540. DOI: 10.1021/acsnano.8b07038
280. Lee, S., Wang, H., Peh, W. Y., He, T., Yen, S. C., Thakor, N. V., Lee, C., Mechano-neuromodulation of autonomic pelvic nerve for underactive bladder: A triboelectric neurostimulator integrated with flexible neural clip interface. *Nano Energy* 2019;60:449–456. DOI: 10.1016/j.nanoen.2019.03.082
281. Hwang, G. -T., Kim, Y., Lee, J. H., Oh, S., Jeong, C. K., Park, D. Y., Ryu, J., Kwon, H., Lee, S.G., Joung, B., Kim, D., Lee, K. J., Self-powered deep brain stimulation via a flexible PIMNT energy harvester. *Energy Environ Sci* 2015;8(9):2677–2684. DOI: 10.1039/C5EE01593F
282. Liao, X., Zhang, Z., Kang, Z., Gao, F., Liao, Q., Zhang, Y., Ultrasensitive and stretchable resistive strain sensors designed for wearable electronics. *Mater Horiz* 2017;4(3):502–510. DOI: 10.1039/c7mh00071e
283. Yi, F., Zhang, Z., Kang, Z., Liao, Q., Zhang, Y., Recent advances in triboelectric nanogenerator-based health monitoring. *Adv Funct Mater* 2019;29(41):1808849. DOI: 10.1002/adfm.201808849
284. Zheng, Q., Shi, B., Li, Z., Wang, Z. L., Recent progress on piezoelectric and triboelectric energy harvesters in biomedical systems. *Adv Sci* 2017;4(7):1700029. DOI: 10.1002/advs.201700029
285. Yao, G., Kang, L., Li, J., Long, Y., Wei, H., Ferreira, C.A., Jeffery, J.J., Lin, Y., Cai, W., Wang, X., Effective weight control via an implanted self-powered vagus nerve stimulation device. *Nat Commun* 2018;9(1):5349. DOI: 10.1038/s41467-018-07764-z
286. Liu, Z., Ma, Y., Ouyang, H., Shi, B., Li, N., Jiang, D., Xie, F., Qu, D., Zou, Y., Huang, Y., Li, H., Transcatheter self-powered ultrasensitive endocardial pressure sensor. *Adv Funct Mater* 2019;29(3):1807560. DOI: 10.1002/adfm.201807560
287. Shi, B., Zheng, Q., Jiang, W., Yan, L., Wang, X., Liu, H., Yao, Y., Li, Z., Wang, Z. L., A packaged self-powered system with universal connectors based on hybridized nanogenerators. *Adv Mater* 2016;28(5):846–852. DOI: 10.1002/adma.201503356
288. Huang, T., Zhang, Y., He, P., Wang, G., Xia, X., Ding, G., Tao, T. H., "Self-matched" tribo/piezoelectric nanogenerators using vapor-induced phase-separated poly (vinylidene fluoride) and recombinant spider silk. *Adv Mater* 2020;32(10):1907336. DOI: 10.1002/adma.201907336
289. Li, H., Zhang, X., Zhao, L., Jiang, D., Xu, L., Liu, Z., Wu, Y., Hu, K., Zhang, M.R., Wang, J., Fan, Y., Li, Z, A hybrid biofuel and triboelectric nanogenerator for bioenergy harvesting. *Nano-Micro Lett* 2020;12(1):50. DOI: 10.1007/s40820-020-0376-8
290. Cheng, X., Tang, W., Song, Y., Chen, H., Zhang, H., Wang, Z. L., Power management and effective energy storage of pulsed output from triboelectric nanogenerator. *Nano Energy* 2019;61:517–532. DOI: 10.1016/j.nanoen.2019.04.096
291. Hu, Y., Yang, J., Niu, S., Wu, W., Wang Z. L., Hybridizing triboelectrification and electromagnetic induction effects for high-efficient mechanical energy harvesting. *ACS Nano* 2014;8(7):7442–7450. DOI: 10.1021/nn502684f
292. Seol, M. -L., Han, J. -W., Park, S. -J., Jeon, S. -B., Choi, Y. -K., Hybrid energy harvester with simultaneous triboelectric and electromagnetic generation from an embedded floating oscillator in a single package. *Nano Energy* 2016;23:50–59. DOI: 10.1016/j.nanoen.2016.03.004
293. Liu, L., Shi, Q., Lee, C., A novel hybridized blue energy harvester aiming at all-weather IoT applications. *Nano Energy*. 2020;76:105052. DOI: 10.1016/j.nanoen.2020.105052
294. Cao, R., Zhou, T., Wang, B., Yin, Y., Yuan, Z., Li, C., Wang, Z. L., Rotating-sleeve triboelectric-electromagnetic hybrid nanogenerator for high efficiency of harvesting mechanical energy. *ACS Nano* 2017;11(8):8370–8378. DOI: 10.1021/acsnano.7b03683

295. Chen, Y. -L., Liu, D., Wang, S., Li, Y. -F., Zhang, X. -S., Self-powered smart active RFID tag integrated with wearable hybrid nanogenerator. *Nano Energy* 2019;64:103911. DOI: 10.1016/j.nanoen.2019.103911
297. Rasel, M. S., Maharjan, P., Park, J. Y., Hand clapping inspired integrated multilayer hybrid nanogenerator as a wearable and universal power source for portable electronics. *Nano Energy* 2019;63:103816. DOI: 10.1016/j.nanoen.2019.06.012
296. Ren, Z., Zheng, Q., Wang, H., Guo, H., Miao, L., Wan, J., Xu, C., Cheng, S., Zhang, H., Wearable and self-cleaning hybrid energy harvesting system based on micro/nanostructured haze film. *Nano Energy* 2020;67:104243. DOI: 10.1016/j.nanoen.2019.104243
298. Chen, J., Huang, Y., Zhang, N., Zou, H., Liu, R., Tao, C., Fan, X., Wang, Z. L., Microcable structured textile for simultaneously harvesting solar and mechanical energy. *Nat Energy* 2016;1(10):16138. DOI: 10.1038/nenergy.2016.138
299. Yang, Y., Xie, L., Wen, Z., Chen, C., Chen, X., Wei, A., Cheng, P., Xie, X., Sun, X., Coaxial triboelectric nanogenerator and supercapacitor fiber-based self-charging power fabric. *ACS Appl Mater Interfaces* 2018;10(49):42356–42362. DOI: 10.1021/acsami.8b15104
300. Wang, S., Lin, Z. H., Niu, S., Lin, L., Xie, Y., Pradel, K. C., Wang, Z. L., Motion charged battery as sustainable flexible-power-unit. *ACS Nano* 2013;7(12):11263–11271. DOI: 10.1021/nn4050408
301. Luo, J., Wang, Z. L., Recent advances in triboelectric nanogenerator based self-charging power systems. *Energy Storage Mater* 2019;23:617–628. DOI: 10.1016/j.ensm.2019.03.009
302. Qin, S., Zhang, Q., Yang, X., Liu, M., Sun, Q., Wang, Z. L., Hybrid piezo/triboelectric-driven self-charging electrochromic supercapacitor power package. *Adv Energy Mater* 2018;8(23):1800069. DOI: 10.1002/aenm.201800069
303. Wen, Z., Yeh, M. H., Guo, H., Wang, J., Zi, Y., Xu, W., Deng, J., Zhu, L., Wang, X., Hu, C., Zhu, L., Self-powered textile for wearable electronics by hybridizing fiber-shaped nanogenerators, solar cells, and supercapacitors. *Sci Adv* 2016;2(10):e1600097. DOI: 10.1126/sciadv.1600097
304. Pu, X., Song, W., Liu, M., Sun, C., Du, C., Jiang, C., Huang, X., Zou, D., Hu, W., Wang, Z. L., Wearable power-textiles by integrating fabric triboelectric nanogenerators and fiber-shaped dye-sensitized solar cells. *Adv Energy Mater* 2016;6(20):1601048. DOI: 10.1002/aenm.201601048
305. Zhang, B., Chen, J., Jin, L., Deng, W., Zhang, L., Zhang, H., Zhu, M., Yang, W., Wang, Z. L., Rotating-disk-based hybridized electromagnetic-triboelectric nanogenerator for sustainably powering wireless traffic volume sensors. *ACS Nano* 2016;10(6):6241–6247. DOI: 10.1021/acsnano.6b02384
306. Yang, H., Deng, M., Zeng, Q., Zhang, X., Hu, J., Tang, Q., Yang, H., Hu, C., Xi, Y., Wang, Z. L., Polydirectional microvibration energy collection for self-powered multifunctional systems based on hybridized nanogenerators. *ACS Nano* 2020;14(3):3328–3336. DOI: 10.1021/acsnano.9b08998
307. He, X., Zi, Y., Yu, H., Zhang, S. L., Wang, J., Ding, W., Zou, H., Zhang, W., Lu, C., Wang, Z. L., An ultrathin paper-based self-powered system for portable electronics and wireless human-machine interaction. *Nano Energy* 2017;39:328–336. DOI: 10.1016/j.nanoen.2017.06.046
308. Xie, Y., Long, J., Zhao, P., Chen, J., Luo, J., Zhang, Z., Li, K., Han, Y., Hao, X., Qu, Z., Lu, M., A self-powered radio frequency (RF) transmission system based on the combination of triboelectric nanogenerator (TENG) and piezoelectric element for disaster rescue/relief. *Nano Energy* 2018;54:331–340. DOI: 10.1016/j.nanoen.2018.10.021
309. Wang, H., Zhu, Q., Ding, Z., Li, Z., Zheng, H., Fu, J., Diao, C., Zhang, X., Tian, J., Zi, Y., A fully-packaged ship-shaped hybrid nanogenerator for blue energy harvesting toward seawater self-desalination and self-powered positioning. *Nano Energy* 2019;57:616–624. DOI: 10.1016/j.nanoen.2018.12.078

310. Wang, W., Xu, J., Zheng, H., Chen, F., Jenkins, K., Wu, Y., Wang, H., Zhang, W., Yang, R., A spring-assisted hybrid triboelectric–electromagnetic nanogenerator for harvesting low-frequency vibration energy and creating a self-powered security system. *Nanoscale* 2018;10(30):14747–14754. DOI: 10.1039/c8nr04276d
311. Rahman, M. T., Salauddin, M., Maharjan, P., Rasel, M. S., Cho, H., Park, J. Y. Natural wind-driven ultra-compact and highly efficient hybridized nanogenerator for self-sustained wireless environmental monitoring system. *Nano Energy* 2019;57:256–268. DOI: 10.1016/j.nanoen.2018.12.052
312. Qian, J., Jing, X., Wind-driven hybridized triboelectric-electromagnetic nanogenerator and solar cell as a sustainable power unit for self-powered natural disaster monitoring sensor networks. *Nano Energy* 2018;52:78–87. DOI: 10.1016/j.nanoen.2018.07.035
313. Gao, L., Lu, S., Xie, W., Chen, X., Wu, L., Wang, T., Wang, A., Yue, C., Tong, D., Lei, W., Yu, H., A self-powered and self-functional tracking system based on triboelectric-electromagnetic hybridized blue energy harvesting module. *Nano Energy* 2020;72:104684. DOI: 10.1016/j.nanoen.2020.104684
314. Chen, Y. -L., Liu, D., Wang, S., Li, Y. -F., Zhang, X. -S., Self-powered smart active RFID tag integrated with wearable hybrid nanogenerator. *Nano Energy* 2019;64:103911. DOI: 10.1016/j.nanoen.2019.103911
315. Qiu, C., Wu, F., Lee, C., Yuce, M. R., Self-powered control interface based on Gray code with hybrid triboelectric and photovoltaics energy harvesting for IoT smart home and access control applications. *Nano Energy* 2020;70:104456. DOI: 10.1016/j.nanoen.2020.104456
316. Liu, H., Gudla, S., Hassani, F. A., Heng, C. H.. Lian, Y., Lee, C., Investigation of the nonlinear electromagnetic energy harvesters from hand shaking. *IEEE Sensors J* 2015;15(4):2356–2364. DOI: 10.1109/JSEN.2014.2375354
317. Kim, S. J., Lee, H. E., Choi, H., Kim, Y., We, J. H., Shin, J. S., Lee, K. J., Cho, B. J., High-performance flexible thermoelectric power generator using laser multiscanning lift-off process. *ACS Nano* 2016;10(12):10851–10857. DOI: 10.1021/acsnano.6b05004
318. Roldán-Carmona, C., Malinkiewicz, O., Soriano, A., Espallargas, G. M., Garcia, A., Reinecke, P., Kroyer, T., Dar, M. I., Nazeeruddin, M. K., Bolink, H. J., Flexible high efficiency perovskite solar cells. *Energy Environ Sci* 2014;7(3):994–997. DOI: 10.1039/c3ee43619e
319. Gupta, R. K., Shi, Q., Dhakar, L., Wang, T., Heng, C. H., Lee, C., Broadband energy harvester using non-linear polymer spring and electromagnetic/triboelectric hybrid mechanism. *Sci Rep* 2017;7(1):41396. DOI: 10.1038/srep41396
320. Sun, J. G., Yang, T. N., Wang, C. Y., Chen, L. J., A flexible transparent one-structure tribo-piezo-pyroelectric hybrid energy generator based on bio-inspired silver nanowires network for biomechanical energy harvesting and physiological monitoring. *Nano Energy* 2018;48:383–390. DOI: 10.1016/j.nanoen.2018.03.071
321. Chen, X., Xie, X., Liu, Y., Zhao, C., Wen, M., Wen, Z., Advances in healthcare electronics enabled by triboelectric nanogenerators. *Adv Funct Mater* 2020;30(43):2004673. DOI: 10.1002/adfm.202004673
322. Hassani, F. A., Shi, Q., Wen, F., He, T., Haroun, A., Yang, Y., Feng, Y., Lee, C., Smart materials for smart healthcare– moving from sensors and actuators to self-sustained nanoenergy nanosystems. *Smart Mater Med* 2020;1:92–124. DOI: 10.1016/j.smaim.2020.07.005
323. Zhu, M., Sun, Z., Zhang, Z., Shi, Q., He, T., Liu, H., Chen, T., Lee, C., Haptic-feedback smart glove as a creative human-machine interface (HMI) for virtual/augmented reality applications. *Sci Adv* 2020;6(19):eaaz8693. DOI: 10.1126/sciadv.aaz8693
324. Jin, T., Sun, Z., Li, L., Zhang, Q., Zhu, M., Zhang, Z., Yuan, G., Chen, T., Tian, Y., Hou, X., Lee, C., Triboelectric nanogenerator sensors for soft robotics aiming at digital twin applications. *Nat Commun* 2020;11(1):5381. DOI: 10.1038/s41467-020-19059-3

Index

www.ingramcontent.com/pod-product-compliance
Lightning Source LLC
Chambersburg PA
CBHW060419220326
41598CB00021BA/2222

* 9 7 8 1 0 3 2 5 5 2 8 4 2 *